UNIVERSE

STARS AND GALAXIES

UNIVERSE
STARS AND GALAXIES

Roger A. Freedman

University of California, Santa Barbara

William J. Kaufmann III

Late of San Diego State University

W. H. Freeman and Company
New York

ACQUISITIONS EDITOR: PATRICK FARACE

PUBLISHER: MICHELLE JULET

DIRECTOR OF MARKETING: JOHN BRITCH

EDITORIAL DEVELOPMENT: BARBARA BROOKS

PROJECT EDITOR: DIANE CIMINO DAVIS

MEDIA/SUPPLEMENTS EDITOR: CHARLES VAN WAGNER

ASSISTANT EDITOR: DANIELLE SWEARENGIN

COVER DESIGN: BLAKE LOGAN

TEXT DESIGN: VICKI TOMASELLI AND BLAKE LOGAN

ILLUSTRATION COORDINATOR: BILL PAGE

ILLUSTRATIONS: FINE LINE ILLUSTRATIONS

PHOTO EDITOR: VIKII WONG

PHOTO RESEARCH: INGE KING

PRODUCTION COORDINATOR: SUSAN WEIN

COMPOSITION: SHERIDAN SELLERS,
W. H. FREEMAN AND COMPANY ELECTRONIC PUBLISHING CENTER, AND TSI GRAPHICS

MANUFACTURING: R. R. DONNELLEY & SONS COMPANY

■ ■ ■

Library of Congress Cataloging-in-Publication Data

Freedman, Roger A.

Universe : stars and galaxies / Roger A. Freedman, William J. Kaufmann III.

p. cm.

ISBN 0-7167-4646-8

1. Astronomy. 2. Cosmology. I. Kaufmann, William J. II. Title.

QB43.3 .F744 2001

523.1'1—dc21

2001002561

Guest Essay photo credits:
p. 18, Don Fukuda/National Optical Astronomy Observatories;
p. 42, courtesy of James Randi; p. 177, courtesy of Geoff Marcy;
p. 418, courtesy of John N. Bahcall; p. 665, Ramono Cagnoni/Black Star;
p. 688, courtesy of John Ruhl; p. 702, courtesy of Matthew Golombek

Printed in the United States of America
Second printing 2002

To lee Johnson Kaufmann and
Caroline Robillard-freedman, strong survivors,
and to the memory of
S/Sgt. Ann Kazmierczak freedman, WAC

Contents Overview

Contents

PART I Introducing Astronomy

1 | Astronomy and the Universe   1

2 | Knowing the Heavens   20

3 | Eclipses and the Motion of the Moon   43

p. 5

p. 144

PART II Planets and Moons

p. 412

PART III　**Stars and Stellar Evolution**

18 | Our Star, the Sun　**389**

19 | The Nature of the Stars　**419**

p. 482

p. 519

p. 624

To the Instructor

From its first edition, Bill Kaufmann's *Universe* has been a tremendous help to my introductory astronomy students. They have benefited from Bill's compelling prose, his emphasis on how astronomers learn about the universe, and his clear exposition. I've benefited as well. The in-depth presentation in *Universe*—never "watered down," but always accessible to the beginning student—has made it possible for me to teach a more satisfying course. And so, it's been my special pleasure to carry on Bill's work by developing *Universe* into its new Sixth Edition.

A narrative of scientific discovery underscores each chapter

When I took over the stewardship of *Universe* five years ago, my goals were to continue its emphasis on the process of scientific discovery; to make it easier for students to grasp the concepts of astronomy; to help students develop the skills needed to solve quantitative problems; and to make the book as modern, lively, and interesting as I possibly could. Above all, I've worked to retain and enhance the strengths that have made *Universe* a standard in astronomy education, including its particular focus on *how scientists work and think*.

Many students who take an astronomy course will have no other exposure to the physical sciences in their college or university career. In a real sense, it is their last chance to gain an appreciation for science and the scientific method. Thus, it is essential that students learn not only *what* we know about our fascinating physical universe but also *how* we know it. The scientific way of knowing, based on observation and continual questioning, is at the heart of both the organization and the features of *Universe*.

Astronomy is an intensely visual and graphic subject, and *Universe* was the first astronomy textbook to be in full color throughout. Continuing this tradition of graphic innovation, new **flipbook** images created from sequences of animations, one for each of the four parts of the book, appear in the upper right-hand corners of the odd-numbered pages in this edition. These images help motivate students to flip through the pages of the book, perhaps to run across an image or topic that piques their interest. The flipbook images also stimulate students to peruse the extensive collection of animations and videos on the *Universe* web site and CD-ROM. Perhaps most important of all, the flipbook images help make this book fun!

The flipbook images highlight the text's flexible organization, which has always suited my needs as an instructor. New to this edition, the split versions of *Universe* make it even easier for instructors to structure their courses as they see fit.

p. 3…

p. 153…

p. 391…

p. 565…

Three versions of UNIVERSE make it easy to customize course content

In addition to the complete, 30-chapter textbook, two briefer versions of the Sixth Edition are now available. These versions provide instructors with new options for shorter courses, as well as for courses that focus on either the solar system or stars and galaxies:

- *Universe* contains the full text of Chapters 1–30. (0-7167-4647-6)

- *Universe: The Solar System* contains the full text of Chapters 1–18 and 30. (0-7167-4645-X)

- *Universe: Stars and Galaxies* contains the full text of Chapters 1–7 and 18–30. (0-7167-4646-8)

All three versions include the fundamental material on positional astronomy, orbits, gravitation, light, and telescopes, as well as all Appendices, Glossary, Answers to Selected Questions, and Index. In addition, each version of *Universe* comes with the full CD-ROM, including *Starry Night*™ planetarium software.

UNIVERSE fully integrates the text with electronic media

Universe was the first textbook to use a web site and a CD-ROM. My vision for the Sixth Edition is that students will use the book, web site, and CD-ROM as aspects of a seamless learning experience. On almost every page, new **media icons** point students to specific, relevant information on the CD-ROM and the *Universe* web site:

 ANIMATIONS illustrating key concepts, including animated versions of selected text figures.

 VIDEOS from space missions as well as ground-based observations.

 IN-DEPTH DISCUSSIONS that supplement text topics.

 LINKS to external web sites where students can find additional information and images.

 INTERACTIVE EXERCISES test student understanding of selected images and diagrams in the text.

 ACTIVE INTEGRATED MEDIA MODULES are interactive learning tools that simplify repetitive calculations and allow students to gain quantitative understanding.

 Many end-of-chapter OBSERVING PROJECTS use the award-winning *Starry Night*™ planetarium software provided on the CD-ROM.

Review aids help students get the most out of the book

My students appreciate how the many features of *Universe* help them review what they've learned. As in previous editions, **chapter-opening questions and full-sentence section headings** help guide their reading. Chapters end with lists of **Key Words** referenced by page to where they are discussed in the chapter, as well as summaries of **Key Ideas**. Each chapter has an extensive collection of **Questions**, including **Problem-Solving Tips and Tools** to help students build their skills. **Observing Tips and Tools** provide students with practical hints to help them with the end-of-chapter **Observing Projects**.

A student-oriented introduction, **How to Get the Most from *Universe*,** follows this preface. I've included hints for students who may never have taken a college science course. Following this is a new **Guided Tour of *Universe*** that explains how best to use the book's text and media features, as well as a new self-quiz that helps students confirm their understanding of these features.

In the previous edition of *Universe*, I introduced **Caution paragraphs** that alert students to common misconceptions. I also added **Analogy paragraphs** and **"The Heavens on the Earth" boxes** that relate astronomical phenomena to aspects of everyday life. On the quantitative side, **"Tools of the Astronomer's Trade" boxes** have worked examples that show students how to solve "real" problems. **Links to other chapters** in blue underlined type help students see the connections among chapters and make it easier for them to locate material they need to review. I've worked hard to make these features even more useful in the Sixth Edition (see the Instructors' Guided Tour of *Universe* on page xix).

For this edition, I've added more than 300 entirely new end-of-chapter questions. In addition to traditional-style questions, many of the new questions ask students to interpret astronomical images or to evaluate how the mass media portray the ideas of astronomy. New **Web/CD-ROM Questions** ask them to analyze images or animations on the textbook CD-ROM or web site, or to do research on the World Wide Web. Other questions require them to use the *Starry Night*™ planetarium software provided on the CD-ROM.

A thoroughly up-to-date text shows science in the making

I've brought the book up to date with the latest information from all aspects of the subject we love. In addition to dramatic new color images and enhanced explanatory diagrams, I've included the latest discoveries, including:

- Recent results from *Mars Global Surveyor, NEAR Shoemaker,* the Chandra X-ray Observatory, and the ESO's Very Large Telescope, as well as other missions and observatories.

- The latest information about brown dwarfs, evidence for a nonzero cosmological constant, gamma-ray bursters, and much more.

- New **guest essays** by Matthew Golombek (JPL) on the exploration of Mars and by John Ruhl (University of California, Santa Barbara) on measuring the cosmic background radiation.

p. 693

I have endeavored to make this new edition even more accessible to students, while being careful to maintain the same intellectual level. One of the ways I have done this is by reorganizing several key chapters. Examples include:

- Chapter 4, Gravitation and the Waltz of the Planets, now presents tidal forces immediately after the discussion of Newton's laws and gravitation.

- Chapter 18, Our Star, the Sun, now discusses the Sun "from the inside out," beginning with how the Sun shines and ending with the solar atmosphere.

p. 411

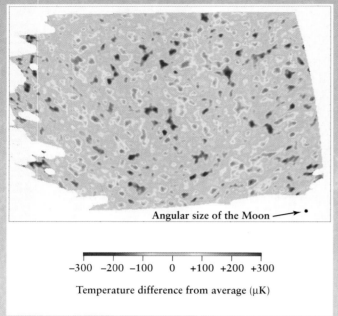

Angular size of the Moon ⟶ •

-300 -200 -100   0   +100 +200 +300

Temperature difference from average (μK)

p. 653

• Chapter 24, Black Holes, has been revised to emphasize observations of black holes as well as black hole theory. New sections describe supermassive black holes and Hawking radiation.

• Chapter 28, Cosmology, has been totally revised in light of recent observations of Type Ia supernovae and of the cosmic microwave background. A new discussion shows how these observations can reveal the presence of dark energy.

• Chapter 30, The Search for Extraterrestrial Life, has been expanded to include the search for extraterrestrial life of all kinds, from single cells to advanced civilizations. A new section describes the ongoing search for life within our solar system.

A constellation of supplements complements this textbook

Students can connect to the universe in two ways: via the web site (**www.whfreeman.com/universe6e**) and via the *Universe 6.0* CD-ROM packaged with every copy of the book. In addition, instructors can choose from a full range of print ancillaries, as well as software tools for course management, assessment, and presentation. For full details on *Universe* supplements for students and instructors, consult the Instructors' Guided Tour, which begins on page xix.

Instructors' Guided Tour of UNIVERSE

UNIVERSE WEB SITE at www.whfreeman.com/universe6e

For instructors, the site offers **Further information** about the topics in the book, **Links** to additional resources, and **a forum** for sharing ideas about using the book.

For students, the web site serves as a study guide. Its features include:

- **Chapter Objectives:** Learning objectives help students formulate study strategies.

- **Q & A:** The self-quizzing feature helps students prepare for exams.

- **Interactive Exercises:** "Drag & drop" exercises that help students understand the vocabulary of astronomy in the context of illustrations from the text.

- **Key Terms:** Flashcard exercises based on vocabulary and definitions.

- *Starry Night*™ **Observing Exercises:** In-depth projects that students can perform using the *Starry Night*™ software provided on the student CD-ROM. These complement the shorter *Starry Night*™ Observing Projects in the textbook.

- **Current Events in Astronomy:** Up-to-date links to information and resources.

- **Animations and Videos,** both original and from sources such as NASA and ESA, are updated regularly.

- **Web Links** lead to World Wide Web sites that contain supplemental material about astronomy.

- **Active Integrated Media Modules (AIMM):** Interactive modules that take students deeper into key topics from the text, including how to determine distances to the stars and beyond, the Doppler effect, the nearest stars, and relativistic redshift.

- **Looking Deeper:** In-depth boxes, thoroughly updated, extend topics discussed in the text, including the search for gravitational waves, how to interpret the shapes of spectral lines, and discoveries about the inner core of the Earth.

UNIVERSE 6.0 CD-ROM featuring STARRY NIGHT™

Packaged with every copy of the textbook, the *Universe 6.0* CD-ROM contains all the content that appears on the web site (except for the Web Links and Current Events in Astronomy). It also includes the easy-to-use, award-winning *Starry Night*™ planetarium software (for both Macintosh and Windows), the ideal electronic accompaniment for any astronomy course. *Starry Night*™ observational activities are provided in each chapter of *Universe*. Others are available on the *Universe* web site.

- **Freeman Observing Guide (ISBN 0-7167-4437-6)**, *William J. F. Wilson and T. Alan Clark, University of Calgary*—This printed supplement has 15 new, comprehensive lab activities for *Starry Night*™ that provide even more opportunities for students to explore the cosmos.

Presentation Tools

INSTRUCTOR'S CD-ROM

Supplied to all instructors who adopt *Universe*, this CD-ROM provides instructors with the tools to create in-class presentations:

- **All text images in JPEG, PICT, and PowerPoint format** for use with any standard presentation software.
- **Presentation Manager Pro software** lets you build classroom presentations using images and video from the *Universe* CD-ROM, *plus* your own digital material (including video) imported from the World Wide Web or other sources.

STARRY NIGHT™ PRO VERSION CD-ROM, FROM SPACE.COM

Available to qualified instructors, the deluxe version of *Starry Night*™ features over 19 million celestial objects (including the complete Hubble Guide Star Catalog), realistically rendered 3D planets, astrophotographs of all 110 Messier objects, a Datamaker application for adding your own custom astronomical databases, QuickTime movie-making abilities, and more.

THE FREEMAN ASTRONOMY IMAGE AND LECTURE GALLERY

- **Images and Diagrams:** This convenient database makes it easy to access electronic figures from the textbook and to use them in web pages or PowerPoint presentations. Registered users can browse, search, and download illustrations from Freeman astronomy titles. They can also create folders on a personalized home page for easy organization of the images.
- **PowerPoint Lecture Presentation:** Instructors adopting *Universe* have access to PowerPoint files for each chapter of the book. These include lecture notes written by Timothy Slater (Montana State University), as well as all the textbook images and their captions. The PowerPoint files can be downloaded and customized to suite the individual instructor's needs.

OVERHEAD TRANSPARENCY SET

100 full-color transparencies of key illustrations, photos, and tables from the text.

Course Management Tools

ASTRONOMY ONLINE, *Timothy Slater, Montana State University*

Available in WebCT, Blackboard, and Metatext formats, **Astronomy Online** combines text content from *Universe* with animations, videos, interactive exercises, images, and quizzing specifically created to accompany the book. It can serve either as the online component of a lecture-based course or as a complete online source for a distance learning course. Throughout **Astronomy Online**, the emphasis is on interactivity and active learning. Students are prompted to answer questions and make observations about the course materials. Upon completion of each module, they receive feedback on their responses. **Astronomy Online** can be used as is or tailored to meet instructors' individual needs. It can be accessed via the *Universe* web site.

ON-LINE COURSE MATERIALS IN WEBCT

Adopters using WebCT can use this service, which provides a fully loaded E-Pack that includes the instructor and student resources for *Universe*. The files can be used as is or customized to fit specific needs. Course outlines, pre-built quizzes, links, activities, and a whole array of materials are included.

INSTRUCTOR'S MANUAL AND RESOURCE GUIDE, *George A. Carlson, Citrus College*

The Manual includes worked-out solutions to all end-of-chapter questions in *Universe*. It also includes detailed chapter outlines as well as classroom-tested teaching strategies and hints. The extensive Resource Guide covers reading materials for students and instructors, audiovisual material, and discussion/paper topics.

Assessment Tools

TEST BANK, *William J. F. Wilson and T. Alan Clark, University of Calgary*

This set of more than 2300 multiple-choice questions is available both in print and on CD-ROM in both Macintosh and Windows formats. All questions are section-referenced and designated as "assignment" or "test" questions. The CD-ROM version makes it easy to add, edit, and re-sequence questions to suit your needs. The CD-ROM is also the access point for Diploma Online Testing.

DIPLOMA ONLINE TESTING, *from the Brownstone Research Group*

Diploma makes it easy to create and administer secure exams over either the Internet or a local area network. Questions (taken from the Test Bank CD-ROM) can incorporate multimedia and interactivity. The program lets you restrict tests to specific computers or time blocks. It includes an impressive suite of gradebook and result-analysis features.

ONLINE QUIZZING, *powered by Questionmark*

Accessed via the *Universe* web site, Questionmark's Perception enables instructors to quiz students online easily and securely using prewritten, multiple-choice questions (not from the Test Bank) for each text chapter. Students receive instant feedback and can take the quizzes multiple times. Instructors can go into a protected web site to view results by quiz, by student, or by question, or can get weekly results via e-mail.

SCIENTIFIC AMERICAN READER FOR UNIVERSE

This collection of 15 astronomy-related articles from *Scientific American* magazine is enhanced with introductions written by Roger Freedman. Each article is accompanied by five questions suitable for homework assignments.

Acknowledgments

I would like to thank my colleagues who carefully scrutinized the manuscript of this edition. This is a stronger and better textbook because of their conscientious efforts:

Robert R. J. Antonucci, *University of California, Santa Barbara*

David Bruning, *University of Wisconsin, Parkside*

Spencer L. Buckner, *Austin Peay State University*

Karen G. Castle, *Diablo Valley College*

Kim Coble, *University of California, Santa Barbara*

Malcolm Coe, *The University of Southampton*

David Dahl, *St. Olaf College*

Steve Danford, *University of North Carolina, Greensboro*

Mary V. Frohne, *Western Illinois University*

Pamela L. Gay, *University of Texas, Austin*

Robert M. Geller, *University of California, Santa Barbara*

Donna Hurlbut Gifford, *Pima Community College*

Owen Gingerich, *Harvard University*

Bernadette Londak Harris, *B.C. Open University*

Dave Kary, *Citrus College*

B. Alexander King III, *Austin Peay State University*

Arthur Kosowsky, *Rutgers University*

David Kriegler, *University of Nebraska, Omaha*

Andrew Lazarewicz, *Boston College*

Ntungwa Maasha, *Coastal Georgia Community College*

Margaret Mazzolini, *Swinburne Astronomy Online*

J. Ward Moody, *Brigham Young University*

Gerald H. Newsom, *The Ohio State University*

Erin O'Connor, *Allan Hancock College*

Fritz Osell, *Leeward Community College*

Robert L. Pompi, *Binghamton University*

Harrison B. Prosper, *Florida State University*

Jim F. Smeltzer, *Northwest Missouri State University*

Larry K. Smith, *Snow College*

Darryl Stanford, *City College of San Francisco*

Donald Terndrup, *The Ohio State University*

George F. Tucker, *The Sage Colleges*

Stephen Walton, *California State University, Northridge*

Joseph C. Wesney, *Greenwich High School*

A. B. Whiting, *U.S. Naval Academy*

My heartfelt thanks also go out to the following instructors, who responded to informal surveys for this edition or examined the fifth edition with particular care with an eye to our plans for revision. Their suggestions were tremendously helpful.

Grant Bazan, *Las Positas Community College*

Richard French, *Wellesley College*

Donald Gudehus, *Georgia State University*

Eric Harpell, *Las Positas Community College*

Paul Hintzen, *California State University, Long Beach*

Michael Kaufman, *San Jose State University*

Paul Marquard, *Casper College*

Donald H. Martins, *University of Alaska, Anchorage*

William Parke, *George Washington University*

Eric R. Peterson, *De Anza College*

Roger Romani, *Stanford University*

Michael Ruiz, *University of North Carolina, Asheville*

James Shea, *University of Wisconsin, Parkside*

Ron Stoner, *Bowling Green State University*

Carol Strong, *University of Alabama, Huntsville*

Colin Terry, *Ventura College*

David Weinberg, *The Ohio State University*

Lynda Williams, *San Francisco State University*

I would also like to thank the many people whose advice on previous editions has had an ongoing influence:

Robert Allen, *University of Wisconsin, La Crosse;* Robert R. J. Antonucci, *University of California, Santa Barbara;* Alice L. Argon, *Harvard-Smithsonian Center for Astrophysics;* Omer Blaes, *University of California, Santa Barbara;* David Van Blerkom, *University of Massachusetts;* John M. Burns, *Mt. San Antonio College;* Bel Campbell, *University of New Mexico;* George A. Carlson, *Citrus College;* Bruce W. Carney, *University of North Carolina;* Bradley W. Carroll, *Weber State University;* George L. Cassiday, *University of Utah;* John J. Cowan, *University of Oklahoma;* John E. Crawford, *McGill University;* Roger B. Culver, *Colorado State University;* Robert Dick, *Carleton University;* James N. Douglas, *University of Texas at Austin;* Robert J. Dukes, Jr., *College of Charleston;* Robert A. Egler, *North Carolina State University;* Debra Meloy Elmegreen, *Vassar College;* David S. Evans, *University of Texas, Austin;* George W. Ficken, Jr., *Cleveland State University;* Andrew Fraknoi, *Astronomical Society of the Pacific;* Juhan Frank, *Louisiana State University;* Steven Giddings, *University of California, Santa Barbara;* Owen Gingerich, *Harvard University;* Paul F. Goldsmith, *University of Massachusetts;* J. Richard Gott III, *Princeton University;* Austin F. Gulliver, *Brandon University;* Buford M. Guy, *Cleveland State Community College;* Carl Gwinn, *University of California, Santa Barbara;* Bruce Hanna, *Old Dominion University;* Charles L. Hartley, *Hartwick College;* Paul A. Heckert, *Western Carolina University;* Bill Herbst, *Wesleyan University;* Paul Hodge, *University of Washington;* Douglas P. Hube, *University of Alberta;* Icko Iben, Jr., *Pennsylvania State University;* Scott B. Johnson, *Idaho State University;* John K. Lawrence, *California State University, Northridge;* Marie E. Machacek, *Northeastern University;*

Laurence A. Marschall, *Gettysburg College;* Raymond C. McNeil, *Northern Kentucky University;* Dimitri Mihalas, *University of Illinois;* Robert M. O'Connell, *College of the Redwoods;* C. Robert O'Dell, *Rice University;* L. D. Opplinger, *Western Michigan University;* Stanton J. Peale, *University of California, Santa Barbara;* John R. Percy, *University of Toronto;* Robert L. Pompi, *State University of New York, Binghamton;* Carlton Pryor, *Rutgers University;* James L. Regas, *California State University, Chico;* Terry Retting, *University of Notre Dame;* Tina Riedinger, *University of Tennessee;* James A. Roberts, *University of North Texas;* Charles W. Rogers, *Southwestern Oklahoma State University;* Kenneth S. Rumstay, *Valdosta State College;* Richard Saenz, *California Polytechnic State University;* Thomas F. Scanlon, *Grossmont College;* Richard L. Sears, *University of Michigan;* Isaac Shlosman, *University of Kentucky;* Alan F. Sill, *Texas Tech University;* Caroline Simpson, *Florida International University;* Michael L. Sitko, *University of Cincinnati;* David B. Slavsky, *Loyola University of Chicago;* Joseph S. Tenn, *Sonoma State University;* Gordon B. Thomson, *Rutgers University;* Charles R. Tolbert, *University of Virginia;* Virginia Trimble, *University of California, Irvine;* Thomas Tsung, *Grossmont College;* Bruce A. Twarog, *University of Kansas;* George Wegner, *Dartmouth College;* Donat G. Wentzel, *University of Maryland;* Nicholas Wheeler, *Reed College;* Raymond E. White, *University of Arizona;* Louis Winkler, *The University of Pennsylvania;* and Robert L. Zimmerman, *University of Oregon.*

I'm particularly grateful to Matthew Golombek and John Ruhl, whose new essays for this edition greatly enhance its coverage. I also thank Geoff Marcy and John Bahcall, who were kind enough to revise their essays in light of recent developments.

Many others have participated in the preparation of this book, and I thank them for their efforts. I am particularly grateful to Patrick Farace, my acquisitions editor, for keeping me entertained; to Barbara Brooks, my development editor, for keeping me encouraged; to Charlie Van Wagner, my media and supplements editor, for supervising the superb web site and CD-ROM; and to Kent Gardner, Annali Kiers, and their colleagues at Sumanas, for translating my rough ideas into elegant multimedia. Special thanks go to Liz Widdicombe, president; Louise B. Ketz, copy editor and indexer; Diane Cimino Davis, project editor; Susan Wein, production coordinator; Sheridan Sellers, W. H. Freeman and Company Electronic Publishing Center, and TSI Graphics, compositors; Inge King and Vikii Wong, photo researchers; John Britch, marketing manager; and the ever-charming Danielle Swearengin, assistant editor.

On a personal note, I would like to thank my father, Richard Freedman, for first cultivating my interest in space many years ago, and for his personal contributions to the exploration of the universe as an engineer for the Atlas and Centaur launch vehicle programs. Most of all, I thank my charming wife, Caroline, for putting up with my long nights slaving over the computer!

Although I have made a concerted effort to make this edition error free, some mistakes may have crept in unbidden. I would appreciate hearing from anyone who finds an error or wishes to comment on the text. You may e-mail or write me.

Roger A. Freedman
Department of Physics
University of California, Santa Barbara
Santa Barbara, CA 93106
airboy@physics.ucsb.edu

To the Student: How to Get the Most from UNIVERSE

If you're like most students just opening this textbook, you're enrolled in one of the few science courses you'll take in college. As you study astronomy, you'll probably do relatively little reading compared with a literature or history course—at least in terms of the number of pages. But your readings will be packed with information, much of it new to you and (I hope) exciting. You can't read this textbook like a novel and expect to learn much from it. Don't worry, though. I wrote this book with you in mind. In this section, I'll suggest how *Universe* can help you succeed in astronomy and take you on a guided tour of the book and media.

Apply these techniques to studying astronomy

- **Read before a lecture** You'll get the most out of your astronomy course if you read each chapter *before* hearing a lecture about its subject matter. That way, many of the topics will already be clear in your mind, and you'll understand the lecture better. You'll be able to spend more of your listening and note-taking time on the more challenging ideas presented in the lecture.

- **Take notes as you read and make use of office hours** Keep a notebook handy as you read, and write down the key points of each section so that you can review them later. If any parts of the section don't seem clear on first reading, make a note of them, too, including the page numbers. Once you've gone through the chapter, reread it with special emphasis on the ideas that gave you trouble the first time. If you're still unsure after the lecture, consult your instructor, either during office hours or after class. Bring your notes with you so your instructor can see which concepts are giving you trouble. Once your instructor has helped clarify things for you, revise your notes so you'll remember your new-found insights. You'll end up with a chapter summary in your own words. This will be a tremendous help when studying for exams!

- **Make use of your fellow students** Many students find it useful to form study groups for astronomy. You can hash out challenging topics with one another and have a good time while you're doing it. But make sure that you write up your homework by yourself, because the penalties for copying or plagiarizing other students' work can be severe in the extreme. Some students find individual assistance useful, too. If you think a tutor will be helpful, link up with one early. Getting a tutor late in the course, in the belief that you'll be able to catch up with what you missed earlier on, is almost always a pointless exercise.

- **Take advantage of the web site and CD-ROM** Take some time to explore the *Universe* web site (**www.whfreeman.com/universe6e**) and the CD-ROM that comes packaged with this book. On both of these you'll find review materials, animations, videos, interactive exercises, flashcards, and many other features keyed to chapters in *Universe*. All these features are designed to help you learn and enjoy astronomy, so make sure to take full advantage of them.

- **Try astronomy for yourself with your star charts and *Starry Night*™** At the back of this book you'll find a set of star charts for the each month of the year in the northern hemisphere. (For a set of southern hemisphere star charts, visit the *Universe* web site.) Star charts can get you started with your own observations of the universe. Hold the chart overhead in the same orientation as the compass points, with "southern horizon" toward the south and "western horizon" toward the west. (To save strain on your arms, you may want to photocopy these pages.) The CD-ROM packaged with this textbook includes the easy-to-use *Starry Night*™ planetarium program, developed in Canada, which you can use to view the sky on any date and time as seen from anywhere on Earth.

Before you study UNIVERSE, take this quiz

Universe has many features designed to help you succeed in your study of astronomy. To get the most from it, understanding these features and knowing how to use them are essential. To make sure that you do, read through **A Guided Tour of Universe**, which begins on the next page. Then take this brief quiz. If you can answer all the questions, you're ready to begin studying astronomy! (You can check your answers on page Q-1.)

1. Which specially labeled paragraphs alert you to common misconceptions and conceptual pitfalls?

2. Which specially labeled paragraphs draw analogies between ideas in astronomy and aspects of everyday life?

3. What is the significance of text that appears in <u>blue underlined type</u>?

4. In many chapters, in addition to the numbered sections, you will also find material set off in boxes. Which type of box provides extra help with solving mathematical problems? Which type relates astronomical principles to phenomena here on Earth?

5. Throughout the book you will encounter icons labeled "Web Link," "Animation," "Video," "AIMM," or "Looking Deeper." Where should you look to find the information to which these icons refer?

6. What is the *Starry Night*™ program? Where can you find a copy to install on your own computer? How much does it cost?

7. Many of the figures in this book are accompanied by the letters R I V U X G, with one of the letters highlighted. For instance, Figure 6-34*a* has R I **V** U X G, while Figure 6-34*d* has R I V U **X** G. What is the significance of the highlighted letter?

8. Where can you find self-tests and review material for each chapter of *Universe*?

9. Refer to the Appendices at the back of this book. On which page(s) of *Universe* would you look to find the following? (a) the value of the Stefan-Boltzmann constant; (b) the average orbital speed of Mars; (c) the average distance from the center of the planet Pluto to its moon, Charon; (d) the distance in light-years to the star Proxima Centauri.

10. Refer to the Index at the back of this book. On which page(s) of *Universe* would you find the following terms described? (a) spicule; (b) refraction; (c) tidal force; (d) aphelion.

11. Refer to the Answers to Selected Questions at the back of this book. What is the answer to Question 35 of Chapter 5? (*NOTE:* Your instructor may assign as homework some of the questions whose answers can be found in the Answers to Selected Questions. If so, your instructor will expect you to write out and explain your calculations to show how this answer is obtained.)

12. Where in this book can you find northern hemisphere star charts for each month of the year?

Here's the most important advice of all

I haven't mentioned the most important thing you should do when studying astronomy: Have fun! Of all the different kinds of scientists, astronomers are among the most excited about what they do and what they study. Let some of that excitement about the universe rub off on you, and you'll have a great time with this course and with this textbook.

In preparing this edition of *Universe*, I tried very hard to make it the kind of textbook that a student like you would find useful. I'm very interested in your comments and opinions! Please feel free to write me, or send me e-mail, and I will respond personally.

Best wishes for success in your studies!

Roger A. Freedman
Department of Physics
University of California, Santa Barbara
Santa Barbara, CA 93106
airboy@physics.ucsb.edu

A Guided Tour of UNIVERSE

This book is designed to help you get the most from your astronomy course

Our Star, the Sun   18

(Institute of Space and Astronomical Science, Yohkoh Project, SXT Group/NASA)

R I V U **X** G

The Sun is by far the brightest object in the sky. By earthly standards, the temperature of its glowing surface is remarkably high, about ... have found regions of the

Sun shines because at its core hundreds of millions of tons of hydrogen are converted to helium every second. We have found the by-products of this transmutation-...

The Sun's size also helps us explain its tremendous energy output. Because the Sun is so large, the total number of square meters of radiating surface—that is, its surface area—is immense. Hence, the total amount of energy emitted by the Sun each second, called its luminosity, is very large indeed: about 3.9×10^{26} watts, or 3.9×10^{26} joules of energy emitted per second. (We discussed the relation between the Sun's surface temperature, radius, and luminosity in Box 5-2.) Astronomers denote the Sun's luminosity by the symbol $L_{\odot}$. A circle with a dot in the center is the astronomical symbol for the Sun and was also used by ancient astrologers.

Chapter openers A brief overview and question set guide you into each chapter's topics. Use the questions as a guide to your reading. You'll find the answer to each question within the corresponding section of the chapter.

Wavelength tabs In addition to using ordinary telescopes, astronomers rely heavily on special telescopes that are sensitive to nonvisible forms of light. To help you appreciate what astronomers learn from these different kinds of observations, wavelength tabs appear with all the images in this textbook. The highlighted letter on each tab shows whether the image was made with **R**adio waves, **I**nfrared radiation, **V**isible light, **U**ltraviolet light, **X** rays, or **G**amma rays. (We'll discuss these different kinds of light in Chapter 5.) The image of the Sun shown here was made with an X-ray telescope, as indicated by the highlighted **X** on the tab.

Key Words and Glossary Every new term that has a special importance in astronomy appears in **boldface type** when it's introduced and whenever it's discussed in depth. Be sure that you understand the meaning of each term and can explain it in your own words. Don't just memorize the definitions! The **Glossary**, which you'll find near the back of the book, gives brief definitions of all the **Key Words** in the book.

Links to other chapters References to topics from previous chapters appear in blue underlined type. The links help you find and review important ideas that apply to topics in different parts of the textbook. For example, the laws that govern the radiation from heated objects (discussed in Chapter 5) help us understand the light we receive from the Sun (Chapter 18).

Flipbook images You can create an animation—an eclipse, for example—when you flip through

the images in the upper corner of each right-hand page. These flipbook images show a different animation for each of the four parts of the book, so they help you locate information and have some fun in the process!

Confronting misconceptions You may think that the phases of the Moon are caused by the Earth's shadow falling on the Moon and that the Earth is closer to the Sun in summer than in winter. But in fact, these "commonsense" ideas are incorrect! (You'll learn the correct explanations in Chapters 3 and 4.) Paragraphs signaled by the caution icon alert you to conceptual pitfalls like these.

Bringing astronomy down to Earth To learn astronomical ideas, it can be helpful to relate them to your experience on Earth. Throughout the book you'll find analogy paragraphs that relate, for example, the motions of the planets to children on a merry-go-round and the bending of light in a telescope lens to a car driving on sand.

---

box 18-1 | Tools of the Astronomer's Trade

Converting Mass into Energy

Figure 18-2 shows the steps involved in the thermonuclear fusion of hydrogen at the Sun's center. In these steps, four protons are converted into a single nucleus of ⁴He, an isotope of helium with two protons and two neutrons. The reaction depicted in Figure 18-2*a* also produces a neutral, nearly massless particle called the *neutrino*. Neutrinos respond hardly at all to ordinary matter, so they travel almost unimpeded through the Sun's massive bulk. Hence, the energy that neutrinos carry is quickly lost into space. This loss is not great, however, because the neutrinos carry relatively little energy. (See Section 18-4 for more about these

4 hydrogen atoms	=	6.693×10^{-27} kg
-1 helium atom	=	-6.645×10^{-27} kg
Mass lost	=	0.048×10^{-27} kg

Thus, a small fraction (0.7%) of the mass of the hydrogen going into the nuclear reaction does not show up in the mass of the helium. This lost mass is converted into an amount of energy $E = mc^2$:

$$E = mc^2 = (0.048 \times 10^{-27} \text{ kg}) (3 \times 10^8 \text{ m/s})^2$$
$$= 4.3 \times 10^{-12} \text{ joule}$$

The Heavens on the Earth boxes show how the principles that astronomers use can explain everyday phenomena, from why diet soft drink cans float to why the daytime sky is blue.

Tools of the Astronomer's Trade boxes with highlighted, easy to follow, Worked Examples help you learn how to perform the calculations you'll probably be asked to work out on exams. Don't just read them; work the examples out on a piece of paper.

Take advantage of the end-of-chapter reviews

Key Words This list appears at the end of each chapter, along with a page number telling you where the word is discussed.

Key Ideas To get the most out of these brief summaries, use their outline format in conjunction with the notes you take while reading. Make sure you fully understand each description in the summary. If any point is unclear, go back and review the text.

Questions and Projects Your instructor may assign items from these sections as homework. The questions help you review and apply text concepts. Some ask you to analyze images in the text or to evaluate how the mass media portray concepts in astronomy. *Web/CD-ROM Questions* challenge you to work with animations and physical concepts on the *Universe* CD-ROM or web site, or to research topics on the World Wide Web. Look for *Problem-Solving Tips and Tools* that provide guidance for the more advanced questions. (Near the end of the book, you'll find a section of *Answers to Selected Questions*.) *Observing Tips and Tools* accompany many *Observing Projects*, which make frequent use of the *Starry Night*™ planetarium software provided on the CD-ROM.

KEY IDEAS

Hydrogen Burning in the Sun's Core: The Sun's energy is produced by hydrogen burning, a thermonuclear fusion process in which four hydrogen nuclei combine to produce a single helium nucleus.

• The energy released in a nuclear reaction corresponds to a slight reduction of mass according to Einstein's equation $E = mc^2$.

• Thermonuclear fusion occurs only at very high temperatures; for example, hydrogen burning occurs only at temperatures in excess of about 10^7 K. In the Sun, fusion occurs only in the dense, hot core.

Models of the Sun's Interior: A theoretical description of a star's interior can be calculated using the laws of physics.

• The standard model of the Sun suggests that hydrogen burning ... extending from the Sun's center ...

WEB/CD-ROM QUESTIONS

47. Search the World Wide Web for the latest information from the neutrino detectors at the Super-Kamiokande Observatory and the Sudbury Neutrino Observatory. What are the most recent results from these detectors? What is the current thinking about the solar neutrino problem? What is the status of a new detector called Borexino?

48. Search the World Wide Web for information about features in the solar atmosphere called *sigmoids*. What are they? What causes them? How might they provide a way to predict coronal mass ejections?

49. **Determining the Lifetime of a Solar Granule.** Access and view the video "Granules on the Sun's Surface" in Chapter 18 of the *Universe* Web site or CD-ROM. Your task is to determine the approximate lifetime of a solar granule on the photosphere. Select an area, then slowly and rhythmically repeat "start, stop, start, stop" until you can consistently predict the appearance and disappearance of granules. While keeping your rhythm, move to a different area of the video and continue monitoring the appearance and disappearance of granules. When you are confident you have the timing right, move your eyes (or use a partner) to the clock shown in the video. Determine the length of time between the

Guest Essays Several chapters in the textbook end with essays by scientists at the cutting edge of astronomy. They include John Bahcall of the Institute for Advanced Study, writing on the physics of the Sun in an essay that follows Chapter 18, and Matthew Golombek of the Jet Propulsion Laboratory, who contributed the essay "Exploring Mars" at the end of Chapter 30. Their essays give you a sense of their excitement about astronomy. We hope you catch this excitement! For a full list of essays, see the Contents Overview on page vii.

The book, CD–ROM, and web site interact

Throughout each chapter, media icons point you to specific, relevant information on the CD-ROM and the *Universe* web site at **www.whfreeman.com/universe6e**

 Web Links, which lead to Internet sites with supplemental material about astronomy, are updated regularly.

 Animations and Videos, both original and from sources such as NASA and ESA.

 Interactive Exercises include "drag & drop" exercises that help you understand the vocabulary of astronomy in the context of illustrations from the text.

 Active Integrated Media Modules take you deeper into key topics from the text.

The UNIVERSE web site also serves as a study guide

- **Chapter Objectives:** Learning objectives help you formulate study strategies.

- **Q & A:** The self-quizzing feature helps you prepare for exams.

- **Key Terms:** Flashcard exercises based on vocabulary and definitions.

- **Current Events in Astronomy:** Information about the latest astronomical developments.

UNIVERSE 6.0 CD–ROM featuring STARRY NIGHT™

 Packaged with every copy of the textbook, the CD-ROM contains all the content that appears on the web site—except for the Web Links and Current Events in Astronomy—and features the easy-to-use, award-winning *Starry Night™* planetarium software.

MATTHEW GOLOMBEK

Exploring Mars

Dr. Matt Golombek, chief scientist of the *Mars Pathfinder* mission, is a research geologist at the Jet Propulsion Laboratory, California Institute of Technology, NASA's lead center for planetary exploration. He received his undergraduate degree from Rutgers University and his master's degree and Ph.D. from the University of Massachusetts, Amherst. Dr. Golombek was a postdoctoral fellow at the Lunar and Planetary Institute, Houston, before joining JPL.

Dr. Golombek has conducted research in the structural geology of the Earth, planets, and satellites. He is presently NASA's Mars Exploration Program Landing Site Scientist, leading the effort to select landing sites for Mars missions.

Dr. Golombek cowrote the book *Mars: Uncovering the Secrets of the Red Planet* for the National Geographic Society and has served as editor and associate editor of professional journals. He has received numerous awards, including the NASA Exceptional Scientific Achievement Medal. In his honor, asteroid 6456 was named Golombek.

 Looking Deeper, in-depth coverage that extends topics discussed in the text, including the search for gravitational waves, how to interpret the shapes of spectral lines, and discoveries about the inner core of the Earth.

 Starry Night™ Observing Exercises are in-depth lab projects that you can perform using *Starry Night™* software provided on the CD-ROM. These complement the shorter *Starry Night™* Observing Projects in the textbook. To use *Starry Night™*, just follow the instructions on the CD-ROM to install the program on your computer.

Astronomy and the Universe

Imagine yourself in the desert on a clear, dark, moonless night, far from the glare of city lights. As you gaze upward, you see a panorama that no poet's words can truly describe and that no artist's brush could truly capture. Literally thousands of stars are scattered from horizon to horizon. The delicate glow of the Milky Way—part of which is visible at the upper right of the photograph, taken from a remote location in northern Mexico—traces a luminous path across the sky. As you watch, the entire spectacle swings slowly overhead from east to west as the night progresses. If you are very lucky indeed, you may even see a rare celestial event such as the comet shown here.

For thousands of years people have looked up at the heavens and contemplated the universe. Like our ancestors, we find our thoughts turning to profound questions as we gaze at the stars. How was the universe created? Where did the Earth, Moon, and Sun come from? What are the planets and stars made of? And how do we fit in? What is our place in the cosmic scope of space and time?

To wonder about the universe is a particularly human endeavor. Our curiosity, our desire to explore and discover, and, most important, our ability to reason about what we have discovered are qualities that distinguish us from other animals. The study of the stars transcends all boundaries of culture, geography, and politics. In a literal sense, astronomy is a universal subject—its subject is the entire universe.

(Courtesy of D. L. Mammana)

R I **V** U X G

1-1 To understand the universe, astronomers use the laws of physics to construct testable theories and models

Astronomy has a rich heritage that dates back to the myths and legends of antiquity. Centuries ago, the heavens were thought to be populated with demons and heroes, gods and goddesses. Astronomical phenomena were explained as the result of supernatural forces and divine intervention.

The course of civilization was greatly affected by a profound realization: *The universe is comprehensible.* This awareness is one of the great gifts to come to us from ancient Greece. Greek astronomers discovered that by observing the heavens and carefully reasoning about what they saw, they could learn something about how the universe operates. For example, as we shall see in Chapter 3, they measured the size of the Earth and understood and predicted eclipses. Modern science is a direct descendant of the intellectual endeavors of these ancient Greek pioneers.

Like art, music, or any other human creative activity, science makes use of intuition and experience. But the approach used by scientists to explore physical reality differs from other forms of intellectual endeavor in that it is based fundamentally on *observation*, *logic*, and *skepticism*. This approach, called the **scientific method**, requires that our ideas about the world around us be consistent with what we actually observe.

The scientific method goes something like this: A scientist trying to understand some phenomenon proposes a **hypothesis**, which is a collection of ideas that seems to explain the phenomenon. It is in developing hypotheses that scientists are at their most creative, imaginative, and intuitive. But their hypotheses must always agree with existing observations and experiments, because a discrepancy with what is observed implies that the hypothesis is wrong. (The exception is if the scientist thinks that the existing results are wrong and can give compelling evidence to show that they are wrong.) The scientist then uses logic to work out the implications of the hypothesis and to make predictions that can be tested. A hypothesis is on firm ground only after it has accurately forecast the results of new experiments or observations. (In prac-

tice, scientists typically go through these steps in a less linear fashion than we have described.)

Scientists describe reality in terms of **models**, which are hypotheses that have withstood observational or experimental tests. A model tells us about the properties and behavior of some object or phenomenon. A familiar example is a model of the atom, which scientists picture as electrons orbiting a central nucleus. Another example, which we will encounter in Chapter 18, is a model that tells us about physical conditions (for example, temperature, pressure, density) in the interior of the Sun. A well-developed model uses mathematics—one of the most powerful tools for logical thinking—to make detailed predictions. For example, a successful model of the Sun's interior should describe in detail how the temperature changes as you look deeper into the Sun. For this reason, mathematics is one of the most important tools used by scientists.

A body of related hypotheses can be pieced together into a self-consistent description of nature called a **theory**. An example from Chapter 5 is the theory of electromagnetism, which in just four equations describes all electric and magnetic phenomena. Without models and theories there is no understanding and no science, only collections of facts.

 In everyday language the word "theory" is often used to mean an idea that looks good on paper, but has little to do with reality. In science, however, a good theory is one that explains reality very well. An excellent example is the theory of electromagnetism, which does a superb job of describing a host of electric and magnetic phenomena. For instance, it predicts that a wave made of oscillating electric and magnetic fields should travel through empty space at the speed of light. Experiment shows that this assertion is exactly correct, and that light itself is a wave of just this type. (We will learn more about the nature of light in Chapter 5.)

Skepticism is an essential part of the scientific method. New hypotheses must be able to withstand the close scrutiny of other scientists. The more radical the hypothesis, the more skepticism and critical evaluation it will receive from the scientific community, because the general rule in science is that extraordinary claims require extraordinary evidence. That is why scientists as a rule do not accept claims that people have been abducted by aliens and taken aboard UFOs. The evidence presented for these claims is flimsy, secondhand, and unverifiable.

At the same time, scientists must be open-minded. They must be willing to discard long-held ideas if these ideas fail to agree with new observations and experiments, provided the new data have survived evaluation. (If an alien spacecraft really did land on Earth, scientists would be the first to accept that aliens existed—provided they could take a careful look at the spacecraft and its occupants.) That is why scientific knowledge is always provisional. As you go through this book, you will encounter many instances where new observations have transformed our understanding of Earth, the planets, the Sun and stars, and indeed the very structure of the universe.

Theories that accurately describe the workings of physical reality have a significant effect on civilization. For example, basing his conclusions in part on observations of how the planets orbit the Sun, the seventeenth-century scientist Isaac Newton deduced a set of fundamental mathematical laws that describe how *all* objects move. These laws, which we will encounter in Chapter 4, work equally well on Earth as in the most distant corner of the universe. They represent our first complete, coherent description of the behavior of the physical universe. **Newtonian mechanics** had an immediate practical application in the construction of machines, buildings, and bridges. It is no coincidence that the Industrial Revolution followed hard on the heels of these theoretical and mathematical advances inspired by astronomy.

Newtonian mechanics and other physical theories have stood the test of time and been shown to have great and general validity. These proven principles are collectively referred to as the **laws of physics**. Astronomers use these laws to interpret and understand their observations of the universe. The laws governing light and its relationship to matter are of particular importance, because the only information we can gather about distant stars and galaxies is in the light that we receive from them. Using the physical laws that describe how objects absorb and emit light, astronomers have measured the temperature of the Sun and even learned what the Sun is made of. By analyzing starlight in the same way, they have discovered that our own Sun is a rather ordinary star, and that the observable universe may contain billions of stars just like the Sun.

An important part of science is the development of new tools for research and new techniques of observation. As an example, until fairly recently everything we knew about the distant universe was based on visible light. Astronomers would peer through telescopes to observe and analyze visible starlight. By the end of the nineteenth century, however, scientists had begun discovering forms of light invisible to the human eye: X rays, gamma rays, radio waves, microwaves, and ultraviolet and infrared radiation.

As we will see in Chapter 6, in recent years astronomers have constructed telescopes that can detect such nonvisible forms of light (Figure 1-1). These instruments give us views of the universe vastly different from anything our eyes can see. These new views have allowed us to see through the atmo-

figure 1-1   R I **V** U X G

A Telescope in Space Because it orbits high above Earth in the vacuum of space, the Hubble Space Telescope (HST) can detect not only visible light but also ultraviolet and near-infrared light. These forms of nonvisible light are absorbed by our atmosphere and hence are difficult or impossible to detect with a telescope on the Earth's surface. This photo of HST was taken by the crew of the space shuttle *Discovery* after a servicing mission in 1997. (NASA)

spheres of distant planets, to study the thin but incredibly violent gas that surrounds our Sun, and even to observe new solar systems being formed around distant stars. Aided by high-technology telescopes, today's astronomers carry on the program of careful observation and logical analysis begun thousands of years ago by their ancient Greek predecessors.

1-2   By exploring the planets, astronomers uncover clues about the formation of the solar system

The science of astronomy is a way to let our intellects voyage across the cosmos. We can think of three stages in this voyage: from the Earth to the solar system, from the solar system to the stars, and from stars to galaxies and the grand scheme of the universe.

The star we call the Sun and all the celestial bodies that orbit the Sun—including Earth, the other eight planets, all their various moons, and smaller bodies such as asteroids and comets—make up the **solar system.** Since the 1960s a series of unmanned spacecraft has visited and explored all the planets except Pluto (Figure 1-2). Using the remote "eyes" of such spacecraft, we have flown over Mercury's cratered surface, peered beneath Venus's poisonous cloud cover, and discovered enormous canyons and extinct volcanoes on Mars. We have found active volcanoes on a moon of Jupiter, seen the rings of

figure 1-2 R I **V** U X G

The Sun and Planets to Scale
This montage of images from various spacecraft and ground-based telescopes shows the relative sizes of the nine planets and the Sun. The Sun is so large compared to the planets that only a portion of it fits into this illustration. The distances from the Sun to each planet are *not* shown to scale; the actual distance from the Sun to the Earth, for instance, is 12,000 times greater than the Earth's diameter. (Calvin J. Hamilton and NASA/JPL)

Saturn and Uranus up close, and looked down on the active atmosphere of Neptune.

Along with rocks brought back by the Apollo astronauts from the Moon (the only world beyond Earth yet visited by humans), this new information has revolutionized our understanding of the origin and evolution of the solar system. We have come to realize that many of the planets and their satellites were shaped by collisions with other objects. Craters on the Moon and on many other worlds are the relics of innumerable impacts by bits of interplanetary rock. The Moon may itself be the result of a catastrophic collision between the Earth and a planet-sized object shortly after the solar system was formed. Such a collision could have torn sufficient material from the primordial Earth to create the Moon.

The discoveries that we have made in our journeys across the solar system are directly relevant to the quality of human life on our own planet. Until recently, our understanding of geology, weather, and climate was based solely on data from the Earth. Since the advent of space exploration, however, we have been able to compare and contrast other worlds with our own. This new knowledge gives us valuable insight into our origins, the nature of our planetary home, and the limits of our natural resources.

1-3 By studying stars and nebulae, astronomers discover how stars are born, grow old, and die

The nearest of all stars to Earth is the Sun. Although humans have used the Sun's warmth since the dawn of our species, it was not until the 1920s and 1930s that physicists figured out how the Sun shines. At the center of the Sun, thermonuclear reactions convert hydrogen (the Sun's primary constituent) into helium. This violent process releases a vast amount of energy, which eventually makes its way to the Sun's surface and escapes as light (Figure 1-3). By 1950 physicists could reproduce such thermonuclear reactions here on Earth in the form of a hydrogen bomb (Figure 1-4). While such thermonuclear weapons are capable of destroy-

ing life on our planet, peaceful applications of this same process may provide a clean source of energy sometime in the next several decades.

All the stars you can see in the nighttime sky also shine by thermonuclear reactions. These reactions consume the material of which stars are made, which means that stars cannot last forever. Rather, they must form, evolve, and eventually die.

Much of what we know about the life stories of stars comes from the study of huge clouds of interstellar gas, called **nebulae** (singular **nebula**), that are found scattered across the sky. Within some nebulae, such as the Orion Nebula shown in Figure 1-5, stars are born from the material of the nebula

figure 1-3 R I **V** U X G

Our Star, the Sun The Sun is a typical star. Its diameter is about 1.39 million kilometers (roughly a million miles), and its surface temperature is about 5500°C (10,000°F). The Sun draws its energy from thermonuclear reactions occurring at its center, where the temperature is about 15 million degrees Celsius. (Celestron International)

Figure 1-4 R I **V** U X G

A Thermonuclear Explosion A hydrogen bomb uses the same physical principle as the thermonuclear reactions at the Sun's center: the conversion of matter into energy by nuclear reactions. This thermonuclear detonation on October 31, 1952, had an energy output equivalent to 10.4 million tons of TNT. This is more than 500 times the output of the bombs exploded over Hiroshima and Nagasaki in 1945, but a mere ten-billionth of the amount of energy released by the Sun in one second. (Defense Nuclear Agency)

itself. Other nebulae reveal what happens when a star dies after millions or billions of years. Some stars that are far more massive than the Sun end their lives with a spectacular detonation called a **supernova** (plural **supernovae**) that blows the star apart. The Crab Nebula (Figure 1-6) is a striking example of a remnant left behind by a supernova.

Dying stars can produce some of the strangest objects in the sky. Some dead stars become **pulsars**, which emit pulses of radio waves, or **X-ray bursters**, which emit powerful bursts of X rays. And some stars end their lives as almost inconceivably dense objects called **black holes**, which are surrounded by gravity so powerful that nothing—not even light—can escape. Even though a black hole itself emits essentially no radiation, a number of black holes have been discovered by Earth-orbiting telescopes. This is done by detecting the X rays emitted by gases falling toward a black hole.

 Astronomers often use biological terms such as "birth" and "death" to describe stages in the evolution of inanimate objects like stars. Keep in mind that such terms are used only as *analogies*, which help us visualize these stages. They are not to be taken literally!

During their death throes, stars return the gas of which they are made to interstellar space. (Figure 1-6 shows these expelled gases expanding away from the site of a supernova explosion.) This gas contains heavy elements created during the star's lifetime by thermonuclear reactions in its interior.

WEB LINK 1.4

Figure 1-5 R I **V** U X G

The Orion Nebula—Birthplace of Stars This beautiful nebula is a stellar "nursery" where stars are formed out of the nebula's gas. Intense ultraviolet light from newborn stars excites the surrounding gas and causes it to glow. Many of the stars embedded in this nebula are less than a million years old, a brief interval in the lifetime of a typical star. The Orion Nebula is some 1500 light-years from Earth and is about 30 light-years across. (© Anglo-Australian Observatory; photograph by David Malin)

WEB LINK 1.5

Figure 1-6 R I **V** U X G

The Crab Nebula—Wreckage of an Exploded Star When a dying star exploded in a supernova, it left behind this elegant funeral shroud of glowing gases blasted violently into space. A thousand years after the explosion these gases are still moving outward at about 1800 km/s (roughly 4 million miles per hour). The Crab Nebula is 6500 light-years from Earth and about 10 light-years across. (The FORS Team, VLT, European Southern Observatory)

Interstellar space thus becomes enriched with newly manufactured atoms and molecules. The Sun and its planets were formed from interstellar material that was enriched in this way. This means that the atoms of iron and nickel that make up the body of the Earth, as well as the carbon in our bodies and the oxygen we breathe, were created deep inside ancient stars. By studying stars and their evolution, we are really studying our own origins.

1-4 By observing galaxies, astronomers learn about the origin and fate of the universe

Stars are not spread uniformly across the universe but are grouped together in huge assemblages called **galaxies.** Galaxies come in a wide range of shapes and sizes. A typical galaxy, like the Milky Way, of which our Sun is part, contains several hundred billion stars. Some galaxies are much smaller, containing only a few million stars. Others are monstrosities that devour neighboring galaxies in a process called "galactic cannibalism."

Our Milky Way Galaxy has arching spiral arms like those of the galaxy shown in Figure 1-7. These arms are particularly active sites of star formation. In recent years, astronomers have discovered a mysterious object at the center of the Milky Way with a mass millions of times greater than that of our Sun. It now seems certain that this curious object is an enormous black hole.

Some of the most intriguing galaxies appear to be in the throes of violent convulsions and are rapidly expelling matter. The centers of these strange galaxies, which may harbor even more massive black holes, are often powerful sources of X rays and radio waves.

Even more awesome sources of energy are found still deeper in space. At distances so great that their light takes billions of years to reach Earth, we find the mysterious **quasars.** Although quasars look like nearby stars (Figure 1-8), they are among the most distant and most luminous objects in the sky. A typical quasar shines with the brilliance of a hundred galaxies. Recent observations of quasars imply that they draw their energy from material falling into enormous black holes.

The motions of distant clusters of galaxies reveal that we live in an expanding universe. Extrapolating into the past, we learn that the universe must have been born from an incredibly dense state (perhaps infinitely dense) some 10 to 15 billion years ago. A variety of evidence indicates that at that moment—the beginning of time—the universe began with a cosmic explosion, known as the **Big Bang,** which occurred

Figure 1-7 R I **V** U X G

A Galaxy This spectacular galaxy, called M100, contains several hundred billion stars. M100 has a diameter of about 120,000 light-years and is located about 56 million light-years from Earth. Along this galaxy's spiral arms you can see a number of glowing clumps. Like the Orion Nebula in our own Milky Way galaxy (Figure 1-5), these are sites of active star formation. (© Anglo-Australian Observatory; photograph by David Malin)

Figure 1-8 R I **V** U X G

A Quasar The two bright starlike objects in this image look almost identical, but they are dramatically different. The object on the left is indeed a star that lies a few hundred light-years from Earth. But the "star" on the right is actually a quasar about 9 billion light-years away. To appear so bright even though they are so distant, quasars like this one must be some of the most luminous objects in the universe. The other objects in this image are galaxies like that in Figure 1-7. (Charles Steidel, California Institute of Technology; and NASA) throughout all space. Events shortly after the Big Bang dictated the present nature of the universe.

Thanks to the combined efforts of astronomers and physicists, we are making steady advances in understanding these cosmic events. This understanding may reveal the origin of some of the most basic properties of physical reality. Studying the most remote galaxies may also answer questions about the ultimate fate of the universe. Will it continue expanding forever, or will it someday stop and collapse back in on itself? The work of unraveling the deepest mysteries of the universe requires specialized tools, including telescopes, spacecraft, and computers. But for many purposes the most useful device for studying the universe is the human brain itself. Our goal in this book is to help you use *your* brain to share in the excitement of scientific discovery.

In the remainder of this chapter we introduce some of the key concepts and mathematics that we will use in subsequent chapters. Study these carefully, for you will use them over and over again throughout your own study of astronomy.

1-5 Astronomers use angles to denote the positions and apparent sizes of objects in the sky

Whether they study planets, stars, galaxies, or the very origins of the universe, astronomers must know where to point their telescopes. For this reason, an important part of astronomy is keeping track of the positions of objects in the sky. *Angles* and a system of *angular measure* are essential tools for this aspect of astronomy.

An **angle** is the opening between two lines that meet at a point. **Angular measure** describes the size of an angle exactly. The basic unit of angular measure is the **degree**, designated by the symbol °. A full circle is divided into 360°, and a right angle measures 90° (Figure 1-9*a*). As Figure 1-9*b* shows, the angle between the two "pointer stars" in the Big Dipper—that is, the **angular distance** between these stars—is about 5°. (These two stars "point" to Polaris, or the North Star, as described in Chapter 2.) The angular distance between the stars that make up the top and bottom of the Southern Cross, which is visible from south of the equator, is about 6° (Figure 1-9*c*).

Astronomers also use angular measure to describe the apparent size of a celestial object—that is, what fraction of the sky that object seems to cover. For example, the angle covered by the diameter of the full moon is about ½° (Figure 1-9*a*). We therefore say that the **angular diameter** (or **angular size**) of the Moon is ½°. Alternatively, astronomers say that the Moon **subtends** an angle of ½°. Ten full moons could fit side by side between the two pointer stars in the Big Dipper.

The adult human hand held at arm's length provides a means of estimating angles, as Figure 1-10 shows. For example, your fist covers an angle of 10°, whereas the tip of your finger is about 1° wide. You can use various segments of your index finger extended to arm's length to estimate angles a few degrees across.

a

b

c

ϝigure 1-9

Angles and Angular Measure **(a)** Angles are measured in degrees (°). There are 360° in a complete circle and 90° in a right angle. For example, the angle between the vertical direction (directly above you) and the horizontal direction (toward the horizon) is 90°. The angular diameter of the full moon in the sky is about ½°. **(b)** The Big Dipper is an easily recognized grouping of seven bright stars, visible from anywhere in the northern hemisphere. The angular distance between the two pointer stars at the front of the Big Dipper is about 5°, roughly ten times the angular diameter of the Moon. **(c)** The four bright stars that make up the Southern Cross can be seen from anywhere in the southern hemisphere. The angular distance between the stars at the top and bottom of the cross is about 6°.

figure 1-10

Estimating Angles with Your Hand Various parts of the adult human hand extended to arm's length can be used to estimate angular distances and angular sizes in the sky.

To talk about smaller angles, we subdivide the degree into 60 **arcminutes** (also called minutes of arc), which is commonly abbreviated as 60 arcmin or 60′. An arcminute is further sub-divided into 60 **arcseconds** (or seconds of arc), usually written as 60 arcsec or 60″. Thus,

$$1° = 60 \text{ arcmin} = 60'$$
$$1' = 60 \text{ arcsec} = 60''$$

For example, on January 1, 2001, the planet Saturn had an angular diameter of 19.7 arcsec as viewed from Earth. That is a convenient, precise statement of how big the planet appeared in Earth's sky on that date. (Because this angular diameter is so small, to the naked eye Saturn appears simply as a point of light. To see any detail on Saturn, such as the planet's rings, requires a telescope.)

If we know the angular size of an object as well as the distance to that object, we can determine the actual linear size of the object (measured in kilometers or miles, for example). Box 1-1 describes how this is done.

1-6 Powers-of-ten notation is a useful shorthand system for writing numbers

Astronomy is a subject of extremes. Astronomers investigate the largest structures in the universe, including galaxies and clusters of galaxies. But they must also study atoms and

box 1-1 Tools of the Astronomer's Trade

The Small-Angle Formula

You can estimate the angular sizes of objects in the sky with your hand and fingers (Figure 1-10). Using rather more sophisticated equipment, astronomers can measure angular sizes to a fraction of an arcsecond. Keep in mind, however, that *angular* size is not the same as *actual* size. As an example, if you extend your arm while looking at a full moon, you can completely cover the Moon with your thumb. That's because from your perspective, your thumb has a larger angular size (that is, it subtends a larger angle) than the Moon. But the actual size of your thumb (about 2 centimeters) is much less than the actual diameter of the Moon (more than 3000 kilometers).

The accompanying figure shows how the angular size of an object is related to its linear size. Part **a** of the figure shows that for a given angular size, the more distant the object, the larger its actual size. For example, your thumb held at arm's length just covers the full moon; the angular size of your thumb and the Moon are about the same, but the Moon is much farther away and is far larger in linear size. Part **b** shows that for a given linear

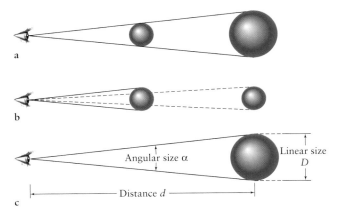

(a) Two objects that have the same angular size may have different linear sizes if they are at different distances from the observer. **(b)** For an object of a given linear size, the angular size is larger the closer the object is to the observer. **(c)** The small-angle formula relates the linear size D of an object to its angular size α and its distance d from the observer.

atomic nuclei, among the smallest objects in the universe, in order to explain how and why stars shine. They also study conditions ranging from the incredibly hot and dense centers of stars to the frigid near-vacuum of interstellar space. To describe such a wide range of phenomena, we need an equally wide range of both large and small numbers.

Astronomers avoid such confusing terms as "a million billion billion" by using a standard shorthand system called **powers-of-ten notation.** All the cumbersome zeros that accompany a large number are consolidated into one term consisting of 10 followed by an **exponent,** which is written as a superscript. The exponent indicates how many zeros you would need to write out the long form of the number. Thus,

$$10^0 = 1$$
$$10^1 = 10$$
$$10^2 = 100$$
$$10^3 = 1000$$
$$10^4 = 10,000$$

and so forth. The exponent also tells you how many tens must be multiplied together to give the desired number, which is why the exponent is also called the **power of ten.** For example, ten thousand can be written as 10^4 ("ten to the fourth" or "ten to the fourth power") because $10^4 = 10 \times 10 \times 10 \times 10 = 10,000$.

In powers-of-ten notation, numbers are written as a figure between one and ten multiplied by the appropriate power

of ten. The approximate distance between Earth and the Sun, for example, can be written as 1.5×10^8 kilometers (or 1.5×10^8 km for short). Once you get used to it, this is more convenient than writing "150,000,000 kilometers" or "one hundred and fifty million kilometers." (The same number could also be written as 15×10^7 or 0.15×10^9, but the preferred form is *always* to have the first figure be between 1 and 10.)

Note that most electronic calculators use a shorthand for powers-of-ten notation. To enter the number 1.5×10^8, you first enter 1.5, then press a key labeled "EXP" or "EE," then enter the exponent 8. (The EXP or EE key takes care of the "× 10" part of the expression.) The number will then appear on your calculator's display as "1.5 E 8," "1.5 8," or some variation of this; typically the "× 10" is not displayed as such. There are some variations from one kind of calculator to another, so you should spend a few minutes reading over your calculator's instruction manual to make sure you know the correct procedure for working with numbers in powers-of-ten notation. You will be using this notation continually in your study of astronomy, so this is time well spent.

 Confusion can result from the way that calculators display powers-of-ten notation. Since 1.5×10^8 is displayed as "1.5 8" or "1.5 E 8," it is not uncommon to think that 1.5×10^8 is the same as 1.5^8. That is not

---

size, the angular size decreases the farther away the object. This is why a car looks smaller and smaller as it drives away from you.

We can put these relationships together into a single mathematical expression called the **small-angle formula.** Suppose that an object subtends an angle α (the Greek letter alpha) and is at a distance d from the observer, as in part **c** of the figure. If the angle α is small, as is almost always the case for objects in the sky, the linear size (D) of the object is given by the following expression:

The small-angle formula

$$D = \frac{\alpha d}{206,265}$$

D = linear size of an object
α = angular size of the object, in arcsec
d = distance to the object

The number 206,265 is required in the formula. Mathematically, it is equal to the number of arcseconds in a complete circle (that is, 360°) divided by the number 2π (the ratio of the circumference of a circle to that circle's radius).

EXAMPLE: On November 28, 2000, Jupiter was 609 million kilometers from Earth. Jupiter's angular diameter on that date was 48.6 arcseconds. Using the small-angle formula, we can calculate Jupiter's diameter as follows:

$$D = \frac{48.6 \times 609,000,000 \text{ km}}{206,265} = 143,000 \text{ km}$$

This answer agrees very well with an equatorial diameter of 142,984 km determined from spacecraft flybys.

EXAMPLE: Under excellent conditions, a telescope on Earth can see details with an angular size as small as 1 arcsec. We can use the small-angle formula to determine the greatest distance at which you could see details as small as 1.7 m (the height of a typical person) under these conditions. We rewrite the formula to solve for the distance d:

$$d = \frac{206,265 \ D}{\alpha} = \frac{206,265 \times 1.7 \text{ m}}{1} = 350,000 \text{ m} = 350 \text{ km}$$

This is much less than the distance to the Moon, which is 384,000 km. Thus, even the best telescope on Earth could not be used to see an astronaut walking on the surface of the Moon.

correct, however; 1.5^8 is equal to 1.5 multiplied by itself eight times, or 25.63, which is not even close to $150,000,000 = 1.5 \times 10^8$. Another, not uncommon, mistake is to write 1.5×10^8 as 15^8. If you are inclined to do this, perhaps you are thinking that you can multiply 1.5 by 10, then tack on the exponent later. This also does not work; 15^8 is equal to 15 multiplied by itself eight times, or 2,562,890,625, which again is nowhere near 1.5×10^8. Reading over the manual for your calculator will help you to avoid these common errors.

You can use powers-of-ten notation for numbers that are less than one by using a minus sign in front of the exponent. A negative exponent tells you to *divide* by the appropriate number of tens. For example, 10^{-2} ("ten to the minus two") means to divide by 10 twice, so $10^{-2} = \frac{1}{10} \times \frac{1}{10} = \frac{1}{100} = 0.01$. This same idea tells us how to interpret other negative powers of ten:

$$10^0 = 1$$
$$10^{-1} = \frac{1}{10} = 0.1$$
$$10^{-2} = \frac{1}{10} \times \frac{1}{10} = \frac{1}{10^2} = 0.01$$
$$10^{-3} = \frac{1}{10} \times \frac{1}{10} \times \frac{1}{10} = \frac{1}{10^3} = 0.001$$
$$10^{-4} = \frac{1}{10} \times \frac{1}{10} \times \frac{1}{10} \times \frac{1}{10} = \frac{1}{10^4} = 0.0001$$

and so forth.

As these examples show, negative exponents tell you how many tenths must be multiplied together to give the desired number. For example, one ten–thousandth, or 0.0001, can be written as 10^{-4} ("ten to the minus four") because $10^{-4} = \frac{1}{10} \times \frac{1}{10} \times \frac{1}{10} \times \frac{1}{10} = 0.0001$.

A useful shortcut in converting a decimal to powers-of-ten notation is to notice where the decimal point is. For example, the decimal point in 0.0001 is four places to the left of the "1," so the exponent is –4, that is, $0.0001 = 10^{-4}$.

You can also use powers-of-ten notation to express a number like 0.00245, which is not a multiple of $\frac{1}{10}$. For example, $0.00245 = 2.45 \times 0.001 = 2.45 \times 10^{-3}$. (Again, the standard for powers-of-ten notation is that the first figure is a number between one and ten.) This notation is particularly useful when dealing with very small numbers. A good example is the diameter of a hydrogen atom, which is much more convenient to state in powers-of-ten notation (1.1×10^{-10} meter, or 1.1×10^{-10} m) than as a decimal (0.00000000011 m) or a fraction (110 trillionths of a meter.)

Powers-of-ten notation lets us write familiar numerical terms in a compact way. For example,

$$\text{one hundred} = 100 = 10^2$$
$$\text{one thousand} = 1000 = 10^3$$
$$\text{one million} = 1{,}000{,}000 = 10^6$$
$$\text{one billion} = 1{,}000{,}000{,}000 = 10^9$$
$$\text{one trillion} = 1{,}000{,}000{,}000{,}000 = 10^{12}$$

and also

$$\text{one one-hundredth} = 0.01 = 10^{-2}$$
$$\text{one one-thousandth} = 0.001 = 10^{-3}$$
$$\text{one one-millionth} = 0.000001 = 10^{-6}$$
$$\text{one one-billionth} = 0.000000001 = 10^{-9}$$
$$\text{one one-trillionth} = 0.000000000001 = 10^{-12}$$

Because it bypasses all the awkward zeros, powers-of-ten notation is ideal for describing the size of objects as small as atoms or as big as galaxies (Figure 1-11). Box 1-2 explains

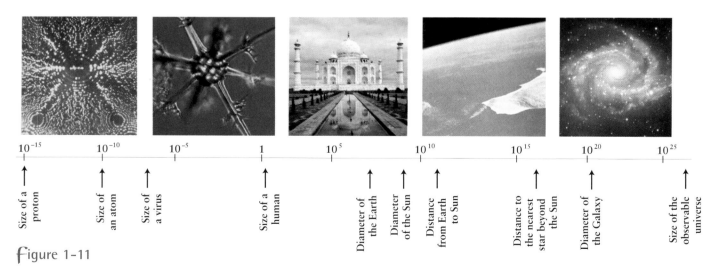

figure 1-11

Examples of Powers-of-Ten Notation The scale gives the sizes of objects in meters, ranging from subatomic particles at the left to the entire observable universe on the right. The photograph at the left shows tungsten atoms, 10^{-10} meter in diameter. Second from left is the crystalline skeleton of a diatom (a single-celled organism), 10^{-4} meter (0.1 millimeter) in size. At the center is the Taj Mahal, about 60 meters tall and within reach of our unaided senses. On the right, looking across the Indian Ocean toward the South Pole, we see the curvature of the Earth, about 10^7 meters in diameter. At the far right is a galaxy, 10^{21} meters (100,000 light-years) in diameter. (Courtesy of Scientific American Books; NASA; and photograph by David Malin from the Anglo-Australian Observatory)

box 1-2 | Tools of the Astronomer's Trade

Arithmetic with Powers-of-Ten Notation

Using powers-of-ten notation makes it easy to multiply numbers. For example, suppose you want to multiply 100 by 1000. If you use ordinary notation, you have to write a lot of zeros:

$$100 \times 1000 = 100,000 \text{ (one hundred thousand)}$$

By converting these numbers to powers-of-ten notation, we can write this same multiplication more compactly as

$$10^2 \times 10^3 = 10^5$$

Because 2 + 3 = 5, we are led to the following general rule for *multiplying* numbers expressed in terms of powers of ten: Simply *add* the exponents.

EXAMPLE:

$$10^4 \times 10^3 = 10^{4+3} = 10^7$$

To *divide* numbers expressed in terms of powers of ten, remember that $10^{-1} = \frac{1}{10}$, $10^{-2} = \frac{1}{100}$, and so on. The general rule for any exponent n is

$$10^{-n} = \frac{1}{10^n}$$

In other words, dividing by 10^n is the same as multiplying by 10^{-n}. To carry out a division, you first transform it into multiplication by changing the sign of the exponent, and then carry out the multiplication by adding the exponents.

EXAMPLE:

$$\frac{10^4}{10^6} = 10^4 \times 10^{-6} = 10^{4+(-6)} = 10^{4-6} = 10^{-2}$$

Usually a computation involves numbers like 3.0×10^{10}, that is, an ordinary number multiplied by a factor of 10 with an exponent. In such cases, to perform multiplication or division, you can treat the numbers separately from the factors of 10^n.

EXAMPLE: We can redo the first numerical example from Box 1-1 in a straightforward manner by using exponents:

$$D = \frac{48.6 \times 609,000,000 \text{ km}}{206,265}$$

$$= \frac{4.86 \times 10 \times 6.09 \times 10^8}{2.06265 \times 10^5} \text{ km}$$

$$= \frac{4.86 \times 6.09 \times 10^{1+8-5}}{2.06265} \text{ km} = 14.3 \times 10^4 \text{ km}$$

$$= 1.43 \times 10 \times 10^4 \text{ km} = 1.43 \times 10^5 \text{ km}$$

how powers-of-ten notation also makes it easy to multiply and divide numbers that are very large or very small.

1-7 Astronomical distances are often measured in astronomical units, parsecs, or light-years

Astronomers use many of the same units of measurement as do other scientists. They often measure lengths in meters (abbreviated m), masses in kilograms (kg), and time in seconds (s). (You can read more about these units of measurement, as well as techniques for converting between different sets of units, in Box 1-3.) Like other scientists, astronomers often find it useful to combine these units with powers of ten and create new units using prefixes. As an example, the number 1000 ($= 10^3$) is represented by the prefix "kilo," and so a distance of 1000 meters is the same as 1 kilometer (1 km).

Here are some of the most common prefixes, with examples of how they are used:

$$\text{one-billionth meter} = 10^{-9} \text{ m} = 1 \text{ nanometer}$$
$$\text{one-millionth second} = 10^{-6} \text{ s} = 1 \text{ microsecond}$$
$$\text{one-thousandth arcsecond} = 10^{-3} \text{ arcsec} = 1 \text{ milliarcsecond}$$
$$\text{one-hundredth meter} = 10^{-2} \text{ m} = 1 \text{ centimeter}$$
$$\text{one thousand meters} = 10^3 \text{ m} = 1 \text{ kilometer}$$
$$\text{one million tons} = 10^6 \text{ tons} = 1 \text{ megaton}$$

In principle, we could express all sizes and distances in astronomy using units based on the meter. Indeed, we will use kilometers to give the diameters of the Earth and Moon, as well as the Earth-Moon distance. But, while a kilometer (roughly equal to three-fifths of a mile) is an easy distance for humans to visualize, a megameter (10^6 m) is not. For this reason, astronomers have devised units of measure that are more appropriate for the tremendous distances between the planets and the far greater distances between the stars.

box 1-3 Tools of the Astronomer's Trade

Units of Length, Time, and Mass

To understand and appreciate the universe, we need to describe phenomena not only on the large scales of galaxies but also on the submicroscopic scale of the atom. Astronomers generally use units that are best suited to the topic at hand. For example, interstellar distances are conveniently expressed in either light-years or parsecs, whereas the diameters of the planets are more comfortably presented in kilometers.

Most scientists prefer to use a version of the metric system called the International System of Units, abbreviated **SI** (after the French name Système International). In SI units, length is measured in meters (m), time is measured in seconds (s), and mass (a measure of the amount of material in an object) is measured in kilograms (kg). How are these basic units related to other measures?

When discussing objects on a human scale, sizes and distances are usually expressed in millimeters (mm), centimeters (cm), and kilometers (km). These units of length are related to the meter as follows:

$$1 \text{ millimeter} = 0.001 \text{ m} = 10^{-3} \text{ m}$$
$$1 \text{ centimeter} = 0.01 \text{ m} = 10^{-2} \text{ m}$$
$$1 \text{ kilometer} = 1000 \text{ m} = 10^{3} \text{ m}$$

Although the English system of inches (in.), feet (ft), and miles (mi) is much older than SI, today the English system is actually based on the SI system: The inch is defined to be exactly 2.54 cm. A useful set of conversions is

$$1 \text{ in.} = 2.54 \text{ cm}$$
$$1 \text{ ft} = 0.3048 \text{ m}$$
$$1 \text{ mi} = 1.609 \text{ km}$$

Each of these equalities can also be written as a fraction equal to 1. For example, you can write

$$\frac{0.3048 \text{ m}}{1 \text{ ft}} = 1$$

Fractions like this are useful for converting a quantity from one set of units to another. For example, the *Saturn V* rocket used to send astronauts to the Moon stands about 363 feet tall. How can we convert this height to meters? The trick is to remember that a quantity does not change if you multiply it by 1. Expressing the number 1 by the fraction (0.3048 m/1 ft), we can write the height of the rocket as

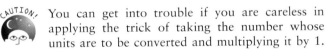

$$363 \text{ ft} \times 1 = 363 \text{ ft} \times \frac{0.3048 \text{ m}}{1 \text{ ft}} = 111 \frac{\text{ft} \times \text{m}}{\text{ft}} = 111 \text{ m}$$

EXAMPLE: The diameter of Mars is 6794 km. Let's try expressing this in miles.

CAUTION! You can get into trouble if you are careless in applying the trick of taking the number whose units are to be converted and multiplying it by 1. For example, if we multiply the diameter by 1 expressed as (1.609 km)/(1 mi), we get

$$6794 \text{ km} \times 1 = 6794 \text{ km} \times \frac{1.609 \text{ km}}{1 \text{ mi}} = 10{,}930 \frac{\text{km}^2}{\text{mi}}$$

The unwanted units of km did not cancel, so this cannot be right. Furthermore, a mile is larger than a

---

When discussing distances across the solar system, astronomers use a unit of length called the **astronomical unit** (abbreviated AU). This is the average distance between Earth and the Sun:

$$1 \text{ AU} = 1.496 \times 10^{8} \text{ km} = 92.96 \text{ million miles}$$

Thus, the average distance between the Sun and Jupiter can be conveniently stated as 5.2 AU.

 To talk about distances to the stars, astronomers use two different units of length. The **light-year** (abbreviated ly) is the distance that light travels in one year. This is a useful concept because the speed of light in empty space always has the same value, 3.00×10^{5} km/s (kilometers per second) or 1.86×10^{5} mi/s (miles per second).

In terms of kilometers or astronomical units, one light-year is given by

$$1 \text{ ly} = 9.46 \times 10^{12} \text{ km} = 63{,}240 \text{ AU}$$

This distance is roughly equal to 6 trillion miles.

CAUTION! Keep in mind that despite its name, the light-year is a unit of distance and not a unit of time. As an example, Proxima Centauri, the nearest star other than the Sun, is a distance of 4.2 light-years from Earth. This means that light takes 4.2 years to travel to us from Proxima Centauri.

Physicists often measure interstellar distances in light-years because the speed of light is one of nature's most imporkilometer, so the diameter expressed in miles should be a smaller number than when expressed in kilometers.

The correct approach is to write the number 1 so that the unwanted units *will* cancel. The number we are starting with is in kilometers, so we must write the number 1 with kilometers in the denominator ("downstairs" in the fraction). Thus, we express 1 as (1 mi)/(1.609 km):

$$6794 \text{ km} \times 1 = 6794 \text{ km} \times \frac{1 \text{ mi}}{1.609 \text{ km}}$$

$$= 4222 \text{ km} \times \frac{\text{mi}}{\text{km}} = 4222 \text{ mi}$$

Now the units of km cancel as they should, and the distance in miles is a smaller number than in kilometers (as it must be).

When discussing very small distances such as the size of an atom, astronomers often use the micrometer (μm) or the nanometer (nm). These are related to the meter as follows:

$$1 \text{ micrometer} = 1 \text{ μm} = 10^{-6} \text{ m}$$
$$1 \text{ nanometer} = 1 \text{ nm} = 10^{-9} \text{ m}$$

Thus, 1 μm = 10^3 nm. (Note that the micrometer is often called the micron.)

The basic unit of time is the second (s). It is related to other units of time as follows:

$$1 \text{ minute (min)} = 60 \text{ s}$$
$$1 \text{ hour (h)} = 3600 \text{ s}$$
$$1 \text{ day (d)} = 86,400 \text{ s}$$
$$1 \text{ year (y)} = 3.156 \times 10^7 \text{ s}$$

In the SI system, speed is properly measured in meters per second (m/s). Quite commonly, however, speed is also expressed in km/s and mi/h:

$$1 \text{ km/s} = 10^3 \text{ m/s}$$
$$1 \text{ km/s} = 2237 \text{ mi/h}$$
$$1 \text{ mi/h} = 0.447 \text{ m/s}$$
$$1 \text{ mi/h} = 1.47 \text{ ft/s}$$

In addition to using kilograms, astronomers sometimes express mass in grams (g) and in solar masses ($M_\odot$), where the subscript ⊙ is the symbol denoting the Sun. It is especially convenient to use solar masses when discussing the masses of stars and galaxies. These units are related to each other as follows:

$$1 \text{ kg} = 1000 \text{ g}$$
$$1 \text{ M}_\odot = 1.99 \times 10^{30} \text{ kg}$$

 CAUTION! You may be wondering why we have not given a conversion between kilograms and pounds. The reason is that these units do not refer to the same physical quantity! A kilogram is a unit of *mass*, which is a measure of the amount of material in an object. By contrast, a pound is a unit of *weight*, which tells you how strongly gravity pulls on that object's material. Consider a person who weighs 110 pounds on Earth, corresponding to a mass of 50 kg. Gravity is only about one-sixth as strong on the Moon as it is on Earth, so on the Moon this person would weigh only one-sixth of 110 pounds, or about 18 pounds. But that person's mass of 50 kg is the same on the Moon; wherever you go in the universe, you take all of your material along with you. We will explore the relationship between mass and weight in Chapter 4.

tant numbers. But many astronomers prefer to use another unit of length, the *parsec*, because its definition is closely related to a method of measuring distances to the stars.

Imagine taking a journey far into space, beyond the orbits of the outer planets. As you look back toward the Sun, Earth's orbit subtends a smaller angle in the sky the farther you are from the Sun. As Figure 1-12 shows, the distance at which 1 AU subtends an angle of 1 arcsec is defined as 1 **parsec** (abbreviated pc):

$$1 \text{ pc} = 3.09 \times 10^{13} \text{ km} = 3.26 \text{ ly}$$

The distance to Proxima Centauri can be stated as 1.3 pc as well as 4.2 ly. Whether you choose to use parsecs or light-years is a matter of personal taste.

For even greater distances, astronomers commonly use **kiloparsecs** and **megaparsecs** (abbreviated kpc and Mpc). As we saw before, these prefixes simply mean "thousand" and "million," respectively:

$$1 \text{ kiloparsec} = 1 \text{ kpc} = 1000 \text{ pc} = 10^3 \text{ pc}$$
$$1 \text{ megaparsec} = 1 \text{ Mpc} = 1,000,000 \text{ pc} = 10^6 \text{ pc}$$

For example, the distance from Earth to the center of our Milky Way Galaxy is about 8 kpc, and the galaxy shown in Figure 1-7 is about 17 Mpc away.

Some astronomers prefer to talk about thousands or millions of light-years rather than kiloparsecs and megaparsecs. Once again, the choice is a matter of personal taste. As a general rule, astronomers use whatever yardsticks seem best

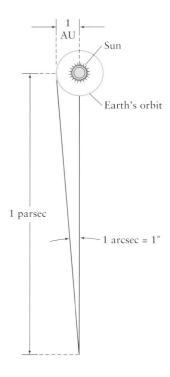

figure 1-12

A Parsec The parsec, a unit of length commonly used by astronomers, is equal to 3.26 light-years. The parsec is defined as the distance at which 1 AU perpendicular to the observer's line of sight subtends an angle of 1 arcsec.

suited for the issue at hand and do not restrict themselves to one system of measurement. For example, an astronomer might say that the supergiant star Antares has a diameter of 860 million kilometers and is located at a distance of 185 parsecs from Earth.

1-8 Astronomy is an adventure of the human mind

An underlying theme of this book is that the universe is rational. It is not a hodgepodge of unrelated things behaving in unpredictable ways. Rather, we find strong evidence that fundamental laws of physics govern the nature of the universe and the behavior of everything in it. These unifying concepts enable us to explore realms far removed from our earthly experience. Thus, a scientist can do experiments in a laboratory to determine the properties of light or the behavior of atoms and then use this knowledge to investigate the structure of the universe.

The discovery of fundamental laws of nature has had a profound influence on humanity. These laws have led to an immense number of practical applications that have fundamentally transformed commerce, medicine, entertainment, transportation, and other aspects of our lives. In particular,

space technology has given us instant contact with any point on the globe through communication satellites. Space technology has also made possible accurate weather forecasts from meteorological satellites, precise navigation to any point on Earth using signals from the satellites of the Global Positioning System (GPS), and long-term monitoring of Earth's climate and environment from orbit (Figure 1-13).

As important as the applications of science are, the pursuit of scientific knowledge for its own sake is no less important. We are fortunate to live in an age in which this pursuit is in full flower. Just as explorers such as Columbus and Magellan discovered the true size of our planet in the fifteenth and sixteenth centuries, astronomers of the twenty-first century are exploring the universe to an extent that is unparalleled in human history. Indeed, even the voyages into space imagined by such great science fiction writers as Jules Verne and H. G. Wells pale in comparison to today's reality. Over a few short

figure 1-13

The Earth's Oceans from Space The *TOPEX/Poseidon* satellite (a joint project of the United States and France) measures variations in the height of the oceans. Areas of warm water bulge upward by several centimeters, so these measurements track changes in ocean temperature. In this map made from *TOPEX/Poseidon* data, elevated areas of warm water appear in white and red while depressed areas of cool water appear in blue and purple. Although the temperature variations are no more than a few degrees above or below normal, they have a profound effect on Earth's global climate. By monitoring these variations, scientists can give early warning of the potential for floods, droughts, and other severe weather. The savings in lives and property that result from these early warnings more than pay for the cost of the satellite. (NASA/JPL) decades, humans have walked on the Moon, sent robot spacecraft to dig into the Martian soil and explore the satellites of Jupiter, and used the most powerful telescopes ever built to probe the limits of the observable universe. Never before has so much been revealed in so short a time.

As you proceed through this book, you will learn about the tools that scientists use to explore the natural world, as well as what they observe with these tools. But, most important, you will see how astronomers build from their observations an understanding of the universe in which we live. It is this search for understanding that makes science more than merely a collection of data and elevates it to one of the great adventures of the human mind. It is an adventure that will continue as long as there are mysteries in the universe—an adventure we hope you will come to appreciate and share.

KEY WORDS

Terms preceded by an asterisk (*) are discussed in the Boxes.

angle, p. 7

angular diameter (angular size), p. 7

angular distance, p. 7

angular measure, p. 7

arcminute (′, minute of arc), p. 8

arcsecond (″, second of arc), p. 8

astronomical unit (AU), p. 12

Big Bang, p. 6

black hole, p. 5

degree (°), p. 7

exponent, p. 9

galaxy, p. 6

hypothesis, p. 2

kiloparsec (kpc), p. 13

laws of physics, p. 3

light-year (ly), p. 12

megaparsec (Mpc), p. 13

model, p. 2

nebula (*plural* nebulae), p. 4

Newtonian mechanics, p. 3

parsec (pc), p. 13

power of ten, p. 9

powers-of-ten notation, p. 9

pulsar, p. 5

quasar, p. 6

scientific method, p. 2

*SI units, p. 12

*small-angle formula, p. 9

solar system, p. 3

subtend (an angle), p. 7

supernova (*plural* supernovae), p. 5

theory, p. 2

X-ray burster, p. 5

KEY IDEAS

Astronomy, Science, and the Nature of the Universe: The universe is comprehensible. The scientific method is a procedure for formulating hypotheses about the universe. These are tested by observation or experimentation in order to build consistent models or theories that accurately describe phenomena in nature.

• Observations of the heavens have helped scientists discover some of the fundamental laws of physics. The laws of physics are in turn used by astronomers to interpret their observations.

The Solar System: Exploration of the planets provides information about the origin and evolution of the solar system, as well as the history and resources of Earth.

Stars and Nebulae: Study of the stars and nebulae helps us learn about the origin and history of the Sun and the solar system.

Galaxies: Observations of galaxies tell us about the origin and history of the universe.

Angular Measure: Astronomers use angles to denote the positions and sizes of objects in the sky. The size of an angle is measured in degrees, arcminutes, and arcseconds.

Powers-of-ten notation is a convenient shorthand system for writing numbers. It allows very large and very small numbers to be expressed in a compact form.

Units of Distance: A variety of distance units are used by astronomers. These include the astronomical unit (the average distance from Earth to the Sun), the light-year (the distance that light travels in one year), and the parsec.

REVIEW QUESTIONS

1. Describe the role that skepticism plays in science.

2. What is the difference between a theory and a law of physics?

3. What caused the craters on the Moon?

4. What role do nebulae like the Orion Nebula play in the life stories of stars?

5. What is the difference between a solar system and a galaxy?

6. What are degrees, arcminutes, and arcseconds used for? What are the relationships among these units of measure?

7. How many arcseconds equal 1°?

8. With the aid of a diagram, explain what it means to say that the Moon subtends an angle of ½°.

9. What is an exponent? How are exponents used in powers-of-ten notation?

10. What are the advantages of using powers-of-ten notation?

11. Write the following numbers using powers-of-ten notation: (a) one hundred million, (b) forty thousand, (c) three one-thousandths, (d) twenty-two billion, (e) your age in years.

12. How is an astronomical unit (AU) defined? Give an example of a situation in which this unit of measure would be convenient to use.

13. What is the advantage to the astronomer of using the light-year as a unit of distance?

14. What is a parsec? How is it related to a kiloparsec and a megaparsec?

15. Give the word or phrase that corresponds to the following standard abbreviations: (a) km, (b) cm, (c) s, (d) km/s, (e) mi/h, (f) m, (g) m/s, (h) h, (i) y, (j) g, (k) kg. Which of these are units of speed? (*Hint:* You may have to refer to a dictionary. All of these abbreviations should be part of your working vocabulary.)

16. In the original (1977) *Star Wars* movie, Han Solo praises the speed of his spaceship by saying, "It's the ship that made the Kessel run in less than 12 parsecs!" Explain why this statement is obvious misinformation.

17. A reporter once described a light-year as "the time it takes light to reach us traveling at the speed of light." How would you correct this statement?

ADVANCED QUESTIONS

Questions preceded by an asterisk (*) involve topics discussed in the Boxes.

Problem-solving tips and tools

The small-angle formula, given in Box 1-1, relates the size of an astronomical object to the angle it subtends. Box 1-3 illustrates how to convert from one unit of measure to another. An object traveling at speed v for a time t covers a distance d given by $d = vt$; for example, a car traveling at 90 km/h (v) for 3 hours (t) covers a distance $d = (90 \text{ km/h})(3 \text{ h}) = 270$ km. Similarly, the time t required to cover a given distance d at speed v is $t = d/v$; for example, if $d = 270$ km and $v = 90$ km/h, then $t = (270 \text{ km})/(90 \text{ km/h}) = 3$ hours.

18. What is the meaning of the letters R I V U X G that appear under some of the figures in this chapter? Why in each case is one of the letters highlighted? (*Hint:* See the section "How to Use This Textbook" that precedes Chapter 1.)

19. A hydrogen atom has a radius of about 5×10^{-9} cm. The radius of the observable universe is about 15 billion light-years. How many times larger than a hydrogen atom is the observable universe? Use powers-of-ten notation.

20. The Sun's mass is 1.99×10^{30} kg, three-quarters of which is hydrogen. The mass of a hydrogen atom is 1.67×10^{-27} kg. How many hydrogen atoms does the Sun contain? Use powers-of-ten notation.

21. The diameter of the Sun is 1.4×10^{11} cm, and the distance to the nearest star, Proxima Centauri, is 4.2 ly. Suppose you want to build an exact scale model of the Sun and Proxima Centauri, and you are using a basketball 30 cm in diameter to represent the Sun. In your scale model, how far away would Proxima Centauri be from the Sun? Give your answer in kilometers, using powers-of-ten notation.

22. How many Suns would it take, laid side by side, to reach the nearest star? Use powers-of-ten notation. (*Hint:* See the preceding question.)

23. The average distance from the Earth to the Sun is 1.496×10^8 km. Express this distance (a) in light-years and (b) in parsecs. Use powers-of-ten notation. (c) Are light-years or parsecs useful units for describing distances of this size? Explain.

24. The speed of light is 3.00×10^8 m/s. How long does it take light to travel from the Sun to Earth? Give your answer in seconds, using powers-of-ten notation.

25. When *Voyager 2* sent back pictures of Neptune during its historic flyby of that planet in 1989, the spacecraft's radio signals traveled for 4 hours at the speed of light to reach Earth. How far away was the spacecraft? Give your answer in kilometers, using powers-of-ten notation. (*Hint:* See the preceding question.)

26. The star Epsilon Eridani is 3.22 pc from Earth. (a) What is the distance to Epsilon Eridani in kilometers? Use powers-of-ten notation. (b) How long does it take for light emanating from Epsilon Eridani to reach Earth? Give your answer in years. (*Hint:* You do not need to know the value of the speed of light.)

27. The age of the universe is about 15 billion years. What is this age in seconds? Use powers-of-ten notation.

*28. Explain where the number 206,265 in the small-angle formula comes from.

*29. At what distance would a person have to hold a nickel (which has a diameter of about 2.0 cm) in order for the nickel to subtend an angle of (a) 1°? (b) 1 arcmin? (c) 1 arcsec? Give your answers in meters.

Object	Distance (km)	Angular size (″)
Sun	1.5×10^8	1800
Saturn	1.5×10^9	16.5
Pluto	6.3×10^9	0.06

*30. The average distance to the Moon is 384,000 km, and the Moon subtends an angle of ½°. Use this information to calculate the diameter of the Moon in kilometers.

*31. Suppose your telescope can give you a clear view of objects and features that subtend angles of at least 2 arcsec. What is the diameter of the smallest crater you can see on the Moon? (*Hint:* See the preceding question.)

*32. A person with good vision can see details that subtend an angle of as small as 1 arcminute. If two dark lines on an eye chart are 2 millimeters apart, how far can such a person be from the chart and still be able to tell that there are two distinct lines? Give your answer in meters.

*33. On December 11, 2000, the planet Venus was at a distance of 0.951 AU from Earth. The diameter of Venus is 12,104 km. What was the angular size of Venus as seen from Earth on December 11, 2000? Give your answer in arcminutes.

DISCUSSION QUESTIONS

34. Scientists assume that "reality is rational." Discuss what this means and the thinking behind it.

35. How do astronomical observations differ from those of other sciences?

WEB/CD-ROM QUESTIONS

36. Use the links given in the *Universe* web site, Chapter 1, to learn about the Orion Nebula (Figure 1-5). Can the nebula be seen with the naked eye? Does the nebula stand alone, or is it part of a larger cloud of interstellar material? What has been learned by examining the Orion Nebula with telescopes sensitive to infrared light?

37. Use the links given in the *Universe* web site, Chapter 1, to learn more about the Crab Nebula (Figure 1-6). When did observers on Earth see the supernova that created this nebula? Does the nebula emit any radiation other than visible light? What kind of object is at the center of the nebula?

38. Use the links given in the *Universe* web site, Chapter 1, to learn more about *TOPEX/Poseidon* (see Figure 1-13) and other space missions that study the Earth. What are an "El Niño" and a "La Niña"? How can they be detected from space? How can the speed of winds over the oceans be measured by an orbiting satellite?

39. Access the AIMM (Active Integrated Media Module) called "Small-Angle Toolbox" in Chapter 1 of the *Universe* CD-ROM or web site. Use this to determine the diameters in kilometers of the Sun, Saturn, and Pluto given the following distances and angular sizes:

OBSERVING PROJECTS

40. On a dark, clear, moonless night, can you see the Milky Way from where you live? If so, briefly describe its appearance. If not, what seems to be interfering with your ability to see the Milky Way?

41. Look up at the sky on a clear, cloud-free night. Is the Moon in the sky? If so, does it interfere with your ability to see the fainter stars? Why do you suppose astronomers prefer to schedule their observations on nights when the Moon is not in the sky?

42. Look up at the sky on a clear, cloud-free night and note the positions of a few prominent stars relative to such reference markers as rooftops, telephone poles, and treetops. Also note the location from where you make your observations. A few hours later, return to that location and again note the positions of the same bright stars that you observed earlier. How have their positions changed? From these changes, can you deduce the general direction in which the stars appear to be moving?

43. Use the CD-ROM that accompanies this book to install the *Starry Night* planetarium software on your computer. Use *Starry Night* to determine when the Moon is visible today during the day and when it is visible tonight. Determine which, if any, of the following planets are visible tonight: Mercury, Venus, Mars, Jupiter, and Saturn. *Hints:* (1) Feel free to experiment with *Starry Night*. You can always return to your starting screen by clicking on the "Home" button in the Control Panel at the top of the main window. To see the sky from your actual location on Earth, select **Set Home Location...** in the **Go** menu and click on the **Lookup...** button to find your city or town. (2) To change your viewing direction, move the mouse until the cursor changes into a little hand. Then, hold down the mouse button (on a Windows computer, the left button) as you move the mouse and you will move the sky. (3) Use the Control Panel at the top of the main window to change the time and date that is displayed as well as how rapidly time appears to change. (4) Use the **Find...** command in the **Edit** menu to locate specific planets or stars by name. (5) To get more information about any object in the sky, point the cursor at the object and double-click the mouse (on a Windows computer, click the right button).

You can find even more information about the program at the *Starry Night* web site.

SANDRA M. FABER

Why Astronomy?

Sandra M. Faber, professor of astronomy at the University of California, Santa Cruz, and astronomer at Lick Observatory, was intrigued by the origins of the universe when she entered Swarthmore College. She completed her doctorate in astronomy at Harvard University but did her dissertation at the Carnegie Institution's Department of Terrestrial Magnetism, where she was influenced by the distinguished astronomer Vera Rubin. Dr. Faber chaired the now legendary group of astronomers called the Seven Samurai, who surveyed the nearest 400 elliptical galaxies and discovered a new mass concentration, the Great Attractor. She is the recipient of many honors and awards for her research,

including the Bok Prize from Harvard University in 1978. She was elected to the National Academy of Sciences in 1985 and in 1986 won the coveted Dannie Heineman Prize from the American Astronomical Society, given in recognition of a sustained body of especially influential astronomical research. She is also on the Board of Trustees of the Carnegie Institution.

As you study astronomy, it is appropriate for you to ask, "Why am I studying this subject? What good is it for human beings in general and for me in particular?" Astronomy, it must be admitted, does not offer the same practical benefits as other sciences. So how can astronomy be important to your life?

On the most basic level, I like to think of astronomy as providing the introductory chapters for the ultimate textbook on human history. Ordinary texts start with recorded history, going back some 3000 years. For events before that, we consult archeologists and anthropologists, who tell us about the early history of our species. For knowledge of the time before that, we consult paleontologists, biologists, and geologists about the origin and evolution of life and the evolution of our planet, altogether going back some five billion years. Astronomy tells us about the vast stretch of time before the origin of the Earth, the ten billion years or so that saw the formation of the Sun and solar system, the Milky Way Galaxy, and the origin of the universe in the Big Bang. Knowledge of astronomy is essential for a well-educated person's view of history.

Astronomy challenges our belief system and impels us to put our "philosophical house" in order. Take the origin of the world, for example. According to Genesis in the Bible, the world and all in it were created in six days by the hand of God. However, the ancient Egyptians believed that the Earth arose spontaneously from the infinite waters of the eternal universe, Nun, and Alaskan legends taught that the world was created by the conscious imaginings of a creator deity named Father Raven.

The modern astronomical story of the creation of the Earth differs from all these in that it claims to be buttressed by physics and by observational fact. The Sun, it is asserted, formed via gravitational collapse from a dense cloud of interstellar dust and gas about five billion years ago. The planets coalesced at the same time as condensates within the swirling solar nebula. This process took not weeks or days but several hundred thousand years. Confidence in this theory stems from our looking out into the Galaxy and seeing young stars actually form in this way.

At issue here, really, is the deep question of how we are to gain information about the nature of the physical world—whether by revelation and intuition or by logic and observation. Where science stops and faith begins is a thorny issue for all human beings, but particularly so for astronomers—and for astronomy students.

Astronomy cultivates our notions about cosmic time and cosmic evolution. It is all too easy, given the short span of human life, to overlook the fact that the universe is a dynamic place. When I first entered astronomy, I found myself handicapped by a conservative mind-set that assumed, subtly, that celestial objects were unchanging. This might seem strange for someone whose avowed interest was in learning how galaxies formed. Such are the vagaries of the human mind! Fortunately, I lost this bias as I grew older, mainly, I think, by observing evolution all around me. Many things that had seemed immutable to me as a child—social structures, customs, even the physical environment—turned out not to be so. With that realization, the veil fell from my eyes, and I was able to accept cosmic evolution as a core concept in my thinking about the universe.

Following closely on the concept of evolution is the concept of fragility—if something can change, it might even actually disappear someday. The most obvious example is the limited lifetime of our Sun. In another five billion years or so, the Sun will enter its death throes, during which it will swell up and brighten to 1000 times its present luminosity and, in the process, incinerate the Earth.

Five billion years is far enough in the future that neither you nor I feel any personal responsibility for preparing the human race to meet this challenge. However, there are many other cosmic catastrophes that will befall us in the meantime. The Earth will be hit by sizable pieces of space junk; eroded impact craters show this happens all the time—every few million years or so. Debris thrown up into the atmosphere from these events could be so extensive as to totally alter the climate for an extended period of time, possibly leading to mass extinctions. There will be enormous volcanic eruptions that will make Mount St. Helens look like a small firecracker. Another Ice Age, which will totally disrupt modern agriculture, is virtually certain within the next 20,000 years, unless we cook the Earth ourselves first by burning too much fossil fuel.

Closer to the human sphere, such common notions as the inevitability of human progress, the desirability of endless economic growth, the ability of the Earth to support its growing human population, are all based on limited experience and may not—and probably *will* not— prove viable in the long run. Consequently, we must rethink who we are as a species and what is our proper activity on Earth. Contemplating cosmic time puts us into the appropriate frame of mind for grappling with these issues. These are problems that are very long range and involve the whole human race yet are totally under human control and are vital to our long-term survival and well-being.

Astronomy is essential to developing a perspective on human existence and its relation to the cosmos. What question could be grander and at the same time more relevant to us? We can hardly contemplate human existence in general without taking at least a brief look at our own lives.

Earth is a mote in the vast cosmic sea, and our lives, in the evolutionary scheme of the universe, would seem at face value to be insignificant. Should that thought terrify us, depriving our lives of any meaning? Should we seek to provide meaning by introducing an intelligent creator? Or should we be fully comforted merely to know that this great universe has given birth to us in an entirely natural, perhaps even inevitable, chain of events?

Not long ago I visited the Southern Hemisphere for the first time to observe the sky from Chile. At that latitude, the center of the Milky Way passes overhead, where it makes a grand show, not the miserable, fuzzy patch that is visible, low on the horizon, from North America. Stepping out on the observatory catwalk one morning before dawn, I saw for the first time the galactic center in all its glory soaring overhead. At that instant, my perceptions underwent a profound alteration. One moment I was standing in an earthbound "room" with the Galaxy painted on the "ceiling." The next instant, the walls of this earthly room fell away, and I was floating freely in outer space, viewing the center of the Galaxy from an outpost on its periphery. The sense of depth was breathtaking: Sparkling star clusters and dense dust clouds were etched against the blazing inner bulge of the Galaxy many thousands of light-years distant. I saw the Galaxy as an immense, three-dimensional object in space for the first time. It became "real." At that moment, my sense of place in the universe underwent a fundamental and irreversible change. I remained a citizen of the Earth, but I also became a citizen of the Galaxy.

Many astronomers believe that the ultimate, proper concept of "home" for the human race is our universe. It seems more and more likely that there are a very large number of other universes. Virtually none of them would appear to resemble ours; they may have different numbers of space and time dimensions—perhaps even no space or time—and very different physical laws. The vast majority probably are incapable of harboring intelligent life as we know it.

The parallel with Earth is striking. Among the solar system planets, only Earth can support human life. Among the great number of planets that likely exist in our Galaxy, only a small fraction are apt to be such that we could call them home. The fraction of hospitable universes is likely to be smaller still. Our universe would seem, then, to be the ultimate instance of "home": a sanctuary in a vast sea of inhospitable universes.

I began this essay by talking about history and ended with questions that border on the ethical and religious. Astronomy is like that: It offers a modern-day version of Genesis—and of Apocalypse, too. Astronomy stimulates and challenges us on our deepest, most human levels. I hope that during this course you will be able to take time out to contemplate the broader implications of what you are studying. This is one of the rare opportunities in life to think about who you are and where you and the human race are going. Don't miss it.

Knowing the Heavens

(Luke Dodd/Science Photo Library/Photo Researchers)

R I V U X G

Hundreds of years ago, the peoples of Polynesia and Micronesia were among the most skilled navigators in the world. They were able to guide their wooden vessels accurately across vast stretches of the Pacific, far beyond sight of land, without compasses, and even by night. Their secret was to watch the positions of certain stars, such as Alpha and Beta Centauri (shown on the left in the photograph) and the Southern Cross (in the center). Each of these stars rises and sets at the same points on the horizon every night. Pacific navigators used these rising and setting points much as a modern sailor uses the points of a compass.

This is but one example of the many uses of *naked-eye astronomy*. You do this same sort of astronomy when you notice where the Moon is in the sky or when you trace the pattern of the Big Dipper.

Naked-eye astronomy cannot tell us what the Sun is made of or how far away the stars are. We need telescopes, and their attendant equipment, to answer those questions. But, as you will learn in this chapter, naked-eye astronomy is nonetheless tremendously important and useful. Before astronomers can train a telescope on a distant star, they must know where in the sky that star can be found. The cycle of day and night and the rhythm of the seasons are directly related to the apparent motions of the Sun across the sky. By studying naked-eye astronomy, you will learn how the Earth moves through space. In this way you will come to understand our true place in the cosmos.

QUESTIONS TO GUIDE YOUR READING

As you read the sections of this chapter, look for the answers to the following questions.

2-1 What role did astronomy play in ancient civilizations?

2-2 Are the stars that make up a constellation actually close to one another?

2-3 Are the same stars visible every night of the year? What is so special about the North Star?

2-4 Are the same stars visible from any location on Earth?

2-5 What causes the seasons? Why are they opposite in the northern and southern hemispheres?

2-6 Has the same star always been the North Star?

2-7 Can we use the rising and setting of the Sun as the basis of our system of keeping time?

2-8 Why are there leap years?

2-1 Naked-eye astronomy had an important place in ancient civilizations

Positional astronomy—the study of the positions of objects in the sky and how these positions change—has roots that extend far back in time. Four to five thousand years ago, the inhabitants of the British Isles erected stone structures, such as Stonehenge, that suggest a preoccupation with the motions of the sky. Alignments of these stones appear to show where the Sun rose and set at key times during the year. A similar structure in the New World is the Medicine Wheel, a circular ring of stones constructed by the Plains Indians atop a windswept plateau in Wyoming. Certain stones in this ring mark the rising points of the Sun and certain bright stars on the first day of summer.

Aztec, Mayan, and Incan architects in Central and South America also designed buildings with astronomical orientations. At the ruined city of Tiahuanaco in Bolivia, the walls of the Temple of the Sun were aligned north-south and east-west with an accuracy of better than one degree. The great Egyptian pyramids, built around 3000 B.C., are likewise oriented north-south and east-west with remarkable precision.

In addition to temples and tombs, some ancient buildings appear to have been dedicated expressly to astronomy. One of the best examples was built a thousand years ago in the Mayan city of Chichén Itzá, on the Yucatán Peninsula (Figure 2-1). The Caracol's cylindrical tower contains windows aligned with the northernmost and southernmost rising and setting points of both the Sun and the planet Venus. A similar four-story adobe building, probably constructed during the fourteenth century, is located at the Casa Grande site in Arizona.

These structures bear witness to an awareness of naked-eye astronomy by the peoples of many cultures. Many of the concepts of modern positional astronomy come to us from these ancients, including the idea of dividing the sky into constellations.

2-2 Eighty-eight constellations cover the entire sky

Looking at the sky on a clear, dark night, you might think that you can see millions of stars. Actually, the unaided human eye can detect only about 6000 stars. Because half of the sky is below the horizon at any one time, you can see at most about 3000 stars. When ancient peoples looked at these thousands of stars, they imagined that groupings of stars traced out pictures in the sky.

figure 2-1 R I **V** U X G

The Caracol at Chichén Itzá This ancient Mayan observatory in the Yucatán was built about A.D. 1000. Its architecture is based on alignments with important celestial events. Mayan astronomers developed a very accurate calendar and measured the motions of celestial bodies with great precision. Special significance was associated with the planet Venus, which inspired sacrificial rites and other ceremonies. (Courtesy of E. C. Krupp)

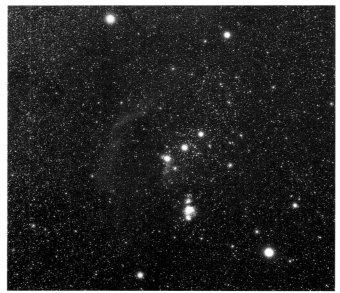

a   R I V U X G

Figure 2-2
Three Views of Orion Orion is a prominent constellation from December through March. During these months, Orion is easily seen high above the southern horizon from North America or Europe. From Australia or South America, it appears high above the northern horizon. **(a)** This time-exposure photograph of Orion shows that stars come in different colors, some red and others blue. You can see these colors with the naked eye; they are even more pronounced as seen through binoculars or a telescope. **(b)** A portion of a modern star atlas shows the names of some of the brightest stars in Orion, along with the distances in light-years (ly) to those stars. The yellow lines show the borders between Orion and its neighboring constellations (labeled in capitals), as defined by modern astronomers. **(c)** This fanciful drawing from a star atlas published in 1835 shows Orion the Hunter as well as other celestial creatures. (a: Luke Dodd/Science Photo Library/Photo Researchers; c: Courtesy of Janus Publications)

b

c

Astronomers still refer to these groupings, called **constellations** (from the Latin for "group of stars").

You may already be familiar with some of these pictures or patterns in the sky, such as the Big Dipper, which is actually part of the large constellation Ursa Major (the Great Bear). Many constellations, such as Orion in Figure 2-2, have names derived from the myths and legends of antiquity. Although some star groupings vaguely resemble the figures they are supposed to represent (see Figure 2-2c), most do not.

The term "constellation" has a broader definition in present-day astronomy. On modern star charts, the entire sky is divided into 88 regions, each of which is called a constellation. For example, the constellation Orion is now defined to be an irregular patch of sky whose borders are shown in Figure 2-2b. When astronomers refer to the "Great Nebula" M42 in Orion, they mean that as seen from Earth this nebula appears to be within Orion's patch of sky. Some constellations cover large areas of the sky (Ursa Major being one of the biggest) and others very small areas

(Crux, the Southern Cross, being the smallest). But because the modern constellations cover the entire sky, every star lies in one constellation or another.

 When you look at a constellation's star pattern, it is tempting to conclude that you are seeing a group of stars that are all relatively close together. In fact, most of these stars are nowhere near one another. As an example, Figure 2-2b shows the distances in light-years to four stars in Orion. Although Bellatrix (Arabic for "the Amazon") and Mintaka ("the belt") appear to be close to each other, Mintaka is actually about a thousand light-years farther away from us. They merely appear close to each other because they are in nearly the same direction as seen from Earth. The same illusion often appears when you see an airliner's lights at night. It is very difficult to tell how far away a single bright light is, which is why you can mistake an airliner a few kilometers away for a star trillions of times more distant.

The star names shown in Figure 2-2b are from the Arabic. For example, Betelgeuse means "armpit," which makes sense when you look at the star atlas drawing in Figure 2-2c. Other types of names are also used for stars. For example, Betelgeuse is also known as α Orionis (α is the Greek letter alpha) and as HD 39801.

A number of unscrupulous commercial firms offer to name a star for you for a fee. The money that they charge you for this "service" is real, but the star names are not; none of these names are recognized by astronomers. If you want to use astronomy to commemorate your name or the name of a friend or relative, consider making a donation to your local planetarium or science museum. The money will be put to much better use!

2-3 The appearance of the sky changes during the course of the night and from one night to the next

Go outdoors soon after dark, find a spot away from bright lights, and note the patterns of stars in the sky. Do the same a few hours later. You will find that the entire pattern of stars (as well as the Moon, if it is visible) has shifted its position. New constellations will have risen above the eastern horizon, and some will have disappeared below the western horizon. If you look again before dawn, you will see that the stars that were just rising in the east when the night began are now low in the western sky. This daily motion, or **diurnal motion,** of the stars is apparent in time-exposure photographs (Figure 2-3). We can understand this apparent motion of the sky by considering the rotation of the Earth.

At any given moment, it is daytime on the half of the Earth illuminated by the Sun and nighttime on the other half. The Earth rotates from west to east, making one complete rotation every 24 hours, which is why there is a daily cycle of day and night. Because of this rotation, stars appear to us to rise in the east and set in the west, as do the Sun and Moon.

Figure 2-4 helps to further explain diurnal motion. It shows two views of the Earth as seen from a point above the North Pole. At the instant shown in Figure 2-4a, it is day in Asia but night in most of North America and Europe. Figure 2-4b shows the Earth 4 hours later. Four hours is one-sixth of a complete 24-hour day, so the Earth has made one-sixth of a rotation between Figures 2-4a and 2-4b. Europe is now in the illuminated half of the Earth (the Sun has risen in Europe), while Alaska has moved from the illuminated to the dark half of the Earth (the Sun has set in Alaska). For a person in

Figure 2-3 R I **V** U X G

Diurnal Motion of the Night Sky This photograph is a time exposure of the night sky over the Kitt Peak National Observatory near Tucson, Arizona. The view is to the northwest. Because the Earth rotates from west to east, the sky appears to rotate from east to west. The stars thus appear as long trails in this photograph. (National Optical Astronomy Observatories)

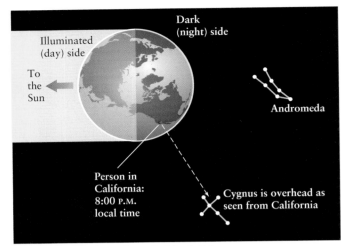

a   Earth as seen from above the North Pole

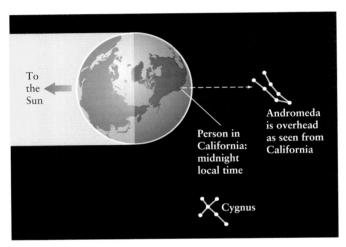

b   4 hours later

figure 2-4

Why Diurnal Motion Happens The diurnal (daily) motion of the stars, the Sun, and the Moon is a consequence of the Earth's rotation. **(a)** This drawing shows the Earth from a vantage point above the North Pole. It is day on the side of the Earth that is illuminated by the Sun, and night on the dark, unilluminated side. In this drawing, the local time in California is 8:00 P.M. and a person in California sees the constellation Cygnus directly overhead. **(b)** Four hours later, the Earth has rotated to the east by one-sixth of a complete rotation. From the perspective of a person on Earth, the entire sky appears to have rotated to the west by one-sixth of a complete rotation. It is now midnight in California, and the constellation directly over California is Andromeda.

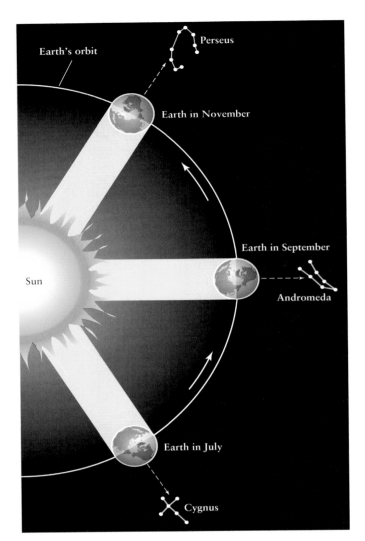

figure 2-5

Why the Night Sky Changes During the Year The particular stars that you see in the night sky are different at different times of the year. This is a consequence of the Earth's orbital motion; as we move around the Sun, the nighttime side of the Earth gradually turns toward different parts of the sky. This figure shows which constellation is overhead at midnight local time—when the Sun is on the opposite side of the Earth from your location—during different months for observers at midnorthern latitudes (including the United States). If you want to view the constellation Andromeda, the best time of the year to do it is in late September, when Andromeda is nearly overhead at midnight.

California, in Figure 2-4*a* the time is 8:00 P.M. and the constellation Cygnus (the Swan) is directly overhead. Four hours later, the constellation over California is Andromeda (named for a mythological princess). Because the Earth rotates from west to east, it appears to us on Earth that the entire sky is rotating around us in the opposite direction, from east to west.

In addition to the diurnal motion of the sky, the constellations visible in the night sky also change slowly over the

course of a year. This happens because the Earth orbits, or revolves around, the Sun (Figure 2-5). Over the course of a year, the Earth makes one complete orbit, and the darkened, nighttime side of the Earth gradually turns toward different parts of the heavens. For example, as seen from the northern hemisphere, at midnight in late July the constellation Cygnus is close to overhead; at midnight in late September the constellation Andromeda is close to overhead; and at midnight in

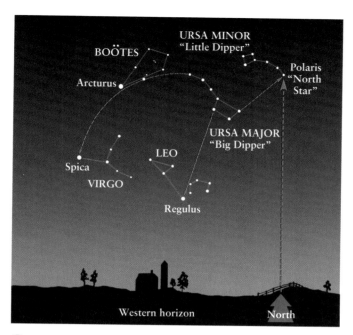

figure 2-6

The Big Dipper as a Guide The North Star can be seen from anywhere in the northern hemisphere on any night of the year. This star chart shows how the Big Dipper can be used to point out the North Star as well as the brightest stars in three other constellations. The chart shows the sky at around 2 A.M. (daylight saving time) on May 1, around midnight on June 1, or around 10 P.M. on July 1. Because of the Earth's rotation and orbital motion around the Sun, the view will be different on different dates and at different times, but the relative positions of the stars will be the same. The angular distance from Polaris to Spica is about 101°, so this chart covers a large fraction of the sky.

late November the constellation Perseus (commemorating a mythological hero) is close to overhead. If you follow a particular star on successive evenings, you will find that it rises approximately 4 minutes earlier each night, or 2 hours earlier each month.

Constellations can help you find your way around the sky. For example, if you live in the northern hemisphere, you can use the Big Dipper in Ursa Major to find the north direction by drawing a straight line through the two stars at the front of the Big Dipper's bowl (Figure 2-6). The first moderately bright star you come to is Polaris, also called the North Star because it is located almost directly over the Earth's north pole. If you draw a line from Polaris straight down to the horizon, you will find the north direction.

By drawing a line through the two stars at the rear of the bowl of the Big Dipper, you can find Leo (the Lion). As Figure 2-6 shows, that line points toward Regulus, the brightest star in the "sickle" tracing the lion's mane. By following the handle of the Big Dipper, you can locate the bright reddish star Arcturus in Boötes (the Shepherd) and the prominent bluish star Spica in Virgo (the Virgin). The saying "Follow the arc to Arcturus and speed to Spica" may help you remember these

stars, which are conspicuous in the evening sky during the spring and summer.

During winter in the northern hemisphere, you can see some of the brightest stars in the sky. Many of them are in the vicinity of the "winter triangle" (Figure 2-7), which connects bright stars in the constellations of Orion (the Hunter), Canis Major (the Large Dog), and Canis Minor (the Small Dog). The "winter triangle" is high above the eastern horizon during the middle of winter at midnight.

A similar feature, the "summer triangle," graces the summer sky in the northern hemisphere. This triangle connects the brightest stars in Lyra (the Harp), Cygnus (the Swan), and Aquila (the Eagle) (Figure 2-8). A conspicuous portion of the Milky Way forms a beautiful background for these constellations, which are nearly overhead during the middle of summer at midnight.

A wonderful tool to help you find your way around the night sky is the planetarium program *Starry Night*, which you will find on the CD-ROM that accompanies this book. In addition, at the end of this book you will find a set of selected star charts for the evening

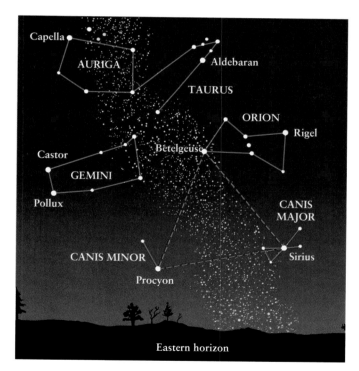

figure 2-7

The "Winter Triangle" This star chart shows the eastern sky as it appears on a late fall or early winter evening in the northern hemisphere (around midnight on November 1, around 10 P.M. on December 1, or around 8 P.M. on January 1). Three of the brightest stars in the sky make up the "winter triangle," which is about 26° on a side. In addition to the constellations involved in the triangle, the chart shows the prominent constellations Gemini (the Twins), Auriga (the Charioteer), and Taurus (the Bull).

figure 2-8

The "Summer Triangle" This star chart shows the eastern sky as it appears in the evening during spring and summer in the northern hemisphere (around 1 A.M.—daylight saving time—on June 1, around 11 P.M. on July 1, and around 9 P.M. on August 1). The angular distance from Deneb to Altair is about 38°. The constellations Sagitta (the Arrow) and Delphinus (the Dolphin) are much fainter than the three constellations that make up the triangle.

hours of all 12 months of the year. You may find stargazing an enjoyable experience, and *Starry Night* and the star charts will help you identify many well-known constellations.

Note that all the star charts in this section and at the end of this book are drawn for an observer in the northern hemisphere. If you live in the southern hemisphere, you can see constellations that are not visible from the northern hemisphere, and vice versa. In the next section we will see why this is so.

2-4 It is convenient to imagine that the stars are located on a celestial sphere

Many ancient societies believed that all the stars are the same distance from the Earth. They imagined the stars to be bits of fire imbedded into the inner surface of an immense hollow sphere, called the **celestial sphere**, with the Earth at its center. In this picture of the universe, the Earth was fixed and did not rotate. Instead, the entire celestial sphere rotated once a day around the Earth from east to west, thereby causing the diurnal motion of the sky. The picture of a rotating celestial sphere fit well with naked-eye observations, and for its time was a

useful model of how the universe works. (We discussed the role of models in science in <u>Section 1-1</u>).

We now know that this simple model of the universe is not correct. Diurnal motion is due to the rotation of the Earth, not the rest of the universe. Furthermore, as we learned when discussing the constellations in Section 2-2, it is impossible to tell with the naked eye how far away the stars are. Indeed, the brightest stars that you can see with the naked eye range from 4.2 to more than 1000 light-years away (see Figure 2-2), and telescopes allow us to see objects at distances of billions of light-years.

Thus, astronomers now recognize that the celestial sphere is an *imaginary* object that has no basis in physical reality. Nonetheless, the celestial sphere model remains a useful tool of positional astronomy. If we imagine, as did the ancients, that the Earth is stationary and that the celestial sphere rotates around us, it is relatively easy to specify the directions to different objects in the sky and to visualize the motions of these objects.

Figure 2-9 depicts the celestial sphere, with the Earth at its center. (A truly proportional drawing would show the celestial sphere as being millions of times larger than the Earth.) If we project the Earth's equator out into space, we obtain the

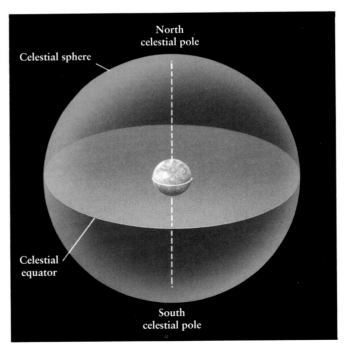

figure 2-9

The Celestial Sphere The celestial sphere is the apparent sphere of the sky. The view in this figure is from the outside of this (wholly imaginary) sphere. Because the Earth is at the center of the celestial sphere, however, our view is always of the *inside* of the sphere. The celestial equator and poles are the projections of the Earth's equator and axis of rotation out into space. The celestial poles are therefore located directly over the Earth's poles.

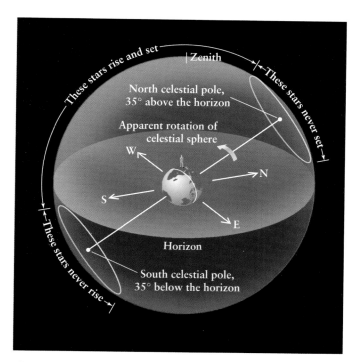

figure 2-10

The View from 35° North Latitude To an observer at 35° north latitude (roughly the latitude of Los Angeles, Atlanta, Tel Aviv, and Tokyo), the north celestial pole is always 35° above the horizon. Stars within 35° of the *north* celestial pole are always above the horizon on any night of the year. These stars are circumpolar; they trace out circles around the north celestial pole during the course of the night. Stars within 35° of the *south* celestial pole are always below the horizon and can never be seen from this latitude. Stars that lie between these two extremes rise in the east and set in the west.

figure 2-11   R I **V** U X G

Circumpolar Star Trails This long exposure, taken from Australia's Siding Spring Mountain and aimed at the south celestial pole, shows the rotation of the sky above the Anglo-Australian Telescope. During the exposure, someone carrying a flashlight walked along the catwalk on the outside of the telescope dome. Another flashlight made the wavy trail at ground level. (© Anglo-Australian Observatory, photograph by David Malin)

celestial equator. The celestial equator divides the sky into northern and southern hemispheres, just as the Earth's equator divides the Earth into two hemispheres.

If we project the Earth's north and south poles into space, we obtain the **north celestial pole** and the **south celestial pole.** The two celestial poles are where the Earth's axis of rotation (extended out into space) intersects the celestial sphere (see Figure 2-9). The star Polaris is less than 1° away from the north celestial pole, which is why it is called the North Star or the Pole Star.

The point in the sky directly overhead an observer anywhere on Earth is called that observer's **zenith.** The zenith and celestial sphere are shown in Figure 2-10 for an observer located at 35° north latitude (that is, at a location on the Earth's surface 35° north of the equator). The zenith is shown at the top of Figure 2-10, so the Earth and the celestial sphere appear "tipped" compared to Figure 2-9. At any time, an observer can see only half of the celestial sphere; the other half is below the horizon, hidden by the body of the Earth. The hidden half of the celestial sphere is darkly shaded in Figure 2-10.

For an observer anywhere in the northern hemisphere, including the observer in Figure 2-10, the north celestial pole is always above the horizon. As the Earth turns from west to east—or, equivalently, as the celestial sphere appears to turn from east to west—stars sufficiently near the north celestial pole revolve around the pole, never rising or setting. Such stars are called **circumpolar.** For example, as seen from North America, Polaris is a circumpolar star and can be seen at any time of night on any night of the year. Similarly, stars near the south celestial pole revolve around that pole but always remain below the horizon of an observer in the northern hemisphere. Hence, these stars can never be seen by the observer in Figure 2-10. Stars at intermediate positions on the celestial sphere rise in the east and set in the west, as in Figure 2-3.

For an observer at 35° south latitude (the latitude of Sydney, Australia), the roles of the north and south celestial poles are reversed: Objects close to the *south* celestial pole are circumpolar, that is, they revolve around that pole and never rise or set. The photograph in Figure 2-11 shows the circular trails of stars around the south celestial pole as seen from Australia.

Stars close to the *north* celestial pole are always below the horizon and can never be seen by an observer in the southern hemisphere. Hence, Australian astronomers never see the North Star but are able to see other stars that are forever hidden from North American observers.

Using the celestial equator and poles, we can define a coordinate system to specify the position of any star on the celestial sphere. As described in Box 2-1, the most commonly used coordinate system uses two angles, *right ascension* and *declination*, that are quite like longitude and latitude on Earth. These coordinates tell us in what direction we should look to see the star. To locate the star's true position in three-dimensional space, we must also know the distance to the star.

Box 2-1 | Tools of the Astronomer's Trade

Celestial Coordinates

To denote the positions of objects in the sky, astronomers use a system based on *right ascension* and *declination*. Declination corresponds to latitude. As shown in the illustration, the **declination** of an object is its angular distance north or south of the celestial equator, measured along a circle passing through both celestial poles. It is measured in degrees, arcminutes, and arcseconds (see Section 1-5).

Right ascension corresponds to longitude. Astronomers measure right ascension from a specific point, called the vernal equinox, on the celestial equator. This point is one of two locations where the Sun crosses the celestial equator during its apparent annual motion, as we discuss in Section 2-5. In the Earth's northern hemisphere, spring officially begins when the Sun reaches the vernal equinox in late March. The **right ascension** of an object in the sky is the angular distance from the vernal equinox eastward along the celestial equator to the circle used in measuring its declination (see illustration). Astronomers do not measure this angular distance in degrees; instead, they use time units (hours, minutes, and seconds) corresponding to the time required for the celestial sphere to rotate through this angle. For example, suppose there is a star at your zenith right now with right ascension 6^h 0^m 0^s. Two hours and 30 minutes from now, there will be a different object at your zenith with right ascension 8^h 30^m 0^s.

 Right ascension and declination are very useful because they tell an astronomer precisely where in the sky an object is located. For example, the coordinates of the bright star Rigel for the year 2000 are R.A. = 5^h 14^m 32.2^s, Decl. = $-8°$ $12'$ $06''$. (R.A. and Decl. are abbreviations for right ascension and declination.) A minus sign on the declination indicates that the star is south of the celestial equator; a plus sign (or no sign at all) indicates that an object is north of the celestial equator. As we discuss in Box 2-2, right ascension is especially useful for determining the best time to observe a particular object.

It is important to state the year for which a star's right ascension and declination are valid. This is so because of precession, which we discuss in Section 2-6.

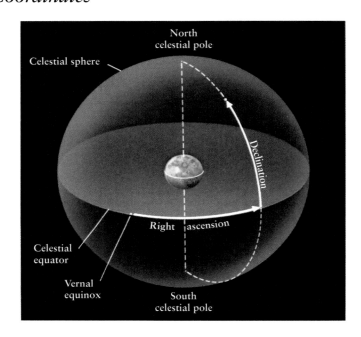

EXAMPLE: What are the coordinates of a star that lies exactly halfway between the vernal equinox and the south celestial pole? Since the circle used to measure this star's declination passes through the vernal equinox, this star's right ascension is R.A. = 0^h 0^m 0^s. The angle between the celestial equator and south celestial pole is 90° 0′ 0″, so the declination of this star is Decl. = $-45°$ $0'$ $0''$. The declination is negative because the star is in the southern half of the celestial sphere.

EXAMPLE: At midnight local time you see a star with R.A. = 2^h 30^m 0^s at your zenith. When will you see a star at your zenith with R.A. = 21^h 0^m 0^s? The time required for the sky to rotate through the angle between the stars is the difference in their right ascensions: 21^h 0^m 0^s − 2^h 30^m 0^s = 18^h 30^m 0^s. So, the second star will be at your zenith 18½ hours after the first one, or at 6:30 P.M. the following evening.

2-5 The seasons are caused by the tilt of Earth's axis of rotation

In addition to rotating on its axis every 24 hours, the Earth revolves around the Sun in about 365¼ days. As we travel with the Earth around its orbit, we experience the annual cycle of seasons. But why *are* there seasons? Furthermore, the seasons are opposite in the northern and southern hemispheres. For example, February is midwinter in North America but midsummer in Australia. Why should this be?

The key fact that explains why we have seasons, and why they are different in different hemispheres, is that the Earth's axis of rotation is not perpendicular to the plane of the Earth's orbit. Instead, as Figure 2-12 shows, it is tilted about 23½° away from the perpendicular. The Earth maintains this tilt as it orbits the Sun, with the Earth's north pole pointing toward the north celestial pole. (This stability is a hallmark of all rotating objects. A top will not fall over as long as it is spinning, and the rotating wheels of a motorcycle help to keep the rider upright.)

During part of the year, when the Earth is in the part of its orbit shown on the left side of Figure 2-12, the northern hemisphere is tilted toward the Sun. As the Earth spins on its axis, a point in the northern hemisphere spends more than 12 hours in the sunlight. Thus, the days there are long and the nights are short, and it is summer in the northern hemisphere. The summer is hot not only because of the extended daylight hours but

also because the Sun is high in the northern hemisphere's sky. As a result, sunlight strikes the ground at a nearly perpendicular angle that heats the ground efficiently (Figure 2-13*a*). During this same time of year in the southern hemisphere, the days are short and the nights are long, because a point in this hemisphere spends fewer than 12 hours a day in the sunlight. The Sun is low in the sky, so sunlight strikes the surface at a grazing angle that causes little heating (Figure 2-13*b*), and it is winter in the southern hemisphere.

Half a year later, the Earth is in the part of its orbit shown on the right side of Figure 2-12. Now the situation is reversed, with winter in the northern hemisphere (which is now tilted away from the Sun) and summer in the southern hemisphere. During spring and autumn, the two hemispheres receive roughly equal amounts of illumination from the Sun, and daytime and nighttime are of equal length everywhere on Earth.

 A common misconception is that the seasons are caused by variations in the distance from the Earth to the Sun. According to this idea, the Earth is closer to the Sun in summer and farther away in winter. But in fact, the Earth's orbit around the Sun is very nearly circular, and the Earth-Sun distance varies only about 3% over the course of a year. (The Earth's orbit only *looks* elongated in Figure 2-12 because this illustration shows an oblique view.) We are slightly closer to the Sun in January than in July, but this small variation has little influence on the cycle of the seasons. Also, if the seasons were really caused by variations in the Earth-Sun distance, the seasons would be the same in both hemispheres!

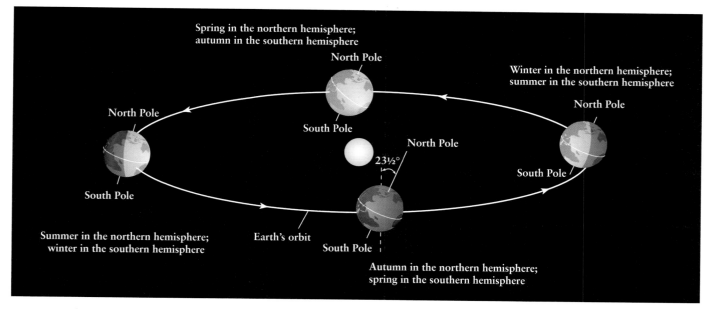

figure 2-12

The Seasons The Earth's axis of rotation is inclined 23½° away from the perpendicular to the plane of the Earth's orbit. The Earth maintains this orientation (with its north pole aimed at the north celestial pole, near the star Polaris) throughout the year as it orbits the Sun. Consequently, the amount of solar illumination and the number of daylight hours at any location on Earth vary in a regular pattern throughout the year. This is the origin of the seasons.

a   b

figure 2-13
Solar Energy in Summer and Winter At different times of the year, sunlight strikes the ground at different angles. **(a)** In summer the Sun is high in the sky. The energy in a shaft of light falls onto a small, nearly circular area. This concentrated solar energy heats the ground effectively and makes the days warm. The days are also longest in summer, which further increases the heating. **(b)** In winter the Sun is low in the sky even at midday. The same shaft of light in (a) now falls on a larger, oval area. Because the sunlight is less concentrated and the days are short, little heating of the ground takes place. This accounts for the low temperatures in winter.

As the Earth orbits around the Sun, the Sun's position (as seen from the Earth) gradually shifts with respect to the background stars. As Figure 2-14 shows, the Sun therefore appears to trace out a circular path, called the **ecliptic,** on the celestial sphere. (This word suggests that the path traced out by the Sun has something to do with eclipses. We will discuss the connection in Chapter 3.) Because there are 365¼ days in a year and 360° in a circle, the Sun appears to move along the ecliptic at a rate of about 1° per day. This motion is from west to east, that is, in the direction opposite to the apparent motion of the celestial sphere.

ANALOGY It may help to envision the celestial sphere as a merry-go-round rotating clockwise, and the Sun as a restless child who is walking slowly around the merry-go-round's rim in the counterclockwise direction. During the time it takes the child to make a round trip, the merry-go-round rotates 365¼ times.

The ecliptic can also be thought of as the projection onto the celestial sphere of the plane of the Earth's orbit. This is *not* the same as the plane of the Earth's equator, thanks to the 23½° tilt of the Earth's rotation axis shown in Figure 2-12. Instead, the ecliptic and the celestial equator are inclined to each other by that same 23½° angle.

The ecliptic and the celestial equator intersect at only two points, which are exactly opposite each other on the celestial sphere. Each point is called an **equinox** (from the Latin for "equal night"), because when the Sun appears at either of these points, day and night are each about 12 hours long at all locations on Earth. The term "equinox" is also used to refer to the date on which the Sun passes through one of these special points on the ecliptic.

On about March 21 of each year, the Sun crosses northward across the celestial equator at the **vernal equinox.** This marks the beginning of spring in the northern hemisphere ("vernal" is from the Latin for "spring"). On about September 22 the Sun moves southward across the celestial equator at the **autumnal equinox,** marking the moment when fall

begins in the northern hemisphere. Since the seasons are opposite in the northern and southern hemispheres, for Australians and South Africans the vernal equinox actually marks the beginning of autumn. The names of the equinoxes come from astronomers of the past who lived north of the equator.

Between the vernal and autumnal equinoxes lie two other significant locations along the ecliptic. The point on the ecliptic farthest north of the celestial equator is called the **summer**

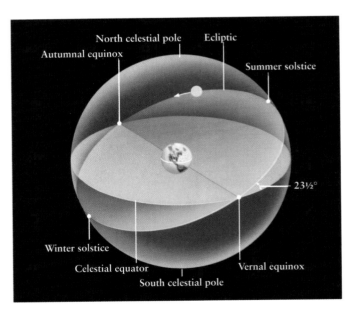

figure 2-14
The Ecliptic, Equinoxes, and Solstices The ecliptic is the apparent annual path of the Sun as projected onto the celestial sphere. Inclined to the celestial equator by 23½° because of the tilt of the Earth's axis of rotation, the ecliptic intersects the celestial equator at two points, called equinoxes. The northernmost point on the ecliptic is the summer solstice, and the corresponding southernmost point is the winter solstice. The Sun is shown in its approximate position for August 1.

solstice. "Solstice" is from the Latin for "solar standstill," and it is at the summer solstice that the Sun stops moving northward on the celestial sphere. At this point, the Sun is as far north of the celestial equator as it can get. It marks the location of the Sun at the moment summer begins in the northern hemisphere (about June 21). At the beginning of the northern hemisphere's winter, the Sun is farthest south of the celestial equator, at a point called the **winter solstice** (about December 21).

Because the Sun's position on the celestial sphere varies slowly over the course of a year, its daily path across the sky (due to the Earth's rotation) also varies with the seasons (Figure 2-15). On the first day of spring or the first day of fall, when the Sun is at one of the equinoxes, the Sun rises directly in the east and sets directly in the west.

When the northern hemisphere is tilted away from the Sun and it is winter in the northern hemisphere, the Sun rises in the southeast. Daylight lasts for fewer than 12 hours as the Sun skims low over the southern horizon and sets in the southwest. Northern hemisphere nights are longest when the Sun is at the winter solstice.

The closer you get to the north pole, the shorter the winter days and the longer the winter nights. In fact, anywhere within 23½° of the north pole (that is, north of latitude 90° − 23½° = 66½° N) the Sun is below the horizon for 24 continuous hours at least one day of the year. The circle around the Earth at 66½° north latitude is called the **Arctic Circle** (Figure 2-16a). The corresponding region around the south pole is bounded by the

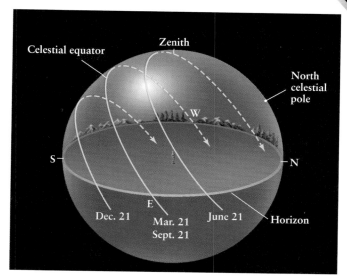

figure 2-15

The Sun's Daily Path Across the Sky Like Figure 2-10, this drawing is for an observer at 35° north latitude. On the first day of spring and the first day of fall, the Sun rises precisely in the east and sets precisely in the west. During summer in the northern hemisphere, the Sun rises in the northeast and sets in the northwest. In winter in the northern hemisphere, the Sun rises in the southeast and sets in the southwest.

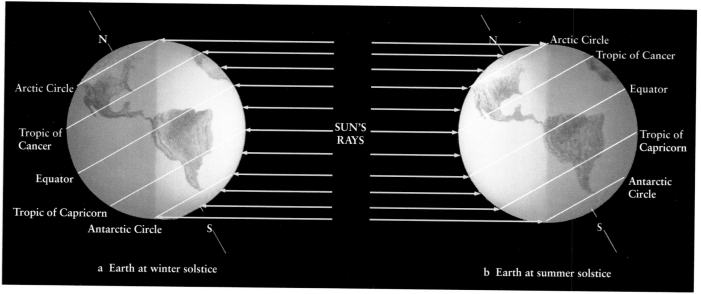

figure 2-16

Tropics and Circles (a) On the day when the Sun is at the winter solstice, the Sun does not rise at all at locations north of the Arctic Circle (66½° north latitude). On this same day the Sun is directly overhead at high noon anywhere on the Tropic of Capricorn (23½° south latitude), and is in the sky for 24 consecutive hours anywhere south of the Antarctic Circle (66½° south latitude). **(b)** The situation is reversed on the day when the Sun is at the summer solstice. The Sun does not set anywhere north of the Arctic Circle and does not rise anywhere south of the Antarctic Circle. The Sun is directly overhead at high noon anywhere on the Tropic of Cancer, at 23½° north latitude.

Antarctic Circle at 66½° south latitude. At the time of the winter solstice, explorers south of the Antarctic Circle enjoy "the midnight sun," or 24 hours of continuous daylight.

During summer in the northern hemisphere, when the northern hemisphere is tilted toward the Sun, the Sun rises in the northeast and sets in the northwest. The Sun is at its northernmost position at the summer solstice, giving the northern hemisphere the greatest number of daylight hours. At the summer solstice the Sun does not set at all north of the Arctic Circle, and does not rise at all south of the Antarctic Circle (Figure 2-16b).

The variations of the seasons are much less pronounced close to the equator. Between the **Tropic of Capricorn** at 23½° south latitude and the **Tropic of Cancer** at 23½° north latitude, the Sun is directly overhead—that is, at the zenith—at high noon at least one day a year. Outside of the tropics, the Sun is never directly overhead, but is always either south of the zenith (as seen from locations north of the Tropic of Cancer) or north of the zenith (as seen from south of the Tropic of Capricorn).

2-6 The Moon helps to cause precession, a slow, conical motion of Earth's axis of rotation

The Moon is by far the brightest and most obvious naked-eye object in the nighttime sky. Like the Sun, the Moon slowly changes its position relative to the background stars; unlike the Sun, the Moon makes a complete trip around the celestial sphere in only about four weeks, or about a month. (The word "month" comes from the same Old English root as the word "moon.") Ancient astronomers realized that this motion occurs because the Moon orbits the Earth in roughly four weeks. In one hour the Moon moves on the celestial sphere by about ½°, or roughly its own angular size.

The Moon's path on the celestial sphere is never far from the Sun's path (that is, the ecliptic). This is because the plane of the Moon's orbit around the Earth is inclined only slightly from the plane of the Earth's orbit around the Sun. The Moon's path varies somewhat from one month to the next, but always remains within a band called the **zodiac** that extends about 8° on either side of the ecliptic. Twelve famous constellations—Aries, Taurus, Gemini, Cancer, Leo, Virgo, Libra, Scorpius, Sagittarius, Capricornus, Aquarius, and Pisces—lie along the zodiac. The Moon is generally found in one of these 12 constellations. As it moves along its orbit, the Moon appears north of the celestial equator for about two weeks and then south of the celestial equator for about the next two weeks. We will learn more about the Moon's motion, as well as why the Moon goes through phases, in Chapter 3.

The Moon not only moves around the Earth but, in concert with the Sun, also causes a slow change in the Earth's rotation. This is because both the Sun and the Moon exert a gravitational pull on the Earth. We will learn much more about gravity in Chapter 4; for now, all we need is the idea that gravity is a universal attraction of matter for other matter.

The gravitational pull of the Sun and the Moon affects the Earth's rotation because the Earth is slightly fatter across the equator than it is from pole to pole: Its equatorial diameter is 43 kilometers (27 miles) larger than the diameter measured from pole to pole. The Earth is therefore said to have an "equatorial bulge." Because of the gravitational pull of the Moon and the Sun on this bulge, the orientation of the Earth's axis of rotation gradually changes.

The Earth behaves somewhat like a spinning top, as illustrated in Figure 2-17. If the top were not spinning, gravity would pull the top over on its side. When the top is spinning, gravity causes the top's axis of rotation to trace out a circle, producing a motion called **precession.**

As the Sun and Moon move along the zodiac, each spends half its time north of the Earth's equatorial bulge and half its

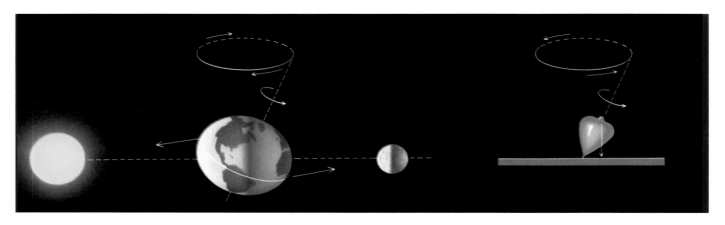

ƒigure 2-17

Precession The gravitational pull of the Moon and the Sun on the Earth's equatorial bulge causes the Earth to precess. As the Earth precesses, its axis of rotation slowly traces out a circle in the sky, like the shifting axis of a spinning top.

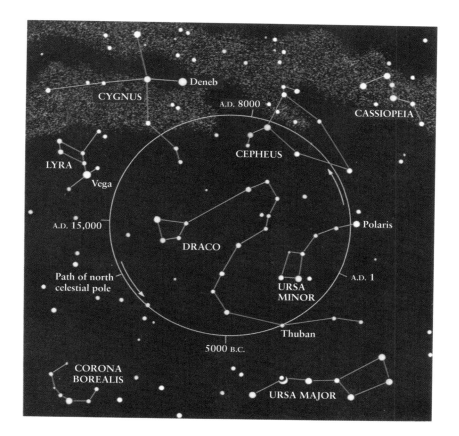

figure 2-18

Precession and the Path of the North Celestial Pole As the Earth precesses, the north celestial pole slowly traces out a circle among the northern constellations. At present, the north celestial pole is near the moderately bright star Polaris, which serves as the North Star. Twelve thousand years from now the bright star Vega will be the North Star.

time south of it. The gravitational pull of the Sun and Moon tugging on the equatorial bulge tries to twist the Earth's axis of rotation to be perpendicular to the plane of the ecliptic. But because the Earth is spinning, the combined actions of gravity and rotation cause the Earth's axis to trace out a circle in the sky, much like what happens to the toy top. As the axis precesses, it remains tilted about 23½° to the perpendicular.

As the Earth's axis of rotation slowly changes its orientation, the north and south celestial poles—which are the projections of that axis onto the celestial sphere—change their positions relative to the stars. At present, the north celestial pole lies within 1° of the star Polaris, which is why Polaris is the North Star. But 5000 years ago, the north celestial pole was closest to the star Thuban in the constellation of Draco (the Dragon). Thus, that star and not Polaris was the North Star. And 12,000 years from now, the North Star will be the bright star Vega in Lyra (the Harp). It takes 26,000 years for the north celestial pole to complete one full precessional circle around the sky (Figure 2-18). The south celestial pole executes a similar circle in the southern sky.

Precession also causes the Earth's equatorial plane to change its orientation. Because this plane defines the location of the celestial equator in the sky, the celestial equator precesses as well. The intersections of the celestial equator and the ecliptic define the equinoxes (see Figure 2-14), so these key locations in the sky also shift slowly from year to year. For this reason, the precession of the Earth is also called the **precession of the equinoxes**. The first person to detect the precession of

the equinoxes, in the second century B.C., was the Greek astronomer Hipparchus, who compared his own observations with those of Babylonian astronomers three centuries earlier. Today, the vernal equinox is located in the constellation Pisces (the Fishes). Two thousand years ago, it was in Aries (the Ram). Around the year A.D. 2600, the vernal equinox will move into Aquarius (the Water Bearer).

 Astrological terms like the "Age of Aquarius" involve boundaries in the sky that are not recognized by astronomers and are generally not even related to the positions of the constellations. Indeed, astrology is *not* a science at all, but merely a collection of superstitions and hokum. Its practitioners use some of the terminology of astronomy but reject the logical thinking that is at the heart of science. James Randi has more to say about astrology and other pseudosciences in his essay "Why Astrology Is Not Science" at the end of this chapter.

The astronomer's system of locating heavenly bodies by their right ascension and declination, discussed in Box 2-1, is tied to the positions of the celestial equator and the vernal equinox. Because of precession, these positions are changing, and thus the coordinates of stars in the sky are also constantly changing. These changes are very small and gradual, but they add up over the years. To cope with this difficulty, astronomers always make note of the date (called the **epoch**) for which a particular set of coordinates is precisely correct.

Consequently, star catalogs and star charts are periodically updated. Most current catalogs and star charts are prepared for the epoch 2000. The coordinates in these reference books, which are precise for January 1, 2000, will require very little correction over the next few decades.

2-7 Positional astronomy plays an important role in keeping track of time

Astronomers have traditionally been responsible for telling time. This is because we want the system of timekeeping used in everyday life to reflect the position of the Sun in the sky. Thousands of years ago, the sundial was invented to keep track of **apparent solar time.** To obtain more accurate measurements astronomers use the **meridian.** As Figure 2-19 shows, this is a north-south circle on the celestial sphere that passes through the zenith (the point directly overhead) and both celestial poles. *Local noon* is defined to be when the Sun crosses the **upper meridian,** which is the half of the meridian above the horizon. At *local midnight,* the Sun crosses the **lower meridian,** the half of the meridian below the horizon; this crossing cannot be observed directly.

The crossing of the meridian by any object in the sky is called a **meridian transit** of that object. If the crossing occurs above the horizon, it is an *upper* meridian transit. An **apparent solar day** is the interval between two successive upper meridian transits of the Sun as observed from any fixed spot on the Earth. Stated less formally, an apparent solar day is the time from one local noon to the next local noon, or from when the Sun is highest in the sky to when it is again highest in the sky.

Unfortunately, the Sun is not a good timekeeper. The length of an apparent solar day (as measured by a device such as an hourglass) varies from one time of year to another. There are two main reasons why this is so, both having to do with the way in which the Earth orbits the Sun.

The first reason is that the Earth's orbit is not a perfect circle but rather an ellipse, as Figure 2-20*a* shows in exaggerated

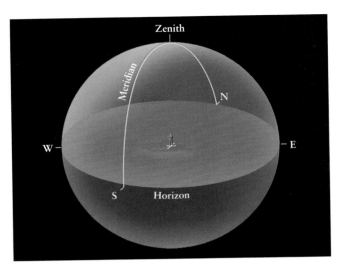

figure 2-19

The Meridian The meridian is a circle on the celestial sphere that passes through the observer's zenith (the point directly overhead) and the north and south points on the observer's horizon. The passing of celestial objects across the meridian can be used to measure time. The upper meridian is the part above the horizon, and the lower meridian (not shown) is the part below the horizon.

form. As we will learn in Chapter 4, the Earth moves more rapidly along its orbit when it is near the Sun than when it is farther away. Hence, the Sun appears to us to move more than 1° per day along the ecliptic in January, when the Earth is nearest the Sun, and less than 1° per day in July, when the Earth is farthest from the Sun. By itself, this effect would cause the apparent solar day to be longer in January than in July.

The second reason why the Sun is not a good timekeeper is the 23½° angle between the ecliptic and the celestial equator (see Figure 2-14). As Figure 2-20*b* shows, this causes a significant part of the Sun's apparent motion when near the equinoxes to be in a north-south direction. The net daily eastward progress in the sky is then somewhat foreshortened. At the summer and winter solstices, by contrast, the Sun's motion

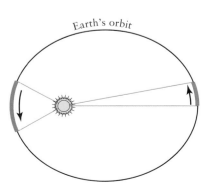

a **A month's motion of the Earth along its orbit**

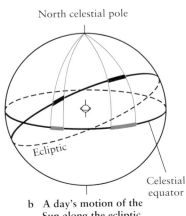

b **A day's motion of the Sun along the ecliptic**

figure 2-20

Why the Sun Is a Poor Timekeeper There are two main reasons that the Sun is a poor timekeeper. **(a)** The Earth's speed along its orbit varies during the year. It moves fastest when closest to the Sun in January and slowest when farthest from the Sun in July. Hence, the apparent speed of the Sun along the ecliptic is not constant. **(b)** Because of the tilt of the Earth's rotation axis, the ecliptic is inclined with respect to the celestial equator. Therefore, the projection of the Sun's daily progress along the ecliptic onto the celestial equator (shown in blue) varies during the year. This causes further variations in the length of the apparent solar day.

is parallel to the celestial equator. Thus, there is no comparable foreshortening around the beginning of summer or winter. This effect by itself would make the apparent solar day shorter in March and September than in June or December. Combining these effects with those due to the Earth's noncircular orbit, we find that the length of the apparent solar day varies in a complicated fashion over the course of a year.

To avoid these difficulties, astronomers invented an imaginary object called the **mean sun** that moves along the celestial equator at a uniform rate. (In science and mathematics, "mean" is a synonym for "average.") The mean sun is sometimes slightly ahead of the real Sun in the sky, sometimes behind. As a result, mean solar time and apparent solar time can differ by as much as a quarter of an hour at certain times of the year.

Because the mean sun moves at a constant rate, it serves as a fine timekeeper. A **mean solar day** is the interval between successive upper meridian transits of the mean sun. It is exactly 24 hours long, the average length of an apparent solar day. One 24-hour day as measured by your alarm clock or wristwatch is a mean solar day.

Time zones were invented for convenience in commerce, transportation, and communication. In a time zone, all clocks and watches are set to the mean solar time for a meridian of longitude that runs approximately through the center of the zone. Time zones around the world are generally centered on meridians of longitude at 15° intervals. In most cases, going from one time zone to the next requires you to change the time on your wristwatch by exactly one hour. The time zones for most of North America are shown in Figure 2-21.

In order to coordinate their observations with colleagues elsewhere around the globe, astronomers often keep track of time using Coordinated Universal Time, somewhat confusingly abbreviated UTC or UT. This is the time in a zone that includes Greenwich, England, a seaport just outside of London where the first internationally accepted time standard was kept. In former times UTC was known as Greenwich Mean Time. In North America, Eastern Standard Time (EST) is 5 hours different from UTC; 9:00 A.M. EST is 14:00 UTC. Coordinated Universal Time is also used by aviators and sailors, who regularly travel from one time zone to another.

Although it is natural to want our clocks and method of timekeeping to be related to the Sun, astronomers often use a system that is based on the apparent motion of the stars. This system, called **sidereal time,** is useful when aiming a telescope. Most observatories are therefore equipped with a clock that measures sidereal time, as discussed in Box 2-2.

2-8 Astronomical observations led to the development of the modern calendar

Just as the day is a natural unit of time based on the Earth's rotation, the year is a natural unit of time based on the Earth's revolution about the Sun. However, nature has not arranged things for our convenience. The year does not divide into

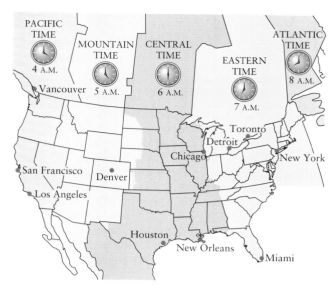

figure 2-21

Time Zones in North America For convenience, the Earth is divided into 24 time zones, generally centered on 15° intervals of longitude around the globe. There are four time zones across the continental United States, making for a 3-hour time difference between New York and California.

exactly 365 whole days. Ancient astronomers realized that the length of a year is approximately 365¼ days, so the Roman emperor Julius Caesar established the system of "leap years" to account for this extra quarter of a day. By adding an extra day to the calendar every four years, he hoped to ensure that seasonal astronomical events, such as the beginning of spring, would occur on the same date year after year.

Caesar's system would have been perfect if the year were exactly 365¼ days long and if there were no precession. Unfortunately, this is not the case. To be more accurate, astronomers now use several different types of years. For example, the **sidereal year** is defined to be the time required for the Sun to return to the same position with respect to the stars. It is equal to 365.2564 mean solar days, or $365^d\ 6^h\ 9^m\ 10^s$.

The sidereal year is the orbital period of the Earth around the Sun, but it is not the year on which we base our calendar. Like Caesar, most people want annual events—in particular, the first days of the seasons—to fall on the same date each year. For example, we want the first day of spring to occur on March 21. But spring begins when the Sun is at the vernal equinox, and the vernal equinox moves slowly against the background stars because of precession. Therefore, to set up a calendar we use the **tropical year,** which is equal to the time needed for the Sun to return to the vernal equinox. This period is equal to 365.2422 mean solar days, or $365^d\ 5^h\ 48^m\ 46^s$. Because of precession, the tropical year is 20 minutes and 24 seconds shorter than the sidereal year.

Caesar's assumption that the tropical year equals 365¼ days was off by 11 minutes and 14 seconds. This tiny error

box 2-2 | Tools of the Astronomer's Trade

Sidereal Time

If you want to observe a particular object in the heavens, the ideal time to do so is when the object is high in the sky, on or close to the upper meridian. This minimizes the distorting effects of the Earth's atmosphere, which increase as you view closer to the horizon. For astronomers who study the Sun, this means making observations at local noon, which is not too different from noon as determined using mean solar time. For astronomers who observe planets, stars, or galaxies, however, the optimum time to observe depends on the particular object to be studied. The problem is this: Given the location of a given object on the celestial sphere, when will that object be on the upper meridian?

To answer this question, astronomers use *sidereal time* rather than solar time. It is different from the time on your wristwatch. In fact, a sidereal clock and an ordinary clock even tick at different rates, because they are based on different astronomical objects. Ordinary clocks are related to the position of the Sun, while sidereal clocks are based on the position of the vernal equinox, the location from which right ascension is measured (see Box 2-1 for a discussion of right ascension).

Regardless of where the Sun is, midnight sidereal time at your location is defined to be when the vernal equinox crosses your upper meridian. (Like solar time, sidereal time depends on where you are on Earth.) A **sidereal day** is the time between two successive upper meridian passages of the vernal equinox. By contrast, an apparent solar day is the time between two successive upper meridian crossings of the Sun. The illustration shows why these two kinds of day are not equal. Because the Earth orbits the Sun, the Earth must make one complete rotation plus about 1° to get from one local solar noon to the next. This extra 1° of rotation corresponds to 4 minutes of time, which is the amount by which a solar day exceeds a sidereal day. To be precise:

$$1 \text{ sidereal day} = 23^{\text{h}} 56^{\text{m}} 4.091^{\text{s}}$$

where the hours, minutes, and seconds are in mean solar time.

One day according to your wristwatch is one mean solar day, which is exactly 24 hours of solar time long. A **sidereal clock** measures sidereal time in terms of sidereal hours, minutes, and seconds, where one sidereal day is divided into 24 sidereal hours. This explains why a sidereal clock ticks at a slightly different rate than your wristwatch. As a result, at some times of the year a sidereal clock will show a very different time than an ordinary clock. (On local noon on March 21, when the Sun is at the vernal equinox, a sidereal clock will say that it is midnight. Do you see why?)

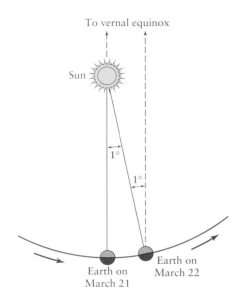

We can now answer the question in the opening paragraph. The vernal equinox, whose celestial coordinates are R.A. = 0^{h} 0^{m} 0^{s}, Decl. = 0° 0′ 0″, crosses the upper meridian at midnight sidereal time (0:00). The autumnal equinox, which is on the opposite side of the celestial sphere at R.A. = 12^{h} 0^{m} 0^{s}, Decl. = 0° 0′ 0″, crosses the upper meridian 12 sidereal hours later at noon sidereal time (12:00). As these examples illustrate, *any* object crosses the upper meridian when the sidereal time is equal to the object's right ascension. That is why astronomers measure right ascension in units of time rather than degrees, and why right ascension is always given in sidereal hours, minutes, and seconds.

EXAMPLE: Suppose you want to observe the bright star Regulus (Figure 2-6), which has epoch 2000 coordinates R.A. = 10^{h} 08^{m} 22.2^{s}, Decl. = +11° 58′ 02″. What is the best time to do this? Regulus will be on the upper meridian and thus ideally suited for observation when the sidereal time at your location is equal to its right ascension, or about 10:08. As this example shows, it is useful to keep track of sidereal time, which is why most observatories are equipped with a sidereal clock.

While sidereal time is extremely useful in astronomy, mean solar time is still the best method of timekeeping for most earthbound purposes. All time measurements in this book are expressed in mean solar time unless otherwise stated.

adds up to about three days every four centuries. Although Caesar's astronomical advisers were aware of the discrepancy, they felt that it was too small to matter. However, by the sixteenth century the first day of spring was occurring on March 11.

The Roman Catholic Church became concerned because Easter kept shifting to progressively earlier dates. To straighten things out, Pope Gregory XIII instituted a calendar reform in 1582. He began by dropping ten days (October 4, 1582, was followed by October 15, 1582), which brought the first day of spring back to March 21. Next, he modified Caesar's system of leap years.

Caesar had added February 29 to every calendar year that is evenly divisible by four. Thus, for example, 1988, 1992, and 1996 were all leap years with 366 days. But we have seen that this system produces an error of about three days every four centuries. To solve the problem, Pope Gregory decreed that only the century years evenly divisible by 400 should be leap years. For example, the years 1700, 1800, and 1900 (which would have been leap years according to Caesar) were not leap years in the improved Gregorian system, but the year 2000, which can be divided evenly by 400, *was* a leap year.

We use the Gregorian system today. It assumes that the year is 365.2425 mean solar days long, which is very close to the true length of the tropical year. In fact, the error is only one day in every 3300 years. That won't cause any problems for a long time.

KEY WORDS

Terms preceded by an asterisk (*) are discussed in the Boxes.

Antarctic Circle, p. 32

apparent solar day, p. 34

apparent solar time, p. 34

Arctic Circle, p. 31

autumnal equinox, p. 30

celestial equator, p. 27

celestial sphere, p. 26

circumpolar, p. 27

constellation, p. 22

*declination, p. 28

diurnal motion, p. 23

ecliptic, p. 30

epoch, p. 33

equinox, p. 30

lower meridian, p. 34

mean solar day, p. 35

mean sun, p. 35

meridian, p. 34

meridian transit, p. 34

north celestial pole, p. 27

positional astronomy, p. 21

precession, p. 32

precession of the equinoxes, p. 33

*right ascension, p. 28

*sidereal clock, p. 36

*sidereal day, p. 36

sidereal time, p. 35

sidereal year, p. 35

south celestial pole, p. 27

summer solstice, pp. 30–31

time zone, p. 35

tropical year, p. 35

Tropic of Cancer, p. 32

Tropic of Capricorn, p. 32

upper meridian, p. 34

vernal equinox, p. 30

winter solstice, p. 31

zenith, p. 27

zodiac, p. 32

KEY IDEAS

Ideas preceded by an asterisk (*) are discussed in the Boxes.

Constellations and the Celestial Sphere: It is convenient to imagine the stars fixed to the celestial sphere with the Earth at its center.

• The surface of the celestial sphere is divided into 88 regions called constellations.

Diurnal (Daily) Motion of the Celestial Sphere: The celestial sphere appears to rotate around the Earth once in each 24-hour period. In fact, it is actually the Earth that is rotating.

• The poles and equator of the celestial sphere are determined by extending the axis of rotation and the equatorial plane of the Earth out to the celestial sphere.

• *The positions of objects on the celestial sphere are described by specifying their right ascension (in time units) and declination (in angular measure).

Seasons and the Tilt of the Earth's Axis: The Earth's axis of rotation is tilted at an angle of about 23½° from the perpendicular to the plane of the Earth's orbit.

• The seasons are caused by the tilt of the Earth's axis.

• Over the course of a year, the Sun appears to move around the celestial sphere along a path called the ecliptic. The ecliptic is inclined to the celestial equator by about 23½°.

• The ecliptic crosses the celestial equator at two points in the sky, the vernal and autumnal equinoxes. The northernmost point that the Sun reaches on the celestial sphere is the summer solstice, and the southernmost point is the winter solstice.

Precession: The orientation of the Earth's axis of rotation changes slowly, a phenomenon called precession.

• Precession is caused by the gravitational pull of the Sun and Moon on the Earth's equatorial bulge.

• Precession of the Earth's axis causes the positions of the equinoxes and celestial poles to shift slowly.

• *Because the system of right ascension and declination is tied to the position of the vernal equinox, the date (or epoch) of observation must be specified when giving the position of an object in the sky.

Timekeeping: Astronomers use several different means of keeping time.

• Apparent solar time is based on the apparent motion of the Sun across the celestial sphere, which varies over the course of the year.

• Mean solar time is based on the motion of an imaginary mean sun along the celestial equator, which produces a uniform mean solar day of 24 hours. Ordinary watches and clocks measure mean solar time.

• *Sidereal time is based on the apparent motion of the celestial sphere.

The Calendar: The tropical year is the period between two passages of the Sun across the vernal equinox. Leap year corrections are needed because the tropical year is not exactly 365 days. The sidereal year is the actual orbital period of the Earth.

REVIEW QUESTIONS

1. A fellow student tells you that only the stars in Figure 2-2b that are connected by blue lines are part of the constellation Orion. How would you respond?

2. How are constellations useful to astronomers? How many stars are not part of any constellation?

3. Why can't a person in Antarctica use the Big Dipper to find the north direction?

4. What is the celestial sphere? Why is this ancient concept still useful today?

5. Imagine that someone suggests sending a spacecraft to land on the surface of the celestial sphere. How would you respond to such a suggestion?

6. Is there any place on Earth where you could see the north celestial pole on the northern horizon? If so, where? Is there any place on Earth where you could see the north celestial pole on the western horizon? If so, where? Explain your answers.

7. What is the celestial equator? How is it related to the Earth's equator? How are the north and south celestial poles related to the Earth's axis of rotation? Where on Earth would you have to be for the celestial equator to pass through your zenith?

8. How many degrees is the angle from the horizon to the zenith? Does your answer depend on what point on the horizon you choose?

9. Using a diagram, explain why the tilt of the Earth's axis relative to the Earth's orbit causes the seasons as we orbit the Sun.

10. Give two reasons why it's warmer in summer than in winter.

11. What is the ecliptic? Why is it tilted with respect to the celestial equator? Does the Sun appear to move along the ecliptic, the celestial equator, or neither? By about how many degrees does the Sun appear to move on the celestial sphere each day?

12. Where on Earth do you have to be in order to see the north celestial pole directly overhead? What is the maximum possible elevation of the Sun above the horizon at that location? On what date can this maximum elevation be observed?

13. What are the vernal and the autumnal equinoxes? What are the summer and winter solstices? How are these four points related to the ecliptic and the celestial equator?

14. At what point on the horizon does the vernal equinox rise? Where on the horizon does it set? (*Hint:* See Figure 2-15.)

15. How does the daily path of the Sun across the sky change with the seasons? Why does it change?

16. Where on Earth do you have to be in order to see the Sun at the zenith? Will it be at the zenith every day? Explain.

17. What causes precession of the equinoxes? How long does it take for the vernal equinox to move 1° along the ecliptic?

18. What is the (fictitious) mean sun? What path does it follow on the celestial sphere? Why is it a better timekeeper than the actual Sun in the sky?

19. Why is it convenient to divide the Earth into time zones?

20. Why is the time given by a sundial not necessarily the same as the time on your wristwatch?

21. What is the difference between the sidereal year and the tropical year? Why are these two kinds of year slightly different in length? Why are calendars based on the tropical year?

22. When is the next leap year? Was 2000 a leap year? Will 2100 be a leap year?

ADVANCED QUESTIONS

Questions preceded by an asterisk (*) involve topics discussed in the Boxes.

> ### Problem-solving tips and tools
>
> To help you visualize the heavens, it is worth taking the time to become familiar with various types of star charts. These include the simple star charts at the end of this book, the monthly star charts published in such magazines as *Sky & Telescope* and *Astronomy*, and the more detailed maps of the heavens found in star atlases. One of the best ways to understand the sky and its motions is to use the *Starry Night* computer program on the CD-ROM that accompanies this book. This easy-to-use program allows you to view the sky on any date and at any time, as seen from any point on Earth, and to animate the sky to visualize its diurnal and annual motions.
>
> You may also find it useful to examine a planisphere, a device consisting of two rotatable disks. The bottom disk shows all the stars in the sky (for a particular latitude), and the top one is opaque with a transparent oval window through which only some of the stars can be seen. By rotating the top disk, you can immediately see which constellations are above the horizon at any time of the year. A planisphere is a convenient tool to carry with you when you are out observing the night sky.

23. On November 1 at 11:00 P.M. you look toward the eastern horizon and see the bright star Mintaka (shown in Figure 2-2b) rising. At approximately what time will Mintaka rise one week later, on November 8?

24. Figure 2-4 shows the situation on September 21, when Cygnus is highest in the sky at 8:00 P.M. local time and Andromeda is highest in the sky at midnight. But as Figure 2-5 shows, on July 21 Cygnus is highest in the sky at midnight. On July 21, at approximately what local time is Andromeda highest in the sky? Explain your reasoning.

25. Figure 2-5 shows which constellations are high in the sky (for observers in the northern hemisphere) in the months of July, September, and November. From this figure, would you be able to see Perseus at midnight on May 15? Draw a picture to justify your answer.

26. Figure 2-6 shows the appearance of Polaris, the Little Dipper, and the Big Dipper at midnight on June 1. Sketch how these objects would appear on this same date at (**a**) 9 P.M. and (**b**) 3 A.M. Include the horizon in your sketches, and indicate the north direction.

27. (**a**) Redraw Figure 2-10 for an observer at the North Pole. (*Hint:* The north celestial pole is directly above this observer.) (**b**) Redraw Figure 2-10 for an observer at the

equator. (*Hint:* The celestial equator passes through this observer's zenith.) (**c**) Using Figure 2-10 and your drawings from (a) and (b), justify the following rule, long used by navigators: The latitude of an observer in the northern hemisphere is equal to the angle in the sky between that observer's horizon and the north celestial pole. (**d**) State the rule that corresponds to (c) for an observer in the southern hemisphere.

28. The time-exposure photograph in Figure 2-11 shows the trails made by individual stars as the celestial sphere appears to rotate around the Earth. (**a**) If you were standing in Australia next to the camera that took this picture and looking in the same direction that the camera was pointing, would you see the stars moving clockwise or counterclockwise? (**b**) For approximately what length of time was the camera shutter left open to take this photograph?

29. Is there any place on Earth where all the visible stars are circumpolar? If so, where? Is there any place on Earth where none of the visible stars is circumpolar? If so, where? Explain your answers.

30. In the northern hemisphere, houses are designed to have "southern exposure," that is, with the largest windows on the southern side of the house. But in the southern hemisphere houses are designed to have "northern exposure." Why are houses designed this way, and why is there a difference between the hemispheres?

31. Figure 2-15 shows the daily path of the Sun across the sky on March 21, June 21, September 21, and December 21 for an observer at 35° north latitude. Sketch a drawing of this kind for (**a**) an observer at 35° south latitude; (**b**) an observer at the equator; and (**c**) an observer at the North Pole.

32. Suppose that you live at a latitude of 40° N. What is the elevation of the Sun above the southern horizon at noon at the time of the winter solstice? Explain your reasoning. Include a drawing as part of your explanation.

33. The city of Mumbai (formerly Bombay) in India is 19° north of the equator. On how many days of the year, if any, is the Sun at the zenith at midday as seen from Mumbai? Explain your answer.

34. The Great Pyramid at Giza has a tunnel that points toward the north celestial pole. At the time the pyramid was built, around 2600 B.C., toward which star did it point? Toward which star does this same tunnel point today? (See Figure 2-18.)

35. Ancient records show that 2000 years ago, the stars of the Southern Cross were visible in the southern sky from Greece. Today, however, these stars cannot be seen from Greece. What accounts for this change?

36. Unlike western Europe, Imperial Russia did not use the revised calendar instituted by Pope Gregory XIII. Explain why the Russian Revolution, which started on November 7, 1917, according to the modern calendar is called "the

October revolution" in Russia. What was this date according to the Russian calendar at the time? Explain.

*37. What is the right ascension of a star that is on the meridian at midnight at the time of the autumnal equinox? Explain.

*38. The coordinates on the celestial sphere of the summer solstice are R.A. = 6^h 0^m 0^s, Decl. = +23° 27′. What are the right ascension and declination of the winter solstice? Explain your answer.

*39. Right ascension is measured in hours, minutes, and seconds. Because 24 hours of right ascension takes you all the way around the celestial equator, it follows that 24^h = 360°. What is the angle in the sky (measured in degrees) between a star with R.A. = 8^h 0^m 0^s, Decl. = 0° 0′ 0″ and a second star with R.A. = 11^h 20^m 0^s, Decl. = 0° 0′ 0″? Explain your answer.

*40. At local noon on March 21, when the Sun is at the vernal equinox, a sidereal clock will say that it is midnight. Explain why.

*41. (a) What is the sidereal time when the vernal equinox rises? (b) On what date is the sidereal time nearly equal to the solar time? Explain.

*42. On a certain night, the first star in Advanced Question 39 passes through the zenith at 12:30 A.M. local time. At what time will the second star pass through the zenith? Explain your answer.

*43. How would the sidereal and solar days change (a) if the Earth's rate of rotation increased, (b) if the Earth's rate of rotation decreased, and (c) if the Earth's rotation were retrograde (that is, if the Earth rotated about its axis opposite to the direction in which it revolves about the Sun)?

44. The image of the Earth shown below was made by the *Galileo* spacecraft while en route to Jupiter. South America

is at the center of the image and Antarctica is at the bottom of the image. (a) In which month of the year was this image made? Explain your reasoning. (b) When this image was made, was the Earth closer to perihelion or to aphelion? Explain your reasoning.

45. The photograph in Figure 2-11 was taken at the Anglo-Australian Observatory (AAO) on Siding Springs Mountain, Australia. Siding Springs is at longitude 149° 03′ 58″ east and latitude 31° 16′ 37″ south. (a) By making measurements on the photograph, find the approximate angular width and angular height of the photo. (b) How far (in degrees) from the south celestial pole can a star be and still be circumpolar as seen from the AAO?

DISCUSSION QUESTIONS

46. Examine a list of the 88 constellations. Are there any constellations whose names obviously date from modern times? Where are these constellations located? Why do you suppose they do not have archaic names?

47. Why is it useful to astronomers to have telescopes in both the southern hemisphere and the northern hemisphere? (See Figure 2-10.)

48. Describe how the seasons would be different if the Earth's axis of rotation, rather than having its present 23½° tilt, were tilted (a) by 0° or (b) by 90°.

49. In William Shakespeare's *Julius Caesar* (act 3, scene 1), Caesar says:

> But I am constant as the northern star,
> Of whose true-fix'd and resting quality
> There is no fellow in the firmament.

Translate Caesar's statement about the "northern star" into modern astronomical language. Is the northern star truly "constant"? Was the northern star the same in Shakespeare's time (1564–1616) as it is today?

WEB/CD-ROM QUESTIONS

50. Search the World Wide Web for information about the national flags of Australia, New Zealand, and Brazil and the state flag of Alaska. Which stars are depicted on these flags? Explain any similarities or differences among these flags.

51. Some people say that on the date that the Sun is at the vernal equinox, and only on this date, you can stand a raw egg on end. Others say that there is nothing special about the vernal equinox, and that with patience you can stand a raw egg on end on any day of the year. Search the World Wide Web for information about this story and for hints about how to stand an egg on end. Use these hints to try the experiment yourself on a day when the Sun is *not* at the

(NASA/JPL)   R I **V** U X G

vernal equinox. What do you conclude about the connection between eggs and equinoxes?

 52. Use the U.S. Naval Observatory web site to find the times of sunset and sunrise on (**a**) your next birthday and (**b**) the date this assignment is due. (**c**) Are the times the same for the two dates? Explain why or why not.

OBSERVING PROJECTS

Observing tips and tools

 Moonlight is so bright that it interferes with seeing the stars. For the best view of the constellations, do your observing when the Moon is below the horizon. You can find the times of moonrise and moonset in your local newspaper or on the World Wide Web. Each monthly issue of the magazines *Sky & Telescope* and *Astronomy* includes much additional observing information.

53. On a clear, cloud-free night, use the star charts at the end of this book to see how many constellations of the zodiac you can identify. Which ones were easy to find? Which were difficult? Are the zodiacal constellations the most prominent ones in the sky?

54. Examine the star charts that are published monthly in such popular astronomy magazines as *Sky & Telescope* and *Astronomy*. How do they differ from the star charts at the back of this book? On a clear, cloud-free night, use one of these star charts to locate the celestial equator and the ecliptic. Note the inclination of the Milky Way to the ecliptic and celestial equator. The Milky Way traces out the plane of our galaxy. What do your observations tell you about the orientation of the Earth and its orbit relative to the galaxy's plane?

55. Suppose you wake up before dawn and want to see which constellations are in the sky. Explain how the star

charts at the end of this book can be quite useful, even though chart times are given only for the evening hours. Which chart most closely depicts the sky at 4:00 A.M. tomorrow morning? Set your alarm clock for 4:00 A.M. to see if you are correct.

 56. Use the *Starry Night* program to observe the diurnal motion of the sky. (**a**) First set *Starry Night* to display the sky as seen from where you live. Select **Set Home Location...** in the **Go** menu and click on the **Lookup** button to find your city or town. Then, using the hand cursor, center your field of view on the northern horizon (if you live in the northern hemisphere) or the southern horizon (if you live in the southern hemisphere). In the Control Panel at the top of the main window, set the time step to 3 minutes and click on the "Forward" button (a triangle that points to the right). To see the stars during the daytime, turn off daylight (select **Daylight** in the **Sky** menu). Do the stars appear to rotate clockwise or counterclockwise? Explain in terms of the Earth's rotation. Are any of the stars circumpolar? (**b**) Recenter your field of view on the southern horizon (if you live in the northern hemisphere) or the northern horizon (if you live in the southern hemisphere). Describe what you see. Are any of these stars circumpolar?

 57. Use the *Starry Night* program to observe the Sun's motion on the celestial sphere. Select **Atlas** in the **Go** menu to see the entire celestial sphere, including the part below the horizon. Center on the Sun by using the **Find...** command in the **Edit** menu. (**a**) Select **Auto Identify** in the **Constellations** menu to display the constellation at the center of the screen. In which constellation is the Sun located today? Is this the same as the astrological sign for today's date? Explain in terms of precession. (**b**) In the Control Panel at the top of the main window, set the time step to 3 days and click on the "Forward" button (a triangle that points to the right). Observe the Sun for a full year of simulated time. How does the Sun appear to move against the constellations? What path does it follow? Does it ever change direction?

JAMES RANDI

Why Astrology Is Not Science

James ("the Amazing") Randi works tirelessly to expose trickery, so that others can relish the greater wonder of science. As a magician, he has had his own television show and an enormous public following. As a lecturer, he addresses teachers, students, and others worldwide. His newsletter and column for *The Skeptic* are key resources for educators. His many books include *Flim-Flam!*, *The Faith Healers*, and *The Mask of Nostradamus*, about a legendary con man with secrets of his own.

Mr. Randi helped found the Committee for the Scientific Investigation of Claims of the Paranormal, and his $10,000 prize for "the performance of any paranormal event . . . under proper observing conditions" has gone unclaimed for more than 25 years. An amateur archeologist and astronomer as well, he lives in Florida with several untalented parrots and the occasional visiting magus.

I'm involved in the strange business of telling folks what they should already know. I meet audiences who believe in all sorts of impossible things, often despite their education and intelligence. My job is to explain how science differs from the unproven, illogical assumptions of pseudoscience—and why it matters. Perhaps my best example is the difference between astronomy and astrology.

Both astrology and astronomy arose from the wonders of the night sky, from the stars to comets, planets, the Sun, and the Moon. Surely, humans have long reasoned, there must be some meaning in their motions. Surely the Moon's effect on tides hints at hidden "causes" for strange events. *Judiciary* (literally "judging") astrology therefore attempted to foretell the future—our earthly future. To serve it, *horary* (literally "hourly") astrology carefully tracked the heavens.

It is the latter that has become astronomy. Thanks to its process of careful measurement and testing, we now understand more about the true nature of the starry universe than astrologers could ever have imagined. With the birth of a new science, astronomers had a logical framework based on physical causes and systematic observations.

Astrology remains a popular delusion. Far too many believe today that patterns in the sky govern our lives. They accept the vague tendencies and portents of seers who cast horoscopes. They shouldn't. Just a glance at the tenets of astrology provides ample evidence of its absurdity.

An individual is said to be born under a sign. To the astrologer, the Sun was located "in" that sign at the moment of birth. (Stars are not seen in the daytime, but no matter—a calculation tells where the Sun is.) Each sign takes its name from a constellation, a totally imaginary figure invented for our convenience in referring to stars. Different cultures have different mythical figures up there, and so different schools of astrology assign different meanings to the signs they use.

In the spirit of equal-opportunity swindling, astrologers divide up the year fairly, ignoring variations in the size of constellations. Since Libra is tiny, while Virgo is huge, they chop some of the sky off Virgo and add it—along with bits of Scorpio—to bring Libra up to size. The Sun could well be declared "in" Libra when it is actually outside that constellation.

It gets worse. Science constantly challenges itself and changes. The rules of astrology could not, although they were made up thousands of years ago, and since then the "fixed" stars have moved. In particular, precession of the equinoxes has shifted objects in the sky relative to our calendar. The constellations have changed but astrology has not. If you were born August 7, you are said to be a Leo, but the Sun that day was really in the same part of the sky as the constellation Cancer.

With a theory like this to back it up, we should not be surprised at the bottom line: *A pseudoscience does not work*. Test after test has checked its predictions, and the result is always the same. One such investigator is Shawn Carlson of the University of California, San Diego. As he put it in *Nature* magazine, astrology is "a hopeless cause." Johannes Kepler, the pioneering astronomer, himself cast horoscopes, but they are little remembered today. Owen Gingerich, a historian of science at Harvard, puts it well: Kepler was the astrologer who destroyed astrology.

Astronomy works, and it works very well indeed. That isn't easy. Because we humans tend to find what we want in any body of data, it takes science's careful process of observation, creative insight, and critical thinking to understand and predict changes in nature. As I write, a transit of Ganymede is due next Thursday at 21:47:20. At exactly that time, the satellite of Jupiter will cross in front of its planet as seen from Earth, and yet most of us will never know it. Still other moons of Jupiter may hold fresh clues to the formation of our entire solar system and the conditions for life elsewhere.

For most people, astronomy has too little fantasy or money in it, and they will never experience the beauty in its predictions. The dedicated labors of generations of scientists have enabled us to perform a genuine wonder.

Eclipses and the Motion of the Moon

R I V U X G

WEB LINK 3.1

On August 11, 1999, millions of humans stood along a narrow corridor of land that extended from England through Europe and the Middle East to India. For a few brief minutes, they saw the rare cosmic spectacle shown here: a total solar eclipse. As the Moon covered the disk of the Sun, the stars and planets became visible in the darkening sky, and the solar corona—the Sun's thin outer atmosphere, which glows with an unearthly pearlescent light—was revealed.

Not everyone will ever see the Moon cover the Sun in this way. But anyone can find the Moon in the sky and observe how its appearance changes from night to night, from new moon to full moon and back again. The times when the Moon rises and sets also differ noticeably from one night to the next.

In this chapter our subject is how the Moon moves as seen from the Earth. We will explore why the Moon goes through a regular cycle of phases, and how the Moon's orbit around the Earth leads to solar eclipses as well as lunar eclipses. We will also see how ancient astronomers used their observations of the Moon to determine the size and shape of the Earth, as well as other features of the solar system. Thus, the Moon—which has always loomed large in the minds of poets, lovers, and dreamers—has also played a key role in the development of our modern picture of the universe.

As you read the sections of this chapter, look for the answers to the following questions.

3-1   Why does the Moon go through phases?

3-2   Is there such a thing as the "dark side of the Moon"?

3-3   What is the difference between a lunar eclipse and a solar eclipse?

3-4   How often do lunar eclipses happen? When one is taking place, where do you have to be to see it?

3-5   How often do solar eclipses happen? Why are they visible only from certain special locations on Earth?

3-6   How did ancient astronomers deduce the sizes of the Earth, the Moon, and the Sun?

3-1   The phases of the Moon are caused by its orbital motion

As seen from Earth, both the Sun and the Moon appear to move from west to east on the celestial sphere—that is, relative to the background of stars. They move at very different rates, however. The Sun takes one year to make a complete trip around the imaginary celestial sphere along the path we call the *ecliptic* (Section 2-5). By comparison, the Moon takes only about four weeks. In the past, these similar motions led people to believe that both the Sun and the Moon orbit around the Earth. We now know that only the Moon orbits the Earth, while the Earth-Moon system as a whole (Figure 3-1) orbits the Sun. (In Chapter 4 we will learn how this was discovered.)

One key difference between the Sun and the Moon is the nature of the light that we receive from them. The Sun emits its own light. So do the stars, which are objects like the Sun but much farther away, and so does an ordinary lightbulb. By contrast, the light that we see from the Moon is reflected light. This is sunlight that has struck the Moon's surface, bounced off, and ended up in our eyes here on Earth.

 You probably associate *reflection* with shiny objects like a mirror or the surface of a still lake. In science, however, the term refers to light bouncing off any object. You see most objects around you by reflected light. When you look at your hand, for example, you are seeing light from the Sun (or from a light fixture) that has been reflected from the skin of your hand and into your eye. In the same way, moonlight is really sunlight that has been reflected by the Moon's surface.

Figure 3-1 shows both the Moon and the Earth as seen from a spacecraft. When this picture was recorded, the Sun was far off to the right. Hence, only the right-hand hemispheres of both worlds were illuminated by the Sun; the left-hand hemispheres were in darkness and are not visible in the picture. In the same way, when we view the Moon from the Earth, we see only the half of the Moon that faces the Sun and

is illuminated. However, not all of the illuminated half of the Moon is necessarily facing us. As the Moon moves around the Earth, from one night to the next we see different amounts of the illuminated half of the Moon. These different appearances of the Moon are called **lunar phases.**

ƒigure 3-1   R I **V** U X G

The Earth and the Moon This picture of the Earth and the Moon was taken in 1992 by the *Galileo* spacecraft on its way toward Jupiter. The Sun, which provides the illumination for both the Earth and the Moon, was far to the right and out of the camera's field of view when this photograph was taken. (NASA/JPL)

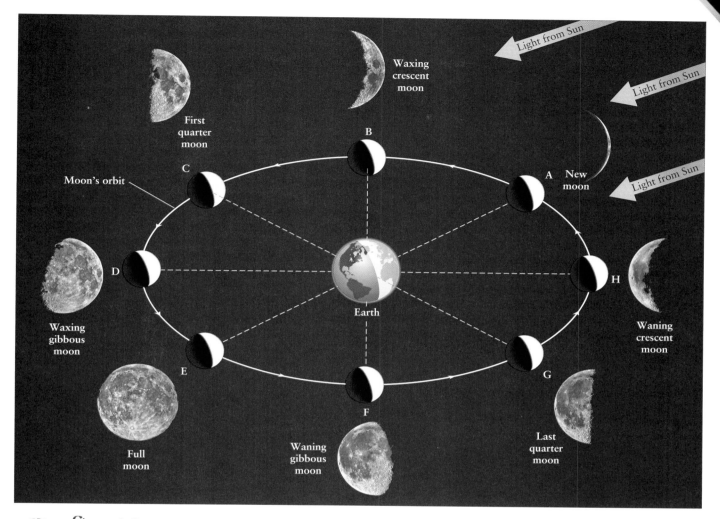

figure 3-2

Why the Moon Goes Through Phases This figure shows the Moon at eight positions on its orbit, along with photographs of what the Moon looks like at each position as seen from Earth. The changes in phase occur because light from the Sun illuminates one half of the Moon, and as the Moon orbits the Earth we see varying amounts of the Moon's illuminated half. It takes about 29½ days for the Moon to go through a complete cycle of phases. (Photographs from Lick Observatory)

Figure 3-2 shows the relationship between the lunar phase visible from Earth and the position of the Moon in its orbit. For example, when the Moon is at position A, we see it in roughly the same direction in the sky as the Sun. Hence, the dark hemisphere of the Moon faces the Earth. This phase, in which the Moon is not visible, is called **new moon.**

As the Moon continues around its orbit from position A in Figure 3-2, more of its illuminated half becomes exposed to our view. The result, shown at position B, is a phase called **waxing crescent moon** ("waxing" is a synonym for "increasing"). About a week after new moon, the Moon is at position C; we then see half of the Moon's illuminated hemisphere and half of the dark hemisphere. This phase is called **first quarter moon.**

Despite the name, a first quarter moon appears to be *half* illuminated, not one-quarter illuminated! The name means that this phase is one-quarter of the way through the complete cycle of lunar phases.

About four days later, the Moon reaches position D in Figure 3-2. Still more of the illuminated hemisphere can now be seen from Earth, giving us the phase called **waxing gibbous moon** ("gibbous" is another word for "swollen"). When you look at the Moon in this phase, as in the waxing crescent and first quarter phases, the illuminated part of the Moon is on the right (toward the west). Two weeks after new moon, when the Moon stands opposite the Sun in the sky (position E), we see the fully illuminated hemisphere. This phase is called **full moon.**

box 3-1   The Heavens on the Earth

Phases and Shadows

figure 3-2 shows how the relative positions of the Earth, Moon, and Sun explain the phases of the Moon. You can visualize lunar phases more clearly by doing a simple experiment here on Earth. All you need are a small round object, such as an orange or a baseball, and a bright source of light, such as a street lamp or the Sun.

In this experiment, you play the role of an observer on the Earth looking at the Moon, and the round object plays the role of the Moon. The light source plays the role of the Sun. Hold the object in your right hand with your right arm stretched straight out in front of you, with the object directly between you and the light source (position A in the accompanying illustration). In this orientation, the illuminated half of the object faces away from you, like the Moon when it is its new phase (position A in Figure 3-2). By analogy, the round object in your hand is in its "new" phase when you hold it between your eyes and the light source.

Now, slowly turn your body to the left so that the object in your hand "orbits" around you (toward positions C, E, and G in the illustration). As you turn, more and more of the illuminated side of the "moon" in your hand will become visible, and it will appear to go through the same cycle of phases—waxing crescent, first quarter, and waxing gibbous—as does the real Moon. When you have rotated

through half a turn so that the light source is directly behind you, you will be looking face on at the illuminated side of the object in your hand. This corresponds to a full moon (position E in Figure 3-2). Make sure your body does not cast a shadow on the "moon" in your hand—that would correspond to a lunar eclipse!

As you continue turning to the left, more of the unilluminated half of the object will become visible as its phase moves through waning gibbous, last quarter, and waning crescent. When your body has rotated back to the same orientation that you were in originally, the unilluminated half of your hand-held "moon" will again be facing toward you, and its phase will again be new. If you continued to rotate, the object in your hand would repeat the cycle of "phases," just as the Moon does as it orbits around the Earth.

The experiment works best when there is just one light source around. If there are several light sources, such as in a room with several lamps turned on, the different sources will create multiple shadows, and it will be difficult to see the phases of your hand-held "moon." If you do the experiment outdoors using sunlight, you may find that it is best to perform it in the early morning or late afternoon when shadows are most pronounced and the Sun's rays are nearly horizontal.

figure 3-3   R I V U X G

The Moon During the Day The Moon can be seen during the daytime as well as at night. The time of day or night when it is visible depends on its phase. When the Moon is full, as shown in this photograph, it is in the part of the sky directly opposite the Sun (see position E in Figure 3-2). Thus, a full moon rises on the eastern horizon when the Sun is setting in the west, and sets on the western horizon when the Sun is rising in the east. (Art Wolfe)

Over the following two weeks, we see less and less of the Moon's illuminated hemisphere as it continues along its orbit. The Moon is said to be *waning*, or decreasing in illumination as seen from Earth. While the Moon is waning, its illuminated side is on the left (toward the east). The phases are called **waning gibbous moon** (position F), **last quarter moon** (position G), and **waning crescent moon** (position H). The Moon takes about four weeks to complete one orbit around the Earth, so it likewise takes about four weeks for a complete cycle of phases from new moon to full moon and back to new moon.

Figure 3-2 also explains why the Moon is often visible in the daytime, as shown in Figure 3-3. From any location on Earth, about half of the Moon's orbit is visible at any time. For example, if it is midnight at your location, you are in the middle of the dark side of the Earth that faces away from the Sun. At that time you can easily see the Moon at positions C, D, E, F, or G. If it is midday at your location, you are in the middle of the Earth's illuminated side, and the Moon will be easily visible if it is at positions A, B, C, G, or H. (The Moon is so bright that it can be seen even against the bright blue sky.) You can see that the Moon is prominent in the midnight sky for about half of its orbit, and prominent in the midday sky for the other half.

CAUTION! A very common misconception about lunar phases is that they are caused by the shadow of the *Earth* falling on the Moon. As Figure 3-2 shows, this is not the case at all. Instead, phases are simply the result of our seeing the illuminated half of the Moon at different angles as the Moon moves around its orbit. To help you better visualize how this works, Box 3-1 describes how you can simulate the cycle shown in Figure 3-2 using ordinary objects on Earth. (As we will learn in Section 3-3, the Earth's shadow does indeed fall on the Moon on rare occasions. When this happens, we see a lunar eclipse.)

3-2   The Moon's rotation always keeps the same face toward the Earth

The phase of the Moon is continuously changing. But one constant aspect of the Moon is that it always keeps essentially the same hemisphere, or face, toward the Earth. Thus, you will always see the same craters and mountains on the Moon, no matter when you look at it; the only difference will be the angle at which these surface features are illuminated by the Sun. (You can verify this by carefully examining the photographs of the Moon in Figure 3-2.)

CAUTION! Why is it that we only ever see one face of the Moon? You might think that it is because the Moon does not rotate (unlike the Earth, which rotates around an axis that passes from its North Pole to its South Pole). To see that this *cannot* be the case, consider Figure 3-4. This figure shows the Earth and the orbiting Moon from a vantage point far above the Earth's North Pole. In this figure two craters on the lunar surface have been colored, one in red and one in blue. If the Moon did not rotate on its axis, as in Figure 3-4a, at some times the red crater would be visible from Earth, while at other times the blue crater would be visible. Thus, we would see different parts of the lunar surface over time, which does not happen in reality.

In fact, the Moon always keeps the same face toward us because it *is* rotating, but in a very special way: It takes exactly as long to rotate on its axis as it does to make one orbit around the Earth. This situation is called **synchronous rotation**. As Figure 3-4b shows, this keeps the crater shown in red always facing the Earth, so that we always see the same face

figure 3-5

The Sidereal and Synodic Months
The sidereal month is the time the
Moon takes to complete one full revolution
around the Earth with respect to the background
stars. However, because the Earth is constantly
moving along its orbit about the Sun, the Moon
must travel through slightly more than 360° to
get from one new moon to the next. Thus, the
synodic month—the time from one new moon to
the next—is longer than the sidereal month.

3-3 Eclipses occur only when the Sun and Moon are both on the line of nodes

From time to time the Sun, Earth, and Moon all happen to lie along a straight line. When this occurs, the shadow of the Earth can fall on the Moon or the shadow of the Moon can fall on the Earth. Such phenomena are called **eclipses.** They are perhaps the most dramatic astronomical events that can be seen with the naked eye.

A **lunar eclipse** occurs when the Moon passes through the Earth's shadow. This occurs when the Sun, Earth, and Moon are in a straight line, with the Earth between the Sun and Moon so that the Moon is at full phase (position E in Figure 3-2). At this point in the Moon's orbit, the face of the Moon seen from Earth would normally be fully illuminated by the Sun. Instead, it appears quite dim because the Earth casts a shadow on the Moon.

A **solar eclipse** occurs when the Earth passes through the Moon's shadow. As seen from Earth, the Moon moves in front of the Sun. Once again, this can happen only when the Sun, Moon, and Earth are in a straight line. However, for a solar eclipse to occur, the Moon must be between the Earth and the Sun. Therefore, a solar eclipse can occur only at new moon (position A in Figure 3-2).

CAUTION! Both new moon and full moon occur at intervals of 29½ days. Hence, you might expect that there would be a solar eclipse every 29½ days, followed by a lunar eclipse about two weeks (half a lunar orbit) later. But in fact, there are only a few solar eclipses and lunar eclipses per year. Solar and lunar eclipses are so infrequent because the plane of the Moon's orbit and the plane of the Earth's orbit are not exactly aligned, as Figure 3-6 shows. The angle between the plane of the Earth's orbit and the plane of the Moon's orbit is about 5°. Because of this tilt, new moon and full moon usually occur when the Moon is either above or below the plane of the Earth's orbit. When the Moon is not in the plane of the Earth's orbit, the Sun, Moon, and Earth cannot align perfectly, and an eclipse cannot occur.

In order for the Sun, Earth, and Moon to be lined up for an eclipse, the Moon must lie in the same plane as the Earth's orbit around the Sun. This is called the **plane of the ecliptic** because its projection onto the celestial sphere is the same as the path that the Sun appears to us to follow among the stars—that is, the ecliptic (see Figure 2-14). Thus, whenever an eclipse occurs, the Moon appears from Earth to be on the ecliptic (which is how the ecliptic gets its name).

The planes of the Earth's orbit and the Moon's orbit intersect along a line called the **line of nodes,** shown in Figure 3-6. The line of nodes passes through the Earth and is pointed in a particular direction in space. Eclipses can occur only if the line of nodes is pointed toward the Sun—that is, if the Sun lies on or near the line of nodes—and if, at the same time, the Moon lies on or very near the line of nodes. Only then do the Sun,

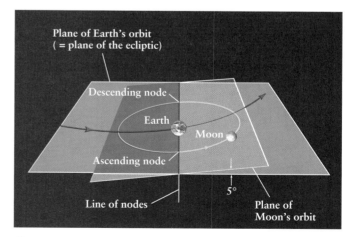

figure 3-6

The Inclination of the Moon's Orbit This drawing shows the Moon's orbit around the Earth (in yellow) and part of the Earth's orbit around the Sun (in red). The plane of the Moon's orbit (shown in brown) is tilted by about 5° with respect to the plane of the Earth's orbit, also called the plane of the ecliptic (shown in blue). These two planes intersect along a line called the line of nodes.

figure 3-7

Conditions for Eclipses Eclipses can take place only if the Sun and Moon are both very near or on the line of nodes. Only then can the Sun, Earth, and Moon all lie along a straight line. A solar eclipse occurs only if the Moon is very near the line of nodes at new moon; a lunar eclipse occurs only if the Moon is very near the line of nodes at full moon. If the Sun and Moon are not near the line of nodes, the Moon's shadow cannot fall on the Earth and the Earth's shadow cannot fall on the Moon.

Earth, and Moon lie in a line straight enough for an eclipse to occur (Figure 3-7).

Anyone who wants to predict eclipses must know the orientation of the line of nodes. But the line of nodes is gradually shifting because of the gravitational pull of the Sun on the Moon. As a result, the line of nodes rotates slowly westward. Astronomers calculate such details to fix the dates and times of upcoming eclipses.

There are at least two—but never more than five—solar eclipses each year. The last year in which five solar eclipses occurred was 1935. The least number of eclipses possible (two solar, zero lunar) happened in 1969. Lunar eclipses occur just about as frequently as solar eclipses, but the maximum possible number of eclipses (lunar and solar combined) in a single year is seven.

3-4 Lunar eclipses can be either total, partial, or penumbral, depending on the alignment of the Sun, Earth, and Moon

The character of a lunar eclipse depends on exactly how the Moon travels through the Earth's shadow. As Figure 3-8 shows, the shadow of the Earth has two distinct parts. In the **umbra,** the darkest part of the shadow, no portion of the

Sun's surface can be seen. A portion of the Sun's surface is visible in the **penumbra,** which therefore is not quite as dark. Most people notice a lunar eclipse only if the Moon passes into the Earth's umbra. As this umbral phase of the eclipse begins, a bite seems to be taken out of the Moon.

The inset in Figure 3-8 shows the different ways in which the Moon can pass into the Earth's shadow. When the Moon passes through only the Earth's penumbra (Path 1), we see a **penumbral eclipse.** During a penumbral eclipse, the Earth blocks only part of the Sun's light and so none of the lunar surface is completely shaded. Because the Moon still looks full but only a little dimmer than usual, penumbral eclipses are easy to miss. If the Moon travels completely into the umbra (Path 2), a **total lunar eclipse** occurs. If only part of the Moon passes through the umbra (Path 3), we see a **partial lunar eclipse.**

If you were on the Moon during a total lunar eclipse, the Sun would be hidden behind the Earth. But some sunlight would be visible through the thin ring of atmosphere around the Earth, just as you can see sunlight through a person's hair if they stand with their head between your eyes and the Sun. As a result, a small amount of light reaches the Moon during a total lunar eclipse, and so the Moon does not completely disappear from the sky as seen from Earth. Most of the light that passes through the Earth's atmosphere is red, and thus the eclipsed Moon glows faintly in reddish hues (Figure 3-9).

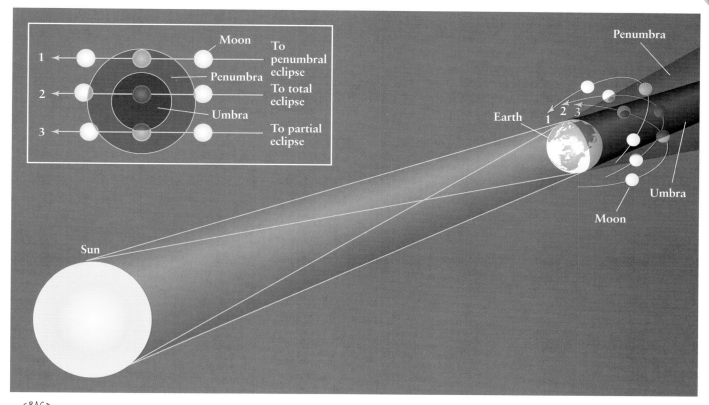

Figure 3-8

Three Types of Lunar Eclipse People on the nighttime side of the Earth see a lunar eclipse when the Moon moves through the Earth's shadow. In the umbra, the darkest part of the shadow, the Sun is completely covered by the Earth. The penumbra is less dark because only part of the Sun is covered by the Earth. The three paths show the motion of the Moon if the lunar eclipse is penumbral (Path 1), total (Path 2), or partial (Path 3). The inset shows these same paths, along with the umbra and penumbra, as viewed from the Earth.

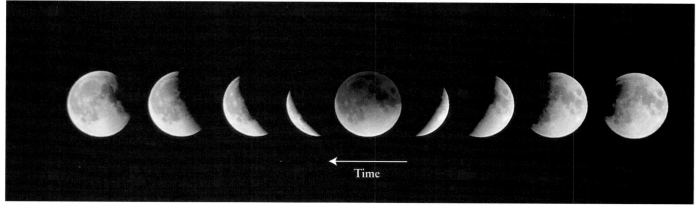

Figure 3-9   R I V U X G

A Total Lunar Eclipse This sequence of nine photographs was taken over a 3-hour period during the lunar eclipse of January 20, 2000. The sequence, which runs from right to left, shows the Moon moving through the Earth's umbra. During the total phase of the eclipse (shown in the center), the Moon has a distinct reddish color. (Fred Espenak, NASA/Goddard Space Flight Center; ©2000 Fred Espenak, MrEclipse.com)

table 3-1	Lunar Eclipses, 2001–2005		
Date	Type	Where visible	Duration of totality (h = hours, m = minutes)
2001 January 9	Total	Eastern Americas, Europe, Africa, Asia	1h 2m
2001 July 5	Partial	Eastern Africa, Asia, Australia, Pacific	—
2001 December 30	Penumbral	Eastern Asia, Australia, Americas	—
2002 May 26	Penumbral	Eastern Asia, Australia, Pacific, western Americas	—
2002 June 24	Penumbral	South America, Europe, Africa, central Asia, Australia	—
2002 November 20	Penumbral	Americas, Europe, Africa, eastern Asia	—
2003 May 16	Total	Central Pacific, Americas, Europe, Africa	53m
2003 November 9	Total	Americas, Europe, Africa, central Asia	24m
2004 May 4	Total	South America, Europe, Africa, Asia, Australia	1h 16m
2004 October 28	Total	Americas, Europe, Africa, central Asia	1h 21m
2005 April 24	Penumbral	Eastern Asia, Australia, Pacific, Americas	—
2005 October 17	Partial	Asia, Australia, Pacific, North America	—

Eclipse predictions by Fred Espenak, NASA/Goddard Space Flight Center. All dates are given in standard astronomical format: year, month, day.

Lunar eclipses occur at full moon, when the Moon is directly opposite the Sun in the sky. Hence, a lunar eclipse can be seen at any place on Earth where the Sun is below the horizon (that is, where it is nighttime). A lunar eclipse has the maximum possible duration if the Moon travels directly through the center of the umbra. The Moon's speed through the Earth's shadow is roughly 1 kilometer per second (3600 kilometers per hour, or 2280 miles per hour), which means that **totality**—the period when the Moon is completely within the Earth's umbra—can last for as long as 1 hour and 42 minutes.

On average, two or three lunar eclipses occur in a year. Table 3-1 lists all twelve lunar eclipses from 2001 to 2005. As the table shows, about half of all lunar eclipses are penumbral.

WEB LINK 3.2

3-5 Solar eclipses can be either total, partial, or annular, depending on the alignment of the Sun, Earth, and Moon

As seen from Earth, the angular diameter of the Moon is almost exactly the same as the angular diameter of the far larger but more distant Sun—about 0.5°. Thanks to this coincidence of nature, the Moon just "fits" over the Sun during a **total solar eclipse.**

A total solar eclipse is a dramatic event. The sky begins to darken, the air temperature falls, and winds increase as the Moon gradually covers more and more of the Sun's disk (Figure 3-10). All nature responds: Birds go to roost, flowers close their petals, and crickets begin to chirp as if evening had arrived. As the last few rays of sunlight peek out from behind the edge of the Moon and the eclipse becomes total,

figure 3-10 R I V U X G

A Total Solar Eclipse This multiple-exposure photograph shows the stages of the total solar eclipse of July 11, 1991, as seen from La Paz, Baja California. The individual exposures were taken at 5-minute intervals as the Sun and Moon moved across the sky from left to right. Notice how the Moon progressively covers then uncovers the Sun. At the center of the photograph, the Sun is totally eclipsed and the solar corona can be seen surrounding the Moon. (Akira Fujii, Hiroyuki Tomioka, and Yonematsu Shiono) the landscape around you is bathed in an eerie gray or, less frequently, in shimmering bands of light and dark. Finally, for a few minutes the Moon completely blocks out the dazzling solar disk and not much else. The **solar corona**—the Sun's thin, hot outer atmosphere, which is normally too dim to be seen—blazes forth in the darkened daytime sky (see the photo that opens this chapter). It is an awe-inspiring sight.

 If you are fortunate enough to see a solar eclipse, keep in mind that the only time when it is safe to look at the Sun is during **totality,** when the solar disk is blocked by the Moon and only the solar corona is visible. Viewing this magnificent spectacle cannot harm you in any way. But you must *never* look directly at the Sun when even a portion of its intensely brilliant disk is exposed. *If you look directly at the Sun at any time without a special filter approved for solar viewing, you will suffer permanent eye damage or blindness.*

To see the remarkable spectacle of a total solar eclipse, you must be inside the darkest part of the Moon's shadow, also called the umbra, where the Moon completely blocks the Sun. Because the Sun and the Moon have nearly the same

angular diameter as seen from Earth, only the tip of the Moon's umbra reaches the Earth's surface (Figure 3-11). As the Earth rotates, the tip of the umbra traces an **eclipse path** across the Earth's surface. Only those locations within the eclipse path are treated to the spectacle of a total solar eclipse. The inset in Figure 3-11 shows the dark spot on the Earth's surface produced by the Moon's umbra.

Immediately surrounding the Moon's umbra is the region of partial shadow called the penumbra. As seen from this area, the Sun's surface appears only partially covered by the Moon. During a solar eclipse, the Moon's penumbra covers a large portion of the Earth's surface, and anyone standing inside the penumbra sees a **partial solar eclipse.** Such eclipses are much less interesting events than total solar eclipses, which is why astronomy enthusiasts strive to be inside the eclipse path. If you are within the eclipse path, you will see a partial eclipse before and after the brief period of totality (see Figure 3-10).

The width of the eclipse path depends primarily on the Earth-Moon distance during totality. The eclipse path is widest if the Moon happens to be at **perigee,** the point in its orbit nearest the Earth. In this case the width of the eclipse

 figure 3-11 R I **V** U X G
The Geometry of a Total Solar Eclipse During a total solar eclipse, the tip of the Moon's umbra reaches the Earth's surface. As the Earth and Moon move along their orbits, this tip traces an eclipse path across the Earth's surface. People within the eclipse path see a total solar eclipse as the tip moves over them. Anyone within the penumbra sees only a partial eclipse. The inset photograph was taken from the *Mir* space station during the August 11, 1999, total solar eclipse (the same eclipse shown in the photo that opens this chapter). The tip of the umbra appears as a black spot on the Earth's surface. At the time the photograph was taken, this spot was 105 km (65 mi) wide and was crossing the English Channel at 3000 km/h (1900 mi/h). (Photograph by Jean-Pierre Haigneré, Centre National d'Etudes Spatiales, France/GSFS)

path can be as great as 270 kilometers (170 miles). In most eclipses, however, the path is much narrower.

In some eclipses the Moon's umbra does not reach all the way to the Earth's surface. This can happen if the Moon is at or near **apogee,** its farthest position from Earth. In this case, the Moon appears too small to cover the Sun completely. The result is a third type of solar eclipse, called an **annular eclipse.** During an annular eclipse, a thin ring of the Sun is seen around the edge of the Moon (Figure 3-12). The length of the Moon's umbra is nearly 5000 kilometers (3100 miles) less than the average distance between the Moon and the Earth's surface. Thus, the Moon's shadow often fails to reach the Earth even when the Sun, Moon, and Earth are properly aligned for an eclipse. Hence, annular eclipses are slightly more common—as well as far less dramatic—than total eclipses.

Even during a total eclipse, most people along the eclipse path observe totality for only a few moments. The Earth's rotation, coupled with the orbital motion of the Moon, causes the umbra to race eastward along the eclipse path at speeds in excess of 1700 kilometers per hour (1060 miles per hour). Because of the umbra's high speed, totality never lasts for more than 7½ minutes. In a typical total solar eclipse, the Sun-Moon-Earth alignment and the Earth-Moon distance are such that totality lasts much less than this maximum.

 The details of solar eclipses are calculated well in advance. They are published in such reference books as the *Astronomical Almanac* and are available on the World Wide Web. Figure 3-13 shows the eclipse paths for all total solar eclipses from 1997 to 2020. Table 3-2 lists all the total, annular, and partial eclipses from 2001 to 2005, including the maximum duration of totality for total eclipses.

figure 3-12   R I **V** U X G

An Annular Solar Eclipse This composite of six photographs taken at sunrise in Costa Rica shows the progress of an annular eclipse of the Sun on December 24, 1973. (Five photographs were made of the Sun, plus one of the hills and sky.) Note that at mideclipse the limb, or outer edge, of the Sun is visible around the Moon. (Courtesy of Dennis di Cicco)

table 3-2	Solar Eclipses, 2001–2005		
Date	**Type**	**Where visible**	**Notes**
2001 June 21	Total	Eastern South America, southern Atlantic, Africa	Maximum duration of totality 4 m 57 s
2001 December 14	Annular	North and Central America, northwest South America	—
2002 June 10	Annular	Eastern Asia, Australia, western North America	—
2002 December 4	Total	Southern Africa, Antarctica, Indonesia, Australia	Maximum duration of totality 2 m 4 s
2003 May 31	Annular	Europe, Asia, northwest North America	—
2003 November 23	Total	Australia, New Zealand, Antarctica, southern South America	Maximum duration of totality 1 m 57 s
2004 April 19	Partial	Antarctica, southern Africa	74% eclipsed
2004 October 14	Partial	Northeast Asia, Hawaii, Alaska	93% eclipsed
2005 April 8	Annular and Total	New Zealand, North and South America	Annular along part of path; maximum duration of totality 0 m 42 s
2005 October 3	Annular	Europe, Africa, southern Asia	—

Eclipse predictions by Fred Espenak, NASA/Goddard Space Flight Center. All dates are given in standard astronomical format: year, month, day.

Eclipse Paths for Total Eclipses, 1997–2020 This map shows the eclipse paths for all 18 total solar eclipses occurring from 1997 through 2020. In each eclipse, the Moon's shadow travels along the eclipse path in a generally eastward direction across the Earth's surface. (Courtesy of Fred Espenak, NASA/Goddard Space Flight Center)

Ancient astronomers achieved a limited ability to predict eclipses. In those times, religious and political leaders who were able to predict such awe-inspiring events as eclipses must have made a tremendous impression on their followers. One of three priceless manuscripts to survive the devastating Spanish Conquest shows that the Mayan astronomers of Mexico and Guatemala had a fairly reliable method for predicting eclipses. The great Greek astronomer Thales of Miletus is said to have predicted the famous eclipse of 585 B.C., which occurred during the middle of a war. The sight was so unnerving that the soldiers put down their arms and declared peace.

In retrospect, it seems that what ancient astronomers actually produced were eclipse "warnings" of various degrees of reliability rather than true predictions. Working with historical records, these astronomers generally sought to discover cycles and regularities from which future eclipses could be anticipated. To see how you might produce eclipse warnings yourself, read Box 3-2.

3-6 Ancient astronomers measured the size of the Earth and attempted to determine distances to the Sun and Moon

The prediction of eclipses was not the only problem attacked by ancient astronomers. More than 2000 years ago, centuries before sailors of Columbus's era crossed the oceans, Greek astronomers were fully aware that the Earth is not flat. They had come to this conclusion using a combination of observation and logical deduction, much like modern scientists. The Greeks noted that during lunar eclipses, when the Moon passes through the Earth's shadow, the edge of the shadow is always circular. Because a sphere is the only shape that always casts a circular shadow from any angle, they concluded that the Earth is spherical.

Around 200 B.C., the Greek astronomer Eratosthenes devised a way to measure the circumference of the spherical

ƀox 3-2	Tools of the Astronomer's Trade

Predicting Solar Eclipses

Suppose that you observe a solar eclipse in your hometown and want to figure out when you and your neighbors might see another eclipse. How would you begin?

First, remember that a solar eclipse can occur only if the line of nodes points toward the Sun at the same time that there is a new moon (see Figure 3-7). Second, you must know that it takes 29.53 days (one synodic month) to go from one new moon to the next. Because solar eclipses occur only during new moon, you must wait several whole lunar months for the proper alignment to occur again.

 However, there is a complication: The line of nodes gradually shifts its position with respect to the background stars. It takes 346.6 days to move from one alignment of the line of nodes pointing toward the Sun to the next identical alignment. This period is called the **eclipse year.**

Therefore, to predict when you will see another solar eclipse, you need to know how many whole lunar months equal some whole number of eclipse years. This will tell you how long you will have to wait for the next virtually identical alignment of the Sun, the Moon, and the line of nodes. By trial and error, you find that 223 lunar months is the same length of time as 19 eclipse years, because

$$223 \times 29.53 \text{ days} = 19 \times 346.6 \text{ days} = 6585 \text{ days}$$

This calculation is accurate to within a few hours. A more accurate calculation gives an interval, called the **saros,** that is about one-third day longer, or 6585.3 days (18 years, 11.3 days). Eclipses separated by the saros interval are said to form an *eclipse series.*

You might think that you and your neighbors would simply have to wait one full saros interval to go from one solar eclipse to the next. However, because of the extra one-third day, the Earth will have rotated by an extra 120°

(one-third of a complete rotation) when the next solar eclipse of a particular series occurs. The eclipse path will thus be one-third of the way around the world from you. Therefore, you must wait three full saros intervals (54 years, 34 days) before the eclipse path comes back around to your part of the Earth. The illustration shows a series of six solar eclipse paths, each separated from the next by one saros interval.

There is evidence that ancient Babylonian astronomers knew about the saros interval. However, the discovery of the saros is more likely to have come from lunar eclipses than solar eclipses. If you are far from the eclipse path, there is a good chance that you could fail to notice a solar eclipse. Even if half the Sun is covered by the Moon, the remaining solar surface provides enough sunlight for the outdoor illumination not to be greatly diminished. By contrast, anyone on the nighttime side of the Earth can see an eclipse of the Moon unless clouds block the view.

Earth. It was known that on the date of the summer solstice (the first day of summer; see Section 2-5) in the town of Syene in Egypt, near present-day Aswan, the Sun shone directly down the vertical shafts of water wells. Hence, at local noon on that day the Sun was at the zenith (see Section 2-4) as seen from Syene. Eratosthenes knew that the Sun never appeared at the zenith at his home in the Egyptian city of Alexandria, which is on the Mediterranean Sea almost due north of Syene. Rather, on the summer solstice in Alexandria, the posi-

tion of the Sun at local noon was about 7° south of the zenith (Figure 3-14). This angle is about one-fiftieth of a complete circle, so he concluded that the distance from Alexandria to Syene must be about one-fiftieth of the Earth's circumference.

In Eratosthenes's day, the distance from Alexandria to Syene was said to be 5000 stades. Therefore, Eratosthenes found the Earth's circumference to be

$$50 \times 5000 \text{ stades} = 250,000 \text{ stades}$$

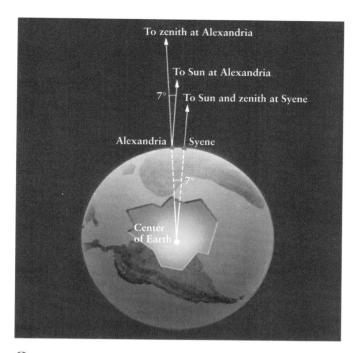

figure 3-14

Eratosthenes's Method of Determining the Diameter of the Earth Around 200 B.C., Eratosthenes noticed that the Sun is about 7° south of the zenith at Alexandria when it is directly overhead at Syene. This angle is about one-fiftieth of a circle, so the distance between Alexandria and Syene must be about one-fiftieth of the Earth's circumference.

Unfortunately, no one today is sure of the exact length of the Greek unit called the stade. One guess is that the stade was about one-sixth of a kilometer, which would mean that Eratosthenes obtained a circumference for the Earth of about 42,000 kilometers. This is remarkably close to the modern value of 40,000 kilometers.

Eratosthenes was only one of several brilliant astronomers to emerge from the so-called Alexandrian school, which by his time had a distinguished tradition. One of the first Alexandrian astronomers, Aristarchus of Samos, had proposed a method of determining the relative distances to the Sun and Moon, perhaps as long ago as 280 B.C.

Aristarchus knew that the Sun, Moon, and Earth form a right triangle at the moment of first or last quarter moon, with the right angle at the location of the Moon (Figure 3-15). He estimated that, as seen from Earth, the angle between the Moon and the Sun at first and last quarters is 87°, or 3° less than a right angle. Using the rules of geometry, Aristarchus concluded that the Sun is about 20 times farther from us than is the Moon. We now know that Aristarchus erred in measuring angles and that the average distance to the Sun is about 390 times larger than the average distance to the Moon. It is nevertheless impressive that people were trying to measure distances across the solar system more than 2000 years ago.

Aristarchus also made an equally bold attempt to determine the relative sizes of the Earth, Moon, and Sun. From his observations of how long the Moon takes to move through the Earth's shadow during a lunar eclipse, Aristarchus estimated the diameter of the Earth to be about 3 times larger

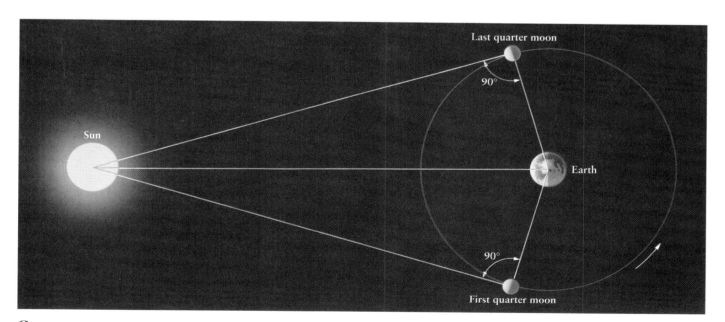

figure 3-15

Aristarchus's Method of Determining Distances to the Sun and Moon Aristarchus knew that the Sun, Moon, and Earth form a right triangle at first and last quarter phases. Using geometrical arguments, he calculated the relative lengths of the sides of these triangles, thereby obtaining the distances to the Sun and Moon.

than the diameter of the Moon. To determine the diameter of the Sun, Aristarchus simply pointed out that the Sun and the Moon have the same angular size in the sky. Therefore, their diameters must be in the same proportion as their distances (see part *a* of the figure in Box 1-1). In other words, because Aristarchus thought the Sun to be 20 times farther from the Earth than the Moon, he concluded that the Sun must be 20 times larger than the Moon. Once Eratosthenes had measured the Earth's circumference, astronomers of the Alexandrian school could estimate the diameters of the Sun and Moon as well as their distances from Earth.

Table 3-3 summarizes some ancient and modern measurements of the sizes of Earth, the Moon, and the Sun and the distances between them. Some of these ancient measurements are far from the modern values. Yet the achievements of our ancestors still stand as impressive applications of observation and reasoning and important steps toward the development of the scientific method.

Table 3-3	Comparison of Ancient and Modern Astronomical Measurements	
	Ancient measure (km)	Modern measure (km)
Earth's diameter	13,000	12,756
Moon's diameter	4,300	3,476
Sun's diameter	9×10^4	1.39×10^6
Earth-Moon distance	4×10^5	3.84×10^5
Earth-Sun distance	10^7	1.50×10^8

KEY WORDS

Terms preceded by an asterisk (*) are discussed in the Boxes.

annular eclipse, p. 54

apogee, p. 54

eclipse, p. 49

eclipse path, p. 53

*eclipse year, p. 56

first quarter moon, p. 45

full moon, p. 45

last quarter moon, p. 47

line of nodes, p. 49

lunar eclipse, p. 49

lunar month, p. 48

lunar phases, p. 44

new moon, p. 45

partial lunar eclipse, p. 50

partial solar eclipse, p. 53

penumbra (*plural* penumbrae), p. 50

penumbral eclipse, p. 50

perigee, p. 53

plane of the ecliptic, p. 49

*saros, p. 56

sidereal month, p. 48

solar corona, p. 53

solar eclipse, p. 49

synchronous rotation, p. 47

synodic month, p. 48

totality (lunar eclipse), p. 52

totality (solar eclipse), p. 53

total lunar eclipse, p. 50

total solar eclipse, p. 52

umbra (*plural* umbrae), p. 50

waning crescent moon, p. 47

waning gibbous moon, p. 47

waxing crescent moon, p. 45

waxing gibbous moon, p. 45

KEY IDEAS

Lunar Phases: The phases of the Moon occur because light from the Moon is actually reflected sunlight. As the relative positions of the Earth, the Moon, and the Sun change, we see more or less of the illuminated half of the Moon.

Length of the Month: Two types of months are used in describing the motion of the Moon.

• With respect to the stars, the Moon completes one orbit around the Earth in a sidereal month, averaging 27.32 days.

• The Moon completes one cycle of phases (one orbit around the Earth with respect to the Sun) in a synodic month (also called a lunar month), averaging 29.53 days.

The Moon's Orbit: The plane of the Moon's orbit is tilted by about 5° from the plane of the Earth's orbit, or ecliptic.

• The line of nodes is the line where the planes of the Moon's orbit and the Earth's orbit intersect. The gravitational pull of the Sun gradually shifts the orientation of the line of nodes with respect to the stars.

Conditions for Eclipses: During a lunar eclipse, the Moon moves through the Earth's shadow. During a solar eclipse, the Earth passes through the Moon's shadow.

• Lunar eclipses occur at full moon, while solar eclipses occur at new moon.

• Either type of eclipse can occur only when the Sun and Moon are both on or very near the line of nodes. If this condition is not met, the Earth's shadow cannot fall on the Moon and the Moon's shadow cannot fall on the Earth.

Umbra and Penumbra: The shadow of an object has two parts: the umbra, within which the light source is completely blocked, and the penumbra, where the light source is only partially blocked.

Lunar Eclipses: Depending on the relative positions of the Sun, Moon, and Earth, lunar eclipses may be total (the Moon passes completely into the Earth's umbra), partial (only part of the Moon passes into the Earth's umbra), or penumbral (the Moon passes only into the Earth's penumbra).

Solar Eclipses: Solar eclipses may be total, partial, or annular.

• During a total solar eclipse, the Moon's umbra traces out an eclipse path over the Earth's surface as the Earth rotates. Observers outside the eclipse path but within the penumbra see only a partial solar eclipse.

• During an annular eclipse, the umbra falls short of the Earth, and the outer edge of the Sun's disk is visible around the Moon at mideclipse.

The Moon and Ancient Astronomers: Ancient astronomers such as Aristarchus and Eratosthenes made great progress in determining the sizes and relative distances of the Earth, the Moon, and the Sun.

REVIEW QUESTIONS

Questions preceded by an asterisk (*) involve topics discussed in the Boxes.

1. (a) Explain why the Moon exhibits phases. (b) A common misconception about the Moon's phases is that they are caused by the Earth's shadow. Use Figure 3-2 to explain why this not correct.

2. Suppose it is the first day of autumn in the northern hemisphere. What is the phase of the Moon if the Moon is located at (a) the vernal equinox? (b) the summer solstice? (c) the autumnal equinox? (d) the winter solstice? (*Hint:* Make a drawing showing the relative positions of the Sun, Earth, and Moon. Compare with Figure 3-2.)

3. How would the sequence and timing of lunar phases be affected if the Moon moved around its orbit (a) in the same direction, but at twice the speed; (b) at the same speed, but in the opposite direction? Explain your answers.

4. Is the far side of the Moon (the side that can never be seen from Earth) the same as the dark side of the Moon? Explain.

5. What is the difference between a sidereal month and a synodic month? Which is longer? Why?

6. (a) If you lived on the Moon, would you see the Sun rise and set, or would it always be in the same place in the sky? Explain. (b) Would you see the Earth rise and set, or would it always be in the same place in the sky? Explain.

7. Astronomers sometimes refer to lunar phases in terms of the age of the Moon. This is the time that has elapsed since new moon phase. Thus, the age of a full moon is half of a 29½-day synodic period, or approximately 15 days. Find the approximate age of (a) a waxing crescent moon; (b) a last quarter moon; (c) a waning gibbous moon.

8. What is the difference between the umbra and the penumbra of a shadow?

9. Why doesn't a lunar eclipse occur at every full moon and a solar eclipse at every new moon?

10. What is the line of nodes? Why is it important to the subject of eclipses?

11. What is a penumbral eclipse of the Moon? Why do you suppose that it is easy to overlook such an eclipse?

12. The maximum duration of totality of a lunar eclipse is 1 hour, 42 minutes. But none of the lunar eclipses listed in Table 3-1 lasts this long. Why is this?

13. Can one ever observe an annular eclipse of the Moon? Why or why not?

14. If you were looking at the Earth from the side of the Moon that faces the Earth, what would you see during (a) a total lunar eclipse? (b) a total solar eclipse? Explain your answers.

15. If there is a total eclipse of the Sun in April, can there be a lunar eclipse three months later in July? Why or why not?

16. Which type of eclipse—lunar or solar—do you think most people on Earth have seen? Why?

17. How is an annular eclipse of the Sun different from a total eclipse of the Sun? What causes this difference?

*18. What is the saros? How did ancient astronomers use it to predict eclipses?

19. How did Eratosthenes measure the size of the Earth?

20. How did Aristarchus try to estimate the distance from the Earth to the Sun and Moon?

ADVANCED QUESTIONS

Problem-solving tips and tools

To estimate the average angular speed of the Moon along its orbit (that is, how many degrees around its orbit the Moon travels per day), divide 360° by the length of a sidereal month. It is helpful to know that the saros interval of 6585.3 days equals 18 years and 11⅓ days if the interval includes four leap years, but is 18 years and 10⅓ days if it includes five leap years.

21. The dividing line between the illuminated and unilluminated halves of the Moon is called the *terminator*. The terminator appears curved when there is a crescent or gibbous moon, but appears straight when there is a first quarter or last quarter moon (see Figure 3-2). Describe how you could use these facts to explain to a friend why lunar phases cannot be caused by the Earth's shadow falling on the Moon.

22. This photograph of the Earth was taken by the crew of the *Apollo 8* spacecraft as they orbited the Moon. A portion of the lunar surface is visible at the right-hand side of the photo. In this photo, the Earth is oriented with its North Pole approximately at the top. When this photo was taken, was the Moon waxing or waning as seen from Earth? Explain your answer with a diagram.

(NASA/JSC)   R I **V** U X G

23. (a) The Moon moves noticeably over the space of a single night. To show this, calculate how long it takes the Moon to move through an angle equal to its own angular diameter (½°) against the background of stars. Give your answer in hours. **(b)** Through what angle (in degrees) does the Moon move during a 12-hour night? Can you notice an angle of this size? (*Hint:* See Figure 1-10.)

24. During an occultation, or "covering up," of Jupiter by the Moon, an astronomer notices that it takes the Moon's edge 90 seconds to cover Jupiter's disk completely. If the Moon's motion is assumed to be uniform and the occultation was "central" (that is, center over center), find the angular diameter of Jupiter. (*Hint:* Assume that Jupiter does not appear to move in the sky during this brief 90-second interval. You will need to convert the Moon's angular speed from degrees per day to arcseconds per second.)

25. How many more sidereal months than synodic months are there in a year? Explain.

26. Suppose the Earth moved a little faster around the Sun, so that it took a bit less than one year to make a complete orbit. If the speed of the Moon's orbit around the Earth were unchanged, would the length of the sidereal month be the same, longer, or shorter than it is now? What about the synodic month? Explain your answers.

27. If the Moon revolved about the Earth in the same orbit but in the opposite direction, would the synodic month be longer or shorter than the sidereal month? Explain your reasoning.

28. A "blue moon" is the second full moon within the same calendar month. There is usually only one full moon within a calendar month, which is why the phrase "once in a blue moon" means "hardly ever." Why are blue moons so rare? Are there any months of the year in which it would be impossible to have two full moons? Explain your answer.

29. The total lunar eclipse of November 29, 1993, was visible from North America. The duration of totality was 48 minutes. Was this total eclipse also visible from India, on the opposite side of the Earth? Explain your reasoning.

30. You are watching a lunar eclipse from some place on the Earth's night side. Will you see the Moon enter the Earth's shadow from the east or from the west? Explain your reasoning.

31. During a total solar eclipse, the Moon's umbra moves in a generally eastward direction across the Earth's surface. Use a drawing like Figure 3-11 to explain why the motion is eastward, not westward.

32. A total solar eclipse was visible from southern Germany on August 11, 1999. (See the photo that opens this chapter.) Draw what the eclipse would have looked like as seen from Switzerland, to the south of the path of totality. Explain the reasoning behind your drawing.

33. Just as the distance from the Earth to the Moon varies somewhat as the Moon orbits the Earth, the distance from the Sun to the Earth changes as the Earth orbits the Sun. The Earth is closest to the Sun at its *perihelion;* it is farthest from the Sun at its *aphelion.* In order for a total solar eclipse to have the maximum duration of totality, should the Earth be at perihelion or aphelion? Assume that the Earth-Moon distance is the same in both situations. As part of your explanation, draw two pictures like Figure 3-11, one with the Earth relatively close to the Sun and one with the Earth relatively far from the Sun.

34. (a) Suppose the diameter of the Moon were doubled, but the orbit of the Moon remained the same. Would total solar eclipses be more common, less common, or just as common as they are now? Explain. (b) Suppose the diameter of the Moon were halved, but the orbit of the Moon remained the same. Explain why there would be *no* total solar eclipses.

*35. On June 21, 2001, residents of southern Africa were treated to a total solar eclipse. (a) On what date and over what part of the world will the next total eclipse of that series occur? Explain. (b) On what date might you next expect a total eclipse of that series to be visible from southern Africa? Explain.

DISCUSSION QUESTIONS

36. Describe the cycle of lunar phases that would be observed if the Moon moved around the Earth in an orbit perpendicular to the plane of the Earth's orbit. Would it be possible for both solar and lunar eclipses to occur under these circumstances? Explain your reasoning.

37. How would a lunar eclipse look if the Earth had no atmosphere? Explain your reasoning.

38. Why do you suppose that total solar eclipse paths fall more frequently on oceans than on land? (You may find it useful to look at Figure 3-13.)

39. In his 1885 novel *King Solomon's Mines*, H. Rider Haggard described a total solar eclipse that was seen in both South Africa and in the British Isles. Is such an eclipse possible? Why or why not?

40. Examine Figure 3-13, which shows all of the total solar eclipses from 1997 to 2020. What are the chances that you might be able to travel to one of the eclipse paths? Do you think you might go through your entire life without ever seeing a total eclipse of the Sun?

 ## WEB/CD-ROM QUESTIONS

41. Search the World Wide Web for information about the next total solar eclipse. Through which major cities, if any, does the path of totality pass? What is the maximum duration of totality? At what location is this maximum duration observed? Will this eclipse be visible (even as a partial eclipse) from your location? Draw a picture showing the Sun, Earth, and Moon when the totality is at its maximum duration, and indicate your location on the drawing of the Earth.

42. Search the World Wide Web for information about the next total lunar eclipse. Will the total phase of the eclipse be visible from your location? If not, will the penumbral phase be visible? Draw a picture showing the Sun, Earth, and Moon when the totality is at its maximum duration, and indicate your location on the drawing of the Earth.

 43. Access the animation "The Moon's Phases" in Chapter 3 of the *Universe* web site or CD-ROM. This shows the Earth-Moon system as seen from a vantage point looking down onto the North Pole. (a) Describe where you would be on the diagram if you are on the equator and the time is 6:00 P.M. (b) If it is 6:00 P.M. and you are standing on Earth's equator, would a third-quarter moon be visible? Why or why not? If it would be visible, describe its appearance.

 44. Access the animation "A Solar Eclipse Viewed from the Moon" in Chapter 3 of the *Universe* web site or CD-ROM. This shows the solar eclipse of August 11, 1999, as viewed from the Moon. Using a diagram, explain why the stars and the Moon's shadow move in the directions shown in this animation.

OBSERVING PROJECTS

45. Observe the Moon on each clear night over the course of a month. On each night, note the Moon's location among the constellations and record that location on a star chart that also shows the ecliptic. After a few weeks, your observations will begin to trace the Moon's orbit. Identify the orientation of the line of nodes by marking the points where the Moon's orbit and the ecliptic intersect. On what dates is the Sun near the nodes marked on your star chart? Compare these dates with the dates of the next solar and lunar eclipses.

46. It is quite possible that a lunar eclipse will occur while you are taking this course. Look up the date of the next lunar eclipse in Table 3-1, in the current issue of a reference such as the *Astronomical Almanac* or *Astronomical Phenomena*, or on the World Wide Web. Then make arrangements to observe this lunar eclipse. You can observe the eclipse with the naked eye, but binoculars or a small telescope will enhance your viewing experience. If the eclipse is partial or total, note the times at which the Moon enters and exits the Earth's umbra. If the eclipse is penumbral, can you see any changes in the Moon's brightness as the eclipse progresses?

47. Use the *Starry Night* program to observe the motion of the Moon. (a) Display the entire celestial sphere, including the part below the horizon (select **Atlas** in the **Go** menu). Center on the Moon by using the **Find...** command in the **Edit** menu. In the Control Panel at the top of the main window, set the time step to 1 day and click on the "Forward" button (a triangle that ponts to the right). How does the Moon appear to move against the background of stars? Does it ever change direction? (b) Determine how many days elapse between successive times when the Moon is on the ecliptic. (If you don't see a green line representing the ecliptic, select **The Ecliptic** in the **Guides** menu.) Then move forward in time to a date when the Moon is on the ecliptic *and* either full or new. What type of eclipse will occur on that date? Confirm your answer by comparing with Tables 3-1 and 3-2 or with lists of eclipses on the World Wide Web.

Gravitation and the Waltz of the Planets

4

I n July 1969 humans first journeyed from the Earth to the surface of the Moon. By studying the rocks that the astronauts collected on the Moon during this and later missions, geologists learned how the surface of our nearest celestial neighbor has evolved over the past several billion years. These discoveries revolutionized our understanding of the very nature of the Moon.

But several other revolutions in human understanding had to take place before these discoveries could be made. One revolution overthrew the ancient idea that the Earth is an immovable object at the center of the universe, around which moved the Sun, the Moon, and the planets. In this chapter we will learn how Nicolaus Copernicus, Galileo Galilei, Tycho Brahe, and Johannes Kepler helped us understand that the Earth is itself one of several planets orbiting the Sun.

We will learn, too, about Isaac Newton's revolutionary discovery of why the planets move in the way that they do. This was just one aspect of Newton's immense body of work, which included formulating the fundamental laws of physics and developing a precise mathematical description of the force of gravity—the force that holds the planets in their orbits.

Newton's laws apply on Earth as well as in the heavens. They also govern the flight of spacecraft, which made missions to the Moon possible. That is why astronaut William Anders radioed the following while en route from the Moon to the Earth: "I think Isaac Newton is doing most of the driving right now."

(Michael Collins, *Apollo 11*, NASA)

R I V U X G

QUESTIONS TO GUIDE YOUR READING

As you read the sections of this chapter, look for the answers to the following questions.

4-1 How did ancient astronomers explain the motions of the planets?

4-2 Why did Copernicus think that the Earth and the other planets go around the Sun?

4-3 What did Galileo see in his telescope that confirmed that the planets orbit the Sun?

4-4 How did Tycho Brahe attempt to test the ideas of Copernicus?

4-5 What paths do the planets follow as they move around the Sun?

4-6 What fundamental laws of nature explain the motions of objects on Earth as well as the motions of the planets?

4-7 Why don't the planets fall into the Sun?

4-8 What keeps the same face of the Moon always pointed toward the Earth?

4-1 Ancient astronomers invented geocentric models to explain planetary motions

Since the dawn of civilization, scholars have attempted to explain the nature of the universe. The ancient Greeks were the first to use the principle that still guides scientists today: The universe can be described and understood logically. For example, more than 2500 years ago Pythagoras and his followers put forth the idea that nature can be described with mathematics. About 200 years later, Aristotle asserted that the universe is governed by physical laws.

As they attempted to create a model (see Section 1-1) of how the universe works, most Greeks assumed that the Sun, the Moon, the stars, and the planets revolve about a stationary Earth. A model of this kind, in which the Earth is at the center of the universe, is called a **geocentric model.** Similar ideas were held by the scholars of ancient China.

Today we recognize that the stars are not merely points of light on an immense celestial sphere. But this is just how the ancient Greeks regarded the stars in their geocentric model of the universe. To explain the diurnal motions of the stars, they assumed that the entire celestial sphere rotated around the stationary Earth once a day. The Sun and Moon both participated in this daily rotation of the sky, which explained their rising and setting motions. To explain why the Sun and Moon both move slowly with respect to the stars, the ancient Greeks imagined that both of these objects orbit around the Earth.

ANALOGY Imagine a merry-go-round that rotates clockwise as seen from above, as in Figure 4-1*a*. As it rotates, two children walk slowly counterclockwise at different speeds around the merry-go-round's platform. Thus, the children rotate along with the merry-go-round and also change

---

Merry-go-round rotates clockwise

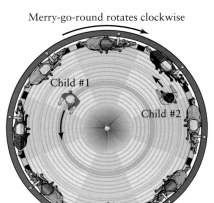

Wooden horses fixed on merry-go-round

a A rotating merry-go-round

Celestial sphere rotates to the west

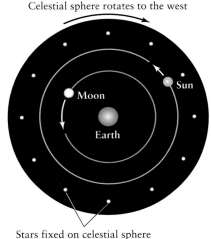

Stars fixed on celestial sphere

b The Greek geocentric model

ANIMATION 4.1 **figure 4-1**
A Merry-Go-Round Analogy
(a) Two children walk at different speeds around a rotating merry-go-round with its wooden horses. **(b)** In an analogous way, the ancient Greeks imagined that the Sun and Moon move at different speeds around the rotating celestial sphere with its fixed stars. Thus, the Sun and Moon move from east to west across the sky every day and also move slowly eastward from one night to the next relative to the background of stars.

their positions with respect to the merry-go-round's wooden horses. This scene is analogous to the way the ancient Greeks pictured the motions of the stars, Sun, and Moon. In their model, the celestial sphere rotated to the west around a stationary Earth (Figure 4-1b). The stars rotate along with the celestial sphere just as the wooden horses rotate along with the merry-go-round in Figure 4-1a. The Sun and Moon are analogous to the two children; they both turn westward with the celestial sphere, making one complete turn each day, and also move slowly eastward at different speeds with respect to the stars.

The geocentric model of the heavens also had to explain the motions of the planets. The ancient Greeks and other cultures of that time knew of five planets: Mercury, Venus, Mars, Jupiter, and Saturn. These planets are quite obvious because they are bright objects in the night sky. For example, when Venus is at its maximum brilliancy, it is 16 times brighter than the brightest star. (By contrast, Uranus, Neptune, and Pluto are quite dim and were not discovered until after the invention of the telescope.)

Like the Sun and Moon, all of the planets rise in the east and set in the west once a day. And like the Sun and Moon, from night to night the planets slowly move on the celestial sphere, that is, with respect to the background of stars. However, the character of this motion on the celestial sphere is quite different for the planets. Both the Sun and the Moon always move from west to east on the celestial sphere, that is, opposite the direction in which the celestial sphere appears to rotate. Furthermore, the Sun and the Moon each move at relatively constant speeds around the celestial sphere. (The Moon's speed is faster than that of the Sun, because it travels all the way around the celestial sphere in about a month while the Sun takes an entire year). By contrast, each of the planets appears to wander back and forth on the celestial sphere, moving from west to east at most times but at other times from east to west. (The name *planet* is well deserved; it comes from a Greek word meaning "wanderer.") As a planet wanders, its speed on the celestial sphere can vary substantially.

You can see this distinctive behavior in Figure 4-2, which shows the motion of Mars with respect to the background of stars during 2005 and 2006. In the language of the merry-go-round analogy in Figure 4-1, the Greeks imagined the planets as children walking around the rotating merry-go-round but who keep changing their minds about which direction to walk!

As seen from Earth, the planets wander primarily across the 12 constellations of the zodiac. (Figure 4-2 shows that during 2005–2006, Mars moves through Pisces, Aries, and Taurus.) As we learned in Section 2-5, these constellations encircle the sky in a continuous band centered on the ecliptic, which is the apparent path of the Sun across the celestial sphere.

If you observe a planet as it travels across the zodiac from night to night, you will find that the planet usually moves

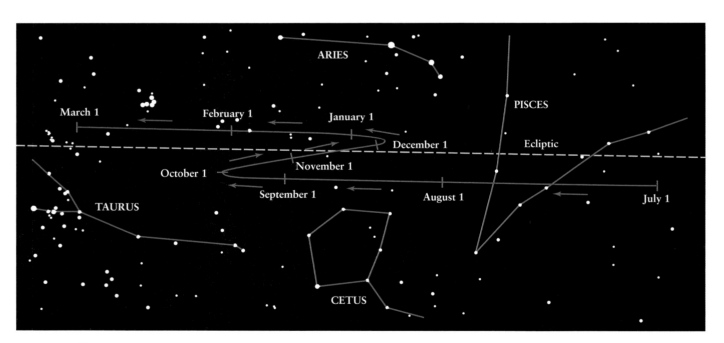

figure 4-2

The Path of Mars in 2005–2006 From July 2005 through February 2006, Mars will move across the constellations Pisces, Aries, and Taurus. Mars's motion will be direct (from west to east, or from right to left in this figure) most of the time but will be retrograde (from east to west, or from left to right in this figure) during October and November 2005. Notice that the speed of Mars relative to the stars is not constant: The planet moves faster and travels farther across the sky from August 1 to September 1 than it does from September 1 to October 1.

a

b

c

figure 4-3

ANIMATION 4.3

A Geocentric Explanation of Retrograde Motion

(a) The ancient Greeks imagined that each planet moves along an epicycle, which in turn moves along a deferent centered approximately on the Earth. The planet moves along the epicycle more rapidly than the epicycle moves along the deferent. (b) At most times the eastward motion of the planet on the epicycle adds to the eastward motion of the epicycle on the deferent. Then the planet moves eastward in direct motion as seen from Earth. (c) When the planet is on the inside of the deferent, its motion along the epicycle is westward. Because this motion is faster than the eastward motion of the epicycle on the deferent, the planet appears from Earth to be moving westward in retrograde motion.

slowly eastward against the background stars. This eastward progress is called **direct motion.** Occasionally, however, the planet will seem to stop and then back up for several weeks or months. This occasional westward movement is called **retrograde motion.** Both direct and retrograde motions are much slower than the apparent daily rotation of the sky caused by the Earth's rotation. Hence, they are best detected by mapping the position of a planet against the background stars from night to night over a long period. Figure 4-2 is a map of just this sort. Mars undergoes retrograde motion about every 22½ months; all the other planets go through retrograde motion, but at different intervals.

Explaining the nonuniform motions of the five planets was one of the main challenges facing the astronomers of antiquity. The Greeks developed many theories to account for retrograde motion and the loops that the planets trace out against the background stars. One of the most successful and enduring models was originated by Hipparchus in the second century B.C. and expanded upon by Ptolemy, the last of the great Greek astronomers, during the second century A.D. Fig-

ure 4-3a sketches the basic concept, usually called the **Ptolemaic system.** Each planet is assumed to move in a small circle called an **epicycle,** whose center in turn moves in a larger circle, called a **deferent,** which is centered approximately on the Earth. Both the epicycle and deferent rotate in the same direction, shown as counterclockwise in Figure 4-3a.

As viewed from Earth, the epicycle moves eastward along the deferent. Most of the time the eastward motion of the planet on its epicycle adds to the eastward motion of the epicycle on the deferent (Figure 4-3b). Then the planet is seen to be in direct (eastward) motion against the background stars. However, when the planet is on the part of its epicycle nearest Earth, the motion of the planet along the epicycle is opposite to the motion of the epicycle along the deferent. The planet therefore appears to slow down and halt its usual eastward movement among the constellations, and actually goes backward in retrograde (westward) motion for a few weeks or months (Figure 4-3c). Thus, the concept of epicycles and deferents enabled Greek astronomers to explain the retrograde loops of the planets.

Using the wealth of astronomical data in the library at Alexandria, including records of planetary positions for hundreds of years, Ptolemy deduced the sizes and rotation rates of the epicycles and deferents needed to reproduce the recorded paths of the planets. After years of tedious work, Ptolemy assembled his calculations into 13 volumes, collectively called the *Almagest*. His work was used to predict the positions and paths of the Sun, Moon, and planets with unprecedented accuracy. In fact, the *Almagest* was so successful that it became the astronomer's bible, and for more than 1000 years, the Ptolemaic system endured as a useful description of the workings of the heavens.

A major problem posed by the Ptolemaic system was a philosophical one: It treated each planet independent of the others. There was no rule in the *Almagest* that related the size and rotation speed of one planet's epicycle and deferent to the corresponding sizes and speeds for other planets. This problem made the Ptolemaic system very unsatisfying to many Islamic and European astronomers of the Middle Ages. They felt that a correct model of the universe should be based on a simple set of underlying principles that applied to all of the planets.

 The idea that simple, straightforward explanations of phenomena are most likely to be correct is called **Occam's razor,** after William of Occam (or Ockham), the fourteenth-century English philosopher who first expressed it. (The "razor" refers to shaving extraneous details from an argument or explanation.) Although Occam's razor has no proof or verification, it appeals to the scientist's sense of beauty and elegance, and it has helped lead to the simple and powerful laws of nature that scientists use today. In the centuries that followed, Occam's razor would help motivate a new and revolutionary view of the universe.

4-2 Nicolaus Copernicus devised the first comprehensive heliocentric model

During the first half of the sixteenth century, a Polish lawyer, physician, canon of the church, and gifted mathematician named Nicolaus Copernicus (Figure 4-4) began to construct a new model of the universe. His model, which placed the Sun at the center, explained the motions of the planets in a more natural way than the Ptolemaic system. As we will see, it also helped lay the foundations of modern physical science. But Copernicus was not the first to conceive a Sun-centered model of planetary motion.

Imagine riding on a fast racehorse. As you pass a slowly walking pedestrian, he appears to move backward, even though he is traveling in the same direction as you and your horse. In the third century B.C., this sort of simple observation inspired the Greek astronomer Aristarchus to suggest a straightforward explanation of retrograde motion.

In Aristarchus's **heliocentric** (Sun-centered) **model,** all the planets, including Earth, revolve about the Sun. Different planets take different lengths of time to complete an orbit, so from time to time one planet will overtake another, just as a fast-moving horse overtakes a person on foot. When the Earth

figure 4-4

Nicolaus Copernicus (1473–1543) Copernicus was the first person to work out the details of a heliocentric system in which the planets, including the Earth, orbit the Sun. (E. Lessing/Magnum)

overtakes Mars, for example, Mars appears to move backward in retrograde motion, as Figure 4-5 shows. Thus, in the heliocentric picture, the occasional retrograde motion of a planet is merely the result of the Earth's motion.

As we saw in Section 3-6, Aristarchus demonstrated that the Sun is bigger than the Earth (see Table 3-3). This made it sensible to imagine the Earth orbiting the larger Sun. He also imagined that the Earth rotated on its axis once a day, which explained the daily rising and setting of the Sun, Moon, and planets and the diurnal motions of the stars. To explain why the apparent motions of the planets never take them far from the ecliptic, Aristarchus proposed that the orbits of the Earth and all the planets must lie in nearly the same plane. (Recall from Section 2-5 that the ecliptic is the projection onto the celestial sphere of the plane of the Earth's orbit.)

This heliocentric model is conceptually much simpler than an Earth-centered system, such as that of Ptolemy, with all its "circles upon circles." In Aristarchus's day, however, the idea of an orbiting, rotating Earth seemed inconceivable, given the Earth's apparent stillness and immobility. Nearly 2000 years would pass before a heliocentric model found broad acceptance.

In the years after 1500, Copernicus came to realize that a heliocentric model has several advantages beyond providing a natural explanation of retrograde motion. In the Ptolemaic system, the arrangement of the planets—that is, which are close to the Earth and which are far away—was chosen

ANIMATION 4.4

figure 4-5

A Heliocentric Explanation of Retrograde Motion In the heliocentric model of Aristarchus, the Earth and the other planets orbit the Sun. The Earth travels around the Sun more rapidly than Mars. Consequently, as the Earth overtakes and passes this slower-moving planet, Mars appears for a few months (from points 4 through 6) to fall behind and move backward with respect to the background of stars.

in large part by guesswork. But using a heliocentric model, Copernicus could determine the arrangement of the planets without ambiguity.

Copernicus realized that because Mercury and Venus are always observed fairly near the Sun in the sky, their orbits must be smaller than the Earth's. Planets in such orbits are called **inferior planets** (Figure 4-6). The other visible planets—Mars, Jupiter, and Saturn—are sometimes seen on the side of the celestial sphere opposite the Sun, so these planets appear high above the horizon at midnight (when the Sun is far below the horizon). When this happens, the Earth must lie between the Sun and these planets. Copernicus therefore concluded that the orbits of Mars, Jupiter, and Saturn must be larger than the Earth's orbit. Hence, these planets are called **superior planets.**

Three additional planets, Uranus, Neptune, and Pluto, were discovered after the telescope was invented (and after the death of Copernicus). All three can be seen at times in the midnight sky, so these are also superior planets with orbits larger than the Earth's.

The heliocentric model also explains why planets appear in different parts of the sky on different dates. Both inferior planets (Mercury and Venus) go through cycles: The planet is seen in the west after sunset for several weeks or months, then for several weeks or months in the east before sunrise, and then in the west after sunset again.

Figure 4-6 shows the reason for this cycle. When Mercury or Venus is visible after sunset, it is near **greatest eastern elongation.** (The angle between the Sun and a planet as viewed from Earth is called the planet's **elongation.**) The planet's position in the sky is as far east of the Sun as possible, so it appears above the western horizon after sunset (that is, to the east of the Sun) and is often called an "evening star." At **greatest western elongation,** Mercury or Venus is as far west of the Sun as it can possibly be. It then rises before the Sun, gracing the predawn sky as a "morning star" in the east. When Mercury or Venus is at **inferior conjunction,** it is between us and the Sun, and it is moving from the evening sky into the morning sky. At **superior conjunction,** when the planet is on the opposite side of the Sun, it is moving back into the evening sky.

A superior planet such as Mars, whose orbit is larger than the Earth's, is best seen in the night sky when it is at **opposi-**

tion. At this point the planet is in the part of the sky opposite the Sun and is highest in the sky at midnight. This is also when the planet appears brightest, because it is closest to us. But when a superior planet like Mars is located behind the Sun at **conjunction,** it is above the horizon during the daytime and thus is not well placed for nighttime viewing.

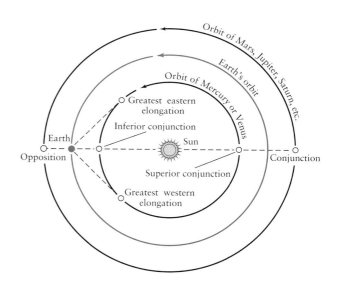

INTERACTIVE EXERCISE 4.2

figure 4-6

Planetary Orbits and Configurations When and where in the sky a planet can be seen from Earth depends on the size of its orbit and its location on that orbit. The inferior planets Mercury and Venus, whose orbits are smaller than the Earth's, are visible after sunset when at or near greatest eastern elongation, and before sunrise when at or near greatest western elongation. The best view of a superior planet (one whose orbit is larger than the Earth's) happens when it is at opposition, so that the planet is highest in the sky at midnight. When an inferior planet is at inferior conjunction or superior conjunction, or when a superior planet is at conjunction, it is only up during the daytime and cannot be seen at night. (Note that in this figure you are looking down onto the solar system from a point far above the Earth's northern hemisphere.)

Table 4-1	Synodic and Sidereal Periods of the Planets	
Planet	Synodic period	Sidereal period
Mercury	116 days	88 days
Venus	584 days	225 days
Earth	—	1.0 year
Mars	780 days	1.9 years
Jupiter	399 days	11.9 years
Saturn	378 days	29.5 years
Uranus	370 days	84.0 years
Neptune	368 days	164.8 years
Pluto	367 days	248.5 years

The Ptolemaic system has no simple rules relating the motion of one planet to another. But Copernicus showed that there are such rules in a heliocentric model. In particular, he found a correspondence between the time a planet takes to complete one orbit—that is, its **period**—and the size of the orbit.

Determining the period of a planet takes some care, because the Earth, from which we must make the observations, is also moving. Realizing this, Copernicus was careful to distinguish between two different periods of each planet. The **synodic period** is the time that elapses between two successive identical configurations as seen from Earth—from one opposition to the next, for example, or from one conjunction to the next. The **sidereal period** is the true orbital period of a planet, the time it takes the planet to complete one full orbit of the Sun relative to the stars.

The synodic period of a planet can be determined by observing the sky, but the sidereal period has to be found by

calculation. Copernicus figured out how to do this (Box 4-1). Table 4-1 shows the results for all of the planets.

To find a relationship between the sidereal period of a planet and the size of its orbit, Copernicus still had to determine the relative distances of the planets from the Sun. He devised a straightforward geometric method of determining the relative distances of the planets from the Sun using trigonometry. His answers turned out to be remarkably close to the modern values, as shown in Table 4-2.

The distances in Table 4-2 are given in terms of the astronomical unit, which is the average distance from Earth to the Sun (Section 1-7). Copernicus did not know the precise value of this distance, so he could only determine the *relative* sizes of the orbits of the planets. One method used by modern astronomers to determine the astronomical unit is to measure the Earth-Venus distance very accurately using radar. At the same time, they measure the angle in the sky between Venus and the Sun, and then calculate the Earth-Sun distance using

Table 4-2	Average Distances of the Planets from the Sun	
Planet	Copernican value (AU*)	Modern value (AU)
Mercury	0.38	0.39
Venus	0.72	0.72
Earth	1.00	1.00
Mars	1.52	1.52
Jupiter	5.22	5.20
Saturn	9.07	9.54
Uranus	—	19.19
Neptune	—	30.06
Pluto	—	39.53

1 AU = 1 astronomical unit = average distance from the Earth to the Sun.

box 4-1

Relating Synodic and Sidereal Periods

We can derive a mathematical formula that relates a planet's sidereal period (the time required for the planet to complete one orbit) to its synodic period (the time between two successive identical configurations). To start with, let's consider an inferior planet (Mercury or Venus) orbiting the Sun as shown in the diagram. Let P be the planet's sidereal period, S the planet's synodic period, and E the sidereal period of Earth or sidereal year (see Section 2-8), which Copernicus knew to be nearly 365¼ days.

The rate at which the Earth moves around its orbit is the number of degrees around the orbit divided by the time to complete the orbit, or 360°/E (equal to a little less than 1° per day). Similarly, the rate at which the inferior planet moves along its orbit is 360°/P.

During a given time interval, the angular distance that the Earth moves around its orbit is its rate, (360°/E), multiplied by the length of the time interval. Thus, during a time S, or one synodic period of the inferior planet, the Earth covers an angular distance of (360°/E)S around its orbit. In that same time, the inferior planet covers an angular distance of (360°/P)S. Note, however, that the inferior planet has gained one full lap on the Earth, and hence has covered 360° more than the Earth has (see the diagram). Thus, (360°/P)S = (360°/E)S + 360°. Dividing each term of this equation by 360° S gives

For an inferior planet:

$$\frac{1}{P} = \frac{1}{E} + \frac{1}{S}$$

P = inferior planet's sidereal period
E = Earth's sidereal period = 1 year
S = inferior planet's synodic period

A similar analysis for a superior planet (for example, Mars, Jupiter, or Saturn) yields

For a superior planet:

$$\frac{1}{P} = \frac{1}{E} - \frac{1}{S}$$

P = superior planet's sidereal period
E = Earth's sidereal period = 1 year
S = superior planet's synodic period

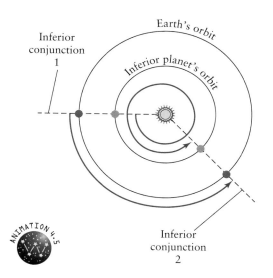

Earth's orbit

Inferior conjunction 1

Inferior planet's orbit

Inferior conjunction 2

ANIMATION 4.5

Using these formulas, we can calculate a planet's sidereal period P from its synodic period S. Often astronomers express P, E, and S in terms of years, by which they mean Earth years of approximately 365.26 days.

EXAMPLE: Jupiter has an observed synodic period S of 398.9 days, or 1.092 years. Because Jupiter is a superior planet and because $E = 1$ year exactly, we determine Jupiter's sidereal period using the second of the two equations above:

$$\frac{1}{P} = \frac{1}{1} - \frac{1}{1.092} = 0.08425,$$

so

$$P = \frac{1}{0.08425} = 11.87 \text{ years}$$

It takes 11.87 years for Jupiter to complete one full orbit of the Sun. The much shorter synodic period of 1.092 years is the time from one opposition to the next, or the time that elapses from when the Earth overtakes Jupiter to when it next overtakes Jupiter. The reason why this is so much shorter than the sidereal period is that Jupiter moves quite slowly around its orbit. The Earth overtakes it a little less often than once per Earth orbit, that is, at intervals of a little bit more than a year.

trigonometry. In this way, the astronomical unit is found to be 1.496×10^8 km (92.96 million miles). Once the astronomical unit is known, the average distances from the Sun to each of the planets in kilometers can be determined from Table 4-2. A table of these distances is given in Appendix 2.

By comparing Tables 4-1 and 4-2, you can see the unifying relationship between planetary orbits in the Copernican model: The farther a planet is from the Sun, the longer it takes to travel around its orbit (that is, the longer its sidereal period). That is not simply because a planet must travel farther to complete a larger orbit; in addition, a planet moves more slowly in a larger orbit. For example, Mercury, with its small orbit, moves at an average speed of 47.9 km/s (107,000 mi/h). Saturn travels around its large orbit much more slowly, at an average speed of 9.64 km/s (21,600 mi/h). The older Ptolemaic model offers no such simple relations between the motions of different planets.

At first, Copernicus assumed that Earth travels around the Sun along a circular path. He found that perfectly circular orbits could not accurately describe the paths of the other planets, so he had to add an epicycle to each planet. (This was *not* to explain retrograde motion, which Copernicus realized was because of the differences in orbital speeds of different planets, as shown in Figure 4-4. Rather, the small epicycles helped Copernicus account for slight variations in each planet's speed along its orbit.) Even though he clung to the old notion that orbits must be made up of circles, Copernicus had shown that a heliocentric model could explain the motions of the planets. He compiled his ideas and calculations into a book entitled *De revolutionibus orbium coelestium* (On the Revolutions of the Celestial Spheres), which was published in 1543, the year of his death.

For several decades after Copernicus, most astronomers saw little reason to change their allegiance from the older geocentric model of Ptolemy. The predictions that the Copernican model makes for the apparent positions of the planets are, on average, no better or worse than those of the Ptolemaic model. The test of Occam's razor does not really favor either model, because both use a combination of circles to describe each planet's motion.

More concrete evidence was needed to convince scholars to abandon the old, comfortable idea of a stationary Earth at the center of the universe. Obtaining this evidence required a quantum leap in the technology of astronomical observation—the invention of the telescope.

4-3 Galileo's discoveries with a telescope strongly supported a heliocentric model

When Dutch opticians invented the telescope during the first decade of the seventeenth century, astronomy was changed forever. The scholar who used this new tool to amass convincing evidence that the planets orbit the Sun, not the Earth, was the Italian mathematician and physical scientist Galileo Galilei (Figure 4-7).

figure 4-7

WEB LINK 4-4

Galileo Galilei (1564–1642) Galileo was one of the first people to use a telescope to observe the heavens. He discovered craters on the Moon, sunspots on the Sun, the phases of Venus, and four moons orbiting Jupiter. His observations strongly suggested that the Earth orbits the Sun, not vice versa. (Art Resource)

While Galileo did not invent the telescope, he was the first to point one of these new devices toward the sky and to publish his observations. Beginning in 1610, he saw sights of which no one had ever dreamed. He discovered mountains on the Moon, sunspots on the Sun, and the rings of Saturn, and he was the first to see that the Milky Way is not a featureless band of light but rather "a mass of innumerable stars."

One of Galileo's most important discoveries with the telescope was that Venus exhibits phases like those of the Moon (Figure 4-8). Galileo also noticed that the apparent size of Venus as seen through his telescope was related to the planet's phase. Venus appears small at gibbous phase and largest at crescent phase. There is also a correlation between the phases of Venus and the planet's angular distance from the Sun.

Figure 4-9 shows that these relationships are entirely compatible with a heliocentric model in which the Earth and Venus both go around the Sun. They are also completely *incompatible* with the Ptolemaic system, in which the Sun and Venus both orbit the Earth. To explain why Venus is never seen very far from the Sun, the Ptolemaic model had to assume that the deferents of Venus and of the Sun move together in lockstep, with the epicycle of Venus centered on a

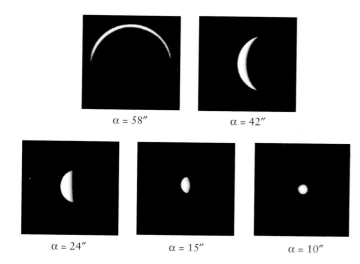

α = 58"   α = 42"

α = 24"   α = 15"   α = 10"

ƒigure 4-8   R I **V** U X G

The Phases of Venus This series of photographs shows how the appearance of Venus changes as it moves along its orbit. The number below each view is the angular diameter α of the planet in arcseconds. Venus has the largest angular diameter when it is a crescent, and the smallest angular diameter when it is gibbous (nearly full). (New Mexico State University Observatory)

straight line between the Earth and the Sun (Figure 4-10). In this model, Venus was never on the opposite side of the Sun from the Earth, and so it could never have shown the gibbous phases that Galileo observed.

Galileo also found more unexpected evidence for the ideas of Copernicus. In 1610 Galileo discovered four moons, now called the Galilean satellites, orbiting Jupiter (Figure 4-11). He realized that they were orbiting Jupiter because they appeared to move back and forth from one side of the planet to the other. Figure 4-12 shows confirming observations made by Jesuit observers in 1620. Astronomers soon realized that the larger the orbit of one of the moons around Jupiter, the slower that moon moves and the longer it takes that moon to travel around its orbit. These are the same relationships that Copernicus deduced for the motions of the planets around the Sun. Thus, the moons of Jupiter behave like a Copernican system in miniature.

Galileo's telescopic observations constituted the first fundamentally new astronomical data in almost 2000 years. Contradicting prevailing opinion, his discoveries strongly suggested a heliocentric structure of the universe. At the time, the Roman Catholic Church attacked Galileo's ideas, because they could not be reconciled with certain passages in the Bible or with the writings of Aristotle and Plato. Galileo was condemned to spend his latter years under house arrest "for

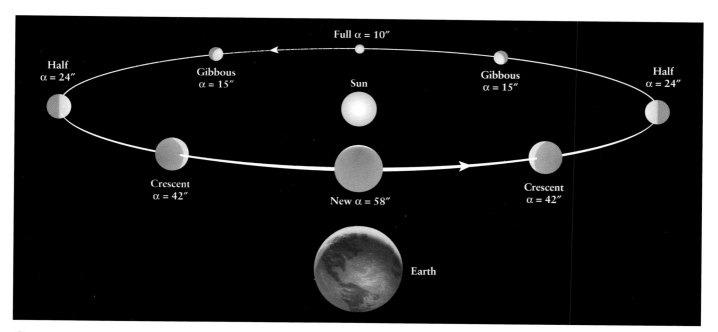

ƒigure 4-9

The Changing Appearance of Venus Explained in a Heliocentric Model A heliocentric model, in which the Earth and Venus both orbit the Sun, provides a natural explanation for the changing appearance of Venus shown in Figure 4-8. When Venus is on the opposite side of the Sun from the Earth, it appears full and has a small angular size. When Venus is on the same side of the Sun as the Earth, we see it in a "new" phase and with a larger angular size. When Venus is at the greatest angle to either side of the Sun (at greatest eastern or western elongation), it appears half illuminated and has an intermediate angular size.

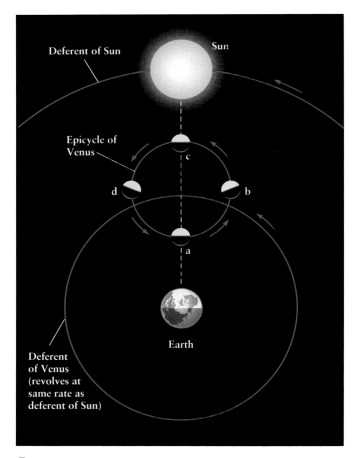

figure 4-10

The Appearance of Venus in the Ptolemaic Model In the geocentric Ptolemaic model, the deferents of Venus and the Sun rotate together, with the epicycle of Venus centered on a line (shown dashed) that connects the Sun and the Earth. In this model an Earth observer would never see Venus as more than half illuminated. (At positions *a* and *c*, Venus appears in a "new" phase; at positions *b* and *d*, it appears as a crescent. Compare with Figure 3-2, which shows the phases of the Moon.) Because Galileo saw Venus in nearly fully illuminated phases, he concluded that the Ptolemaic model must be incorrect.

vehement suspicion of heresy." Nevertheless, there was no turning back.

Galileo's observations showed convincingly that the Ptolemaic model was entirely wrong and that a heliocentric model is the more nearly correct one. Galileo did not, however, prove that the particular heliocentric model proposed by Copernicus is *the* correct one. What was needed was a heliocentric description of planetary motion that makes accurate predictions for the apparent positions of planets in the sky.

As it happened, the details of such a model were being worked out at about the same time that Galileo was making his telescopic discoveries. To construct such a model, it was essential to have an extensive set of data on planetary positions against which the predictions of the model could be tested.

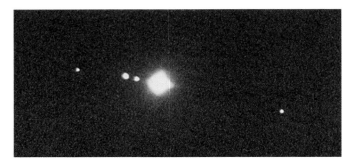

figure 4-11 R I **V** U X G

Jupiter and Its Largest Moons This photograph, taken by an amateur astronomer with a small telescope, shows the four Galilean satellites alongside an overexposed image of Jupiter. Each satellite is bright enough to be seen with the unaided eye, were it not overwhelmed by the glare of Jupiter. (Courtesy of C. Holmes)

figure 4-12

Early Observations of Jupiter's Moons In 1610 Galileo discovered four "stars" that move back and forth across Jupiter from one night to the next. He concluded that these are four moons that orbit Jupiter, much as our Moon orbits the Earth. This drawing shows notations made by Jesuit observers on successive nights in 1620. The circle represents Jupiter and the stars its moons. Compare the drawing numbered 13 with the photograph in Figure 4-11. (Yerkes Observatory)

The story of how these data came to be collected begins nearly 40 years before Galileo's observations, when a young Danish astronomer pondered the nature of a new star in the heavens.

4-4   Tycho Brahe's astronomical observations disproved ancient ideas about the heavens

On November 11, 1572, a bright star suddenly appeared in the constellation of Cassiopeia. At first, it was even brighter than Venus, but then it began to grow dim. After 18 months, it faded from view. Modern astronomers recognize this event as a supernova explosion, the violent death of a massive star.

In the sixteenth century, however, the vast majority of scholars held with the ancient teachings of Aristotle and Plato, who had argued that the heavens are permanent and unalterable. Consequently, the "new star" of 1572 could not really be a star at all, because the heavens do not change; it must instead be some sort of bright object quite near the Earth, perhaps not much farther away than the clouds overhead.

The 25-year-old Danish astronomer Tycho Brahe (1546–1601) realized that straightforward observations might reveal the distance to the new star. It is common experience that when you walk from one place to another, nearby objects appear to change position against the background of more distant objects. This phenomenon, whereby the apparent position of an object changes because of the motion of the observer, is called **parallax.** If the new star was nearby, then its position should shift against the background stars over the course of a single night because Earth's rotation changes our viewpoint. Figure 4-13 shows this predicted shift. (Actually, Tycho believed that the heavens rotate about the Earth, as in the Ptolemaic model, but the net effect is the same.)

Tycho's careful observations failed to disclose any parallax. The farther away an object is, the less it appears to shift against the background as we change our viewpoint, and the smaller the parallax. Hence, the new star had to be quite far away, farther from Earth than anyone had imagined. Tycho also attempted to measure the parallax of a bright comet that appeared in 1577 and, again, found it too small to measure. Thus, the comet also had to be far beyond the Earth. With these observations, Tycho showed that the heavens are by no means pristine and unchanging. This discovery flew in the face of nearly 2000 years of astronomical thought.

Tycho's observations of the supernova of 1572 attracted the attention of the king of Denmark, who was so impressed that he financed the construction of two magnificent observatories for Tycho on the island of Hven, just off the Danish coast. The bequest for these two observatories—Uraniborg ("heavenly castle") and Stjerneborg ("star castle")—allowed Tycho to design and have built a set of astronomical instruments vastly superior in quality to any earlier instruments (Figure 4-14). With this state-of-the-art equipment, Tycho proceeded to measure the positions of stars and planets with unprecedented accuracy. In addition, he and his assistants were careful to make several observations of the same star or planet

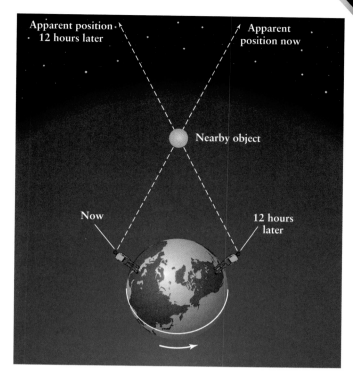

ƒigure 4-13

The Parallax of a Nearby Object Tycho Brahe argued that if an object is near the Earth, its position relative to the background stars should change over the course of a night. Tycho failed to measure such changes for a supernova in 1572 and a comet in 1577. He therefore concluded that these objects were far from the Earth.

with different instruments, in order to identify any errors that might be caused by the instruments themselves. This painstaking approach to mapping the heavens revolutionized the practice of astronomy and is used by astronomers today.

A key goal of Tycho's observations during this period was to test the ideas Copernicus had proposed decades earlier about the Earth going around the Sun. Tycho argued that if the Earth was in motion, then nearby stars should appear to shift their positions with respect to background stars as we orbit the Sun. Tycho failed to detect any such parallax, and he concluded that the Earth was at rest and the Copernican system was wrong.

On this point Tycho was in error, for nearby stars do in fact shift their positions as he had suggested. But even the nearest stars are so far away that the shifts in their positions are less than an arcsecond, too small to be seen with the naked eye. Tycho would have needed a telescope to detect the parallax that he was looking for, but the telescope was not invented until after his death in 1601. Indeed, the first accurate determination of stellar parallax was not made until 1838.

Although he remained convinced that the Earth was at the center of the universe, Tycho nonetheless made a tremendous contribution toward putting the heliocentric model on a solid foundation. From 1576 to 1597, he used his instruments to

figure 4-14

Tycho Brahe (1546–1601) Observing Tycho Brahe measured the positions of stars and planets with greater accuracy than ever before. This contemporary illustration shows Tycho with some of the measuring apparatus at Uraniborg, one of the two observatories that he built under the patronage of Frederik II of Denmark. (This magnificent observatory lacked a telescope, which had not yet been invented.) The data that Tycho collected were crucial to the development of astronomy in the years after his death. (Photo Researchers, Inc.)

figure 4-15

Johannes Kepler (1571–1630) By analyzing Tycho Brahe's detailed records of planetary positions, Kepler developed three general principles, called Kepler's laws, that describe how the planets move about the Sun. Kepler was the first to realize that the orbits of the planets are ellipses and not circles. (E. Lessing/Magnum)

make comprehensive measurements of the positions of the planets with an accuracy of 1 arcminute. This is as well as can be done with the naked eye and was far superior to any earlier measurements. Within the reams of data that Tycho compiled lay the truth about the motions of the planets. The person who would extract this truth was a German mathematician who became Tycho's assistant in 1600, a year before the great astronomer's death. His name was Johannes Kepler (Figure 4-15).

4-5 Johannes Kepler proposed elliptical paths for the planets about the Sun

The task that Kepler took on at the beginning of the seventeenth century was to find a model of planetary motion that agreed completely with Tycho's extensive and very accurate observations of planetary positions. To do this, Kepler found that he had to break with an ancient prejudice about planetary motions.

Astronomers had long assumed that heavenly objects move in circles, which were considered the most perfect and harmonious of all geometric shapes. They believed that if a perfect God resided in heaven along with the stars and planets, then the motions of these bodies must be perfect too. Against this context, Kepler dared to try to explain planetary motions with noncircular curves. In particular, he found that he had the best success with a particular kind of curve called an **ellipse.**

You can draw an ellipse by using a loop of string, two thumbtacks, and a pencil, as shown in Figure 4-16a. Each thumbtack in the figure is at a **focus** (plural **foci**) of the ellipse; an ellipse has two foci. The longest diameter of an ellipse, called the **major axis,** passes through both foci. Half of that distance is called the **semimajor axis** and is usually designated by the letter *a*. A circle is a special case of an ellipse in which the two foci are at the same point (this corresponds to using only a single thumbtack in Figure 4-16a). The semimajor axis of a circle is equal to its radius.

By assuming that planetary orbits were ellipses, Kepler found, to his delight, that he could make his theoretical calculations match precisely to Tycho's observations. This impor-

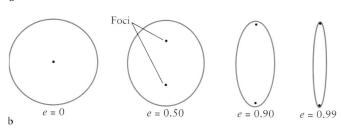

b

figure 4-16

Ellipses **(a)** To draw an ellipse, use two thumbtacks to secure the ends of a piece of string, then use a pencil to pull the string taut. If you move the pencil while keeping the string taut, the pencil traces out an ellipse. The thumbtacks are located at the two foci of the ellipse. The major axis is the greatest distance across the ellipse; the semimajor axis is half of this distance. The minor axis is perpendicular to the major axis and passes through the center of the ellipse. **(b)** A series of ellipses with the same major axis but different eccentricities. An ellipse can have any eccentricity from $e = 0$ (a circle) to just under $e = 1$ (virtually a straight line).

The orbit of a planet about the Sun is an ellipse with the Sun at one focus.

There is no object at the other focus of a planet's elliptical orbit. This "empty focus" has geometrical significance, because it helps to define the shape of the ellipse, but plays no other role. The semimajor axis a of a planet's orbit is the average distance between the planet and the Sun.

Ellipses come in different shapes, depending on the elongation of the ellipse. The shape of an ellipse is described by its **eccentricity,** designated by the letter e. The value of e can range from 0 (a circle) to just under 1 (nearly a straight line). The greater the eccentricity, the more elongated the ellipse. Figure 4-16*b* shows a few examples of ellipses with different eccentricities. Because a circle is a special case of an ellipse, it is possible to have a perfectly circular orbit. But all of the planets have orbits that are at least slightly elliptical, with eccentricities that range from 0.007 for Venus (the most nearly circular of any planetary orbit) to 0.248 for Pluto (the most elongated orbit).

Once he knew the shape of a planet's orbit, Kepler was ready to describe exactly *how* it moves on that orbit. As a planet travels in an elliptical orbit, its distance from the Sun varies. Kepler realized that the speed of a planet also varies along its orbit. A planet moves most rapidly when it is nearest the Sun, at a point on its orbit called **perihelion.** Conversely, a planet moves most slowly when it is farthest from the Sun, at a point called **aphelion** (Figure 4-17).

After much trial and error, Kepler found a way to describe just how a planet's speed varies as it moves along its orbit. This discovery, also published in 1609, is illustrated in Figure 4-17. Suppose that it takes 30 days for a planet to go from point A to point B. During that time, an imaginary line joining the Sun and the planet sweeps out a nearly triangular

figure 4-17

Kepler's First and Second Laws

According to Kepler's first law, a planet travels around the Sun along an elliptical orbit with the Sun at one focus. According to his second law, a planet moves fastest when closest to the Sun (at perihelion) and slowest when farthest from the Sun (at aphelion). As the planet moves, an imaginary line joining the planet and the Sun sweeps out equal areas in equal intervals of time (from A to B or from C to D). Kepler found that by using these laws, his calculations gave a perfect fit to the apparent motions of the planets.

area. Kepler discovered that a line joining the Sun and the planet also sweeps out exactly the same area during any other 30-day interval. In other words, if the planet also takes 30 days to go from point C to point D, then the two shaded segments in Figure 4-17 are equal in area. **Kepler's second law** can be stated thus:

A line joining a planet and the Sun sweeps out equal areas in equal intervals of time.

This relationship is also called the **law of equal areas.** In the idealized case of a circular orbit, a planet would have to move at a constant speed around the orbit in order to satisfy Kepler's second law.

A good analogy for Kepler's second law is a twirling ice skater. If the skater pulls her arms straight in to her body, she spins faster; if she lets her arms extend away from her body, her rate of spin decreases. In the same way, a planet in an elliptical orbit travels at a higher speed when it moves closer to the Sun (toward perihelion) and travels at a lower speed when it moves away from the Sun (toward aphelion).

Kepler's second law describes how the speed of a given planet changes as it orbits the Sun. Kepler also deduced from Tycho's data a relationship that can be used to compare the motions of *different* planets. Published in 1618 and now called **Kepler's third law,** it states a relationship between the size of a planet's orbit and the time the planet takes to go once around the Sun:

The square of the sidereal period of a planet is directly proportional to the cube of the semimajor axis of the orbit.

Kepler's third law says that the larger a planet's orbit—that is, the larger the semimajor axis, or average distance from the planet to the Sun—the longer the sidereal period, which is the time it takes the planet to complete an orbit. From Kepler's third law one can show that the larger the semimajor axis, the slower the average speed at which the planet moves around its orbit. (By contrast, Kepler's *second* law describes how the speed of a given planet is sometimes faster and sometimes slower than its average speed.) This qualitative relationship between orbital size and orbital speed is just what Aristarchus and Copernicus used to explain retrograde motion, as we saw in Section 4-2. Kepler's great contribution was to make this relationship a quantitative one.

It is useful to restate Kepler's third law as an equation. If a planet's sidereal period P is measured in years and the length of its semimajor axis a is measured in astronomical units (AU), where 1 AU is the average distance from the Earth to the Sun (Section 1-7), then Kepler's third law is

Kepler's third law

$$P^2 = a^3$$

P = planet's sidereal period, in years
a = planet's semimajor axis, in AU

If you know either the sidereal period of a planet or the semimajor axis of its orbit, you can find the other quantity using this equation. Box 4-2 gives some examples of how this is done.

We can verify Kepler's third law for all of the planets, including those that were discovered after Kepler's death, using data from Tables 4-1 and 4-2. If Kepler's third law is correct, for each planet the numerical values of P^2 and a^3 should be equal. This is indeed true to very high accuracy, as Table 4-3 shows.

Kepler's laws are a landmark in the history of astronomy. They made it possible to calculate the motions of the planets with far better accuracy than any geocentric model ever had, and they helped to justify the idea of a heliocentric model.

Table 4-3	A Demonstration of Kepler's Third Law			
Planet	Sidereal period P (years)	Semimajor axis a (AU)	P^2	a^3
Mercury	0.24	0.39	0.06	0.06
Venus	0.61	0.72	0.37	0.37
Earth	1.00	1.00	1.00	1.00
Mars	1.88	1.52	3.53	3.51
Jupiter	11.86	5.20	140.7	140.6
Saturn	29.46	9.54	867.9	868.3
Uranus	84.01	19.19	7,058	7,067
Neptune	64.79	30.06	27,160	27,160
Pluto	248.54	39.53	61,770	61,770

box 4-2 | Tools of the Astronomer's Trade

Using Kepler's Third Law

Kepler's third law relates the sidereal period P of an object orbiting the Sun to the semimajor axis a of its orbit:

$$P^2 = a^3$$

You must keep two essential points in mind when working with this equation:

1. The period P *must* be measured in years, and the semimajor axis a *must* be measured in astronomical units (AU). Otherwise you will get nonsensical results.

2. This equation applies *only* to the special case of an object, like a planet, that orbits the Sun. If you want to analyze the orbit of the Moon around the Earth, of a spacecraft around Mars, or of a planet around a distant star, you must use a different, generalized form of Kepler's third law. We discuss this alternative equation in Section 4-7 and Box 4-4.

EXAMPLE: The average distance from Venus to the Sun—that is, the semimajor axis a of Venus's orbit—is 0.72 AU. To determine the sidereal period of Venus, we first cube the semimajor axis (multiply it by itself twice):

$$a^3 = (0.72)^3 = 0.72 \times 0.72 \times 0.72 = 0.373$$

According to Kepler's third law this is also equal to P^2, the square of the sidereal period. So, to find P, we have to "undo" the square, that is, take the square root. Using a calculator, we find

$$P = \sqrt{P^2} = \sqrt{0.373} = 0.61$$

The sidereal period of Venus is 0.61 years, or a bit more than seven Earth months.

EXAMPLE: Consider a small asteroid (a rocky body a few tens of kilometers across) that takes eight years to complete one orbit around the Sun. We can use Kepler's third law to determine the semimajor axis a of the asteroid's orbit. We first square the period:

$$P^2 = 8^2 = 8 \times 8 = 64$$

From Kepler's third law, this is also equal to a^3. To determine a, we must take the *cube root* of a^3, that is, find the number whose cube is 64. If your calculator has a cube root function, denoted by the symbol $\sqrt[3]{}$, you can use it to find that the cube root of 64 is 4: $\sqrt[3]{64} = 4$. Otherwise, you can determine by trial and error that the cube of 4 is 64:

$$4^3 = 4 \times 4 \times 4 = 64$$

Because the cube of 4 is 64, it follows that the cube root of 64 is 4 (taking the cube root "undoes" the cube).

With either technique you find that the orbit of this asteroid has semimajor axis $a = 4$ AU. This is intermediate between the orbits of Mars and Jupiter. Many asteroids are known with semimajor axes in this range, forming a region in the solar system called the asteroid belt.

Kepler's laws also pass the test of Occam's razor, for they are simpler in every way than the schemes of Ptolemy or Copernicus, both of which used a complicated combination of circles.

But the significance of Kepler's laws goes beyond understanding planetary orbits. These same laws are also obeyed by spacecraft orbiting the Earth, by two stars revolving about each other in a binary star system, and even by galaxies in their orbits about each other. Throughout this book, we shall use Kepler's laws in a wide range of situations.

What Kepler was not able to explain is *why* planets move in accordance with his three laws. The first person who was able to provide such an explanation was the Englishman Isaac Newton, born on Christmas Day of 1642, a dozen years after the death of Kepler and the same year that Galileo died. While

Galileo and Kepler revolutionized our understanding of planetary motions, Newton's contribution was far greater: He deduced the basic laws that govern all motions on Earth as well as in the heavens.

4-6 Isaac Newton formulated three laws that describe fundamental properties of physical reality

Until the mid-seventeenth century, virtually all attempts to describe the heavens mathematically were *empirical*, or based directly on data and observations. From Ptolemy to Kepler, astronomers would adjust their ideas and calculations by trial and error until they ended up with answers that agreed with observation.

Isaac Newton (Figure 4-18) introduced a new approach. He began with three quite general statements, now called **Newton's laws of motion.** These laws, deduced from experimental observation, apply to all forces and all bodies. Newton then showed that Kepler's three laws follow logically from these laws of motion and from a formula for the force of gravity that he derived from observation.

In other words, Kepler's laws are not just an empirical description of the motions of the planets, but a direct consequence of the fundamental laws of physical matter. Using this deeper insight into the nature of motions in the heavens, Newton and his successors were able to describe accurately not just the orbits of the planets but also the orbits of the Moon and comets.

Newton's laws of motion describe objects on Earth as well as in the heavens. Thus, we can understand each of these laws by considering the motions of objects around us. We begin with **Newton's first law of motion,** or **law of inertia:**

A body remains at rest, or moves in a straight line at a constant speed, unless acted upon by a net outside force.

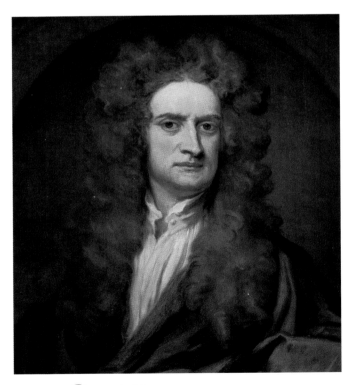

figure 4-18

Isaac Newton (1642–1727) Using mathematical techniques that he devised, Isaac Newton formulated the law of universal gravitation and demonstrated that the planets orbit the Sun according to simple mechanical rules. (National Portrait Gallery, London)

By **force** we mean any push or pull that acts on the body. An *outside* force is one that is exerted on the body by something other than the body itself. The net, or total, outside force is the combined effect of all of the individual outside forces that act on the body.

Right now, you are demonstrating the first part of Newton's first law. As you sit in your chair reading this, there are two outside forces acting on you: The force of gravity pulls you downward, and the chair pushes up on you. These two forces are of equal strength but of opposite direction, so their effects cancel—there is no *net* outside force. Hence, your body remains at rest as stated in Newton's first law. If you try to lift yourself out of your chair by grabbing your knees and pulling up, you will remain at rest because this force is not an outside force.

 The second part of Newton's first law, about objects in motion, may seem to go against common sense. If you want to make this book move across the floor in a straight line at a constant speed, you must continually push on it. You might therefore think that there *is* a net outside force, the force of your push. But another force also acts on the book—the force of friction as the book rubs across the floor. As you push the book across the floor, the force of your push exactly balances the force of friction, so again there is no net outside force. The effect is to make the book move in a straight line at constant speed, just as Newton's first law says. If you stop pushing, there will be nothing to balance the effects of friction. Then there will be a net outside force and the book will slow to a stop.

Newton's first law tells us that a force must be acting on the planets. A planet moving through empty space encounters no friction, so it would tend to fly off into space along a straight line if there were no other outside force acting on it. Because this does not happen, Newton concluded that there must be a force that acts continuously on the planets to keep them in their elliptical orbits.

Newton's second law describes how the motion of an object changes if there is a net outside force acting on it. To appreciate Newton's second law, we must first understand three quantities that describe motion—speed, velocity, and acceleration.

Speed is a measure of how fast an object is moving. Speed and direction of motion together constitute an object's **velocity.** Compared with a car driving north at 100 km/h (62 mi/h), a car driving east at 100 km/h has the same speed but a different velocity.

Acceleration is the rate at which velocity changes. Because velocity involves both speed and direction, acceleration can result from changes in either. Contrary to popular use of the term, acceleration does not simply mean speeding up. A car is accelerating if it is speeding up, and it is also accelerating if it is slowing down or turning (that is, changing the direction in which it is moving).

You can verify these statements about acceleration if you think about the sensations of riding in a car. If the car is moving with a constant velocity (in a straight line at a constant speed), you feel the same as if the car were not moving at all.

box 4-3 | The Heavens on the Earth

Newton's Laws in Everyday Life

In our study of astronomy, we use Newton's three laws of motion to help us understand the motions of objects in the heavens. But you can see applications of Newton's laws every day in the world around you. By considering these everyday applications, we can gain insight into how Newton's laws apply to celestial events that are far removed from ordinary human experience.

Newton's *first* law, or principle of inertia, says that an object at rest naturally tends to remain at rest and that an object in motion naturally tends to remain in motion. This explains the sensations that you feel when riding in an automobile. When you are waiting at a red light, your car and your body are both at rest. When the light turns green and you press on the gas pedal, the car accelerates forward but your body attempts to stay where it was. Hence, the seat of the accelerating car pushes forward into your body, and it feels as though you are being pushed back in your seat.

Once the car is up to cruising speed, your body wants to keep moving in a straight line at this cruising speed. If the car makes a sharp turn to the left, the right side of the car will move toward you. Thus, you will feel as though you are being thrown to the car's right side (the side on the outside of the turn). If you bring the car to a sudden stop by pressing on the brakes, your body will continue moving forward until the seat belt stops you. In this case, it feels as though you are being thrown toward the front of the car.

Newton's *second* law states that the net outside force on an object equals the product of the object's mass and its acceleration. You can accelerate a crumpled-up piece of paper to a pretty good speed by throwing it with a moderate force. But if you try to throw a backpack full of books by using the same force, the acceleration will be much less because the backpack has much more mass than the crumpled paper. Because of the reduced acceleration during the throw, the backpack will leave your hand moving at only a slow speed.

Automobile airbags are based on the relationship between force and acceleration. It takes a large force to bring a fast-moving object suddenly to rest because this requires a large acceleration. In a collision, the driver of a car not equipped with airbags is jerked to a sudden stop and the large forces that act can cause major injuries. But if the car has airbags that deploy in an accident, the driver's body will slow down more gradually as it contacts the airbag, and the driver's acceleration will be less. (Remember that *acceleration* can refer to slowing down as well as to speeding up.) Hence, the force on the driver and the chance of injury will both be greatly reduced.

Newton's *third* law, the principle of action and reaction, explains how a car can accelerate at all. It is not correct to say that the engine pushes the car forward, because Newton's second law tells us that it takes a force acting from outside the car to make the car accelerate. Rather, the engine makes the wheels and tires turn, and the tires push backward on the ground. (You can see this backward force acting when a car drives through mud and sprays mud backward from the spinning tires.) From Newton's third law, the ground must exert an equally large forward force on the car, and this is the force that pushes the car forward.

You use the same principles when you walk: You push backward on the ground with your foot, and the ground pushes forward on you. Icy pavement or a freshly waxed floor have greatly reduced friction. In these situations, your feet and the surface under you can exert only weak forces on each other, and it is much harder to walk.

But you can feel it when the car accelerates in any way: You feel thrown back in your seat if the car speeds up, thrown forward if the car slows down, and thrown sideways if the car changes direction in a tight turn. In Box 4-3 we discuss the reasons for these sensations, along with other applications of Newton's laws to everyday life.

An apple falling from a tree is a good example of acceleration that involves only an increase in speed. Initially, at the moment the stem breaks, the apple's speed is zero. After 1 second, its downward speed is 9.8 meters per second, or 9.8 m/s (32 feet per second, or 32 ft/s). After 2 seconds, the apple's speed is twice this, or 19.6 m/s. After 3 seconds, the speed is 29.4 m/s. Because the apple's speed increases by 9.8 m/s for each second of free fall, the rate of acceleration is 9.8 meters per second per second, or 9.8 m/s^2 (32 ft/s^2). Thus, the Earth's gravity gives the apple a constant acceleration of 9.8 m/s^2 downward, toward the center of the Earth.

A planet revolving about the Sun along a perfectly circular orbit is an example of acceleration that involves change of direction only. As the planet moves along its orbit, its speed remains constant. Nevertheless, the planet is continuously being accelerated because its direction of motion is continuously changing.

Newton's second law of motion says that the acceleration of an object is proportional to the net outside force acting on the object. In other words, the harder you push on an object, the greater the resulting acceleration. This law can be succinctly stated as an equation. If a net outside force *F* acts on an object of mass *m*, the object will experience an acceleration *a* such that

Newton's second law

$$F = ma$$

F = net outside force on an object
m = mass of object
a = acceleration of object

The **mass** of an object is a measure of the total amount of material in the object. It is usually expressed in kilograms (kg) or grams (g). For example, the mass of the Sun is 2×10^{30} kg, the mass of a hydrogen atom is 1.7×10^{-27} kg, and the mass of an average adult is 75 kg. The Sun, a hydrogen atom, and a person have these masses regardless of where they happen to be in the universe.

It is important not to confuse the concepts of mass and weight. **Weight** is the force of gravity that acts on a body and, like any force, is usually expressed in pounds or newtons (1 newton = 0.225 pound).

We can use Newton's second law to relate mass and weight. We have seen that the acceleration caused by the Earth's gravity is 9.8 m/s². When a 50-kg swimmer falls from a diving board, the only outside force acting on her as she falls is her weight. Thus, from Newton's second law ($F = ma$), her weight is equal to her mass multiplied by the acceleration due to gravity:

$$50 \text{ kg} \times 9.8 \text{ m/s}^2 = 490 \text{ newtons} = 110 \text{ pounds}$$

Note that this answer is correct only when the swimmer is on Earth. She would weigh less on the Moon, where the pull of gravity is weaker, and more on Jupiter, where the gravitational pull is stronger. Floating deep in space, she would have no weight at all; she would be "weightless." Nevertheless, in all these circumstances, she would always have exactly the same mass, because mass is an inherent property of matter unaffected by details of the environment. Whenever we describe the properties of planets, stars, or galaxies, we speak of their masses, never of their weights.

We have seen that a planet is continually accelerating as it orbits the Sun. From Newton's second law, this means that there must be a net outside force that acts continually on each of the planets. As we will see in the next section, this force is the gravitational attraction of the Sun.

The last of Newton's general laws of motion, called **Newton's third law of motion**, is the famous statement about action and reaction:

Whenever one body exerts a force on a second body, the second body exerts an equal and opposite force on the first body.

For example, if you weigh 110 pounds, when you are standing up you are pressing down on the floor with a force of 110 pounds. Newton's third law tells us that the floor is also pushing up against your feet with an equal force of 110 pounds.

Newton realized that because the Sun is exerting a force on each planet to keep it in orbit, each planet must also be exert-ing an equal and opposite force on the Sun. However, the planets are much less massive than the Sun (for example, the Earth has only 1/300,000 of the Sun's mass). Therefore, although the Sun's force on a planet is the same as the planet's force on the Sun, the planet's much smaller mass gives it a much larger acceleration, according to Newton's second law. This is why the planets circle the Sun instead of vice versa. Thus, Newton's laws reveal the reason for our heliocentric solar system.

4-7 Newton's description of gravity accounts for Kepler's laws and explains the motions of the planets

Tie a ball to one end of a piece of string, hold the other end of the string in your hand, and whirl the ball around in a circle. As the ball "orbits" your hand, it is continuously accelerating because its velocity is changing. (Even if its speed is constant, its direction of motion is changing.) In accordance with Newton's second law, this can happen only if the ball is continuously acted on by an outside force—the pull of the string. The pull is directed along the string toward your hand. In the same way, Newton saw, the force that keeps a planet in orbit around the Sun is a pull that always acts toward the Sun. That pull is **gravity**, or **gravitational force**.

Newton's discovery about the forces that act on planets led him to suspect that the force of gravity pulling a falling apple straight down to the ground is fundamentally the same as the force on a planet that is always directed straight at the Sun. In other words, gravity is the force that shapes the orbits of the planets. What is more, he was able to determine how the force of gravity depends on the distance between the Sun and the planet. His result was a law of gravitation that could apply to the motion of distant planets as well as to the flight of a football on Earth. Using this law, Newton achieved the remarkable goal of deducing Kepler's laws from fundamental principles of nature.

To see how Newton reasoned, think again about a ball attached to a string. If you use a short string, so that the ball orbits in a small circle, and whirl the ball around your hand at a high speed, you will find that you have to pull fairly hard on the string (Figure 4-19*a*). But if you use a longer string, so that the ball moves in a larger orbit, and if you make the ball orbit your hand at a slow speed, you only have to exert a light tug on the string (Figure 4-19*b*). The orbits of the planets behave in the same way; the larger the size of the orbit, the slower the planet's speed. By analogy to the force of the string on the orbiting ball, Newton concluded that the force that attracts a planet toward the Sun must decrease with increasing distance between the Sun and the planet.

Using his own three laws and Kepler's three laws, Newton succeeded in formulating a general statement that describes the nature of the gravitational force. Newton's **law of universal gravitation** is as follows:

Two bodies attract each other with a force that is directly proportional to the mass of each body and inversely proportional to the square of the distance between them.

Strong pull required

a   Ball moves at a high speed
    in a small circle

Only a weak pull
required

b   Ball moves at a low speed
    in a large circle

Figure 4-19

A Ball on a String **(a)** To make a ball on a string move at high speed around a small circle, you have to exert a substantial pull on the string. Similarly, a planet that orbits close to the Sun moves at high speed and requires a substantial gravitational force from the Sun. **(b)** If you lengthen the string and make the same ball move at low speed around a large circle, much less pull is required. In the same way, a planet in a large orbit moves at low speed and requires less gravitational force to stay in orbit. This example shows that the Sun's gravitational pull must decrease with increased distance.

This law states that *any* two objects exert gravitational pulls on each other. Normally, you notice only the gravitational force that the Earth exerts on you, otherwise known as your weight. In fact, you feel gravitational attractions to *all* the objects around you. For example, this book is exerting a gravitational force on you as you read it. But because the force exerted on you by this book is proportional to the book's mass, which is very small compared to the Earth's mass, the force is too small to notice. (It can actually be measured with sensitive equipment.)

Consider two 1-kg objects separated by a distance of 1 meter. Newton's law of universal gravitation says that the force is directly proportional to the mass, so if we double the mass of one object to 2 kg, the force between the objects will double. If we double both masses so that we have two 2-kg objects separated by 1 meter, the force will be $2 \times 2 = 4$ times what it was originally (the force is directly proportional to the mass of *each* object). If we go back to two 1-kg masses, but double their separation to 2 meters, the force will be only one-quarter its original value. This is because the force is inversely proportional to the distance: If we double the distance, the force is multiplied by a factor of

$$\frac{1}{2^2} = \frac{1}{4}$$

Newton's law of universal gravitation can be stated more succinctly as an equation. If two objects have masses m_1 and m_2 and are separated by a distance r, then the gravitational force F between these two objects is given by the equation

Newton's law of universal gravitation

$$F = G\left(\frac{m_1 m_2}{r^2}\right)$$

F = gravitational force between two objects
m_1 = mass of first object
m_2 = mass of second object
r = distance between objects
G = universal constant of gravitation

If the masses are measured in kilograms and the distance between them in meters, then the force is measured in new-

tons. In this formula, G is a number called the **universal constant of gravitation.** Laboratory experiments have yielded a value for G of

$$G = 6.67 \times 10^{-11} \text{ newton m}^2/\text{kg}^2$$

We can use Newton's law of universal gravitation to calculate the force with which any two bodies attract each other. For example, to compute the gravitational force that the Sun exerts on the Earth, we substitute values for the Earth's mass ($m_1 = 5.98 \times 10^{24}$ kg), the Sun's mass ($m_2 = 1.99 \times 10^{30}$ kg), the distance between them ($r = 1$ AU $= 1.5 \times 10^{11}$ m), and the value of G into Newton's equation. We get

$$F_{\text{Sun-Earth}} = 6.67 \times 10^{-11}\left[\frac{(5.98 \times 10^{24}) \times (1.99 \times 10^{30})}{(1.50 \times 10^{11})^2}\right]$$

$$= 3.53 \times 10^{22} \text{ newtons}$$

If we calculate the force that the Earth exerts on the Sun, we get exactly the same result. (Mathematically, we just let m_1 be the Sun's mass and m_2 be the Earth's mass instead of the other way around. The product of the two numbers is the same, so the force is the same.) This is in accordance with Newton's third law: Any two objects exert *equal* gravitational forces on each other.

Because there is a gravitational force between any two objects, Newton concluded that gravity is also the force that keeps the Moon in orbit around the Earth. It is also the force that keeps artificial satellites in orbit. But if the force of gravity attracts two objects to each other, why don't satellites immediately fall to Earth? Why doesn't the Moon fall into the Earth? And, for that matter, why don't the planets fall into the Sun?

To see the answer, imagine (as Newton did) dropping a ball from a great height above the Earth's surface, as in Figure 4-20. After you drop the ball, it, of course, falls straight down (path A in Figure 4-20). But if you *throw* the ball horizontally, it travels some distance across the Earth's surface before hitting the ground (path B). If you throw the ball harder, it travels a greater distance (path C). If you could throw at just the right speed, the curvature of the

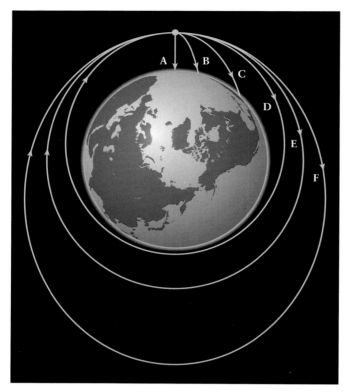

WEB LINK 4.8

figure 4-20

An Explanation of Orbits If a ball is dropped from a great height above the Earth's surface, it falls straight down (A). If the ball is thrown with some horizontal speed, it follows a curved path before hitting the ground (B, C). If thrown with just the right speed (E), the ball goes into circular orbit; the ball's path curves but it never gets any closer to the Earth's surface. If the ball is thrown with a speed that is slightly less (D) or slightly more (F) than the speed for a circular orbit, the ball's orbit is an ellipse.

ball's path will exactly match the curvature of the Earth's surface (path E). Although the Earth's gravity is making the ball fall, the Earth's surface is falling away under the ball at the same rate. Hence, the ball does not get any closer to the surface, and the ball is in circular orbit. So the ball in path E is in fact falling, but it is falling *around* the Earth rather than *toward* the Earth.

A spacecraft is launched into orbit in just this way—by throwing it fast enough. The thrust of a rocket is used to give the spacecraft the necessary orbital speed. Once the spacecraft is in orbit, the rocket turns off and the spacecraft falls continually around the Earth.

CAUTION! An astronaut on board an orbiting spacecraft feels "weightless," but this is *not* because she is "beyond the pull of gravity." The astronaut is herself an independent satellite of the Earth, and the Earth's gravitational pull is what holds her in orbit. She feels "weightless" because

she and her spacecraft are falling *together* around the Earth, so there is nothing pushing her against any of the spacecraft walls. You feel the same "weightless" sensation whenever you are falling, such as when you jump off a diving board or ride the free-fall ride at an amusement park.

If the ball in Figure 4-20 is thrown with a slightly slower speed than that required for a circular orbit, its orbit will be an ellipse (path D). An elliptical orbit also results if instead the ball is thrown a bit too fast (path F). In this way, spacecraft can be placed into any desired orbit around the Earth by adjusting the thrust of the rockets.

Just as the ball in Figure 4-20 will not fall to Earth if given enough speed, the Moon does not fall to Earth and the planets do not fall into the Sun. The planets acquired their initial speeds around the Sun when the solar system first formed 4.6 billion years ago. Figure 4-20 shows that a circular orbit is a very special case, so it is no surprise that the orbits of the planets are not precisely circular.

CAUTION! Orbiting satellites do sometimes fall out of orbit and crash back to Earth. When this happens, however, the real culprit is not gravity but air resistance. A satellite in a relatively low orbit is actually flying through the tenuous outer wisps of the Earth's atmosphere. The resistance of the atmosphere slows the satellite and changes a circular orbit like E in Figure 4-20 to an elliptical one like D. As the satellite sinks to lower altitude, it encounters more air resistance and sinks even lower. Eventually, it either strikes the Earth or burns up in flight due to air friction. By contrast, the Moon and planets orbit in the near-vacuum of interplanetary space. Hence, they are unaffected by this kind of air resistance, and their orbits are much more long-lasting.

Using his three laws of motion and his law of gravity, Newton found that he could prove Kepler's three laws mathematically. Kepler's first law, concerning the elliptical shape of planetary orbits, proved to be a direct consequence of the $1/r^2$ factor in the law of universal gravitation. (Had the nature of gravity in our universe been different, so that this factor was given by a different function such as $1/r$ or $1/r^3$, elliptical orbits would not have been possible.) The law of equal areas, or Kepler's second law, turns out to be a consequence of the Sun's gravitational force on a planet being directed straight toward the Sun.

Newton also demonstrated that Kepler's third law follows logically from his law of gravity. Specifically, he proved that if two objects with masses m_1 and m_2 orbit each other, the period P of their orbit and the semimajor axis a of their orbit (that is, the average distance between the two objects) are related by the equation

$$P^2 = \left[\frac{4\pi^2}{G(m_1 + m_2)} \right] a^3$$

This equation is called **Newton's form of Kepler's third law.** It is valid whenever two objects orbit each other because of their mutual gravitational attraction. It is invaluable in the

box 4-4 | **T**ools of the Astronomer's Trade

Newton's Form of Kepler's Third Law

Kepler's original statement of his third law, $P^2 = a^3$, is valid only for objects that orbit the Sun. (Box 4-2 shows how to use this equation.) But Newton's form of Kepler's third law is much more general: It can be used in *any* situation where two bodies of masses m_1 and m_2 orbit each other. For example, Newton's form is the equation to use for a moon orbiting a planet or a satellite orbiting the Earth. This equation is

Newton's form of Kepler's third law

$$P^2 = \left[\frac{4\pi^2}{G(m_1 + m_2)} \right] a^3$$

P = sidereal period of orbit, in seconds
a = semimajor axis of orbit, in meters
m_1 = mass of first object, in kilograms
m_2 = mass of second object, in kilograms
G = universal constant of gravitation = 6.67×10^{-11}

Notice that P, a, m_1, and m_2 *must* be expressed in these particular units. If you fail to use the correct units, your answer will be incorrect.

EXAMPLE: Io is one of the four large moons of Jupiter discovered by Galileo. It orbits at a distance of 421,600 km from the center of Jupiter and has an orbital period of 1.77 days. Because this is not an orbit around the Sun, we must use Newton's form of Kepler's third law. We can use this form to determine the combined mass of Jupiter (m_1) and Io (m_2). We first rewrite the equation in the form

$$m_1 + m_2 = \frac{4\pi^2 a^3}{GP^2}$$

To use this equation, we have to convert the distance a from kilometers to meters and convert the period P from days to seconds. There are 1000 meters in 1 kilometer and 86,400 seconds in 1 day, so

$$a = (421{,}600 \text{ km}) \times \frac{1000 \text{ m}}{1 \text{ km}} = 4.216 \times 10^8 \text{ m}$$

$$P = (1.77 \text{ days}) \times \frac{86{,}400 \text{ s}}{1 \text{ day}} = 1.529 \times 10^5 \text{ s}$$

We can now put these values and the value of G into the above equation:

$$m_1 + m_2 = \frac{4\pi^2 (4.216 \times 10^8)^3}{(6.67 \times 10^{-11})(1.529 \times 10^5)^2} = 1.90 \times 10^{27} \text{ kg}$$

Io is very much smaller than Jupiter, so its mass is only a small fraction of the mass of Jupiter. Thus, $m_1 + m_2$ is very nearly the mass of Jupiter alone. We conclude that Jupiter has a mass of 1.90×10^{27} kg, or about 300 times the mass of Earth. This technique can be used to determine the mass of any object that has a second, much smaller object orbiting around it. Astronomers use this technique to find the masses of stars, black holes, and entire galaxies of stars.

study of binary star systems, in which two stars orbit each other. If the orbital period P and semimajor axis a of the two stars in a binary system are known, astronomers can use this formula to calculate the sum $m_1 + m_2$ of the masses of the two stars. Within our own solar system, Newton's form of Kepler's third law makes it possible to learn about the masses of planets. By measuring the period and semimajor axis for a satellite, astronomers can determine the sum of the masses of the planet and the satellite. (The satellite can be a moon of the planet or a spacecraft that we place in orbit around the planet. Newton's laws apply in either case.) Box 4-4 gives some examples using Newton's form of Kepler's third law.

Newton also discovered new features of orbits around the Sun. For example, his equations soon led him to conclude that the orbit of an object around the Sun need not be an ellipse. It could be any one of a family of curves called conic sections.

A **conic section** is any curve that you get by cutting a cone with a plane, as shown in Figure 4-21. You can get circles and ellipses by slicing all the way through the cone. You can also get two types of open curves called **parabolas** and **hyperbolas.** If you were to throw the ball in Figure 4-20 with a fast enough speed, it would follow a parabolic or hyperbolic orbit and would fly off into space, never to return. Comets hurtling toward the Sun from the depths of space sometimes follow hyperbolic orbits.

Newton's ideas turned out to be applicable to an incredibly wide range of situations. Using his laws of motion, Newton himself proved that the Earth's axis of rotation must precess because of the gravitational pull of the Moon and the Sun on the Earth's equatorial bulge (see Figure 2-17). In fact, all the details of the orbits of the planets and their satellites could be explained mathematically with a body of knowledge built on Newton's work that is today called **Newtonian mechanics.**

figure 4-21

Conic Sections A conic section is any one of a family of curves obtained by slicing a cone with a plane. The orbit of one body about another can be any one of these curves: a circle, an ellipse, a parabola, or a hyperbola.

Not only could Newtonian mechanics explain a variety of known phenomena in detail, but it could also predict new phenomena. For example, one of Newton's friends, Edmund Halley, was intrigued by three similar historical records of a comet that had been sighted at intervals of 76 years. Assuming these records to be accounts of the same comet, Halley used Newton's methods to work out the details of the comet's orbit and predicted its return in 1758. It was first sighted on Christmas night of 1757, a fitting memorial to Newton's birthday, and to this day the comet bears Halley's name (Figure 4-22).

Perhaps the most dramatic success of Newton's ideas was their role in the discovery of the eighth planet from the Sun. The seventh planet, Uranus, had been discovered accidentally by William Herschel in 1781 during a telescopic survey of the sky. Fifty years later, however, it was clear that Uranus was not following its predicted orbit. Two mathematicians, John Couch Adams in England and Urbain Le Verrier in France,

independently calculated that the gravitational pull of a yet unknown, more distant planet could explain the deviations of Uranus from its orbit. They each predicted that the planet would be found at a certain location in the constellation of Aquarius. A brief telescopic search on September 23, 1846, revealed the planet Neptune within 1° of the calculated position. Before it was sighted with a telescope, Neptune was actually predicted with pencil and paper.

Because it has been so successful in explaining and predicting many important phenomena, Newtonian mechanics has become the cornerstone of modern physical science. Even today, as we send astronauts into Earth orbit and spacecraft to the outer planets, Newton's equations are used to calculate the orbits and trajectories of these spacecraft.

In the twentieth century, scientists found that Newton's laws do not apply in all situations. A new theory called *quantum mechanics* had to be developed to explain the behavior of matter on the very smallest of scales, such as within the atom and within the atomic nucleus. Albert Einstein developed the *theory of relativity* to explain what happens at very high speeds approaching the speed of light and in places where gravitational forces are very strong. For many purposes in astronomy, however, Newton's laws are as useful today as when Newton formulated them more than three centuries ago.

figure 4-22 R I **V** U X G

Comet Halley This most famous of all comets orbits the Sun with an average period of about 76 years. During the twentieth century, the comet passed near the Sun in 1910 and again in 1986. It will next be prominent in the sky in 2061. This photograph shows how the comet looked in 1986. (David Malin, Anglo-Australian Observatory/Royal Observatory, Edinburgh)

4-8   Gravitational forces between the Earth and Moon produce tides

We have seen how Newtonian mechanics explains why the Moon stays in orbit around the Earth. It also explains why there are ocean tides, as well as why the Moon always keeps the same face toward the Earth. Both of these are consequences of *tidal forces*—an aspect of gravity that deforms planets and reshapes galaxies.

Tidal forces are differences in the gravitational pull at different points in an object. As an illustration, imagine that three billiard balls are lined up in space at some distance from a planet, as in Figure 4-23a. According to Newton's law of universal gravitation, the force of attraction between two

a

b

c

figure 4-23

The Origin of Tidal Forces
(a) Imagine three identical billiard balls placed some distance from a planet and released. (b) The closer a ball is to the planet, the more gravitational force the planet exerts on it. Thus, a short time after the balls are released, the blue 2-ball has moved farther toward the planet than the yellow 1-ball, and the red 3-ball has moved farther still. (c) From the perspective of the 2-ball in the center, it appears that forces have pushed the 1-ball away from the planet and pulled the 3-ball toward the planet. These forces are called tidal forces.

objects is greater the closer the two objects are to each other. Thus, the planet exerts more force on the 3-ball (in red) than on the 2-ball (in blue), and, in turn, the planet exerts more force on the 2-ball than on the 1-ball (in yellow). Now, imagine that the three balls are released and allowed to fall toward the planet. Figure 4-23b shows the situation a short time later. Because of the differences in gravitational pull, a short time later the 3-ball will have moved farther than the 2-ball, which will in turn have moved farther than the 1-ball. But now imagine that same motion from the perspective of the 2-ball. From this perspective, it appears as though the 3-ball is pulled toward the planet while the 1-ball is pushed away from the planet (Figure 4-23c). These apparent pushes and pulls are called tidal forces.

The Moon has a similar effect on the Earth as the planet in Figure 4-23 has on the three billiard balls. The arrows in Figure 4-24a indicate the strength and direction of the gravitational force of the Moon at several locations on the Earth. The side of the Earth closest to the Moon feels a greater gravitational pull than the Earth's center does, and the side of the Earth that faces away from the Moon feels less gravitational pull than does the Earth's center. Like the billiard balls in Figure 4-23, the result is that there are tidal forces acting on the Earth (Figure 4-24b). These forces try to elongate the

Earth along a line connecting the centers of the Earth and the Moon and try to squeeze the Earth inward in the direction perpendicular to that axis. The net effect is that the Moon's gravity tries to deform the Earth into a football shape.

Because the body of the Earth is largely rigid, it cannot deform very much in response to the Moon's tidal forces. But the water in the oceans can and does deform into a football shape, as Figure 4-25a shows. As the Earth rotates, a point on its surface goes from where the water is shallow to where the water is deep and back again. This is the origin of low and high ocean tides. (In this simplified description we have assumed that the Earth is completely covered with water. The full story of the tides is much more complex, because the shapes of the continents and the effects of winds must also be taken into account.)

Actually, the Sun also exerts tidal forces on the Earth's oceans. (The tidal effects of the Sun are about half as great as those of the Moon.) When the Sun, Moon, and the Earth are aligned, which happens at either new moon or full moon, the tidal effects of the Sun and Moon reinforce each other and the tidal distortion of the oceans is greatest. This produces large shifts in water level called **spring tides** (Figure 4-25b). At first quarter and last quarter, when the Sun and Moon form a right angle with the Earth, the tidal effects of the Sun and Moon

a

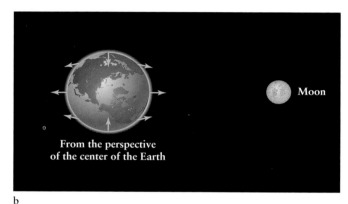

b

figure 4-24

Tidal Forces on the Earth **(a)** These arrows indicate the strength and direction of the Moon's gravitational pull at selected points on the Earth. **(b)** These arrows indicate the strength and direction of the tidal forces acting on the Earth. At any location, the tidal force equals the Moon's gravitational pull at that point minus the gravitational pull of the Moon at the center of the Earth. These tidal forces tend to deform the Earth into a nonspherical shape.

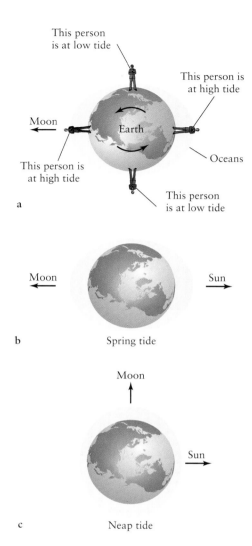

figure 4-25

High and Low Tides **(a)** The gravitational forces of the Moon and Sun deform the oceans. As the Earth rotates underneath the oceans, a given location experiences alternating low tides and high tides. **(b)** The greatest deformation (spring tides) occurs when the Sun, Earth, and Moon are aligned and the tidal effects of the Sun and Moon reinforce each other. **(c)** The least deformation (neap tides) occurs when the Sun, Earth, and Moon form a right angle. In this situation the tidal effects of the Sun and Moon partially cancel each other.

partially cancel each other. Hence, the tidal distortion of the oceans is the least pronounced, producing smaller tidal shifts called **neap tides** (Figure 4-25c).

CAUTION! Note that spring tides have nothing to do with the season of the year called spring. Instead, the name refers to the way that the ocean level "springs up" to a greater than normal height. Spring tides occur whenever there is a new moon or full moon, no matter what the season of the year.

Just as the Moon exerts tidal forces on the Earth, the Earth exerts tidal forces on the Moon. Soon after the Moon formed some 4.6 billion years ago, it was molten throughout its volume. The Earth's tidal forces deformed the molten Moon into a slightly elongated shape, with the long axis of the Moon pointed toward the Earth. The Moon retained this shape and orientation when it cooled and solidified. To keep its long axis

pointed toward the Earth, the Moon spins once on its axis as it makes one orbit around the Earth—that is, it is in synchronous rotation (Section 3-2). Hence, the same side of the Moon always faces the Earth, and this is the side that we see. For the same reason, most of the satellites in the solar system are in synchronous rotation, and thus always keep the same side facing their planet.

Tidal forces are also important on scales much larger than the solar system. Figure 4-26 shows two spiral galaxies, like

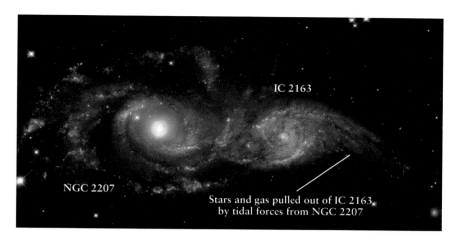

Stars and gas pulled out of IC 2163 by tidal forces from NGC 2207

figure 4-26 R I **V** U X G
Tidal Forces on a Galaxy For millions of years the galaxies NGC 2207 and IC 2163 have been moving ponderously past each other. The larger galaxy's tremendous tidal forces have drawn a streamer of material a hundred thousand light-years long out of IC 2163. If you lived on a planet orbiting a star within this streamer, you would have a magnificent view of both galaxies. NGC 2207 and IC 2163 are respectively 143,000 light-years and 101,000 light-years in diameter. Both galaxies are 114 million light-years away in the constellation Canis Major. (NASA and the Hubble Heritage Team, AURA/STScI)

the one in Figure 1-7, undergoing a near-collision. During the millions of years that this close encounter has been taking place, the tidal forces of the larger galaxy have pulled an immense streamer of stars and interstellar gas out of the smaller galaxy.

Many galaxies, including our own Milky Way Galaxy, show signs of having been disturbed at some time by tidal interactions with other galaxies. By their effect on the interstellar gas from which stars are formed, tidal interactions can actually trigger the birth of new stars. Our own Sun and solar system may have been formed as a result of tidal interactions of this kind. Hence we may owe our very existence to tidal forces. In this and many other ways, the laws of motion and of universal gravitation shape our universe and our destinies.

KEY WORDS

acceleration, p. 78

aphelion, p. 75

conic section, p. 83

conjunction, p. 67

deferent, p. 65

direct motion, p. 65

eccentricity, p. 75

ellipse, p. 74

elongation, p. 67

epicycle, p. 65

focus (of an ellipse; *plural* foci), p. 74

force, p. 78

geocentric model, p. 63

gravitational force, p. 80

gravity, p. 80

greatest eastern elongation, p. 67

greatest western elongation, p. 67

heliocentric model, p. 66

hyperbola, p. 83

inferior conjunction, p. 67

inferior planet, p. 67

Kepler's first law, p. 75

Kepler's second law, p. 76

Kepler's third law, p. 76

law of equal areas, p. 76

law of inertia, p. 78

law of universal gravitation, p. 80

major axis (of an ellipse), p. 74

mass, p. 80

neap tides, p. 86

Newtonian mechanics, p. 83

Newton's first law of motion, p. 78

Newton's form of Kepler's third law, p. 82

Newton's second law of motion, p. 79

Newton's third law of motion, p. 80

Occam's razor, p. 66

opposition, p. 67

parabola, p. 83

parallax, p. 73

perihelion, p. 75

period (of a planet), p. 68

Ptolemaic system, p. 65

retrograde motion, p. 65

semimajor axis (of an ellipse), p. 74

sidereal period, p. 68

speed, p. 78

spring tides, p. 85

superior conjunction, p. 67

superior planet, p. 67

synodic period, p. 68

tidal forces, p. 84

universal constant of gravitation, p. 81

velocity, p. 78

weight, p. 80

KEY IDEAS

Apparent Motions of the Planets: Like the Sun and Moon, the planets move on the celestial sphere with respect to the background of stars. Most of the time a planet moves eastward in direct motion, in the same direction as the Sun and the Moon, but from time to time it moves westward in retrograde motion.

Geocentric Model: Ancient astronomers believed the Earth to be at the center of the universe. They invented a complex system of epicycles and deferents to explain the direct and retrograde motions of the planets on the celestial sphere.

Heliocentric Model: Copernicus's heliocentric (Sun-centered) theory simplified the general explanation of planetary motions.

• In a heliocentric system, the Earth is one of the planets orbiting the Sun.

• A planet undergoes retrograde motion as seen from Earth when the Earth and the planet pass each other.

• The sidereal period of a planet, its true orbital period, is measured with respect to the stars. Its synodic period is measured with respect to the Earth and the Sun (for example, from one opposition to the next).

Evidence for the Heliocentric Model: The invention of the telescope led Galileo to new discoveries that supported a heliocentric model. These included his observations of the phases of Venus and of the motions of four moons around Jupiter.

Elliptical Orbits and Kepler's Laws: Copernicus thought that the orbits of the planets were combinations of circles. Using data collected by Tycho Brahe, Kepler deduced three laws of planetary motion: (1) the orbits are in fact ellipses; (2) a planet's speed varies as it moves around its elliptical orbit; and (3) the orbital period of a planet is related to the size of its orbit.

Newton's Laws of Motion: Isaac Newton developed three principles, called the laws of motion, that apply to the motions of objects on Earth as well as in space. These are (1) the law of inertia, (2) the relationship between the net outside force on an object and the object's acceleration, and (3) the principle of action and reaction. These laws and Newton's law of universal gravitation can be used to deduce Kepler's laws. They lead to extremely accurate descriptions of planetary motions.

• The mass of an object is a measure of the amount of matter in the object. Its weight is a measure of the force with which the gravity of some other object pulls on it.

• In general, the path of one object about another, such as that of a planet or comet about the Sun, is one of the curves called conic sections: circle, ellipse, parabola, or hyperbola.

Tidal Forces: Tidal forces are caused by differences in the gravitational pull that one object exerts on different parts of a second object.

• The tidal forces of the Moon and Sun produce tides in the Earth's oceans.

• The tidal forces of the Earth have locked the Moon into synchronous rotation.

REVIEW QUESTIONS

1. In what direction does a planet move relative to the stars when it is in direct motion? When it is in retrograde motion? How do these compare with the direction in which we see the Sun move relative to the stars?

2. **(a)** In what direction does a planet move relative to the horizon over the course of one night? **(b)** The answer to (a) is the same whether the planet is in direct motion or retrograde motion. What does this tell you about the speed at which planets move on the celestial sphere?

3. What is an epicycle? How is it important in Ptolemy's explanation of the retrograde motions of the planets?

4. What is the significance of Occam's razor as a tool for analyzing theories?

5. How did Copernicus explain the retrograde motion of the planets?

6. At what configuration (for example, superior conjunction, greatest eastern elongation, and so on) would it be best to observe Mercury or Venus with an Earth-based telescope? At what configuration would it be best to observe Mars, Jupiter, or Saturn? Explain your answers.

7. Which planets can never be seen at opposition? Which planets can never be seen at inferior conjunction? Explain your answers.

8. What is the difference between the synodic period and the sidereal period of a planet?

9. What observations did Galileo make that reinforced the heliocentric model? Why did these observations contradict the older model of Ptolemy? Why could these observations not have been made before Galileo's time?

10. What did Tycho Brahe's observations of a supernova and a comet suggest about the ancient Greek model of the universe?

11. What observations did Tycho Brahe make in an attempt to test the heliocentric model? What were his results? Explain why modern astronomers get different results.

12. What are Kepler's three laws? Why are they important?

13. What are the foci of an ellipse? If the Sun is at one focus of a planet's orbit, what is at the other focus?

14. A line joining the Sun and an asteroid is found to sweep out 6.3 AU^2 of space during 2004. How much area is swept out during 2005? Over a period of five years?

15. The orbit of a spacecraft about the Sun has a perihelion distance of 0.1 AU and an aphelion distance of 0.4 AU. What is the semimajor axis of the spacecraft's orbit? What is its orbital period?

16. A comet with a period of 125 years moves in a highly elongated orbit about the Sun. At perihelion, the comet comes very close to the Sun's surface. What is the comet's average distance from the Sun? What is the farthest it can get from the Sun?

17. What are Newton's three laws? Give an everyday example of each law.

18. How much force do you have to exert on a 4-kg brick to give it an acceleration of 3 m/s^2? If you double this force, what is the brick's acceleration? Explain.

19. What is your weight in pounds and in newtons? What is your mass in kilograms?

20. What is the difference between weight and mass?

21. Suppose that the Earth were moved to a distance of 0.25 AU from the Sun. How much stronger or weaker would the Sun's gravitational pull be on the Earth? Explain.

22. The mass of the Moon is 7.35×10^{22} kg, while that of the Earth is 5.98×10^{24} kg. The average distance from the center of the Moon to the center of the Earth is 384,400 km. What is the size of the gravitational force that the Earth exerts on the Moon? What is the size of the gravitational force that the Moon exerts on the Earth? How do your answers compare with the force between the Sun and the Earth calculated in the text?

23. How far would you have to go from Earth to be completely beyond the pull of its gravity? Explain.

24. What are conic sections? In what way are they related to the orbits of planets in the solar system?

25. Why was the discovery of Neptune an important confirmation of Newton's law of universal gravitation?

26. What is a tidal force? How do tidal forces produce tides in the Earth's oceans?

27. What is the difference between spring tides and neap tides?

ADVANCED QUESTIONS

Questions preceded by an asterisk (*) involve topics discussed in the Boxes.

Problem-solving tips and tools

Box 4-1 explains sidereal and synodic periods in detail. The semimajor axis of an ellipse is half the length of the long, or major, axis of the ellipse. For data about the planets and their satellites, see Appendices 1, 2, and 3 at the back of this book. If you want to calculate the gravitational force that you feel on the surface of a planet, the distance r to use is the planet's radius (the distance between you and the center of the planet). Boxes 4-2 and 4-4 show how to use Kepler's third law in its original form and in Newton's form.

28. Figure 4-2 shows the retrograde motion of Mars as seen from Earth. Sketch a similar figure that shows how Earth would appear to move against the background of stars during this same time period as seen by an observer on Mars.

*29. The synodic period of Mercury (an inferior planet) is 115.88 days. Calculate its sidereal period in days.

*30. Table 4-1 shows that the synodic period is *greater* than the sidereal period for the inferior planets Mercury and Venus, and that the synodic period is *less* than the sidereal period for all the superior planets. Draw diagrams like the one in Box 4-1 to explain why this is so.

*31. A general rule for superior planets is that the greater the average distance from the planet to the Sun, the more frequently that planet will be at opposition. Explain how this rule comes about.

32. In 2000, Mercury was at greatest western elongation on March 28, July 27, and November 15. It was at greatest eastern elongation on February 15, June 9, and October 6. Does Mercury take longer to go from eastern to western elongation, or vice versa? Why do you suppose this is the case?

33. A certain asteroid is 2 AU from the Sun at perihelion and 6 AU from the Sun at aphelion. (a) Find the semimajor axis of the asteroid's orbit. (b) Find the sidereal period of the orbit.

34. A comet orbits the Sun with a sidereal period of 27.0 years. (a) Find the semimajor axis of the orbit. (b) At aphelion, the comet is 17.5 AU from the Sun. How far is it from the Sun at perihelion?

35. One trajectory that can be used to send spacecraft from the Earth to Venus is an elliptical orbit that has the Sun at

one focus, its aphelion at the Earth, and its perihelion at Venus. The spacecraft is launched from Earth and coasts along this ellipse until it reaches Venus, when a rocket is fired either to put the spacecraft into orbit around Venus or to cause it to land on Venus. (a) Find the semimajor axis of the ellipse. (*Hint:* Draw a picture showing the Sun and the orbits of the Earth, Venus, and the spacecraft. Treat the orbits of the Earth and Venus as circles.) (b) Calculate how long (in days) such a one-way trip to Venus would take.

36. Suppose that you traveled to a planet with 3 times the mass and 3 times the diameter of the Earth. Would you weigh more or less on that planet than on Earth? By what factor?

37. The mass of Saturn is approximately 100 times that of Earth, and the semimajor axis of Saturn's orbit is approximately 10 AU. To this approximation, how does the gravitational force that the Sun exerts on Saturn compare to the gravitational force that the Sun exerts on the Earth? How do the accelerations of Saturn and the Earth compare?

38. On Earth, a 50-kg astronaut weighs 490 newtons. What would she weigh if she landed on Jupiter's moon Europa? What fraction is this of her weight on Earth? See Appendix 3 for relevant data about Europa.

39. A satellite is said to be in a "geosynchronous" orbit if it appears always to remain over the exact same spot on Earth. (a) What is the period of this orbit? (b) At what distance from the center of the Earth must such a satellite be placed into orbit? (c) Explain why the orbit must be in the plane of the Earth's equator.

40. Imagine a planet like the Earth orbiting a star with 4 times the mass of the Sun. If the semimajor axis of the planet's orbit is 1 AU, what would be the planet's sidereal period? (*Hint:* Use Newton's form of Kepler's third law. Compared with the case of the Earth orbiting the Sun, by what factor has the quantity $m_1 + m_2$ changed? Has a changed? By what factor must P^2 change?)

41. The photograph that opens this chapter shows the lunar module *Eagle* in orbit around the Moon after completing the first successful lunar landing in July 1969. (The photograph was taken from the command module *Columbia,* in which the astronauts returned to Earth.) The spacecraft orbited 111 km above the surface of the Moon. Calculate the period of the spacecraft's orbit. See Appendix 3 for relevant data about the Moon.

*42. In Box 4-4 we analyzed the orbit of Jupiter's moon Io. Look up information about the orbits of Jupiter's three other large moons (Europa, Ganymede, and Callisto) in the appendices. Demonstrate that these data are in agreement with Newton's form of Kepler's third law.

*43. Suppose a newly discovered asteroid is in a circular orbit with synodic period 1.25 years. The asteroid lies between the orbits of Mars and Jupiter. (a) Find the sidereal period of the orbit. (b) Find the distance from the asteroid to the Sun.

44. The average distance from the Moon to the center of the Earth is 384,400 km, and the diameter of the Earth is 12,756 km. Calculate the gravitational force that the Moon exerts (a) on a 1-kg rock at the point on the Earth's surface closest to the Moon and (b) on a 1-kg rock at the point on the Earth's surface farthest from the Moon. (c) Find the difference between the two forces you calculated in parts (a) and (b). This difference is the tidal force pulling these two rocks away from each other, like the 1-ball and 3-ball in Figure 4-23. Explain why tidal forces cause only a very small deformation of the Earth.

DISCUSSION QUESTIONS

45. Which planet would you expect to exhibit the greatest variation in apparent brightness as seen from Earth? Which planet would you expect to exhibit the greatest variation in angular diameter? Explain your answers.

46. Use two thumbtacks, a loop of string, and a pencil to draw several ellipses. Describe how the shape of an ellipse varies as the distance between the thumbtacks changes.

 ## WEB/CD-ROM QUESTIONS

47. (a) Search the World Wide Web for information about Galileo. What were his contributions to physics? Which of Galileo's new ideas were later used by Newton to construct his laws of motion? (b) Search the World Wide Web for information about Kepler. Before he realized that the planets move on elliptical paths, what other models of planetary motion did he consider? What was Kepler's idea of "the music of the spheres?" (c) Search the World Wide Web for information about Newton. What were some of the contributions that he made to physics other than developing his laws of motion? What contributions did he make to mathematics?

 48. Monitoring the Retrograde Motion of Mars. Watching Mars night after night reveals that it changes its position with respect to the background stars. To track its motion, access and view the animation "The Path of Mars in 2004–2005" in Chapter 4 of the *Universe* web site or CD-ROM. (a) Through which two constellations does Mars move? (b) On approximately what date does Mars stop its direct (west-to-east) motion and begin its retrograde motion? (*Hint:* Use the "Stop" function on your animation controls.) (c) Over how many days does Mars move retrograde?

OBSERVING PROJECTS

49. It is quite probable that within a few weeks of your reading this chapter one of the planets will be near opposition or greatest eastern elongation, making it readily visible in the evening sky. Select a planet that is at or near such a configuration by searching the World Wide Web or by consulting a reference book, such as the current issue of the *Astronomical Almanac* or the pamphlet entitled *Astronomical Phenomena* (both published by the U.S. government). At that configuration, would you expect the planet to be moving rapidly or slowly from night to night against the background stars? Verify your expectations by observing the planet once a week for a month, recording your observations on a star chart.

50. If Jupiter happens to be visible in the evening sky, observe the planet with a small telescope on five consecutive clear nights. Record the positions of the four Galilean satellites by making nightly drawings, just as the Jesuit priests did in 1620 (see Figure 4-12). From your drawings, can you tell which moon orbits closest to Jupiter and which orbits farthest? Was there a night when you could see only three of the moons? What do you suppose happened to the fourth moon on that night?

51. If Venus happens to be visible in the evening sky, observe the planet with a small telescope once a week for a month. On each night, make a drawing of the crescent that you see. From your drawings, can you determine if the planet is nearer or farther from the Earth than the Sun is?

Do your drawings show any changes in the shape of the crescent from one week to the next? If so, can you deduce if Venus is coming toward us or moving away from us?

52. Use the *Starry Night* program to observe the moons of Jupiter. Turn off daylight (select **Daylight** in the **Sky** menu) and show the entire celestial sphere (select **Atlas** in the **Go** menu). Center on the planet Jupiter (use **Find...** in the **Edit** menu), then use the controls at the right-hand end of the Control Panel to zoom in or out until you can see the planet and its four Galilean satellites. (**a**) Step forward in increments of 1 hour and draw the positions of the moons at each time step. (**b**) From your drawings, can you tell which moon orbits closest to Jupiter and which orbits farthest away? Explain your reasoning. (**c**) Are there times when only three of the satellites are visible? What happens to the fourth moon at those times?

53. Use the *Starry Night* program to observe the changing appearance of Mercury. Turn off daylight (select **Daylight** in the **Sky** menu) and show the entire celestial sphere (select **Atlas** in the **Go** menu). Center on the planet Mercury (use **Find...** in the **Edit** menu), then use the controls at the right-hand end of the Control Panel to zoom in or out until you can clearly see details on the planet's surface. (**a**) Step forward in increments of one day and record the changes in Mercury's phase and apparent size. (**b**) Explain why the phase and apparent size change in this way.

The Nature of Light

Molecular hydrogen

Neon

Lithium

Iron

Barium

Calcium

The Sun

Incandescent lamp

Fluorescent lamp

(Courtesy of Bausch and Lomb)

R I V U X G

In the early 1800s, the French philosopher Auguste Comte argued that because the stars are so far away, humanity would never know their nature and composition. But the means to learn about the stars was already there for anyone to see—starlight. Just a few years after Comte's bold pronouncement, scientists began analyzing starlight to learn the very things that he had deemed unknowable.

We now know that atoms of each chemical element emit and absorb light at a unique set of wavelengths characteristic of that element alone. Some of these atomic "fingerprints" are shown in the accompanying photo, along with the spectra of light from the Sun and from ordinary lamps. By looking for these wavelengths in the light from celestial objects, astronomers have learned that the same chemical elements found on Earth are also found in nearby planets, distant stars, and remote galaxies.

In this chapter we learn about the basic properties of light. Light is a form of energy with the properties of both waves and particles. Because the light emitted by an object depends upon the object's temperature, we can determine the surface temperatures of stars. By studying the structure of atoms, we will learn why each element emits and absorbs light only at specific wavelengths, and see how astronomers determine what the atmospheres of planets and stars are made of. The motion of a light source also affects wavelengths, permitting us to deduce how fast stars and other objects are approaching or receding. These are but a few of the reasons why understanding light is a prerequisite to understanding the universe.

As you read the sections of this chapter, look for the answers to the following questions.

5-1 How fast does light travel? How can this speed be measured?

5-2 Why do we think light is a wave? What kind of wave is it?

5-3 How is the light from an ordinary lightbulb different from the light emitted by a neon sign?

5-4 How can astronomers measure the temperatures of the Sun and stars?

5-5 What is a photon? How does an understanding of photons help explain why ultraviolet light causes sunburns?

5-6 How can astronomers tell what distant celestial objects are made of?

5-7 What are atoms made of?

5-8 How does the structure of atoms explain what kind of light those atoms can emit or absorb?

5-9 How can we tell if a star is approaching us or receding from us?

5-1 Light travels through empty space at a speed of 300,000 km/s

Galileo Galilei and Isaac Newton were among the first to ask basic questions about light. Does light travel instantaneously from one place to another, or does it move with a measurable speed? Whatever the nature of light, it does seem to travel swiftly from a source to our eyes. We see a distant event before we hear the accompanying sound. (For example, we see a flash of lightning before we hear the thunderclap.)

In the early 1600s, Galileo tried to measure the speed of light. He and an assistant stood at night on two hilltops a known distance apart, each holding a shuttered lantern. First, Galileo opened the shutter of his lantern; as soon as his assistant saw the flash of light, he opened his own. Galileo used his pulse as a timer to try to measure the time between opening his lantern and seeing the light from his assistant's lantern. From the distance and time, he hoped to compute the speed at which the light had traveled to the distant hilltop and back.

Galileo found that the measured time failed to increase noticeably, no matter how distant the assistant was stationed. Galileo therefore concluded that the speed of light is too high to be measured by slow human reactions. Thus, he was unable to tell whether or not light travels instantaneously.

The first evidence that light does *not* travel instantaneously was presented in 1676 by Olaus Rømer, a Danish astronomer. Rømer had been studying the orbits of the moons of Jupiter by carefully timing the moments when they passed into or out of Jupiter's shadow. To Rømer's surprise, the timing of these eclipses of Jupiter's moons seemed to depend on the relative positions of Jupiter and the Earth. When the Earth was far from Jupiter (that is, near conjunction; see Figure 4-6), the eclipses occurred several minutes later than when the Earth was close to Jupiter (near opposition).

Rømer realized that this puzzling effect could be explained if light needs time to travel from Jupiter to the Earth. When the Earth is closest to Jupiter, the image of a satellite disappearing into Jupiter's shadow arrives at our telescopes a little sooner than it does when Jupiter and the Earth are farther apart (Figure 5-1). The range of variation in the times at which such eclipses are observed is about 16.6 minutes, which Rømer interpreted as the length of time required for light to travel across the diameter of the Earth's orbit (a distance of 2 AU). The size of the Earth's orbit was not accurately known in Rømer's day, and he never actually calculated the speed of light. Today, using the modern value of 150 million kilometers for the astronomical unit, Rømer's method yields a value for the speed of light equal to roughly 300,000 km/s (186,000 mi/s).

Almost two centuries after Rømer, the speed of light was measured very precisely in an experiment carried out on Earth.

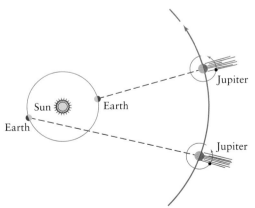

ƒigure 5-1

Rømer's Proof that Light Does Not Travel Instantaneously
When the Earth is near Jupiter, eclipses of Jupiter's moons occur slightly earlier than anticipated. When the Earth-Jupiter distance is large, the eclipses occur slightly later than anticipated. Rømer correctly attributed this effect to variations in the time required for light to travel from Jupiter to the Earth.

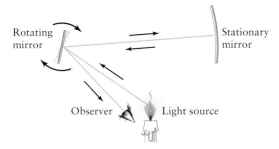

ƒigure 5-2

The Fizeau-Foucault Method of Measuring the Speed of Light Light from a light source is reflected off a rotating mirror to a stationary mirror and from there back to the rotating mirror. The ray that reaches the observer is deflected away from the path of the initial beam because the rotating mirror has moved slightly while the light was making the round trip. The speed of light is calculated from the deflection angle and the dimensions of the apparatus.

In 1850, the French physicists Armand-Hippolyte Fizeau and Jean Foucault built the apparatus sketched in Figure 5-2. Light from a light source reflects from a rotating mirror toward a stationary mirror 20 meters away. The rotating mirror moves slightly while the light is making the round trip, so the returning light ray is deflected away from the source by a small angle. By measuring this angle and knowing the dimensions of their apparatus, Fizeau and Foucault could deduce the speed of light. Once again, the answer was very nearly 300,000 km/s.

The speed of light in a vacuum is usually designated by the letter c (from the Latin *celeritas*, meaning "speed"). The modern accepted value is $c = 299,792.458$ km/s (186,282.397 mi/s). In most calculations you can use

$$c = 3.0 \times 10^5 \text{ km/s} = 3.0 \times 10^8 \text{ m/s}$$

The most convenient set of units to use for c is different in different situations. The value in kilometers per second (km/s) is often most useful when comparing c to the speeds of objects in space, while the value in meters per second (m/s) is preferred when doing calculations involving the wave nature of light (which we will discuss in Section 5-2).

Note that the quantity c is the speed of light *in a vacuum*. Light travels more slowly through air, water, glass, or any other transparent substance than it does in a vacuum. In our study of astronomy, however, we will almost always consider light traveling through the vacuum (or near vacuum) of space.

The speed of light in empty space is one of the most important numbers in modern physical science. This value appears in many equations that describe atoms, gravity, electricity, and magnetism. According to Einstein's special theory of relativity, nothing can travel faster than the speed of light.

5-2 Light is electromagnetic radiation and is characterized by its wavelength

Light is energy. This fact is apparent to anyone who has felt the warmth of the sunshine on a summer's day. But what exactly is light? How is it produced? What is it made of? How does it move through space? Scholars have struggled with these questions throughout history.

The first major breakthrough in understanding light came from a simple experiment performed by Isaac Newton around 1670. Newton was familiar with what he called the "celebrated Phenomenon of Colours," in which a beam of sunlight passing through a glass prism spreads out into the colors of the rainbow (Figure 5-3). This rainbow is called a **spectrum** (plural **spectra**).

 Until Newton's time, it was thought that a prism somehow added colors to white light. To test this idea, Newton placed a second prism so that just one color of the spectrum passed through it (Figure 5-4). According to the old theory, this should have caused a further

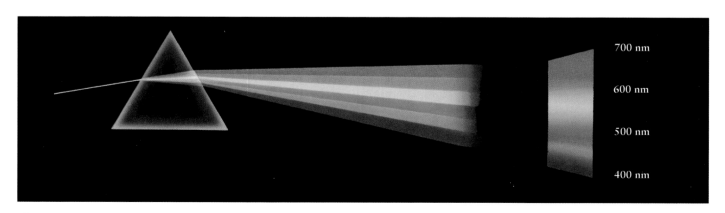

ƒigure 5-3

A Prism and a Spectrum When a beam of sunlight passes through a glass prism, the light is broken into a rainbow-colored band called a spectrum. The numbers on the right side of the spectrum indicate the wavelengths of different colors of light.

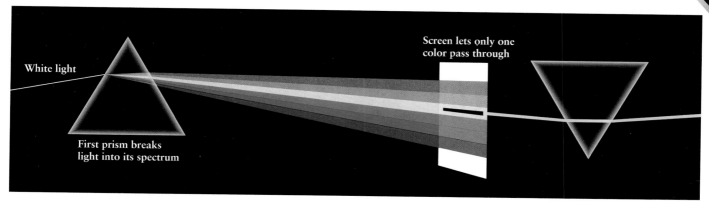

figure 5-4

Newton's Experiment on the Nature of Light In a crucial experiment, Newton took sunlight that had passed through a prism and sent it through a second prism. Between the two prisms was a screen with a hole in it that allowed only one color of the spectrum to pass through. This same color emerged from the second prism. Newton's experiment proved that prisms do not add color to light but merely bend different colors through different angles. It also proved that white light, such as sunlight, is actually a combination of all the colors that appear in its spectrum.

change in the color of the light. But Newton found that each color of the spectrum was unchanged by the second prism; red remained red, blue remained blue, and so on. He concluded that a prism merely separates colors and does not add color. Hence, the spectrum produced by the first prism shows that sunlight is a mixture of all the colors of the rainbow.

Newton suggested that light is composed of particles too small to detect individually. (The modern theory of quantum mechanics, developed in the early 1900s, takes a similar viewpoint and postulates that light is composed of tiny packets of energy. We will explore these ideas in Section 5-5.) In 1678, however, the Dutch physicist and astronomer Christiaan Huygens proposed a rival explanation. He suggested that light travels in the form of waves rather than particles. Scientists now realize that light possesses both particlelike and wavelike characteristics.

Around 1801, Thomas Young in England carried out an experiment that convincingly demonstrated the wavelike aspect of light. He passed a beam of light through two thin, parallel slits in an opaque screen, as shown in Figure 5-5a. On a white surface some distance beyond the slits, the light formed a pattern of alternating bright and dark bands. Young reasoned that if a beam of light was a stream of particles (as Newton had suggested), the two beams of light from the slits should simply form bright images of the slits on the white surface. The pattern of bright and dark bands he observed is just what would be expected, however, if light had wavelike properties. An analogy with water waves demonstrates why.

ANALOGY Imagine ocean waves pounding against a reef or breakwater that has two openings (Figure 5-5b). A pattern of ripples is formed on the other side of the barrier as the waves come through the two openings and interfere with each other. At certain points, wave crests arrive simultaneously from the two openings. These reinforce each

other and produce high waves. At other points, a crest from one opening meets a trough from the other opening. These cancel each other out, leaving areas of still water. This process of combining two waves also takes place in Young's double-slit experiment: The bright bands are regions where waves from the two slits reinforce each other, while the dark bands appear where waves from the two slits cancel each other.

The discovery of the wave nature of light posed some obvious questions. What exactly is "waving" in light? That is, what is it about light that goes up and down like water waves on the ocean? Because we can see light from the Sun, planets, and stars, light waves must be able to travel across empty space. Hence, whatever is "waving" cannot be any material substance. What, then, is it?

The answer came from a seemingly unlikely source—a comprehensive theory that described electricity and magnetism. Numerous experiments during the first half of the nineteenth century demonstrated an intimate connection between electric and magnetic forces. A central idea to emerge from these experiments is the concept of a *field*, an immaterial yet measurable disturbance of any region of space in which electric or magnetic forces are felt. Thus, an electric charge is surrounded by an electric field, and a magnet is surrounded by a magnetic field. Experiments in the early 1800s demonstrated that moving an electric charge produces a magnetic field; conversely, moving a magnet gives rise to an electric field.

In the 1860s, the Scottish mathematician and physicist James Clerk Maxwell succeeded in describing all the basic properties of electricity and magnetism in four equations. This mathematical achievement demonstrated that electric and magnetic forces are really two aspects of the same phenomenon, which we now call **electromagnetism.**

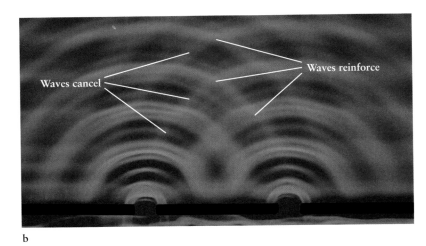

Waves cancel

Waves reinforce

a   b

Young's Double-Slit Experiment **(a)** Thomas Young's classic double-slit experiment can easily be repeated in the modern laboratory by shining light from a laser onto two closely spaced parallel slits. Alternating dark and light bands appear on a screen beyond the slits. **(b)** The intensity of light on the screen in (a) is analogous to the height of water waves that pass through a barrier with two openings. (The photograph shows this experiment with water waves in a small tank.) In certain locations, wave crests from both openings reinforce each other to produce extra high waves. At other locations a crest from one opening meets a trough from the other. The crest and trough cancel each other, producing still water. (© Eric Schrempp/Photo Researchers)

By combining his four equations, Maxwell showed that electric and magnetic fields should travel through space in the form of waves at a speed of 3.0×10^5 km/s—a value exactly equal to the best available value for the speed of light. Maxwell's suggestion that these waves do exist and are observed as light was soon confirmed by experiments. Because of its electric and magnetic properties, light is also called **electromagnetic radiation.**

CAUTION! You may associate the term *radiation* with radioactive materials like uranium, but this term refers to anything that radiates, or spreads away, from its source. For example, scientists sometimes refer to sound waves as "acoustic radiation." Radiation does not have to be related to radioactivity!

Electromagnetic radiation consists of oscillating electric and magnetic fields, as shown in Figure 5-6. The distance between two successive wave crests is called the **wavelength** of the light, usually designated by the Greek letter λ (lambda). No matter what the wavelength, electromagnetic radiation always travels at the same speed $c = 3.0 \times 10^5$ km/s $= 3.0 \times 10^8$ m/s in a vacuum.

More than a century elapsed between Newton's experiments with a prism and the confirmation of the wave nature of light. One reason for this delay is that **visible light,** the light to which the human eye is sensitive, has extremely short wavelengths—less than a thousandth of a millimeter—that are not easily detectable. To express such tiny distances conve-

niently, scientists use a unit of length called the **nanometer** (abbreviated nm), where 1 nm $= 10^{-9}$ m. Experiments demonstrated that visible light has wavelengths covering the range from about 400 nm for violet light to about 700 nm for red light. Intermediate colors of the rainbow like yellow (550 nm) have intermediate wavelengths, as shown in Figure 5-7. (Some astronomers prefer to measure wavelengths in *angstroms*. One angstrom, abbreviated Å, is one-tenth of a nanometer: 1 Å $= 0.1$ nm $= 10^{-10}$ m. In these units, the wavelengths of visible light extend from about 4000 Å to about 7000 Å. We will not use these units in this book, however.)

Maxwell's equations place no restrictions on the wavelength of electromagnetic radiation. Hence, electromagnetic waves could and should exist with wavelengths both longer and shorter than the 400–700 nm range of visible light. Consequently, researchers began to look for *invisible* forms of light. These are forms of electromagnetic radiation to which the cells of the human retina do not respond.

The first kind of invisible radiation to be discovered actually preceded Maxwell's work by more than a half century. Around 1800 the British astronomer William Herschel passed sunlight through a prism and held a thermometer just beyond the red end of the visible spectrum. The thermometer registered a temperature increase, indicating that it was being exposed to an invisible form of energy. This invisible energy, now called **infrared radiation,** was later realized to be electromagnetic radiation with wavelengths somewhat longer than those of visible light.

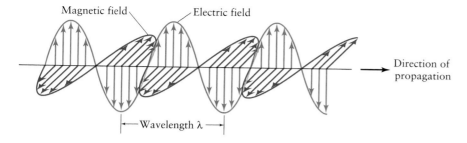

figure 5-6

Electromagnetic Radiation All forms of light consist of oscillating electric and magnetic fields that move through space at a speed of 3.0×10^5 km/s $= 3.0 \times 10^8$ m/s. This figure shows a "snapshot" of these fields at one instant. The distance between two successive crests, called the wavelength of the light, is usually designated by the Greek letter λ (lambda).

In experiments with electric sparks in 1888, the German physicist Heinrich Hertz succeeded in producing electromagnetic radiation with even longer wavelengths of a few centimeters or more. These are now known as **radio waves.** In 1895 another German physicist, Wilhelm Röntgen, invented a machine that produces electromagnetic radiation with wavelengths shorter than 10 nm, now known as **X rays.** The X-ray machines in modern medical and dental offices are direct descendants of Röntgen's invention. Over the years radiation has been discovered with many other wavelengths.

Thus, visible light occupies only a tiny fraction of the full range of possible wavelengths, collectively called the **electromagnetic spectrum.** As Figure 5-7 shows, the electromagnetic spectrum stretches from the longest-wavelength radio waves to the shortest-wavelength gamma rays.

On the long-wavelength side of the visible spectrum, infrared radiation covers the range from about 700 nm to 1 mm. Astronomers interested in infrared radiation often express wavelength in *micrometers* or *microns,* abbreviated μm, where 1 μm $= 10^{-3}$ mm $= 10^{-6}$ m. **Microwaves** have wavelengths from roughly 1 mm to 10 cm, while radio waves have even longer wavelengths.

At wavelengths shorter than those of visible light, **ultraviolet radiation** extends from about 400 nm down to 10 nm. Next are X rays, which have wavelengths between about 10 and 0.01 nm, and beyond them at even shorter wavelengths are **gamma rays.** Note that these rough boundaries are simply arbitrary divisions in the electromagnetic spectrum.

Astronomers who work with radio telescopes often prefer to speak of *frequency* rather than wavelength. The **frequency** of a wave is the number of wave crests that pass a given point in one second. Equivalently, it is the number of complete *cycles* of the wave that pass per second (a complete cycle is from one crest to the next). Frequency is usually denoted by the Greek letter ν (nu). The unit of frequency is the cycle per second, also called the hertz (abbreviated Hz) in honor of Heinrich Hertz, the physicist who first produced radio waves. For example, if 500 crests of a wave pass you in one second, the frequency of the wave is 500 cycles per second or 500 Hz.

In working with frequencies, it is often convenient to use the prefix *mega-* (meaning "million," or 10^6, and abbreviated M) or *kilo-* (meaning "thousand," or 10^3, and abbreviated k). For example, AM radio stations broadcast at frequencies between 535 and 1605 kHz (kilohertz), while FM radio

stations broadcast at frequencies in the range from 88 to 108 MHz (megahertz).

The relationship between the frequency and wavelength of an electromagnetic wave is a simple one. Because light moves at a constant speed $c = 3 \times 10^8$ m/s, if the wavelength (distance from one crest to the next) is made shorter, the frequency must increase (more of those closely spaced crests pass

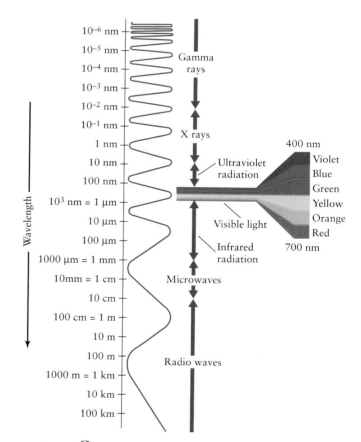

figure 5-7

The Electromagnetic Spectrum The full array of all types of electromagnetic radiation is called the electromagnetic spectrum. It extends from the longest-wavelength radio waves to the shortest-wavelength gamma rays. Visible light occupies only a tiny portion of the full electromagnetic spectrum.

you each second). Mathematically, the frequency ν of light is related to its wavelength λ by

Frequency and wavelength of an electromagnetic wave

$$\nu = \frac{c}{\lambda}$$

ν = frequency of an electromagnetic wave (in Hz)

c = speed of light = 3×10^8 m/s

λ = wavelength of the wave (in meters)

For example, hydrogen atoms in space emit radio waves with a wavelength of 21.12 cm. To calculate the frequency of this radiation, we must first express the wavelength in meters rather than centimeters: λ = 0.2112 m. Then we can use the above formula to find the frequency ν:

$$\nu = \frac{c}{\lambda} = \frac{3 \times 10^8 \text{ m/s}}{0.2112 \text{ m/s}} = 1.42 \times 10^9 \text{ Hz} = 1420 \text{ MHz}$$

Visible light has a much shorter wavelength and higher frequency than radio waves. You can use the above formula to show that for yellow-orange light of wavelength 600 nm, the frequency is 5×10^{14} Hz or 500 *million* megahertz!

5-3 A dense object emits electromagnetic radiation according to its temperature

To learn about objects in the heavens, astronomers study the character of the electromagnetic radiation coming from those objects. Such studies can be very revealing because different kinds of electromagnetic radiation are typically produced in different ways. As an example, on Earth the most common way to generate radio waves is to make an electric current oscillate back and forth (as is done in the broadcast antenna of a radio station). By contrast, X rays for medical and dental purposes are usually produced by bombarding atoms with fast-moving particles extracted from within other atoms. Our own Sun emits radio waves from near its glowing surface and X rays from its corona (see the photo that opens Chapter 3). Hence, these observations indicate the presence of electric currents near the Sun's surface, and of fast-moving particles in the Sun's outermost regions. (We will discuss the Sun at length in Chapter 18.)

The simplest and most common way to produce electromagnetic radiation, either on or off the Earth, is to heat an object. The hot filament of wire inside an ordinary lightbulb emits white light, and a neon sign has a characteristic red glow because neon gas within the tube is heated by an electric current. In like fashion, almost all the visible light that we receive from space comes from hot objects like the Sun and the stars. The kind and amount of light emitted by a hot object tell us not only how hot it is but also about other properties of the object.

We can tell whether the hot object is made of relatively dense or relatively thin material. Consider the difference between a lightbulb and a neon sign. The dense, solid filament of a lightbulb makes white light, which is a mixture of all different visible wavelengths, while the thin neon gas produces light of a rather definite red color and, hence, a rather definite wavelength. For now we will concentrate our attention on the light produced by dense objects. (We will return to gases in Section 5-6.) Even though the Sun and stars are gaseous, not solid, it turns out that they emit light with many of the same properties as light emitted by a hot, glowing, solid object.

Imagine a welder or blacksmith heating a bar of iron. As the bar becomes hot, it begins to glow deep red, as shown in Figure 5-8a. (You can see this same glow from the coils of a toaster, or from an electric range turned on "high.") As the temperature rises further, the bar begins to give off a brighter, reddish-orange light (Figure 5-8b). At still higher temperatures, it shines with a brilliant yellowish-white light (Figure 5-8c). If the bar could be prevented from melting and vaporizing, at extremely high temperatures it would emit a dazzling blue-white light.

As this example shows, the amount of energy emitted by the hot, dense object and the dominant wavelength of the emitted radiation both depend on the temperature of the object. The hotter the object, the more energy it emits and the shorter the wavelength at which most of the energy is

a

b

c

ƒigure 5-8

Heating a Bar of Iron This sequence of drawings shows how the appearance of a heated bar of iron changes with temperature. As the temperature increases, the bar glows more brightly because it radiates more energy. The color of the bar also changes because the dominant wavelength of light emitted by the bar decreases as the temperature goes up.

figure 5-9 R **I** V U X G

An Infrared Boy and His Dog A camera sensitive to infrared radiation took this picture of a boy and his dog. The different colors in the image represent regions of different temperature; white areas are the warmest and emit the most infrared light, while black areas (including the dog's cold nose) are at the lowest temperatures and emit the least radiation. (R. P. Clark and M. Goff)

If something is hot, its atoms are moving at high speeds; if it is cold, its atoms are moving slowly. Scientists usually prefer to use the Kelvin temperature scale, on which temperature is measured in **kelvins** (K) upward from **absolute zero.** This is the coldest possible temperature, at which atoms move as slowly as possible (they can never quite stop completely). On the more familiar Celsius and Fahrenheit temperature scales, absolute zero (0 K) is −273°C and −460°F. Ordinary room temperature is 293 K, 20°C, or 68°F. Box 5-1 discusses the relationships among the Kelvin, Celsius, and Fahrenheit temperature scales.

Figure 5-10 depicts quantitatively how the radiation from a dense object depends on its Kelvin temperature. Each curve in this figure shows the intensity of light emitted at each wavelength by a dense object at a given temperature. In other words, the curves show the spectrum of light emitted by such an object. At any temperature, a hot, dense object emits at all wavelengths, so its spectrum is a smooth, continuous curve with no gaps in it.

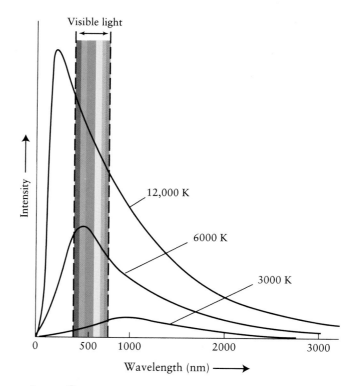

figure 5-10

Blackbody Curves Each of these curves shows the intensity of light at every wavelength that is emitted by a blackbody (an idealized case of a dense object) at a particular temperature. The higher the temperature, the shorter the wavelength of maximum emission (where the curve peaks) and the greater the amount of light emitted at every wavelength. The rainbow-colored band shows the range of visible wavelengths. The vertical scale has been compressed so that all three curves can be seen; the peak intensity for the 12,000-K curve is actually about 1000 times greater than the peak intensity for the 3000-K curve.

emitted. Colder objects emit relatively little energy, and this emission is primarily at long wavelengths.

These observations explain why you can't see in the dark. The temperatures of people, animals, and furniture are rather less than even that of the iron bar in Figure 5-8a. So, while these objects emit radiation even in a darkened room, most of this emission is at wavelengths greater than those of red light, in the infrared part of the spectrum (see Figure 5-7). Your eye is not sensitive to infrared, and you thus cannot see ordinary objects in a darkened room. But you can detect this radiation by using a camera that is sensitive to infrared light (Figure 5-9).

To better understand the relationship between the temperature of a dense object and the radiation it emits, it is helpful to know just what "temperature" means. The temperature of a substance is directly related to the average speed of the tiny **atoms**—the building blocks that come in distinct forms for each distinct chemical element—that make up the substance. (Typical atoms are about 10^{-10} m = 0.1 nm in diameter, or about 1/5000 as large as a typical wavelength of visible light.)

box 5-1 | Tools of the Astronomer's Trade

Temperatures and Temperature Scales

 Three temperature scales are in common use. Throughout most of the world, temperatures are expressed in **degrees Celsius** (°C). The Celsius temperature scale is based on the behavior of water, which freezes at 0°C and boils at 100°C at sea level on Earth. This scale is named in honor of the Swedish astronomer Anders Celsius, who proposed it in 1742.

Astronomers usually prefer the Kelvin temperature scale. This is named after the nineteenth-century British physicist Lord Kelvin, who made many important contributions to our understanding of heat and temperature. Absolute zero, the temperature at which atomic motion is at the absolute minimum, is −273°C in the Celsius scale but 0 K in the Kelvin scale. Because it is impossible for atomic motion to be any less than the minimum, nothing can be colder than 0 K; hence, there are no negative temperatures on the Kelvin scale. Note that we do *not* use degree (°) with the Kelvin temperature scale.

A temperature expressed in kelvins is always equal to the temperature in degrees Celsius plus 273. On the Kelvin scale, water freezes at 273 K and boils at 373 K. Water must be heated through a change of 100 K or 100°C to go from its freezing point to its boiling point. Thus, the "size" of a kelvin is the same as the "size" of a Celsius degree. When considering temperature changes, measurements in kelvins and Celsius degrees are the same.

The now-archaic Fahrenheit scale, which expresses temperature in **degrees Fahrenheit** (°F), is used only in the United States. When the German physicist Gabriel Fahrenheit introduced this scale in the early 1700s, he intended 100°F to represent the temperature of a healthy human body. On the Fahrenheit scale, water freezes at 32°F and boils at 212°F. There are 180 Fahrenheit degrees between the freezing and boiling points of water, so a degree Fahrenheit is only 100/180 = 5/9 as large as either a Celsius degree or a kelvin.

Two simple equations allow you to convert a temperature from the Celsius scale to the Fahrenheit scale and from Fahrenheit to Celsius:

$$T_F = \frac{9}{5}T_c + 32$$

$$T_C = \frac{5}{9}(T_F - 32)$$

T_F = temperature in degrees Fahrenheit

T_C = temperature in degrees Celsius

EXAMPLE: A typical room temperature is 68°F. We can convert this to the Celsius scale using the second equation:

$$T_C = \frac{5}{9}(68 - 32) = 20°C$$

To convert this to the Kelvin scale, we simply add 273 to the Celsius temperature. Thus,

$$68°F = 20°C = 293 \text{ K}$$

The diagram displays the relationships among these three temperature scales.

The shape of the spectrum depends on temperature, however. An object at relatively low temperature (say, 3000 K) has a low curve, indicating a low intensity of radiation. The **wavelength of maximum emission,** at which the curve has its peak and the emission of energy is strongest, is at a long wavelength. The higher the temperature, the higher the curve (indicating greater intensity) and the shorter the wavelength of maximum emission.

 Figure 5-10 shows that for a dense object at a temperature of 3000 K, the wavelength of maximum emission is around 1000 nm (1 μm). Because this is an infrared wavelength well outside the visible range, you might think that you cannot see the radiation from an object at this temperature. In fact, the glow from such an object *is* visible; the curve shows that this object emits plenty of light within the visible range, as well as at even shorter wavelengths. The 3000-K curve is quite a bit higher at the red end of the visible spectrum than at the violet end, so a dense object at this temperature will appear red in color. Similarly, the 12,000-K curve has its wavelength of maximum emission in the ultraviolet part of the spectrum, at a wavelength shorter than visible light. But such a hot, dense object also emits copious amounts of visible light (much more than at 6000 K or 3000 K, for which the curves are lower) and thus will have a very visible glow. The curve for this temperature is higher for blue light than for red light, and so the color of a dense object at 12,000 K is a brilliant blue or blue-white. These conclusions agree with the color changes of a heated rod shown in Figure 5-8. The same principles apply to stars: A star that looks blue has a high surface temperature, while a red star has a relatively cool surface.

The curves in Figure 5-10 are drawn for an idealized type of dense object called a **blackbody.** A perfect blackbody does not reflect any light at all; instead, it absorbs all radiation falling on it. Because it reflects no electromagnetic radiation, the radiation that it does emit is entirely the result of its temperature. Ordinary objects, like tables, textbooks, and people, are not perfect blackbodies; they reflect light, which is why they are visible. A star such as the Sun, however, behaves very much like a perfect blackbody, because it absorbs almost completely any radiation falling on it from outside. The light emitted by a blackbody is called **blackbody radiation,** and the curves in Figure 5-10 are often called **blackbody curves.**

 Despite its name, a blackbody does not necessarily look black. The Sun, for instance, does not look black because its temperature is high (around 5800 K), and so it glows brightly. But a room-temperature (around 300 K) blackbody would appear very black indeed. Even if it were as large as the Sun, it would emit only about 1/100,000 as much energy. (Its blackbody curve is far too low to graph in Figure 5-10.) Furthermore, most of this radiation would be at wavelengths that are too long for our eyes to perceive.

Figure 5-11 illustrates how accurately a star like the Sun can be modeled as a blackbody. The graph shows how the intensity of sunlight varies with wavelength, as measured from above the Earth's atmosphere. (This is necessary because the Earth's atmosphere absorbs certain wavelengths.) The peak of the curve is at a wavelength of about 500 nm, near the middle of the visible spectrum.

Figure 5-11 also shows the blackbody curve for a temperature of 5800 K. Note how closely the observed intensity curve for the Sun matches the blackbody curve. This is a strong indication that the temperature of the Sun's glowing surface is about 5800 K—a temperature that we can measure across a distance of 150 million kilometers! The close correlation between blackbody curves and the observed intensity curves for most stars is a key reason why astronomers are interested in the physics of blackbody radiation.

Blackbody radiation depends only on the temperature of the object emitting the radiation, not on the chemical composition of the object. The light emitted by molten gold at

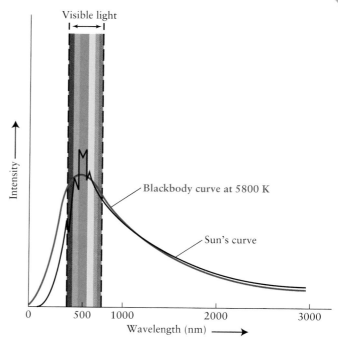

figure 5-11

The Sun as a Blackbody This graph compares the intensity of sunlight over a wide range of wavelengths (black curve) with the intensity of radiation coming from a blackbody at a temperature of 5800 K (red curve). The measurements of the Sun's intensity were made above the Earth's atmosphere. The Sun mimics a blackbody remarkably well. Note that the Sun's spectrum peaks in the visible spectrum. This is not surprising, because the human eye evolved to take advantage of the most plentiful light available.

2000 K is very nearly the same as that emitted by molten lead at 2000 K. Therefore, it might seem that analyzing the light from the Sun or from a star can tell astronomers the object's temperature but not what the star is made of. As Figure 5-11 shows, however, the intensity curve for the Sun (a typical star) is not precisely that of a blackbody. We will see later in this chapter that the *differences* between a star's spectrum and that of a blackbody allow us to determine the chemical composition of the star.

5-4 Wien's law and the Stefan-Boltzmann law are useful tools for analyzing glowing objects like stars

The mathematical formula that describes the blackbody curves in Figure 5-10 is a rather complicated one. But there are two simpler formulas for blackbody radiation that prove to be very useful in many branches of astronomy. They are used by astronomers who investigate the stars as well as by those who study the planets (which are dense, relatively cool objects that emit infrared radiation). One of these formulas relates the temperature of a blackbody to its wavelength of maximum emission, and the other relates the temperature to the amount of energy that the blackbody emits.

Figure 5-10 shows that the higher the temperature (T) of a blackbody, the shorter its wavelength of maximum emission (λ_{max}). In 1893 the German physicist Wilhelm Wien used ideas about both heat and electromagnetism to make this relationship quantitative. The formula that he derived, which today is called **Wien's law**, is

Wien's law for a blackbody

$$\lambda_{max} = \frac{0.0029}{T}$$

λ_{max} = wavelength of maximum emission of the object (in meters)

T = temperature of object (in kelvins)

 According to Wien's law, the wavelength of maximum emission of a blackbody is inversely proportional to its temperature in kelvins. In other words, if the temperature of the blackbody doubles, its wavelength of maximum emission is halved, and vice versa. For example, Figure 5-10 shows blackbody curves for temperatures of 3000 K, 6000 K, and 12,000 K. From Wien's law, a blackbody with a temperature of 6000 K has a wavelength of maximum emission λ_{max} = (0.0029)/(6000) = 4.8 $\times$ 10^{-7} m = 480 nm, in the visible part of the electromagnetic spectrum. At 12,000 K, or twice the temperature, the blackbody has a wavelength of maximum emission half as great, or λ_{max} = 240 nm; this is in the ultraviolet. At 3000 K, just half our original temperature, the value of λ_{max} is twice the original value—960 nm, which is an infrared wavelength. You can see that these wavelengths agree with the peaks of the curves in Figure 5-10.

 Remember that Wien's law involves the wavelength of maximum emission in *meters*. If you want to convert the wavelength to nanometers, you must multiply the wavelength in meters by (10^9 nm)/(1 m).

Wien's law is very useful for determining the surface temperatures of stars. It is not necessary to know how far away the star is, how large it is, or how much energy it radiates into space. All we need to know is the dominant wavelength of the star's electromagnetic radiation.

The other useful formula for the radiation from a blackbody involves the total amount of energy the blackbody radiates at all wavelengths. (By contrast, the curves in Figure 5-10 show how much energy a blackbody radiates at each individual wavelength.)

Energy is usually measured in **joules** (J), named after the nineteenth-century English physicist James Joule. A joule is the amount of energy contained in the motion of a 2-kilogram mass moving at a speed of 1 meter per second. The joule is a convenient unit of energy because it is closely related to the familiar **watt** (W): 1 watt is 1 joule per second, or 1 W = 1 J/s = 1 J s^{-1}. (The superscript -1 means you are dividing by that quantity.) For example, a 100-watt lightbulb uses energy at a rate

of 100 joules per second, or 100 J/s. The energy content of food is also often measured in joules; in most of the world, diet soft drinks are labeled as "low joule" rather than "low calorie."

The amount of energy emitted by a blackbody depends both on its temperature and on its surface area. This makes sense: A large burning log radiates much more heat than a burning match, even though the temperatures are the same. To consider the effects of temperature alone, it is convenient to look at the amount of energy emitted from each square meter of an object's surface in a second. This quantity is called the **energy flux** (F). Flux means "rate of flow," and thus F is a measure of how rapidly energy is flowing out of the object. It is measured in joules per square meter per second, usually written as J/m^2/s or J m^{-2} s^{-1}. Alternatively, because 1 watt equals 1 joule per second, we can express flux in watts per square meter (W/m^2, or W m^{-2}).

In 1879 the Austrian physicist Josef Stefan reported experimental evidence that the flux from a blackbody is proportional to the fourth power of the object's temperature (measured in kelvins). Five years after Stefan announced his law, another Austrian physicist, Ludwig Boltzmann, showed how it could be derived mathematically from basic assumptions about atoms and molecules. For this reason, Stefan's law is commonly known as the **Stefan-Boltzmann law**. Written as an equation, the Stefan-Boltzmann law is

Stefan-Boltzmann law for a blackbody

$$F = \sigma T^4$$

F = energy flux, in joules per square meter of surface per second

σ = a constant = 5.67 $\times$ 10^{-8} W m^{-2} K^{-4}

T = object's temperature, in kelvins

The value of the constant σ (the Greek letter sigma) is known from laboratory experiments.

The Stefan-Boltzmann law says that if you double the temperature of an object (for example, from 300 K to 600 K), then the energy emitted from the object's surface each second increases by a factor of 2^4 = 16. If you increase the temperature by a factor of 10 (for example, from 300 to 3000 K), the rate of energy emission increases by a factor of 10^4 = 10,000. Thus, a chunk of iron at room temperature (around 300 K) emits very little electromagnetic radiation (and essentially no visible light), but an iron bar heated to 3000 K glows quite intensely.

Box 5-2 gives several examples of applying Wien's law and the Stefan-Boltzmann law to typical astronomical problems.

5-5 Light has properties of both waves and particles

At the end of the nineteenth century, physicists mounted a valiant effort to explain all the characteristics of blackbody radiation. To this end they constructed theories based on

Using the Laws of Blackbody Radiation

The Sun and stars behave like nearly perfect blackbodies. Wien's law and the Stefan-Boltzmann law can therefore be used to relate the surface temperature of the Sun or a distant star to the energy flux and wavelength of maximum emission of its radiation. The following examples show how to do this.

EXAMPLE: The surface temperature of the Sun can be determined using Wien's law. The Sun emits energy over a wide range of wavelengths, but the maximum intensity of sunlight is at a wavelength of roughly 500 nm = 5.0×10^{-7} m. From Wien's law, we find the Sun's surface temperature $T_\odot$ to be

$$T_\odot = \frac{0.0029}{\lambda_{max}} = \frac{0.0029}{5.0 \times 10^{-7}} = 5800 \text{ K}$$

This is about the same temperature as an iron welding arc. (The symbol $\odot$ is the standard astronomical symbol for the Sun.)

EXAMPLE: We can also find the Sun's surface temperature using the Stefan-Boltzmann law. Using detectors above the Earth's atmosphere, astronomers have measured the average flux of solar energy arriving at Earth; this value, called the **solar constant,** is equal to 1370 W m^{-2}. However, the quantity F in the Stefan-Boltzmann law refers to the flux measured at the Sun's surface, not at the Earth.

To determine the value of F, we first imagine a huge sphere of radius 1 AU with the Sun at its center, as shown in the figure. Each square meter of that sphere receives 1370 watts of power from the Sun, so the total energy radiated by the Sun per second is equal to the solar constant multiplied by the sphere's surface area. The result, called the **luminosity** of the Sun and denoted by the symbol $L_\odot$, is $L_\odot = 3.90 \times 10^{26}$ W. That is, in 1 second the Sun radiates 3.90×10^{26} joules of energy into space. Because we know the size of the Sun, we can compute the energy flux (energy emitted per square meter per second) at its surface. The radius of the Sun is $R_\odot = 6.96 \times 10^{8}$ m, and the Sun's surface area is $4\pi R_\odot^2$. Therefore, its energy flux is the luminosity (total energy emitted by the Sun per second) divided by the surface area (the number of square meters of surface):

$$F_\odot = \frac{L_\odot}{4\pi R_\odot^2} = \frac{3.90 \times 10^{26} \text{ W}}{4\pi(6.96 \times 10^8 \text{ m})^2} = 6.41 \times 10^7 \text{ W m}^{-2}$$

Notice that the solar constant of 1370 W m^{-2} is very much less than this. By the time the Sun's radiation reaches Earth, it is spread over a greatly increased area.

Once we have the Sun's energy flux $F_\odot$, we can use the Stefan-Boltzmann law to find the Sun's surface temperature $T_\odot$:

$$T_\odot^4 = \frac{F_\odot}{\sigma} = 1.13 \times 10^{15} \text{ K}^4$$

Taking the fourth root (the square root of the square root) of this value, we find the surface temperature of the Sun to be $T_\odot = 5800$ K. This result agrees with the value we computed in the previous example using Wien's law.

EXAMPLE: Sirius, the brightest star in the night sky, has a surface temperature of about 10,000 K. We can use Wien's law to calculate the wavelength (λ_{max}) at which Sirius emits most intensely:

$$\lambda_{max} = \frac{0.0029}{T} = \frac{0.0029}{10,000} = 2.9 \times 10^{-7} \text{ m} = 290 \text{ nm}$$

Sirius therefore emits light most intensely in the ultraviolet. In the visible part of the spectrum, it emits more blue light than red light (see the curve for 12,000 K in Figure 5-10), so Sirius has a distinct blue color.

(continued on following page)

box 5-2 *(continued)*

EXAMPLE: We can use the Stefan-Boltzmann law to compare the energy flux from Sirius with that from the Sun. For the Sun, the Stefan-Boltzmann law is $F_\odot = \sigma T_\odot^4$, and for Sirius we can likewise write $F_* = \sigma T_*^4$, where the subscripts $\odot$ and $*$ refer to the Sun and Sirius, respectively. If we divide one equation by the other, the Stefan-Boltzmann con°stants cancel out, giving us an expression in terms of ratios:

$$\frac{F_*}{F_\odot} = \frac{T_*^4}{F_\odot^4}$$

In the preceding examples, we found that the temperatures of Sirius and the Sun are 10,000 K and 5800 K, respectively. Thus, we get

$$\frac{F_*}{F_\odot} = \frac{(10{,}000\ \text{K})^4}{(5800\ \text{K})^4} = \left(\frac{10{,}000}{5800}\right)^4 = 8.8$$

In other words, each square meter of Sirius's surface emits 8.8 times more energy per second than a square meter of the Sun's surface.

Maxwell's description of light as electromagnetic waves. But all such theories failed to explain the characteristic shapes of blackbody curves shown in Figure 5-10.

In 1900, however, the German physicist Max Planck discovered that he could derive a formula that correctly described blackbody curves if he assumed that electromagnetic energy is emitted in discrete, particlelike packets. He further had to assume that the energy of each such packet, today called a **photon,** is related to the wavelength of light: The greater the wavelength, the lower the energy of a photon associated with that wavelength. Thus, a photon of red light (wavelength $\lambda = 700$ nm) has less energy than a photon of violet light ($\lambda = 400$ nm). In this picture, light has a dual personality; it behaves as a stream of particlelike photons, but each photon has wavelike properties. In this sense, the best answer to the question "Is light a wave or a stream of particles?" is "Yes!"

It was soon realized that Planck's photon hypothesis explains more than just the detailed shape of blackbody curves. For example, it explains why only ultraviolet light causes suntans and sunburns. The reason is that tanning or burning involves a chemical reaction in the skin. High-energy, short-wavelength ultraviolet photons can trigger these reactions, but the lower-energy, longer-wavelength photons of visible light cannot. Similarly, normal photographic film is sensitive to visible light but not to infrared light; a long-wavelength infrared photon does not have enough energy to cause the chemical change that occurs when film is exposed to the higher-energy photons of visible light. Another phenomenon explained by the photon hypothesis is the **photoelectric effect.** In this effect, a metal plate is illuminated by a light beam. If ultraviolet light is used, tiny negatively charged particles called **electrons** are emitted from the metal plate. (We will see in Section 5-7 that the electron is one of the basic particles of the atom.) But if visible light is used, no matter how bright, no electrons are emitted.

In 1905 the great German-born physicist Albert Einstein explained this behavior by noting that a certain minimum amount of energy is required to

remove an electron from the metal plate. The energy of a short-wavelength ultraviolet photon is greater than this minimum value, so an electron that absorbs a photon of ultraviolet light will have enough energy to escape from the plate. But an electron that absorbs a photon of visible light, with its longer wavelength and lower energy, does not gain enough energy to escape and so remains within the metal. Einstein and Planck both won Nobel prizes for their contributions to understanding the nature of light.

The relationship between the energy E of a single photon and the wavelength of the electromagnetic radiation can be expressed in a simple equation:

Energy of a photon (in terms of wavelength)

$$E = \frac{hc}{\lambda}$$

E = energy of a photon

h = Planck's constant

c = speed of light

λ = wavelength of light

The value of the constant h in this equation, now called *Planck's constant,* has been shown in laboratory experiments to be

$$h = 6.625 \times 10^{-34}\ \text{J s}$$

The units of h are joules multiplied by seconds, called "joule-seconds" and abbreviated J s.

Because the value of h is so tiny, a single photon carries a very small amount of energy. For example, a photon of red light with wavelength 633 nm has an energy of only 3.14×10^{-19} J (Box 5-3). This is why we ordinarily do not notice that light comes in the form of photons; even a dim light source emits so many photons per second that it seems to be radiating a continuous stream of energy.

Photons at the Supermarket

A beam of light can be regarded as a stream of tiny packets of energy called photons. The Planck relationships $E = hc/\lambda$ and $E = h\nu$ can be used to relate the energy E carried by a photon to its wavelength λ and frequency ν.

As an example, the laser bar-code scanners used at stores and supermarkets emit orange-red light of wavelength 633 nm. To calculate the energy of a single photon of this light, we must first express the wavelength in meters. A nanometer (nm) is equal to 10^{-9} m, so the wavelength is

$$\lambda = (633 \text{ nm})\left(\frac{10^{-9} \text{ m}}{1 \text{ nm}}\right)$$

$$= 633 \times 10^{-9} \text{ m}$$

$$= 6.33 \times 10^{-7} \text{ m}$$

Then, using the Planck formula $E = hc/\lambda$, we find that the energy of a single photon is

$$E = \frac{hc}{\lambda} = \frac{(6.625 \times 10^{-34} \text{ J s})(3 \times 10^8 \text{ m/s})}{6.33 \times 10^{-7} \text{ m}} = 3.14 \times 10^{-19} \text{ J}$$

This is a very small amount of energy. The laser in a typical bar-code scanner emits 10^{-3} joule of light energy per second, so the number of photons emitted per second is

$$\frac{10^{-3} \text{ joule per second}}{3.14 \times 10^{-19} \text{ joule per photon}} = 3.2 \times 10^{15} \text{ photons per second}$$

This is such a large number that the laser beam seems like a continuous flow of energy rather than a stream of little energy packets.

The energies of photons are sometimes expressed in terms of a small unit of energy called the **electron volt** (eV). One electron volt is equal to 1.602×10^{-19} J, so a 633-nm photon has an energy of 1.96 eV. If energy is expressed in electron volts, Planck's constant is best expressed in electron volts multiplied by seconds, abbreviated eV s:

$$h = 4.135 \times 10^{-15} \text{ eV s}$$

Because the frequency ν of light is related to the wavelength λ by $\nu = c/\lambda$, we can rewrite the equation for the energy of a photon as

Energy of a photon (in terms of frequency)

$$E = h\nu$$

E = energy of a photon

h = Planck's constant

ν = frequency of light

The equations $E = hc/\lambda$ and $E = h\nu$ are together called **Planck's law**. Both equations express a relationship between a particlelike property of light (the energy E of a photon) and a wavelike property (the wavelength λ or frequency ν).

The photon picture of light is essential for understanding the detailed shapes of blackbody curves. As we will see, it also helps to explain how and why the spectra of the Sun and stars differ from those of perfect blackbodies.

5-6 Each chemical element produces its own unique set of spectral lines

In 1814 the German master optician Joseph von Fraunhofer repeated Newton's classic experiment of shining a beam of sunlight through a prism (see Figure 5-3). But this time Fraunhofer subjected the resulting rainbow-colored spectrum to intense magnification. To his surprise, he discovered that the solar spectrum contains hundreds of fine, dark lines, now called **spectral lines.** By contrast, if the light from a perfect blackbody were sent through a prism, it would produce a smooth, continuous spectrum with no dark lines. Fraunhofer counted more than 600 dark lines in the Sun's spectrum; today we know of more than 30,000. The photograph of the Sun's spectrum in Figure 5-12 shows hundreds of these spectral lines.

Half a century later, chemists discovered that they could produce spectral lines in the laboratory and use these spectral lines to analyze what kinds of atoms different substances are made of. Chemists had long known that many substances emit distinctive colors when sprinkled into a flame. To facilitate study of these colors, around 1857 the German chemist Robert Bunsen invented a gas burner (today called a Bunsen burner) that produces a clean flame with no color of its own. Bunsen's colleague, the Prussian-born physicist Gustav Kirchhoff, suggested that the colored light produced by substances in a flame might best be studied by passing the light through a prism (Figure 5-13). The two scientists promptly discovered that the spectrum from a flame consists of a pattern of thin, bright

ƒigure 5-12   R I **V** U X G

The Sun's Spectrum Numerous dark spectral lines are seen in this photograph of the Sun's spectrum. The spectrum is spread out so much that it had to be cut into segments to fit on this page. (National Optical Astronomy Observatories)

spectral lines against a dark background. The same kind of spectrum is produced by heated gases such as neon or argon.

Kirchhoff and Bunsen then found that each chemical element produces its own unique pattern of spectral lines. Thus was born in 1859 the technique of **spectral analysis,** the identification of chemical substances by their unique patterns of spectral lines.

A chemical **element** is a fundamental substance that cannot be broken down into more basic chemicals. Some examples are hydrogen, oxygen, carbon, iron, gold, and silver. After Kirchhoff and Bunsen had recorded the prominent spectral lines of all the then-known elements, they soon began to discover other spectral lines in the spectra of vaporized mineral samples. In this way they discovered elements whose presence had never before been suspected. In 1860 Kirchhoff and Bunsen found a new line in the blue portion of the spectrum of a sample of mineral water. After isolating the previously unknown element responsible for making the line, they named it cesium (from the Latin *caesium,* "gray-blue"). The next year, a new line in the red portion of the spectrum of a mineral sample led them to discover the element rubidium (Latin *rubidium,* "red").

Spectral analysis even allowed the discovery of new elements outside Earth. During the solar eclipse of 1868, astronomers found a new spectral line in light coming from the hot gases at the upper surface of the Sun while the main body of the Sun was hidden by the Moon. This line was attributed to a new element that was named helium (from the Greek *helios,* "sun"). Helium was not discovered on Earth until 1895, when it was found in gases obtained from a uranium mineral.

The spectrum of the Sun, with its dark spectral lines superimposed on a bright background (see Figure 5-12), may seem to

be unrelated to the spectra of bright lines against a dark background produced by substances in a flame (see Figure 5-13). But by the early 1860s Kirchhoff's experiments had revealed a direct connection between these two types of spectra. His conclusions are summarized in three important statements about spectra that are today called **Kirchhoff's laws.** These laws, which are illustrated in Figure 5-14, are as follows:

ƒigure 5-13

The Kirchhoff-Bunsen Experiment In the mid-1850s, Gustav Kirchhoff and Robert Bunsen discovered that when a chemical substance is heated and vaporized, the spectrum of the emitted light exhibits a series of bright spectral lines. They also found that each chemical element produces its own characteristic pattern of spectral lines. (In an actual laboratory experiment, lenses would be needed to focus the image of the slit onto the screen.)

Law 1 A hot opaque body, such as a perfect blackbody, or a hot, dense gas produces a **continuous spectrum**—a complete rainbow of colors without any spectral lines.

Law 2 A hot, transparent gas produces an **emission line spectrum**—a series of bright spectral lines against a dark background.

Law 3 A cool, transparent gas in front of a source of a continuous spectrum produces an **absorption line spectrum**—a series of dark spectral lines among the colors of the continuous spectrum. Furthermore, the dark lines in the absorption spectrum of a particular gas occur at exactly the *same* wavelengths as the bright lines in the emission spectrum of that same gas.

Kirchhoff's laws imply that if a beam of white light is passed through a gas, the atoms of the gas somehow extract light of very specific wavelengths from the white light. Hence, an observer who looks straight through the gas at the white-light source (the blackbody in Figure 5-14) will receive light whose spectrum has dark absorption lines superimposed on the continuous spectrum of the white light. The gas atoms then radiate light of precisely these same wavelengths in all directions. An observer at an oblique angle (that is, one who is not sighting directly through the cloud toward the blackbody) will receive only this light radiated by the gas cloud; the spectrum of this light is bright emission lines on a dark background.

 Figure 5-14 shows that light can either pass through a cloud of gas or be absorbed by the gas. But there is also a third possibility: The light can simply bounce off the atoms or molecules that make up the gas, a phenomenon called **light scattering**. In other words, photons passing through a gas cloud can miss the gas atoms altogether, be swallowed whole by the atoms (absorption), or bounce off the atoms like billiard balls colliding (scattering). Box 5-4 describes how light scattering explains the blue color of the sky and the red color of sunsets.

Whether an emission line spectrum or an absorption line spectrum is observed from a gas cloud depends on the relative temperatures of the gas cloud and its background. Absorption lines are seen if the background is hotter than the gas, and emission lines are seen if the background is cooler.

For example, if sodium is placed in the flame of a Bunsen burner in a darkened room, the flame will emit a characteristic

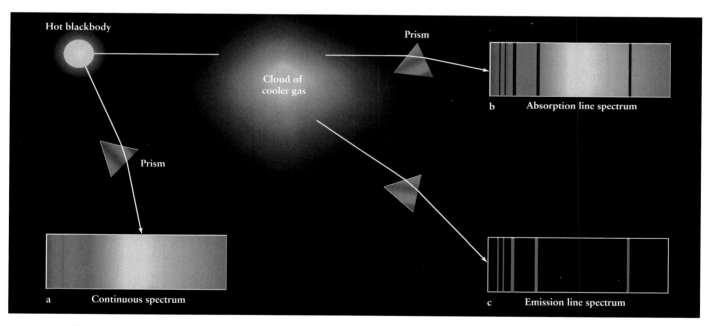

figure 5-14

Continuous, Absorption Line, and Emission Line Spectra A hot, glowing object such as a blackbody emits a continuous spectrum of light (spectrum *a*). If this light is passed through a cloud of a cooler gas, the cloud absorbs light of certain specific wavelengths. Hence, the light from the hot, glowing object that passes directly through the cloud is depleted in these wavelengths, and the spectrum of this light has dark absorption lines (spectrum *b*). The cloud does not retain all the light energy that it absorbs but radiates it outward in all directions. The spectrum of this reradiated light contains bright emission lines (spectrum *c*). The dark absorption lines in spectrum *b* have exactly the same wavelengths as the bright emission lines in spectrum *c*. The specific wavelengths observed depend on the chemical composition of the cloud.

box 5-4 **T**he Heavens on the Earth

Light Scattering

Light scattering is the process whereby photons bounce off particles in their path. These particles can be atoms, molecules, or clumps of molecules. You are reading these words using photons from the Sun or a lamp that bounced off the page—that is, were scattered by the particles that make up the page.

An important fact about light scattering is that very small particles—ones that are smaller than a wavelength of visible light—are quite effective at scattering short-wavelength photons of blue light, but less effective at scattering long-wavelength photons of red light. This fact explains a number of phenomena that you can see here on Earth.

The light that comes from the daytime sky is sunlight that has been scattered by the molecules that make up our atmosphere (see part *a* of the accompanying figure). Air molecules are less than 1 nm across, far smaller than the wavelength of visible light, so they scatter blue light more than red light—which is why the sky looks blue. Smoke particles are also quite small, which explains why the smoke from a cigarette or a fire has a bluish color.

Distant mountains often appear blue thanks to sunlight's being scattered from the atmosphere between the mountains and your eyes. (The Blue Ridge Mountains, which extend from Pennsylvania to Georgia, and Australia's Blue Mountains derive their names from this effect.) Sunglasses often have a red or orange tint, which blocks out blue light. This cuts down on the amount of scattered light from the sky reaching your eyes and allows you to see distant objects more clearly.

Light scattering also explains why sunsets are red. The light from the Sun contains photons of all visible wavelengths, but as this light passes through our atmosphere the blue photons are scattered away from the straight-line path from the Sun to your eye. Red photons undergo relatively little scattering, so the Sun always looks a bit redder than it really is. When you look toward the setting sun, the sunlight that reaches your eye has had to pass through a relatively thick layer of atmosphere (part *b* of the accompanying figure). Hence, a large fraction of the blue light from the Sun has been scattered, and the Sun appears quite red.

The same effect also applies to sunrises, but sunrises seldom look as red as sunsets do. The reason is that dust is lifted into the atmosphere during the day by the wind (which is typically stronger in the daytime than at night), and dust particles in the atmosphere help to scatter even more blue light.

If the small particles that scatter light are sufficiently concentrated, there will be almost as much scattering of red light as of blue light, and the scattered light will appear white. This explains the white color of clouds, fog, and haze, in which the scattering particles are ice crystals or water droplets. Whole milk looks white because of light scattering from tiny fat globules; nonfat milk has only a very few of these globules and so has a slight bluish cast.

Light scattering has many applications to astronomy. For example, it explains why very distant stars in our Galaxy appear surprisingly red. The reason is that there is a very thin gas in the space between the stars, and that gas scatters blue photons. By studying how much scattering takes place, astronomers have learned about the tenuous material that fills interstellar space.

a

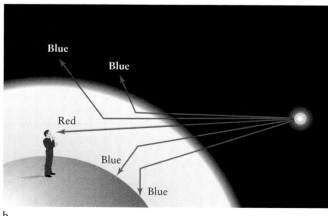

b

orange-yellow glow. (This same glow is produced if we use ordinary table salt, which is a compound of sodium and chlorine.) If we pass the light from the flame through a prism, it displays an emission line spectrum with two closely spaced spectral lines at wavelengths of 588.99 and 589.59 nm, in the orange-yellow part of the spectrum. We now turn on a lightbulb whose filament is hotter than the flame and shine the bulb's white light through the flame. The spectrum of this light after it passes through the flame's sodium vapor is the continuous spectrum from the lightbulb, but with two closely spaced *dark* lines at 588.99 and 589.59 nm. Thus, the chemical composition of the gas is revealed by either bright emission lines or dark absorption lines.

Spectroscopy is the systematic study of spectra and spectral lines. Spectral lines are tremendously important in astronomy, because they provide reliable evidence about the chemical composition of distant objects. As an example, the spectrum of the Sun shown in Figure 5-12 is an absorption line spectrum. The continuous spectrum comes from the hot surface of the Sun, which acts like a blackbody. The dark absorption lines are caused by this light passing through a cooler gas; this gas is the atmosphere that surrounds the Sun. Therefore, by identifying the spectral lines present in the solar spectrum, we can determine the chemical composition of the Sun's atmosphere.

Figure 5-15 shows both a portion of the Sun's absorption line spectrum and the emission line spectrum of iron vapor over the same wavelength range. This pattern of bright spectral lines in the lower spectrum is iron's own distinctive "fingerprint," which no other substance can imitate. Because some absorption lines in the Sun's spectrum coincide with the iron lines, some vaporized iron must exist in the Sun's atmosphere.

Spectroscopy can also help us analyze gas clouds in space, such as the nebula NGC 2363 shown in Figure 5-16. Such glowing clouds have emission line spectra, because we see them against the black background of space. The dominant red color of NGC 2363 is due to an emission line at a wavelength near 656 nm. This is one of the characteristic wavelengths emitted by hydrogen gas, so we can conclude that this nebula contains hydrogen. More detailed analyses of this kind show that hydrogen is the most common element in gaseous nebulae, and indeed in the universe as a whole.

What is truly remarkable about spectroscopy is that it can determine chemical composition at any distance. The 656-nm red light produced by a sample of heated hydrogen gas on

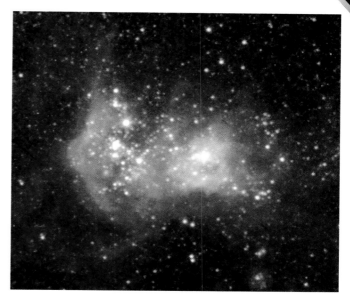

figure 5-16 R I **V** U X G

The Nebula NGC 2363 The glowing gas cloud in this Hubble Space Telescope image lies within a galaxy some 10 million light-years away in the constellation Camelopardis (the Camel). Hot stars within the nebula emit high-energy, ultraviolet photons, which are absorbed by the surrounding gas and heat the gas to high temperature. This heated gas produces light with an emission line spectrum. The particular wavelength of red light emitted by the nebula is 656 nm, characteristic of hydrogen gas. (L. Drissen, J.-R. Roy, and C. Robert/Département de Physique and Observatoire du Mont Mégantic, Université Laval; and NASA)

Earth is the same as that observed coming from NGC 2363 in Figure 5-16, located about 10 million light-years away. By using the basic principles outlined by Kirchhoff, astronomers have the tools to make chemical assays of objects that are almost inconceivably distant.

To make full use of Kirchhoff's laws, it is helpful to understand why they work. Why does an atom absorb light of only particular wavelengths? And why does it then emit light of only these same wavelengths? Maxwell's theory of electromagnetism

figure 5-15 R I **V** U X G

Iron in the Sun The upper part of this figure is a portion of the Sun's absorption line spectrum from 420 to 430 nm. Numerous dark spectral lines are visible. The lower part of the figure is a corresponding portion of the emission line spectrum of vaporized iron. Several bright spectral lines are seen against a black background. The iron lines coincide with some of the solar lines, which proves that there is some iron (albeit a very tiny amount) in the Sun's atmosphere. (Carnegie Observatories)

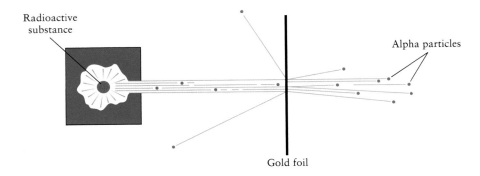

Radioactive substance

Alpha particles

Gold foil

Rutherford's Experiment Alpha particles from a radioactive source are directed at a thin metal foil. Most alpha particles pass through the foil with very little deflection. Occasionally, however, an alpha particle will recoil, indicating that it has collided with the massive nucleus of an atom. This experiment provided the first evidence that the nuclei of atoms are relatively massive and compact.

(Section 5-2) could not answer these questions. The answers did not come until early in the twentieth century, when scientists began to discover the structure and properties of atoms.

5-7 An atom consists of a small, dense nucleus surrounded by electrons

The first important clue about the internal structure of atoms came from an experiment conducted in 1910 by Ernest Rutherford, a gifted chemist and physicist from New Zealand. Rutherford and his colleagues at the University of Manchester in England had been investigating the recently discovered phenomenon of radioactivity. Certain radioactive elements, such as uranium and radium, were known to emit particles of various types. One type, the alpha particle, has about the same mass as a helium atom and is emitted from some radioactive substances with considerable speed.

In one series of experiments, Rutherford and his colleagues were using alpha particles as projectiles to probe the structure of solid matter. They directed a beam of these particles at a thin sheet of metal (Figure 5-17). Almost all the alpha particles passed through the metal sheet with little or no deflection from their straight-line paths. To the surprise of the experimenters, however, an occasional alpha particle bounced back from the metal sheet as though it had struck something quite dense. Rutherford later remarked, "It was almost as incredible as if you fired a fifteen-inch shell at a piece of tissue paper and it came back and hit you."

Rutherford concluded from this experiment that most of the mass of an atom is concentrated in a compact, massive lump of matter that occupies only a small part of the atom's volume. Most of the alpha particles pass freely through the nearly empty space that makes up most of the atom, but a few particles happen to strike the dense mass at the center of the atom and bounce back.

Rutherford proposed a new model for the structure of an atom, shown in Figure 5-18. According to this model, a massive, positively charged **nucleus** at the center of the atom is orbited by tiny, negatively charged electrons. Rutherford concluded that at least 99.98% of the mass of an atom must be concentrated in its nucleus, whose diameter is only about

10^{-14} m. (The diameter of a typical atom is far larger, about 10^{-10} m.)

 To appreciate just how tiny the nucleus is, imagine expanding an atom by a factor of 10^{12}, so that it was 100 meters across, about the size of a sports stadium. To this scale, the nucleus would be just a centimeter across— no larger than your thumbnail.

We know today that the nucleus of an atom contains two types of particles, **protons** and **neutrons**. A proton has a positive electric charge, equal and opposite to that of an electron. As its name suggests, a neutron has no electric charge—it is electrically neutral. As an example, an alpha particle (such as those Rutherford's team used) is actually a nucleus of the

Rutherford's Model of the Atom Electrons orbit the atom's nucleus, which contains most of the atom's mass. The nucleus contains two types of particles, protons and neutrons.

helium atom, with two protons and two neutrons. Protons and neutrons are held together in a nucleus by the so-called strong nuclear force, whose great strength overcomes the electric repulsion between the positively charged protons. A proton and a neutron have almost the same mass, 1.7×10^{-27} kg, and each has about 2000 times as much mass as an electron (9.1×10^{-31} kg). In an ordinary atom there are as many positive protons as there are negative electrons, so the atom has no net electric charge. Because the mass of the electron is so small, the mass of an atom is not much greater than the mass of its nucleus. That is why an alpha particle has nearly the same mass as an atom of helium.

While the solar system is held together by gravitational forces, atoms are held together by electrical forces. The electric forces attracting the positively charged protons and the negatively charged electrons keep the atom from coming apart. Box 5-5 describes more about the connection between the structure of atoms and the chemical and physical properties of substances made of those atoms.

Rutherford's experiments clarified the structure of the atom, but they did not explain how these tiny particles within the atom give rise to spectral lines. The task of reconciling Rutherford's atomic model with Kirchhoff's laws of spectral analysis was undertaken by the young Danish physicist Niels Bohr, who joined Rutherford's group at Manchester in 1911.

5-8 Spectral lines are produced when an electron jumps from one energy level to another within an atom

 Niels Bohr began his study of the connection between atomic spectra and atomic structure by trying to understand the structure of hydrogen, the simplest and lightest of the elements. (As we discussed in Section 5-6, hydrogen is also the most common element in the universe.) When Bohr was done, he had not only found a way to explain this atom's spectrum but had also found a justification for Kirchhoff's laws in terms of atomic physics.

The most common type of hydrogen atom consists of a single electron and a single proton. Hydrogen has a simple

visible-light spectrum consisting of a pattern of lines that begins at a wavelength of 656.3 nm and ends at 364.6 nm. The first spectral line is called H_α (H-alpha), the second spectral line is called H_β (H-beta), the third is H_γ (H-gamma), and so forth. The closer you get to the short-wavelength end of the spectrum at 364.6 nm, the more spectral lines you see.

The regularity in this spectral pattern was described mathematically in 1885 by Johann Jakob Balmer, a Swiss schoolteacher. The spectral lines of hydrogen at visible wavelengths are today called **Balmer lines,** and the entire pattern from H_α onward is called the **Balmer series.** More than two dozen Balmer lines are seen in the spectrum of the star shown in Figure 5-19. Stars in general, including the Sun, have Balmer absorption lines in their spectra, which shows they have atmospheres that contain hydrogen.

Using trial and error, Balmer discovered a formula from which the wavelengths (λ) of hydrogen's spectral lines can be calculated. Balmer's formula is usually written

$$\frac{1}{\lambda} = R\left(\frac{1}{4} - \frac{1}{n^2}\right)$$

where n can be any integer (whole number) greater than 2. Here R is the *Rydberg constant* ($R = 1.097 \times 10^7$ m^{-1}), named in honor of the Swedish spectroscopist Johannes Rydberg. To get the wavelength λ_α of the spectral line H_α, you first put $n = 3$ into Balmer's formula:

$$\frac{1}{\lambda_\alpha} = (1.097 \times 10^7 \text{ m}^{-1})\left(\frac{1}{4} - \frac{1}{3^2}\right) = 1.524 \times 10^6 \text{ m}^{-1}$$

Then take the reciprocal:

$$\lambda_\alpha = \frac{1}{1.524 \times 10^6 \text{ m}^{-1}} = 6.563 \times 10^{-7} \text{ m} = 656.3 \text{ nm}$$

To get the wavelength of H_β, use $n = 4$, and to get the wavelength of H_γ, use $n = 5$. If you use $n = \infty$ (the symbol ∞ stands for infinity), you get the short-wavelength end of the hydrogen spectrum at 364.6 nm. (Note that 1 divided by infinity equals zero.)

Bohr realized that to fully understand the structure of the hydrogen atom, he had to be able to derive Balmer's formula

ƒigure 5-19 R I V U X G

Balmer Lines in the Spectrum of a Star
This portion of the spectrum of the star HD 193182 shows nearly two dozen Balmer lines, from H_{13} through H_{40}. (Numbers are used beyond the first few Balmer lines, so you do not have to memorize the entire Greek alphabet!) The series converges at 364.6 nm, just to the left of H_{40}. This star's spectrum also contains the first 12 Balmer lines (H_α through H_{12}), but they are located beyond the right edge of this photograph. (Carnegie Observatories)

box 5-5 | Tools of the Astronomer's Trade

Atoms, the Periodic Table, and Isotopes

Each different chemical element is made of a specific type of atom. Each specific atom has a characteristic number of protons in its nucleus. For example, a hydrogen atom has 1 proton in its nucleus, an oxygen atom has 8 protons in its nucleus, and so on.

The number of protons in an atom's nucleus is the **atomic number** for that particular element. The chemical elements are most conveniently listed in the form of a **periodic table** (shown in the figure). Elements are arranged in the periodic table in order of increasing atomic number. With only a few exceptions, this sequence also corresponds to increasing average mass of the atoms of the elements. Thus, hydrogen (symbol H), with atomic number 1, is the lightest element. Iron (symbol Fe) has atomic number 26 and is a relatively heavy element.

All the elements listed in a single vertical column of the periodic table have similar chemical properties. For example, the elements in the far right column are all gases under the conditions of temperature and pressure found at the Earth's surface, and they are all very reluctant to react chemically with other elements.

In addition to nearly 100 naturally occurring elements, the periodic table includes a number of artificially produced elements. All these elements are heavier than uranium (symbol U) and are highly radioactive, which means that they decay into lighter elements within a short

time of being created in laboratory experiments. Scientists have succeeded in creating only a few atoms of elements 110 through 118 in this way, and these "superheavy" elements have not yet been given official names.

The number of protons in the nucleus of an atom determines which element that atom is. Nevertheless, the same element may have different numbers of neutrons in its nucleus. For example, oxygen (O) has atomic number 8, so every oxygen nucleus has exactly 8 protons. But oxygen nuclei can have 8, 9, or 10 neutrons. These three slightly different kinds of oxygen are called **isotopes.** The isotope with 8 neutrons is by far the most abundant variety. It is written as ^{16}O, or oxygen-16. The rarer isotopes with 9 and 10 neutrons are designated as ^{17}O and ^{18}O, respectively.

The superscript that precedes the chemical symbol for an element equals the total number of protons and neutrons in a nucleus of that particular isotope. For example, a nucleus of the most common isotope of iron, ^{56}Fe or iron-56, contains a total of 56 protons and neutrons. From the periodic table, the atomic number of iron is 26, so every iron atom has 26 protons in its nucleus. Therefore, the number of neutrons in an iron-56 nucleus is $56 - 26 = 30$. (Most nuclei have more neutrons than protons, especially in the case of the heaviest elements.)

It is extremely difficult to distinguish chemically between the various isotopes of a particular element. Ordinary

---

using the laws of physics. He first made the rather wild assumption that the electron in a hydrogen atom can orbit the nucleus only in certain specific orbits. (This was a significant break with the ideas of Newton, in whose mechanics any orbit should be possible.) Figure 5-20 shows the four smallest of these **Bohr orbits**, labeled by the numbers $n = 1$, $n = 2$, $n = 3$, and so on.

Although confined to one of these allowed orbits while circling the nucleus, an electron can jump from one Bohr orbit to another. For an electron to do this, the hydrogen atom must gain or lose a specific amount of energy. The atom must absorb energy for the electron to go from an inner to an outer orbit; the atom must release energy for the electron to go from an outer to an inner orbit. As an example, Figure 5-21 shows an electron jumping between the $n = 2$ and $n = 3$ orbits of a hydrogen atom as the atom absorbs or emits an H_α photon.

When the electron jumps from one orbit to another, the energy of the photon that is emitted or absorbed equals the difference in energy between these two orbits. This energy difference, and hence the photon energy, is the same whether the jump is from a low orbit to a high orbit (Figure 5-21*a*) or from the high orbit back to the low one (Figure 5-21*b*). According

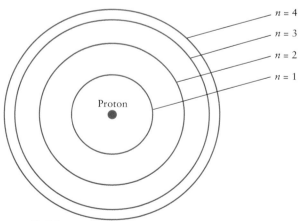

figure 5-20

The Bohr Model of the Hydrogen Atom According to the Bohr model of the hydrogen atom, an electron circles the nucleus (a proton) only in allowed orbits $n = 1, 2, 3$, and so forth. The first four Bohr orbits are shown here. This figure is not drawn to scale; in the Bohr model, the $n = 2, 3$, and 4 orbits are respectively 4, 9, and 16 times larger than the $n = 1$ orbit.

chemical reactions involve only the electrons that orbit the atom, never the neutrons buried in its nucleus. But there are small differences in the wavelengths of the spectral lines for different isotopes of the same element. For example, the spectral line wavelengths of the hydrogen isotope ^{2}H are about 0.03% greater than the wavelengths for the most common hydrogen isotope, ^{1}H. Thus, different isotopes can be distinguished by careful spectroscopic analysis.

Isotopes are important in astronomy for a variety of reasons. By measuring the relative amounts of different isotopes of a given element in a Moon rock or meteorite, the age of that sample can be determined. The mixture of isotopes left behind when a star explodes into a supernova (see Section 1-3) tells astronomers about the processes that led to the explosion. And knowing the properties of different isotopes of hydrogen and helium is crucial to understanding the nuclear reactions that make the Sun shine. Look for these and other applications of the idea of isotopes in later chapters.

Periodic Table of the Elements

to Planck and Einstein, if two photons have the same energy E, the relationship $E = hc/\lambda$ tells us that they must also have the same wavelength λ. It follows that if an atom can emit photons of a given energy and wavelength, it can also absorb photons of precisely the same energy and wavelength. Thus, Bohr's picture explains Kirchhoff's observation that atoms emit and absorb the same wavelengths of light.

The Bohr picture also helps us visualize what happens to produce an emission line spectrum. When a gas is heated, its atoms move around rapidly and can collide forcefully with each other. These energetic collisions excite the atoms' electrons into high orbits. The electrons then cascade back down to the innermost possible orbit, emitting photons whose energies are equal to the energy differences between different Bohr orbits. In this

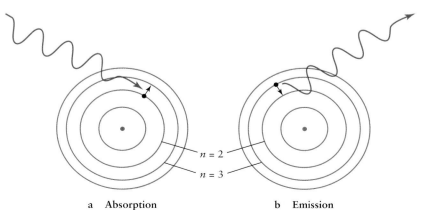

a   Absorption   b   Emission

figure 5-21

The Absorption and Emission of an H_{α} Photon This schematic diagram, drawn according to the Bohr model of the atom, shows what happens when a hydrogen atom absorbs or emits a photon whose wavelength is 656.3 nm. **(a)** The photon is absorbed by the atom, causing the electron to jump from the $n = 2$ orbit up to $n = 3$. **(b)** The photon is emitted by the atom as the electron falls from the $n = 3$ orbit down to the $n = 2$ orbit.

fashion, a hot gas produces an emission line spectrum with a variety of different wavelengths.

To produce an absorption line spectrum, begin with a relatively cool gas, so that the electrons in most of the atoms are in inner, low-energy orbits. If a beam of light with a continuous spectrum is shone through the gas, most wavelengths will pass through undisturbed. Only those photons will be absorbed whose energies are just right to excite an electron to an allowed outer orbit. Hence, only certain wavelengths will be absorbed, and dark lines will appear in the spectrum at those wavelengths.

Using his picture of allowed orbits and the formula $E = hc/\lambda$, Bohr was able to prove mathematically that the wavelength λ of the photon emitted or absorbed as an electron jumps between an inner orbit N and an outer orbit n is

Bohr formula for hydrogen wavelengths

$$\frac{1}{\lambda} = R\left(\frac{1}{N^2} - \frac{1}{n^2}\right)$$

N = number of inner orbit

n = number of outer orbit

R = Rydberg constant = 1.097×10^7 m^{-1}

λ = wavelength (in meters) of emitted or absorbed photon

If Bohr let $N = 2$ in this formula, he got back the formula that Balmer discovered by trial and error. Hence, Bohr deduced the meaning of the Balmer series: All the Balmer lines are produced by electrons jumping between the second Bohr orbit ($N = 2$) and higher orbits ($n = 3, 4, 5,$ and so on).

Bohr's formula also correctly predicts the wavelengths of other series of spectral lines that occur at nonvisible wavelengths. Using $N = 1$ gives the **Lyman series,** which is entirely in the ultraviolet. All the spectral lines in this series involve electron transitions between the lowest Bohr orbit and all higher orbits ($n = 2, 3, 4,$ and so on). This pattern of spectral lines begins with L_α (Lyman alpha) at 122 nm and converges on L_∞ at 91.2 nm. Using $N = 3$ gives a series of infrared wavelengths called the **Paschen series.** This series, which involves transitions between the third Bohr orbit and all higher orbits, begins with P_α (Paschen alpha) at 1875 nm and converges on P_∞ at 822 nm. Additional series exist at still longer wavelengths. Figure 5-22 shows some examples of electron transitions in the Bohr atom for these series, along with the wavelength for each transition.

Today's view of the atom owes much to the Bohr model, but is different in certain ways. The modern picture is based on **quantum mechanics,** a branch of physics dealing with photons and subatomic particles that was developed during the 1920s. As a result of this work, physicists no longer picture electrons as moving in specific orbits about the nucleus. Instead, electrons are now known to have both particle and wave properties and are said to occupy only certain **energy levels** in the atom.

figure 5-22

Electron Transitions in the Hydrogen Atom
Electron transitions between the lowest orbit ($n = 1$) and higher orbits ($n = 2, 3, 4, \ldots$) in the hydrogen atom give rise to spectral lines in the ultraviolet called the Lyman series. Transitions between the second orbit ($n = 2$) and higher orbits ($n = 3, 4, 5, \ldots$) cause visible spectral lines in the Balmer series, and transitions between the third orbit ($n = 3$) and higher orbits ($n = 4, 5, 6, \ldots$) create the Paschen series of spectral lines in the infrared. The same wavelengths occur when the electron drops from a high orbit to a low one, emitting a photon, and when the electron jumps from a low orbit to a high one, absorbing a photon. The orbits are not shown to scale.

An extremely useful way of displaying the structure of an atom is with an **energy-level diagram.** Figure 5-23 shows such a diagram for hydrogen. The lowest energy level, called the **ground state,** corresponds to the $n = 1$ Bohr orbit. Higher energy levels, called **excited states,** correspond to successively larger Bohr orbits.

An electron can jump from the ground state up to the $n = 2$ level only if the atom absorbs a Lyman-alpha photon with a wavelength of 122 nm. Such a photon has energy $E = hc/\lambda = 10.2$ eV (electron volts; see Section 5-5). That's why the energy level of $n = 2$ is shown in Figure 5-23 as having an energy 10.2 eV above that of the ground state (which is usually assigned a value of 0 eV). Similarly, the $n = 3$ level is 12.1 eV above the ground state, and so forth. Electrons can make transitions to higher energy levels by absorbing a photon or in a collision between atoms; they can make transitions to lower energy levels by emitting a photon.

On the energy-level diagram for hydrogen, the $n = \infty$ level has an energy of 13.6 eV. (This corresponds to an infinitely large orbit in the Bohr model.) If the electron is initially in the ground state and the atom absorbs a photon of any energy greater than 13.6 eV, the electron will be removed completely from the atom. This process is called **ionization**. A 13.6-eV photon has a wavelength of 91.2 nm, equal to the shortest wavelength in the ultraviolet Lyman series (L_∞). So any photon with a wavelength of 91.2 nm or less can ionize hydrogen. (The Planck formula $E = hc/\lambda$ tells us that the higher the photon energy, the shorter the wavelength.)

As an example, the gaseous nebula NGC 2363 shown in Figure 5-16 has hot stars in its neighborhood that produce copious amounts of ultraviolet photons with wavelengths less than 91.2 nm. Hydrogen atoms in the nebula that absorb these photons become ionized and lose their electrons. When the electrons recombine with the nuclei, they cascade down the energy levels to the ground state and emit visible light in the process. This is what makes the nebula glow.

The same basic principles that explain the hydrogen spectrum also apply to the atoms of other elements. Electrons in each kind of atom can be only in certain energy levels, so only photons of certain wavelengths can be emitted or absorbed. Because each kind of atom has its own unique arrangement of electron levels, the pattern of spectral lines is likewise unique to that particular type of atom (see the photograph that opens this chapter). These patterns are in general much more complicated than for the hydrogen atom. Hence, there is no simple relationship analogous to the Bohr formula that applies to the spectra of all atoms.

The idea of energy levels explains the emission line spectra and absorption line spectra of gases. But what about the continuous spectra produced by dense objects like the filament of a lightbulb or the coils of a toaster? These objects are made of atoms, so why don't they emit light with an emission line spectrum characteristic of the particular atoms of which they are made?

The reason is directly related to the difference between a gas on the one hand and a liquid or solid on the other. In a gas, atoms are widely separated and can emit photons without interference from other atoms. But in a liquid or a solid, atoms are so close that they almost touch, and thus these atoms interact strongly with each other. These interactions interfere with the process of emitting photons. As a result, the pattern of distinctive bright spectral lines that the atoms would emit in isolation becomes "smeared out" into a continuous spectrum.

Think of atoms like tuning forks. If you strike a single tuning fork, it produces a sound wave with a single clear frequency and wavelength, just as an isolated atom emits light of definite wavelengths. But if you shake a box packed full of tuning forks, you will hear a clanging noise that is a mixture of sounds of all different frequencies and wavelengths. This is directly analogous to the continuous spectrum of light emitted by a dense object with closely packed atoms.

With the work of such people as Planck, Einstein, Rutherford, and Bohr, the interchange between astronomy and physics came full circle. Modern physics was born when Newton set out to understand the motions of the planets. Two and a half centuries later, physicists in their laboratories probed the properties of light and the structure of atoms. Their labors had immediate applications in astronomy. Armed with this new understanding of light and matter, astronomers were able to probe in detail the chemical and physical properties of planets, stars, and galaxies.

5-9 The wavelength of a spectral line is affected by the relative motion between the source and the observer

In addition to telling us about temperature and chemical composition, the spectrum of a planet, star, or galaxy can also reveal something about that object's motion through space. This idea dates from 1842, when Christian Doppler, a professor of mathematics in Prague, pointed out that the observed wavelength of light must be affected by motion. Figure 5-24 shows why. In this figure, a light source is moving from right to left, and the circles represent the crests of waves emitted from the moving source at various positions. Each successive wave crest is emitted from a position slightly closer to the observer on the left, so she sees a shorter wavelength—the distance from one crest to the next—than she would if the source were stationary. All the lines in the spectrum of an approaching source are shifted toward the

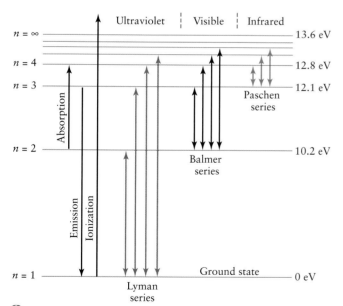

Figure 5-23

Energy-Level Diagram of Hydrogen A convenient way to display the structure of the hydrogen atom is in a diagram like this, which shows the allowed energy levels. The diagram shows a number of possible electron jumps, or transitions, between energy levels. An upward transition occurs when the atom absorbs a photon; a downward transition occurs when the atom emits a photon. (Compare with Figure 5-22.)

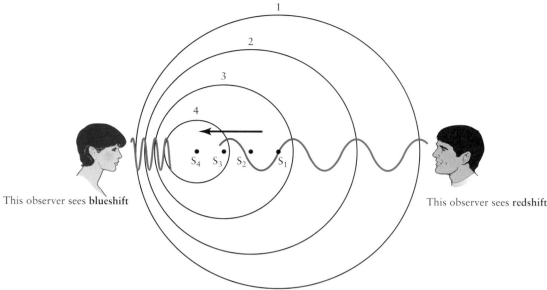

figure 5-24

The Doppler Effect The wavelength of light is affected by motion between the light source and an observer. The light source shown here is moving toward the left. Circles 1 through 4 indicate the crests of light waves that were emitted by the source when it was at points S_1 through S_4, respectively. The waves are crowded together in front of the moving source but are spread out behind it. Consequently, wavelengths appear shortened (blueshifted) if the source is moving toward the observer. They appear lengthened (redshifted) if the source is moving away from the observer. Motion perpendicular to an observer's line of sight does not affect wavelength.

short-wavelength (blue) end of the spectrum. This phenomenon is called a **blueshift.**

The source is receding from the observer on the right in Figure 5-24. The wave crests that reach him are stretched apart, so that he sees a longer wavelength than he would if the source were stationary. All the lines in the spectrum of a receding source are shifted toward the longer-wavelength (red) end of the spectrum, producing a **redshift.** In general, the effect of relative motion on wavelength is called the **Doppler effect.**

You have probably noticed a similar Doppler effect for sound waves. When a police car is approaching, the sound waves from its siren have a shorter wavelength and higher frequency than if the siren were at rest, and hence you hear a higher pitch. After the police car passes you and is moving away, you hear a lower pitch from the siren because the sound waves have a longer wavelength and a lower frequency.

Suppose that λ_0 is the wavelength of a particular spectral line from a light source that is not moving. It is the wavelength that you might look up in a reference book or determine in a laboratory experiment for this spectral line. If the source is moving, this particular spectral line is shifted to a different wavelength λ. The size of the wavelength shift is usually written as $\Delta\lambda$, where $\Delta\lambda = \lambda - \lambda_0$. Thus, $\Delta\lambda$ is the difference between the

wavelength listed in reference books and the wavelength that you actually observe in the spectrum of a star or galaxy.

Doppler proved that the wavelength shift ($\Delta\lambda$) is governed by the following simple equation:

Doppler shift equation

$$\frac{\Delta\lambda}{\lambda_0} = \frac{v}{c}$$

$\Delta\lambda$ = wavelength shift

λ_0 = wavelength if source is not moving

v = velocity of the source measured along the line of sight

c = speed of light = 3.0×10^5 km/s

The capital Greek letter Δ, or "delta," is commonly used to denote a change in the value of a quantity. Thus, $\Delta\lambda$ is the change in the wavelength λ due to the Doppler effect. It is *not* equal to a quantity Δ multiplied by a second quantity λ!

The velocity determined from the Doppler effect is called **radial velocity,** because v is the component of the star's motion parallel to our line of sight, or along the "radius" drawn from Earth to the star. Of course, a sizable fraction of a star's motion may be perpendicular to our line of sight. The speed

Box 5-6 | Tools of the Astronomer's Trade

Applications of the Doppler Effect

Doppler's formula relates the radial velocity of an astronomical object to the wavelength shift of its spectral lines. Here are two examples that show how to use this remarkably powerful formula.

EXAMPLE: As measured in the laboratory, the prominent H_α spectral line of hydrogen has a wavelength $\lambda_0 = 656.285$ nm. But in the spectrum of the star Vega, the brightest star in the constellation Lyra (the Harp), this line has a wavelength $\lambda = 656.255$ nm. The wavelength shift is

$$\Delta\lambda = \lambda - \lambda_0 = 656.255 \text{ nm} - 656.285 \text{ nm} = -0.030 \text{ nm}$$

The negative value means that we see the light from Vega shifted to shorter wavelengths—that is, there is a blueshift. (Note that the shift is very tiny and can be measured only using specialized equipment.) From Doppler's formula, the star's radial velocity is:

$$v = c\frac{\Delta\lambda}{\lambda_0} = (3.0 \times 10^5 \text{ km/s})\left(\frac{-0.030 \text{ nm}}{656.285 \text{ nm}}\right) = -14 \text{ km/s}$$

The minus sign indicates that the star is coming toward us at 14 km/s.

By plotting the motions of different stars toward and away from us, astronomers have been able to learn how the Milky Way Galaxy (of which our Sun is a part) is rotating. From this knowledge, and aided by Newton's universal law of gravitation (Section 4-7), they have made the surprising discovery that the Milky Way contains roughly 10 times more matter than had once been thought! The nature of this unseen *dark matter* is still a subject of debate.

EXAMPLE: In the radio region of the electromagnetic spectrum, hydrogen atoms emit and absorb photons with a wavelength of 21.12 cm, giving rise to a spectral feature commonly called the *21-centimeter line*. The galaxy NGC 3840 in the constellation Leo (the Lion) is receding from us at a speed of 7370 km/s, or about 2.5% of the speed of light. Using the Doppler formula, we can predict the wavelength at which we expect to detect the 21-cm line from this galaxy. The wavelength shift is

$$\Delta\lambda = \lambda_0\left(\frac{v}{c}\right) = (21.12 \text{ cm})\left(\frac{7370 \text{ km/s}}{3.0 \times 10^5 \text{ km/s}}\right) = 0.52 \text{ cm}$$

Therefore, we will detect the 21-cm line of hydrogen from this galaxy at a wavelength of

$$\lambda = \lambda_0 + \Delta\lambda = 21.12 \text{ cm} + 0.52 \text{ cm} = 21.64 \text{ cm}$$

The 21-cm line has been redshifted to a longer wavelength because the galaxy is receding from us. In fact, most galaxies are receding from us. This observation is one of the key pieces of evidence that the universe is expanding outward from a Big Bang that took place some 10 to 15 billion years ago.

---

of this transverse movement across the sky does not affect wavelengths if the speed is small compared with c. Box 5-6 includes two examples of calculations with radial velocity using the Doppler formula.

CAUTION! The redshifts and blueshifts of stars visible to the naked eye, or even through a small telescope, are only a small fraction of a nanometer. These tiny wavelength changes are far too small to detect visually. (Astronomers were able to detect the tiny Doppler shifts of starlight only after they had developed highly sensitive equipment for measuring wavelengths. This was done around 1890, a half-century after Doppler's original proposal.) So if you see a star with a red color, it means that the star really is red; it does *not* mean that it is moving rapidly away from us.

The Doppler effect is an important tool in astronomy because it uncovers basic information about the motions of planets, stars, and galaxies. For example, the rotation of the planet Venus was deduced from the Doppler shift of radar waves reflected from its surface. Small Doppler shifts in the spectrum of sunlight have shown that the entire Sun is vibrating like an immense gong. The back-and-forth Doppler shifting of the spectral lines of certain stars reveals that these stars are being orbited by unseen companions; from this astronomers have discovered planets around other stars and massive objects that may be black holes. Astronomers also use the Doppler effect along with Kepler's third law to measure the masses of galaxies. These are but a few examples of how Doppler's discovery has empowered astronomers in their quest to understand the universe.

In this chapter we have glimpsed how much can be learned by analyzing light from the heavens. To analyze this light, however, it is first necessary to collect as much of it as possible, because most light sources in space are very dim. Collecting the faint light from distant objects is a key purpose of telescopes. In the next chapter we will describe both how telescopes work and how they are used.

KEY WORDS

Terms preceded by an asterisk (*) are discussed in the Boxes.

absolute zero, p. 99

absorption line spectrum, p. 107

atom, p. 99

*atomic number, p. 112

Balmer line, p. 111

Balmer series, p. 111

blackbody, p. 101

blackbody curves, p. 101

blackbody radiation, p. 101

blueshift, p. 116

Bohr orbits, p. 112

continuous spectrum, p. 107

*degrees Celsius, p. 100

*degrees Fahrenheit, p. 100

Doppler effect, p. 116

electromagnetic radiation, p. 96

electromagnetic spectrum, p. 97

electromagnetism, p. 95

electron, p. 104

electron volt, p. 105

element, p. 106

emission line spectrum, p. 107

energy flux, p. 102

energy level, p. 114

energy-level diagram, p. 114

excited state, p. 114

frequency, p. 97

gamma rays, p. 97

ground state, p. 114

infrared radiation, p. 96

ionization, p. 115

*isotope, p. 112

joule, p. 102

kelvin, p. 99

Kirchhoff's laws, p. 106

light scattering, p. 107

*luminosity, p. 103

Lyman series, p. 114

microwaves, p. 97

nanometer, p. 96

neutron, p. 110

nucleus, p. 110

Paschen series, p. 114

*periodic table, p. 112

photoelectric effect, p. 104

photon, p. 104

Planck's law, p. 105

proton, p. 110

quantum mechanics, p. 114

radial velocity, p. 116

radio waves, p. 97

redshift, p. 116

*solar constant, p. 103

spectral analysis, p. 106

spectral line, p. 105

spectroscopy, p. 109

spectrum (*plural* spectra), p. 94

Stefan-Boltzmann law, p. 102

ultraviolet radiation, p. 97

visible light, p. 96

watt, p. 102

wavelength, p. 96

wavelength of maximum emission, p. 100

Wien's law, p. 102

X rays, p. 97

KEY IDEAS

The Nature of Light: Light is electromagnetic radiation. It has wavelike properties described by its wavelength λ and frequency ν, and travels through empty space at the constant speed $c = 3.0 \times 10^8$ m/s $= 3.0 \times 10^5$ km/s.

Blackbody Radiation: A blackbody is a hypothetical object that is a perfect absorber of electromagnetic radiation at all wavelengths. Stars closely approximate the behavior of blackbodies, as do other hot, dense objects.

• The intensities of radiation emitted at various wavelengths by a blackbody at a given temperature are shown by a blackbody curve.

• Wien's law states that the dominant wavelength at which a blackbody emits electromagnetic radiation is inversely proportional to the Kelvin temperature of the object: λ_{max} (in meters) $= (0.0029)/T$.

• The Stefan-Boltzmann law states that a blackbody radiates electromagnetic waves with a total energy flux F directly proportional to the fourth power of the Kelvin temperature T of the object: $F = \sigma T^4$.

Photons: An explanation of blackbody curves led to the discovery that light has particlelike properties. The particles of light are called photons.

• Planck's law relates the energy E of a photon to its frequency ν or wavelength λ: $E = h\nu = hc/\lambda$, where h is Planck's constant.

Kirchhoff's Laws: Kirchhoff's three laws of spectral analysis describe conditions under which different kinds of spectra are produced.

• A hot, dense object such as a blackbody emits a continuous spectrum covering all wavelengths.

• A hot, transparent gas produces a spectrum that contains bright (emission) lines.

• A cool, transparent gas in front of a light source that itself has a continuous spectrum produces dark (absorption) lines in the continuous spectrum.

Atomic Structure: An atom has a small dense nucleus composed of protons and neutrons. The nucleus is surrounded by electrons that occupy only certain orbits or energy levels.

• When an electron jumps from one energy level to another, it emits or absorbs a photon of appropriate energy (and hence of a specific wavelength).

• The spectral lines of a particular element correspond to the various electron transitions between energy levels in atoms of that element.

• Bohr's model of the atom correctly predicts the wavelengths of hydrogen's spectral lines.

The Doppler Shift: The Doppler shift enables us to determine the radial velocity of a light source from the displacement of its spectral lines.

• The spectral lines of an approaching light source are shifted toward short wavelengths (a blueshift); the spectral lines of a receding light source are shifted toward long wavelengths (a redshift).

• The size of a wavelength shift is proportional to the radial velocity between the light source and the observer.

REVIEW QUESTIONS

1. How long does it take light to travel from the Moon to the Earth, a distance of 384,000 km?

2. Approximately how many times around the world could a beam of light travel in one second?

3. (a) Describe an experiment in which light behaves like a wave. (b) Describe an experiment in which light behaves like a particle.

4. What is meant by the frequency of light? How is frequency related to wavelength?

5. A light source emits infrared radiation at a wavelength of 1060 nm. What is the frequency of this radiation?

6. A cellular phone is actually a radio transmitter and receiver. You receive an incoming call in the form of a radio wave of frequency 893.07 MHz. What is the wavelength (in meters) of this wave?

7. Using Wien's law and the Stefan-Boltzmann law, explain the color and intensity changes that are observed as the temperature of a hot, glowing object increases.

8. What is a blackbody? In what way is a blackbody black? If a blackbody is black, how can it emit light? If you were to shine a flashlight beam on a perfect blackbody, what would happen to the light?

9. Explain why astronomers are interested in blackbody radiation.

10. Why do astronomers find it convenient to use the Kelvin temperature scale in their work rather than the Celsius or Fahrenheit scale?

11. The bright star Aldebaran in the constellation Taurus (the Bull) has a surface temperature of 3850 K. What is its

wavelength of maximum emission in nanometers? What color is this star?

12. The bright star Hadar in the constellation Centaurus (the Centaur) emits the greatest intensity of radiation at a wavelength of 121 nm. What is the surface temperature of the star? What color is this star?

13. If you double the Kelvin temperature of a hot piece of steel, how much more energy will it radiate per second?

14. How is the energy of a photon related to its wavelength? What kind of photons carry the most energy? What kind of photons carry the least energy?

15. How do we know that atoms have massive, compact nuclei?

16. Describe the spectrum of hydrogen at visible wavelengths and explain how Bohr's model of the atom accounts for the Balmer lines.

17. Why do different elements display different patterns of lines in their spectra?

18. What is the Doppler effect? Why is it important to astronomers?

19. If you see a blue star, what does its color tell you about how the star is moving through space? Explain your answer.

ADVANCED QUESTIONS

Questions preceded by an asterisk (*) involve topics discussed in the Boxes.

Problem-solving tips and tools

You can find formulas in Box 5-1 for converting between temperature scales. Box 5-2 discusses how a star's radius, luminosity, and surface temperature are related. Box 5-3 shows how to use Planck's law to calculate the energy of a photon. To learn how to do calculations using the Doppler effect, see Box 5-6.

20. What is the temperature of the Sun's surface in degrees Fahrenheit?

21. Your normal body temperature is 98.6°F. What kind of radiation do you predominantly emit? At what wavelength (in nm) do you emit the most radiation?

22. What wavelength of electromagnetic radiation is emitted with greatest intensity by this book? To what region of the electromagnetic spectrum does this wavelength correspond?

*23. The bright star Deneb in the constellation Cygnus (the Swan) has a surface temperature of 8700 K. How much more energy is emitted each second from each square meter of Deneb's surface than from each square meter of the Sun's surface?

*24. Jupiter's moon Io has an active volcano named Pele whose temperature can be as high as 320°C. (a) What is the wavelength of maximum emission for the volcano at this temperature? In what part of the electromagnetic spectrum is this? (b) The average temperature of Io's surface is −150°C. Compared with a square meter of surface at this temperature, how much more energy is emitted per second from each square meter of Pele's surface?

*25. The bright star Sirius in the constellation of Canis Major (the Large Dog) has a radius of 1.67 $R_\odot$ and a luminosity of 25 $L_\odot$. (a) What is the energy flux at the surface of Sirius? (b) What is the star's surface temperature?

26. Black holes are objects whose gravity is so strong that not even an object moving at the speed of light can escape from their surface. Hence, black holes do not themselves emit light. But it is possible to detect radiation from material falling *toward* a black hole. Calculations suggest that as this matter falls, it is compressed and heated to temperatures around 10^6 K. Calculate the wavelength of maximum emission for this temperature. In what part of the electromagnetic spectrum does this wavelength lie?

27. In Figure 5-12 you can see two distinct dark lines at the boundary between the orange and yellow parts of the Sun's spectrum (in the center of the third colored band from the top of the figure). The wavelengths of these dark lines are 588.99 and 589.59 nm. What do you conclude from this about the chemical composition of the Sun's atmosphere?

28. Since the 1970s, instruments on board balloons and spacecraft have detected 511-keV photons coming from the direction of the center of our Galaxy. (The prefix k means *kilo*, or thousand, so 1 keV = 10^3 eV.) What is the wavelength of these photons? To what part of the electromagnetic spectrum do these photons belong?

29. (a) Calculate the wavelength of P_δ, the fourth wavelength in the Paschen series. (b) Draw a schematic diagram of the hydrogen atom and indicate the electron transition that gives rise to this spectral line. (c) In what part of the electromagnetic spectrum does this wavelength lie?

30. (a) Calculate the wavelength of H_{13}, the spectral line for an electron transition between the $n = 14$ and $n = 2$ orbits of hydrogen. (b) In what part of the electromagnetic spectrum does this wavelength lie? Use this to explain why Figure 5-19 is labeled R I V **U** X G.

31. (a) Can a hydrogen atom in the ground state absorb an H-alpha (H_α) photon? Explain why or why not. (b) Can a hydrogen atom in the $n = 2$ state absorb a Lyman-alpha (L_α) photon? Explain why or why not.

32. The spaces between the galaxies are filled with a very cold, very thin gas of hydrogen atoms. Ultraviolet radiation with any wavelength shorter than 91.2 nm cannot pass through this gas; instead, it is absorbed. Explain why.

33. An imaginary atom has just 3 energy levels: 0 eV, 1 eV, and 3 eV. Draw an energy-level diagram for this atom. Show all possible transitions between these energy levels. For each transition, determine the photon energy and the photon wavelength. Which transitions involve the emission or absorption of visible light?

34. The wavelength of H_β in the spectrum of the star Megrez in the Big Dipper (part of the constellation Ursa Major, the Great Bear) is 486.112 nm. Laboratory measurements demonstrate that the normal wavelength of this spectral line is 486.133 nm. Is the star coming toward us or moving away from us? At what speed?

35. The nebula NGC 2363 shown in Figure 5-16 is located within the galaxy NGC 2366 in the constellation Camelopardis (the Camel). This galaxy and the nebula within it are moving away from us at 252 km/s. At what wavelength does the red H_α line of hydrogen (which causes the color of the nebula) appear in the nebula's spectrum?

36. You are given a traffic ticket for going through a red light (wavelength 700 nm). You tell the police officer that because you were approaching the light, the Doppler effect caused a blueshift that made the light appear green (wavelength 500 nm). How fast would you have had to be going for this to be true? Would the speeding ticket be justified? Explain.

DISCUSSION QUESTIONS

37. The equation that relates the frequency, wavelength, and speed of a light wave, $\nu = c/\lambda$, can be rewritten as $c = \nu\lambda$. A friend who has studied mathematics but not much astronomy or physics might look at this equation and say: "This equation tells me that the higher the frequency ν, the greater the wave speed c. Since visible light has a higher frequency than radio waves, this means that visible light goes faster than radio waves." How would you respond to your friend?

38. Why do you suppose that ultraviolet light can cause skin cancer but ordinary visible light does not?

39. The accompanying visible-light image shows the star cluster NGC 3293 in the constellation Carina (the Ship's Keel). What can you say about the surface temperatures of most of the bright stars in this cluster? In what part of the electromagnetic spectrum do these stars emit most intensely? Are your eyes sensitive to this type of radiation? If not, how is it possible to see these stars at all? There is at least one bright star in this cluster with a distinctly

(David Malin/Anglo-Australian Observatory)   R I **V** U X G

different color from the others; what can you conclude about its surface temperature?

40. (a) If you could see ultraviolet radiation, how might the night sky appear different? Would ordinary objects appear different in the daytime? **(b)** What differences might there be in the appearance of the night sky and in the appearance of ordinary objects in the daytime if you could see infrared radiation?

41. The human eye is most sensitive over the same wavelength range at which the Sun emits the greatest intensity of radiation. Suppose creatures were to evolve on a planet orbiting a star somewhat hotter than the Sun. To what wavelengths would their vision most likely be sensitive?

 WEB/CD-ROM QUESTIONS

42. Search the World Wide Web for information about rainbows. Why do rainbows form? Why do they appear as circular arcs? Why can you see different colors?

 43. Measuring Stellar Temperatures. Access the Active Integrated Media Module "Blackbody Curves" in Chapter 5 of the *Universe* web site or CD-ROM. **(a)** Use the module to determine the range of temperatures over which a star's peak wavelength is in the visible spectrum. **(b)** Determine if any of the following stars have a peak wavelength in the visible spectrum: Rigel, $T = 14{,}000$ K; Deneb, $T = 9500$ K; Arcturus, $T = 4500$ K; Vega, $T = 11{,}500$ K; Betelgeuse, $T = 3100$ K.

OBSERVING PROJECTS

44. Turn on an electric stove or toaster oven and carefully observe the heating elements as they warm up. Relate your observations to Wien's law and the Stefan-Boltzmann law.

45. Obtain a glass prism (or a diffraction grating, which is probably more readily available and is discussed in the next chapter) and look through it at various light sources, such as an ordinary incandescent light, a neon sign, and a mercury vapor street lamp. **Do not look at the sun! Looking directly at the Sun causes permanent eye damage or blindness.** Do you have any trouble seeing spectra? What do you have to do to see a spectrum? Describe the differences in the spectra of the various light sources you observed.

 46. Use the *Starry Night* program to examine some distant celestial objects. First turn off daylight (select **Daylight** in the **Sky** menu) and show the entire celestial sphere (select **Atlas** in the **Go** menu). Then search for objects (i), (ii), and (iii) listed below (select **Find...** in the **Edit** menu). For each object, zoom in until you can see it in detail. For each object, state whether it has a continuous spectrum, an absorption line spectrum, or an emission line spectrum, and explain your reasoning. (i) The Trifid Nebula in Sagittarius. (*Hint:* See Figure 5-16.) (ii) M31, the great galaxy in the constellation Andromeda. (*Hint:* The light coming from this galaxy is the combined light of hundreds of billions of individual stars.) (iii) The Moon. (*Hint:* Recall from Section 3-1 that moonlight is simply reflected sunlight.)

Optics and Telescopes

University of Hawaii
0.6-m Telescope

Caltech
Submillimeter
Observatory

James Clerk Maxwell
Telescope

Submillimeter
Array

Subaru

Keck I

Keck II

United Kingdom
Infrared Telescope

Gemini North

University of Hawaii
2.2-meter Telescope

Canada-France-Hawaii Telescope

NASA Infrared
Telescope Facility

(Richard J. Wainscoat, University of Hawaii)

R I V U X G

Just a few kilometers from the resorts of Hawaii's Kona Coast, in a place where few tourists ever go, is one of the best locations on Earth to view the heavens. This is the summit of Mauna Kea, an extinct volcano that reaches some 4200 m (14,000 ft) above the waters of the Pacific. Astronomers prize Mauna Kea because the skies there are unusually clear, still, and dark. And so it is that Mauna Kea has become the home of nearly a dozen powerful telescopes (see the accompanying photo).

Large telescopes such as those atop Mauna Kea can make images that are far brighter and sharper than the images formed by our eyes. They can also produce finely detailed spectra that reveal the chemical composition of nearby planets as well as distant galaxies.

A number of modern telescopes can even compensate for the "twinkling" of starlight, which occurs when light passes through turbulent air. Some telescopes deal with this problem by avoiding it altogether; the most famous example is the Hubble Space Telescope, which orbits high above the blurring effects of the atmosphere.

Special telescopes on Earth and in orbit explore the electromagnetic spectrum at wavelengths outside the visible range. Radio telescopes have mapped out the structure of our Milky Way Galaxy; with infrared telescopes, astronomers have peered at stars in the process of formation; and X-ray telescopes may have located black holes in orbit around ordinary stars. The telescope, in all its variations, is by far the most useful tool that astronomers have for collecting data about the universe.

As you read the sections of this chapter, look for the answers to the following questions.

6-1 Why is it important that telescopes be large?

6-2 Why do most modern telescopes use a large mirror rather than a large lens?

6-3 Why are observatories in such remote locations?

6-4 Do astronomers use ordinary photographic film to take pictures of the sky? Do they actually look through large telescopes?

6-5 How do astronomers use telescopes to measure the spectra of distant objects?

6-6 Why do astronomers need telescopes that detect radio waves and other nonvisible forms of light?

6-7 Why is it useful to put telescopes in orbit?

6-1 A refracting telescope uses a lens to concentrate incoming light at a focus

The **optical telescope**—that is, a telescope designed for use with visible light—was invented in the Netherlands in the early seventeenth century. Soon after, Galileo used one of these new inventions for his groundbreaking astronomical observations (see Section 4-3). These first telescopes used *lenses* to make distant objects appear larger and brighter. Telescopes of this same basic design are used today by many amateur astronomers. To understand telescopes of this kind, we need to understand how lenses work.

All lenses, including those used in telescopes, make use of the same physical principle: *Light travels at a slower speed in a dense substance.* Thus, although the speed of light in a vacuum is 3.0×10^8 m/s, its speed in glass is less than 2×10^8 m/s. Just as a woman's walking pace slows suddenly when she walks from a boardwalk onto a sandy beach, so light slows abruptly as it enters a piece of glass. Upon exiting the glass, light resumes its original speed, just as a woman stepping back onto a boardwalk easily resumes her original pace.

A material through which light travels is called a **medium** (plural **media**). As a beam of light passes from one transparent medium into another—say, from air into glass, or from glass back into air—the direction of the light can change. This phenomenon, called **refraction**, is caused by the change in the speed of light.

ANALOGY Imagine driving a car from a smooth pavement onto a sandy beach (Figure 6-1*a*). If the car approaches the beach head-on, it slows down when it enters the sand but keeps moving straight ahead. If the car approaches the beach at an angle, however, one of the front wheels will be slowed by the sand before the other is, and the car will veer from its original direction. In the same way, a beam of light changes direction when it enters a piece of glass at an angle (Figure 6-1*b*).

Figure 6-2*a* shows the refraction of a beam of light passing through a piece of flat glass. As the beam enters the

a

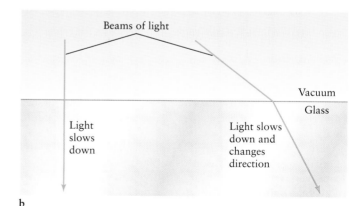
b

Figure 6-1

Refraction (a) When a car drives from smooth pavement into soft sand, it slows down. If it enters the sand at an angle, the front wheel on one side feels the drag of the sand before the other wheel, causing the car to veer to the side and change direction. (b) Similarly, light slows down when it passes from a vacuum into glass. The direction of the light beam changes if it enters the glass at an angle.

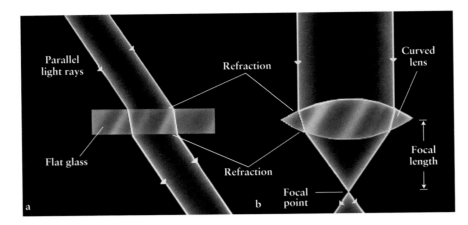

ƒigure 6-2

Refraction and Lenses **(a)** Refraction is the change in direction of a light ray when it passes into or out of a transparent medium such as glass. When light rays pass through a flat piece of glass, the two refractions bend the rays in opposite directions. There is no overall change in the direction in which the light travels. **(b)** If the glass is in the shape of a convex lens, parallel light rays converge to a focus at a special point called the focal point. The distance from the lens to the focal point is called the focal length of the lens.

upper surface of the glass, refraction takes place, and the beam is bent to a direction more nearly perpendicular to the surface of the glass. As the beam exits from the glass back into the surrounding air, a second refraction takes place, and the beam bends in the opposite sense. (The amount of bending depends on the speed of light in the glass, so different kinds of glass produce slightly different amounts of refraction.) Because the two surfaces of the glass are parallel, the beam emerges from the glass traveling in the same direction in which it entered.

Something more useful happens if the glass is curved into a convex shape (one that is fatter in the middle than at the edges), like the lens in Figure 6-2*b*. When a beam of light rays passes through the lens, refraction causes all the rays to converge at a point called the **focus.** If the light rays entering the lens are all parallel, the focus occurs at a special point called the **focal point.** The distance from the lens to the focal point is called the **focal length** of the lens.

The case of parallel light rays, shown in Figure 6-2*b*, is not merely a theoretical ideal. The stars are so far away that light rays from them are essentially parallel, as Figure 6-3 shows. Consequently, a lens always focuses light from an astronomical object to the focal point. If the object has a very small angular size, like a distant star, all the light entering the lens from that object converges onto the focal point. The resulting image is just a single bright dot.

But if the object is *extended*—that is, has a relatively large angular size, like the Moon or a planet—then light coming from each point on the object is brought to a focus at its own individual point. The result is an extended image that lies in the **focal plane** of the lens (Figure 6-4), which is a plane that includes the focal point. You can use an ordinary magnifying glass in this way to make an image of the Sun on the ground.

To use a lens to make a photograph of an astronomical object, you would place a piece of film in the focal plane. An ordinary camera works in the same way for taking photographs of relatively close objects here on Earth. But if you instead want to view the image visually, you would add a second lens to magnify the image formed in the focal plane. Such an arrangement of two lenses is called a **refracting telescope,**

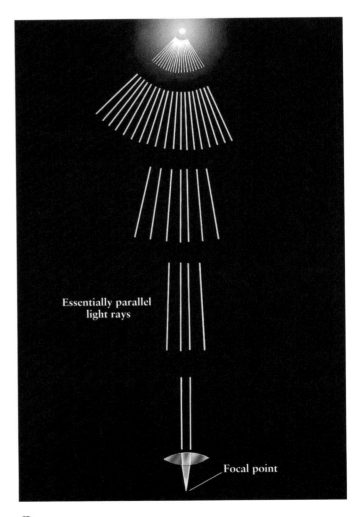

ƒigure 6-3

Light Rays from Distant Objects Are Parallel Light rays travel away in all directions from an ordinary light source. If a lens is located very far from the light source, only a few of the light rays will enter the lens, and these rays will be essentially parallel. This is why we drew parallel rays entering the lens in Figure 6-2*b*.

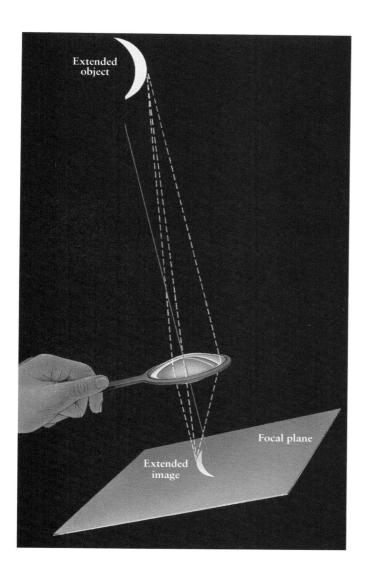

A Lens Creates an Extended Image of an Extended Object
Light coming from each point on an extended object passes
through a lens and produces an image of that point. All of these
tiny images put together make an extended image of the entire
object. The image of a very distant object is formed in a plane
called the focal plane. The distance from the lens to the focal plane
is called the focal length.

or **refractor** (Figure 6-5). The large-diameter, long-focal-length
lens at the front of the telescope, called the **objective lens,**
forms the image; the smaller, shorter-focal-length lens at the
rear of the telescope, called the **eyepiece lens,** magnifies the
image for the observer.

In addition to the focal length, the other important dimension of the objective lens of a refractor is the diameter. Compared with a small-diameter lens, a large-diameter lens captures more light, produces brighter images, and allows
astronomers to detect fainter objects. (For the same reason,
the iris of your eye opens when you go into a darkened room
to allow you to see dimly lit objects.)

The **light-gathering power** of a telescope is directly proportional to the area of the objective lens, which in turn is
proportional to the square of the lens diameter (Figure 6-6).
Thus, if you double the diameter of the lens, the light-
gathering power increases by a factor of $2^2 = 2 \times 2 = 4$.
Box 6-1 describes how to compare the light-gathering power
of different telescopes.

Because light-gathering power is so important for seeing
faint objects, the lens diameter is almost always given in
describing a telescope. For example, the Lick telescope on
Mount Hamilton in California is a 90-cm refractor, which

ƒigure 6-5

**A Refracting
Telescope**
A refracting telescope consists of a
large-diameter objective lens with
a long focal length and a small
eyepiece lens of short focal length.
The eyepiece lens magnifies the
image formed by the objective lens
in its focal plane (shown as a
dashed line).

ƒigure 6-6 R I U X G

Light-Gathering Power These two photographs of the galaxy M31 in Andromeda were taken using the same exposure time and at the same magnification, but with two different telescopes with objective lenses of different diameters. The right-hand photograph is brighter and shows more detail because it was made using the larger-diameter lens, which intercepts more starlight than a small-diameter lens. This same principle applies to telescopes that use curved mirrors rather than lenses to collect light (see Section 6-2). (Association of Universities for Research in Astronomy)

means that it is a refracting telescope whose objective lens is 90 cm in diameter. By comparison, Galileo's telescope of 1610 was a 3-cm refractor. The Lick telescope has an objective lens 30 times larger in diameter, and so has $30 \times 30 = 900$ times the light-gathering power of Galileo's instrument.

In addition to their light-gathering power, telescopes are useful because they magnify distant objects. As an example, the angular diameter of the Moon as viewed with the naked eye is about 0.5°. But when Galileo viewed the Moon through his telescope, its apparent angular diameter was 10°, large enough so that he could identify craters and mountain ranges. The **magnification,** or **magnifying power,** of a telescope is the ratio of an object's angular diameter seen through the telescope to its naked-eye angular diameter. Thus, the magnification of Galileo's telescope was 10°/0.5° = 20 times, usually written as 20×.

 The magnification of a refracting telescope depends on the focal lengths of both of its lenses:

$$\text{Magnification} = \frac{\text{focal length of objective lens}}{\text{focal length of eyepiece lens}}$$

This formula shows that using a long-focal-length objective lens with a short-focal-length eyepiece gives a large magnification. Box 6-1 illustrates how this formula is used.

Many people think that the primary purpose of a telescope is to magnify images. While magnification is certainly important, it is *not* the most important aspect of a telescope. The reason is that there is a limit to how sharp any astronomical image can be. For telescopes that view the heavens from Earth's surface, this limit is usually due to the blurring caused by having to view through Earth's atmosphere. (We will discuss this phenomenon in more detail in Section 6-3.) Magnifying a blurred image may make it look bigger but will not make it any clearer. Thus, beyond a certain point, there is nothing to be gained by further magnification. Astronomers put much more store in the light-gathering power of a telescope than in its magnification. Greater light-gathering power means brighter images, which makes it easier to see faint details.

If you were to build a telescope like that in Figure 6-5 using only the instructions given so far, you would probably be disappointed with the results. The problem is that a lens bends different colors of light through different angles, just as a prism does (recall Figure 5-3). As a result, different colors do not focus at the same point, and stars viewed through a telescope that uses a simple lens are surrounded by fuzzy, rainbow-colored halos. Figure 6-7a shows this optical defect, called **chromatic aberration.**

box 6-1 | Tools of the Astronomer's Trade

Magnification and Light-Gathering Power

The magnification of a telescope is equal to the focal length of the objective divided by the focal length of the eyepiece. Telescopic eyepieces are usually interchangeable, so the magnification of a telescope can be changed by using eyepieces of different focal lengths.

EXAMPLE: A small refracting telescope has an objective of focal length 120 cm. If the eyepiece has a focal length of 4.0 cm, the magnification of the telescope is

$$\text{Magnification} = \frac{120}{4.0} = 30 \text{ (usually written as } 30\times)$$

As viewed through this telescope, a large lunar crater that subtends an angle of 1 arcminute to the naked eye will appear to subtend an angle 30 times greater, or 30 arcminutes (one-half of a degree). This makes the details of the crater much easier to see.

If a 2.0-cm-focal-length eyepiece is used instead, the magnification will be 120/2.0 = 60×. The shorter the focal length of the eyepiece, the greater the magnification.

The light-gathering power of a telescope depends on the diameter of the objective lens; it does not depend on the focal length. The light-gathering power is proportional to the square of the diameter.

EXAMPLE: A fully dark adapted human eye has a pupil diameter of about 5 mm. By comparison, a small telescope whose objective lens is 5 cm in diameter has 10 times the diameter and $10^2 = 100$ times the light-gathering power of the eye. (Recall that there are 10 mm in 1 cm.) Hence, this telescope allows you to see objects 100 times fainter than you can see without a telescope.

The same relationships apply to reflecting telescopes, discussed in Section 6-2. Each of the two Keck telescopes on Mauna Kea in Hawaii (see the chapter opening photograph) uses a concave mirror 10 m in diameter to bring starlight to a focus. Because there are 1000 mm in one meter, this diameter can also be expressed as

$$10 \text{ m} \times \frac{1000 \text{ mm}}{1 \text{ m}} = 10,000 \text{ mm}$$

Thus, the light-gathering power of either of the Keck telescopes is greater than that of the human eye by a factor of

$$\frac{(10,000 \text{ mm})^2}{(5 \text{ mm})^2} = (2000)^2 = 4 \times 10^6, \text{ or } 4,000,000$$

When it comes to light-gathering power, the bigger the telescope, the better!

figure 6-7

Chromatic Aberration **(a)** A single lens suffers from a defect called chromatic aberration, in which different colors of light are brought to a focus at different distances from the lens. **(b)** This problem can be corrected by adding a second lens made from a different kind of glass.

a

b

ℱigure 6-8   R I 𝐕 U X G

A Large Refracting Telescope (a) This giant refractor, built at the end of the 1800s, is housed at Yerkes Observatory near Chicago. The objective lens is at the upper end of the telescope tube, and the eyepiece is at the lower end. The telescope tube is 19.5 m (64 ft) long; it has to be this long because the focal length of the objective is just under 19.5 m (see Figure 6-5). The entire observatory floor can be raised to bring the eyepiece within reach. **(b)** This historical photograph shows the astronomer George van Biesbrock with the objective lens of the Yerkes refractor. This lens, the largest ever made, is 102 cm (40 in.) in diameter. (Yerkes Observatory)

One way to correct for chromatic aberration is to use an objective lens that is not just a single piece of glass. Different types of glass can be manufactured by adding small amounts of chemicals to the glass when it is molten. Because of these chemicals, the speed of light varies slightly from one kind of glass to another, and the refractive properties vary as well. If a thin lens is mounted just behind the main objective lens of a telescope, as shown in Figure 6-7*b*, and if the telescope designer carefully chooses two different kinds of glass for these two lenses, different colors of light can be brought to a focus at the same point.

Chromatic aberration is only the most severe of a host of optical problems that must be solved in designing a high-quality refracting telescope. Master opticians of the nineteenth century devoted their careers to solving these problems, and several magnificent refractors were constructed in the late 1800s. Figure 6-8 shows the world's largest refracting telescope, completed in 1897.

Unfortunately, there are several negative aspects of refractors that not even the finest optician can overcome. First, because faint light must readily pass through the objective lens, the glass from which the lens is made must be totally free of defects, such as the bubbles that frequently form when molten glass is poured into a mold. Such defect-free glass is extremely expensive. Second, glass is opaque to certain kinds of light. Ultraviolet light is absorbed almost completely, and even visible light is dimmed substantially as it passes through the thick slab of glass that makes up the objective lens. Third, it is impossible to produce a large lens that is entirely free of chromatic aberration. Fourth, because the lens can be supported only around its edges, it tends to sag and distort under its own weight. This has adverse effects on the image clarity.

For these reasons and more, few major refractors have been built since the beginning of the twentieth century. Instead, astronomers have avoided all of the limitations of refractors by building telescopes that use a mirror instead of a lens to form an image.

6-2   A reflecting telescope uses a mirror to concentrate incoming light at a focus

Almost all modern telescopes form an image using the principle of **reflection**. To understand reflection, imagine drawing a dashed line perpendicular to the surface of a flat mirror at the point where a light ray strikes the mirror (Figure 6-9). The angle *i* between the *incident* (arriving) light ray and the perpendicular is always equal to the angle *r* between the *reflected* ray and the perpendicular.

In 1663, the Scottish mathematician James Gregory first proposed a telescope using reflection from a concave mirror—

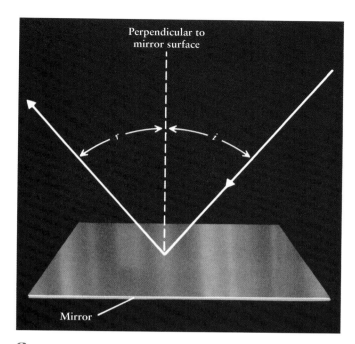

figure 6-9

Reflection by a Flat Mirror The angle at which a beam of light approaches a mirror, called the angle of incidence (*i*), is always equal to the angle at which the beam is reflected from the mirror, called the angle of reflection (*r*). That is, *i* = *r*.

one that is fatter at the edges than at the middle. Such a mirror makes parallel light rays converge to a focus (Figure 6-10). The distance between the reflecting surface and the focus is the focal length of the mirror. A telescope that uses a curved mirror to make an image of a distant object is called a **reflecting telescope,** or **reflector.** Using terminology similar to that used for refractors, the mirror that forms the image is called the **objective mirror** or **primary mirror.**

To make a reflector, an optician grinds and polishes a large slab of glass into the appropriate concave shape. The glass is then coated with silver, aluminum, or a similar highly reflective substance. Because light reflects off the surface of the glass rather than passing through it, defects within the glass—which would have very negative consequences for the objective lens of a refracting telescope—have no effect on the optical quality of a reflecting telescope.

Another advantage of reflectors is that they do not suffer from the chromatic aberration that plagues refractors. This is because reflection is not affected by the wavelength of the incoming light, so all wavelengths are reflected to the same focus. (A small amount of chromatic aberration may arise if the image is viewed using an eyepiece lens.) Furthermore, the mirror can be fully supported by braces on its back, so that a large, heavy mirror can be mounted without much danger of breakage or surface distortion.

Although a reflecting telescope has many advantages over a refractor, the arrangement shown in Figure 6-10 is not ideal.

One problem is that the focal point is in front of the objective mirror. If you try to view the image formed at the focal point, your head will block part or all of the light from reaching the mirror.

To get around this problem, in 1668 Isaac Newton simply placed a small, flat mirror at a 45° angle in front of the focal point, as sketched in Figure 6-11*a*. This secondary mirror deflects the light rays to one side, where Newton placed an eyepiece lens to magnify the image. A reflecting telescope with this optical design is appropriately called a **Newtonian reflector.** The magnifying power of a Newtonian reflector is calculated in the same way as for a refractor: The focal length of the objective mirror is divided by the focal length of the eyepiece (see Box 6-1).

Later astronomers modified Newton's original design. The objective mirrors of some modern reflectors are so large that the astronomer could actually sit in an "observing cage" at the undeflected focal point directly in front of the objective mirror. (In practice, riding in this cage on a winter's night is a remarkably cold and uncomfortable experience.) This arrangement is called a **prime focus** (Figure 6-11*b*). It usually provides the highest-quality image, because there is no need for a secondary mirror (which might have imperfections).

Another popular optical design, called a **Cassegrain focus** after the French contemporary of Newton who first proposed it, also has a convenient, accessible focal point. A hole is drilled directly through the center of the primary mirror, and a convex secondary mirror placed in front of the original focal point reflects the light rays back through the hole (Figure 6-11*c*).

A fourth design is useful when there is optical equipment too heavy or bulky to mount directly on the telescope. Instead, a series of mirrors channels the light rays away from the telescope to a remote focal point where the equipment is located.

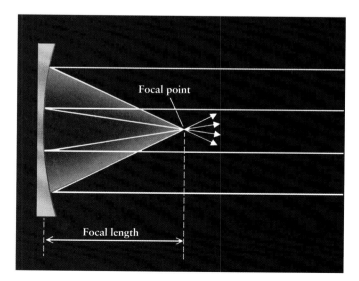

figure 6-10

Reflection by a Concave Mirror A concave mirror causes parallel light rays to converge to a focus at the focal point. The distance between the mirror and the focus is the focal length of the mirror.

a Newtonian focus   b Prime focus   c Cassegrain focus   d Coudé focus

ƒigure 6-11

Designs for Reflecting Telescopes
Four of the most popular optical designs for reflecting telescopes are shown here. **(a)** The Newtonian focus is used today on some small reflecting telescopes for amateur astronomers. **(b)** Prime focus is used only on some large telescopes; an observer or instrument is placed directly at the focal point, within the barrel of the telescope. **(c)** The Cassegrain focus is used on reflecting telescopes of all sizes, from 90-mm (3.5-in.) reflectors used by amateur astronomers to the giant 10-m Keck telescopes on Mauna Kea (see the chapter opening figure). **(d)** The coudé focus is useful when large and heavy optical apparatus is to be used at the focal point. Light reflects off the objective mirror to a secondary mirror, then back down to an angled tertiary mirror. With this arrangement, the heavy apparatus at the focal point does not have to move when the telescope is repositioned.

This design is called a **coudé focus,** from a French word meaning "bent like an elbow" (Figure 6-11d).

You might think that the secondary mirror in the Newtonian, Cassegrain, and coudé designs shown in Figure 6-11 would cause a black spot or hole in the center of the telescope image. But this does not happen. The reason is that light from every part of the object lands on every part of the primary, objective mirror. Hence, any portion of the mirror can itself produce an image of the distant object, as Figure 6-12 shows. The only effect of the secondary mirror is that it prevents part of the light from reaching the objective mirror, which reduces somewhat the light-gathering power of the telescope.

A reflecting telescope must be designed to minimize a defect called **spherical aberration.** At issue is the precise shape of a mirror's concave surface. A spherical surface is easy to grind and polish, but different parts of a spherical mirror have slightly different focal lengths (Figure 6-13a). This results in a fuzzy image.

One common way to eliminate spherical aberration is to polish the mirror's surface to a parabolic shape, because a parabola reflects parallel light rays to a common focus (Figure 6-13b). Unfortunately, the astronomer then no longer has a wide-angle view. Furthermore, unlike spherical mirrors, parabolic mirrors suffer from a defect called **coma,** wherein star images far from the center of the field of view are elongated to look like tiny teardrops. A different approach is to use a spherical mirror, thus minimizing coma, and to place a thin correcting lens at the front of the telescope to eliminate spherical aberration (Figure 6-13c). This approach is only used on relatively small reflecting telescopes for amateur astronomers.

At the time of this writing, there are eleven optical reflectors around the world with primary mirrors more than 8 meters (26.2 feet) in diameter. These are listed in Table 6-1. The largest full-capability optical telescopes are the twin 10-meter reflectors of the W. M. Keck Observatory, located atop the dormant volcano Mauna Kea in Hawaii (see the photograph that opens this chapter). Figure 6-14 shows the objective mirror of the Keck I telescope. Even larger is the 11-meter Hobby-Eberly Telescope, located in the Davis Mountains of western Texas. It cannot be pointed to all portions of the sky, but it cost only one-seventh as much as the $94 million price of either Keck telescope.

A matching pair of 8.1-meter telescopes, called Gemini, have recently gone into operation atop Mauna Kea and Cerro Pachón, Chile. These twins will allow astronomers to observe both the northern and southern parts of the celestial sphere with essentially the same state-of-the-art instrument. Two other "twins" are the side-by-side 8.4-m objective mirrors of the Large Binocular Telescope in Arizona. Combining the light from these two mirrors gives double the light-gathering power, equivalent to a single 11.8-m mirror. In the following section we will learn about the special characteristics of the other telescopes listed in Table 6-1.

figure 6-12

The Secondary Mirror Does Not Cause a Hole in the Image
This illustration shows how even a small portion of the primary (objective) mirror of a reflecting telescope can make a complete image of the Moon. Thus, the secondary mirror does not cause a black spot or hole in the image. (It does, however, make the image a bit dimmer by reducing the total amount of light that reaches the primary mirror.)

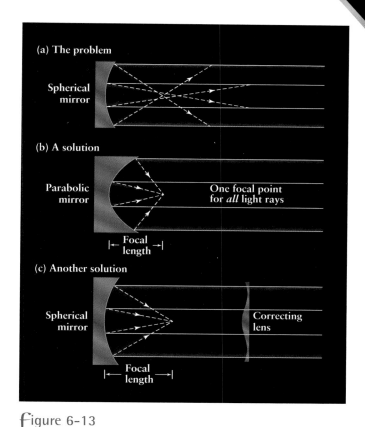

figure 6-13

Spherical Aberration (a) Different parts of a spherically concave mirror reflect light to slightly different focal points. This effect, called spherical aberration, causes image blurring. This difficulty can be corrected by either (b) using a parabolic mirror or (c) using a correcting lens in front of the mirror.

table 6-1	The World's Largest Optical Telescopes		
Telescope	**Location**	**Year completed**	**Mirror diameter (m)**
Hobby-Eberly Telescope	McDonald Observatory, Texas	1998	11.0
Keck II	Mauna Kea, Hawaii	1996	10.0
Keck I	Mauna Kea, Hawaii	1993	10.0
Large Binocular Telescope	Mount Graham, Arizona	2002	Two 8.4
Subaru	Mauna Kea, Hawaii	1999	8.3
VLT UT1–Antu	Cerro Paranal, Chile	1998	8.2
VLT UT 2–Kueyen	Cerro Paranal, Chile	1999	8.2
VLT UT 3–Melipal	Cerro Paranal, Chile	2000	8.2
VLT UT 4–Yepun	Cerro Paranal, Chile	2000	8.2
Gemini North	Mauna Kea, Hawaii	1999	8.1
Gemini South	Cerro Pachón, Chile	2000	8.1

Figure 6-14 R I **V** U X G

The Keck I Telescope on Mauna Kea This photograph is a view down the Keck I telescope, looking toward the 10-meter objective mirror. (Note the astronomers standing on platforms to either side of the telescope.) The objective mirror is actually an arrangement of 36 hexagonal mirrors, each of which is 1.8 m (6 ft) across. Sensors and mechanical actuators on the back of each hexagonal mirror keep the arrangement in perfect alignment. The entire system is much lighter than a single 10-meter mirror would be. The hole in the center of the mirror arrangement is for the Cassegrain focus (see Figure 6-11c); the 1.4-m secondary mirror is housed in the dark hexagonal structure above the center of the photograph. (Roger Ressmeyer-Starlight)

In addition to the "giants" listed in Table 6-1, several other reflectors have objective mirrors between 3 and 6 meters in diameter, and dozens of smaller but still powerful telescopes have mirrors in the range of 1 to 3 meters. There are thousands of professional astronomers, each of whom has several ongoing research projects, and thus the demand for all of these telescopes is high. On any night of the year, nearly every research telescope in the world is being used to explore the universe.

6-3 Telescope images are degraded by the blurring effects of the atmosphere and by light pollution

In addition to providing a brighter image, a large telescope also helps achieve a second major goal: It produces star images that are sharp and crisp. A quantity called **angular resolution** gauges how well fine details can be seen. Poor angular resolution causes star images to be fuzzy and blurred together.

To determine the angular resolution of a telescope, pick out two adjacent stars whose separate images are just barely discernible (Figure 6-15). The angle θ (the Greek letter theta) between these stars is the telescope's angular resolution; the *smaller* that angle, the finer the details that can be seen and the sharper the image.

When you are asked to read the letters on an eye chart, what's being measured is the angular resolution of your eye. If you have 20/20 vision, the angular resolution θ of your eye is about 1 arcminute, or 60 arcseconds. (You may want to review the definitions of these angular measures in Section 1-5.) Hence, with the naked eye it is impossible to distinguish two stars less than 1 arcminute apart or to see details on the Moon with an angular size smaller than this. All the planets have angular sizes (as seen from Earth) of 1 arcminute or less, which is why they appear as featureless points of light to the naked eye.

One factor limiting angular resolution is **diffraction,** which is the tendency of light waves to spread out when they are confined to a small area like the lens or mirror of a telescope. (A rough analogy is the way water exiting a garden hose sprays out in a wider angle when you cover part of the end of the hose with your thumb.) As a result of diffraction, a narrow beam of light tends to spread out within a telescope's optics, thus blurring the image. If diffraction were the only limit, the angular resolution of a telescope would be given by the formula

Angular resolution

$$\theta = 2.5 \times 10^5 \frac{\lambda}{D}$$

θ = diffraction-limited angular resolution of a telescope, in arcseconds

λ = wavelength of light, in meters

D = diameter of telescope objective, in meters

For a given wavelength of light, using a telescope with an objective of *larger* diameter D *reduces* the amount of diffraction and makes the angular resolution θ *smaller* (and hence better). For example, with visible light with wavelength 600 nm, or 6×10^{-7} m, the diffraction-limited resolution of a 20-cm (8-in.) telescope of the sort used by amateur astronomers would be θ = $(2.5 \times 10^5) (6 \times 10^{-7}$ m$)/(0.20$ m$) = 0.75$ arcsec. With the 10-meter Keck telescope in Figure 6-14, the angular resolution would be 0.015 arcsec, or 50 times better.

a

b

ꬵigure 6-15 R I **V** U X G

Angular Resolution The angular resolution of a telescope indicates the sharpness of the telescope's images. **(a)** This telescope view shows two sources of light whose angular separation is greater than the angular resolution. The two sources can easily be distinguished. **(b)** The light sources have been moved together so that their angular separation is equal to the angular resolution. In this case it is just barely possible to tell that there are two sources. If the sources were moved any closer together, the telescope image would show them as a single source.

In practice, however, ordinary optical telescopes cannot achieve such exceptionally fine angular resolution. The problem is that turbulence in the air causes star images to jiggle around and twinkle. Even through the largest telescopes, a star still looks like a tiny blob rather than a pinpoint of light. A measure of the limit that atmospheric turbulence places on a telescope's resolution is called the **seeing disk**. This disk is the angular diameter of a star's image broadened by turbulence. The size of the seeing disk varies from one observatory site to another and from one night to another. At the observatories on Kitt Peak in Arizona and Cerro Tololo in Chile, the seeing disk is typically around 1 arcsec. Some of the very best conditions in the world can be found at the observatories atop Mauna Kea in Hawaii, where the seeing disk is often as small as 0.5 arcsec. This is one reason why so many telescopes have been built there (see the photograph that opens this chapter).

In many cases the angular resolution of a telescope will be even worse than the limit imposed by the seeing disk. This will be the case if the objective mirror deforms even slightly due to variations in air temperature or flexing of the telescope mount. To combat this, most of the large telescopes listed in Table 6-1 are equipped with an **active optics** system. Such a system adjusts the mirror shape every few seconds to help keep the telescope in optimum focus and properly aimed at its target.

Changing the mirror shape is also at the heart of a more refined technique called **adaptive optics.** The goal of this technique is to compensate for atmospheric turbulence, so that the angular resolution can be smaller than the size of the seeing disk and can even approach the theoretical limit set by diffraction. Turbulence causes the image of a star to "dance" around erratically. Optical sensors monitor this dancing motion 10 to 100 times per second, and a powerful computer rapidly calculates the mirror shape needed to compensate. Fast-acting mechanical devices called *actuators* then deform the mirror accordingly, at a much faster rate than in an active optics system. (In some adaptive-optics systems, the actuators deform a small secondary mirror rather than the large objective mirror.)

Figure 6-16 shows the dramatic improvement in angular resolution possible with adaptive optics. Images made with adaptive optics are nearly as sharp as if the telescope were in the vacuum of space, where there is no atmospheric distortion whatsoever and the only limit on angular resolution is

a

b

ꬵigure 6-16 R ■ V U X G

Using Adaptive Optics to "Unblur" Telescope Images Both these images of the star κ (Kappa) Pegasi (in the constellation Pegasus, the Winged Horse) were made with the same 3.5-m telescope. **(a)** Without adaptive optics, the star's image is blurred. Here, the star's image is blurred across a seeing disk of more than 1 arcsecond by turbulence in the atmosphere. **(b)** This image was made with an adaptive optics system turned on. The angular resolution is much finer, revealing that κ Pegasi is actually a binary star (two stars orbiting each other). The angular separation between the stars is only 0.3 arcsec. (R. Q. Fugate, USAF Research Laboratory, Starfire Optical Range)

WEB LINK 6.5

figure 6-17   R I **V** U X G

The Very Large Telescope This photograph shows the
four domes of the European Southern Observatory's Very
Large Telescope (VLT). The VLT is atop a 2640-m (8660-ft) mountain in
the Atacama desert of Chile, where the skies are very dark and
cloudless and the seeing is excellent. The four 8.2-m telescopes bear
the names of celestial objects in the local Mapuche language: Antu
(Sun), Kueyen (Moon), Melipal (Southern Cross), and Yepun (Venus, the
evening star). (European Southern Observatory)

diffraction. Several of the telescopes listed in Table 6-1, including the Japanese telescope Subaru (the Japanese term for the
Pleiades star cluster), are designed to make use of adaptive
optics. At present, however, adaptive optics can be used only
with rather bright objects.

Perhaps the finest angular resolution will be obtained with
the Very Large Telescope (VLT), a project of the multinational
European Southern Observatory. The VLT is actually four
reflecting telescopes located in adjacent buildings at Cerro
Paranal, Chile (Figure 6-17). Each telescope has a relatively
flexible 8.2-meter objective mirror equipped with 214 active-
optics actuators for controlling the mirror's shape. The four
reflectors can be used individually to look at different objects,
but by 2006 it will be possible to observe the same object
simultaneously with all four telescopes and to combine the light
received by all four. This will give the same light-gathering
power as a single 16.4-m (53.8-ft) mirror. But because the
reflectors are spread out over a distance of 200 m (670 ft, or
more than two football fields), the angular resolution will be
that of a 200-m telescope. In this way it will be possible to
achieve the amazing resolution of 0.001 arcsec, which corresponds to being able to distinguish the two headlights on a car
located on the Moon!

Light from city street lamps and from buildings
also degrades telescope images. This **light pollution**
illuminates the sky, making it more difficult to see

the stars. You can appreciate the problem if you have ever
looked at the night sky from a major city. Only a few of the
very brightest stars can be seen, as against the thousands that
can be seen with the naked eye in the desert or the mountains.
To avoid light pollution, observatories are built in remote
locations far from any city lights.

Unfortunately, the expansion of cities has brought light
pollution to observatories that in former times had none. As
an example, the growth of Tucson, Arizona, has had deleteri-
ous effects on observations at the nearby Kitt Peak National
Observatory. Efforts have been made to have cities adopt light
fixtures that provide safe illumination for their citizens but
produce little light pollution. These efforts have met with only
mixed success.

One factor over which astronomers have absolutely no
control is the weather. Optical telescopes cannot see through
clouds, so it is important to build observatories where the
weather is usually good. One advantage of mountaintop observatories such as Mauna Kea or Cerro Paranal is that most
clouds form at altitudes below the observatory, giving astronomers a better chance of having clear skies.

In many ways the best location for a telescope is in orbit
around Earth, where it is unaffected by weather, light pollution, or atmospheric turbulence. We will discuss orbiting telescopes in Section 6-7.

6-4 An electronic device is commonly used to record the image at a telescope's focus

Telescopes provide astronomers with detailed pictures of distant
objects. The task of recording these pictures is called **imaging.**

Astronomical imaging really began in the nineteenth century with the invention of photography. It was soon realized
that this new invention was a boon to astronomy. By taking
long exposures with a camera mounted at the focus of a telescope, an astronomer can record features too faint to be seen
by simply looking through the telescope. Such long exposures
can reveal details in galaxies, star clusters, and nebulae that
would not be visible to an astronomer looking through a telescope. Indeed, most large, modern telescopes do not have eyepieces at all.

Unfortunately, photographic film is not a very efficient
light detector. Only about 1 out of every 50 photons striking
photographic film triggers the chemical reaction needed to
produce an image. Thus, roughly 98% of the light falling onto
photographic film is wasted.

The most sensitive light detector currently available to
astronomers is the **charge-coupled device (CCD).** At the heart of
a CCD is a semiconductor wafer divided into an array of small
light-sensitive squares called picture elements or, more commonly, **pixels** (Figure 6-18). For example, some state-of-the-art
CCDs for astronomy have more than 16 million pixels arranged
in 4096 rows by 4096 columns. They have about a thousand
times more pixels per square centimeter than on a typical computer screen, which means that a CCD of this type can record

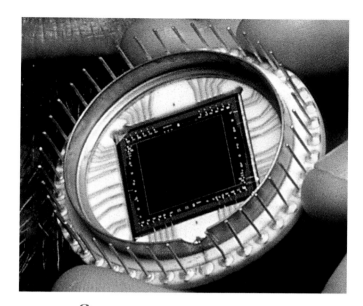

very fine image details. CCDs with smaller numbers of pixels are used in digital cameras, scanners, and fax machines.

When an image from a telescope is focused on the CCD, an electric charge builds up in each pixel in proportion to the number of photons falling on that pixel. When the exposure is finished, the amount of charge on each pixel is read into a computer, where the resulting image can be stored in digital form and either viewed on a monitor or printed out. Compared with photographic film, CCDs are some 35 times more sensitive to light (they commonly respond to 70% of the light falling on them, versus 2% for film), can record much finer details, and respond more uniformly to light of different colors. Figure 6-19 shows the dramatic difference between photographic and CCD images. The great sensitivity of CCDs also makes them useful for **photometry,** which is the measurement of the brightnesses of stars and other astronomical objects.

In the modern world of CCD astronomy, astronomers need no longer spend the night in the unheated dome of a telescope. Instead, they operate the telescope electronically from a separate control room, where the electronic CCD images can be viewed on a computer monitor. The control room need not even be adjacent to the telescope. Although the Keck telescopes (see Figure 6-14) are at an altitude of 4200 m (13,800 feet), astronomers can now make observations from a facility elsewhere on the island of Hawaii that is much closer to sea level. This saves the laborious drive to the summit of Mauna Kea and eliminates the need for astronomers to acclimate to the high altitude.

figure 6-18   R I **V** U X G

A Charge-Coupled Device (CCD) This tiny silicon square records images in its 16,777,216 light-sensitive pixels. At the end of each exposure, additional circuits control the transfer and readout of the data to a waiting computer. Somewhat less elaborate CCDs are used by amateur astronomers and in consumer electronics. (Institute for Astronomy, University of Hawaii, and Roger Ressmeyer ©1993 Corbis)

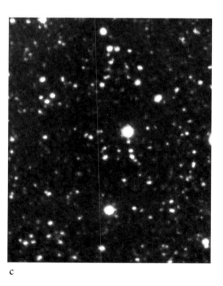

a   b   c

figure 6-19   R I **V** U X G

Imaging: Photographic Film Versus CCD **(a)** This is a negative print (black stars and white sky) of a portion of the sky as imaged with a 4-meter telescope and photographic film. **(b)** This negative image of the same area as in (a) was made using the same telescope, but with the photographic film replaced by a CCD. You can see dozens of stars and galaxies that did not appear at all in the photograph. **(c)** To produce this color positive view of the same region, a series of CCD images were made using different color filters. These were then combined using a computer image-processing program. (Patrick Seitzer, National Optical Astronomy Observatories)

Most of the images that you will see in this book were made with CCDs. Because of their extraordinary sensitivity and their ability to be used in conjunction with computers, CCDs have attained a role of central importance in astronomy.

6-5 Spectrographs record the spectra of astronomical objects

We saw in Section 5-6 how the spectrum of an astronomical object provides a tremendous amount of information about that object. This is why measuring spectra, or **spectroscopy,** is one of the most important uses of telescopes. (The giant 11-meter Hobby-Eberly Telescope described in Section 6-2 is designed solely for measuring the spectra of distant, faint objects; it will never be used for imaging.) An essential tool of spectroscopy is the **spectrograph,** a device that records spectra on a CCD or photographic film. This optical device is mounted at the focus of a telescope.

An older design for a spectrograph, shown in Figure 6-20, uses a prism to form the spectrum of a planet, star, or galaxy. A slit, two lenses, and the prism are arranged to focus the spectrum onto a small piece of photographic film. Next, the exposed portion of the photographic plate is covered, and light from a hot gas (usually an element like iron or argon) is focused on the spectrograph slit. This result is a *comparison spectrum* above and below the spectrum of the star or galaxy (Figure 6-21). The wavelengths of the bright spectral lines of the comparison spectrum are already known from laboratory experiments and can therefore serve as reference markers.

There are drawbacks to this old-fashioned spectrograph. A prism does not disperse the colors of the rainbow evenly: Blue and violet portions of the spectrum are spread out more than the red portion. In addition, because the blue and violet wavelengths must pass through more glass than the red wavelengths (examine Figure 5-3), light is absorbed unevenly across the spectrum. Indeed, a glass prism is opaque to near-ultraviolet light.

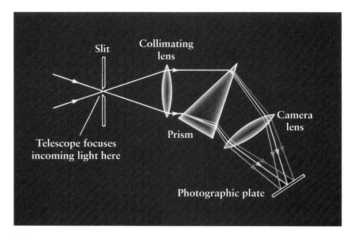

figure 6-20

A Prism Spectrograph This optical device uses a prism to break up the light from a source into a spectrum. The collimating lens directs incoming light rays so that they enter the prism parallel to one another. A second lens, called a camera lens, then focuses the spectrum onto a photographic plate.

A better device for breaking starlight into the colors of the rainbow is a **diffraction grating,** or **grating** for short. This is a piece of glass on which thousands of closely spaced parallel lines have been cut. Some of the finest diffraction gratings have more than 10,000 lines per centimeter, which are usually cut by drawing a diamond back and forth across the glass. The spacing of the lines must be very regular, and the best results are obtained when the grooves are beveled. When light is shone on a diffraction grating, a spectrum is produced by the way in which light waves leaving different parts of the grating interfere with each other. (This same effect produces the rainbow of colors you see reflected from a compact disc or CD-ROM. Information is stored on the disc in a series of closely spaced pits, which act as a diffraction grating.) Figure 6-22 shows the design of a modern grating spectrograph.

figure 6-21   R I **V** U X G

A Spectrogram The photographic record of a spectrum is called a spectrogram. The spectrogram running across the middle of this figure shows the spectrum of a star. Note the many dark absorption lines, which are characteristic of the light from stars (see Section 5-6). Above and below the star's spectrum are emission lines produced by an iron arc at the observatory. These emission lines serve as a comparison spectrum. (Palomar Observatory)

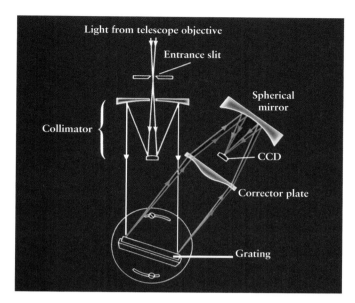

figure 6-22

A Grating Spectrograph This optical device uses a diffraction grating to break up the light from a source into a spectrum. The collimator ensures that light rays striking the grating are parallel. A corrector lens and mirror then focus the spectrum onto a CCD, which is far more sensitive than a photographic plate.

Labels in figure: Light from telescope objective; Entrance slit; Collimator; Spherical mirror; CCD; Corrector plate; Grating

figure 6-23

Two Representations of a Spectrum This drawing compares two ways of displaying a spectrum. When a photographic plate is placed at the focus of a spectrograph, a rainbow-colored spectrum is obtained. The dark absorption lines in this example are the Balmer lines of hydrogen (see Section 5-8). When a CCD is placed at the focus of a spectrograph, a computer program can be used to convert the CCD's data into a graph of intensity versus wavelength. Note that absorption lines appear as dips on such a graph.

In Figure 6-22 a CCD, instead of a photographic plate, is placed at the focus of the spectrograph. When the exposure is finished, electronic equipment measures the charge that has accumulated in each pixel. These data are used to graph light intensity against wavelength. Dark absorption lines in the spectrum appear as depressions or valleys on the graph, while bright emission lines appear as peaks. Figure 6-23 compares two ways of exhibiting a spectrum in which several absorption lines appear. Later in this book we shall see spectra presented in both these ways.

6-6 A radio telescope uses a large concave dish to reflect radio waves to a focus

For thousands of years, all the information that astronomers gathered about the universe was based on ordinary visible light. In the twentieth century, however, astronomers first began to explore the nonvisible electromagnetic radiation coming from astronomical objects. In this way they have discovered aspects of the cosmos that are forever hidden to optical telescopes.

Astronomers have used ultraviolet light to map the outer regions of the Sun and the clouds of Venus, and used infrared radiation to see new stars and perhaps new planetary systems in the process of formation. By detecting radio waves from Jupiter and Saturn, they have mapped the intense magnetic fields that surround those giant planets; by detecting curious bursts of X rays from space, they have learned about the utterly alien conditions in the vicinity of a black hole. It is no exaggeration to say that today's astronomers learn as much about the universe using telescopes for nonvisible wavelengths as they do using visible light.

Radio waves were the first part of the electromagnetic spectrum beyond the visible to be exploited for astronomy. This happened as a result of a research project seemingly unrelated to astronomy. In the early 1930s, Karl Jansky, a young electrical engineer at Bell Telephone Laboratories, was trying to locate what was causing interference with the then-new transatlantic radio link. By 1932 he realized that one kind of radio noise is strongest when the constellation Sagittarius is high in the sky. The center of our Galaxy is located in the direction of Sagittarius, and Jansky concluded that he was detecting radio waves from an astronomical source.

At first only Grote Reber, a radio engineer living in Illinois, took up Jansky's research. In 1936 Reber built in his backyard the first **radio telescope,** a radio-wave detector dedicated to astronomy. He modeled his design after an ordinary reflecting telescope, with a parabolic metal "dish" (reflecting antenna) measuring 31 ft (10 m) in diameter and a radio receiver at the focal point of the dish.

Reber spent the years from 1938 to 1944 mapping radio emissions from the sky at wavelengths of 1.9 m and 0.63 m. He found radio waves coming from the entire Milky Way, with the greatest emission from the center of the Galaxy. These results, together with the development of improved radio technology during World War II, encouraged the growth

Figure 6-24 R I V U X G

A Radio Telescope The dish of this radio telescope, silhouetted by the setting Sun, is 64 m (210 ft) in diameter. It is located at the Parkes Observatory in New South Wales, Australia. (Roger Ressmeyer-Starlight)

of radio astronomy and the construction of new radio telescopes around the world. Radio observatories are as common today as major optical observatories.

Like Reber's prototype, a typical modern radio telescope has a large parabolic dish (Figure 6-24). A small antenna tuned to the desired frequency is located at the focus, and the incoming signal is relayed to amplifiers and recording instruments, typically located in a room at the base of the telescope's pier.

 The radio telescope in Figure 6-24 looks like a radar dish but is used in a different way. In radar, the dish is used to send out a narrow beam of radio waves. If this beam encounters an object like an airplane, some of the radio waves will be reflected back to the radar dish and detected by a receiver at the focus of the dish. Thus, a radar dish looks for radio waves *reflected* by distant objects. By contrast, a radio telescope is designed to receive radio waves *emitted* by objects in space.

Many radio telescope dishes, like the one in Figure 6-24, have visible gaps in them like a wire mesh. This does not affect their reflecting power because the holes are much smaller than the wavelengths of the radio waves. The same idea is used in the design of microwave ovens. The glass window in the oven door would allow the microwaves to leak out, so the window is covered by a metal screen with small holes. These holes are much smaller than the 12.2-cm (4.8-in.) wavelength of the microwaves, so the screen reflects the microwaves back into the oven.

One great drawback of early radio telescopes was their very poor angular resolution. Recall from Section 6-3 that angular resolution is the smallest angular separation between two stars that can just barely be distinguished as separate objects. For radio telescopes, the limitation on angular resolution is not atmospheric turbulence, as it is for optical telescopes. Rather, the problem is that angular resolution is

directly proportional to the wavelength being observed (see the formula for the angular resolution θ in Section 6-3). The longer the wavelength, the larger (and hence worse) the angular resolution and the fuzzier the picture. As an example, a 1-m radio telescope detecting radio waves of 5-cm wavelength has 100,000 times poorer angular resolution than a 1-m optical telescope. Because radio radiation has very long wavelengths, astronomers thought that radio telescopes could produce only blurry, indistinct images.

 A very large radio telescope can produce a somewhat sharper radio image, because as the diameter of the telescope increases, the angular resolution decreases. In other words, the bigger the dish, the better the resolution. For this reason, most modern radio telescopes have dishes more than 30 m (100 ft) in diameter. This is also useful for increasing light-gathering power, because radio signals from astronomical objects are typically very weak in comparison with the intensity of visible light. But even the largest single radio dish in existence, the 305-m (1000-ft) Arecibo radio telescope in Puerto Rico, cannot come close to the resolution of the best optical instruments.

A very clever technique makes it possible to produce radio images with excellent resolution. Unlike ordinary light, radio signals can be carried over electrical wires. Consequently, two radio telescopes observing the same astronomical object can be hooked together, even if they are separated by many kilometers. This technique is called **interferometry**, because the incoming radio signals are made to "interfere," or blend together. This makes the combined signal sharp and clear. The effective angular resolution of two such radio telescopes is equivalent to that of one gigantic dish with a diameter equal to the **baseline**, or distance between the two telescopes.

One of the largest arrangements of radio telescopes for interferometry is the Very Large Array (VLA), located in the desert near Socorro, New Mexico (Figure 6-25). The VLA consists of 27 parabolic dishes, each 25 m (82 ft) in diameter. These 27 telescopes are arranged along the arms of a gigantic **Y** that covers an area 27 km (17 mi) in diameter. By pointing all 27 telescopes at the same object and combining the 27 radio signals, this system can produce radio views of the sky with an angular resolution comparable to that of the very best optical telescopes. (The Very Large Telescope, shown in Figure 6-17, will combine the images from its four optical reflectors to do interferometry with *visible* light. In this way the VLT may achieve even better angular resolution with visible light than the VLA does using radio waves.)

 Dramatically better angular resolution can be obtained by combining the signals from radio telescopes at different observatories thousands of kilometers apart. This technique is called **very-long-baseline interferometry (VLBI)**. VLBI is used by a system called the Very Long Baseline Array (VLBA), which consists of ten 25-meter dishes at different locations between Hawaii and the Caribbean. Although the ten dishes are not physically connected, they are all used to observe the same object at the same time. The data from each telescope are recorded electronically and processed later. By carefully synchronizing the ten recorded

WEB LINK 6.9

Figure 6-25 R I **V** U X G

The Very Large Array (VLA) The 27 radio telescopes of the VLA are arranged along the arms of a **Y** in central New Mexico. (Only 13 of the 27 telescopes can be seen in this photograph.) The north arm of the array is 19 km long; the southwest and southeast arms are each 21 km long. By spreading the telescopes out along the legs and combining the signals received, the VLA can give the same angular resolution as a single dish many kilometers in radius. (National Radio Astronomy Observatory)

signals, they can be combined just as if the telescopes had been linked together during the observation. With VLBA, features smaller than 0.001 arcsec can be distinguished at radio wavelengths. This angular resolution is 100 times better than a large optical telescope with adaptive optics.

Even better angular resolutions can be obtained by adding radio telescopes in space; the baseline is then the distance from the VLBA to the orbiting telescope. The first such space radio telescope, the Japanese HALCA spacecraft, went into operation in 1997. Orbiting at a maximum distance of 21,400 km (13,300 mi) above the Earth's surface, HALCA gives a baseline 3 times longer—and thus an angular resolution 3 times better—than can be obtained with Earthbound telescopes alone.

Figure 6-26 shows how optical and radio images of the same object can give different and complementary kinds of information. The visible-light image of Saturn (Figure 6-26a) shows clouds in the planet's atmosphere and the structure of the rings. Like the visible light from the Moon, the light used to make this image is just reflected sunlight. By contrast, the radio image of Saturn (Figure 6-26b) is a record of waves emitted by electrically charged particles outside the planet. These particles emit because of their motion in Saturn's magnetic field, so this radio image provides clues about the conditions in the interior of Saturn where the magnetic field is produced. This information could never be obtained from a visible-light image such as Figure 6-26a.

Figure 6-26b is also an example of how radio astronomers often use "false color" to display their views of astronomical objects. In this figure the most intense radio emission is shown in red, the least intense in blue. Intermediate colors of the rainbow represent intermediate levels of radio intensity. Black indicates no detectable radio emission. Astronomers working at other nonvisible wavelength ranges also frequently use false-color techniques to display images obtained from their instruments. You will find such false-color images throughout this book.

a   R I **V** U X G

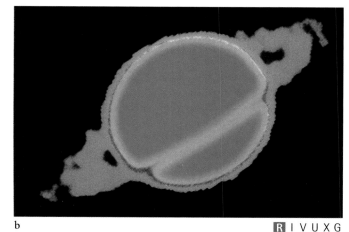

b   **R** I V U X G

Figure 6-26

Optical and Radio Views of Saturn (a) This picture was taken by a spacecraft 18 million kilometers from Saturn. The view was produced by sunlight reflecting from the planet's cloudtops and rings. (b) This VLA image shows radio emission from Saturn at a wavelength of 2 cm. This emission is caused by electrically charged particles moving within Saturn's strong magnetic field and acting as tiny radio transmitters. In this image, color represents the intensity of radio emission, ranging from blue for the weakest to red for the strongest. Note the blue radio "shadow" caused by Saturn's rings where they lie in front of the planet. (a: NASA; b: National Radio Astronomy Observatory)

6-7 Telescopes in orbit around the Earth detect radiation that does not penetrate the atmosphere

The many successes of radio astronomy show the value of observations at nonvisible wavelengths. But the Earth's atmosphere is opaque to many wavelengths. Other than visible light and radio waves, very little radiation from space manages to penetrate the air we breathe. To overcome this, astronomers have placed a variety of telescopes in orbit.

Figure 6-27 shows the transparency of the Earth's atmosphere to different wavelengths of electromagnetic radiation. The atmosphere is most transparent in two wavelength regions, the **optical window** (which includes the entire visible spectrum) and the **radio window** (which includes part, but not all, of the radio spectrum). There are also several relatively transparent regions at infrared wavelengths between 1 and 40 μm. Infrared radiation within these wavelength intervals can penetrate the Earth's atmosphere somewhat and can be detected with ground-based telescopes. This wavelength range is called the *near-infrared*, because it lies just beyond the red end of the visible spectrum.

Water vapor is the main absorber of infrared radiation from space, which is why infrared observatories are located at sites with exceptionally low humidity. The site must also be at high altitude to get above as much of the atmosphere's water vapor as possible. One site that meets both criteria is the summit of Mauna Kea in Hawaii, shown in the photograph that opens this chapter. (The complete lack of vegetation on the summit attests to its extreme dryness.) Some of the telescopes on Mauna Kea are designed exclusively for detecting infrared radiation. Others, such as the two

Keck telescopes, are used for both visible and near-infrared observations.

 Even at the elevation of Mauna Kea, water vapor in the atmosphere restricts the kinds of infrared observations that astronomers can make. This situation can be improved by carrying telescopes on board high-altitude balloons or aircraft. But the ultimate solution is to place a telescope in Earth orbit and radio its data back to astronomers on the ground. The first such orbiting infrared observatory, the Infrared Astronomical Satellite (IRAS), was launched in 1983. During its nine-month mission, IRAS used its 57-cm (22-in.) telescope to map almost the entire sky at wavelengths from 12 to 100 μm.

The IRAS data revealed the presence of dust disks around nearby stars. Planets are thought to coalesce from disks of this kind, so this was the first concrete (if indirect) evidence that there might be planets orbiting other stars. The dust that IRAS detected is warm enough to emit infrared radiation but too cold to emit much visible light, so it remained undetected by ordinary optical telescopes. IRAS also discovered distant, ultraluminous galaxies that emit almost all their radiation at infrared wavelengths.

In 1995 the Infrared Space Observatory (ISO), a more advanced 60-cm reflector with better light detectors, was launched into orbit by the European Space Agency (Figure 6-28). During its 2½-year mission, ISO made a number of groundbreaking observations of very distant galaxies and of the thin, cold material between the stars of our own Galaxy. Like IRAS, ISO had to be cooled by liquid helium to temperatures just a few degrees above absolute zero. Had this not been done, the infrared blackbody radiation from the telescope itself would have outshone the infrared radiation from astronomical objects. The ISO mission came to an end when the last of the helium evaporated into space.

figure 6-27

The Transparency of the Earth's Atmosphere This graph shows the percentage of radiation that can penetrate the Earth's atmosphere at different wavelengths. At wavelengths where the curve of transparency is high, such as in the optical and radio windows, the atmosphere is relatively transparent; at wavelengths where the curve is low, the atmosphere is opaque. Absorption by oxygen and nitrogen makes the atmosphere completely opaque at wavelengths less than about 290 nm. The opaque region between the optical and radio windows is due to absorption by water vapor and carbon dioxide. At wavelengths longer than about 20 m, radiation is reflected back into space by ionized gases in the upper atmosphere.

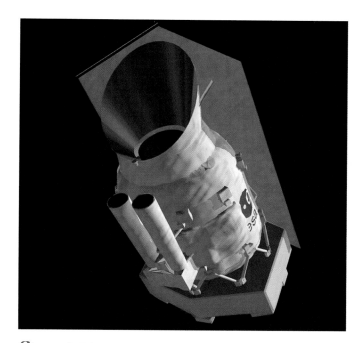

Figure 6-28

The Infrared Space Observatory (ISO) Between 1995 and 1998, this spacecraft observed the sky at wavelengths between 2.5 and 240 μm. It provided astronomers with more than 26,000 infrared images and other data about planets, comets, gas and dust around stars and in interstellar space, galaxies, and the large-scale distribution of matter in the universe. (European Space Agency)

The next great orbiting infrared observatory is SIRTF (Space Infrared Telescope Facility), an 85-cm infrared telescope that will survey the infrared sky with unprecedented resolution. SIRTF will be used to study brown dwarfs (exotic objects larger than planets but smaller than stars), search for more direct evidence of planets forming around other stars, and probe the interiors of galaxies billions of light-years away. As of this writing NASA plans to launch SIRTF in 2002 on a mission lasting from 2½ to 5 years.

Astronomers are also very interested in observing at ultraviolet wavelengths. These observations can reveal a great deal about hot stars, ionized clouds of gas between the stars, and the Sun's high-temperature corona (see Section 3-5), all of which emit copious amounts of ultraviolet light. The spectrum of ultraviolet sunlight reflected from a planet can also reveal the composition of the planet's atmosphere. However, Earth's atmosphere is opaque to ultraviolet light except for the narrow *near-ultraviolet* range, which extends from about 400 nm (the violet end of the visible spectrum) down to 300 nm.

To see shorter-wavelength *far-ultraviolet* light, astronomers must again make their observations from space. The first ultraviolet telescope was placed in orbit in 1962, and several others have since followed it into space. Small rockets have also lifted ultraviolet cameras briefly above the Earth's atmosphere. Figure 6-29 shows an ultraviolet view of the constellation Orion, along with infrared and visible views.

The Far Ultraviolet Spectroscopic Explorer (FUSE), which went into orbit in 1999, specializes in measuring spectra at wavelengths from 90 to 120 nm.

a   R I V [U] X G

b   R [I] V U X G

c   R I [V] U X G

d

Figure 6-29

Orion Seen at Ultraviolet, Infrared, and Visible Wavelengths **(a)** An ultraviolet view of the constellation of Orion was obtained during a brief rocket flight in 1975. This 100-s exposure covers the wavelength range 125–200 nm. **(b)** The false-color view from the Infrared Astronomical Satellite displays emission at different wavelengths in different colors: red for 100-μm radiation, green for 60-μm radiation, and blue for 12-μm radiation. Compare these images with **(c)** an ordinary optical photograph and **(d)** a star chart of Orion. (a: G. R. Carruthers, Naval Research Laboratory; b: NASA; c: R. C. Mitchell, Central Washington University)

Figure 6-30 R I V U X G

The Hubble Space Telescope (HST) This photograph of the HST in orbit was taken from the space shuttle *Discovery* during a servicing mission in 1999. Light enters the telescope through the open aperture cover on the top. The large, dark-colored solar panels on either side of the telescope provide electric power. The telescope is 13.1 m (43.0 ft) long and has a mass of 11,600 kg (13 tons). (NASA/Space Telescope Science Institute)

Highly ionized oxygen atoms, which can exist only in an extremely high-temperature gas, have a characteristic spectral line in this range. By looking for this spectral line in various parts of the sky, FUSE has confirmed that our Milky Way Galaxy (Section 1-4) is surrounded by an immense "halo" of gas at temperatures in excess of 200,000 K. Only an ultraviolet telescope could have detected this "halo," which is thought to have been produced by exploding stars called supernovae (Section 1-3).

Infrared and ultraviolet satellites give excellent views of the heavens at selected wavelengths. But since the 1940s astronomers had dreamed of having one large telescope that could be operated at any wavelength from the near-infrared through the visible range and out into the ultraviolet. This is the mission of the Hubble Space Telescope (HST), which was placed in a 600-km-high orbit by the space shuttle *Discovery* in 1990 (Figure 6-30). HST has a 2.4-meter (7.9-ft) objective mirror and

was designed to observe at wavelengths from 115 nm to 1 μm. Like most ground-based telescopes, HST uses a CCD to record images. (In fact, the development of HST helped drive advances in CCD technology.) The images are then radioed back to Earth in digital form.

The great promise of HST was that from its vantage point high above the atmosphere, its angular resolution would be limited only by diffraction. But soon after HST was placed in orbit, astronomers discovered that a manufacturing error had caused the telescope's 2.4-m primary mirror to suffer from spherical aberration. The mirror should have been able to concentrate 70% of a star's light into an image with an angular diameter of 0.1 arcsec. Instead, only 20% of the light was focused into this small area. The remainder was smeared out over an area about 1 arcsec wide, giving images little better than those achieved at major ground-based observatories.

On an interim basis, astronomers used only the 20% of incoming starlight that was properly focused and, with computer processing, discarded the remaining poorly focused 80%. This was practical only for brighter objects on which astronomers could afford to waste light. But many of the observing projects scheduled for HST involved extremely dim galaxies and nebulae.

These problems were resolved by a second space shuttle mission in 1993. Astronauts installed a set of small secondary mirrors whose curvature exactly compensated for the error in curvature of the primary mirror. Once these were in place, HST was able to make truly sharp images of extremely faint objects (Figure 6-31). Astronomers have used the repaired HST to make discoveries about the nature of planets, the evolution of stars, the inner workings of galaxies, and the expansion of the universe. You will see many HST images in later chapters.

The success of HST has inspired plans for its sequel, the Next Generation Space Telescope (NGST). This 8-m reflector, the largest ever placed in orbit, will observe at visible and infrared wavelengths with more than 10 times the light-gathering power of HST. Planned for a 2007 launch, NGST will study faint objects such as planetary systems forming around other stars and galaxies near the limit of the observable universe.

Space telescopes have also made it possible to explore objects whose temperatures reach the almost inconceivable values of 10^6 to 10^8 K. Atoms in such a high-temperature gas move so fast that when they collide, they emit X-ray photons of very high energy and very short wavelengths less than 10 nm. X-ray telescopes designed to detect these photons must be placed in orbit, since Earth's atmosphere is totally opaque at these wavelengths.

CAUTION! X-ray telescopes work on a very different principle from the X-ray devices used in medicine and dentistry. If you have your foot "X-rayed" to check for a broken bone, a piece of photographic film sensitive to X rays is placed under your foot and an X-ray beam is directed at your foot from above. The radiation penetrates through soft

a   b

ƒigure 6-31   R I ▣ U X G

HST Images: Before and After Repair These Hubble Space Telescope images show the central region of the galaxy M100 in the constellation Coma Berenices (Berenice's Hair), some 70 million light-years from Earth. Each image shows an area about 25 arcseconds wide. **(a)** Spherical aberration in the original, uncorrected HST gave this blurred image. **(b)** This image, made after a 1993 space shuttle mission to install corrective optics, shows details that are less than a tenth of an arcsecond across. (Space Telescope Science Institute)

tissue but not through bone, so the bones cast an "X-ray shadow" on the film. A fracture will show as a break in the shadow. X-ray telescopes, by contrast, do *not* send beams of X rays toward astronomical objects in an attempt to see inside them. Rather, these telescopes detect X rays that the objects emit on their own.

Astronomers got their first quick look at the X-ray sky from brief rocket flights during the late 1940s. These observations confirmed that the Sun's corona (see Section 3-5) is a source of X rays, and must therefore be at a temperature of millions of kelvins. In 1962 a rocket experiment revealed that objects beyond the solar system also emit X rays.

Since 1970, a series of increasingly sensitive and sophisticated X-ray observatories have been placed in orbit, including NASA's Einstein Observatory, the European Space Agency's Exosat, and the German-British-American ROSAT. These telescopes have shown that other stars also have high-temperature coronae, and have found hot, X-ray-emitting gas clouds so immense that hundreds of galaxies fit inside them. They also discovered unusual stars that emit X rays in erratic bursts. These bursts are now thought to be coming from heated gas swirling around a small but massive object—possibly a black hole.

X-ray astronomy took a quantum leap forward in 1999 with the launch of NASA's Chandra X-ray Observatory and the European Space Agency's XMM-Newton. Named for the Indian-American astrophysicist Subrahmanyan Chandrasekhar, Chandra can view the X-ray sky with an angular resolution of 0.5 arcsec (Figure 6-32*a*). This is comparable to the best ground-based optical telescopes, and more than a thousand times better than the resolution of the first orbiting X-ray telescope. Chandra can also measure X-ray spectra 100 times more precisely than any previous spacecraft, and can detect variations in X-ray emissions on time scales as short as 16 microseconds. This latter capability is essential for understanding how X-ray bursts are produced around black holes.

XMM-Newton (for *X*-ray *M*ulti-*m*irror *M*ission) is actually three X-ray telescopes that all point in the same direction (Figure 6-32*b*). Their combined light-gathering power is 5 times greater than that of Chandra, which makes XMM-Newton able to observe fainter objects. (For reasons of economy, the mirrors were not ground as precisely as those on Chandra, so the angular resolution of XMM-Newton is only about 6 arcseconds.) It also carries a small but highly capable telescope for ultraviolet and visible observations. Hot X-ray sources are usually accompanied by

a

b

figure 6-32

Two Orbiting X-Ray Observatories **(a)** X rays are absorbed by ordinary mirrors like those used in optical reflectors, but they can be reflected if they graze the mirror surface at a very shallow angle. In the Chandra X-ray Observatory, X rays (shown in red) enter the telescope from the left and graze the inner, polished surface of a gently curved cylinder (shown in blue). The reflected rays come together in the focal plane (shown in yellow) 10 m (33 ft) behind the mirror. **(b)** XMM-Newton is about the same size as Chandra and forms images in the same way. This artist's rendering shows the large apertures for the three X-ray telescopes. (a: NASA/Chandra X-ray Observatory Center/Smithsonian Astrophysical Observatory; b: D. Ducros/European Space Agency)

cooler material that radiates at these longer wavelengths, so XMM-Newton will be able to observe these hot and cool regions simultaneously.

Gamma rays, the shortest-wavelength photons of all, help us to understand phenomena even more energetic than those that produce X rays. As an example, when a massive star explodes into a supernova, it produces radioactive atomic nuclei that are strewn across interstellar space. Observing the gamma rays emitted by these nuclei helps astronomers understand the nature of supernova explosions.

Like X rays, gamma rays do not penetrate Earth's atmosphere, so space telescopes are required. The most powerful gamma-ray telescope placed in orbit to date is the Compton Gamma Ray Observatory (CGRO), shown in Figure 6-33. One particularly important task for CGRO was the study of gamma-ray bursts, which are brief, unpredictable, and very intense flashes of gamma rays that occur unpredictably from all parts of the sky. Data from CGRO and other orbiting observatories have shown that the sources of these gamma-ray bursts are billions of light-years away. For these bursts to be visible across such great distances, their sources must be among the most energetic objects in the universe.

The advantages and benefits of Earth-orbiting observatories cannot be overemphasized. We are no longer limited to the narrow ranges of whatever wavelengths manage to leak through our shimmering, hazy atmosphere (Figure 6-34). For the first time, we are really *seeing* the universe.

figure 6-33   R I **V** U X G

The Compton Gamma Ray Observatory (CGRO)
This photograph shows the Compton Observatory being deployed from the space shuttle *Atlantis* in 1991. Named in honor of Arthur Holly Compton, an American scientist who made important discoveries about gamma rays, CGRO carried four different gamma-ray detectors. When its mission ended in 2000, CGRO disintegrated as it re-entered our atmosphere. (NASA)

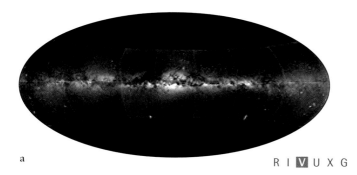

a   R **V** U X G

b   **R** I V U X G

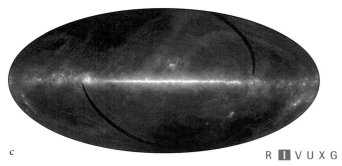

c   R **I** V U X G

d   R I V U **X** G

e   R I V U X **G**

Figure 6-34

WEB LINK 6.18

The Entire Sky at Five Wavelength Ranges These five views show the entire sky at visible, radio, infrared, X-ray, and gamma-ray wavelengths. The entire celestial sphere is mapped onto an oval, with the Milky Way stretching horizontally across the center. The black crescents in the infrared and X-ray images are where data are missing. (GFSC/NASA)

(a) In the visible view the constellation Orion is at the right, Sagittarius in the middle, and Cygnus toward the left. Many of the dark areas along the Milky Way are locations where interstellar dust is sufficiently thick to block visible light.

(b) The radio view shows the sky at a wavelength of 21 cm. This wavelength is emitted by hydrogen atoms in interstellar space. The brightest regions (shown in red) are in the plane of the Milky Way, where the hydrogen is most concentrated.

(c) The infrared view from IRAS shows emission at 100 μm, 60 μm, and 12 μm. Most of the emission is from dust particles in the plane of the Milky Way that have been warmed by starlight.

(d) The X-ray view from ROSAT shows wavelengths of 0.8 nm (blue), 1.7 nm (green), and 5.0 nm (red), corresponding to photon energies of 1500, 750, and 250 eV. Extremely high temperature gas emits these X rays. The white regions, which emit strongly at all X-ray wavelengths, are remnants of supernovae.

(e) The gamma-ray view from the Compton Gamma Ray Observatory includes all wavelengths less than about 1.2×10^{-5} nm (photon energies greater than 10^8 eV). The diffuse radiation from the Milky Way is emitted when fast-moving subatomic particles collide with the nuclei of atoms in interstellar gas clouds. The bright spots above and below the Milky Way are distant, extremely energetic galaxies.

KEY WORDS

active optics, p. 133

adaptive optics, p. 133

angular resolution, p. 132

baseline, p. 138

Cassegrain focus, p. 129

charge-coupled device (CCD), p. 134

chromatic aberration, p. 126

coma, p. 130

coudé focus, p. 130

diffraction, p. 132

diffraction grating, p. 136

eyepiece lens, p. 125

focal length, p. 124

focal plane, p. 124

focal point, p. 124

focus (of a lens or mirror), p. 124

grating, p. 136

imaging, p. 134

interferometry, p. 138

light-gathering power, p. 125

light pollution, p. 134

magnification (magnifying power), p. 126

medium (*plural* media), p. 123

Newtonian reflector, p. 129

objective lens, p. 125

objective mirror (primary mirror), p. 129

optical telescope, p. 123

optical window (in the Earth's atmosphere), p. 140

photometry, p. 135

pixel, p. 134

prime focus, p. 129

radio telescope, p. 137

radio window (in the Earth's atmosphere), p. 140

reflecting telescope (reflector), p. 129

reflection, p. 128

refracting telescope (refractor), pp. 124–125

refraction, p. 123

seeing disk, p. 133

spectrograph, p. 136

spectroscopy, p. 136

spherical aberration, p. 130

very-long-baseline interferometry (VLBI), p. 138

KEY IDEAS

Refracting Telescopes: Refracting telescopes, or refractors, produce images by bending light rays as they pass through glass lenses.

• Chromatic aberration is an optical defect whereby light of different wavelengths is bent in different amounts by a lens.

• Glass impurities, chromatic aberration, opacity to certain wavelengths, and structural difficulties make it inadvisable to build extremely large refractors.

Reflecting Telescopes: Reflecting telescopes, or reflectors, produce images by reflecting light rays to a focus point from curved mirrors.

• Reflectors are not subject to most of the problems that limit the useful size of refractors.

Angular Resolution: A telescope's angular resolution, which indicates ability to see fine details, is limited by two key factors.

• Diffraction is an intrinsic property of light waves. Its effects can be minimized by using a larger objective lens or mirror.

• The blurring effects of atmospheric turbulence can be minimized by placing the telescope atop a tall mountain with very smooth air. They can be dramatically reduced by the use of adaptive optics and can be eliminated entirely by placing the telescope in orbit.

Charge-Coupled Devices: Sensitive light detectors called charge-coupled devices (CCDs) are often used at a telescope's focus to record faint images.

Spectrographs: A spectrograph uses a diffraction grating and lenses to form the spectrum of an astronomical object.

Radio Telescopes: Radio telescopes use large reflecting antennas or dishes to focus radio waves.

• Very large dishes provide reasonably sharp radio images. Higher resolution is achieved with interferometry techniques that link smaller dishes together.

Transparency of the Earth's Atmosphere: The Earth's atmosphere absorbs much of the radiation that arrives from space.

• The atmosphere is transparent chiefly in two wavelength ranges known as the optical window and the radio window. A few wavelengths in the near-infrared also reach the ground.

Telescopes in Space: For observations at wavelengths to which the Earth's atmosphere is opaque, astronomers depend on telescopes carried above the atmosphere by rockets or spacecraft.

• Satellite-based observatories provide new information about the universe and permit coordinated observation of the sky at all wavelengths.

REVIEW QUESTIONS

1. Describe refraction and reflection. Explain how these processes enable astronomers to build telescopes.

2. With the aid of a diagram, describe a refracting telescope. Which dimensions of the telescope determine its light-gathering power? Which dimensions determine the magnification?

3. What is the purpose of a telescope eyepiece? What aspect of the eyepiece determines the magnification of the image? In what circumstances would the eyepiece not be used?

4. Do most professional astronomers actually look through their telescopes? Why or why not?

5. What is chromatic aberration? For what kinds of telescopes does it occur? How can it be corrected?

6. Quite often advertisements appear for telescopes that extol their magnifying power. Is this a good criterion for evaluating telescopes? Explain your answer.

7. With the aid of a diagram, describe a reflecting telescope. Describe four different ways in which an astronomer can access the focal plane.

8. Explain some of the advantages of reflecting telescopes over refracting telescopes.

9. What is spherical aberration? How can it be corrected?

10. What kind of telescope would you use if you wanted to take a color photograph entirely free of chromatic aberration? Explain your answer.

11. Explain why a Cassegrain reflector can be substantially shorter than a refractor of the same focal length.

12. No major observatory has a Newtonian reflector as its primary instrument, whereas Newtonian reflectors are extremely popular among amateur astronomers. Explain why this is so.

13. What is active optics? What is adaptive optics? Why are they useful? Would either of these be a good feature to include on a telescope to be placed in orbit?

14. Compare an optical reflecting telescope and a radio telescope. What do they have in common? How are they different?

15. Why can radio astronomers make observations at any time during the day, whereas optical astronomers are mostly limited to observing at night? (*Hint:* Does your radio work any better or worse in the daytime than at night?)

16. What are the optical window and the radio window? Why isn't there an X-ray window or an ultraviolet window?

17. How are the images made by an X-ray telescope different from those made by a medical X-ray machine?

18. Why must astronomers use satellites and Earth-orbiting observatories to study the heavens at X-ray and gamma-ray wavelengths?

ADVANCED QUESTIONS

Problem-solving tips and tools

You may find it useful to review the small-angle formula discussed in Box 1-1. The area of a circle is proportional to the square of its diameter. Data on the planets can be found in the appendices at the end of this book. Section 5-2 discusses the relationship between frequency and wavelength. Box 6-1 gives examples of how to calculate magnifying power and light-gathering power.

19. Ordinary photographs made with a telephoto lens make distant objects appear close. How does the focal length of a telephoto lens compare with that of a normal lens? Explain your reasoning.

20. Show by means of a diagram why the image formed by a simple refracting telescope is upside down.

21. The observing cage in which an astronomer sits at the prime focus of the 5-m telescope on Palomar Mountain is about 1 m in diameter. Calculate what fraction of the incoming starlight is blocked by the cage.

22. Compare the light-gathering power of the Subaru 8.3-m telescope with that of the Hubble Space Telescope (HST), which has a 2.4-m objective mirror. What advantages does HST have over Subaru?

23. Suppose your Newtonian reflector has an objective mirror 20 cm (8 in.) in diameter with a focal length of 2 m. What magnification do you get with eyepieces whose focal lengths are (**a**) 9 mm, (**b**) 20 mm, and (**c**) 55 mm? What is the telescope's diffraction-limited angular resolution when used with orange light of wavelength 600 nm?

24. The Hobby-Eberly Telescope (HET) has a spherical mirror (the least expensive shape to grind). Consequently, the telescope has spherical aberration. Explain why this doesn't affect the usefulness of HET for spectroscopy.

25. The four largest moons of Jupiter are roughly the same size as our Moon and are about 628 million (6.28×10^8) kilometers from Earth at opposition. What is the size in kilometers of the smallest surface features that the Hubble Space Telescope (resolution of 0.1 arcsec) can detect? How does this compare with the smallest features that can be seen on the Moon with the unaided human eye (resolution of 1 arcmin)?

26. Figure 6-31*b* shows the center of the galaxy M100. **(a)** If the angular resolution of this Hubble Space Telescope image is 0.1 arcsec, what is the diameter in light-years of the smallest detail that can be discerned in the image? Use the small-angle formula and the information given in the caption to Figure 6-31. **(b)** At what distance would a U.S. dime (diameter 1.8 cm) have an angular size of 0.1 arcsec? Give your answer in kilometers.

27. Can the Hubble Space Telescope distinguish any features on Pluto? Justify your answer using calculations.

28. The Russian Space Agency plans to place a radio telescope into an even higher orbit than the Japanese HALCA telescope. Using this telescope in concert with the VLBA, baselines as long as 77,000 km may be obtainable. Astronomers want to use this combination to study radio emission at a frequency of 1665 MHz from distant objects called quasars. **(a)** What is the wavelength of this emission? **(b)** Taking the baseline to be the effective diameter of this radio-telescope array, what angular resolution can be achieved?

29. The mission of the Submillimeter Wave Astronomy Satellite (SWAS), launched in 1998, is to investigate interstellar clouds within which stars form. One of the frequencies at which it observes these clouds is 557 GHz (1 GHz = 1 gigahertz = 10^9 Hz), characteristic of the emission from interstellar water molecules. **(a)** What is the wavelength (in meters) of this emission? In what part of the electromagnetic spectrum is this? **(b)** Why is it necessary to use a satellite for these observations? **(c)** SWAS has an angular resolution of 4 arcminutes. What is the diameter of its primary mirror?

30. To search for ionized oxygen gas surrounding our Milky Way Galaxy, astronomers aimed the ultraviolet telescope of the FUSE spacecraft at a distant galaxy far beyond the Milky Way. They then looked for an ultraviolet spectral line of ionized oxygen in that galaxy's spectrum. Were they looking for an emission line or an absorption line? Explain.

31. A sufficiently thick interstellar cloud of cool gas can absorb low-energy X rays but is transparent to high-energy X rays and gamma rays. Explain why both part *b* and part *d* of Figure 6-34 reveal the presence of cool gas in the Milky Way. Could you infer the presence of this gas from the visible-light image in Figure 6-34*a*? Explain.

DISCUSSION QUESTIONS

32. If you were in charge of selecting a site for a new observatory, what factors would you consider important?

33. Discuss the advantages and disadvantages of using a small telescope in Earth orbit versus a large telescope on a mountaintop.

WEB/CD-ROM QUESTIONS

34. Several telescope manufacturers build telescopes with a design called a "Schmidt-Cassegrain." These use a correcting lens in an arrangement like that shown in Figure 6-13*b*. Consult advertisements on the World Wide Web to see the appearance of these telescopes and find out their cost. Why do you suppose they are very popular among amateur astronomers?

35. Projects are under way to build large optical reflectors in South Africa (SALT, the Southern African Large Telescope) and in the Canary Islands (GTC, the Gran Telescopio Canarias). Search for current information about these telescopes on the World Wide Web. What will be the sizes of the primary mirrors? Are the telescopes designed for imaging, for spectroscopy, or both? Will they observe only at visible wavelengths? In what ways do they complement or surpass existing telescopes?

36. Three of the telescopes shown in the photo that opens this chapter— the James Clerk Maxwell Telescope (JCMT), the Caltech Submillimeter Observatory (CSO), and the Submillimeter Array (SMA)—are designed to detect radiation with wavelengths close to 1 mm. Search for current information about JCMT, CSO, and SMA on the World Wide Web. What kinds of celestial objects emit radiation at these wavelengths? What can astronomers see using JCMT, CSO, and SMA that cannot be observed at other wavelengths? Why is it important that they be at high altitude? How large are the primary mirrors used in JCMT, CSO, and SMA? What are the differences among the three telescopes? Which can be used in the daytime? What recent discoveries have been made using JCMT, CSO, or SMA?

37. The Hubble Space Telescope was originally intended to have a 15-year lifetime. At the time of this writing, NASA has decided to extend this to 20 years. Consult the Space Telescope Science Institute web site to learn about plans for HST's final years of operation. Are future space shuttle missions planned to service HST? If so, what changes will be made to HST on such missions? What will become of HST at the end of its mission lifetime?

38. NASA and the European Space Agency (ESA) have plans to launch a number of advanced space telescopes. These include Constellation-X, the Far Infrared and Submillimeter Telescope (FIRST), the Gamma-ray Large Area Space Telescope (GLAST), and the Space Interferometry Mission (SIM). Search the World Wide Web for information about at least two of these. What are the scientific goals of these projects? What is unique about each telescope? What advantages would they have over existing ground-based or orbiting telescopes? What kind of orbit will each telescope be in? When will each be launched and placed in operation?

39. Telescope Magnification. Access the Active Integrated Media Module "Telescope Magnification" in Chapter 6 of the *Universe* web site or CD-ROM. A common telescope found in department stores is a 3-inch (76-mm) diameter refractor that boasts a magnification of 300 times. Use the magnification calculator to determine the magnifications that are achieved by using each of the following commonly found eyepieces on that telescope: Eyepiece A with focal length 40 mm; Eyepiece B with focal length 25 mm; Eyepiece C with focal length 12 mm; and Eyepiece D with focal length 2.5 mm.

OBSERVING PROJECTS

40. Obtain a telescope during the daytime along with several eyepieces of various focal lengths. If you can determine the telescope's focal length, calculate the magnifying powers of the eyepieces. Focus the telescope on some familiar object, such as a distant lamppost or tree. **DO NOT FOCUS ON THE SUN! Looking directly at the Sun can cause blindness.** Describe the image you see through the telescope. Is it upside down? How does the image move as you slowly and gently shift the telescope left and right or up and down? Examine the eyepieces, noting their focal lengths. By changing the eyepieces, examine the distant object under different magnifications. How do the field of view and the quality of the image change as you go from low power to high power?

41. On a clear night, view the Moon, a planet, and a star through a telescope using eyepieces of various focal lengths and known magnifying powers. (To determine the locations in the sky of the Moon and planets, you may want to use the *Starry Night* program on the CD-ROM that comes with this book. You may also want to consult such magazines as *Sky & Telescope* and *Astronomy* or their web sites.) In what way does the image seem to degrade as you view with increasingly higher magnification? Do you see any chromatic aberration? If so, with which object and which eyepiece is it most noticeable?

42. Many towns and cities have amateur astronomy clubs. If you are so inclined, attend a "star party" hosted by your local club. People who bring their telescopes to such gatherings are delighted to show you their instruments and take you on a telescopic tour of the heavens. Such an experience can lead to a very enjoyable, lifelong hobby.

43. The field of view of a typical small telescope for amateur astronomers is about 30 arcminutes (30′). Use the *Starry Night* program to simulate the view that such a telescope provides of various celestial objects. First turn off daylight (select **Daylight** in the **Sky** menu) and show the entire celestial sphere (select **Atlas** in the **Go** menu). Then search for objects (i), (ii), and (iii) listed below (select **Find…** in the **Edit** menu). For each object, use the controls at the right-hand end of the Control Panel to adjust the field of view to approximately 30′. Describe how much detail you can see on each object. (i) Jupiter. (ii) Saturn. (iii) M31, the great galaxy in the constellation Andromeda.

Our Solar System

(GSFC/NASA)

R I V U X G

For a good view of the planets, go to a dark location far from city lights. But for even better views, leave the Earth altogether and view the sky from space—like this view from the *Clementine* spacecraft, which shows (from left to right) Saturn, Mars, and Mercury, as well as the Sun and the Moon. What are the planets like? Are they larger or smaller than the Earth? What are they made of? And where did they come from?

Fifty years ago, astronomers knew precious little about the solar system. Even the best telescopes provided images of the planets that were frustratingly hazy and indistinct. Of asteroids, comets, and the satellites of the planets, we knew even less.

Today, our knowledge of the solar system has grown many thousandfold. Spacecraft such as *Clementine* have flown at close range past all the planets save Pluto, showing us details that astronomers of an earlier generation could only dream about. We have landed spacecraft on the Moon, Venus, and Mars and dropped a probe into the immense atmosphere of Jupiter. This is truly the golden age of solar system exploration.

In this chapter we paint in broad outline our present understanding of the solar system. We will see that the planets come in a variety of sizes and chemical compositions. There is also rich variety among the moons of the planets and among smaller bodies we call asteroids and comets. By studying this diverse cast of characters, we will come to understand how the solar system formed some 4.6×10^9 years ago. We will also learn about planets beyond our solar system and how even more planets may be forming around other stars.

As you read the sections of this chapter, look for the answers to the following questions.

7-1   Are all the other planets similar to Earth, or are they very different?

7-2   Do other planets have moons like Earth's Moon?

7-3   How do astronomers know what the other planets are made of?

7-4   Are all the planets made of basically the same material?

7-5   What is the difference between an asteroid and a comet?

7-6   Why are some elements (like gold) quite rare, while others (like carbon) are more common?

7-7   How do astronomers think the solar system formed?

7-8   Did all of the planets form in the same way?

7-9   Are there planets orbiting other stars? How do astronomers search for such planets?

7-1   There are two broad categories of planets: Earthlike and Jupiterlike

Each of the nine planets that orbit the Sun is unique. Only Earth has liquid water and an atmosphere breathable by humans; only Venus has a perpetual cloud layer made of sulfuric acid droplets; only Jupiter has immense storm systems that persist for centuries; and only Pluto always keeps the same face toward its moon. But there are also striking similarities among planets. Volcanoes exist not only on Earth but also on Venus and Mars; there are rings around Jupiter, Saturn, Uranus, and Neptune; and craters on Mercury, Venus, Earth, and Mars show that all of these planets have been bombarded by interplanetary debris.

How can we make sense of the many similarities and differences among the planets? An important step is to organize our knowledge of the planets in a systematic way. There are two useful ways to do this. First, we can contrast the orbits of different planets around the Sun; and second, we can compare the planets' physical properties such as diameter, mass, average density, and chemical composition.

The planets fall naturally into two classes according to the sizes of their orbits. As Figure 7-1 shows, the orbits of the four inner planets (Mercury, Venus, Earth, and Mars) are crowded in close to the Sun. In contrast, the orbits of the next four planets (Jupiter, Saturn, Uranus, and Neptune) are widely spaced at great distances from the Sun. Table 7-1 lists the orbital characteristics of all the planets.

 Note that while Figure 7-1 shows the orbits of the planets, it does not show the planets themselves. The reason is simple: If Jupiter, the largest of the planets, were to be drawn to the same scale as the rest of this figure, it would be a dot just 0.0002 cm across—about $1/300$ of the width of a human hair and far too small to see without a microscope. The planets themselves are *very* small compared to the distances between the planets. Indeed, while an airliner traveling at 1000 km/h (620 mi/h) can fly around the Earth in less than two days, at this speed it would take 17 *years* to fly from the Earth to the Sun. The solar system is a very large and very empty place!

Most of the planets have orbits that are nearly circular. As we learned in Section 4-5, Kepler discovered in the sixteenth century that these orbits are actually ellipses. Astronomers denote the elongation of an ellipse by its *eccentricity* (see Figure 4-16b). The eccentricity of a circle is zero, and indeed most planets have orbital eccentricities that are very close to zero. The exceptions are Mercury and Pluto. In fact, Pluto's noncircular orbit sometimes takes it nearer the Sun than its neighbor, Neptune. (Happily, the orbits of Pluto and Neptune are such that the two planets will never collide.)

If you could observe the solar system from a point several astronomical units (AU) above the Earth's North Pole, you would see that all the planets orbit the Sun in the same counterclockwise direction. Furthermore, the orbits of the planets all lie in nearly the same plane. In other words, the orbits of the planets are inclined at only slight angles to the plane of the ecliptic, which is the plane of the Earth's orbit around the Sun. (Again, however, Pluto is an exception; the plane of its orbit is tilted at about 17° to the ecliptic, as Figure 7-1 shows.) Furthermore, the plane of the Sun's equator is very closely aligned with the orbital planes of the planets. As we will learn, these near-alignments are not a coincidence. They provide important clues about the origin of the planets.

When we compare the physical properties of the planets, we again find that they fall naturally into two classes—the four inner planets and the four outer ones—with Pluto as an exception. The four inner planets are called **terrestrial planets** because they resemble the Earth (in Latin, *terra*). They all have hard, rocky surfaces with mountains, craters, valleys, and volcanoes. You could stand on the surface of any one of them, although you would need a protective spacesuit on Mercury, Venus, or Mars. The outer four planets are called **Jovian planets** because they resemble Jupiter. (Jove was another name for the Roman god Jupiter.) An attempt to land a spacecraft on the surface of any of the Jovian planets would be futile, because the materials of which these planets are made are mostly gaseous or liquid. The visible "surface" features of a Jovian planet are actually cloud formations in the planet's atmosphere. The photographs in Table 7-1 show the distinctive appearances of the two classes of planets.

The most apparent difference between the terrestrial and Jovian planets is their diameters. You can compute the

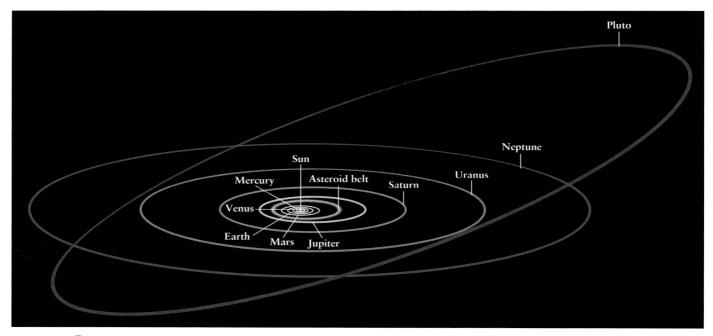

figure 7-1

The Solar System to Scale This scale drawing shows a perspective view of the orbits of the planets around the Sun. The four inner planets are crowded in close to the Sun, while the five outer planets orbit the Sun at much greater distances. Many smaller objects called asteroids orbit within a belt between the orbits of Mars and Jupiter. On the scale of this drawing, the planets themselves would be much smaller than the diameter of a human hair and too small to see.

diameter of a planet if you know its angular diameter as seen from Earth and its distance from Earth. For example, on December 11, 2000, Venus was 1.42×10^8 km from Earth and had an angular diameter of 17.6 arcsec. Using the small-angle formula from Box 1-1, we can calculate the diameter of Venus to be 12,100 km (7520 mi). Similar calculations demonstrate that the Earth, with its diameter of about 12,756 km (7926 mi), is the largest of the four inner, terrestrial planets. In sharp contrast, the four outer, Jovian planets are much larger than the terrestrial planets. First place goes to Jupiter, whose equa-

torial diameter is more than 11 times that of the Earth. Pluto, always the exception, is even smaller than the inner planets, despite being the outermost planet. Its diameter is less than one-fifth that of the Earth. Figure 7-2 shows the Sun and the planets drawn to the same scale. The diameters of the planets are given in Table 7-1.

The masses of the terrestrial and Jovian planets are also dramatically different. If a planet has a satellite, you can calculate the planet's mass from the satellite's period and semi-major axis by using Newton's form of Kepler's third law (see

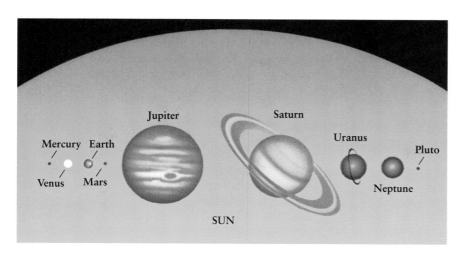

figure 7-2

The Sun and the Planets to Scale This drawing shows the nine planets in front of the disk of the Sun, with all ten bodies drawn to the same scale. The four terrestrial planets (Mercury, Venus, Earth, and Mars) have orbits nearest the Sun. They are small and dense, and made of rocky materials. The Jovian planets are the next four planets from the Sun (Jupiter, Saturn, Uranus, and Neptune). They are large and of low density and are composed primarily of hydrogen and helium. Pluto, an exceptional case, is a mixture of ice and rock.

table 7-1	Characteristics of the Planets			

	The Inner Planets			
	Mercury	**Venus**	**Earth**	**Mars**
Average distance from Sun (10^6 km)	57.91	108.2	149.60	227.93
Average distance from Sun (AU)	0.3871	0.7233	1.0000	1.5236
Orbital period (years)	0.2408	0.6152	1.0000	1.8808
Orbital eccentricity	0.206	0.007	0.017	0.093
Inclination of orbit to the ecliptic	7.00°	3.39°	0.00°	1.85°
Equatorial diameter (km)	4880	12,104	12,756	6794
Equatorial diameter (Earth = 1)	0.383	0.949	1.000	0.533
Mass (kg)	3.302×10^{23}	4.868×10^{24}	5.974×10^{24}	6.418×10^{23}
Mass (Earth = 1)	0.0553	0.8150	1.0000	0.1074
Average density (kg/m³)	5430	5243	5515	3934

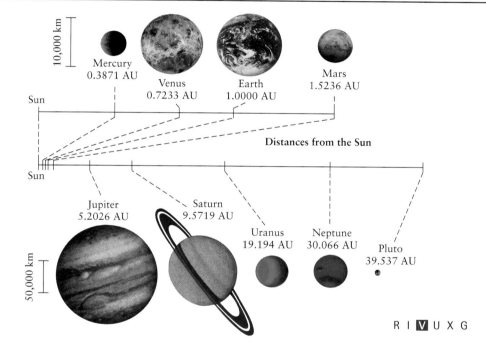

10,000 km

Mercury
0.3871 AU

Venus
0.7233 AU

Earth
1.0000 AU

Mars
1.5236 AU

Sun

Distances from the Sun

Sun

Jupiter
5.2026 AU

Saturn
9.5719 AU

Uranus
19.194 AU

Neptune
30.066 AU

Pluto
39.537 AU

50,000 km

(NASA)

R I **V** U X G

	The Outer Planets				
	Jupiter	**Saturn**	**Uranus**	**Neptune**	**Pluto**
Average distance from Sun (10^6 km)	778.30	1431.9	2877.4	4497.8	5914.7
Average distance from Sun (AU)	5.2026	9.5719	19.194	30.066	39.537
Orbital period (years)	11.856	29.369	84.099	164.86	248.60
Orbital eccentricity	0.048	0.053	0.043	0.010	0.250
Inclination of orbit to the ecliptic	1.30°	2.48°	0.77°	1.77°	17.12°
Equatorial diameter (km)	142,984	120,536	51,118	49,528	2300
Equatorial diameter (Earth = 1)	11.209	9.449	4.007	3.883	0.180
Mass (kg)	1.899×10^{27}	5.685×10^{26}	8.682×10^{25}	1.024×10^{26}	1.31×10^{22}
Mass (Earth = 1)	317.8	95.16	14.53	17.15	0.002
Average density (kg/m³)	1326	687	1318	1638	2000

Box 7-1 The Heavens on the Earth

Average Density

Average density—the mass of an object divided by that object's volume—is a useful quantity for describing the differences between planets in our solar system. This same quantity has many applications here on Earth.

A rock tossed into a lake sinks to the bottom, while an air bubble produced at the bottom of a lake (for example, by the air tanks of a scuba diver) rises to the top. These are examples of a general principle: An object sinks in a fluid if its average density is greater than that of the fluid, but rises if its average density is less than that of the fluid. The average density of water is 1000 kg/m^3, which is why a typical rock (with an average density of about 3000 kg/m^3) sinks, while an air bubble (average density of about 1.2 kg/m^3) rises.

At many summer barbecues, cans of soft drinks are kept cold by putting them in a container full of ice. When the ice melts, the cans of diet soda always rise to the top, while the cans of regular soda sink to the bottom. Why is this? The average density of a can of diet soda—which includes water, flavoring, artificial sweetener, and the trapped gas that makes the drink fizzy—is slightly less than the density of water, and so the can floats. A can of regular soda contains sugar instead of artificial sweetener, and the sugar is a bit heavier than the sweetener. The extra weight is just enough to make the average density of a can of regular soda slightly more than that of water, making the can sink. (You can test these statements by putting unopened cans of diet soda and regular soda in a sink or bathtub full of water.)

The concept of average density provides geologists with important clues about the early history of the Earth. The average density of surface rocks on Earth, about 3000 kg/m^3, is less than the Earth's average density of 5515 kg/m^3. The simplest explanation is that in the ancient past, the Earth was completely molten throughout its volume, so that low-density materials rose to the surface and high-density materials sank deep into the Earth's interior in a process called *chemical differentiation*. This series of events also suggests that the Earth's core must be made of relatively dense materials, such as iron and nickel. A tremendous amount of other geological evidence has convinced scientists that this picture is correct.

Section 4-7 and Box 4-4). Astronomers have also measured the mass of each planet except Pluto by sending a spacecraft to pass near the planet. The planet's gravitational pull (which is proportional to its mass) deflects the spacecraft's path, and the amount of deflection tells us the planet's mass. Using these techniques, astronomers have found that the four Jovian planets have masses that are tens or hundreds of times greater than the mass of any of the terrestrial planets. Again, first place goes to Jupiter, whose mass is 318 times greater than the Earth's.

Once we know the diameter and mass of a planet, we can learn something about what that planet is made of. The trick is to calculate the planet's **average density,** or mass divided by volume, measured in kilograms per cubic meter (kg/m^3). The average density of any substance depends in part on that substance's composition. For example, air near sea level on Earth has an average density of 1.2 kg/m^3, water's average density is 1000 kg/m^3, and a piece of concrete has an average density of 2000 kg/m^3. Box 7-1 describes some applications of the idea of average density to everyday phenomena on Earth.

The four inner, terrestrial planets have very high average densities (see Table 7-1); the average density of the Earth, for example, is 5515 kg/m^3. By contrast, a typical rock found on the Earth's surface has a lower average density, about 3000 kg/m^3. Thus, the Earth must contain a large amount of material that is denser than rock. This information provides our first clue that terrestrial planets have dense iron cores.

In sharp contrast, the outer, Jovian planets have quite low densities. Saturn has an average density less than that of water. This information strongly suggests that the giant outer planets are composed primarily of light elements such as hydrogen and helium. All four Jovian planets probably have large cores of mixed rock and highly compressed water buried beneath low-density outer layers tens of thousands of kilometers thick.

Pluto is again an oddity. Although it is even smaller than the dense inner planets, its average density is closer to that of the giant outer planets. Because its average density of about 1900 kg/m^3 is intermediate between the densities of ice (that is, frozen water) and of rock, Pluto is probably composed of a mixture of these two substances.

7-2 Seven large satellites are almost as big as the terrestrial planets

All the planets except Mercury and Venus have satellites. More than 60 satellites are known (Jupiter, Saturn, and Uranus each have at least 16), and dozens of other small ones probably remain to be discovered. The known satellites fall into two distinct categories. Seven giant satellites are larger than Pluto

Table 7-2	The Seven Giant Satellites						
	Moon	Io	Europa	Ganymede	Callisto	Titan	Triton
Parent planet	Earth	Jupiter	Jupiter	Jupiter	Jupiter	Saturn	Neptune
Diameter (km)	3476	3642	3130	5268	4806	5150	2706
Mass (kg)	7.35×10^{22}	8.93×10^{22}	4.80×10^{22}	1.48×10^{23}	1.08×10^{23}	1.34×10^{23}	2.15×10^{22}
Average density (kg/m³)	3340	3530	2970	1940	1850	1880	2050
Substantial atmosphere?	No	No	No	No	No	Yes	No

Moon   Io   Europa   Ganymede   Callisto   Titan   Triton

(JPL/NASA)

R I **V** U X G

and roughly as big as the planet Mercury. (Table 7-2 shows these satellites to the same scale.) Like the terrestrial planets, all seven giant satellites have hard surfaces of either rock or ice. All the other satellites also have solid surfaces but are much smaller, with diameters less than 2000 km.

Interplanetary spacecraft have made many surprising and fascinating discoveries about these giant satellites. We now know that Jupiter's satellite Io is the most geologically active world in the solar system, with numerous geyserlike volcanoes that continually belch forth sulfur-rich compounds. The fractured surface of Europa, another of Jupiter's large satellites, suggests that a worldwide ocean of liquid water may lie beneath its icy surface. Saturn's largest satellite, Titan, has an atmosphere nearly twice as dense as the Earth's atmosphere, with a perpetual haze layer that gives it a featureless appearance (see the photographs in Table 7-2).

7-3 Spectroscopy reveals the chemical composition of the planets

The average densities of the planets and satellites give us a crude measure of their **chemical compositions**—that is, what substances they are made of. For example, the low average density of the Moon (3344 kg/m³) compared with the Earth (5515 kg/m³)

tells us that the Moon contains relatively little of dense metals like iron. But to truly understand the nature of the planets and satellites, we need to know their chemical compositions in much greater detail than we can learn from average density alone.

The most accurate way to determine chemical composition is by directly analyzing samples taken from a planet's atmosphere and soil. Unfortunately, we have such direct information only for the Earth and the three worlds on which spacecraft have landed—Venus, the Moon, and Mars. In all other cases, astronomers must analyze sunlight reflected from the distant planets and their satellites. To do that, astronomers bring to bear one of their most powerful tools, **spectroscopy,** the systematic study of spectra and spectral lines. (We discussed spectroscopy in Sections 5-6 and 6-5.)

Spectroscopy is a sensitive probe of the composition of a planet's atmosphere. If a planet has an atmosphere, then sunlight reflected from that planet must have passed through its atmosphere. During this passage, some of the wavelengths of sunlight will have been absorbed. Hence, the spectrum of this reflected sunlight will have dark absorption lines. Astronomers look at the particular wavelengths absorbed and the amount of light absorbed at those wavelengths. Both of these depend on the kinds of chemicals present in the planet's atmosphere and the abundance of those chemicals.

For example, astronomers have used spectroscopy to analyze the atmosphere of Saturn's largest satellite, Titan (see

a

b

figure 7-3

The Spectrum of Saturn's Satellite Titan (a) This graph shows the spectrum of sunlight reflected from Titan, the largest satellite of Saturn and the only satellite in the solar system with a substantial atmosphere. The dips in the curve are due to absorption by atoms of hydrogen (H), molecules of oxygen (O_2), and molecules of methane (CH_4). Of these gases, only methane is actually present in Titan's atmosphere. **(b)** The spectrum in (a) was made by light that was emitted from the Sun's glowing surface, passed through the Sun's atmosphere, was reflected by Titan's atmosphere, and passed through the Earth's atmosphere before reaching a telescope. The absorption lines of hydrogen and oxygen are produced in the atmospheres of the Sun and Earth, respectively. Only the methane dips are caused by absorption in Titan's atmosphere.

Table 7-2 for a photograph of Titan). The graph in Figure 7-3a shows the spectrum of visible sunlight reflected from Titan. (We first saw this method of displaying spectra in Figure 6-23.) The dips in this curve of intensity versus wavelength represent absorption lines. However, not all of these absorption lines are produced in the atmosphere of Titan (Figure 7-3b). Before reaching Titan, light from the Sun's glowing surface must pass through the Sun's hydrogen-rich atmosphere. This produces the hydrogen absorption line in Figure 7-3a at a wavelength of 656 nm. After being reflected from Titan, the light must pass through the Earth's atmosphere before reaching the telescope; this is where the oxygen absorption line in Figure 7-3a is produced. Only the two dips near 620 nm and 730 nm are caused by gases in Titan's atmosphere.

These two absorption lines are caused not by individual atoms in the atmosphere of Titan but by atoms combined to form **molecules.** For example, two hydrogen atoms can combine with an atom of oxygen to form a molecule of water. Water's chemical formula, H_2O, describes the atoms in a water molecule. In a similar way, two oxygen atoms can bond with a carbon atom to produce a molecule of carbon dioxide, whose formula is CO_2. Molecules, like atoms, also produce unique patterns of lines in the spectra of astronomical objects. The absorption lines in Figure 7-3a indicate the presence in Titan's atmosphere of molecules of methane, or CH_4 (a molecule made of one carbon atom and four hydrogen atoms).

This shows that Titan is a curious place indeed, because on Earth, methane is the primary ingredient in natural gas! The spectra of all of the planetary atmospheres in the solar system display absorption lines of molecules of various types.

In addition to visible-light measurements such as those in Figure 7-3a, it is very useful to study the *infrared* and *ultraviolet* spectra of planetary atmospheres. Many molecules have much stronger spectral lines in these nonvisible wavelength bands than in the visible. As an example, the ultraviolet spectrum of Titan shows that nitrogen molecules (N_2) are the dominant constituent of Titan's atmosphere. Furthermore, Titan's infrared spectrum includes spectral lines of a variety of molecules of carbon and hydrogen, indicating that Titan's atmosphere has a very complex chemistry. None of these molecules could have been detected by visible light alone.

Spectroscopy can also provide useful information about the solid surfaces of planets and satellites without atmospheres. When light shines on a solid surface, some wavelengths are absorbed while others are reflected. (For example, a plant leaf absorbs red and violet light but reflects green light—which is why leaves look green.) Unlike a gas, a solid illuminated by sunlight does not produce sharp, definite spectral lines. Instead, only broad absorption features appear in the spectrum. By comparing such a spectrum with the spectra of samples of different substances on Earth, astronomers can infer the chemical composition of the surface of a planet or satellite.

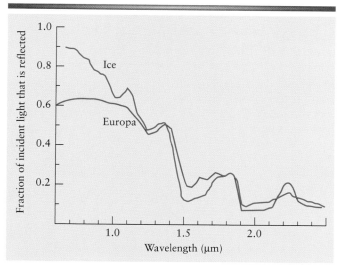

WEB LINK 7.5

figure 7-4

The Spectrum of Jupiter's Moon Europa Infrared light from the Sun that is reflected from the surface of Europa, one of Jupiter's moons, has almost exactly the same spectrum as sunlight reflected from ordinary water ice. This shows that the surface of Europa is made predominantly of ice and not rock.

As an example, Figure 7-4 shows the infrared spectrum of light reflected from the surface of Jupiter's satellite Europa (Table 7-2 includes a photograph of Europa). Because this spectrum is so close to that of ordinary ice (that is, frozen water), astronomers conclude that ice is the dominant constituent of Europa's surface. However, this technique tells us little about what the material is like just below the surface. There is no substitute for sending a spacecraft to a planet and examining its surface directly.

7-4 Hydrogen and helium are abundant on the Jovian planets, whereas the terrestrial planets are composed mostly of heavy elements

Spectroscopic observations from Earth and spacecraft show that the outer layers of the Jovian planets are composed primarily of the lightest gases, hydrogen and helium. In contrast, chemical analysis of soil samples from Venus, Earth, and Mars demonstrate that the terrestrial planets are made mostly of heavy elements, such as iron, silicon, magnesium, sulfur, and nickel. Spacecraft images such as Figures 7-5 and 7-6 only hint at these striking differences in chemical composition.

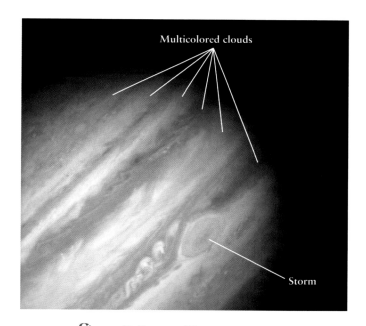

WEB LINK 7.6

figure 7-5   R I **V** U X G

A Jovian Planet This detailed view of Jupiter's cloudtops was made by the Hubble Space Telescope in 1999. Jupiter is composed mostly of the lightest elements, hydrogen and helium. Hydrogen and helium are colorless; the colors in the atmosphere are caused by trace amounts of other substances. The giant storm at lower right, called the Great Red Spot, has been raging for more than 300 years. (Space Telescope Science Institute/JPL/NASA)

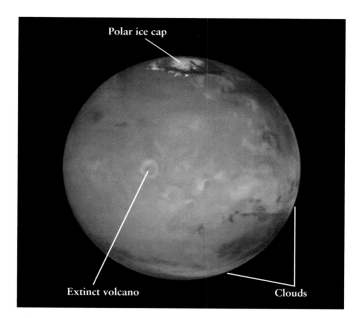

WEB LINK 7.7

figure 7-6   R I **V** U X G

A Terrestrial Planet Mars is composed mostly of heavy elements such as iron, silicon, magnesium, and sulfur. The planet's red surface can be seen clearly in this Hubble Space Telescope image because the Martian atmosphere is thin and nearly cloudless. Olympus Mons, the extinct volcano to the left of center, is nearly 3 times the height of Mount Everest. (Space Telescope Science Institute/JPL/NASA)

Temperature plays a major role in determining whether the materials of which planets are made exist as solids, liquids, or gases. Hydrogen and helium are gaseous except at extremely low temperatures and extraordinarily high pressures. By contrast, rock-forming compounds such as iron and silicon are solids except at temperatures exceeding 1000 K. (You may want to review the discussion of temperature scales in Box 5-1.) Between these two extremes are substances such as water (H_2O), carbon dioxide (CO_2), methane (CH_4), and ammonia (NH_3). At low temperatures, typically below 100 to 300 K, these common chemicals solidify into solids called **ices**. (In astronomy, frozen water is just one kind of "ice.") At somewhat higher temperatures, they can exist as liquids or gases. For example, clouds of ammonia ice crystals are found in the cold upper atmosphere of Jupiter, but within Jupiter's warmer interior, ammonia exists primarily as a liquid.

As you might expect, a planet's surface temperature is related to its distance from the Sun. The four inner planets are quite warm. For example, midday temperatures on Mercury may climb to 700 K (= 427°C = 801°F), and during midsummer on Mars, it is sometimes as warm as 290 K (= 17°C = 63°F). The outer planets, which receive much less solar radiation, are cooler. Typical temperatures range from about 125 K (= −148°C = −234°F) in Jupiter's upper atmosphere to about 40 K (= −233°C = −387°F) on Pluto.

The higher surface temperatures of the terrestrial planets help to explain the following observation: The atmospheres of the terrestrial planets contain virtually no hydrogen or helium. Instead, the atmospheres of Venus, Earth, and Mars are composed of heavier molecules such as nitrogen (N_2, 14 times more massive than a hydrogen molecule), oxygen (O_2, 16 times more massive), and carbon dioxide (22 times more massive). To understand the connection between surface temperature and the absence of hydrogen and helium, we need to know a few basic facts about gases.

 The temperature of a gas is directly related to the speeds at which the atoms or molecules of the gas move: The higher the gas temperature, the greater the speed of its atoms or molecules. Furthermore, for a given temperature, lightweight atoms and molecules move more rapidly than heavy ones. On the four inner, terrestrial planets, where atmospheric temperatures are high, low-mass hydrogen molecules and helium atoms move so swiftly that they can escape from the relatively weak gravity of these planets. Hence, the atmospheres that surround the terrestrial planets are composed primarily of more massive, slower-moving molecules such as CO_2, N_2, O_2, and water vapor (H_2O). On the four Jovian planets, low temperatures and relatively strong gravity prevent even lightweight hydrogen and helium gases from escaping into space, and so their atmospheres are much more extensive. The combined mass of Jupiter's atmosphere, for example, is about a million (10^6) times greater than that of the Earth's atmosphere. This is comparable to the mass of the entire Earth! Box 7-2 describes more about the ability of a planet's gravity to retain gases.

7-5 Small chunks of rock and ice also orbit the Sun

In addition to the nine planets, many smaller objects orbit the Sun. Between the orbits of Mars and Jupiter are thousands of rocky objects called **asteroids**. There is no sharp dividing line between planets and asteroids, which is why asteroids are also called **minor planets**. The largest asteroid, Ceres, has a diameter of about 900 km. The next largest, Pallas and Vesta, are each about 500 km in diameter. Still smaller ones, like the asteroid shown in close-up in Figure 7-7, are increasingly numerous. There are thousands of kilometer-sized asteroids and millions of asteroids that are boulder-sized or smaller.

Most (though not all) asteroids orbit the Sun at distances of 2 to 3.5 AU. This region of the solar system between the orbits of Mars and Jupiter is called the **asteroid belt**.

 One common misconception about asteroids is that they are the remnants of an ancient planet that somehow broke apart or exploded, like the fictional planet Krypton in the comic book adventures of Superman. In fact, the asteroids were probably never part of any planet-sized body. The early solar system is thought to have been filled with asteroidlike objects, most of which coalesced to form the planets. The "leftover" objects that missed out on this process make up our present-day population of asteroids.

 Figure 7-7   R I **V** U X G

An Asteroid The asteroid shown in this image, 433 Eros, is only 33 km (21 mi) long and 13 km (8 mi) wide. Because Eros is so small, its gravity is too weak to have pulled it into a spherical shape. This image was taken in March 2000 by *NEAR Shoemaker*, the first spacecraft to go into orbit around an asteroid. (*NEAR* Project, Johns Hopkins University, Applied Physics Laboratory, and NASA)

Kinetic Energy, Temperature, and Whether Planets Have Atmospheres

A moving object possesses energy as a result of its motion. The faster it moves, the more energy it has. Energy of this type is called **kinetic energy**. If an object of mass m is moving with a speed v, its kinetic energy E_k is given by

Kinetic energy

$$E_k = \frac{1}{2}mv^2$$

E_k = kinetic energy of an object

m = mass of object

v = speed of object

This expression for kinetic energy is valid for all objects, both big and small, from atoms and molecules to planets and stars, as long as their speed is slow in relation to the speed of light. If the mass is expressed in kilograms and the speed in meters per second, the kinetic energy is expressed in joules (J).

EXAMPLE: An automobile of mass 1000 kg driving at a typical freeway speed of 30 m/s (= 108 km/h = 67 mi/h) has a kinetic energy of

$$E_k = \frac{1}{2}(1000 \times 30^2) = 450,000 \text{ J} = 4.5 \times 10^5 \text{ J}$$

Consider a gas, such as the atmosphere of a star or planet. Some of the gas atoms or molecules will be moving slowly, with little kinetic energy, while others will be moving faster and have more kinetic energy. The temperature of the gas is a direct measure of the *average* amount of kinetic energy per atom or molecule. The hotter the gas, the faster atoms or molecules move, on average, and the greater the average kinetic energy of an atom or molecule.

If the gas temperature is sufficiently high, typically several thousand kelvins, molecules move so fast that when they collide with one another, the energy of the collision can break the molecules apart into their constituent atoms. Thus, the Sun's atmosphere, where the temperature is 5800 K, consists primarily of individual hydrogen atoms rather than hydrogen molecules. By contrast, the hydrogen atoms in the Earth's atmosphere (temperature 290 K) are combined with oxygen atoms into molecules of water vapor (H_2O).

The physics of gases tells us that in a gas of temperature T (in kelvins), the average kinetic energy of an atom or molecule is

Kinetic energy of a gas atom or molecule

$$E_k = \frac{3}{2}kT$$

E_k = average kinetic energy of a gas atom or molecule, in joules

k = 1.38×10^{-23} J/K

T = temperature of gas, in kelvins

The quantity k is called the Boltzmann constant. Note that the higher the gas temperature, the greater the average kinetic energy of an atom or molecule of the gas. This average kinetic energy becomes zero at absolute zero, or $T = 0$, the temperature at which molecular motion is at a minimum.

At a given temperature, all kinds of atoms and molecules will have the same average kinetic energy. But the average *speed* of a given kind of atom or molecule depends on its mass. To see this, note that the average kinetic energy of a gas atom or molecule can be written in two equivalent ways:

$$E_k = \frac{1}{2}mv^2 = \frac{3}{2}kT$$

where v represents the average speed of an atom or molecule in a gas with temperature T. Rearranging this equation, we obtain

Average speed of a gas atom or molecule

$$v = \sqrt{\frac{3kT}{m}}$$

v = average speed of a gas atom or molecule, in m/s

k = 1.38×10^{-23} J/K

T = temperature of gas, in kelvins

m = mass of the atom or molecule, in kilograms

For a given gas temperature, the greater the mass of a given type of gas atom or molecule, the slower its average speed. (The value of v given by this equation is actually slightly higher than the average speed of the atoms or molecules in the gas, but it is close enough for our purposes here. If you are studying physics, you may know that v is actually the root-mean-square speed.)

(continued on following page)

box 7-2 *(continued)*

EXAMPLE: Suppose you want to know the average speed of the oxygen molecules that you breathe at a room temperature of 20°C (= 68°F = 293 K). From a reference book, you can find that the mass of an oxygen atom is 2.66×10^{-26} kg. The mass of an oxygen molecule (O_2) is twice the mass of an oxygen atom, or $2(2.66 \times 10^{-26}$ kg) $= 5.32 \times 10^{-26}$ kg. Thus, the average speed is

$$v = \sqrt{\frac{3(1.38 \times 10^{-23})(293)}{5.32 \times 10^{-26}}} = 478 \text{ m/s} = 0.478 \text{ km/s}$$

This is more than 1700 kilometers per hour (1100 miles per hour). Hence, atoms and molecules move rapidly in even a moderate-temperature gas.

In some situations, atoms and molecules in a gas may be moving so fast that they can overcome the attractive force of a planet's gravity and escape into interplanetary space. The minimum speed that an object at a planet's surface must have in order to permanently leave the planet is called the planet's **escape speed**. The escape speed for a planet of mass M and radius R is given by

$$v_{escape} = \sqrt{\frac{2GM}{R}}$$

where $G = 6.67 \times 10^{-11}$ N m²/kg² is the universal constant of gravitation.

The accompanying table gives the escape speed for various objects in the solar system. For example, to get to another planet, a spacecraft must leave Earth with a speed greater than 11.2 km/s (25,100 mi/h).

A good rule of thumb is that a planet can retain a gas if the escape speed is at least 6 times greater than the average speed of the molecules in the gas. (Some molecules are moving slower than average, and others are moving faster, but very few are moving more than 6 times faster than average.) In such a case, very few molecules will be moving fast enough to escape from the planet's gravity.

EXAMPLE: Consider the Earth's atmosphere. We saw that the average speed of oxygen molecules is 0.478 km/s at room temperature. The escape speed from the Earth (11.2 km/s) is much more than 6 times the average speed of the oxygen molecules, so the Earth has no trouble keeping oxygen in its atmosphere.

A similar calculation for hydrogen molecules (H_2) gives a different result, however. At 293 K, the average speed of a hydrogen molecule is 1.9 km/s. Six times this speed is 11.4 km/s, which is slightly higher than the escape speed from the Earth. Thus, the Earth does not retain hydrogen in its atmosphere. Any hydrogen released into the air slowly leaks away into space.

Planet	Escape speed (km/s)
Mercury	4.3
Venus	10.4
Earth	11.2
Moon	2.4
Mars	5.0
Jupiter	59.5
Saturn	35.5
Uranus	21.3
Neptune	23.5
Pluto	1.3

Quite far from the Sun, beyond the orbit of Neptune, are chunks of very dirty ice called **comets**. Many comets have highly elongated orbits that occasionally bring them close to the Sun. When this happens, the Sun's radiation vaporizes some of the comet's ices, producing long flowing tails of gas and dust particles (Figure 7-8).

 Science-fiction movies and television programs sometimes show comets tearing across the night sky like a rocket. That would be a pretty impressive sight—but that is not what comets look like. Like the planets, comets orbit the Sun. And like the planets, comets move hardly at all against the background of stars over the course of a single night (see Section 4-1). If you are lucky enough to see a bright comet, it will not zoom dramatically from horizon to horizon. Instead, it will seem to hang majestically among the stars, so you can admire it at your leisure.

Both asteroids and comets are thought to be debris left over from the formation of the solar system. In the inner regions of the solar system, rocky fragments have been able to endure continuous exposure to the Sun's heat, but any ice originally present would have evaporated. Far from the Sun, chunks of ice have survived for billions of years. Thus, debris in the solar system naturally divides into two families (asteroids and comets), which can be arranged according to dis-

figure 7-8   R I **V** U X G

A Comet This photograph shows Comet Hale-Bopp as it appeared in April 1997. The solid part of a comet like this is a chunk of dirty ice a few tens of kilometers in diameter. When a comet passes near the Sun, solar radiation vaporizes some of the icy material. The released material forms a bluish tail of gas and a white tail of dust, both of which can extend for tens of millions of kilometers. (Richard J. Wainscoat, University of Hawaii)

tance from the Sun, just like the two categories of planets (terrestrial and Jovian).

These ideas show that to understand the solar system, we must understand how it formed and how it evolved over time. To do this, we need to understand the origin of the basic ingredients of the solar system—the particular mix of chemical elements out of which planets, satellites, asteroids, comets, and the Sun itself are made. The answers to many of these questions lie in the stars, for this is where most of the chemical elements are manufactured.

7-6   The relative abundances of the elements are the result of cosmic processes

Some chemical elements are very common in our solar system, but others are very rare. Hydrogen is by far the most abundant substance in the solar system: It makes up nearly three-quarters of the combined mass of the Sun and planets. (We have seen that Jupiter, whose mass is greater than that of all the other planets combined, is composed mostly of hydrogen.) Helium is the second most abundant element. Together, hydrogen and helium account for about 98% of the mass of all the material in the solar system. All of the other chemical elements combined make up the remaining 2%.

The dominance of hydrogen and helium is not merely a characteristic of our local part of the universe. By analyzing the spectra of stars and galaxies, astronomers have found the same pattern of chemical abundances out to the farthest dis-

tance attainable by the most powerful telescopes. It is fair to say that the vast majority of the atoms in the universe are hydrogen and helium atoms. The elements that make up the bulk of the Earth (mostly iron, oxygen, and silicon) are relatively rare in the universe as a whole, as are the elements of which living organisms are made (carbon, oxygen, nitrogen, and phosphorus, among others).

There is a good reason for this overwhelming abundance of hydrogen and helium. A wealth of evidence has led astronomers to conclude that the universe began some 10 to 15 billion years ago with a violent event called the Big Bang (see Section 1-4). Only the lightest elements—hydrogen, helium, and tiny amounts of lithium and beryllium—emerged from the enormously high temperatures following this cosmic event. All the heavier elements were manufactured by stars later, either by nuclear fusion reactions deep in their interiors or by the violent explosions that mark the end of massive stars. Were it not for these processes that take place only in stars, there would be no heavy elements in the universe, no planets like Earth, no living creatures, and no one to contemplate the nature of the cosmos.

Because our solar system contains heavy elements, it must be that at least some of its material was once inside other stars. But how did this material become available to help build our solar system? The answer is that near the ends of their lives, stars cast much of their matter back out into space. For most stars this process is a comparatively gentle one, in which a star's outer layers are gradually expelled. Figure 7-9 shows a star which is losing material in this fashion. This ejected

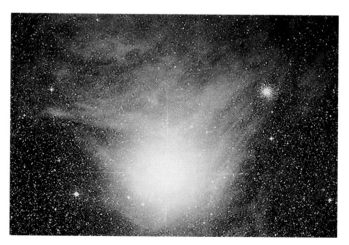

figure 7-9   R I **V** U X G

A Mass-Loss Star The star Antares is shedding material from its outer layers that forms a cloud around the star. Although this cloud is very thin, we can see it because some of the ejected material has condensed into tiny grains of dust that reflect the star's light. Dust particles in the air around you reflect light in the same way, which is why you can see them within a shaft of sunlight in a darkened room. Antares lies some 600 light-years from Earth in the constellation Scorpio. (David Malin/Anglo-Australian Observatory) material appears as the cloudy region, or **nebulosity** (from *nubes,* Latin for "cloud"), that surrounds the star and is illuminated by it. A few stars eject matter much more dramatically at the very end their lives, in a spectacular detonation called a *supernova,* which blows the star apart (see Figure 1-6).

No matter how it escapes, the ejected material contains heavy elements dredged up from the star's interior, where they were formed. This material becomes part of the **interstellar medium,** a tenuous collection of gas and dust that pervades the spaces between the stars. As different stars die, they increasingly enrich the interstellar medium with heavy elements. Observations show that new stars form as condensations in the interstellar medium (Figure 7-10). Thus, these new stars have an adequate supply of heavy elements from which to develop a system of planets, satellites, comets, and asteroids. Our own solar system must have formed from enriched material in just this way.

Stars create different heavy elements in different amounts. For example, carbon, oxygen, silicon, and iron are readily produced in the interiors of massive stars, whereas gold is created only under special circumstances. Consequently, gold is rare in our solar system and in the universe as a whole, while carbon is relatively abundant (although still much less abundant than hydrogen or helium).

A convenient way to express the relative abundances of the various elements is to say how many atoms of a particular element are found for every trillion (10^{12}) hydrogen atoms. For example, for every 10^{12} hydrogen atoms in space, there are about 70 billion (70×10^9, or 7×10^{10}) helium atoms. From spectral analysis of stars and chemical analysis of Earth rocks, Moon rocks, and meteorites, scientists have determined the relative abundances of the elements in our part of the Galaxy today. Table 7-3 lists the ten most abundant elements, in order of their **atomic number.** An element's atomic number is the number of protons in the nucleus of an atom of that element. It is also equal to the number of electrons orbiting the nucleus (see Box 5-5). In general, the greater the atomic number of an atom, the greater its mass.

figure 7-10 R I **V** U X G

A Dusty Region of Star Formation Unlike Figure 7-9, which depicts an old star that is ejecting material into space, this image shows young stars in the constellation Orion (the Hunter) that have only recently formed from a cloud of gas and dust. The bluish, wispy appearance of the cloud (called NGC 1973-1975-1977) is caused by starlight reflecting off interstellar dust grains within the cloud. The grains are made of heavy elements produced by earlier generations of stars. (David Malin/Anglo-Australian Observatory)

In addition to these ten very common elements, five elements are moderately abundant: sodium, aluminum, argon, calcium, and nickel. These elements have abundances in the range of 10^6 to 10^7 relative to the standard 10^{12} hydrogen atoms. Most of the other elements are much rarer. For exam-

table 7-3	**Abundances of the Most Common Elements**		
Atomic number	Element	Symbol	Relative abundance
1	hydrogen	H	1×10^{12}
2	helium	He	7×10^{10}
6	carbon	C	4×10^8
7	nitrogen	N	9×10^7
8	oxygen	O	7×10^8
10	neon	Ne	1×10^8
12	magnesium	Mg	4×10^7
14	silicon	Si	4×10^7
16	sulfur	S	2×10^7
26	iron	Fe	3×10^7

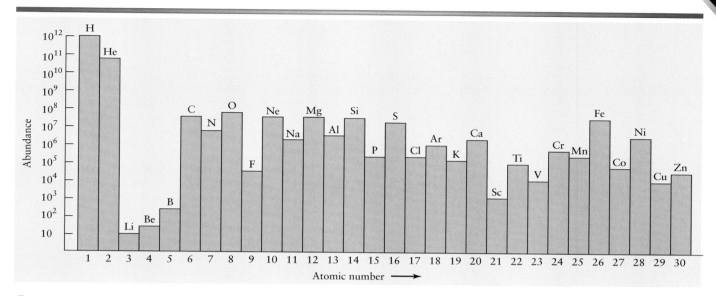

figure 7-11

Abundances of the Lighter Elements This graph shows the abundances of the 30 lightest elements (listed in order of increasing atomic number) compared to a value of 10^{12} for hydrogen. Notice that the vertical scale is not a linear one; each division on the scale corresponds to a tenfold increase in abundance. All elements heavier than zinc (Zn) have abundances less than 1000 atoms per 10^{12} atoms of hydrogen.

ple, for every 10^{12} hydrogen atoms in the solar system, there are only six atoms of gold. Figure 7-11, an extension of Table 7-3, shows the relative abundances of the 30 lightest elements.

Some 4.6 billion years ago, a collection of hydrogen, helium, and heavy elements came together to form the Sun and all the objects that orbit around it. All of those heavy elements, including the carbon atoms in your body and the oxygen atoms that you breathe, were created and cast off by stars that lived and died during the first 5 to 10 billion years of the universe's existence. Thus, by studying the abundances of the elements, we are led to a remarkable insight: We are literally made of star dust.

7-7 The Sun and planets formed from a solar nebula

Processes in the Big Bang and within ancient stars produced the ingredients to make our solar system. But given these ingredients, how did the Sun and planets form? In other words, where did the solar system come from?

This question has tantalized astronomers for centuries. While we do not yet have a wholly complete answer, we think we have identified the most likely series of events that led to the present-day system of Sun and planets.

A key piece of evidence about the origin of the solar system is that all the planets orbit the Sun in the same direction and in nearly the same plane, as we saw in Section 7-1. In the eighteenth century, the German philosopher Immanuel

Kant and the French scientist Pierre-Simon de Laplace independently suggested that this state of affairs could not be a coincidence. They proposed that our entire solar system— the Sun as well as all of its planets and satellites—formed from a vast, rotating cloud of gas and dust called the **solar nebula** (Figure 7-12*a*).

The consensus among today's astronomers is that Kant and Laplace were exactly right. In the modern version of their theory, at the outset the solar nebula was similar in character to the nebulosity shown in Figure 7-10 and had a mass somewhat greater than that of our present-day Sun.

Each part of the nebula exerted a gravitational attraction on the other parts, and these mutual gravitational pulls tended to make the nebula contract. As it contracted, the greatest concentration of matter occurred at the center of the nebula, forming a relatively dense region called the **protosun**. As its name suggests, this part of the solar nebula eventually developed into the Sun. The planets formed from the much sparser material in the outer regions of the solar nebula. Indeed, the mass of all the planets together is only 0.1% of the Sun's mass.

When you drop a ball, the gravitational attraction of the Earth makes the ball fall faster and faster as it falls; in the same way, material falling inward toward the protosun would have gained speed. As this fast-moving material ran into the protosun, the energy of the collision was converted into thermal energy, causing the temperature deep inside the solar nebula to climb. This process, in which the gravitational energy of a contracting gas cloud is converted into thermal energy, is called **Kelvin-Helmholtz contraction**, after the nineteenth-century physicists who first described it.

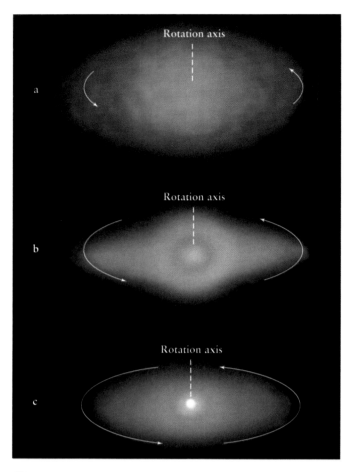

figure 7-12

The Birth of the Solar System This sequence of drawings shows three stages in the formation of the solar system. **(a)** A slowly rotating cloud of interstellar gas and dust begins to contract because of its own gravity. **(b)** A central condensation forms as the cloud flattens and spins more rapidly around its rotation axis. **(c)** Roughly 10^5 years after the cloud begins to contract, a glowing protosun has formed at its center. Surrounding this is a flattened disk of gas and dust, out of which the planets will form.

Temperatures within the newly created protosun soon climbed to several thousand kelvins. While the protosun's surface temperature stayed roughly constant, the temperature inside the protosun increased even more by means of further contraction. Eventually, after perhaps 10^7 (10 million) years had passed since the solar nebula first began to contract, the central temperature of the protosun reached a few million kelvins (that is, a few times 10^6 K), and nuclear reactions that convert hydrogen into helium began in the protosun's interior. When this happened, the contraction stopped and a true star was born. Nuclear reactions in the interior of the present-day Sun are the source of all the energy that the Sun radiates into space.

If the solar nebula had not been rotating at all, everything would have fallen directly into the protosun, leaving nothing behind to form the planets. Instead, the solar nebula must have had an overall slight rotation, which caused its evolution to follow a different path. As the slowly rotating nebula collapsed inward, it would naturally have tended to rotate faster. This relationship between the size of an object and its rotation speed is an example of a general principle called the **conservation of angular momentum**.

ANALOGY Figure skaters make use of the conservation of angular momentum. When a spinning skater pulls her arms and legs in close to her body, the rate at which she spins automatically increases. (If you are not a figure skater, you can demonstrate this yourself by sitting on an office chair. Sit with your arms outstretched and hold a weight, like a shoe or a full water bottle, in either hand. Now use your feet to start yourself rotating, lift your feet off the ground, and then pull your arms inward. Your rotation will speed up quite noticeably.)

As the solar nebula began to rotate more rapidly, it also tended to flatten out (Figure 7-12*b*). From the perspective of a particle rotating along with the nebula, it felt as though there were a force pushing the particle away from the nebula's axis of rotation. (Likewise, passengers on a spinning carnival ride seem to feel a force pushing them outward and away from the ride's axis of rotation.) This apparent force was directed opposite to the inward pull of gravity, and so it tended to slow the contraction of material toward the nebula's rotation axis. But there was no such effect opposing contraction in a direction parallel to the rotation axis. Some 10^5 (100,000) years after the solar nebula first began to contract, it had developed the structure shown in Figure 7-12*c*, with a rotating, flattened disk surrounding the protosun. The planets formed from the material in this disk, which explains why their orbits all lie in essentially the same plane and why they all orbit the Sun in the same direction.

There were no humans to observe these processes taking place during the formation of the solar system. But Earth astronomers have seen disks of material surrounding other stars that formed only recently. These are called **protoplanetary disks**, or **proplyds**, because it is thought that planets can eventually form from these disks around other stars. Thus, proplyds are planetary systems that are still "under construction." By studying these proplyds, astronomers are able to examine what our solar nebula may have been like some 4.6×10^9 years ago.

 Figure 7-13 shows a number of proplyds in the Orion Nebula, a region of active star formation. A star is visible at the center of each proplyd, which reinforces the idea that our Sun began to shine before the planets were fully formed. A study of 110 stars in the Orion Nebula detected proplyds around 56 of them, which suggests that systems of planets may form around a substantial fraction of stars. Later in this chapter we will see direct evidence for planets that have formed around stars other than the Sun.

a

b

figure 7-13   R I **V** U X G

Protoplanetary Disks (Proplyds) in the Orion Nebula **(a)** The Orion Nebula is a star-forming region located some 1500 light-years from Earth. To the naked eye, it is the middle star in Orion's "sword" (see Figure 2-2). The smaller, bluish nebula is the object shown in close-up in Figure 7-13. **(b)** This view of the center of the Orion Nebula is a mosaic of Hubble Space Telescope images. The four insets are extreme close-ups of four proplyds, or protoplanetary disks, that lie within the nebula. False colors are used in the insets to show brightness variations within each proplyd. The inset at the lower left shows the size of our own solar system for comparison. A young, recently formed star is at the center of each proplyd. (The proplyd at upper right is seen nearly edge-on.) The material in each proplyd may, with time, form into planets. (a: Anglo-Australian Observatory image by David Malin; b: C. R. O'Dell and S. K. Wong, Rice University; NASA)

7-8   The planets formed by the accretion of planetesimals and the accumulation of gases in the solar nebula

We have seen how the solar nebula would have contracted to form a young Sun with a protoplanetary disk, or proplyd, rotating around it. But how did this disk coalesce into planets? And how did the formation of the inner, high-density, terrestrial planets differ from that of the outer, low-density, Jovian planets?

To understand how the planets, asteroids, and comets formed, we must look at the conditions that prevailed within the solar nebula. The density of material in the nebula was rather low, as was the pressure of the nebula's gas. If the pressure is sufficiently low, a substance cannot remain in the liquid state, but must end up as either a solid or a gas. For a given pressure, what determines whether a certain substance is a solid or a gas is its **condensation temperature.** If the temperature of a substance is above its condensation temperature,

the substance is a gas; if the temperature is below the condensation temperature, the substance solidifies into tiny specks of dust or snowflakes. You can often see similar behavior on a cold morning. The air temperature can be above the condensation temperature of water, while the cold windows of parked cars may have temperatures below the condensation temperature. Thus, water molecules in the air remain as a gas (water vapor) but form solid ice particles on the car windows.

Substances such as water (H_2O), methane (CH_4), and ammonia (NH_3) have low condensation temperatures, ranging from 100 to 300 K. Rock-forming substances have much higher condensation temperatures, in the range from 1300 to 1600 K. The gas cloud from which the solar system formed had an initial temperature near 50 K, so all of these substances could have existed in solid form. Thus, the solar nebula would have been populated by an abundance of small ice particles and dust grains like the one in Figure 7-14. (Recall that "ice" can refer to frozen CO_2, CH_4, or NH_3, as well as frozen water.) But hydrogen and helium, the most abundant elements in the solar nebula, have condensation temperatures so near absolute zero that these substances always existed as gases

Figure 7-14

A Grain of Cosmic Dust This greatly enlarged image shows a tiny dust grain from interplanetary space that entered Earth's upper atmosphere and was collected by a high-flying aircraft. Dust grains of this sort are abundant in star-forming regions like that shown in Figure 7-13, and were almost certainly abundant in the nebula from which our solar system formed. These tiny grains served as the building blocks of the planets. The grain shown here is about 20 μm (0.02 mm) long. (Donald Brownlee, University of Washington)

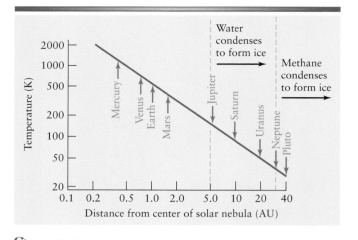

Figure 7-15

Temperature Distribution in the Solar Nebula This graph shows how temperatures probably varied across the solar nebula as the planets were forming. For the inner planets, temperatures ranged from roughly 1200 K at Mercury to 500 K at the orbit of Mars. From the position of Jupiter outward, temperatures were everywhere less than 200 K. Beyond 5 AU from the center of the nebula, temperatures were low enough for water to condense and form ice; beyond 30 AU, methane (CH_4) could also condense.

during the creation of the solar system. You can best visualize the initial state of the solar nebula as a thin gas of hydrogen and helium strewn with tiny dust particles.

This state of affairs changed as the central part of the solar nebula underwent Kelvin-Helmholtz contraction to form the protosun. During this phase, the protosun was actually quite a bit more luminous than the present-day Sun, and it heated the inner part of the solar nebula to temperatures above 2000 K. Meanwhile, temperatures in the outermost regions of the solar nebula remained below 50 K.

Figure 7-15 shows the probable temperature distribution in the solar nebula at this stage in the formation of our solar system. In the inner regions of the solar nebula, water, methane, and ammonia were vaporized by the high temperatures. Only materials with high condensation temperatures could have remained solid. Of these materials, iron, silicon, magnesium, and sulfur were particularly abundant, followed closely by aluminum, calcium, and nickel. (These elements were also present in the form of oxides, which are chemical compounds containing oxygen. These elements also have high condensation temperatures.) In contrast, ice particles and ice-coated dust grains were able to survive in the cooler, outer portions of the solar nebula.

Small chunks would have formed in the inner part of the solar nebula from collisions between neighboring dust grains. Initially electric forces (chemical bonds) held them together. Over a few million years, these chunks coalesced into roughly 10^9 asteroidlike objects called **planetesimals,** with diameters

of a kilometer or so. During the next stage, the gravitational attraction between the planetesimals caused them to collide and coalesce into still-larger objects called **protoplanets,** which were roughly the size and mass of our Moon. This accumulation of material to form larger and larger objects is called **accretion.** During the final stage, these Moon-sized protoplanets collided to form the terrestrial planets. This final episode must have involved some truly spectacular, world-shattering collisions.

Important clues to this stage in the evolution of the inner solar system come from studies of **meteorites,** bits of rocky solar system debris that have collided with the Earth. Many of these are fragments of planetesimals that were never incorporated into the planets. Most meteorites contain not only dust grains but also **chondrules,** which are small, glassy, roughly spherical blobs (Figure 7-16). (The "ch" in "chondrule" is pronounced as in "chord.") Liquids tend to form spherical drops (like drops of water), so the shape of chondrules shows that they were once molten.

Attempts to produce chondrules in the laboratory show that the blobs must have melted suddenly, then solidified over the space of only an hour or so. This means they could not have been melted by the temperature of the inner solar nebula, which would have remained high for hundreds of thousands of years after the formation of the protosun. Instead, chondrules must have been produced by sudden, high-energy events that quickly melted material and then permitted it to cool. It is not clear what these events were, though the sheer number of

|⊢——————— 1 cm ———————⊣|

ⓕigure 7-16   R I **V** U X G

Chondrules This is not a collection of pebbles; it is a cross-section of the interior of a meteorite. The "pebbles" are individual chondrules ranging in size from a few millimeters to a few tenths of a millimeter. Many of the chondrules are coated with dust grains like the one shown in Figure 7-14. (Courtesy of John A. Wood and the Smithsonian Astrophysical Observatory)

chondrules suggests that they happened throughout the inner solar nebula over a period of more than 10^6 years. The words "cloud" and "nebula" may suggest material moving in gentle orbits around the protosun; the existence of chondrules suggests that conditions in the inner solar nebula were at times quite violent.

Astronomers use computer simulations to learn how the inner planets could have formed from planetesimals. A com-

puter is programmed to simulate a large number of particles circling a newborn sun along orbits dictated by Newtonian mechanics. As the simulation proceeds, the particles coalesce to form larger objects, which in turn collide to form planets. By performing a variety of simulations, each beginning with somewhat different numbers of planetesimals in different orbits, it is possible to see what kinds of planetary systems would have been created under different initial conditions. Such studies demonstrate that a wide range of initial conditions ultimately lead to basically the same result in the inner solar system: Accretion continues for roughly 10^8 (100 million) years and typically forms four or five terrestrial planets with orbits between 0.3 and 1.6 AU from the Sun.

Figure 7-17 shows one particular computer simulation. The calculations began with 100 planetesimals, each having a mass of 1.2×10^{23} kg. This choice ensures that the total mass (1.2×10^{25} kg) equals the combined mass of the four terrestrial planets (Mercury through Mars) plus their satellites. The initial orbits of planetesimals are inclined to each other by angles of less than 5° to simulate a thin layer of particles orbiting the protosun.

After an elapsed time simulating 30 million (3×10^7) years, the 100 original planetesimals have coalesced into 22 protoplanets. After 79 million (7.9×10^7) years, 11 larger protoplanets remain. Nearly another 100 million (10^8) years elapse before the total number of growing protoplanets is reduced to six. Figure 7-17c shows four planets following nearly circular orbits after a total elapsed time of 441 million (4.41×10^8) years. In this particular simulation, the fourth planet from the Sun ends up being the most massive; in our own solar system, it is the third planet (Earth) that is the most massive of the terrestrial planets. Note that the four planets in the simulation end up in orbits that are nearly circular, just like the orbits of most of the planets in our solar system.

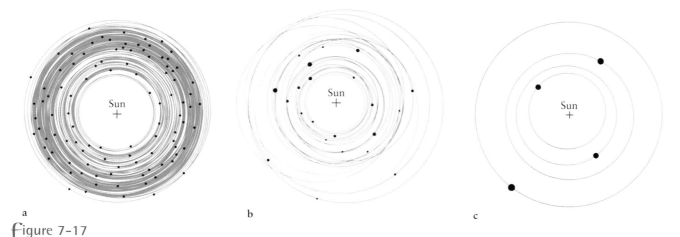

a   b   c

ⓕigure 7-17

Accretion of the Terrestrial Planets These three drawings show the results of a computer simulation of the formation of the inner planets. **(a)** The simulation begins with 100 planetesimals orbiting the Sun. **(b)** After 30 million years, these planetesimals have coalesced into 22 protoplanets. **(c)** This final view shows four planets orbiting the Sun after an elapsed time of 441 million years. In fact, in this simulation the inner planets were essentially formed after 150 million years. (Adapted from George W. Wetherill)

As we have seen, the material from which the inner protoplanets accreted was rich in substances with high condensation temperatures. This material remained largely in solid form despite the high temperatures in the inner solar nebula. But as the protoplanets grew, they were heated by the violent impacts of planetesimals as well as the decay of radioactive elements such as uranium, and this heat caused melting. Thus, the terrestrial planets began their existence as spheres of at least partially molten rock. Material was free to move within these molten spheres, so the denser, iron-rich minerals sank to the centers of the planets while the less dense silicon-rich minerals floated to their surfaces. This process is called **chemical differentiation** (see Box 7-1). In this way the terrestrial planets developed their dense iron cores.

Like the inner planets, the outer planets may have begun to form by the accretion of planetesimals. The key difference is that ices as well as rocky grains were able to survive in the colder outer regions of the solar nebula. The elements of which ices are made are much more abundant than those that form rocky grains. Thus, much more solid material would have been available to form planetesimals in the outer solar nebula than in the inner part. As a result, objects several times larger than any of the terrestrial planets formed in the outer solar nebula. Each such object, made up of a mixture of ices and rock, could have become the core of a Jovian planet and served as a "seed" around which the rest of the planet eventually grew.

Thanks to the lower temperatures in the outer solar system, gas atoms (principally hydrogen and helium) were moving relatively slowly and so could more easily be captured by the strong gravity of these massive cores (see Box 7-2). Thus, the core of a Jovian protoplanet began to capture an envelope of gas as it continued to grow by accretion. Calculations suggest that both rock and gas slowly accumulated for about a million years, until the masses of the core and the envelope became equal. From that critical moment on, the envelope pulled in all the gas it could get, dramatically increasing the protoplanet's mass and size.

This runaway growth of the protoplanet continued until all the available gas was used up. The result was a huge planet with an enormously thick, hydrogen-rich envelope surrounding a rocky core with 5 to 10 times the mass of the Earth. This scenario occurred at four different distances from the Sun, thus creating the four Jovian planets. Figure 7-18 summarizes this story of the formation of the solar system.

Like the terrestrial planets, the Jovian planets were quite a bit hotter during their formation than they are today. A heated gas expands, so these gas-rich planets must also have been much larger than at present. As each planet cooled and contracted, it would have formed a disk like a solar nebula in miniature (see Figure 7-12). The satellites of the Jovian planets are thought to have formed from ice particles and dust grains within these disks.

The Jovian planets may also explain why there are asteroids and comets. Jupiter, the most massive of the planets, was almost certainly the first to form. Jupiter's tremendous mass would have exerted strong gravitational forces on any nearby planetesimals. Some would have been sent crashing into the terrestrial planets or into the Sun, while others would have

a

b

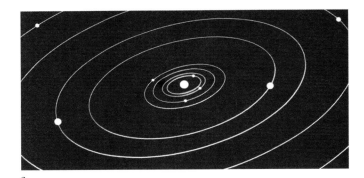

c

figure 7-18

Formation of the Planets These sketches show the major stages in the birth of the solar system. **(a)** The solar nebula in its initial stages. **(b)** Terrestrial planets accrete from rocky material in the warm inner regions of the solar nebula. Meanwhile, the huge, gaseous Jovian planets form in the cold outer regions. **(c)** Planetary formation is nearly complete after some 10^8 (100 million) years.

been ejected completely from the solar system. A relatively few rocky planetesimals survived to produce the present-day population of asteroids. Comets, by contrast, began as icy planetesimals beyond the orbit of Jupiter. The strong gravitational forces from all of the Jovian planets pushed these into orbits even further from the Sun. Only a miniscule fraction of these wander close enough to the Sun to produce a visible tail (see Figure 7-8).

a

b

WEB LINK 7.16

Figure 7-19   R I **V** U X G

Jets and Winds from Young Stars (a) This young star in the southern constellation Vela (the Sails) is ejecting matter at speeds of hundreds of kilometers per second, forming an immense jet. The jet is about 0.2 light-year (12,500 AU) in length. **(b)** An outpouring of particles and radiation from the surfaces of these young stars has carved out a cavity in the surrounding dusty material. The stars lie within the Trifid Nebula in the constellation Sagittarius (the Archer). (a: C. Burrows, STScI; J. Hester, Arizona State University; J. Morse, STScI; and NASA; b: David Malin/Anglo-Australian Observatory)

While the planets, satellites, asteroids, and comets were forming, the protosun was also evolving into a full-fledged star with nuclear reactions occurring in its core (see Section 1-3). The time required for this to occur was about 10^8 years, roughly the same as that required for the formation of the terrestrial planets. Before the onset of nuclear reactions, however, the young Sun probably expelled a substantial portion of its mass into space. Magnetic fields within the solar nebula would have funneled a portion of the nebula's mass into oppositely directed **jets** along the rotation axis of the nebula. Figure 7-19a shows one such jet emanating from a young star.

In addition, instabilities within the young Sun would have caused it to eject its tenuous outermost layers into space. This brief but intense burst of mass loss, observed in many young stars across the sky, is called a **T Tauri wind,** after the star in the constellation Taurus (the Bull) where it was first identified (see Figure 7-19b). Each of the proplyds in Figure 7-13 has a T Tauri star at its center. (The present-day Sun also loses mass from its outer layers in the form of high-speed electrons and protons, a flow called the **solar wind.** But this is miniscule compared with a T Tauri wind, which causes a star to lose mass 10^6 to 10^7 times faster than in the solar wind.)

With the passage of time, the combined effects of jets, the T Tauri wind, and accretion onto the planets would have swept the solar system nearly clean of gas and dust. With no more interplanetary material to gather up, the planets would have stabilized at roughly their present-day sizes, and the formation of the solar system would have been complete.

7-9   Astronomers have discovered planets orbiting other stars

If planets formed around our Sun, have they formed around other stars? That is, are there **extrasolar planets** orbiting stars other than the Sun? Our model for the formation of the planets would seem to suggest so. This model is based on the laws of physics and chemistry, which to the best of our understanding are the same throughout the universe. The discovery of a set of planets orbiting another star, with terrestrial planets in orbit close to the star and Jovian planets orbiting further away, would be a tremendous vindication of our theory of solar system formation. It would also be one of the most remarkable scientific discoveries of all time, for it would tell us that our planetary system is not unique in the universe. Because at least one planet in our solar system—the Earth—has the ability to support life, perhaps other planetary systems could also harbor living organisms.

Since 1995 evidence has accumulated for the existence of dozens of planets orbiting stars other than the Sun. However, none of these is a terrestrial planet. Most of them are more massive than Jupiter, and some are in eccentric, noncircular orbits quite unlike planetary orbits in our solar system. There is even some controversy as to whether these objects orbiting other stars are really planets at all. To understand this controversy, we must look at the process that astronomers go through to search for extrasolar planets.

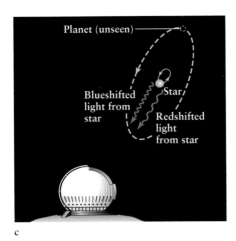

a
b
c

ƒigure 7-20

Detecting Planets Orbiting Other Stars **(a)** A planet and its star both orbit around their common center of mass, always staying on opposite sides of this point. Even if the planet cannot be seen, its presence can be inferred if the star's motion can be detected. **(b)** The astrometric method of detecting the unseen planet involves making direct measurements of the star's orbital motion. **(c)** In the radial velocity method, astronomers measure the Doppler shift of the star's spectrum as it moves alternately toward and away from the Earth. The amount of Doppler shift can be used to determine the size of the star's orbit, which in turn allows a determination of the unseen planet's orbit.

It is very difficult to make direct observations of planets orbiting other stars. The problem is that planets are small and dim compared with stars; at visible wavelengths, the Sun is 10^9 times brighter than Jupiter and 10^{10} times brighter than the Earth. A hypothetical planet orbiting a distant star, even a planet 10 times larger than Jupiter, would be lost in the star's glare as seen through even the largest telescope on Earth.

Instead, indirect methods must be used to search for extrasolar planets. One very powerful method is to search for stars that appear to "wobble." If a star has a planet, it is not quite correct to say that the planet orbits the star. Rather, both the planet and the star move in elliptical orbits around a point called the **center of mass.** Imagine the planet and the star as sitting at opposite ends of a very long seesaw; the center of mass is the point where you would have to place the fulcrum in order to make the seesaw balance. Because of the star's much greater mass, the center of mass is much closer to the star than to the planet. Thus, while the planet may move in an orbit that is hundreds of millions of kilometers across, the star will move in a much smaller orbit (Figure 7-20a).

As an example, the Sun and Jupiter both orbit their common center of mass with an orbital period of 5.2 years. (Jupiter has more mass than the other eight planets put together, so it is a reasonable approximation to consider the Sun's wobble as being due to Jupiter alone.) Jupiter's orbit has a semimajor axis of 7.78×10^8 km, while the Sun's orbit has a much smaller semimajor axis of 742,000 km. The Sun's radius is 696,000 km, so the Sun slowly wobbles around a point not far outside its surface. If astronomers elsewhere in the Galaxy could detect the Sun's wobbling motion, they could tell that there was a large planet (Jupiter) orbiting the Sun. They could

even determine the planet's mass and the size of its orbit, even though the planet itself was unseen.

Detecting the wobble of other stars is not an easy task. One approach to the problem, called the **astrometric method,** involves making very precise measurements of a star's position in the sky relative to other stars. The goal is to find stars whose positions change in a cyclic way (Figure 7-20b). The measurements must be made with very high accuracy (0.001 arcsec or better) and, ideally, over a long enough time to span an entire orbital period of the star's motion.

A different approach to the problem is the **radial velocity method** (Figure 7-20c). This is based on the Doppler effect, which we described in Section 5-9. A wobbling star will alternately move away from and toward the Earth. This will cause the dark absorption lines in the star's spectrum (see Figures 5-12 and 5-18) to change their wavelengths in a periodic fashion. When the star is moving away from us, its spectrum will undergo a redshift to longer wavelengths. When the star is approaching, there will be a blueshift of the spectrum to shorter wavelengths. These wavelength shifts are very small because the star's motion around its orbit is quite slow. As an example, the Sun moves around its small orbit at only 12.5 m/s (45 km/h, or 28 mi/h). If the Sun was moving directly toward an observer at this speed, the hydrogen absorption line at a wavelength of 656 nm in the Sun's spectrum would be shifted by only 2.6×10^{-5} nm, or about 1 part in 25 million. Detecting these tiny shifts requires extraordinarily careful measurements and painstaking data analysis.

In 1995, Michel Mayor and Didier Queloz of the Geneva Observatory in Switzerland announced that they had used the radial velocity method to discover a planet orbiting the

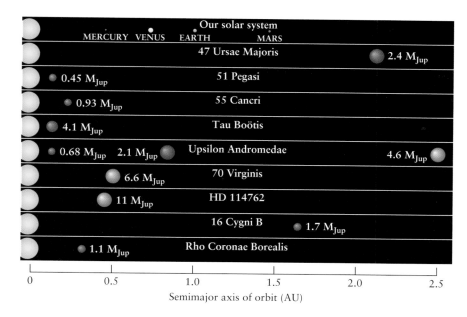

figure 7-21

A Selection of Extrasolar Planets
This figure summarizes what we know about planets orbiting nine other stars. The star name is given at the middle of each line. Each planet is shown at its average distance from its star (equal to the semimajor axis of its orbit). Comparison with our own solar system (at the top of the figure) shows how closely these extrasolar planets orbit their stars. The planets are *not* shown to the same scale as their stars, and the relative sizes of the extrasolar planets are estimates only. The mass of each planet—actually a lower limit—is given as a multiple of Jupiter's mass (M_{Jup}). (Adapted from G. Marcy and R. P. Butler)

star 51 Pegasi, which is 48 light-years from Earth in the constellation Pegasus (the Winged Horse). Their results were soon confirmed by the team of Geoff Marcy of San Francisco State University and Paul Butler of the University of California, Berkeley, using observations made at the University of California's Lick Observatory. For the first time, solid evidence had been found for a planet orbiting a star like our own Sun. Marcy and Butler, along with other astronomers, have since discovered more than 30 planets orbiting other stars by means of the radial velocity method. Figure 7-21 depicts a selection of these discoveries.

The extrasolar planets discovered by the radial velocity method all have masses comparable with or larger than that of Saturn, and thus are presumably Jovian planets made primarily of hydrogen and helium. (This cannot be confirmed directly, because the spectra of the planets are too faint to be detected with present technology.) According to the picture we presented in Section 7-8, such Jovian planets would be expected to orbit relatively far from their stars, where temperatures were low enough to allow the buildup of a massive envelope of hydrogen and helium gas. But as Figure 7-21 shows, many extrasolar planets are in fact found orbiting very *close* to their stars. For example, the planet orbiting 51 Pegasi has a mass at least 0.45 times as great as that of Jupiter but orbits only 0.051 AU from its star with an orbital period of just 4.2 days. In our own solar system, this orbit would lie well inside the orbit of Mercury!

Another surprising result is that about a third of the extrasolar planets found so far have very eccentric orbits. As an example, the planet around 16 Cygni B has an orbital eccentricity of 0.67; its distance from the star varies between 0.6 AU and 2.7 AU. This is quite unlike planetary orbits in our own solar system, where no planet has an orbital eccentricity greater than 0.25.

Do these observations mean that our picture of how planets form is incorrect? If Jupiterlike extrasolar planets such as

that orbiting 51 Pegasi formed close to their stars, the mechanism of their formation must have been very different from that which operated in our solar system. But another possibility is that extrasolar planets actually formed at large distances from their stars and have migrated inward since their formation. If enough gas and dust remains in a disk around a star after its planets form, interactions between the disk material and an orbiting planet will cause the planet to lose energy and to spiral inward toward the star around which it orbits. The planet's inward migration can eventually stop because of subtle gravitational effects from the disk or from the star. Gravitational interactions between the planet and the disk, or between planets in the same planetary system, could also have forced an extrasolar planet into a highly eccentric, noncircular orbit.

Yet a third possibility is that some of these extrasolar planets are not planets at all! The problem is that the radial velocity method cannot give precise values for the masses of planets, only lower limits. (An exact determination of the mass would require knowing how the plane of the planet's orbit is inclined to our line of sight. Unfortunately, this angle is not known because the planets themselves are unseen.) Hence, the actual masses of some of the "planets" shown in Figure 7-21 may be much larger than the values shown in that figure. It is thought that planets more massive than about 13 times the mass of Jupiter are actually **brown dwarfs,** objects that are like the Sun but insufficiently massive to sustain nuclear reactions in their cores (Figure 7-22). Like stars, but unlike true planets, brown dwarfs are thought to form directly from gas by gravitational contraction rather than by accretion of planetesimals.

Happily, some extrasolar planets appear to be of a sort that could fit into our own solar system. The best example is the planet orbiting 47 Ursae Majoris, a star very similar to the Sun. This planet is in a nearly circular orbit with a radius of 2.1 AU, intermediate between the orbits of Mars and Jupiter, and has a mass of at least 2.4 times that of Jupiter. Some

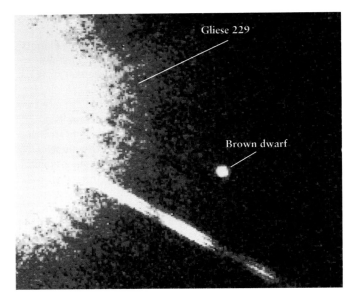

figure 7-22   R **I** V U X G

A Brown Dwarf This false-color infrared image shows a brown dwarf that orbits the star Gliese 229 at an average distance of 32 AU. The brown dwarf is estimated to have a mass of 30 to 55 times the mass of Jupiter (between 0.030 and 0.055 of the Sun's mass) but about the same radius as Jupiter. Its surface temperature is about 1000 K. The spike running diagonally is an illusion caused by the telescope optics. Gliese 229 and its brown dwarf are 19 ly from Earth in the constellation Lepus (the Hare). (S. Kulkarni, Caltech; D. Golimowski, Johns Hopkins University; and NASA)

simulations of planet formation produce planets with orbital radii and masses in this range. Evidence for another solar system-like case has been presented by George Gatewood of the Allegheny Observatory. Applying the astrometric method to the star Lalande 21185, he finds that the wobble of that star indicates the presence of *two* planets, one with 0.9 Jupiter masses at a distance of 2.2 AU from the star and another with 1.1 Jupiter masses at 11 AU from the star. (These masses are actual values, not lower limits, because the astrometric method makes it possible to determine the inclinations of the planets' orbits.) There is also evidence for a system of three planets orbiting Upsilon Andromedae (see Figure 7-21).

The study of extrasolar planets is still in its infancy. We know little about how they formed or how their formation may have differed from that of our own solar system. We do not yet know what extrasolar planets look like or what they are made of. No Earth-sized extrasolar planets have yet been discovered, because their masses are too small to be detected by either the radial velocity method or the astrometric method using current telescope technology. Hence, we do not yet know whether solar systems like our own exist around other stars.

Eventually, however, advances in adaptive optics (see Section 6-3) may make it possible to record images of extrasolar planets the size of Jupiter. Orbiting telescopes are now being planned by both NASA and the European Space Agency to search for terrestrial planets around other stars. The next few decades may tell whether our own solar system is an exception or just one of a host of similar systems throughout the Galaxy.

KEY WORDS

Terms preceded by an asterisk (*) are discussed in the Boxes.

accretion, p. 166

asteroid, p. 158

asteroid belt, p. 158

astrometric method (for detecting extrasolar planets), p. 170

atomic number, p. 162

average density, p. 154

brown dwarf, p. 171

center of mass, p. 170

chemical composition, p. 155

chemical differentiation, p. 168

chondrule, p. 166

comet, p. 160

condensation temperature, p. 165

conservation of angular momentum, p. 164

*escape speed, p. 160

extrasolar planet, p. 169

ices, p. 158

interstellar medium, p. 162

jets, p. 169

Jovian planet, p. 151

Kelvin-Helmholtz contraction, p. 163

*kinetic energy, p. 159

meteorite, p. 166

minor planet, p. 158

molecule, p. 156

nebulosity, p. 162

planetesimal, p. 166

protoplanet, p. 166

protoplanetary disk (proplyd), p. 164

protosun, p. 163

radial velocity method (for detecting extrasolar planets), p. 170

solar nebula, p. 163

solar wind, p. 169

spectroscopy, p. 155

T Tauri wind, p. 169

terrestrial planet, p. 151

KEY IDEAS

Properties of the Planets: All of the planets orbit the Sun in the same direction and in almost the same plane. Most of the planets have nearly circular orbits.

• The four inner planets are called terrestrial planets. They are relatively small (with diameters of 5000 to 13,000 km), have high average densities (4000 to 5500 kg/m^3), and are composed primarily of rock.

• The four giant outer planets are called Jovian planets. They have large diameters (50,000 to 143,000 km) and low average densities (700 to 1700 kg/m^3) and are composed primarily of hydrogen and helium.

• Pluto appears to be a special case. It is smaller than any of the terrestrial planets and has an intermediate average density of about 1900 kg/m^3, suggesting that it is composed of a mixture of ice and rock.

Satellites and Small Bodies in the Solar System: Besides the planets, the solar system includes satellites of the planets, asteroids, and comets.

• Seven large planetary satellites (one of which is the Moon) are comparable in size to the planet Mercury. The remaining satellites of the solar system are much smaller.

• Asteroids are small, rocky objects, while comets are small objects made of dirty ice. Both are remnants left over from the formation of the planets.

Spectroscopy and the Composition of the Planets: Spectroscopy, the study of spectra, provides information about the chemical composition of objects in the solar system.

• By mass, 98% of the ordinary matter in the universe is hydrogen and helium, the two lightest elements. These elements were probably formed shortly after the universe was created. The heavier elements were produced much later, in the centers of stars, and were cast into space when the stars died.

• The atoms of various elements can combine to form many different kinds of molecules.

• The basic planet-forming substances can be classified as gases, ices, or rock, depending on their condensation temperatures. The terrestrial planets are composed primarily of rock, whereas the Jovian planets are composed largely of gas.

Formation of the Solar System: The solar system formed from a disk-shaped cloud of hydrogen and helium that also contained ice and dust particles.

• The four inner planets formed through the accretion of dust particles into planetesimals, then into larger protoplanets.

• The four outer planets probably began as rocky protoplanetary cores, similar in character to the terrestrial planets. Gas then accreted onto these cores in a runaway fashion.

• The Sun formed by gravitational contraction of the center of the nebula. After about 10^8 years, temperatures at the protosun's center became high enough to ignite nuclear reactions that convert hydrogen into helium, thus forming a true star.

Extrasolar Planets: Astronomers have discovered planets orbiting other stars. The planets themselves are not visible; their presence is detected by the "wobble" of the stars around which they orbit.

• Most of the extrasolar planets discovered to date are quite massive and have orbits that are very different from planets in our solar system.

REVIEW QUESTIONS

1. What are the characteristics of a terrestrial planet?

2. What are the characteristics of a Jovian planet?

3. In what ways does Pluto not fit the usual classification of either terrestrial or Jovian planets?

4. In what ways are the largest satellites similar to the terrestrial planets? In what ways are they different? Which satellites are these?

5. What is an asteroid? What is a comet? In what ways are these minor members of the solar system like or unlike the planets?

6. What is meant by the average density of a planet? What does the average density of a planet tell us?

7. If hydrogen and helium account for 98% of the mass of all the material in the universe, why aren't the Earth and Moon composed primarily of these two gases?

8. What is meant by a substance's condensation temperature? What role did condensation temperatures play in the formation of the planets?

9. What is a planetesimal? How did planetesimals give rise to the planets?

10. What is meant by accretion?

11. What is a chondrule? How do we know they were not formed by the ambient heat of the solar nebula?

12. Why did the terrestrial planets form close to the Sun while the Jovian planets formed far from the Sun?

13. Why is the combined mass of all the asteroids so small?

14. Explain how our current understanding of the formation of the solar system can account for the following characteristics of the solar system: (a) All planetary orbits lie in nearly the same plane. (b) All planetary orbits are nearly circular. (c) The planets orbit the Sun in the same direction that the Sun itself rotates.

15. Explain why most of the satellites of Jupiter orbit that planet in the same direction that Jupiter rotates.

16. What techniques are used to detect planets orbiting other stars? Why are these techniques unable to detect planets like Earth?

17. Summarize the differences between the planets of our solar system and those found orbiting other stars.

18. A 1999 news story about the discovery of three planets orbiting the star Upsilon Andromedae (see Figure 7-21) stated that "the newly discovered galaxy, with three large planets orbiting a star known as Upsilon Andromedae, is 44 light-years away from Earth." What is wrong with this statement?

ADVANCED QUESTIONS

Questions preceded by an asterisk (*) involve topics discussed in the Boxes.

Problem-solving tips and tools

The volume of a sphere of radius r is $\frac{4}{3}\pi r^3$, and the volume of a disk of radius r and thickness t is $\pi r^2 t$. The average density of an object is its mass divided by its volume. To calculate escape speeds, you will need to review Box 7-2; to compute the mass of Mars or 70 Virginis, you may have to review Box 4-4. Be sure to use the same system of units (meters, seconds, kilograms) in all your calculations involving escape speeds, orbital speeds, and masses. Appendix 6 gives conversion factors between different sets of units. Box 5-1 has formulas relating various temperature scales, and Box 1-1 explains the relationship between the angular size of an object and its actual size.

19. Mars has two small satellites, Phobos and Deimos. Phobos circles Mars once every 0.31891 day at an average altitude of 5980 km above the planet's surface. The diameter of Mars is 6794 km. Using this information, calculate the mass and average density of Mars.

20. Figure 7-3 shows the spectrum of Saturn's largest satellite, Titan. Can you think of a way that astronomers can tell which absorption lines are due to Titan's atmosphere and which are due to the atmospheres of the Sun and Earth?

21. Table 7-3 shows that carbon, nitrogen, and oxygen are the most abundant elements (after hydrogen and helium). In our solar system, the atoms of these elements are found primarily in the molecules CH_4 (methane), NH_3 (ammonia), and H_2O (water). Why do you suppose this is?

*22. A hypothetical spherical asteroid 2 km in diameter, and composed of rock with an average density of 2500 kg/m^3, strikes the Earth with a speed of 25 km/s. (a) What is the kinetic energy of the asteroid at the moment of impact? (b) How does this energy compare with that released by a 20-kiloton nuclear weapon, like the device that destroyed Hiroshima, Japan, on August 6, 1945? (*Hint*: 1 kiloton of TNT releases 4.2×10^{12} joules of energy.)

*23. (a) Find the speed required to escape from the surface of the asteroid described in Question 22. (b) A typical jogging speed is 3 m/s. What would happen to an astronaut who decided to go for a jog on this asteroid?

*24. A hydrogen atom has a mass of 1.673×10^{-27} kg, and the temperature of the Sun's surface is 5800 K. What is the average speed of hydrogen atoms at the Sun's surface?

*25. The Sun's mass is 1.989×10^{30} kg, and its radius is 6.96×10^8 m. (a) What is the escape speed from the Sun's surface? (b) Using your answer to Question 24, explain why the Sun has lost very little hydrogen over its entire 4.6-billion-year history.

*26. Suppose a spacecraft landed on Jupiter's moon Europa (see Table 7-2), which moves around Jupiter in an orbit of radius 670,900 km. After collecting samples from the satellite's surface, the spacecraft prepares to return to Earth. (a) Calculate the escape speed from Europa. (b) Calculate the escape speed from Jupiter at the distance of Europa's orbit. (c) In order to begin its homeward journey, the spacecraft must leave Europa with a speed greater than either your answer to (a) or your answer to (b). Explain why.

27. (a) If the Earth had retained hydrogen and helium in the same proportion to the heavier elements that exist elsewhere in the universe, what would its mass be? Give your answer as a multiple of the Earth's actual mass, given in Table 7-1. Explain your reasoning. (b) How does your answer to (a) compare with the masses of the Jovian planets? What does your answer imply about how large the cores of the Jovian planets must be?

28. Imagine a planet with roughly the same mass as Earth but located 50 AU from the Sun. (a) What do you think this planet would be made of? Explain your reasoning. (b) On the basis of this speculation, assume a reasonable density for this planet and calculate its diameter. How many times bigger or smaller than Earth would it be?

29. The distance from the asteroid 433 Eros (Figure 7-7) to the Sun varies between 1.13 and 1.78 AU. (a) Find the period of Eros's orbit. (b) Does Eros lie in the asteroid belt? How can you tell?

30. Suppose you were to use the Hubble Space Telescope to monitor one of the protoplanetary disks shown in Figure 7-13*b*. Over the course of ten years, would you expect to see planets forming within the disk? Why or why not?

31. The protoplanetary disk, or proplyd, at the upper right of Figure 7-13*b* is seen edge-on. The diameter of the disk is about 700 AU. (a) Make measurements on this image to

determine the thickness of the disk in AU. (b) Explain why the disk will continue to flatten as time goes by.

32. The accompanying infrared image shows IRAS 04302+2247, a young star that is still surrounded by a disk of gas and dust. The scale bar at the lower right of the image shows that at the distance of IRAS 04302+2247, an angular size of 2 arcseconds corresponds to a linear size of 280 AU. Find the distance to IRAS 04302+2247.

2"
280 AU

(Courtesy of D. Padgett; W. Brandner,   R **I** V U X G
IPAC/Caltech; K. Stapelfeldt, JPL; and NASA)

33. The image accompanying Question 32 shows a dark, opaque disk of material surrounding the young star IRAS 04302+2247. The disk is edge-on to our line of sight, so it appears as a dark band running vertically across this image (north to south in the sky). The material to the left and right of this band is still falling onto the disk. (a) Make measurements on this image to determine the diameter of the disk in AU. Use the scale bar at the lower right of this image. (b) If the thickness of the disk is 50 AU, find its volume in cubic meters. (c) The total mass of the disk is perhaps 2×10^{28} kg (0.01 of the mass of the Sun). How many atoms are in the disk? Assume that the disk is all hydrogen. A single hydrogen atom has a mass of 1.673×10^{-27} kg. (d) Find the number of atoms per cubic meter in the disk. By comparison, the air that you breathe contains about 5.4×10^{25} atoms per cubic meter. Would you describe the disk material as thick or thin by Earth standards?

34. Discuss why the "planet" that orbits the star HD 114762 (see Figure 7-21) could possibly be a brown dwarf, while the planet that orbits 51 Pegasi probably is not.

*****35.** The planet discovered orbiting the star 70 Virginis, 59 light-years from Earth, moves in an orbit with semimajor axis 0.47 AU and eccentricity 0.40. The period of the orbit is 116.6 days. Find the mass of 70 Virginis. Compare your answer with the mass of the Sun.

*****36.** Because of the presence of Jupiter, the Sun moves in a small orbit of radius 742,000 km with a period of

11.86 years. (a) Calculate the Sun's orbital speed in meters per second. (b) An astronomer on a hypothetical planet orbiting the star Vega, 25 light-years from the Sun, wants to use the astrometric method to search for planets orbiting the Sun. What would be the angular diameter of the Sun's orbit as seen by this alien astronomer? Would the Sun's motion be discernible if the alien astronomer could measure positions to an accuracy of 0.001 arcsec? (c) Repeat part (b), but now let the astronomer be located on a hypothetical planet in the Pleiades star cluster, 360 light-years from the Sun. Would the Sun's motion be discernible to this astronomer?

DISCUSSION QUESTIONS

37. Propose an explanation why the Jovian planets are orbited by terrestrial-like satellites.

38. Suppose that a planetary system is now forming around some protostar in the sky. In what ways might this planetary system turn out to be similar to or different from our own solar system? Explain your reasoning.

39. Suppose astronomers discovered a planetary system in which the planets orbit a star along randomly inclined orbits. How might a theory for the formation of that planetary system differ from that for our own?

WEB/CD-ROM QUESTIONS

40. Search the World Wide Web for information about recent observations of protoplanetary disks. What insights have astronomers gained from these observations? Is there any evidence that planets have formed within these disks?

41. In 2000, extrasolar planets with masses comparable to that of Saturn were first detected around the stars HD 16141 (also called 79 Ceti) and HD 46375. (All extrasolar planets discovered prior to these are more massive than Jupiter.) Search the World Wide Web for information about these "lightweight" planets. Do these planets move around their stars in the same kind of orbit as Saturn follows around the Sun? Why do you suppose this is? How does the discovery of these planets reinforce the model of planet formation described in Section 7-8?

42. As of this writing, not even the most powerful telescopes have been able to actually *see* planets orbiting other stars. Nonetheless, astronomers have evidence of having "seen" an extrasolar planet come between its star and us. Search the World Wide Web for information about this planet, which orbits the star HD 209458. If the planet is too small and faint to be seen, how can astronomers tell that it came between HD 209458 and us? What have they learned about the planet from this event that cannot be learned with the radial velocity method?

43. Determining Terrestrial Planet Orbital Periods. Access the animation "Planetary Orbits" in Chapter 6 of the *Universe* web site or CD-ROM. Focus on the motions of the inner planets at the last half of the animation. Using the stop and start buttons, determine how many days it takes Mars, Venus, and Mercury to orbit the Sun once if Earth takes approximately 365 days.

OBSERVING PROJECTS

44. There are many young stars still embedded in the clouds of gas and dust from which they formed. In the winter evening sky, for example, is the famous Orion Nebula (Figure 7-13*a*). In the summer night sky, the Lagoon, Omega, and Trifid nebulae are found in the Milky Way. Examine some of these nebulae with a telescope. Describe their appearance. Can you guess which stars in your field of view are actually associated with the nebulosity? You can use the *Starry Night* program that comes with this book to help you find some of these

nebulae. The table below gives their coordinates (right ascension and declination; see Box 2-1) for the year 2000.

Nebula	Right ascension	Declination
Lagoon	18^h 03.8^m	−24° 23′
Omega	18 20.8	−16 11
Trifid	18 02.3	−23 02
Orion	5 35.4	−5 27

45. Use the *Starry Night* program to observe magnified images of at least four of the planets. First turn off daylight (select **Daylight** in the **Sky Display** menu) and show the entire celestial sphere (select **Atlas** in the **Go** menu). Select **Planet List** in the **Window** menu, and then double-click on the name of the planet you wish to view. Use the controls at the right-hand end of the Control Panel to zoom in until you can see a detailed view of the planet. Describe each planet's appearance. From what you observe, is there any way of knowing whether you are looking at a planet's surface or simply a cloud cover?

GEOFF MARCY

Alien Planets

When Geoff Marcy sat down to write this essay, 50 planets were known to orbit other stars. He leads the team that has found 32 of them. Dr. Marcy became interested in astronomy at age 14, when his parents bought him a small reflecting telescope. Since receiving his Ph.D. from the University of California, Santa Cruz, he has studied stars similar to our Sun. He helped show that magnetic regions on their surfaces cause dark "star spots" and stellar flares. He also showed that brown

dwarfs—stars too small to burn hydrogen—rarely orbit other stars. His work may soon reveal whether a planetary system like our own is common or a quirk of the cosmos. Dr. Marcy is an astronomer at the University of California, Berkeley.

Astronomers have now surveyed 1000 nearby stars in the search for orbiting planets. My team is using the 10-meter Keck telescope in Hawaii to measure the Doppler effect in stars that wobble because of planets orbiting around them. So far, we and other teams have found 50 planets, some more massive than Jupiter and some probably less massive than Saturn.

Two systems of multiple planets orbiting a single star have been found, and many more such planetary systems will be discovered in the next few years. We hope to compare the architecture of these planetary systems to that of our own solar system. Is our solar system a common, garden-variety type? Do all planetary systems have giant planets orbiting far away from the host star, with small, rocky planets orbiting close in?

Remarkably, the planetary systems observed so far seem very different from our own solar system. In some, a giant planet orbits very close to the star. More surprising is that almost all alien planets that orbit beyond 0.1 AU from their star reside in highly elliptical orbits that resemble eccentric comet orbits. Why are these planets in such elliptical orbits, so different from the nearly circular orbits of planets in our own solar system? Astronomers are baffled by this difference.

One possibility is that planets usually form as close-knit families around a star. Over time, the planets may gravitationally pull on one another, yanking themselves out of their original circular orbits. These perturbed planets may then venture near other planets, yanking on them, too. Soon, all the planets may pass close to other planets, gravitationally slinging some of them completely out of the planetary system. These far-flung planets are destined to

roam the darkness of galactic space, cooling to frigid temperatures that are inhospitable to life.

The extrasolar planets cast a mystery back on the Earth. Why is our solar system immune to this chaotic episode that scatters planets into wacky orbits? Imagine the disaster if our Earth were suddenly thrust into a highly elliptical orbit. During half the year, we would be roasting too close to the Sun, with oceans vaporized into an enormous steam bath. The other half of the year our orbit would carry us too far from the Sun, causing all the oceans to freeze over. Life on Earth would be severely challenged to survive in such an elongated orbit.

Why are we humans so lucky as to live in a stable planetary system in which circular orbits ensure that we receive the same warming light intensity from our Sun, all year round? Perhaps the mystery is explained by our very existence. After all, intelligent life might not have evolved here if the Earth suffered from wild temperature swings that would make liquid water rare. We humans, and intelligent species in general, may flourish mostly on worlds that maintain nearly constant temperature. If so, our solar system and the Earth might be a bizarre, rare galactic quirk that just happens to be suitable for life. We wouldn't be here otherwise. If this picture is correct, our Earth is indeed a precious oasis in our Milky Way Galaxy.

In the next decade, we will discover more planetary systems by using the Doppler method. We are working night and day to find Jovian planets that orbit 5 AU from their star, similar to the orbital radius of our Jupiter. Their orbits, circular or elliptical, will shed light on the most intriguing questions of our time: How common are configurations like our Solar System, and how common is advanced life in the universe?

A Star-Forming Region This infrared image shows a swarm of young stars in the Omega Nebula, a star-forming region some 550 light-years from Earth in the constellation Sagittarius. New planetary systems may be in the process of formation within this nebula. (European Southern Observatory)

R **I** V U X G

This version of *Universe* now moves beyond the planets to study the Sun, the stars, and galaxies. For full coverage of our remarkable solar system—including the planets, satellites, asteroids, meteoroids, and comets—see either the full version of *Universe* (ISBN 0-7167-4647-6) or *Universe: The Solar System* (ISBN 0-7167-4645-X).

Our Star, the Sun

(Institute of Space and Astronomical Science, Yohkoh Project, SXT Group/NASA)

R I V U X G

The Sun is by far the brightest object in the sky. By earthly standards, the temperature of its glowing surface is remarkably high, about 5800 K. Yet astronomers have found regions of the Sun that reach far higher temperatures of several million kelvins. Gases at such temperatures emit X rays and can be seen in an image like this one from an X-ray telescope. Bright, X-ray-emitting regions, like those in this image, can spawn solar flares—powerful eruptions that propel solar material across space to reach the Earth and other planets.

In recent decades, astronomers have used tools such as X-ray telescopes to make a wealth of discoveries about the Sun. We have learned that the Sun shines because at its core hundreds of millions of tons of hydrogen are converted to helium every second. We have found the by-products of this transmutation—strange, ethereal particles called neutrinos—streaming outward from the Sun into space. We have discovered that the Sun has a surprisingly violent atmosphere, with a host of features such as sunspots whose numbers rise and fall on a predictable 11-year cycle. By studying the Sun's vibrations, we have begun to probe beneath its surface into hitherto unexplored realms. And we have just begun to investigate how changes in the Sun's activity can affect the Earth's environment as well as our technological society.

As you read the sections of this chapter, look for the answers to the following questions.

18-1   What is the source of the Sun's energy?

18-2   What is the internal structure of the Sun?

18-3   How can astronomers measure the properties of the Sun's interior?

18-4   How can we be sure that thermonuclear reactions are happening in the Sun's core?

18-5   Does the Sun have a solid surface?

18-6   Since the Sun is so bright, how is it possible to see its dim outer atmosphere?

18-7   Where does the solar wind come from?

18-8   What are sunspots? Why do they appear dark?

18-9   What is the connection between sunspots and the Sun's magnetic field?

18-10 What causes eruptions in the Sun's atmosphere?

18-1   The Sun's energy is generated by thermonuclear reactions in its core

The Sun is the largest member of the solar system, with almost a thousand times more mass than all the planets, moons, asteroids, comets, and meteoroids put together. But the Sun is also a star. In fact, it is a remarkably typical star, with a mass, size, surface temperature, and chemical composition that are approximately midway between the extremes exhibited by the myriad other stars in the heavens. Table 18-1 lists essential data about the Sun.

For most people, what matters most about the Sun is the energy that it radiates into space. Without the Sun's warming rays, our atmosphere and oceans would freeze into an icy layer coating a desperately cold planet, and life on Earth would be impossible. To understand why we are here, we must understand the nature of the Sun.

What makes the Sun such an important source of energy? An important part of the answer is that the Sun has a far higher surface temperature than any of the planets or moons. The Sun's spectrum is close to that of an idealized blackbody with a temperature of 5800 K (see Section 5-3 and Section 5-4, especially Figure 5-11). Thanks to this high temperature, each square meter of the Sun's surface emits a tremendous amount of radiation, principally at visible wavelengths. Indeed, the Sun is the only object in the solar system that emits substantial amounts of visible light. The light that we see from the Moon and planets is actually sunlight that struck those worlds and was reflected toward Earth.

The Sun's size also helps us explain its tremendous energy output. Because the Sun is so large, the total number of square meters of radiating surface—that is, its surface area—is immense. Hence, the total amount of energy emitted by the Sun each second, called its **luminosity,** is very large indeed: about 3.9×10^{26} watts, or 3.9×10^{26} joules of energy emitted per second. (We discussed the relation among the Sun's surface temperature, radius, and luminosity in Box 5-2.) Astronomers denote the Sun's luminosity by the symbol $L_\odot$. A circle with a dot in the center is the astronomical symbol for the Sun and was also used by ancient astrologers.

These ideas lead us to a more fundamental question: What keeps the Sun's visible surface so hot? Or, put another way, what is the fundamental source of the tremendous energies that the Sun radiates into space? For centuries, this was one of the greatest mysteries in science. The mystery deepened in the nineteenth century, when geologists and biologists found convincing evidence that life had existed on Earth for at least several hundred million years. (We now know that the Earth is 4.6 billion years old and that life has existed on it for most of its history.) Since life as we know it depends crucially on sunlight, the Sun must be as old. This posed a severe problem for physicists. What source of energy could have kept the Sun shining for so long (Figure 18-1)?

One attempt to explain solar energy was made in the mid-1800s by the English physicist Lord Kelvin (for whom the temperature scale is named) and the German scientist Hermann von Helmholtz. They argued that the tremendous weight of the Sun's outer layers should cause the Sun to contract gradually, compressing its interior gases. Whenever a gas is compressed, its temperature rises. (You can demonstrate this with a bicycle pump: As you pump air into a tire, the temperature of the air increases and the pump becomes warm to the touch.) Kelvin and Helmholtz thus suggested that gravitational contraction could cause the Sun's gases to become hot enough to radiate energy out into space.

This process, called *Kelvin-Helmholtz contraction,* actually does occur during the earliest stages of the birth of a star like the Sun (see Section 7-7). But Kelvin-Helmholtz contraction cannot be the major source of the Sun's energy today. If it were, the Sun would have had to have been much larger in the relatively recent past. Helmholtz's own calculations showed that the Sun could have started its initial collapse from the solar nebula no more than about 25 million years ago. But the geological and fossil record shows that the Earth is far older than that, and so the Sun must be as well. Hence, this model of a Sun that shines because it shrinks cannot be correct.

On Earth, a common way to produce heat and light is by burning fuel, such as a log in a fireplace or coal in a power plant. Is it possible that a similar process explains the energy released by the Sun? The answer is no, because this process could not continue for a long enough time to explain the age

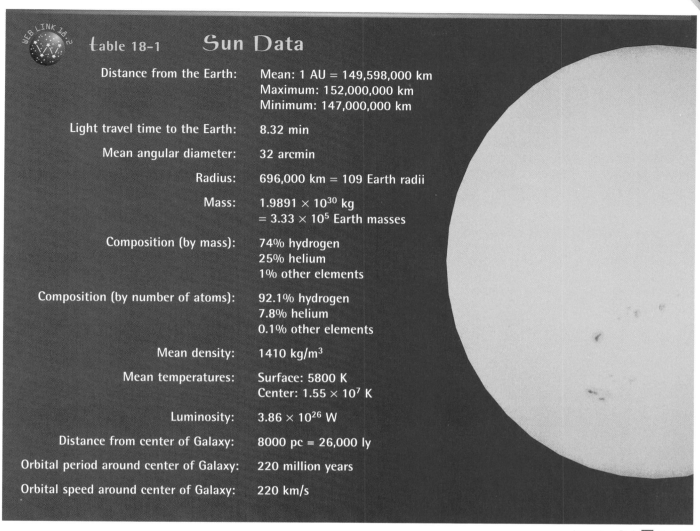

Table 18-1	Sun Data
Distance from the Earth:	Mean: 1 AU = 149,598,000 km
	Maximum: 152,000,000 km
	Minimum: 147,000,000 km
Light travel time to the Earth:	8.32 min
Mean angular diameter:	32 arcmin
Radius:	696,000 km = 109 Earth radii
Mass:	1.9891×10^{30} kg
	$= 3.33 \times 10^5$ Earth masses
Composition (by mass):	74% hydrogen
	25% helium
	1% other elements
Composition (by number of atoms):	92.1% hydrogen
	7.8% helium
	0.1% other elements
Mean density:	1410 kg/m^3
Mean temperatures:	Surface: 5800 K
	Center: 1.55×10^7 K
Luminosity:	3.86×10^{26} W
Distance from center of Galaxy:	8000 pc = 26,000 ly
Orbital period around center of Galaxy:	220 million years
Orbital speed around center of Galaxy:	220 km/s

(NOAO)   R I **V** U X G

of the Earth. The chemical reactions involved in burning release roughly 10^{-19} joule of energy per atom. Therefore, the number of atoms that would have to undergo chemical reactions each second to generate the Sun's luminosity of 3.9×10^{26} joules per second is approximately

$$\frac{3.9 \times 10^{26} \text{ joules per second}}{10^{-19} \text{ joule per atom}} = 3.9 \times 10^{45} \text{ atoms per second}$$

figure 18-1   R I **V** U X G

The Sun The Sun's visible surface has a temperature of about 5800 K. At this temperature, all solids and liquids vaporize to form gases. It was only in the twentieth century that scientists discovered what has kept the Sun so hot for billions of years: the thermonuclear fusion of hydrogen nuclei in the Sun's core.
(Jeremy Woodhouse/PhotoDisc)

From its mass and chemical composition, we know that the Sun contains about 10^{57} atoms. Thus, the length of time required to consume the entire Sun by burning would be

$$\frac{10^{57} \text{ atoms}}{3.9 \times 10^{45} \text{ atoms per second}} = 3 \times 10^{11} \text{ seconds}$$

There are about 3×10^7 seconds in a year. Hence, in this model, the Sun would burn itself out in a mere 10,000 (10^4) years! This is far shorter than the known age of the Earth, so chemical reactions also cannot explain how the Sun shines.

The source of the Sun's luminosity could be explained if there were a process like burning but that released much more energy per atom. Then the rate at which atoms would have to be consumed would be far less, and the lifetime of the Sun could be long enough to be consistent with the known age of the Earth. Albert Einstein discovered the key to such a process in 1905. According to his *special theory of relativity,* a quantity m of mass can in principle be converted into an amount of energy E according to a now-famous equation:

Einstein's mass-energy equation

$$E = mc^2$$

m = quantity of mass, in kg

c = speed of light = 3×10^8 m/s

E = amount of energy into which the mass can be converted, in joules

The speed of light c is a large number, so c^2 is huge. Therefore, a small amount of matter can release an awesome amount of energy.

Inspired by Einstein's ideas, astronomers began to wonder if the Sun's energy output might come from the conversion of matter into energy. The Sun's low density of 1410 kg/m^3 indicates that it must be made of the very lightest atoms, primarily hydrogen and helium. In the 1920s, the British astronomer Arthur Eddington showed that temperatures near the center of the Sun must be so high that atoms become completely ionized. Hence, at the Sun's center we expect to find hydrogen nuclei and electrons flying around independent of each other.

Another British astronomer, Robert Atkinson, suggested that under these conditions hydrogen nuclei could fuse together to produce helium nuclei in a *nuclear reaction* that transforms a tiny amount of mass into a large amount of energy. Experiments in the laboratory using individual nuclei show that such reactions can indeed take place. The process of converting hydrogen into helium is called **hydrogen burning,** even though nothing is actually burned in the conventional sense. (Ordinary burning involves chemical reactions that rearrange the outer electrons of atoms but have no effect on the atoms' nuclei.) Hydrogen burning provides the devastating energy released in a hydrogen bomb (see Figure 1-4).

The fusing together of nuclei is also called **thermonuclear fusion,** because it can take place only at extremely high temperatures. The reason is that all nuclei have a positive electric charge and so tend to repel one another. But in the extreme heat and pressure at the Sun's center, hydrogen nuclei (protons) are moving so fast that they can overcome their electric repulsion and actually touch one another. When that happens, thermonuclear fusion can take place.

ANALOGY You can think of protons as tiny electrically charged spheres that are coated with a very powerful glue. If the spheres are not touching, the repulsion between their charges pushes them apart. But if the spheres are forced into contact, the strength of the glue "fuses" them together.

CAUTION! Be careful not to confuse thermonuclear fusion with the similar-sounding process of *nuclear fission.* In nuclear fusion, energy is released by joining together nuclei of lightweight atoms such as hydrogen. In nuclear fission, by contrast, the nuclei of very massive atoms such as uranium or plutonium release energy by fragmenting into smaller nuclei. Nuclear power plants produce energy using

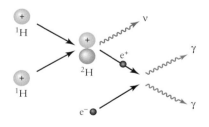

a   Step 1

b   Step 2

c   Step 3

figure 18-2

Fusing Hydrogen into Helium Hydrogen burning in the Sun usually takes place in three energy-releasing steps. (See Box 18-1.) **(a)** Two protons (hydrogen nuclei, ^1H) fuse to form a hydrogen isotope with one proton and one neutron (^2H), a neutral, nearly massless neutrino (ν), and a positively-charged electron, or positron (e^+). This positron encounters an ordinary electron (e^-), annihilating both particles and converting them into gamma-ray photons (γ). **(b)** The ^2H nucleus from the first step combines with a third proton, forming a helium isotope (^3He) and releasing another gamma-ray photon. **(c)** Two ^3He nuclei collide, forming a different helium isotope with two protons and two neutrons (^4He) and releasing two protons.

fission, not fusion. (Generating power using fusion has been a goal of researchers for decades, but no one has yet discovered a commercially viable way to do this.)

We learned in Section 5-8 that the nucleus of a hydrogen atom (H) consists of a single proton. The nucleus of a helium atom (He) consists of two protons and two neutrons. In the nuclear process that Atkinson described, four hydrogen nuclei combine to form one helium nucleus, with a concurrent release of energy:

$$4 \text{ H} \rightarrow \text{He} + \text{energy}$$

In several separate reactions, two of the four protons are changed into neutrons, and eventually combine with the remaining protons to produce a helium nucleus (Figure 18-2). Each time this process takes place, a small fraction (0.7%) of the combined mass of the hydrogen nuclei does not show up in the mass of the helium nucleus. This "lost" mass is converted into energy. Box 18-1 describes how to use Einstein's mass-energy equation to calculate the amount of energy released.

Box 18-1 | Tools of the Astronomer's Trade

Converting Mass into Energy

Figure 18-2 shows the steps involved in the thermonuclear fusion of hydrogen at the Sun's center. In these steps, four protons are converted into a single nucleus of ^{4}He, an isotope of helium with two protons and two neutrons. The reaction depicted in Figure 18-2a also produces a neutral, nearly massless particle called the *neutrino*. Neutrinos respond hardly at all to ordinary matter, so they travel almost unimpeded through the Sun's massive bulk. Hence, the energy that neutrinos carry is quickly lost into space. This loss is not great, however, because the neutrinos carry relatively little energy. (See Section 18-4 for more about these curious particles.)

Most of the energy released by thermonuclear fusion appears in the form of gamma-ray photons. The energy of these photons remains trapped within the Sun for a long time, thus maintaining the Sun's intense internal heat. Some gamma-ray photons are produced by the reaction in Figure 18-2b. Others appear when an electron in the Sun's interior annihilates a positively charged electron, or **positron**, which is a by-product of the reaction in Figure 18-2a. An electron and a positron are respectively matter and antimatter, and they convert entirely into energy when they meet. (You may have thought that "antimatter" was pure science fiction, but it is being created and annihilated in the Sun as you read these words.)

We can summarize the thermonuclear fusion of hydrogen as follows:

$$4 \text{ H} \rightarrow \text{He} + \text{neutrinos} + \text{gamma-ray photons}$$

To calculate how much energy is released in this process, we use Einstein's mass-energy formula: The energy released is equal to the amount of mass consumed multiplied by c^2, where c is the speed of light. To see how much mass is consumed, we compare the combined mass of four hydrogen atoms (the ingredients) to the mass of one helium atom (the product):

$$
\begin{array}{rl}
4 \text{ hydrogen atoms} & = 6.693 \times 10^{-27} \text{ kg} \\
-1 \text{ helium atom} & = -6.645 \times 10^{-27} \text{ kg} \\
\hline
\text{Mass lost} & = 0.048 \times 10^{-27} \text{ kg}
\end{array}
$$

Thus, a small fraction (0.7%) of the mass of the hydrogen going into the nuclear reaction does not show up in the mass of the helium. This lost mass is converted into an amount of energy $E = mc^2$:

$$E = mc^2 = (0.048 \times 10^{-27} \text{ kg}) (3 \times 10^8 \text{ m/s})^2$$
$$= 4.3 \times 10^{-12} \text{ joule}$$

This is the amount of energy released by the formation of a single helium atom. It would light a 10-watt lightbulb for almost one-half of a trillionth of a second.

EXAMPLE: Let's consider the conversion of 1 kg of hydrogen into helium. Although a kilogram of hydrogen goes into this reaction, 0.7% of the mass is lost, and only 0.993 kg of helium comes out. Using Einstein's equation, we find that the missing 0.007 kg of matter has been transformed into an amount of energy equal to

$$E = mc^2 = (0.007 \text{ kg}) (3 \times 10^8 \text{ m/s})^2 = 6.3 \times 10^{14} \text{ joules}$$

For comparison, this equals the energy released by burning 20,000 metric tons (2×10^7 kg) of coal!

The Sun's luminosity is 3.9×10^{26} joules per second. To generate this much power, hydrogen must be consumed at a rate of

$$\frac{3.9 \times 10^{26} \text{ joules per second}}{6.3 \times 10^{14} \text{ joules per kilogram}}$$
$$= 6 \times 10^{11} \text{ kilograms per second}$$

That is, the Sun converts 600 million metric tons of hydrogen into helium within its core every second.

You may have heard the idea that mass is always conserved (that is, it is neither created nor destroyed), or that energy is always conserved in a reaction. Einstein's ideas show that neither of these statements is quite correct, because mass can be converted into energy and vice versa. A more accurate statement is that the total amount of mass *plus* energy is conserved. Hence, the destruction of mass in the Sun does not violate any laws of nature.

For every four hydrogen nuclei converted into a helium nucleus, 4.3×10^{-12} joule of energy is released. This may seem like only a tiny amount of energy, but it is about 10^7 times larger than the amount of energy released in a typical chemical reaction, such as occurs in ordinary burning. Thus, thermonuclear fusion can explain how the Sun could have been shining for billions of years.

To produce the Sun's luminosity of 3.9×10^{26} joules per second, 6×10^{11} kg (600 million metric tons) of hydrogen must be converted into helium each second. This rate is prodigious, but there is a literally astronomical amount of hydrogen in the Sun. In particular, the Sun's core contains enough hydrogen to have been giving off energy at the present rate for as long as the solar system has existed, about 4.6 billion years, and to continue doing so for another 5 billion years.

 The Sun's energy actually comes from a *series* of nuclear reactions, called the **proton-proton chain,** in which hydrogen nuclei combine to form helium (see Figure 18-2 and Box 18-1). The proton-proton chain is also the energy source for many of the stars in the sky. In stars with central temperatures that are much hotter than that of the Sun, however, hydrogen burning proceeds according to a different set of nuclear reactions, called the **CNO cycle,** in which carbon, nitrogen, and oxygen nuclei absorb protons to produce helium nuclei. Still other thermonuclear reactions, such as helium burning, carbon burning, and oxygen burning, occur late in the lives of many stars.

18-2 A theoretical model of the Sun shows how energy gets from its center to its surface

While thermonuclear fusion is the source of the Sun's energy, this process cannot take place everywhere within the Sun. As we have seen, extremely high temperatures—in excess of 10^7 K—are required for atomic nuclei to fuse together to form larger nuclei. The temperature of the Sun's visible surface, about 5800 K, is too low for these reactions to occur there. Hence, thermonuclear fusion can be taking place only within the Sun's interior. But precisely where does it take place? And how does the energy produced by fusion make its way to the surface, where it is emitted into space in the form of photons?

To answer these questions, we must understand conditions in the Sun's interior. Ideally, we would send an exploratory spacecraft to probe deep into the Sun; in practice, the Sun's intense heat would vaporize even the sturdiest

spacecraft. Instead, astronomers use the laws of physics to construct a theoretical model of the Sun. (We discussed the use of models in science in Section 1-1.) Let's see what ingredients go into building a model of this kind.

Note first that the Sun is not undergoing any dramatic changes. The Sun is not exploding or collapsing, nor is it significantly heating or cooling. The Sun is thus said to be in both *hydrostatic equilibrium* and *thermal equilibrium*.

To understand what is meant by **hydrostatic equilibrium,** imagine a slab of material in the solar interior (Figure 18-3a). In equilibrium, the slab on average will move neither up nor down. (In fact, there are upward and downward motions of material inside the Sun, but these motions average out.) Equilibrium is maintained by a balance among three forces that act on this slab:

1. The downward pressure of the layers of solar material above the slab

2. The upward pressure of the hot gases beneath the slab

3. The slab's weight—that is, the downward gravitational pull it feels from the rest of the Sun

The pressure from below must balance both the slab's weight and the pressure from above. Hence, the pressure below the slab must be greater than that above the slab. In other words, pressure has to increase with increasing depth. For the same reason, pressure increases as you dive deeper into the ocean (Figure 18-3b) or as you move toward lower altitudes in our atmosphere.

Hydrostatic equilibrium also tells us about the density of the slab. If the slab is too dense, its weight will be too great and it will sink; if the density is too low, the slab will rise. To prevent this, the density of solar material must have a certain value at each depth within the solar interior. (The same principle applies to objects that float beneath the surface of the ocean. Scuba divers wear weight belts to increase their average density so that they will neither rise nor sink but will stay submerged at the same level.)

Another consideration is that the Sun's interior is so hot that it is completely gaseous. Gases compress and become more dense when you apply greater pressure to them, so density must increase along with pressure as you go to greater depths within the Sun. Furthermore, when you compress a gas, its temperature tends to rise, so the temperature must also increase as you move toward the Sun's center.

While the temperature in the solar interior is different at different depths, the temperature at each depth remains constant in time. This principle is called **thermal equilibrium.** For the Sun to be in thermal equilibrium, all the energy generated by thermonuclear reactions in the Sun's core must be transported to the Sun's glowing surface, where it can be radiated into space. If too much energy flowed from the core to the surface to be radiated away, the Sun's interior would cool down; alternatively, the Sun would heat up if too little energy flowed to the surface.

But exactly how is energy transported from the Sun's center to its surface? There are three methods of energy transport:

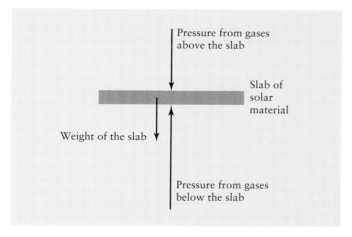

a   In hydrostatic equilibrium, forces balance

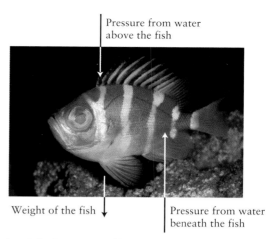

b   In hydrostatic equilibrium, forces balance

figure 18-3

Hydrostatic Equilibrium **(a)** The Sun's interior is in hydrostatic equilibrium, which means that, on average, material tends to move neither up nor down. For any slab of material in the solar interior, the pressure from material below the slab (which tends to push the slab up) must balance the combined effects of the slab's weight and the pressure from material above the slab (both of which tend to push the slab down). Hence, the pressure must increase with increasing depth. **(b)** The same principle applies to a fish in water. In equilibrium, the forces balance and the fish neither rises nor sinks. (Ken Usami/PhotoDisc)

conduction, convection, and *radiative diffusion.* Only the last two are important inside the Sun.

If you heat one end of a metal bar with a blowtorch, energy flows to the other end of the bar so that it too becomes warm. The efficiency of this method of energy transport, called **conduction,** varies significantly from one substance to another. For example, copper is a good conductor of heat, but wood is not (which is why copper pots often have wooden handles). Conduction is not an efficient means of energy transport inside stars like the Sun, for which the average density is relatively low.

Inside stars like our Sun, energy moves from center to surface by two other means: convection and radiative diffusion. **Convection** is the circulation of fluids—gases or liquids—between hot and cool regions. Hot gases rise toward a star's surface, while cool gases sink back down toward the star's center. This physical movement of gases transports heat energy outward in a star, just as the physical movement of water boiling in a pot transports energy from the bottom of the pot (where the heat is applied) to the cooler water at the surface.

In **radiative diffusion,** photons created in the thermonuclear inferno at a star's center diffuse outward toward the star's surface. Individual photons are absorbed and reemitted by atoms and electrons inside the star. The overall result is an outward migration from the hot core, where photons are constantly created, toward the cooler surface, where they escape into space.

To construct a model of a star like the Sun, astrophysicists express the ideas of hydrostatic equilibrium, thermal equilib-

rium, and energy transport as a set of equations. To ensure that the model applies to the particular star under study, they also make use of astronomical observations of the star's surface. (For example, to construct a model of the Sun, they use the data that the Sun's surface temperature is 5800 K, its luminosity is 3.9×10^{26} W, and the gas pressure and density at the surface are almost zero.) The astrophysicists then use a computer to solve their set of equations and calculate conditions layer by layer in toward the star's center. The result is a model of how temperature, pressure, and density increase with increasing depth below the star's surface.

Table 18-2 and Figure 18-4 show a theoretical model of the Sun that was calculated in just this way. Different models of the Sun use slightly different assumptions, but all models give essentially the same results as those shown here. From such computer models we have learned that at the Sun's center the density is 160,000 kg/m³ (14 times the density of lead!), the temperature is 1.55×10^7 K, and the pressure is 3.4×10^{11} atm. (One atmosphere, or 1 atm, is the average atmospheric pressure at sea level on Earth.)

Table 18-2 and Figure 18-4 show that the solar luminosity rises to 100% at about one-quarter of the way from the Sun's center to its surface. In other words, the Sun's energy production occurs within a volume that extends out only to 0.25 $R_\odot$. (The symbol $R_\odot$ denotes the *solar radius,* or radius of the Sun as a whole, equal to 696,000 km.) Outside 0.25 $R_\odot$, the density and temperature are too low for thermonuclear reactions to take place. Also note that 94% of the total mass of the Sun is found within the inner 0.5 $R_\odot$. Hence, the outer 0.5 $R_\odot$ contains only a relatively small amount of material.

Table 18-2 — A Theoretical Model of the Sun

Distance from the Sun's center (solar radii)	Fraction of luminosity	Fraction of mass	Temperature (× 10⁶ K)	Density (kg/m³)	Pressure (relative to pressure at center)
0.0	0.00	0.00	15.5	160,000	1.00
0.1	0.42	0.07	13.0	90,000	0.46
0.2	0.94	0.35	9.5	40,000	0.15
0.3	1.00	0.64	6.7	13,000	0.04
0.4	1.00	0.85	4.8	4,000	0.007
0.5	1.00	0.94	3.4	1,000	0.001
0.6	1.00	0.98	2.2	400	0.0003
0.7	1.00	0.99	1.2	80	4×10^{-5}
0.8	1.00	1.00	0.7	20	5×10^{-6}
0.9	1.00	1.00	0.3	2	3×10^{-7}
1.0	1.00	1.00	0.006	0.00030	4×10^{-13}

Note: The distance from the Sun's center is expressed as a fraction of the Sun's radius ($R_\odot$). Thus, 0.0 is at the center of the Sun and 1.0 is at the surface. The fraction of luminosity is that portion of the Sun's total luminosity produced within each distance from the center; this is equal to 1.00 for distances of 0.25 $R_\odot$ or more, which means that all of the Sun's nuclear reactions occur within 0.25 solar radius from the Sun's center. The fraction of mass is that portion of the Sun's total mass lying within each distance from the Sun's center. The pressure is expressed as a fraction of the pressure at the center of the Sun.

How energy flows from the Sun's center toward its surface depends on how easily photons move through the gas. If the solar gases are comparatively transparent, photons can travel moderate distances before being scattered or absorbed, and energy is thus transported by radiative diffusion. If the gases are comparatively opaque, photons cannot get through the gas easily and heat builds up. Convection then becomes the most efficient means of energy transport. The gases start to churn, with hot gas moving upward and cooler gas sinking downward.

From the center of the Sun out to about 0.71 $R_\odot$, energy is transported by radiative diffusion. Hence, this region is called the **radiative zone**. Beyond about 0.71 $R_\odot$, the temperature is low enough (a mere 2×10^6 K or so) for electrons and hydrogen nuclei to join into hydrogen atoms. These atoms are very effective at absorbing photons, much more so than free electrons or nuclei, and this absorption chokes off the outward flow of photons. Therefore, beyond about

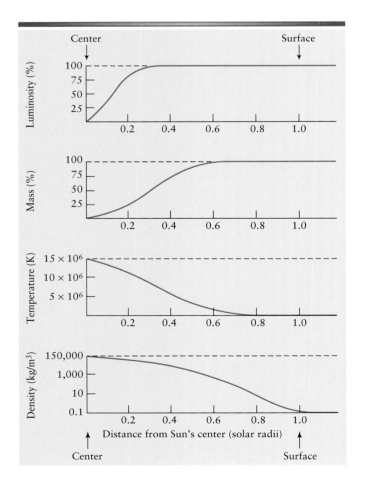

Figure 18-4

A Theoretical Model of the Sun's Interior From top to bottom, these graphs depict what percentage of the Sun's total luminosity is produced within each distance from the center, what percentage of the total mass lies within each distance from the center, the temperature at each distance, and the density at each distance. (See Table 18-2 for a numerical version of this model.)

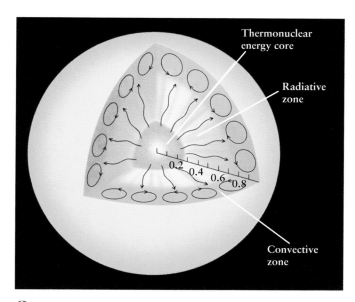

Figure 18-5

The Sun's Internal Structure Thermonuclear reactions occur in the Sun's core, which extends out to a distance of 0.25 $R_\odot$ from the center. Energy is transported outward, via radiative diffusion, to a distance of about 0.71 $R_\odot$. In the outer layers between 0.71 $R_\odot$ and 1.00 $R_\odot$, energy flows outward by convection.

0.71 $R_\odot$, radiative diffusion is not an effective way to transport energy. Instead, convection dominates the energy flow in this outer region, which is why it is called the **convective zone.** Figure 18-5 shows these aspects of the Sun's internal structure.

Although energy travels through the radiative zone in the form of photons, the photons have a difficult time of it. Table 18-2 shows that the material in this zone is extremely dense, so photons from the Sun's core take a long time to diffuse through the radiative zone. As a result, it takes approximately 170,000 years for energy created at the Sun's center to travel 696,000 km to the solar surface and finally escape as sunlight. The energy flows outward at an average rate of 50 centimeters per hour, or about 20 times slower than a snail's pace.

Once the energy escapes from the Sun, it travels much faster—at the speed of light. Thus, solar energy that reaches you today took only 8 minutes to travel the 150 million kilometers from the Sun's surface to the Earth. But this energy was actually produced by thermonuclear reactions that took place in the Sun's core hundreds of thousands of years ago.

18-3 Astronomers probe the solar interior using the Sun's own vibrations

 We have described how astrophysicists construct models of the Sun. But since we cannot see into the Sun's opaque interior, how can we check these models to see if they are accurate? What is needed is a technique for probing the Sun's interior. A very powerful technique of just this kind involves measuring vibrations of the Sun as a whole. This field of solar research is called **helioseismology.**

Vibrations are a useful tool for examining the hidden interiors of all kinds of objects. Food shoppers test whether melons are ripe by tapping on them and listening to the vibrations. Geologists can determine the structure of the Earth's interior by using seismographs to record vibrations during earthquakes.

Although there are no true "sunquakes," the Sun does vibrate at a variety of frequencies, somewhat like a ringing bell. These vibrations were first noticed in 1960 by Robert Leighton of the California Institute of Technology, who made high-precision Doppler shift observations of the solar surface. These measurements revealed that parts of the Sun's surface move up and down about 10 meters every 5 minutes (Figure 18-6). Since the mid-1970s, several astronomers have reported slower vibrations, having periods ranging from 20 to 160 minutes. The detection of extremely slow vibrations has inspired astronomers to organize networks of telescopes around and in orbit above the Earth to monitor the Sun's vibrations on a continuous basis.

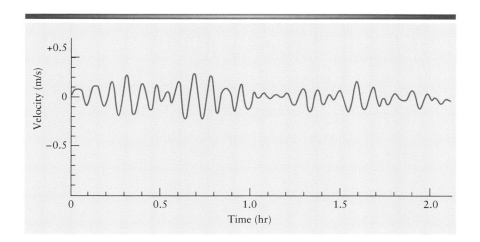

Figure 18-6

Vibrations of the Sun's Surface Doppler shift measurements give the speed at which areas of the Sun's surface bob up and down. The data shown here extend over 2 hours and show numerous oscillations with a period of about 5 minutes. The size of the oscillations varies gradually due to interference between a variety of vibrations with different wavelengths. (Adapted from O. R. White)

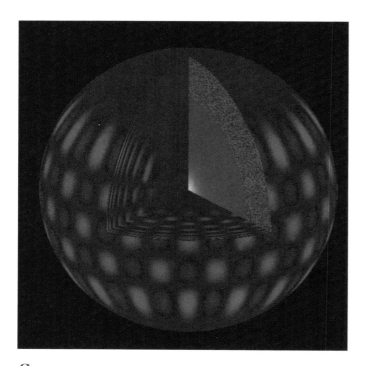

figure 18-7

A Sound Wave Resonating in the Sun This computer-generated image shows one of the millions of ways in which the Sun vibrates because of sound waves resonating in its interior. The regions that are moving outward are colored blue, those moving inward, red. The cutaway shows how deep these oscillations are thought to extend. (National Solar Observatory)

The vibrations of the Sun's surface can be compared with sound waves. If you could somehow survive within the Sun's outermost layers, you would first notice a deafening roar, somewhat like a jet engine, produced by turbulence in the Sun's gases. But superimposed on this noise would be a variety of nearly pure tones. You would need greatly enhanced hearing to detect these tones, however; the strongest has a frequency of just 0.003 hertz, 13 octaves below the lowest frequency audible to humans. (Recall from Section 5-2 that one hertz is one oscillation per second.)

In 1970, Roger Ulrich, at UCLA, pointed out that sound waves moving upward from the solar interior would be reflected back inward after reaching the solar surface. However, as a reflected sound wave descended back into the Sun, the increasing density and pressure would bend the wave severely, turning it around and aiming it back toward the solar surface. In other words, sound waves bounce back and forth between the solar surface and layers deep within the Sun. These sound waves can reinforce each other if their wavelength is the right size, just as sound waves of a particular wavelength resonate inside an organ pipe.

The Sun oscillates in millions of ways as a result of waves resonating in its interior. Figure 18-7 is a computer-generated illustration of one such mode of vibration. Helioseismologists

can deduce information about the solar interior from measurements of these oscillations. For example, they have been able to set limits on the amount of helium in the Sun's core and convective zone and to determine the thickness of the transition region between the radiative zone and convective zone. They have also found that the convective zone is thicker than previously thought.

18-4 Neutrinos provide information about the Sun's core—and have surprises of their own

We have seen circumstantial evidence that thermonuclear fusion is the source of the Sun's power. To be certain, however, we need more definitive evidence. How can we show that thermonuclear fusion really is taking place in the Sun's core? Unfortunately, helioseismology is of little assistance. The vibrations that astronomers see on the Sun's surface do not penetrate that far into the interior.

Although the light energy that we receive from the Sun originates in the core, it provides few clues about conditions there. The problem is that this energy has changed form repeatedly during its passage from the core: It appeared first as photons diffusing through the radiative zone, then as heat transported through the outer layers by convection, and then again as photons emitted from the Sun's glowing surface. As a result of these transformations, much of the information that the Sun's radiated energy once carried about conditions in the core has been lost.

ANALOGY If you make a photocopy of a photocopy of a photocopy of an original document, the final result may be so blurred as to be unreadable. In an analogous way, because solar energy is transformed many times while en route to Earth, the story it could tell us about the Sun's core is hopelessly blurred.

But there is a way for scientists to learn about conditions in the Sun's core and to get direct evidence that thermonuclear fusion really does happen there. The trick is to detect the subatomic by-products of thermonuclear fusion reactions.

As part of the process of hydrogen burning, protons change into neutrons and release **neutrinos** (see Figure 18-2*a* and Box 18-1). Like photons, neutrinos are particles that carry energy, have no electric charge, and have almost no mass. (If neutrinos are truly massless, they must travel through space at the speed of light, just as photons do. Some experiments suggest that neutrinos might have a tiny mass, probably less than one ten-thousandth the mass of an electron, in which case they would travel slightly slower than light.)

Unlike photons, however, neutrinos interact only very weakly with matter. Even the vast bulk of the Sun offers little impediment to their passage, so neutrinos must be streaming out of the core and into space. Indeed, the conversion of hydrogen into helium at the Sun's center produces 10^{38} neutrinos each second. Every second, about 10^{14} neutrinos from

Figure 18-8 R I U X G

The Davis Solar Neutrino Experiment This tank of cleaning fluid lies 1.5 km below the surface in the Homestake Gold Mine in South Dakota. The Earth shields the tank from stray particles, so that only solar neutrinos can convert chlorine atoms in the fluid into radioactive argon atoms. The number of argon atoms created in the tank is a direct measure of the neutrino flux from the Sun. The detected flux was only about one-third the predicted value, a state of affairs called the "solar neutrino problem." (Courtesy of R. Davis, Brookhaven National Laboratory)

the Sun—that is, **solar neutrinos**—must pass through each square meter of the Earth.

If it were possible to detect these solar neutrinos, we would have direct evidence that thermonuclear reactions really do take place in the Sun's core. Beginning in the 1960s, scientists began to build neutrino detectors for precisely this purpose.

The challenge is that neutrinos are exceedingly difficult to detect. Just as neutrinos pass unimpeded through the Sun, they also pass through the Earth *almost* as if it were not there. We stress the word "almost," because neutrinos can and do interact with matter, albeit infrequently.

On rare occasions a neutrino will strike a neutron and convert it into a proton. This effect was the basis of the original solar neutrino detector, designed and built by Raymond Davis of the Brookhaven National Laboratory in the 1960s. This device used 100,000 gallons of perchloroethylene (C_2Cl_4), a fluid used in dry cleaning, in a huge tank buried deep underground (Figure 18-8). Most of the solar neutrinos that entered Davis's tank passed right through it with no effect whatsoever. But occasionally a neutrino struck the nucleus of one of the chlorine atoms (^{37}Cl) in the cleaning fluid and converted one

of its neutrons into a proton, creating a radioactive atom of argon (^{37}Ar).

The rate at which argon is produced is related to the *neutrino flux*—that is, the number of neutrinos from the Sun arriving at the Earth per square meter per second. By counting the number of newly created argon atoms, Davis was able to determine the neutrino flux from the Sun. (Other subatomic particles besides neutrinos can also induce reactions that create radioactive atoms. By placing the experiment deep underground, however, the body of the Earth absorbs essentially all such particles—with the exception of neutrinos.)

Davis and his collaborators found that solar neutrinos created one radioactive argon atom in the tank every three days. But this rate corresponds to only *one-third* of the neutrino flux predicted from standard models of the Sun. This troubling discrepancy between theory and observation is called the **solar neutrino problem.** It has motivated scientists around the world to conduct new experiments to measure the flux of solar neutrinos.

 The various nuclear reactions in the proton-proton chain produce neutrinos of differing energies. The vast majority of the neutrinos from the Sun are created during the first reaction, in which two protons combine to form a heavy isotope of hydrogen (see Figure 18-2a). But these neutrinos have too little energy to convert chlorine into argon. Davis's experiment responded only to high-energy neutrinos produced by reactions that occur only part of the time near the end of the proton-proton chain. (Figure 18-2 does not show these reactions.)

To measure the flux of low-energy neutrinos produced by the first reaction of the proton-proton chain, teams of physicists have constructed two neutrino detectors that use several tons of gallium rather than cleaning fluid. Low-energy neutrinos convert gallium (^{71}Ga) into a radioactive isotope of germanium (^{71}Ge). By chemically separating the germanium from the gallium and counting the radioactive atoms, the physicists have been able to measure the low-energy neutrino flux from the Sun. Both detectors—GALLEX in Italy and SAGE (Soviet-American Gallium Experiment) in Russia—have detected only 50% to 60% of the expected neutrino flux.

An experiment in Japan, called Kamiokande, was designed to detect solar neutrinos with energies even higher than those in the Davis experiment. A large underground tank containing 3000 tons of water was surrounded by 1000 light detectors. From time to time, a high-energy solar neutrino struck an electron in one of the water molecules, dislodging it and sending it flying like a pin hit by a bowling ball. The recoiling electron produced a flash of light, which was sensed by the detectors. Kamiokande detected about 45% of the expected flux of neutrinos. By analyzing the flashes, scientists could tell the direction from which the neutrinos were coming and confirmed that they emanated from the Sun. This gave direct evidence that thermonuclear fusion is indeed occurring in the Sun's core.

All these experiments suggest that the Sun is emitting fewer neutrinos than predicted by standard solar models. This discrepancy between theory and observation suggests that either something is wrong with standard solar models or we do not fully understand neutrinos. Some scientists have proposed nonstandard solar models, while others theorize that neutrinos change their character as they travel through space.

One proposed solution to the neutrino problem is that the Sun's core is cooler than predicted by solar models. If the Sun's central temperature is only 10% less than the current estimate, fewer neutrinos would be produced and the neutrino flux would agree with experiments. However, a lower central temperature would cause other obvious features, such as the Sun's size and surface temperature, to be different from what we observe.

Many astronomers and physicists suspect that the solution to the neutrino problem lies not in how neutrinos are produced, but rather in what happens to them between the Sun's core and detectors on the Earth. Physicists have found that there are actually three types of neutrinos. Only one of these types is produced in the Sun, and it is only this type that can be detected by the experiments we have described. But if some of the solar neutrinos change in flight into a different type of neutrino, our detectors would record only a fraction of the total neutrino flux. This effect is called *neutrino oscillation*. In June 1998, scientists at the Super-Kamiokande neutrino observatory in Japan (a larger and more sensitive device than Kamiokande) revealed evidence that neutrino oscillation does indeed take place.

 Super-Kamiokande and the Sudbury Neutrino Observatory in Canada may help solve the solar neutrino problem. While both detectors are sensitive only to high-energy neutrinos, they can detect neutrinos of all three types and can even determine how many of them have undergone neutrino oscillation. The data from these and other solar neutrino experiments may finally reveal the truth about the Sun's core.

18-5 The photosphere is the lowest of three main layers in the Sun's atmosphere

Although the Sun's core is hidden from our direct view, we can easily see sunlight coming from the high-temperature gases that make up the Sun's atmosphere. These outermost layers of the Sun prove to be the sites of truly dramatic activity, much of which has a direct impact on our planet. By studying these layers, we gain further insight into the character of the Sun as a whole.

In a visible-light photograph of the Sun like Figure 18-9, it appears that the Sun has a definite surface. This is actually an illusion; the Sun is gaseous throughout its volume because of its high internal temperature. As you move farther away from the Sun's center, the gases simply become less and less dense.

Why, then, does the Sun appear to have a sharp, well-defined surface? The reason is that essentially all of the Sun's visible light emanates from a single, thin layer of gas called the **photosphere** ("sphere of light"). Just as you can see only a certain distance through the Earth's atmosphere before objects vanish in the haze, we can see only about 400 km into the

figure 18-9   R I **V** U X G

The Photosphere The photosphere is the layer in the solar atmosphere from which the Sun's visible light is emitted. Note that the Sun appears darker around its limb, or edge; here we are seeing the upper photosphere, which is relatively cool and thus glows less brightly. (The dark sunspots, which we discuss in Section 18-8, are also relatively cool regions.) (Celestron International)

photosphere. This distance is so small compared with the Sun's radius of 696,000 km that the photosphere appears to be a definite surface.

The photosphere is actually the lowest of the three layers that together constitute the solar atmosphere. Above it are the *chromosphere* and the *corona*, both of which are transparent to visible light. We can see them only using special techniques, which we discuss later in this chapter. Everything below the photosphere is called the *solar interior*.

The photosphere is heated from below by energy streaming outward from the solar interior. Hence, temperature should decrease as you go upward in the photosphere, just as in the solar interior (see Table 18-2 and Figure 18-4). We know this is the case because the photosphere appears darker around the edge, or *limb*, of the Sun than it does toward the center of the solar disk, an effect called **limb darkening** (examine Figure 18-9). This happens because when we look near the Sun's limb, we do not see as deeply into the photosphere as we do when we look near the center of the disk (Figure 18-10). The high-altitude gas we observe at the limb is not as hot and thus does not glow as brightly as the deeper, hotter gas seen near the disk center.

The spectrum of the Sun's photosphere confirms how its temperature varies with altitude. As we saw in Section 18-1, the photosphere shines like a nearly perfect blackbody with an average temperature of about 5800 K. However, superimposed on this spectrum are many dark absorption lines (see Figure 5-12). As discussed in Section 5-6, we see an *absorption line spectrum* of this sort whenever we view a hot, glow-

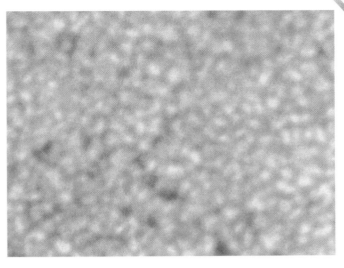

Figure 18-10

The Origin of Limb Darkening Light coming from both the Sun's limb and the center of the Sun's disk travels about the same straight-line distance through the photosphere to reach a telescope on Earth. Because of the Sun's curved shape, light from the limb comes from a greater height within the photosphere, where the temperature is lower and the gases glow less brightly. This gives the limb its darker appearance.

 Figure 18-11   R I 【V】 U X G

Solar Granulation High-resolution photographs of the Sun's surface, such as this one, reveal a blotchy pattern called granulation. The individual yellow blobs in this image are granules, each of which measures about 1000 km (600 mi) across. Granules are actually convection cells in the Sun's outer layers (see Figure 18-12). (NOAO)

ing object through a relatively cool gas. In this case, the hot object is the lower part of the photosphere; the cooler gas is in the upper part of the photosphere, where the temperature declines to about 4400 K. All the absorption lines in the Sun's spectrum are produced in this relatively cool layer, as atoms selectively absorb photons of various wavelengths streaming outward from the hotter layers below.

 You may find it hard to think of 4400 K as "cool." But keep in mind that the ratio of 4400 K to 5800 K, the temperature in the lower photosphere, is the same as the ratio of the temperature on a Siberian winter night to that of a typical day in Hawaii.

We can learn still more about the photosphere by examining it with a telescope—but only when using special dark filters to prevent eye damage. *Looking directly at the Sun without the correct filter, whether with the naked eye or with a telescope, can cause permanent blindness!* Under good observing conditions, astronomers using such filter-equipped telescopes can often see a blotchy pattern in the photosphere, called **granulation** (Figure 18-11). Each light-colored **granule** measures about 1000 km (600 mi) across and is surrounded by a darkish boundary. The difference in brightness between the center and the edge of a granule corresponds to a temperature drop of about 300 K.

Granulation is caused by convection of the gas in the photosphere. Gas from lower levels rises upward in granules, cools off, spills over the edges of the granules, and then plunges back

down into the Sun (Figure 18-12). This can occur only if the gas is heated from below, like a pot of water being heated on a stove (see Section 18-2). Along with limb darkening and the Sun's absorption line spectrum, granulation shows that the upper part of the photosphere must be cooler than the lower part.

 Figure 18-12

Convection in the Photosphere Hot gas rising upward produces the bright granules shown in Figure 18-11. Cooler gas sinks downward along the boundaries between granules; this gas glows less brightly, giving the boundaries their dark appearance. The circulating convective motion shown here transports heat from the Sun's interior outward into the solar atmosphere.

ƒigure 18-13 R I **V** U X G

Supergranules and Large-Scale Convection Supergranules display relatively little contrast between their center and edges, so they are hard to observe in ordinary images. But they can be seen in a Doppler image like this one. Light from gas that is approaching us (that is, rising) is shifted toward shorter wavelengths, while light from receding gas (that is, descending) is shifted toward longer wavelengths (see Section 5-9). The colors in this image are false: areas of rising and falling gas are colored blue and red, respectively. (David Hathaway, MSFC/NASA)

Time-lapse photography reveals more of the photosphere's dynamic activity. Granules form, disappear, and reform in cycles lasting only a few minutes. At any one time, about 4 million granules cover the solar surface. A typical granule occupies about 10^6 square kilometers, equal to the areas of Texas and Oklahoma combined.

Superimposed on the pattern of granulation are even larger convection cells called **supergranules** (Figure 18-13). As in granules, gases rise upward in the middle of a supergranule, move horizontally outward toward its edge, and descend back into the Sun. The difference is that a typical supergranule is about 35,000 km in diameter, large enough to enclose several hundred granules. This large-scale convection moves at only about 0.4 km/s (1400 km/h, or 900 mi/h), about one-tenth the speed of gases churning in a granule. A given supergranule lasts about a day.

ANALOGY Similar patterns of large-scale and small-scale convection can be found in the Earth's atmosphere. On the large scale, air rises gradually at a low-pressure area, then sinks gradually at a high-pressure area, which might be hundreds of kilometers away. This is analogous to the flow in a supergranule. Thunderstorms in our atmo-

sphere are small but intense convection cells within which air moves rapidly up and down. Like granules, they last only a relatively short time before they dissipate.

Although the photosphere is a very active place, it actually contains relatively little material. Careful examination of the spectrum shows that it has a density of only about 10^{-4} kg/m^3, roughly 0.01% the density of the Earth's atmosphere at sea level. The photosphere is made primarily of hydrogen and helium, the most abundant elements in the solar system (see Table 7-3).

Despite being such a thin gas, the photosphere is surprisingly opaque to visible light. If it were not so opaque, we could see into the Sun's interior to a depth of hundreds of thousands of kilometers, instead of the mere 400 km that we can see down into the photosphere. What makes the photosphere so opaque is that its hydrogen atoms sometimes acquire an extra electron, becoming **negative hydrogen ions.** The extra electron is only loosely attached and can be dislodged if it absorbs a photon of any visible wavelength. Hence, negative hydrogen ions are very efficient light absorbers, and there are enough of these light-absorbing ions in the photosphere to make it quite opaque. Because it is so opaque, the photosphere's spectrum is close to that of an ideal blackbody.

18-6 The chromosphere is characterized by spikes of rising gas

An ordinary visible-light photograph such as Figure 18-9 gives the impression that the Sun ends at the top of the photosphere. But during a total solar eclipse, the Moon blocks the photosphere from our view, revealing a glowing, pinkish layer of gas above the photosphere (Figure 18-14). This is the tenuous **chromosphere** ("sphere of color"), the second of the three major levels in the Sun's atmosphere. The chromosphere is only about one ten-thousandth (10^{-4}) as dense as the photosphere, or about 10^{-8} as dense as our own atmosphere. No wonder it is normally invisible!

Unlike the photosphere, which has an absorption line spectrum, the chromosphere has a spectrum dominated by *emission* lines. An emission line spectrum is produced by the atoms of a hot, thin gas (see Section 5-6 and Section 5-8). As their electrons fall from higher to lower energy levels, the atoms emit photons.

One of the strongest emission lines in the chromosphere's spectrum is the H_α line at 656.3 nm, which is emitted by a hydrogen atom when its single electron falls from the $n = 3$ level to the $n = 2$ level (recall Figure 5-21b). This wavelength is in the red part of the spectrum, which gives the chromosphere its characteristic pinkish color. The spectrum also contains emission lines of singly ionized calcium, as well as lines due to ionized helium and ionized metals. In fact, helium was originally discovered in the chromospheric spectrum in 1868, almost thirty years before helium gas was first isolated on Earth.

figure 18-14 R I **V** U X G

The Chromosphere This photograph, taken during a total solar eclipse, shows the Sun's chromosphere glowing around the edge of the Moon. It appears pinkish because its hot gases emit light at only certain discrete wavelengths, principally the H_α emission of hydrogen at a red wavelength of 656.3 nm. The expanded area above shows spicules, jets of chromospheric gas that surge upward into the Sun's outer atmosphere. In making this image, a filter was used that is transparent only to H_α emission. (NOAO)

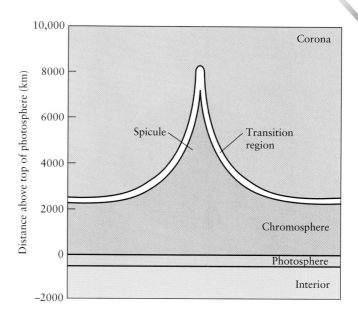

figure 18-15

The Solar Atmosphere This schematic diagram shows the three layers of the solar atmosphere. The lowest, the photosphere, is about 400 km thick. The chromosphere above it extends to an altitude of about 2000 km, with spicules jutting up to nearly 10,000 km above the photosphere. The third, outermost layer is the corona, which we discuss in Section 18-7. It extends many millions of kilometers out into space. (Adapted from J. A. Eddy)

Analysis of the chromospheric spectrum shows that temperature *increases* with increasing height in the chromosphere. This is just the opposite of the situation in the photosphere, where temperature decreases with increasing height. The temperature is about 4400 K at the top of the photosphere; 2000 km higher, at the top of the chromosphere, the temperature is nearly 25,000 K.

The photospheric spectrum is dominated by absorption lines at certain wavelengths, while the spectrum of the chromosphere has emission lines at these same wavelengths. In other words, the photosphere appears dark at the wavelengths at which the chromosphere emits most strongly, such as the H_α wavelength of 656.3 nm. By viewing the Sun through a special filter that is transparent to light only at the wavelength of H_α, astronomers can screen out light from the photosphere and make the chromosphere visible. (The same technique can be used with other wavelengths at which the chromosphere emits strongly, including nonvisible wavelengths.) This makes it possible to see the chromosphere at any time, not just during a solar eclipse.

The top photograph in Figure 18-14 is a high-resolution image of the Sun's chromosphere taken through an H_α filter. This image shows numerous vertical spikes, which are actually jets of rising gas called **spicules**. A typical spicule lasts just 15 minutes or so: It rises at the rate of about 20 km/s

(72,000 km/h, or 45,000 mi/h), reaches a height of several thousand kilometers, then collapses and fades away (Figure 18-15). Approximately 300,000 spicules exist at any one time, covering about 1% of the Sun's surface.

Spicules are generally located directly above the edges of supergranules (see Figure 18-13). This is a surprising result, because chromospheric gases are rising in a spicule while photospheric gases are *descending* at the edge of a supergranule. What, then, is pulling gases upward to form spicules? As we will see in Section 18-10, the answer proves to be the Sun's intense magnetic field. But before we delve into how this happens, let us to complete our tour of the solar atmosphere by exploring its outermost, least dense, most dynamic, and most bizarre layer—the region called the corona.

18-7 The corona ejects mass into space to form the solar wind

The **corona,** or outermost region of the Sun's atmosphere, begins at the top of the chromosphere. It extends out to a distance of several million kilometers. Despite its tremendous extent, the corona is only about one-millionth (10^{-6}) as bright as the photosphere—no brighter than the full moon. Hence,

ƒigure 18-16   R I **V** U X G

The Solar Corona This striking photograph of the corona was taken during the total solar eclipse of July 11, 1991. Numerous streamers extend for millions of kilometers above the solar surface. The unearthly light of the corona is one of the most extraordinary aspects of experiencing a solar eclipse. (Courtesy of R. Christen and M. Christen, Astro-Physics, Inc.)

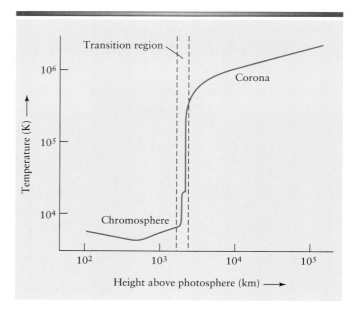

ƒigure 18-17

Temperatures in the Sun's Upper Atmosphere This graph shows how temperature varies with altitude in the Sun's chromosphere and corona. In order to show a large range of values, both the vertical and horizontal scales are nonlinear. In the narrow transition region between the chromosphere and corona, the temperature rises abruptly by a hundredfold, from about 10^4 K to about 10^6 K. (Adapted from A. Gabriel)

the corona can be viewed only when the light from the photosphere is blocked out, either by use of a specially designed telescope or during a total eclipse.

Figure 18-16 is an exceptionally detailed photograph of the Sun's corona taken during a solar eclipse. It shows that the corona is not merely a spherical shell of gas surrounding the Sun. Rather, numerous streamers extend in different directions far above the solar surface. The shapes of these streamers vary on time scales of days or weeks. (For other views of the corona during solar eclipses, see Figure 3-10 and the photograph that opens Chapter 3.)

The corona has an emission line spectrum, characteristic of a hot, thin gas. When this spectrum was first measured in the nineteenth century, astronomers found a number of emission lines at wavelengths that had never been seen in the laboratory. Their explanation was that the corona contained elements that had not yet been detected on the Earth. However, laboratory experiments in the 1930s revealed that these unusual emission lines were in fact caused by the same atoms found elsewhere in the universe—but in highly ionized states. For example, a prominent green line at 530.3 nm is caused by highly ionized iron atoms, each of which has been stripped of 13 of its 26 electrons. In order to strip that many electrons from atoms, temperatures in the corona must reach 2 million kelvins (2×10^6 K) or even higher. Figure 18-17 shows how temperature in both the chromosphere and corona varies with altitude.

 The corona is actually not very "hot"—that is, it contains very little thermal energy. The reason is that the corona is nearly a vacuum. In the corona there are only about 10^{11} atoms per cubic meter, compared with about 10^{23} atoms per cubic meter in the Sun's photosphere and about 10^{25} atoms per cubic meter in the air that we breathe. Because of the corona's high temperature, the atoms there are moving at very high speeds. But because there are so few atoms in the corona, the total amount of energy in these moving atoms (a measure of how "hot" the gas is) is rather low. If you flew a spaceship into the corona, you would have to worry about becoming overheated by the intense light coming from the photosphere, but you would notice hardly any heating from the corona's ultrathin gas.

 The situation in the corona is similar to that inside a conventional oven that is being used for baking. Both the walls of the oven and the air inside the oven are at the same high temperature, but the air contains very few atoms and thus carries little energy. If you put your hand in the oven momentarily, the lion's share of the heat you feel is radiation from the oven walls.

The low density of the corona explains why it is so dim compared with the photosphere. In general, the higher the temperature of a gas, the brighter it glows. But because there are so few atoms in the corona, the net amount of light that it emits is very feeble compared with the light from the much cooler, but also much denser, photosphere.

The Earth's gravity keeps our atmosphere from escaping into space. In the same way, the Sun's powerful gravitational attraction keeps most of the gases of the photosphere, chromosphere, and corona from escaping. But the corona's high temperature means that its atoms and ions are moving at very high speeds, around a million kilometers per hour. As a result, some of the coronal gas does escape. This outflow of gas, which we first encountered in Section 7-8, is called the **solar wind.**

Each second the Sun ejects about a million tons (10^9 kg) of material into the solar wind. But the Sun is so massive that, even over its entire lifetime, it will eject only a few tenths of a percent of its total mass. The solar wind is composed almost entirely of electrons and nuclei of hydrogen and helium. About 0.1% of the solar wind is made up of ions of more massive atoms, such as silicon, sulfur, calcium, chromium, nickel, iron, and argon. The aurorae seen at far northern or southern latitudes on Earth are produced when electrons and ions from the solar wind enter our upper atmosphere.

Special telescopes enable astronomers to see the origin of the solar wind. To appreciate what sort of telescopes are needed, note that because the temperature of the coronal gas is so high, ions in the corona are moving very fast (see Box 7-2). When ions collide, the energy of the impact is so great that the ion's electrons are boosted to very high energy levels. As the electrons fall back to lower levels, they emit high-energy photons in the ultraviolet and X-ray portions of the spectrum—wavelengths at which the photosphere and chromosphere are relatively dim. Hence, telescopes sensitive to these short wavelengths are ideal for studying the corona and the flow of the solar wind.

The Earth's atmosphere is opaque to ultraviolet light and X rays, so telescopes for these wavelengths must be placed on board spacecraft (see Section 6-7, especially Figure 6-27). Figure 18-18 shows a composite ultraviolet view of the corona from the spacecraft *SOHO (Solar and Heliospheric Observatory)*, a joint project of the European Space Agency (ESA) and NASA.

Figure 18-18 reveals that the corona is not uniform in temperature or density. The densest, highest-temperature regions appear bright, while the thinner, lower-temperature regions are dark. Note the large dark area, called a **coronal hole** because it is almost devoid of luminous gas. Particles streaming away from the Sun can most easily flow outward through these particularly thin regions. Therefore, it is thought that coronal holes are the main corridors through which particles of the solar wind escape from the Sun. Evidence in favor of this picture has come from the *Ulysses* spacecraft, another joint ESA/NASA mission. In 1994 and 1995, *Ulysses* became the first spacecraft to fly over the Sun's north and south poles, where there are apparently permanent coronal holes. The spacecraft indeed measured a stronger solar wind emanating from these holes.

The temperatures in the corona and the chromosphere are not at all what we would expect. Just as you feel warm

Coronal hole

igure 18-18   R I V U X G

The Ultraviolet Corona This false-color view of the corona is a composite of two ultraviolet images from *SOHO*. The black ring separates the two images. The outer image shows material from the corona streaming outward to join the solar wind. The inner image shows the origin of these streamers in the lower regions of the corona. The dark feature running across the Sun's disk from the top is a coronal hole, a region where the coronal gases are thinner and at a lower temperature than average. (Inner image: *SOHO* Extreme Ultraviolet Imaging Telescope Consortium. Outer image: Ultraviolet Coronagraph Spectrometer Consortium)

if you stand close to a campfire but cold if you move away, we would expect that the temperature in the corona and chromosphere would *decrease* with increasing altitude and, hence, increasing distance from the warmth of the Sun's photosphere. Why, then, does the temperature in these regions *increase* with increasing altitude? This has been one of the major unsolved mysteries in astronomy for the past half-century. As astronomers have tried to resolve this dilemma, they have found important clues in one of the Sun's most familiar features—sunspots.

18-8 Sunspots are low-temperature regions in the photosphere

Granules, supergranules, spicules, and the solar wind occur continuously. These features are said to be aspects of the *quiet Sun.* But other, more dramatic features appear periodically, including massive eruptions and regions of concentrated magnetic fields. When these are present, astronomers refer to the *active Sun.* One aspect of the active Sun that can easily be seen with even a small telescope (although only with an appropriate filter attached) is the feature known as sunspots.

a

b

Figure 18-19   R I **V** U X G

Sunspots (a) This high-resolution photograph of the photosphere shows a mature sunspot. The dark center of the spot is called the umbra. It is bordered by the penumbra, which is less dark and has a featherlike appearance. (b) In this view of a typical sunspot group, several sunspots are close enough to overlap. In both images you can see granulation in the surrounding, undisturbed photosphere. (NOAO)

Sunspots are irregularly shaped dark regions in the photosphere. Sometimes sunspots appear in isolation (Figure 18-19*a*), but frequently they are found in sunspot groups (Figure 18-19*b*; see also Figure 18-9). Although sunspots vary greatly in size, typical ones measure a few tens of thousands of kilometers across—comparable to the diameter of the Earth. Sunspots are not permanent features of the photosphere but last between a few hours and a few months.

Each sunspot has a dark central core, called the *umbra*, and a brighter border called the *penumbra*. We used these same terms in Section 3-4 to refer to different parts of the Earth's or the Moon's shadow. But a sunspot is not a shadow; it is a region in the photosphere where the temperature is relatively low, which makes it appear darker than its surroundings. If the surrounding photosphere is blocked from view, a sunspot's umbra appears red and the penumbra appears orange. As we saw in Section 5-4, Wien's law relates the color of a blackbody (which depends on the wavelength at which it emits the most light) to the blackbody's temperature. The colors of a sunspot indicate that the temperature of the umbra is typically 4300 K and that of the penumbra is typically 5000 K. While high by earthly standards, these temperatures are quite a bit lower than the average photospheric temperature of 5800 K.

The Stefan-Boltzmann law (see Section 5-4) tells us that the energy flux from a blackbody is proportional to the fourth power of its temperature. This law lets us compare the amounts of light energy emitted by a square meter of a sun-

spot's umbra and by a square meter of undisturbed photosphere. The ratio is

$$\frac{\text{flux from umbra}}{\text{flux from photosphere}} = \left(\frac{4300 \text{ K}}{5800 \text{ K}}\right)^4 = 0.30$$

That is, the umbra emits only 30% as much light as an equally large patch of undisturbed photosphere. This is why sunspots appear so dark.

On rare occasions, a sunspot group is large enough to be seen without a telescope. Chinese astronomers recorded such sightings 2000 years ago, and a huge sunspot group visible to the naked eye (with an appropriate filter) was seen in 1989. But it was not until Galileo introduced the telescope into astronomy (see Section 4-3) that anyone was able to examine sunspots in detail.

Galileo discovered that he could determine the Sun's rotation rate by tracking sunspots as they moved across the solar disk (Figure 18-20). He found that the Sun rotates once in about four weeks. A typical sunspot group lasts about two months, so a specific one can be followed for two solar rotations.

Further observations by the British astronomer Richard Carrington in 1859 demonstrated that the Sun does not rotate as a rigid body. Instead, the equatorial regions rotate more rapidly than the polar regions. This phenomenon is known as **differential rotation**. Thus, while a sunspot near the solar equator takes only 25 days to go once

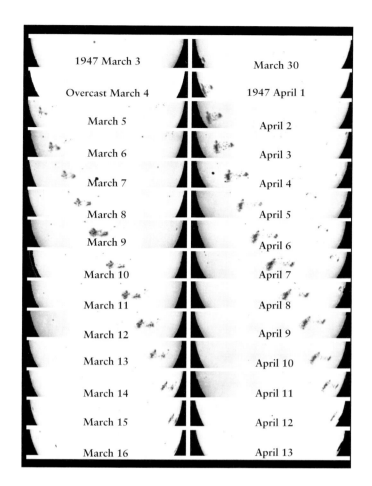

1947 March 3

Overcast March 4

March 5

March 6

March 7

March 8

March 9

March 10

March 11

March 12

March 13

March 14

March 15

March 16

March 30

1947 April 1

April 2

April 3

April 4

April 5

April 6

April 7

April 8

April 9

April 10

April 11

April 12

April 13

figure 18-20   R I **V** U X G

Tracking the Sun's Rotation with Sunspots By observing the same group of sunspots from one day to the next, Galileo found that the Sun rotates once in about four weeks. The equatorial regions of the Sun actually rotate somewhat faster than the polar regions, a phenomenon called differential rotation. This series of photographs shows the same sunspot group over 1½ solar rotations. (The Carnegie Observatories)

around the Sun, a sunspot at 30° north or south of the equator takes 27½ days. The rotation period at 75° north or south is about 33 days, while near the poles it may be as long as 35 days.

The average number of sunspots on the Sun is not constant, but varies in a predictable **sunspot cycle** (Figure 18-21*a*). This phenomenon was first reported by the German astronomer Heinrich Schwabe in 1843 after many years of observing. As Figure 18-21*a* shows, the average number of sunspots varies with a period of about 11 years. A period of exceptionally many sunspots is a **sunspot maximum,** as occurred in 1979, 1989, and 2000. Conversely, the Sun is almost devoid of sunspots at a **sunspot minimum,** as occurred in 1976, 1986, and 1996 and is projected to occur in 2007.

The locations of sunspots also vary with the same 11-year sunspot cycle. At the beginning of a cycle, just after a sunspot minimum, sunspots first appear at latitudes around 30°

a

b   R I **V** U X G

c   R I **V** U X G

figure 18-21

The Sunspot Cycle (a) The number of sunspots on the Sun varies with a period of about 11 years. The most recent sunspot maximum occurred in 2000, and the most recent sunspot minimum occurred in 1996. **(b)** This photograph, taken near sunspot maximum in 1989, shows a number of sunspots and large sunspot groups. The sunspot group visible near the bottom of the Sun's disk has about the same diameter as the planet Jupiter. **(c)** Near sunspot minimum, as in this 1986 photograph, essentially no sunspots are visible. (NOAO)

figure 18-22

Variations in the Average Latitude of Sunspots At the beginning of each sunspot cycle, most sunspots are found about 30° north or south of the Sun's equator. Sunspots that appear later in each cycle typically form closer and closer to the equator.

north and south of the solar equator (Figure 18-22). Over the succeeding years, the sunspots occur closer and closer to the equator, until at the end of the cycle virtually all sunspots lie on the solar equator.

18-9 Sunspots are produced by a 22-year cycle in the Sun's magnetic field

Why should the number of sunspots vary with an 11-year cycle? Why should their average latitude vary over the course of a cycle? And why should sunspots exist at all? The first step

toward answering these questions came in 1908, when the American astronomer George Ellery Hale discovered that sunspots are associated with intense magnetic fields on the Sun. When Hale focused a spectroscope on sunlight coming from a sunspot, he found that many spectral lines appear to be split into several closely spaced lines (Figure 18-23). This "splitting" of spectral lines is called the **Zeeman effect**, after the Dutch physicist Pieter Zeeman, who first observed it in his laboratory in 1896. Zeeman showed that a spectral line splits when the atoms are subjected to an intense magnetic field. The more intense the magnetic field, the wider the separation of the split lines.

Hale's discovery showed that sunspots are places where the hot gases of the photosphere are bathed in a concentrated magnetic field. Because of the high temperature of the Sun's atmosphere, many of the atoms there are ionized. The solar atmosphere is thus a gaseous mixture called a **plasma,** in which electrically charged ions and electrons can move freely. Like any moving, electrically charged objects, they can be deflected by magnetic fields. Figure 18-24 shows how a magnetic field in the laboratory bends a beam of fast-moving electrons into a curved trajectory. Similarly, the paths of moving ions and electrons in the photosphere are deflected by the Sun's magnetic field. In particular, magnetic forces act on the hot plasma that rises from the Sun's interior due to convection. Where the magnetic field is particularly strong, these forces push the hot plasma away. The result is a localized region where the gas is relatively cool and thus glows less brightly—in other words, a sunspot.

a

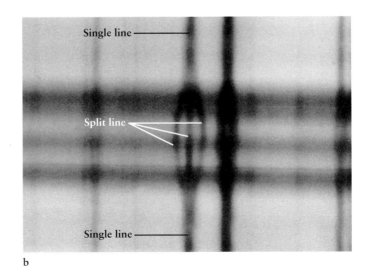

b

figure 18-23 R I **V** U X G

Sunspots Have Strong Magnetic Fields (a) Astronomers measure the magnetic field of a sunspot using a spectrograph. A black line in this image of a sunspot shows where the slit of the spectrograph was aimed. (b) This is a portion of the resulting spectrum, including a dark absorption line caused by iron atoms in the photosphere. Within the sunspot, this line is split into three components. The separation in wavelength between these components can be used to calculate the strength of the magnetic field in the sunspot. Typical sunspot fields are more than 5000 times stronger than the Earth's magnetic field at its north and south poles. (NOAO)

figure 18-24 R I **V** U X G

Magnetic Fields Deflect Moving, Electrically Charged Objects
In this laboratory experiment, a beam of negatively charged electrons (shown by a blue arc) is aimed straight upward. The entire apparatus is inside a large magnet, and the magnetic field deflects the beam into a curved path. This effect is used to focus the electron beam in a television picture tube and to make electrons oscillate so that they produce microwaves in a microwave oven. (Courtesy of Central Scientific Company)

To get a fuller picture of the Sun's magnetic fields, astronomers take images of the Sun at two wavelengths, one just less than and one just greater than the wavelength of a magnetically split spectral line. From the difference between these two images, they can construct a picture called a **magnetogram,** which displays the magnetic fields in the solar atmosphere. Figure 18-25*a* is an ordinary white-light photograph of the Sun taken at the same time as the magnetogram in Figure 18-25*b*. In the magnetogram, dark blue indicates areas of the photosphere with one magnetic polarity (north), and yellow indicates areas with the opposite (south) magnetic polarity. This image shows that many sunspot groups have roughly comparable areas covered by north and south magnetic polarities (see also Figure 18-25*c*). Thus, a sunspot group resembles a giant bar magnet, with a north magnetic pole at one end and a south magnetic pole at the other.

If different sunspot groups were unrelated to one another, their magnetic poles would be randomly oriented, like a bunch of compass needles all pointing in random directions. As Hale discovered, however, there is a striking regularity in the magnetization of sunspot groups. As a given sunspot group moves with the Sun's rotation, the sunspots in front are called the "preceding members" of the group. The spots that follow behind are referred to as the "following members." Hale compared the sunspot groups in the two solar hemispheres, north or south of the Sun's equator. He found that the preceding members in one solar hemisphere all have the same magnetic polarity, while the preceding members in the other hemisphere have the opposite polarity. Furthermore, in the hemisphere where the Sun has its north magnetic pole, the preceding members of all sunspot groups have north magnetic polarity. In the opposite hemisphere, where the Sun has its south magnetic pole, the preceding members all have south magnetic polarity.

Along with his colleague Seth B. Nicholson, Hale also discovered that the Sun's polarity pattern completely reverses itself every 11 years—the same interval as the time from one

a

b

c

figure 18-25 R I **V** U X G

Mapping the Sun's Magnetic Field This visible-light image of the Sun **(a)** was taken at the same time as this magnetogram **(b)**. Dark blue and yellow areas in the magnetogram have north and south magnetic polarity, respectively. Regions with weak magnetic fields, which include much of the Sun's disk, are shown in blue-green. Notice how regions with strong magnetic fields in (b) correlate with the locations of sunspots in (a). **(c)** This close-up magnetogram of a large sunspot group shows that the two ends of the group have opposite magnetic polarities (colored blue and yellow), like the two ends of a giant bar magnet. (NOAO)

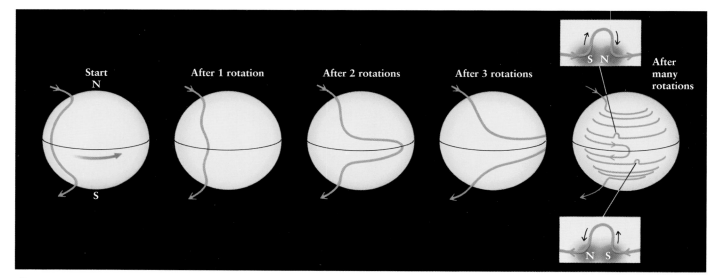

figure 18-26

Babcock's Magnetic Dynamo The sunspot cycle may be due in part to the Sun's differential rotation. Magnetic field lines tend to move along with the plasma in the Sun's outer layers. A field line that starts off running from the Sun's north magnetic pole (N) to its south magnetic pole (S), as in the drawing on the far left, soon becomes distorted because the Sun's equator rotates faster than the poles. After many rotations the magnetic field is wrapped around the Sun like twine wrapped around a ball. The insets on the far right show how sunspot groups appear where the concentrated magnetic field breaks through the solar surface. In each sunspot group, the preceding members—in the insets, the sunspots on the right—have the same polarity as the Sun's magnetic pole in that hemisphere (in this drawing, N in the upper hemisphere and S in the lower hemisphere).

solar maximum to the next. The hemisphere that has preceding north magnetic poles during one 11-year sunspot cycle will have preceding south magnetic poles during the next 11-year cycle, and vice versa. The north and south magnetic poles of the Sun itself also reverse every 11 years. Thus, the Sun's magnetic pattern repeats itself only after two sunspot cycles, which is why astronomers speak of a **22-year solar cycle.**

In 1960, the American astronomer Horace Babcock proposed a description that seems to account for many features of this 22-year solar cycle. Babcock's scenario, called a **magnetic-dynamo model,** makes use of two basic properties of the Sun's photosphere—differential rotation and convection. Differential rotation causes the magnetic field in the photosphere to become wrapped around the Sun (Figure 18-26). As a result, the magnetic field becomes concentrated at certain latitudes on either side of the solar equator. Convection in the photosphere creates tangles in the concentrated magnetic field, and "kinks" erupt through the solar surface. Sunspots appear where the magnetic field protrudes through the photosphere. The theory suggests that sunspots should appear first at northern and southern latitudes and later form nearer to the equator. This is just what is observed (see Figure 18-22). Note also that as shown in Figure 18-26, the preceding member of a sunspot group has the same polarity as the Sun's magnetic pole in that hemisphere, just as Hale observed.

Differential rotation eventually undoes the twisted magnetic field. The preceding members of sunspot groups move toward the Sun's equator, while the following members migrate toward the poles. Because the preceding members from the two hemispheres have opposite magnetic polarities, their magnetic fields cancel each other out when they meet at the equator. The following members in each hemisphere have the opposite polarity to the Sun's pole in that hemisphere; hence, when they converge on the pole, the following members first cancel out and then reverse the Sun's overall magnetic field. The fields are now completely relaxed. Once again, differential rotation begins to twist the Sun's magnetic field, but now with all magnetic polarities reversed. In this way, Babcock's model helps to explain the change in field direction every 11 years.

Recent discoveries in helioseismology (Section 18-3) offer new insights into the Sun's magnetic field. By comparing the speeds of sound waves that travel with and against the Sun's rotation, helioseismologists have been able to determine the Sun's rotation rate at different depths and latitudes. As shown in Figure 18-27, the Sun's surface pattern of differential rotation persists through the convective zone. Farther in, within the radiative zone, the Sun seems to rotate like a rigid object with a period of 27 days at all latitudes. Astronomers suspect that the Sun's magnetic field originates in a relatively thin layer where the radiative and convective zones meet and slide past each other due to their different rotation rates.

Much about sunspots and solar activity remains mysterious. The best calculations predict that a sunspot should break

35 days

25 days

WEB LINK 18·12

Figure 18-27

Rotation Rates in the Solar Interior This cutaway picture of the Sun shows how the solar rotation rate varies with depth and latitude. Colors represent rotation periods according to the scale on the right. The surface rotation pattern, which varies from 25 days at the equator to 35 days near the poles, persists throughout the Sun's convective zone. The radiative zone seems to rotate like a rigid sphere. (Courtesy of K. Libbrecht, Big Bear Solar Observatory)

up and disperse fairly quickly, but, in fact, sunspots can persist for many weeks. There are also perplexing irregularities in the solar cycle. For example, the overall reversal of the Sun's magnetic field is often piecemeal and haphazard. One pole may reverse polarity long before the other. For several weeks the Sun's surface may have two north magnetic poles and no south magnetic pole at all.

What is more, there seem to be times when all traces of sunspots and the sunspot cycle vanish for many years. For example, virtually no sunspots were seen from 1645 through 1715. Curiously, during these same years Europe experienced record low temperatures, often referred to as the Little Ice Age, whereas the western United States was subjected to severe drought. By contrast, there was apparently a period of increased sunspot activity during the eleventh and twelfth centuries, during which the Earth was warmer than it is today. Thus, variations in solar activity appear to affect climates on the Earth. The origin of this Sun-Earth connection is a topic of ongoing research.

18-10 The Sun's magnetic field also produces other forms of solar activity

Astronomers now understand that the Sun's magnetic field does more than just explain the presence of sunspots. It is also responsible for the existence of spicules, as well as a host of other dramatic phenomena in the chromosphere and corona.

In a plasma, magnetic field lines and the material of the plasma tend to move together. This means that as convection pushes material toward the edge of a supergranule, it pushes magnetic field lines as well. The result is that vertical magnetic field lines pile up around a supergranule. Plasma that "sticks" to these magnetic field lines thus ends up lifted upward, forming a spicule (see Figure 18-14).

The tendency of plasma to follow the Sun's magnetic field may also explain why the temperature of the corona is so high. Spacecraft observations show magnetic field arches extending tens of thousands of kilometers into the corona, with streamers of electrically charged particles moving along each arch (Figure 18-28). If two arches come near to each other, their flowing charges can interact to form a gigantic

WEB LINK 18·13

Figure 18-28 R I V **U** X G

Magnetic Heating of the Corona This false-color ultraviolet image from NASA's *TRACE* spacecraft (*Transition Region and Coronal Explorer*) shows magnetic field loops suspended high above the solar surface. The largest loops are about 200,000 km (120,000 mi, or 15 times the Earth's diameter) across. "Short circuits" among the field lines have raised the gas temperature to as high as 2.7×10^6 K, making it glow at ultraviolet wavelengths. (Stanford-Lockheed Institute for Space Research; *TRACE*; and NASA)

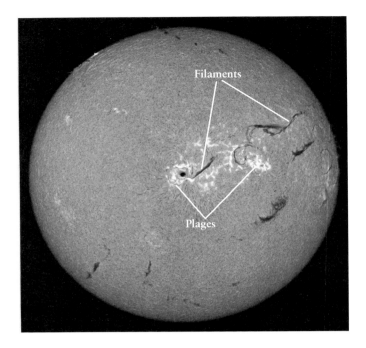

Figure 18-29 R I ☑ U X G

Plages and Filaments in the Chromosphere The bright areas in this H_α image, called plages, are hotter than the surrounding chromosphere. The dark filaments extend far above the normal extent of the chromosphere; when seen against the background of space, they appear as in Figure 18-30. Plages and filaments are associated with strong magnetic fields and are always found near sunspots. (Sacramento Peak Observatory)

Figure 18-30 R I V U ☒ G

A Solar Prominence A huge prominence arches above the solar surface in this X-ray image from the *SOHO* spacecraft. The image was recorded using light at a wavelength of 30.4 nm, emitted by singly ionized helium atoms at a temperature of about 60,000 K. The gas ejected from this 1997 prominence did not strike the Earth. (Solar and Heliospheric Observatory)

"short circuit." Just as a short circuit on the Earth can make sparks fly, these coronal short circuits release a tremendous amount of energy into the corona. (A single arch contains as much energy as a hydroelectric power plant would generate in a million years.) The amount of energy released in this way appears to be more than enough to maintain the corona's temperature.

Magnetic heating can also explain why the parts of the corona that are most prominent in X-ray images are often those that lie on top of sunspots. (Some examples are the bright regions in the image that opens this chapter.) The intense magnetic field of the sunspots helps give the overlying coronal gas such a high temperature that it emits X rays.

Spicules and coronal heating occur even when the Sun is quiet. But magnetic fields can also explain many aspects of the active Sun in addition to sunspots. Figure 18-29 is an image of the chromosphere made with an H_α filter during a sunspot maximum. The bright areas are called **plages** (from the French word for "beach"). These are bright, hot regions in the chromosphere that tend to form just before the appearance of new sunspots. They are probably created by magnetic fields that push upward from the Sun's interior, compressing and heating a portion of the chromosphere. The dark streaks, called **filaments**, are relatively cool and dense parts of the chromo-

sphere that have been pulled along with magnetic field lines as they arch to high altitudes.

When seen from the side, so that they are viewed against the dark background of space, filaments appear as bright, arching columns of gas called **prominences** (Figure 18-30). They can extend for tens of thousands of kilometers above the photosphere. Some prominences last for only a few hours, while others persist for many months. The prominence in Figure 18-30 was so energetic that it broke free of the magnetic fields that confined it and burst into space.

Violent, eruptive events on the Sun, called **solar flares,** occur in complex sunspot groups. Within only a few minutes, temperatures in a compact region may soar to 5×10^6 K and vast quantities of particles and radiation are blasted out into space. These eruptions can also cause disturbances that spread outward in the solar atmosphere, like the ripples that appear when you drop a rock into a pond.

The most energetic flares carry as much as 10^{30} joules of energy, equivalent to 10^{14} one-megaton nuclear weapons being exploded at once! However, the energy of a solar flare does not come from thermonuclear fusion in the solar atmosphere; instead, it appears to be released from the intense magnetic field around a sunspot group.

a   b   c

ƒigure 18-31   R I **V** **U X G**

A Coronal Mass Ejection These visible-light images of a coronal mass ejection are from the *Solar Maximum Mission* spacecraft. The light of the Sun's disk has been blocked to make the corona visible. **(a)** Before the ejection, the corona appears quiet. **(b)** As the ejection begins, it is already hundreds of thousands of kilometers across. **(c)** Just 16 minutes after the image in (b), the leading edge of the ejection has expanded outward by almost a million kilometers. (High Altitude Observatory/*Solar Maximum Mission*)

As energetic as solar flares are, they are dwarfed by **coronal mass ejections** (Figure 18-31). In such an event, more than 10^{12} kilograms (a billion tons) of high-temperature coronal gas is blasted into space at speeds of hundreds of kilometers per second. A typical coronal mass ejection lasts a few hours. These explosive events seem to be related to large-scale alterations in the Sun's magnetic field. Such events occur every few months; smaller eruptions may occur almost daily.

If a solar flare or coronal mass ejection happens to be aimed toward Earth, a stream of high-energy electrons and nuclei reaches us a few days later.

When this plasma arrives, it can interfere with satellites, pose a health hazard to astronauts in orbit, and disrupt electrical and communications equipment on the Earth's surface. Telescopes on Earth and on board spacecraft now monitor the Sun continuously to provide warnings of dangerous levels of solar particles.

The numbers of plages, filaments, solar flares, and coronal mass ejections all vary with the same 11-year cycle as sunspots. But unlike sunspots, coronal mass ejections never completely cease, even when the Sun is at its quietest. Astronomers are devoting substantial effort to understanding these and other aspects of our dynamic Sun.

KEY WORDS

The term preceded by an asterisk (*) is discussed in Box 18-1.

22-year solar cycle, p. 410

chromosphere, p. 402

CNO cycle, p. 394

conduction, p. 395

convection, p. 395

convective zone, p. 397

corona, p. 403

coronal hole, p. 405

coronal mass ejection, p. 413

differential rotation, p. 406

filament, p. 412

granulation, p. 401

granule, p. 401

helioseismology, p. 397

hydrogen burning, p. 392

hydrostatic equilibrium, p. 394

limb darkening, p. 400

luminosity (of the Sun), p. 390

magnetic-dynamo model, p. 410

magnetogram, p. 409

negative hydrogen ion, p. 402

neutrino, p. 398

photosphere, p. 400

plage, p. 412

plasma, p. 408

*positron, p. 393

prominence, p. 412

proton-proton chain, p. 394

KEY IDEAS

Hydrogen Burning in the Sun's Core: The Sun's energy is produced by hydrogen burning, a thermonuclear fusion process in which four hydrogen nuclei combine to produce a single helium nucleus.

• The energy released in a nuclear reaction corresponds to a slight reduction of mass according to Einstein's equation $E = mc^2$.

• Thermonuclear fusion occurs only at very high temperatures; for example, hydrogen burning occurs only at temperatures in excess of about 10^7 K. In the Sun, fusion occurs only in the dense, hot core.

Models of the Sun's Interior: A theoretical description of a star's interior can be calculated using the laws of physics.

• The standard model of the Sun suggests that hydrogen burning takes place in a core extending from the Sun's center to about 0.25 solar radius.

• The core is surrounded by a radiative zone extending to about 0.71 solar radius. In this zone, energy travels outward through radiative diffusion.

• The radiative zone is surrounded by a rather opaque convective zone of gas at relatively low temperature and pressure. In this zone, energy travels outward primarily through convection.

Solar Neutrinos and Helioseismology: Conditions in the solar interior can be inferred from measurements of solar neutrinos and of solar vibrations.

• Neutrinos emitted in thermonuclear reactions in the Sun's core have been detected, but in smaller numbers than expected. Recent experiments may explain why this is so.

• Helioseismology is the study of how the Sun vibrates. These vibrations have been used to infer pressures, densities, chemical compositions, and rotation rates within the Sun.

The Sun's Atmosphere: The Sun's atmosphere has three main layers: the photosphere, the chromosphere, and the corona. Everything below the solar atmosphere is called the solar interior.

• The visible surface of the Sun, the photosphere, is the lowest layer in the solar atmosphere. Its spectrum is similar to that of a blackbody at a temperature of 5800 K. Convection in the photosphere produces granules.

• Above the photosphere is a layer of less dense but higher-temperature gases called the chromosphere. Spicules extend upward from the photosphere into the chromosphere along the boundaries of supergranules.

• The outermost layer of the solar atmosphere, the corona, is made of very high-temperature gases at extremely low density. Activity in the corona includes coronal mass ejections and coronal holes. The solar corona blends into the solar wind at great distances from the Sun.

The Active Sun: The Sun's surface features vary in an 11-year cycle. This is related to a 22-year cycle in which the surface magnetic field increases, decreases, and then increases again with the opposite polarity.

• Sunspots are relatively cool regions produced by local concentrations of the Sun's magnetic field. The average number of sunspots increases and decreases in a regular cycle of approximately 11 years, with reversed magnetic polarities from one 11-year cycle to the next. Two such cycles make up the 22-year solar cycle.

• The magnetic-dynamo model suggests that many features of the solar cycle are due to changes in the Sun's magnetic field. These changes are caused by convection and the Sun's differential rotation.

• A solar flare is a brief eruption of hot, ionized gases from a sunspot group. A coronal mass ejection is a much larger eruption that involves immense amounts of gas from the corona.

REVIEW QUESTIONS

1. What is hydrogen burning? Why is hydrogen burning fundamentally unlike the burning of a log in a fireplace?

2. Why do thermonuclear reactions occur only in the Sun's core?

3. Describe how the net result of the reactions in Figure 18-2 is the conversion of four protons into a single helium nucleus. What other particles are produced in this process? How many of each particle are produced?

4. Give an everyday example of hydrostatic equilibrium. Give an example of thermal equilibrium. Explain how these equilibrium conditions apply to each example.

5. If thermonuclear fusion in the Sun were suddenly to stop, what would eventually happen to the overall radius of

the Sun? Justify your answer using the ideas of hydrostatic equilibrium and thermal equilibrium.

6. Give some everyday examples of conduction, convection, and radiative diffusion.

7. Describe the Sun's interior. Include references to the main physical processes that occur at various depths within the Sun.

8. Suppose thermonuclear fusion in the Sun stopped abruptly. Would the intensity of sunlight decrease just as abruptly? Why or why not?

9. Explain how studying the oscillations of the Sun's surface can give important, detailed information about physical conditions deep within the Sun.

10. What is a neutrino? Why is it useful to study neutrinos coming from the Sun? What do they tell us that cannot be learned from other avenues of research?

11. Unlike all other types of telescopes, neutrino detectors are placed deep underground. Why?

12. Describe the dangers in attempting to observe the Sun. How have astronomers learned to circumvent these observational problems?

13. Briefly describe the three layers that make up the Sun's atmosphere. In what ways do they differ from each other?

14. What is solar granulation? Describe how convection gives rise to granules.

15. High-resolution spectroscopy of the photosphere reveals that absorption lines are blueshifted in the spectrum of the central, bright regions of granules but are redshifted in the spectrum of the dark boundaries between granules. Explain how these observations show that granulation is due to convection.

16. What is the difference between granules and supergranules?

17. What are spicules? Where are they found? How can you observe them? What causes them?

18. How do astronomers know that the temperature of the corona is so high?

19. How do astronomers know when the next sunspot maximum and minimum will occur?

20. Why do astronomers say that the solar cycle is really 22 years long, even though the number of sunspots varies over an 11-year period?

21. Explain how the magnetic-dynamo model accounts for the solar cycle.

22. Explain why the surface of the Sun appears black in Figure 18-28.

23. Why should solar flares and coronal mass ejections be a concern for businesses that use telecommunication satellites?

ADVANCED QUESTIONS

The question preceded by an asterisk (*) involves the topic discussed in Box 18-1.

Problem-solving tips and tools

You may have to review Wien's law and the Stefan-Boltzmann law, which are the subject of Section 5-4. Section 5-5 discusses the properties of photons. As we described in Box 5-2, you can simplify calculations by taking ratios, such as the ratio of the flux from a sunspot to the flux from the undisturbed photosphere. When you do this, all the cumbersome constants cancel out. Figure 5-7 shows the various parts of the electromagnetic spectrum. We introduced the Doppler effect in Section 5-9 and Box 5-6. For information about the planets, see Table 7-1.

24. How much energy would be released if each of the following masses were converted *entirely* into their equivalent energy: (a) a carbon atom with a mass of 2×10^{-26} kg, (b) 1 kilogram, and (c) a planet as massive as the Earth (6×10^{24} kg)?

25. Use the luminosity of the Sun (given in Table 18-1) and the answers to the previous question to calculate how long the Sun must shine in order to release an amount of energy equal to that produced by the complete mass-to-energy conversion of (a) a carbon atom, (b) 1 kilogram, and (c) the Earth.

26. Assuming that the current rate of hydrogen burning in the Sun remains constant, what fraction of the Sun's mass will be converted into helium over the next 5 billion years? How will this affect the overall chemical composition of the Sun?

27. (a) Estimate how many kilograms of hydrogen the Sun has consumed over the past 4.6 billion years, and estimate the amount of mass that the Sun has lost as a result. Assume that the Sun's luminosity has remained constant during that time. (b) In fact, the Sun's luminosity when it first formed was only about 70% of its present value. With this in mind, explain whether your answers to part (a) are an overestimate or an underestimate.

*28. (a) A positron has the same mass as an electron (see Appendix 7). Calculate the amount of energy released by the annihilation of an electron and positron. (b) The products of this annihilation are two photons, each of equal energy. Calculate the wavelength of each photon, and confirm from Figure 5-7 that this wavelength is the gamma-ray range.

29. Sirius is the brightest star in the night sky. It has a luminosity of 23.5 $L_\odot$, that is, it is 23.5 times as luminous as the Sun and burns hydrogen at a rate 23.5 times greater than the Sun. How many kilograms of hydrogen does Sirius convert into helium each second?

30. (Refer to the preceding question.) Sirius has 2.3 times the mass of the Sun. Do you expect that the lifetime of Sirius will be longer, shorter, or the same length as that of the Sun? Explain your reasoning.

31. What would happen if the Sun were not in a state of both hydrostatic and thermal equilibrium? Explain your reasoning.

32. Using the mass and size of the Sun given in Table 18-1, verify that the average density of the Sun is 1410 kg/m^3. Compare your answer with the average densities of the Jovian planets.

33. Use the data in Table 18-2 to calculate the *average* density of material within 0.1 $R_\odot$ of the center of the Sun. (You will need to use the mass and radius of the Sun as a whole, given in Table 18-1.) Explain why your answer is not the same as the density at 0.1 $R_\odot$ given in Table 18-2.

34. In a typical solar oscillation, the Sun's surface moves up or down at a maximum speed of 0.1 m/s. An astronomer sets out to measure this speed by detecting the Doppler shift of an absorption line of iron with wavelength 557.6099 nm. What is the maximum wavelength shift that she will observe?

35. The amount of energy required to dislodge the extra electron from a negative hydrogen ion is 1.2×10^{-19} J. (a) The extra electron can be dislodged if the ion absorbs a photon of sufficiently short wavelength. (Recall from Section 5-5 that the higher the energy of a photon, the shorter its wavelength.) Find the longest wavelength (in nm) that can accomplish this. (b) In what part of the electromagnetic spectrum does this wavelength lie? (c) Would a photon of visible light be able to dislodge the extra electron? Explain. (d) Explain why the photosphere, which contains negative hydrogen ions, is quite opaque to visible light but is less opaque to light with wavelengths longer than the value you calculated in (a).

36. Astronomers often use an H$_\alpha$ filter to view the chromosphere. Explain why this can also be accomplished with filters that are transparent only to the wavelengths of the H and K lines of ionized calcium.

37. Calculate the wavelengths at which the photosphere, chromosphere, and corona emit the most radiation. Explain how the results of your calculations suggest the best way to observe these regions of the solar atmosphere. (*Hint*: Treat each part of the atmosphere as a perfect blackbody. Assume average temperatures of 50,000 K and 1.5×10^6 K for the chromosphere and corona, respectively.)

38. On November 15, 1999, the planet Mercury passed in front of the Sun as seen from Earth. The *TRACE* spacecraft made these time-lapse images of this event using ultraviolet light (top) and visible light (bottom). (Mercury moved from left to right in these images. The time between successive views of Mercury is 6 to 9 minutes.) Explain why the Sun appears somewhat larger in the ultraviolet image than in the visible-light image.

R I V U X G

(K. Schrijver, Stanford-Lockheed Institute for Space Research; *TRACE*; and NASA)

R I V U X G

39. Find the wavelength of maximum emission of the umbra of a sunspot and the wavelength of maximum emission of a sunspot's penumbra. In what part of the electromagnetic spectrum do these wavelengths lie?

40. (a) Find the ratio of the energy flux from a patch of a sunspot's penumbra to the energy flux from an equally large patch of undisturbed photosphere. Which patch is brighter? (b) Find the ratio of the energy flux from a patch of a sunspot's penumbra to the energy flux from an equally large patch of umbra. Again, which patch is brighter?

41. Suppose that you want to determine the Sun's rotation rate by observing its sunspots. Is it necessary to take the Earth's orbital motion into account? Why or why not?

42. The amount of visible light emitted by the Sun varies only a little over the 11-year sunspot cycle. But the amount of X rays emitted by the Sun can be ten times greater at solar maximum than at solar minimum. Explain why these two types of radiation should be so different in their variability.

DISCUSSION QUESTIONS

43. Discuss the extent to which cultures around the world have worshiped the Sun as a deity throughout history. Why do you suppose there has been such widespread veneration?

44. In the movie *Star Trek IV: The Voyage Home*, the starship *Enterprise* flies on a trajectory that passes close to the Sun's surface. What features should a real spaceship have to survive such a flight? Why?

45. Discuss some of the difficulties in correlating solar activity with changes in the Earth's climate.

46. Describe some of the advantages and disadvantages of observing the Sun (**a**) from space and (**b**) from the Earth's South Pole. What kinds of phenomena and issues might solar astronomers want to explore from these locations?

 WEB/CD-ROM QUESTIONS

47. Search the World Wide Web for the latest information from the neutrino detectors at the Super-Kamiokande Observatory and the Sudbury Neutrino Observatory. What are the most recent results from these detectors? What is the current thinking about the solar neutrino problem? What is the status of a new detector called Borexino?

48. Search the World Wide Web for information about features in the solar atmosphere called *sigmoids*. What are they? What causes them? How might they provide a way to predict coronal mass ejections?

49. Determining the Lifetime of a Solar Granule. Access and view the video "Granules on the Sun's Surface" in Chapter 18 of the *Universe* Web site or CD-ROM. Your task is to determine the approximate lifetime of a solar granule on the photosphere. Select an area, then slowly and rhythmically repeat "start, stop, start, stop" until you can consistently predict the appearance and disappearance of granules. While keeping your rhythm, move to a different area of the video and continue monitoring the appearance and disappearance of granules. When you are confident you have the timing right, move your eyes (or use a partner) to the clock shown in the video. Determine the length of time between the appearance and disappearance of the granules and record your answer.

OBSERVING PROJECTS

50. Use a telescope with a solar filter to observe the surface of the Sun. Do you see any sunspots? Sketch their appearance. Can you distinguish between the umbrae and penumbrae of the sunspots? Can you see limb darkening? Can you see any granulation?

51. If you have access to an H_α filter attached to a telescope especially designed for viewing the Sun safely, use this instrument to examine the solar surface. How does the appearance of the Sun differ from that in white light? What do sunspots look like in H_α? Can you see any prominences? Can you see any filaments? Are the filaments in the H_α image near any sunspots seen in white light? (Note that the amount of activity that you see will be much greater at some times during the solar cycle than at others.)

 52. Use the *Starry Night* program to measure the Sun's rotation. Set the time in the Control Panel to 12:00:00 P.M., center on the Sun (select **Find...** in the **Edit** menu), and use the controls in the Control Panel to zoom in until the sunspots are clearly visible. Set the time step to 1 day. Using the single-step time control buttons, step through enough time to determine the rotation rate of the Sun. Which part of the actual Sun's surface rotates at the rate shown in *Starry Night*? (The program does not show the Sun's differential rotation.)

Observing tips and tools

At the risk of repeating ourselves, we remind you to *never look directly at the Sun, because it can easily cause permanent blindness.* You can view the Sun safely without a telescope just by using two pieces of white cardboard. First, use a pin to poke a small hole in one piece of cardboard; this will be your "lens," and the other piece of cardboard will be your "viewing screen." Hold the "lens" piece of cardboard so that it is face-on to the Sun and sunlight can pass through the hole. With your other hand, hold the "viewing screen" so that the sunlight from the "lens" falls on it. Adjust the distance between the two pieces of cardboard so that you see a sharp image of the Sun on the "viewing screen." This image is perfectly safe to view. It is actually possible to see sunspots with this low-tech apparatus.

For a better view, use a telescope with a solar filter that fits on the front of the telescope. A standard solar filter is a piece of glass coated with a thin layer of metal to give it a mirrorlike appearance. This coating reflects almost all the sunlight that falls on it, so that only a tiny, safe amount of sunlight enters the telescope. An H_α filter, which looks like a red piece of glass, keeps the light at a safe level by admitting only a very narrow range of wavelengths. (Filters that fit on the back of the telescope are *not* recommended. The telescope focuses concentrated sunlight on such a filter, heating it and making it susceptible to cracking—and if the filter cracks when you are looking through it, your eye will be ruined instantly and permanently.)

To use a telescope with a solar filter, first aim the telescope away from the Sun, then put on the filter. Keep the lens cap on the telescope's secondary wide-angle "finder scope" (if it has one), because the heat of sunlight can fry the finder scope's optics. Next, aim the telescope toward the Sun, using the telescope's shadow to judge when you are pointed in the right direction. You can then safely look through the telescope's eyepiece. When you are done, make sure you point the telescope away from the Sun before removing the filter and storing the telescope.

Note that the amount of solar activity that you can see (sunspots, filaments, flares, prominences, and so on) will depend on where the Sun is in its 11-year sunspot cycle.

JOHN N. BAHCALL

Searching for Neutrinos Beyond the Textbooks

John Bahcall's wide-ranging expertise includes models of the Galaxy, dark matter, atomic and nuclear physics applied to astronomy, stellar evolution, and quasar spectra. After receiving his Ph.D. in physics from Harvard in 1961, he joined the faculty of the California Institute of Technology. Since 1971 he has been a professor of natural sciences at the Institute for Advanced Study in Princeton, New Jersey.

Dr. Bahcall received the National Medal of Science, the nation's highest scientific award, in 1998. His other awards include the Dannie Heinemann Prize of the American Astronomical Society and the American Institute of Physics for his research on solar neutrinos, and the NASA Distinguished Public Service Medal for his work with the Hubble Space Telescope, He has served as president of the American Astronomical Society and as chair of the Astronomy and Astrophysics Survey Committee of the National Academy of Sciences.

In attempting to understand the Sun, physicists, chemists, and astronomers have been confronted with a mystery—the case of the missing neutrinos. In the early 1960s, Ray Davis and I proposed to test the theory of how the Sun shines. Ray, a chemist at Brookhaven National Laboratory, had developed a neutrino detector that uses a cleaning fluid containing chlorine. Using standard theories of physics and astronomy, I calculated the rate at which neutrinos are produced in the Sun. I could then predict the rate at which neutrinos should be captured in the largest detector Ray could build. If my calculations matched experiment, they would confirm that the Sun shines by nuclear fusion in its interior.

The actual experiment used 100,000 gallons of perchloroethylene, about enough to fill an Olympic-sized swimming pool. Ray and his collaborators put their detector in a deep gold mine, to shield it from other particles that hit the surface of the Earth. To everyone's surprise, Ray's chlorine detector captured many fewer neutrinos than I had predicted. The results were challenged and checked repeatedly over the following three decades, but always with the same result: Many neutrinos appear to be missing! The case of the missing neutrinos has grown stronger with time. Three other experiments, each with a different type of detector, searched for neutrinos from the Sun. They all found fewer than I predicted.

What could be wrong? Where have the neutrinos gone? There are three possibilities. Either the experiments are wrong, the standard model of how the Sun (and other stars) shine is wrong, or something happens to the neutrinos after they are produced. New experiments are now under way in Japan, in Italy, and in Canada to test these hypotheses.

Most people working in the field think that the last of the three explanations is most likely to be correct: Only physics beyond the standard textbooks can describe what has happened to solar neutrinos. Somehow, most physicists think, neutrinos created in the solar interior change into neutrinos that are more difficult to detect as they pass out of the Sun and travel to the Earth. They change their personalities, so to speak! If this is correct, it would be the first experimental demonstration of a process beyond the standard model of particle physics.

So far, evidence for the new physics is only circumstantial. We know only that the results differ markedly from predictions based on our understanding now of how the Sun shines. Future experiments designed to search for a "smoking gun"—unequivocal evidence of processes not in the physics textbooks—will use the fact that neutrinos come in different types. Most easily detected are the so-called *electron-type* neutrinos; more difficult to detect are *muon-type* and *tau-type* neutrinos.

Suppose some of these neutrinos from the Sun convert into neutrinos that are easier to detect as they pass through the Earth at night on their way to the detector. The change would make the Sun appear brighter (in terms of neutrinos) at night than during the day. If that were seen, it would provide a dramatic demonstration that unconventional physics is occurring.

If such a smoking gun is found, it could offer a clue to new laws of particle physics. Recent experiments in Japan using neutrinos produced in the Earth's atmosphere by energetic particles from outside our solar system (cosmic rays) have already shown that neutrinos have a tiny mass, which allows them to convert from one type to another. Observations of solar neutrinos may lead us to further revise the laws that govern the smallest scales of matter.

I do not know the correct explanation for the missing neutrinos. However, the particle physics ("split personality") explanation has a mathematical beauty and simplicity that are very attractive. If the deity has not chosen this solution to the mystery, then he or she has missed an excellent opportunity to enrich the laws of the universe.

However, the real message of the experiments on solar neutrinos is even more remarkable. It is that working on the frontier of science, you may stumble across something that is beautiful and unexpected. We do not yet know precisely what we have discovered with solar neutrinos, but we do know that future experiments will solve the mystery for us. Their outcome might point the way to a better understanding of fundamental physics and of the stars.

The Nature of the Stars

To the unaided eye, the night sky is spangled with several thousand stars, each appearing as a bright pinpoint of light. With a pair of binoculars, you can see some 10,000 other, fainter stars; with a 15-cm (6-in.) telescope, the total rises to more than 2 million. Astronomers now know that there are in excess of 100 billion (10^{11}) stars in our Milky Way Galaxy alone.

But what are these distant pinpoints? To the great thinkers of ancient Greece, the stars were bits of light embedded in a vast sphere with the Earth at its center. They thought the stars were composed of a mysterious "fifth element," quite unlike anything found on Earth.

Today, we know that the stars are made of the same chemical elements found on Earth. We know their sizes, their temperatures, their masses, and something of their internal structures. We understand, too, why the stars in the accompanying image come in a range of beautiful colors: blue stars have high surface temperatures, while the surface temperatures of red stars are relatively low.

How have we learned these things? How can we know the nature of the stars, objects so distant that their light takes years or centuries to reach us? In this chapter, we will learn about the measurements and calculations that astronomers make to determine the properties of stars. We will also take a first look at the

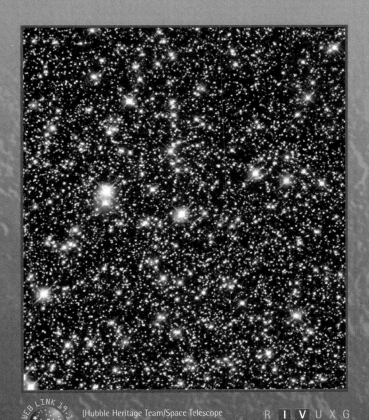

WEB LINK 19.1 (Hubble Heritage Team/Space Telescope Science Institute)

R **I** V U X G

Hertzsprung-Russell diagram, an important tool that helps astronomers systematize the wealth of available information about the stars. In later chapters, we will use this diagram to help us understand how stars are born, evolve, and eventually die.

As you read the sections of this chapter, look for the answers to the following questions.

19-1   How far away are the stars?

19-2   What evidence do astronomers have that the Sun is a typical star?

19-3   What is meant by a "first-magnitude" or "second-magnitude" star?

19-4   Why are some stars red and others blue?

19-5   What are the stars made of?

19-6   As stars go, is our Sun especially large or small?

19-7   What are giant, supergiant, and white dwarf stars?

19-8   How do we know the distances to remote stars?

19-9   Why are binary star systems important in astronomy?

19-10   How can a star's spectrum show whether it is actually a binary system?

19-11   What do astronomers learn from stars that eclipse each other?

19-1   Careful measurements of the parallaxes of stars reveal their distances

 The vast majority of stars are objects very much like the Sun. This understanding followed from the discovery that the stars are tremendously far from us, at distances so great that their light takes years to reach us. Because the stars at night are clearly visible to the naked eye despite these huge distances, it must be that the **luminosity** of the stars—that is, how much energy they emit into space per second—is comparable to or greater than that of the Sun. Just as for the Sun, the only explanation for such tremendous luminosities is that thermonuclear reactions are occurring within the stars (see Section 18-1).

Clearly, then, it is important to know how distant the stars are. But how do we measure these distances? You might think this is done by comparing how bright different stars appear. Perhaps the star Betelgeuse in the constellation Orion appears bright because it is relatively close, while the dimmer and less conspicuous star Polaris (the North Star, in the constellation Ursa Minor) is farther away.

But this line of reasoning is incorrect: Polaris is actually closer to us than Betelgeuse! How bright a star appears is *not* a good indicator of its distance. If you see a light on a darkened road, it could be a motorcycle headlight a kilometer away or a person holding a flashlight just a few meters away. In the same way, a bright star might be extremely far away but have an unusually high luminosity, and a dim star might be relatively close but have a rather low luminosity. Astronomers must use other techniques to determine the distances to the stars.

The most straightforward way of measuring stellar distances uses an effect called **parallax.** This is the apparent displacement of an object because of a change in the observer's point of view (Figure 19-1). To see how parallax works, hold your arm out straight in front of you. Now look at the hand on your outstretched arm, first with your left eye closed, then with your right eye closed. When you close one eye and open the other, your hand appears to shift back and forth against the background of more distant objects.

The closer the object you are viewing, the greater the parallax shift. To see this, repeat the experiment with your hand held closer to your face. Your brain analyzes such parallax shifts constantly as it compares the images from your left and right eyes, and in this way determines the distances to objects around you. This is the origin of depth perception.

To measure the distance to a star, astronomers measure the parallax shift of the star using two points of view that are as far apart as possible—at opposite sides of the Earth's orbit. The direction from Earth to a nearby star changes as our planet orbits the Sun, and the nearby star appears to move back and forth against the background of more distant stars (Figure 19-2). This motion is called **stellar parallax.** The par-

Figure 19-1

Parallax Imagine looking at some nearby object (a tree) against a distant background (mountains). When you move from one location to another, the nearby object appears to shift with respect to the distant background scenery. This familiar phenomenon is called parallax.

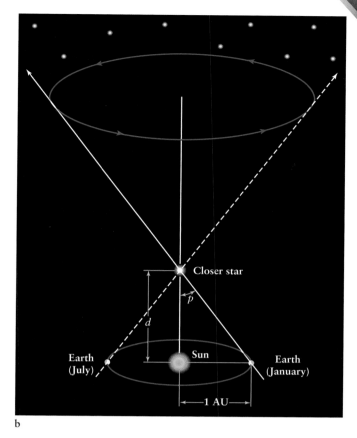

a

b

figure 19-2

Stellar Parallax (a) As the Earth orbits the Sun, a nearby star appears to shift its position against the background of distant stars. The parallax (p) of the star is equal to the angular radius of the Earth's orbit as seen from the star. **(b)** The closer the star is to us, the greater the parallax angle p. The distance d to the star (in parsecs) is equal to the reciprocal of the parallax angle p (in arcseconds): $d = 1/p$.

allax (p) of a star is equal to half the angle through which the star's apparent position shifts as the Earth moves from one side of its orbit to the other. The larger the parallax p, the smaller the distance d to the star (compare Figure 19-2*a* with Figure 19-2*b*).

It is convenient to measure the distance d in **parsecs**. A star with a parallax angle of 1 second of arc ($p = 1$ arcsec) is at a distance of 1 parsec ($d = 1$ pc). (The word "parsec" is a contraction of the phrase "the distance at which a star has a *par*allax of one arc*sec*ond." Recall from Section 1-7 that 1 parsec equals 3.26 light-years, 3.09×10^{13} km, or 206,265 AU; see Figure 1-12.) If the angle p is measured in arcseconds, then the distance d to the star in parsecs is given by the following equation:

Relation between a star's distance and its parallax

$$d = \frac{1}{p}$$

d = distance to a star, in parsecs

p = parallax angle of that star, in arcseconds

This simple relationship between parallax and distance in parsecs is one of the main reasons that astronomers usually measure cosmic distances in parsecs rather than light-years. For example, a star whose parallax is $p = 0.1$ arcsec is at a distance $d = 1/(0.1) = 10$ parsecs from the Earth. Barnard's star, named for the American astronomer Edward E. Barnard, has a parallax of 0.545 arcsec. Hence, the distance to this star is:

$$d = \frac{1}{p} = \frac{1}{0.545} = 1.83 \ \text{pc}$$

Because 1 parsec is 3.26 light-years, this can also be expressed as

$$d = 1.83 \ \text{pc} \times \frac{3.26 \ \text{ly}}{1 \ \text{pc}} = 5.98 \ \text{ly}$$

All known stars have parallax angles less than one arcsecond. In other words, the closest star is more than 1 parsec away. It was not until 1838 that the first successful parallax measurements were made by the German astronomer and mathematician Friedrich Wilhelm Bessel. He found the parallax angle of the star 61 Cygni to be just ⅓ arcsec—equal to the angular diameter of a dime at a distance of 11 kilometers, or 7 miles. He thus determined that this star is about 3 pc from the Earth. (Modern measurements give a slightly smaller parallax angle, which means that 61 Cygni is actually more than 3 pc away.) The star Proxima Centauri has the largest known parallax angle, 0.772 arcsec, and hence is the closest known star (other than the Sun); its distance is 1/(0.772) = 1.30 pc.

Appendix 4 at the back of this book lists all the stars within 4 pc of the Sun, as determined by parallax measurements. Most of these stars are far too dim to be seen with the naked eye, which is why their names are probably unfamiliar to you. By contrast, the majority of the familiar, bright stars in the nighttime sky (listed in Appendix 5) are so far away that their parallaxes cannot be measured from the Earth's surface. They appear bright not because they are close, but because they are far more luminous than the Sun. The brightest stars in the sky are *not* necessarily the nearest stars!

Parallax angles smaller than about 0.01 arcsec are extremely difficult to measure from the Earth, in part because of the blurring effects of the atmosphere. Therefore, the parallax method used with ground-based telescopes can give fairly reliable distances only for stars nearer than about 1/0.01 = 100 pc. But an observatory in space is unhampered by the atmosphere. Observations made from spacecraft therefore permit astronomers to measure even smaller parallax angles and thus determine the distances to more remote stars.

In 1989 the European Space Agency (ESA) launched the satellite Hipparcos, an acronym for *Hi*gh *P*recision *Parallax* *Co*llecting *S*atellite (and a commemoration of the ancient Greek astronomer Hipparchus, who created one of the first star charts). Over more than three years of observations, the telescope aboard Hip-

parcos was used to measure the parallaxes of 118,000 stars with an accuracy of 0.001 arcsecond. This has enabled astronomers to determine stellar distances out to several hundred parsecs, and with much greater precision than has been possible with ground-based observations. In the years to come, astronomers will increasingly turn to space-based observations to determine stellar distances.

Unfortunately, most of the stars in the Galaxy are so far away that their parallax angles are too small to measure even with an orbiting telescope. Later in this chapter, we will discuss a technique that can be used to find the distances to these more remote stars. In Chapters 26 and 28 we will learn about other techniques that astronomers use to determine the much larger distances to galaxies beyond the Milky Way. These techniques also help us understand the overall size, age, and structure of the universe.

Because it can be used only on relatively close stars, stellar parallax might seem to be of limited usefulness. But parallax measurements are the cornerstone for all other methods of finding the distances to remote objects. These other methods require a precise and accurate knowledge of the distances to nearby stars, as determined by stellar parallax. Hence, any inaccuracies in the parallax angles for nearby stars can translate into substantial errors in measurement for the whole universe. For this reason, astronomers are continually trying to perfect their parallax-measuring techniques.

Stellar parallax is an *apparent* motion of stars caused by the Earth's orbital motion around the Sun. But stars are not fixed objects and actually do move through space. As a result, stars change their positions on the celestial sphere (Figure 19-3), and they move either toward or away from the Sun. These motions are sufficiently slow, however, that changes in the positions of the stars are hardly noticeable over a human lifetime. Box 19-1 describes how astronomers study these motions and what insights they gain from these studies.

August 24, 1894

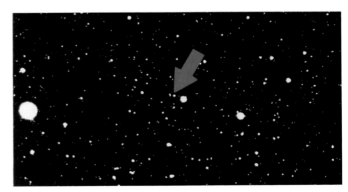
May 30, 1916

ƒigure 19-3   R I [V] U X G

The Motion of Barnard's Star These two photographs of Barnard's star, which lies 1.82 parsecs away in the constellation Ophiuchus, were taken 22 years apart. Over this time interval, Barnard's star moved almost 230 arcseconds on the celestial sphere (about 4 arcminutes, or 0.06°), more than any other star. Although it is close to us, Barnard's star is only about 10^{-4} as luminous as the Sun. As a result, this star is too dim to be seen with the naked eye. (Yerkes Observatory)

box 19-1 | Tools of the Astronomer's Trade

Stellar Motions

Stars can move through space in any direction. The space velocity of a star describes how fast and in what direction it is moving. As the accompanying figure shows, a star's space velocity v can be broken into components parallel and perpendicular to our line of sight.

The component perpendicular to our line of sight—that is, across the plane of the sky—is called the star's **tangential velocity** (v_t). To determine it, astronomers must know the distance to a star (d) and its **proper motion** (μ, the Greek letter mu), which is the number of arcseconds that the star appears to move per year on the celestial sphere. Proper motion does not repeat itself yearly, so it can be distinguished from the apparent back-and-forth motion due to parallax. In terms of a star's distance and proper motion, its tangential velocity (in km/s) is

$$v_t = 4.74\mu d$$

where μ is in arcseconds per year and d is in parsecs. For example, Barnard's star (Figure 19-3) has a proper motion of 10.358 arcseconds per year and a distance of 1.82 pc. Hence, its tangential velocity is

$$v_t = 4.74(10.358)(1.82) = 89.4 \text{ km/s}$$

The component of a star's motion parallel to our line of sight—that is, either directly toward us or directly away from us—is its **radial velocity** (v_r). It can be determined from measurements of the Doppler shifts of the star's spectral lines (see Section 5-9 and Box 5-6). If a star is approaching us, the wavelengths of all of its spectral lines are decreased (blueshifted); if the star is receding from us, the wavelengths are increased (redshifted). The radial velocity is related to the wavelength shift by the equation

$$\frac{\lambda - \lambda_0}{\lambda_0} = \frac{v_r}{c}$$

In this equation, λ is the wavelength of light coming from the star, λ_0 is what the wavelength would be if the star were not moving, and c is the speed of light. As an illustration, a particular spectral line of iron in the spectrum of Barnard's star has a wavelength (λ) of 516.438 nm. As measured in a laboratory on the Earth, the same spectral line has a wavelength (λ_0) of 516.629 nm. Thus, for Barnard's star, our equation becomes

$$\frac{516.48 \text{ nm} - 516.629 \text{ nm}}{516.629 \text{ nm}} = -0.000370 = \frac{v_r}{c}$$

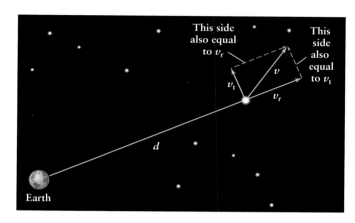

Solving this equation for the radial velocity v_r, we find

$$v_r = (-0.000370)c = (-0.000370)(3.00 \times 10^5 \text{ km/s})$$
$$= -111 \text{ km/s}$$

The negative sign means that Barnard's star is moving toward us. You can check this interpretation by noting that the wavelength $\lambda = 516.438$ nm received from Barnard's star is less than the laboratory wavelength $\lambda_0 = 516.629$ nm; hence, the light from the star is blueshifted, which indeed means that the star is approaching. If the star were receding, its radial velocity would be positive.

The illustration shows that the tangential velocity and radial velocity form two sides of a right triangle. The long side (hypotenuse) of this triangle is the space velocity (v). From the Pythagorean theorem, the space velocity is

$$v = \sqrt{v_t^2 + v_r^2}$$

For Barnard's star, the space velocity is

$$v = \sqrt{(89.4 \text{ km/s})^2 + (-111 \text{ km/s})^2} = 143 \text{ km/s}$$

Therefore, Barnard's star is moving through space at a speed of 143 km/s (515,000 km/h, or 320,000 mi/h) relative to the Sun.

Determining the space velocities of stars is essential for understanding the structure of the Galaxy. Studies show that the stars in our local neighborhood are moving in wide orbits around the center of the Galaxy, which lies some 8000 pc (26,000 light-years) away in the direction of the constellation Sagittarius (the Archer). While many of the orbits are roughly circular and lie in nearly the same plane, others are highly elliptical or steeply inclined to the galactic plane. We will see in Chapter 25 how the orbits of stars and gas clouds reveal the Galaxy's spiral structure.

19-2 If a star's distance is known, its luminosity can be determined from its brightness

All the stars you can see in the nighttime sky shine by thermonuclear fusion, just as the Sun does (see Section 18-1). But they are by no means merely identical copies of the Sun. Stars differ in their luminosity (L), the amount of light energy they emit each second. Luminosity is usually measured either in watts (1 watt, or 1 W, is 1 joule per second) or as a multiple of the Sun's luminosity ($L_\odot$, equal to 3.86×10^{26} W). Most stars are less luminous than the Sun, but some blaze forth with a million times the Sun's luminosity. Knowing a star's luminosity is essential for determining the star's history, its present-day internal structure, and its future evolution.

To determine the luminosity of a star, we first note that as light energy moves away from its source, it spreads out over increasingly larger regions of space. Imagine a sphere of radius d centered on the light source, as in Figure 19-4. The amount of energy that passes each second through a square meter of the sphere's surface area is the total luminosity of the source (L) divided by the total surface area of the sphere (equal to $4\pi d^2$). This quantity is called the **apparent brightness** of the light, or just **brightness** (b), because how bright a light source appears depends on how much light energy per second enters through the area of a light detector (such as your eye). Apparent brightness is measured in watts per square meter (W/m^2):

Inverse-square law relating apparent brightness and luminosity

$$b = \frac{L}{4\pi d^2}$$

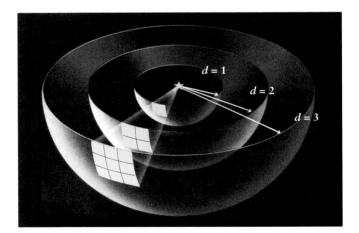

figure 19-4

The Inverse-Square Law This drawing shows how the same amount of radiation from a light source must illuminate an ever-increasing area as distance from the light source increases. This area increases as the square of the distance from the source, so the apparent brightness decreases as the square of the distance. The brightness at $d = 2$ is $1/(2^2) = \frac{1}{4}$ of the brightness at $d = 1$, and the brightness at $d = 3$ is $1/(3^2) = \frac{1}{9}$ of that at $d = 1$.

b = apparent brightness of a star's light, in W/m^2

L = star's luminosity, in W

d = distance to star, in meters

This relationship is called the **inverse-square law,** because the apparent brightness of light that an observer can see or measure is inversely proportional to the square of the observer's distance (d) from the source. If you double your distance from a light source, its radiation is spread out over an area 4 times larger, so the apparent brightness you see is decreased by a factor of 4. Similarly, at triple the distance, the apparent brightness is ⅑ as great (see Figure 19-4).

We can apply the inverse-square law to the Sun, which is 1.50×10^{11} m from Earth. Its apparent brightness ($b_\odot$) is

$$b_\odot = \frac{3.86 \times 10^{26}\,W}{4\pi(1.50 \times 10^{11}\,m)^2} = 1370 \;\; W/m^2$$

That is, a solar panel with an area of 1 square meter receives 1370 watts of power from the Sun.

Astronomers measure the apparent brightness of a star using a telescope with an attached light-sensitive instrument, similar to the light meter in a camera that determines the proper exposure. Measuring a star's apparent brightness is called **photometry.**

The inverse-square law says that we can find a star's luminosity if we know its distance and its apparent brightness. To do this, it is convenient to express this law in a somewhat different form. We first rearrange the above equation:

$$L = 4\pi d^2 b$$

We then apply this equation to the Sun. That is, we write a similar equation relating the Sun's luminosity ($L_\odot$), the distance from the Earth to the Sun ($d_\odot$, equal to 1 AU), and the Sun's apparent brightness ($b_\odot$):

$$L_\odot = 4\pi d_\odot^2 b_\odot$$

If we take the ratio of these two equations, the unpleasant factor of 4π drops out and we are left with the following:

Determining a star's luminosity from its apparent brightness

$$\frac{L}{L_\odot} = \left(\frac{d}{d_\odot}\right)^2 \frac{b}{b_\odot}$$

$L/L_\odot$ = ratio of the star's luminosity to the Sun's luminosity

$d/d_\odot$ = ratio of the star's distance to the Earth-Sun distance

$b/b_\odot$ = ratio of the star's apparent brightness to the Sun's apparent brightness

We need to know just two things to find a star's luminosity: the distance to a star as compared to the Earth-Sun distance (the ratio $d/d_\odot$), and how that star's apparent brightness compares to that of the Sun (the ratio $b/b_\odot$). Then we can use the above equation to find how luminous that star is compared to the Sun (the ratio $L/L_\odot$). Box 19-2 explains how to

box 19-2 | Tools of the Astronomer's Trade

Luminosity, Distance, and Apparent Brightness

The inverse-square law (Section 19-2) relates a star's luminosity, distance, and apparent brightness to the corresponding quantities for the Sun:

$$\frac{L}{L_\odot} = \left(\frac{d}{d_\odot}\right)^2 \frac{b}{b_\odot}$$

We can use a similar equation to relate the luminosities, distances, and apparent brightnesses of any two stars, which we call star 1 and star 2:

$$\frac{L_1}{L_2} = \left(\frac{d_1}{d_2}\right)^2 \frac{b_1}{b_2}$$

EXAMPLE: The star ε Eridani is 3.22 pc from Earth. As seen from Earth, this star appears only 6.73×10^{-13} as bright as the Sun. What is the luminosity of ε Eridani compared with that of the Sun?

Since we are asked to compare this star to the Sun, the first equation (which involves the Sun's luminosity, distance, and brightness) is the one to use. The information given tells us that the ratio of the apparent brightness of ε Eridani to that of the Sun is $b/b_\odot = 6.73 \times 10^{-13}$. To get the distance ratio, we remember that the distance from the Earth to the Sun is $d_\odot = 1$ AU. Furthermore, there are 206,625 AU in 1 parsec, so we can write the distance to ε Eridani as $d = (3.22 \text{ pc})(206{,}625 \text{ AU/pc}) = 6.65 \times 10^5$ AU. Hence, the ratio of distances is $d/d_\odot = (6.65 \times 10^5 \text{ AU})/(1 \text{ AU}) = 6.65 \times 10^5$. Hence, the ratio of the luminosity of ε Eridani (L) to the Sun's luminosity ($L_\odot$) is

$$\frac{L}{L_\odot} = \left(\frac{d}{d_\odot}\right)^2 \frac{b}{b_\odot} = (6.65 \times 10^5)^2 \times (6.73 \times 10^{-13}) = 0.30$$

This means that ε Eridani has only 0.30 the luminosity of the Sun.

EXAMPLE: Suppose star 1 is at half the distance of star 2 (that is, $d_1/d_2 = \frac{1}{2}$) and that star 1 appears twice as bright as star 2 (that is, $b_1/b_2 = 2$). How do the luminosities of these two stars compare?

We can answer this question immediately using the second of our two equations:

$$\frac{L_1}{L_2} = \left(\frac{1}{2}\right)^2 \times 2 = 0.5$$

This says that star 1 has only half the luminosity of star 2. Despite this, star 1 appears brighter than star 2 because it is closer to us.

The two equations above are also useful in the method of *spectroscopic parallax*, which we discuss in Section 19-8. It turns out that a star's luminosity can be determined simply by analyzing the star's spectrum. If the star's apparent brightness is also known, the star's distance can be calculated. The inverse-square law can be rewritten as an expression for the ratio of the star's distance from the Earth (d) to the Earth-Sun distance ($d_\odot$):

$$\frac{d}{d_\odot} = \sqrt{\frac{(L/L_\odot)}{(b/b_\odot)}}$$

We can also this formula as a relation between the properties of two stars, 1 and 2:

$$\frac{d_1}{d_2} = \sqrt{\frac{(L_1/L_2)}{(b_1/b_2)}}$$

EXAMPLE: The star Regulus has 140 times the luminosity of the Sun (so $L/L_\odot = 140$) but appears only 5.2×10^{-12} as bright as the Sun (so $b/b_\odot = 5.2 \times 10^{-12}$). How far is Regulus from Earth?

The first equation above tells us that

$$\frac{d}{d_\odot} = \sqrt{\frac{(L/L_\odot)}{(b/b_\odot)}} = \sqrt{\frac{140}{5.2 \times 10^{-12}}} = \sqrt{2.7 \times 10^{13}} = 5.2 \times 10^6$$

Since $d_\odot = 1$ AU, this means that Regulus is 5.2×10^6 AU from Earth. Using the conversion 206,265 AU = 1 pc, the star's distance is $d = (5.2 \times 10^6 \text{ AU})(1 \text{ pc}/206{,}265 \text{ AU}) = 25$ pc.

EXAMPLE: The star δ (delta) Cephei, which lies 300 pc from the Earth, is thousands of times more luminous than the Sun. Thanks to this great luminosity, stars like δ Cephei can be seen in galaxies millions of parsecs away. As an example, the Hubble Space Telescope has detected stars like δ Cephei within the galaxy NGC 3351, which lies in the direction of the constellation Leo. These stars appear only 9×10^{-10} as bright as δ Cephei. If we let 1 denote a star in NGC 3351 and 2 denote δ Cephei itself, we have $b_1/b_2 = 9 \times 10^{-10}$. Because δ Cephei has the same luminosity as its sister stars in NGC 3351, we also have

(continued on following page)

box 19-2   *(continued from previous page)*

$L_1/L_2 = 1$. Given these ratios, we can find the ratio d_1/d_2 (the distance from the Earth to NGC 3351 divided by the distance from the Earth to δ Cephei):

$$\frac{d_1}{d_2} = \sqrt{\frac{(L_1/L_2)}{(b_1/b_2)}} = \sqrt{\frac{1}{9 \times 10^{-10}}} = \sqrt{1.1 \times 10^9} = 33,000$$

Hence, NGC 3351 is 33,000 times further away than δ Cephei. The distance from the Earth to NGC 3351 is therefore $(33,000)(300 \text{ pc}) = 10^7$ pc, or 10 megaparsecs (10 Mpc). We will learn more about Cepheids in Chapter 21. In Chapter 26 we will explore further how they are used to determine the distances to remote galaxies.

---

use this equation to determine the luminosity of the nearby star ε (epsilon) Eridani, the fifth brightest star in the constellation Eridanus (named for a river in Greek mythology). Parallax measurements indicate that ε Eridani is 3.22 pc away, and photometry shows that the star appears only 6.73×10^{-13} as bright as the Sun; using the above equation, we find that ε Eridani has only 0.30 times the luminosity of the Sun.

Calculations of this kind show that stars come in a wide variety of different luminosities, with values that range from about 10^6 $L_\odot$ (a million times the Sun's luminosity) to only about 10^{-4} $L_\odot$ (a mere ten-thousandth of the Sun's light output). The most luminous star emits roughly 10^{10} times more energy each second than the least luminous! (To put this number in perspective, about 10^{10} human beings have lived on the Earth since our species first evolved.)

As stars go, our Sun is neither extremely luminous nor extremely dim; it is a rather ordinary, garden-variety star. It is somewhat more luminous than most stars, however. Of more than 30 stars within 4 pc of the Sun (Appendix 4), only three (α Centauri, Sirius, and Procyon) have a greater luminosity than the Sun.

To better characterize a typical population of stars, astronomers count the stars out to a certain distance from the Sun and plot the number of stars that have different luminosities. The resulting graph is called the **luminosity function.** Figure 19-5 shows the luminosity function for stars in our part of the Milky Way Galaxy. The curve declines very steeply for the most luminous stars toward the left side of the graph, indicating that they are quite rare. For example, this graph shows that stars like the Sun are about 10,000 times more common than stars like Spica (with a luminosity of 2100 $L_\odot$).

The exact shape of the curve in Figure 19-5 applies only to the vicinity of the Sun and similar regions in our Milky Way Galaxy. Other locations have somewhat different luminosity functions. In stellar populations in general, however, low-luminosity stars are much more common than high-luminosity ones.

19-3   Astronomers often use the magnitude scale to denote brightness

Because astronomy is among the most ancient of sciences, some of the tools used by modern astronomers are actually many centuries old. One such tool is the **magnitude scale,** which astronomers frequently use to denote the brightness of stars. This scale was invented in the second century B.C. by the Greek astronomer Hipparchus, who called the brightest stars first-magnitude stars. Stars about half as bright as first-magnitude stars were called second-magnitude stars, and so forth, down to sixth-magnitude stars, the dimmest ones he

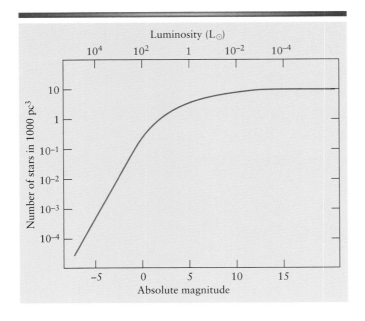

figure 19-5

The Luminosity Function This graph shows how many stars of a given luminosity lie within a representative 1000 cubic-parsec volume. The steep decline of the curve toward the left (high-luminosity) side of the graph shows that very luminous stars are quite rare. The scale at the bottom of the graph shows absolute magnitude, an alternative measure of a star's luminosity (described in Section 19-3). (Adapted from J. Bahcall and R. Soneira) could see. After telescopes came into use, astronomers extended Hipparchus's magnitude scale to include even dimmer stars.

The magnitudes in Hipparchus's scale are properly called **apparent magnitudes**, because they describe how bright an object *appears* to an Earth-based observer. Apparent magnitude is directly related to apparent brightness.

CAUTION! The magnitude scale has a tendency to be confusing because it works "backward." Keep in mind that the *greater* the apparent magnitude, the *dimmer* the star. A star of apparent magnitude +3 (a third-magnitude star) is dimmer than a star of apparent magnitude +2 (a second-magnitude star).

In the nineteenth century, astronomers developed better techniques for measuring the light energy arriving from a star. These measurements showed that a first-magnitude star is about 100 times brighter than a sixth-magnitude star. In other words, it would take 100 stars of magnitude +6 to provide as much light energy as we receive from a single star of magnitude +1. To make computations easier, the magnitude scale was redefined so that a magnitude difference of 5 corresponds exactly to a factor of 100 in brightness. A magnitude difference of 1 corresponds to a factor of 2.512 in brightness, because

$$2.512 \times 2.512 \times 2.512 \times 2.512 \times 2.512 = (2.512)^5 = 100$$

Thus, it takes 2.512 third-magnitude stars to provide as much light as we receive from a single second-magnitude star.

Figure 19-6 illustrates the modern apparent magnitude scale. The dimmest stars visible through a pair of binoculars have an apparent magnitude of +10, and the dimmest stars that can be photographed in a one-hour exposure with the Keck telescopes (see Section 6-2) or the Hubble Space Telescope have apparent magnitude +30. Modern astronomers also use negative numbers to extend Hipparchus's scale to include very bright objects. For example, Sirius, the brightest star in the sky, has an apparent magnitude of −1.44. The Sun, the brightest object in the sky, has an apparent magnitude of −26.7.

Apparent magnitude is a measure of a star's apparent brightness as seen from Earth. A related quantity that measures a star's true energy output—that is, its luminosity—is called **absolute magnitude.** This is the apparent magnitude a star would have if it were located exactly 10 parsecs from Earth.

ANALOGY If you wanted to compare the light output of two different lightbulbs, you would naturally place them side by side so that both bulbs were the same distance from you. In the absolute magnitude scale, we imagine doing the same thing with stars to compare their luminosities.

a

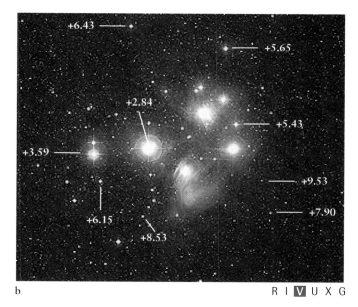

b   R I **V** U X G

figure 19-6

WEB LINK 19.6

The Apparent Magnitude Scale (a) Astronomers denote the apparent brightness of objects in the sky by their apparent magnitudes. The stars visible to the naked eye have magnitudes from −1.44 to about +6. Very long exposure photography using the largest telescopes can reveal stars as faint as magnitude +30. **(b)** This photograph of the Pleiades cluster, located about 110 pc away in the constellation Taurus, shows the apparent magnitudes of some of its stars. Most of these stars are too faint to be seen by the naked eye. (David Malin/Anglo-Australian Observatory)

box 19-3 | Tools of the Astronomer's Trade

Apparent Magnitude and Absolute Magnitude

Astronomers commonly express a star's apparent brightness in terms of apparent magnitude (denoted by a lowercase m), and the star's luminosity in terms of absolute magnitude (denoted by a capital M). While we do not use these quantities extensively in this book, it is useful to know a few simple relationships involving them.

Consider two stars, labeled 1 and 2, with apparent magnitudes m_1 and m_2 and brightnesses b_1 and b_2, respectively. The *ratio* of their apparent brightnesses (b_1/b_2) corresponds to a *difference* in their apparent magnitudes ($m_2 - m_1$). As we learned in Section 19-3, each step in magnitude corresponds to a factor of 2.512 in brightness; we receive 2.512 times more energy per square meter per second from a third-magnitude star than from a fourth-magnitude star. This idea was used to construct the table that follows.

Apparent magnitude difference ($m_2 - m_1$)	Ratio of apparent brightness (b_1/b_2)
1	2.512
2	$(2.512)^2 = 6.31$
3	$(2.512)^3 = 15.85$
4	$(2.512)^4 = 39.82$
5	$(2.512)^5 = 100$
10	$(2.512)^{10} = 10^4$
15	$(2.512)^{15} = 10^6$
20	$(2.512)^{20} = 10^8$

A simple equation relates the difference between two stars' apparent magnitudes to the ratio of their brightnesses:

Magnitude difference related to brightness ratio

$$m_2 - m_1 = 2.5 \log \left(\frac{b_1}{b_2} \right)$$

m_1, m_2 = apparent magnitudes of stars 1 and 2

b_1, b_2 = apparent brightnesses of stars 1 and 2

In this equation, $\log (b_1/b_2)$ is the logarithm of the brightness ratio. The logarithm of $1000 = 10^3$ is 3, the logarithm of $10 = 10^1$ is 1, and the logarithm of $1 = 10^0$ is 0.

EXAMPLE: At its most brilliant, Venus has a magnitude of about −4, while the dimmest stars visible to the naked eye have a magnitude of +6. The magnitude difference between these is $+6 - (-4) = 10$. From the table, this difference corresponds to a brightness ratio of $(2.512)^{10} = 10^4$. It would thus take 10,000 sixth-magnitude stars to shine as brilliantly as Venus.

EXAMPLE: In August 1975 a nova appeared in the constellation Cygnus (the Swan). As we will see in Chapter 23, a nova is a violent outburst of a certain kind of star. The magnitude of this nova changed in just two days from +15 to +2, so the change in magnitude ($m_2 - m_1$) was 13. (Here we use the labels "1" and "2" to refer to two different states of the same star.) From the table, a magnitude difference of 10 corresponds to a brightness ratio of 10^4, and a magnitude difference of 3 corresponds to a brightness ratio of 15.85. Combining these, a magnitude difference of $13 = 10 + 3$ corresponds to a brightness ratio of $10^4 \times 15.85 = 158,500$. That is, the nova's brightness increased in two days by a factor of 158,500.

EXAMPLE: The variable star RR Lyrae in the constellation Lyra (the Harp) periodically doubles its light output. By how much does its apparent magnitude change?

The brightness of the star varies by a factor of 2, so the ratio of its maximum brightness (b_1) to its minimum brightness (b_2) is $b_1/b_2 = 2$. Using a calculator, we find that the apparent magnitude difference is $m_2 - m_1 = 2.5 \log (2) = 2.5 \times 0.30 = 0.75$. RR Lyrae therefore varies

---

If the Sun were moved to a distance of 10 parsecs from the Earth, it would have an apparent magnitude of +4.8. The absolute magnitude of the Sun is thus +4.8. The absolute magnitudes of the stars range from approximately +15 for the least luminous to −10 for the most luminous. (*Note:* Like apparent magnitudes, absolute magnitudes work "backward": The

greater the absolute magnitude, the *less luminous* the star.) The Sun's absolute magnitude is about in the middle of this range.

There is a mathematical relationship between absolute magnitude and luminosity, which astronomers use to convert one to the other as they see fit. It is also possible to rewrite the inverse-square law, which we introduced in Section 19-2, periodically in brightness by 0.75 magnitude. Stated another way, at its dimmest RR Lyrae has an apparent magnitude m_2 that is 0.75 *greater* than its apparent magnitude when it is brightest, m_1. (Remember that a greater value of apparent magnitude means the star is dimmer, not brighter!)

The inverse-square law relating a star's apparent brightness and luminosity can be rewritten in terms of the star's apparent magnitude (m), absolute magnitude (M), and distance from the Earth (d). This can be expressed as an equation:

Relation between a star's apparent magnitude and absolute magnitude

$$m - M = 5 \log d - 5$$

m = star's apparent magnitude

M = star's absolute magnitude

d = distance from the Earth to the star in parsecs

In this expression $m - M$ is called the **distance modulus**, and $\log d$ means the logarithm of the distance d in parsecs. For convenience, the following table gives the values of the distance d corresponding to different values of $m - M$.

Distance modulus $m - M$	Distance d (pc)
−4	1.6
−3	2.5
−2	4.0
−1	6.3
0	10
1	16
2	25
3	40
4	63
5	100
10	10^3
15	10^4
20	10^5

This table shows that if a star is less than 10 pc away, its distance modulus $m - M$ is negative. That is, its apparent magnitude (m) is less than its absolute magnitude (M). If the star is more than 10 pc away, $m - M$ is positive and m is greater than M. As an example, the star Epsilon Indi (in the southern constellation Indus) has apparent magnitude $m = +4.7$; it is 3.6 pc away, which is less than 10 pc, so its apparent magnitude is less than its absolute magnitude.

EXAMPLE: We can determine the absolute magnitude of ε (epsilon) Indi using the formula $m - M = 5 \log d - 5$. Using $d = 3.6$ pc, we find using a calculator that $\log d = 0.56$. Therefore, this star's distance modulus is $m - M = 5(0.56) - 5 = -2.2$; this is less than zero, as it should be for a star less than 10 pc away. The star's absolute magnitude is $M = m - (-2.2) = +4.7 + 2.2 = +6.9$. Hence, ε Indi is less luminous than the Sun, which has absolute magnitude +4.8.

EXAMPLE: Suppose you were viewing the Sun from a planet orbiting another star 100 pc away. The Sun has absolute magnitude $M = +4.8$, and from the table above the distance modulus is $m - M = 5$. So, as seen from this planet, the Sun's apparent magnitude would be $m = M + 5 = +4.8 + 5 = +9.8$. This is too dim to be seen with the naked eye and just barely bright enough to be seen with binoculars. To an inhabitant of such a distant planetary system, our own Sun would be just another insignificant star.

The magnitude system is also used by astronomers to express the colors of stars as seen through different filters, as we describe in Section 19-4. For example, rather than quantifying a star's color by the color ratio b_V/b_B (a star's apparent brightness as seen through a V filter divided by the brightness through a B filter), astronomers commonly use the *color index* B–V, which is the difference in the star's apparent magnitude as measured with these two filters. We will not use this system in this book, however (but see Advanced Questions 44 and 45).

as an equation that relates a star's apparent magnitude (a measure of its apparent brightness), its absolute magnitude (a measure of its luminosity), and its distance. Box 19-3 describes these relationships.

Because the "backward" magnitude scales can be confusing, we will use them only occasionally in this book. We will usually speak of a star's luminosity rather than its absolute magnitude and will describe a star's appearance in terms of apparent brightness rather than apparent magnitude. But if you go on to learn more about astronomy, you will undoubtedly make frequent use of apparent and absolute magnitude.

19-4 A star's color depends on its surface temperature

 The image that opens this chapter shows that stars come in different colors. You can see these colors even with the naked eye. For example, you can easily see the red color of Betelgeuse, the star in the "armpit" of the constellation Orion, and the blue tint of Bellatrix at Orion's other "shoulder" (see Figure 2-2). Colors are most evident for the brightest stars, because your color vision does not work well at low light levels.

 It's true that the light from a star will appear redshifted if the star is moving away from you and blueshifted if it's moving toward you. But for even the fastest stars, these color shifts are so tiny that it takes sensitive instruments to measure them. The red color of Betelgeuse and the blue color of Bellatrix are not due to their motions; they are the actual colors of the stars.

We saw in Section 5-3 that a star's color is directly related to its surface temperature. The intensity of light from a relatively cool star peaks at long wavelengths, making the star look red (Figure 19-7a). A hot star's intensity curve peaks at shorter wavelengths, so the star looks blue (Figure 19-7c). For a star with an intermediate temperature, such as the Sun, the intensity peak is near the middle of the visible spectrum. This gives the star a yellowish color (Figure 19-7b).

Figure 19-7 shows that astronomers can determine the surface temperatures of a star by accurately measuring its color. To do this, the star's light is collected by a telescope and passed through one of a set of differently colored filters. The filtered light is then collected by a light-sensitive device such as a CCD (see Section 6-4). The process is then repeated with each of the filters in the set. The star's image will have a dif-

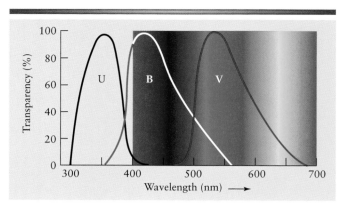

ƒigure 19-8

U, B, and V Filters This graph shows the wavelength ranges over which the standard U, B, and V filters are transparent. The U filter is transparent to near-ultraviolet light. The B filter is transparent to violet, blue, and green light, while the V filter is transparent to green and yellow light. By measuring the apparent brightness of a star with each of these filters and comparing the results, an astronomer can determine the star's surface temperature.

ferent brightness through each colored filter, and by comparing these brightnesses astronomers can find the wavelength at which the star's intensity curve has its peak—and hence the star's temperature.

Let's look at this procedure in more detail. The most commonly used filters are called U, B, and V, and the technique that uses them is called **UBV photometry**. Each filter is transparent in a different band of wavelengths: the ultraviolet (U), the blue (B), and the yellow-green (V, for *visual*) region of the visible spectrum (Figure 19-8). The transparency of the V filter mimics the sensitivity of the human eye.

a **This star looks red**

b **This star looks yellow-white**

c **This star looks blue-white**

ƒigure 19-7

Temperature and Color These graphs show the intensity of light emitted by three hypothetical stars plotted against wavelength (compare with Figure 5-10). The rainbow band indicates the range of visible wavelengths. The star's apparent color depends on whether the intensity curve has larger values at the short-wavelength or long-wavelength end of the visible spectrum.

To determine a star's temperature using UBV photometry, the astronomer first measures the star's brightness through each of the filters individually. This gives three apparent brightnesses for the star, designated b_U, b_B, and b_V. The astronomer then compares the intensity of starlight in neighboring wavelength bands by taking the ratios of these brightnesses: b_V/b_B and b_B/b_U. Table 19-1 gives values for these **color ratios** for several stars with different surface temperatures.

If a star is very hot, its radiation is skewed toward short, ultraviolet wavelengths as in Figure 19-7c. This makes the star dim through the V filter, brighter through the B filter, and brightest through the U filter. Hence, for a hot star b_V is less than b_B, which in turn is less than b_U, and the ratios b_V/b_B and b_B/b_U are both less than 1. One such star is Bellatrix (see Table 19-1), which has a surface temperature of 21,500 K.

In contrast, if a star is cool, its radiation peaks at long wavelengths as in Figure 19-7a. Such a star appears brightest through the V filter, dimmer through the B filter, and dimmest through the U filter (see Figure 19-8). In other words, for a cool star b_V is greater than b_B, which in turn is greater than b_U. Hence, the ratios b_V/b_B and b_B/b_U will both be greater than 1. The star Betelgeuse (surface temperature 3500 K) is an example.

You can see these differences between hot and cool stars in parts *a* and *c* of Figure 6-29, which show the constellation Orion at ultraviolet wavelengths (a bit shorter than those transmitted by the U filter) and at visible wavelengths that approximate the transmission of a V filter. The hot star Bellatrix is brighter in the ultraviolet image (Figure 6-29a) than at visible wavelengths (Figure 6-29c). (Figure 6-29d shows the names of the stars.) The situation is reversed for the cool star Betelgeuse: It is bright at visible wavelengths, but at ultraviolet wavelengths it is too dim to show up in the image.

Figure 19-9 graphs the relationship between a star's b_V/b_B color ratio and its temperature. If you know the value of the b_V/b_B color ratio for a given star, you can use this graph to find the star's surface temperature. As an example, for the Sun b_V/b_B equals 1.87, which corresponds to a surface temperature of 5800 K.

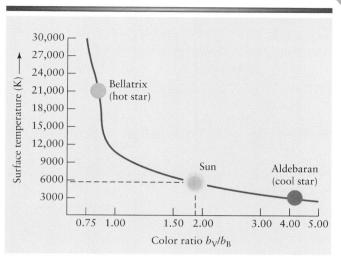

Figure 19-9

Temperature, Color, and Color Ratio The b_V/b_B color ratio is the ratio of a star's apparent brightnesses through a V filter and through a B filter. This ratio is small for hot, blue stars but large for cool, red stars. After measuring a star's brightness with the B and V filters, an astronomer can estimate the star's surface temperature from a graph like this one.

CAUTION! As we will see in Chapter 20, dust and gas that pervade interstellar space cause distant stars to appear redder than they really are. (In the same way, particles in the Earth's atmosphere make the setting Sun look redder; see Box 5-4.) Astronomers must take this reddening into account whenever they attempt to determine a star's surface temperature from its color ratios. A star's spectrum provides a more precise measure of a star's surface temperature, as we will see next. But it is quicker and easier to observe a star's colors with a set of U, B, and V filters than it is to take the star's spectrum with a spectrograph.

Table 19-1   Colors of Selected Stars

Star	Surface temperature (K)	b_V/b_B	b_B/b_U	Apparent color
Bellatrix (γ Orionis)	21,500	0.81	0.45	Blue
Regulus (α Leonis)	12,000	0.90	0.72	Blue-white
Sirius (α Canis Majoris)	9400	1.00	0.96	Blue-white
Megrez (δ Ursae Majoris)	8630	1.07	1.07	White
Altair (α Aquilae)	7800	1.23	1.08	Yellow-white
Sun	5800	1.87	1.17	Yellow-white
Aldebaran (α Tauri)	4000	4.12	5.76	Orange
Betelgeuse (α Orionis)	3500	5.55	6.66	Red

Source: J.-C. Mermilliod, B. Hauck, and M. Mermilliod, University of Lausanne.

19-5 The spectra of stars reveal their chemical compositions as well as surface temperatures

We have seen how the color of a star's light helps astronomers determine its surface temperature. To determine the other properties of a star, astronomers must analyze the spectrum of its light in more detail. This technique of *stellar spectroscopy* began in 1817 when Joseph Fraunhofer, a German instrument maker, attached a spectroscope to a telescope and pointed it toward the stars. Fraunhofer had earlier observed that the Sun has an absorption line spectrum—that is, a continuous spectrum with dark absorption lines (see Section 5-6). He found that stars have the same kind of spectra, which reinforces the idea that our Sun is a rather typical star. But Fraunhofer also found that the pattern of absorption lines is different for different stars.

We see an absorption line spectrum when a cool gas lies between us and a hot, glowing object (recall Figure 5-14). The light from the hot, glowing object itself has a continuous spectrum. In the case of a star, light with a continuous spectrum is produced at low-lying levels of the star's atmosphere where the gases are hot and dense. The absorption lines are created when this light flows outward through the upper layers of the star's atmosphere. Atoms in these cooler, less dense layers absorb radiation at specific wavelengths, which depend on the specific kinds of atoms present—hydrogen, helium, or other elements—and on whether or not the atoms are ionized. Absorption lines in the Sun's spectrum are produced in this same way (see Section 18-5).

Some stars have spectra in which the Balmer absorption lines of hydrogen are prominent. But in the spectra of other stars, including the Sun, the Balmer lines are nearly absent and the dominant absorption lines are those of heavier elements such as calcium, iron, and sodium. Still other stellar spectra are dominated by broad absorption lines caused by molecules, such as titanium oxide, rather than single atoms. To cope with this diversity, astronomers group similar-appearing stellar spectra into **spectral classes.** In a popular classification scheme that emerged in the late 1890s, a star was assigned a letter from A through O according to the strength or weakness of the hydrogen Balmer lines in the star's spectrum.

Nineteenth-century science could not explain why or how the spectral lines of a particular chemical are affected by the temperature and density of the gas. Nevertheless, a team of astronomers at the Harvard College Observatory forged ahead with a monumental project of examining the spectra of hundreds of thousands of stars. Their goal was to develop a system of spectral classification in which all spectral features, not just Balmer lines, change smoothly from one spectral class to the next.

The Harvard project was financed by the estate of Henry Draper, a wealthy New York physician and amateur astronomer who in 1872 became the first person to photograph stellar absorption lines. Researchers on the project included

figure 19-10   R I V U X G

Classifying the Spectra of the Stars The modern scheme of classifying stars by their spectra was developed at the Harvard College Observatory in the late nineteenth century. A team of women astronomers led by Edward C. Pickering and Williamina Fleming (standing) analyzed hundreds of thousands of spectra. Social conventions of the time prevented most women astronomers from using research telescopes or receiving salaries comparable to men's. (Harvard College Observatory)

Edward C. Pickering, Williamina Fleming, Antonia Maury, and Annie Jump Cannon (Figure 19-10). As a result of their efforts, many of the original A-through-O classes were dropped and others were consolidated. The remaining spectral classes were reordered in the sequence **OBAFGKM.** You can remember this sequence with the mnemonic: "*Oh, Be A Fine Girl* (or *Guy*), *Kiss Me!*"

Cannon refined the original OBAFGKM sequence into smaller steps called **spectral types.** These steps are indicated by attaching an integer from 0 through 9 to the original letter. For example, the spectral class F includes spectral types F0, F1, F2, . . . , F8, F9, which are followed by the spectral types G0, G1, G2, . . . , G8, G9, and so on.

Figure 19-11 shows representative spectra of several spectral types. The strengths of spectral lines change gradually from one spectral type to the next. For example, the Balmer absorption lines of hydrogen become increasingly prominent as you go from spectral type B0 to A0. From A0 onward through the F and G classes, the hydrogen lines weaken and almost fade from view. The Sun, whose spectrum is dominated by calcium and iron, is a G2 star.

The Harvard project culminated in the *Henry Draper Catalogue,* published between 1918 and 1924. It listed 225,300 stars, each of which Cannon had personally classified. Meanwhile, physicists had been making important discoveries about the structure of atoms. Ernest Rutherford had shown that atoms have nuclei (recall Figure 5-17), and Niels Bohr made the remarkable hypothesis that electrons circle atomic nuclei along discrete orbits (see Figure 5-20). These advances gave scientists the conceptual and mathematical tools needed to understand stellar spectra.

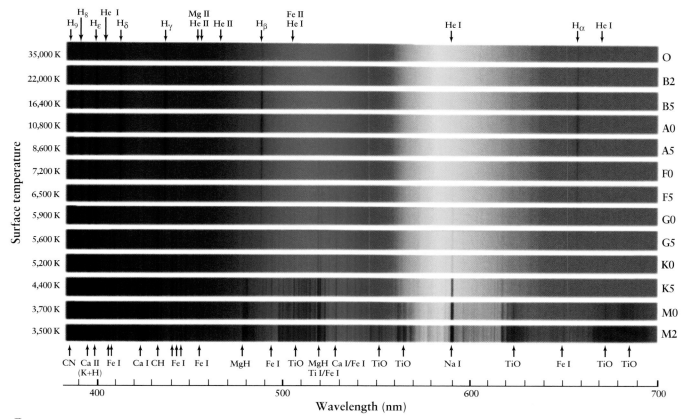

ƒigure 19-11   R I ■V■ U X G

Principal Types of Stellar Spectra This figure shows the spectra for stars of different spectral classes and different surface temperatures. The Balmer lines of hydrogen (H_α, H_β, H_γ, and H_δ) are strongest in stars of spectral class A, which have surface temperatures of about 10,000 K. The spectra of G and K stars (surface temperatures 4000 to 6000 K) have numerous lines due to metals such as iron (Fe) and calcium (Ca). The broad, dark bands in the spectra of M stars are caused by molecules of titanium oxide (TiO), which can exist in a star only when the surface temperature is below about 3500 K. A roman numeral after a chemical symbol shows whether the absorption line is caused by un-ionized atoms (roman numeral I) or by atoms that have lost one electron (roman numeral II). (R. Bell, University of Maryland, and M. Briley, University of Wisconsin at Oshkosh)

In the 1920s, the Harvard astronomer Cecilia Payne and the Indian physicist Meghnad Saha demonstrated that the OBAFGKM spectral sequence is actually a sequence in temperature. The hottest stars are O stars. Their absorption lines can occur only if these stars have surface temperatures above 25,000 K. M stars are the coolest stars. The spectral features of M stars are consistent with stellar surface temperatures of about 3000 K.

To see why the appearance of a star's spectrum is profoundly affected by the star's surface temperature, consider the Balmer lines of hydrogen. Hydrogen is by far the most abundant element in the universe, accounting for about three-quarters of the mass of a typical star. Yet the Balmer lines do not necessarily show up in a star's spectrum. As we saw in Section 5-8, Balmer absorption lines are produced when an electron in the $n = 2$ orbit of hydrogen is lifted into a higher orbit by absorbing a photon with the right amount of energy (see Figure 5-22). If the star is much hotter than

10,000 K, the photons pouring out of the star's interior have such high energy that they easily knock electrons out of hydrogen atoms in the star's atmosphere. This process ionizes the gas. With its only electron torn away, a hydrogen atom cannot produce absorption lines. Hence, the Balmer lines will be relatively weak in the spectra of such hot stars, such as the hot O and B2 stars in Figure 19-11.

Conversely, if the star's atmosphere is much cooler than 10,000 K, almost all the hydrogen atoms are in the lowest ($n = 1$) energy state. Most of the photons passing through the star's atmosphere possess too little energy to boost electrons up from the $n = 1$ to the $n = 2$ orbit of the hydrogen atoms. Hence, very few of these atoms will have electrons in the $n = 2$ orbit, and only these few can absorb the photons characteristic of the Balmer lines. As a result, these lines are nearly absent from the spectrum of a cool star. (You can see this in the spectra of the cool M0 and M2 stars in Figure 19-11.)

figure 19-12

The Strengths of Absorption Lines This graph shows how the relative strengths of different stellar absorption lines depend on the temperature of a star's outer layers. Each curve peaks at the stellar surface temperature for which that chemical's absorption line is strongest. (Roman numeral I denotes neutral, un-ionized atoms; II, III, and IV denote ions that have lost one, two, or three electrons.) For example, hydrogen (H) absorption lines are strongest in A stars with surface temperatures near 10,000 K. The Sun's spectrum has strong absorption lines of singly ionized calcium and iron (Ca II and Fe II), corresponding to a spectral type of G2 and a surface temperature of 5800 K.

For the Balmer lines to be prominent in a star's spectrum, the star must be hot enough to excite the electrons out of the ground state but not so hot that all the hydrogen atoms become ionized. A stellar surface temperature of about 9000 K produces the strongest hydrogen lines; this is the case for the stars of spectral types A0 and A5 in Figure 19-11.

 Every other type of atom or molecule also has a characteristic temperature range in which it produces prominent absorption lines in the observable part of the spectrum. Figure 19-12 shows the relative strengths of absorption lines produced by different chemicals. By measuring the details of these lines in a given star's spectrum, astronomers can accurately determine that star's surface temperature.

For example, the spectral lines of neutral (that is, un-ionized) helium are strong around 25,000 K. At this temperature, photons have enough energy to excite helium atoms without tearing away the electrons altogether. In stars hotter than about 30,000 K, helium atoms become singly ionized, that is, they lose one of their two electrons. The remaining electron produces a set of spectral lines that is recognizably different from those of neutral helium. Hence, when the spectral lines of singly ionized helium appear in a star's spectrum, we know that the star's surface temperature is greater than 30,000 K.

Astronomers use the term **metals** to refer to all elements other than hydrogen and helium. (This idiosyncratic use of the term "metal" is quite different from the definition used by chemists and other scientists. To a chemist, sodium and iron are metals but carbon and oxygen are not; to an astronomer, all of these substances are metals.) In this terminology, metals dominate the spectra of stars cooler than 10,000 K. Ionized metals are prominent for surface temperatures between 6000 and 8000 K, while neutral metals are strongest between approximately 5500 and 4000 K.

Below 4000 K, certain atoms in a star's atmosphere combine to form molecules. (At higher temperatures atoms move so fast that when they collide, they bounce off each other rather than "sticking together" to form molecules.) As these molecules vibrate and rotate, they produce bands of spectral lines that dominate the star's spectrum. Most noticeable are the lines of titanium oxide (TiO), which are strongest for surface temperatures of about 3000 K.

Since 1995 astronomers have found a number of stars with surface temperatures even lower than those of spectral class M. Strictly speaking, these are not stars but **brown dwarfs**, which we introduced in Section 7-9. Brown dwarfs are too small to sustain thermonuclear fusion in their cores. Instead, these "substars" glow primarily from the heat released by Kelvin-Helmholtz contraction (see Section 18-1). (They do undergo fusion reactions for a brief period during their evolution.) Brown dwarfs are so cold that they are best observed with infrared telescopes (see Figure 7-22). Such observations reveal that brown dwarf spectra have a rich variety of absorption lines due to molecules. Some of these molecules actually form into solid grains in a brown dwarf's atmosphere. To describe brown dwarf spectra, astronomers have defined two new spectral classes, L and T. Thus, the modern spectral sequence of stars and brown dwarfs is OBAFGKMLT. (Can you think of a new mnemonic that includes L and T?) Table 19-2 summarizes the relationship between the temperature and spectra of stars and brown dwarfs.

When the effects of temperature are accounted for, astronomers find that *all* stars have essentially the same chemical composition. By mass, almost all stars (including the Sun) and brown dwarfs are about three-quarters hydrogen, one-quarter helium, and 1% or less metals.

19-6 Stars come in a wide variety of sizes

With even the best telescopes, stars appear as nothing more than bright points of light. On a photograph or CCD image, brighter stars appear larger than dim ones (see Figure 19-3 and Figure 19-6b), but these apparent sizes give no indication

Table 19-2	The Spectral Sequence			
Spectral class	Color	Temperature (K)	Spectral lines	Examples
O	Blue-violet	30,000–50,000	Ionized atoms, especially helium	Naos (ζ Puppis), Mintaka (δ Orionis)
B	Blue-white	11,000–30,000	Neutral helium, some hydrogen	Spica (α Virginis), Rigel (β Orionis)
A	White	7500–11,000	Strong hydrogen, some ionized metals	Sirius (α Canis Majoris), Vega (α Lyrae)
F	Yellow-white	5900–7500	Hydrogen and ionized metals such as calcium and iron	Canopus (α Carinae), Procyon (α Canis Minoris)
G	Yellow	5200–5900	Both neutral and ionized metals, especially ionized calcium	Sun, Capella (α Aurigae)
K	Orange	3900–5200	Neutral metals	Arcturus (α Boötis), Aldebaran (α Tauri)
M	Red-orange	2500–3900	Strong titanium oxide and some neutral calcium	Antares (α Scorpii), Betelgeuse (α Orionis)
L	Red	1300–2500	Neutral potassium, rubidium, and cesium, and metal hydrides	Brown dwarf Teide 1
T	Red	below 1300	Strong neutral potassium and some water (H_2O)	Brown dwarf Gliese 229B

of the star's actual size. To determine the size of a star, astronomers combine information about its luminosity (determined from its distance and apparent brightness) and its surface temperature (determined from its spectral type). In this way, they find that some stars are quite a bit smaller than the Sun, while others are a thousand times larger.

The key to finding a star's radius from its luminosity and surface temperature is the Stefan-Boltzmann law (Section 5-4). This law says that the amount of energy radiated per second from a square meter of a blackbody's surface—that is, the energy flux (F)—is proportional to the fourth power of the temperature of that surface (T), as given by the equation $F = \sigma T^4$. This equation applies very well to stars, whose spectra are quite similar to that of a perfect blackbody. (Absorption lines, while important for determining the star's chemical composition and surface temperature, make only relatively small modifications to a star's blackbody spectrum.)

A star's luminosity is the amount of energy emitted per second from its entire surface. This equals the energy flux F multiplied by the total number of square meters on the star's surface (that is, the star's surface area). We expect that most stars are nearly spherical, like the Sun, so we can use the formula for the surface area of a sphere. This is $4\pi R^2$, where R is the star's radius (the distance from its center to its surface). Multiplying together the formulas for energy flux and surface area, we can write the star's luminosity as:

Relationship between a star's luminosity, radius, and surface temperature

$$L = 4\pi R^2 \sigma T^4$$

L = star's luminosity, in watts

R = star's radius, in meters

σ = Stefan-Boltzmann constant = 5.67×10^{-8} W m^{-2} K^{-4}

T = star's surface temperature, in kelvins

This equation says that a relatively cool star (low surface temperature T), for which the energy flux is quite low, can nonetheless be very luminous if it has a large enough radius R. Alternatively, a relatively hot star (large T) can have a very low luminosity if the star has only a little surface area (small R).

ANALOGY In a similar way, a roaring campfire can emit more light than a welder's torch. The campfire is at a lower temperature than the torch, but has a much larger surface area from which it emits light.

Box 19-4 describes how to use the above equation to find a star's radius if its luminosity and surface temperature are known. In this way astronomers find that stars come in a wide

box 19-4   Tools of the Astronomer's Trade

Stellar Radii, Luminosities, and Surface Temperatures

Because stars emit light in almost exactly the same fashion as blackbodies, we can use the Stefan-Boltzmann law to relate a star's luminosity (L), surface temperature (T), and radius (R). The relevant equation is

$$L = 4\pi R^2 \sigma T^4$$

As written, this equation involves the Stefan-Boltzmann constant σ, which is equal to 5.67×10^{-8} W m^{-2} K^{-4}. In many calculations, it is more convenient to relate everything to the Sun, which is a typical star. Specifically, for the Sun we have $L_\odot = 4\pi R_\odot^2 \sigma T_\odot^4$, where $L_\odot$ is the Sun's luminosity, $R_\odot$ is the Sun's radius, and $T_\odot$ is the Sun's surface temperature (equal to 5800 K). Dividing the general equation for L by this specific equation for the Sun, we obtain

$$\frac{L}{L_\odot} = \left(\frac{R}{R_\odot}\right)^2 \left(\frac{T}{T_\odot}\right)^4$$

This is an easier formula to use because the constant σ has cancelled out. We can also rearrange terms to arrive at a useful equation for the radius (R) of a star:

Radius of a star related to its luminosity and surface temperature

$$\frac{R}{R_\odot} = \left(\frac{T_\odot}{T}\right)^2 \sqrt{\frac{L}{L_\odot}}$$

$R/R_\odot$ = ratio of the star's radius to the Sun's radius

$T_\odot/T$ = ratio of the Sun's surface temperature to the star's surface temperature

$L/L_\odot$ = ratio of the star's luminosity to the Sun's luminosity

EXAMPLE: Consider the bright reddish star Betelgeuse in the constellation Orion (see Figure 2-2 or Figure 6-29c). Betelgeuse is 60,000 times more luminous than the Sun (so $L = 60,000 L_\odot$) and has surface temperature $T = 3500$ K. Substituting these data into the previous equation, we get

$$\frac{R}{R_\odot} = \left(\frac{5800\text{ K}}{3500\text{ K}}\right)^2 \sqrt{6 \times 10^4} = 670$$

In other words, Betelgeuse's radius is 670 times larger than that of the Sun. The Sun's radius is 6.96×10^5 km, so the radius of Betelgeuse is $(670)(6.96 \times 10^5$ km$) = 4.7 \times 10^8$ km, or more than 3 AU. If Betelgeuse were located at the center of our solar system, it would extend beyond the orbit of Mars!

EXAMPLE: Sirius, the brightest star in the sky, is actually two stars orbiting each other (a binary star). The less luminous star, Sirius B, is a white dwarf that is too dim to see with the naked eye. Its luminosity is 0.0025 $L_\odot$ and its surface temperature is 10,000 K. Hence, the ratio of its radius to that of the Sun is

$$\frac{R}{R_\odot} = \left(\frac{5800\text{ K}}{10,000\text{ K}}\right)^2 \sqrt{0.0025} = 0.17$$

Sirius B is only 0.017 as large as the Sun: its radius is $(0.017)(6.96 \times 10^5$ km$) = 12,000$ km, or about twice the radius of the Earth.

The radii of some stars have been measured with other techniques (see Section 19-11). These other methods yield values consistent with those calculated by the methods we have just described.

---

range of sizes. The smallest stars visible through ordinary telescopes, called *white dwarfs*, are about the same size as the Earth. Although their surface temperatures can be very high (25,000 K or more), white dwarfs have so little surface area that their luminosities are very low (less than 0.01 $L_\odot$). The largest stars, called *supergiants*, are a thousand times larger in radius than the Sun (and 10^5 times larger than white dwarfs). If our own Sun were replaced by one of these supergiants, the Earth's orbit would lie completely inside the star!

Figure 19-13 summarizes how astronomers determine the distance from Earth, luminosity, surface temperature, chemical composition, and radius of a star close enough to us so that its parallax can be measured. Remarkably, all of these properties

can be deduced from just a few measured quantities: the star's parallax angle, apparent brightness, and spectrum.

19-7   Hertzsprung-Russell (H-R) diagrams reveal the different kinds of stars

Astronomers have collected a wealth of data about the stars, but merely having tables of numerical data is not enough. Astronomers, like all scientists, want to analyze their data to look for trends and underlying principles. One of the best

ƒigure 19-13
Finding Key Properties of Nearby Stars This flowchart shows how astronomers deduce the properties of relatively nearby stars (those close enough that their distance is known from parallax measurements). Given the star's distance, measuring the apparent brightness gives the star's luminosity from the inverse-square law (Section 19-2). The star's spectrum reveals the star's chemical composition and surface temperature (Section 19-5). And from luminosity and surface temperature, the Stefan-Boltzmann law allows the radius to be calculated (Section 19-6). To determine these same properties for more distant stars, different calculations have to be made (see Section 19-8, especially Figure 19-16).

ways to look for trends in any set of data, whether it comes from astronomy, finance, medicine, or meteorology, is to create a graph showing how one quantity depends on another. For example, investors consult graphs of stock market values versus dates, and weather forecasters make graphs of temperature versus altitude to determine whether thunderstorms will form. Astronomers have found that a particular graph of stellar properties shows that stars fall naturally into just a few categories. This graph, one of the most important in all astronomy, will in later chapters help us understand how stars form, evolve, and eventually die.

Which properties of stars should we include in a graph? Most stars have about the same chemical composition, but two properties of stars—their luminosities and surface temperatures—differ substantially from one star to another. Stars also come in a wide range of radii, but a star's radius is a secondary property that can be found from the luminosity and surface temperature (as we saw in Section 19-6 and Box 19-4). We also relegate the positions and space velocities of stars to secondary importance. (In a similar way, a physician is more interested in your weight and blood pressure than in where you live or how fast you drive.) We can then ask the following question: What do we learn when we graph the luminosities of stars versus their surface temperatures?

The first answer to this question was given in 1911 by the Danish astronomer Ejnar Hertzsprung. He pointed out that a regular pattern appears when the absolute magnitudes of stars (which measure their luminosities) are plotted against their colors (which measure their surface temperatures). Two years later, the American astronomer Henry Norris Russell independently discovered a similar regularity in a graph using spectral types (another measure of surface temperature) instead of colors. In recognition of their originators, graphs of this kind are today known as **Hertzsprung-Russell diagrams, or H-R diagrams.**

Figure 19-14 is a typical Hertzsprung-Russell diagram. Each dot represents a star whose spectral type and luminosity have been determined. The most luminous stars are near the top of the diagram, the least luminous stars near the bottom. Hot stars of spectral classes O and B are toward the left side of the graph and cool stars of spectral class M are toward the right.

CAUTION! You are probably accustomed to graphs where the numbers on the horizontal axis increase as you move to the right. (For example, the business section of a newspaper includes a graph of stock market values versus dates, with later dates to the right of earlier ones.) But on an H-R diagram the temperature scale on the horizontal axis increases toward the *left*. This practice stems from the original diagrams of Hertzsprung and Russell, who placed hot O stars on the left and cool M stars on the right. This arrangement is a tradition that no one has seriously tried to change.

The most striking feature of the H-R diagram is that the data points are not scattered randomly over the graph but are grouped in a few distinct regions. The luminosities and surface temperatures of stars do *not* have random values; these two quantities are related!

The band stretching diagonally across the H-R diagram includes about 90% of the stars in the night sky. This band, called the **main sequence**, extends from the hot, luminous, blue stars in the upper left corner of the diagram to the cool,

figure 19-14

A Hertzsprung-Russell (H-R) Diagram On an
H-R diagram, the luminosities of stars are plotted
against their spectral types. Each dot on this graph represents a star
whose luminosity and spectral type have been determined. Included
on the diagram are some well-known stars. The data points are
grouped in just a few regions on the graph, showing that
luminosity and spectral type are correlated. Main-sequence stars lie
along the red curve. Giants like Arcturus are to the right of the
main sequence, supergiants like Rigel and Betelgeuse are above it,
and white dwarfs like Sirius B are below it. The absolute
magnitudes and surface temperatures are also shown at the right
and top of the graph, respectively.

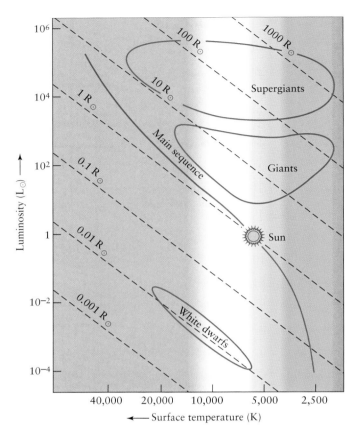

figure 19-15

The Sizes of Stars on an H-R Diagram This H-R diagram plots
stellar luminosities versus surface temperature. The dashed diagonal
lines indicate different radii. These lines were drawn using the equation
$L = 4\pi R^2 \sigma T^4$ relating a star's luminosity (L), radius (R), and surface
temperature (T). For a given stellar radius, as the surface temperature
increases (corresponding to moving from right to left in the
H-R diagram), the star glows more intensely and the luminosity
increases (corresponding to moving upward in the diagram). While
individual stars are not plotted, we show the regions of the diagram in
which main sequence, giant, supergiant, and white dwarf stars are
found. Note that the Sun is intermediate in luminosity, surface
temperature, and radius; it is very much a middle-of-the-road star.

dim, red stars in the lower right corner. A star whose properties place it in this region of an H-R diagram is called a **main-
sequence star.** The Sun (spectral type G2, luminosity 1 L☉,
absolute magnitude +4.8) is such a star. We will find that all
main-sequence stars are like the Sun in that *hydrogen burning*—thermonuclear fusion that converts hydrogen into
helium (Section 18-1)—is taking place in their cores.

The upper right side of the H-R diagram shows a second
major grouping of data points. Stars represented by these
points are both luminous and cool. From the Stefan-Boltzmann law, we know that a cool star radiates much less light

per unit of surface area than a hot star. In order for these stars
to be as luminous as they are, they must be huge, and so they
are called **giants.** These stars are around 10 to 100 times larger
than the Sun (see Box 19-4). You can see this in Figure 19-15,
which is an H-R diagram to which dashed lines have been
added to represent stellar radii. Most giant stars are around
100 to 1000 times more luminous than the Sun and have surface temperatures of about 3000 to 6000 K. Cooler members
of this class of stars (those with surface temperatures from
about 3000 to 4000 K) are often called **red giants** because they
appear reddish. Aldebaran in the constellation Taurus and

Arcturus in Boötes are two red giants that can be easily seen with the naked eye.

A few rare stars are considerably bigger and brighter than typical red giants, with radii up to 1000 $R_\odot$. Appropriately enough, these superluminous stars are called **supergiants.** Betelgeuse in Orion and Antares in Scorpius are two supergiants you can find in the nighttime sky. Together, giants and supergiants make up about 1% of the stars in the sky.

Both giants and supergiants have thermonuclear reactions occurring in their interiors, but the character of those reactions and where in the star they occur can be quite different than for a main-sequence star like the Sun. We will study these stars in more detail in Chapters 21 and 22.

The remaining 9% of stars form a distinct grouping of data points toward the lower left corner of the Hertzsprung-Russell diagram. Although these stars are hot, their luminosities are quite low; hence, they must be small. They are appropriately called **white dwarfs.** These stars, which are so dim that they can be seen only with a telescope, are approximately the same size as the Earth. As we will learn in Chapter 22, no thermonuclear reactions take place within white dwarf stars. Rather, like embers left from a fire, they are the still-glowing remnants of what were once giant stars.

By contrast, *brown* dwarfs (which lie at the extreme lower right of the main sequence, off the bottom and right-hand edge of Figure 19-14 or Figure 19-15) are objects that will never become stars. They are comparable in radius to the planet Jupiter (that is, intermediate in size between the Earth and the Sun; see Figure 7-2). The study of brown dwarfs is still in its infancy, but it appears that there may be twice as many brown dwarfs as there are "real" stars.

 You can think of white dwarfs as "has-been" stars whose days of glory have passed. In this analogy, a brown dwarf is a "never-will-be."

The existence of fundamentally different types of stars is the first important lesson to come from the H-R diagram. In later chapters we will find that these different types represent various stages in the lives of stars. We will use the H-R diagram as an essential tool for understanding how stars evolve.

19-8 Details of a star's spectrum reveal whether it is a giant, a white dwarf, or a main-sequence star

A star's surface temperature largely determines which lines are prominent in its spectrum. Therefore, classifying stars by spectral type is essentially the same as categorizing them by surface temperature. But, as the H-R diagram in Figure 19-15 shows, stars of the same surface temperature can have very different luminosities. As an example, a star with surface temperature 5800 K could be either a white dwarf, a main-sequence star, a giant, or a supergiant, depending on its luminosity. By examining the details of a star's spectrum, however, astronomers can determine to which of these categories a star belongs. This

gives astronomers a tool to determine the distances to stars millions of parsecs away, far beyond the maximum distance that can be measured using stellar parallax.

Figure 19-16 compares the spectra of two stars of the same spectral type but different luminosity: a B8 supergiant and a B8 main-sequence star. Note that the Balmer lines of hydrogen are narrow in the spectrum of the very luminous supergiant but quite broad in the spectrum of the less luminous main-sequence star. In general, for stars of spectral types B through F, the more luminous the star, the narrower its hydrogen lines.

Fundamentally, these differences between stars of different luminosity are due to differences between the stars' atmospheres, where absorption lines are produced. Hydrogen lines in particular are affected by the density and pressure of the gas in a star's atmosphere. The higher the density and pressure, the more frequently hydrogen atoms collide and interact with other atoms and ions in the atmosphere. These collisions shift the energy levels in the hydrogen atoms and thus broaden the hydrogen spectral lines.

In the atmosphere of a luminous giant star, the density and pressure are quite low because the star's mass is spread over a huge volume. Atoms and ions in the atmosphere are relatively far apart; hence, collisions between them are sufficiently infrequent that hydrogen atoms can produce narrow Balmer lines. A main-sequence star, however, is much more compact than a giant or supergiant. In the denser atmosphere of a main-sequence star, frequent interatomic collisions perturb the energy levels in the hydrogen atoms, thereby producing broader Balmer lines.

In the 1930s, W. W. Morgan and P. C. Keenan of the Yerkes Observatory of the University of Chicago developed a system of **luminosity classes** based upon the subtle differences

figure 19-16   R I **V** U X G

How Luminosity Affects a Star's Spectrum These are the spectra of two stars of the same spectral type (B8) and surface temperature (13,400 K) but different luminosities: **(a)** the B8 supergiant Rigel (luminosity 58,000 $L_\odot$), one of the brightest stars in Orion, and **(b)** the B8 main-sequence star Algol (luminosity 100 $L_\odot$), the second brightest star in Perseus. Note the different widths of the hydrogen Balmer lines H_γ and H_δ in the two spectra. (From *An Atlas of Stellar Spectra*, W. W. Morgan, P. C. Keenan, and E. Kellman)

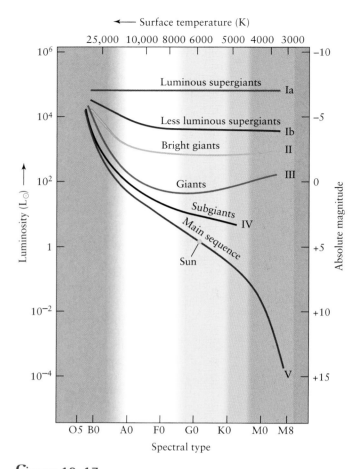

f**igure 19-17**

Luminosity Classes By dividing the H-R diagram into regions corresponding to luminosity classes, finer distinctions can be made between giants and supergiants. Luminosity classes Ia and Ib include the supergiants, luminosity class V encompasses the main-sequence stars, and classes II, III, and IV denote giants of intermediate size. White dwarfs do not have their own luminosity class.

in spectral lines. When these luminosity classes are plotted on an H-R diagram (Figure 19-17), they provide a useful subdivision of the star types in the upper right half of the diagram. Luminosity classes Ia and Ib are composed of supergiants; luminosity class V includes all the main-sequence stars. The intermediate classes distinguish giant stars of various luminosities. As we will see in Chapters 21 and 22, different luminosity classes represent different stages in the evolution of a star. White dwarfs are not given a luminosity class of their own; as we mentioned in Section 19-7, they represent a final stage in stellar evolution in which no thermonuclear reactions take place.

Astronomers commonly use a shorthand description that combines a star's spectral type and its luminosity class. For example, the Sun is said to be a G2 V star. The spectral type indicates the star's surface temperature, and the luminosity

class indicates its luminosity. Thus, an astronomer knows immediately that any G2 V star is a main-sequence star with a luminosity of about 1 $L_\odot$ and a surface temperature of about 5800 K. Similarly, a description of Aldebaran as a K5 III star tells an astronomer that it is a red giant with a luminosity of around 370 $L_\odot$ and a surface temperature of about 4000 K.

A star's spectral type and luminosity class, combined with the information on the H-R diagram, enable astronomers to estimate the star's distance from the Earth. As an example, consider the star Regulus (also called α Leonis, because it is the brightest star in the constellation Leo, the Lion). Its spectrum reveals Regulus to be a B7 V star (a hot, blue, main-sequence star). Using Figure 19-17, we can read off that such a star's luminosity is 140 $L_\odot$. Given the star's luminosity and its apparent brightness—in the case of Regulus, 5.2×10^{-12} of the apparent brightness of the Sun—we can use the inverse-square law to determine its distance from the Earth. The mathematical details are worked out in Box 19-2.

This method for determining distance, in which the luminosity of a star is found using spectroscopy, is called **spectroscopic parallax**. (The name is a bit misleading, because no parallax angle is involved. A better name for this method, although not the one used by astronomers, would be "spectroscopic distance determination.") Figure 19-18 summarizes the method of spectroscopic parallax.

Spectroscopic parallax is an incredibly powerful technique. No matter how remote a star is, this technique allows astronomers to determine its distance, provided only that its spectrum and apparent brightness can be measured. Box 19-2 gives an example of how spectroscopic parallax has been used to find the distance to stars in other galaxies tens of millions of parsecs away. By contrast, we saw in Section 19-1 that "real" stellar parallaxes can be measured only for stars within a few hundred parsecs.

Unfortunately, spectroscopic parallax has its limitations; distances to individual stars determined using this method are only accurate to at best 10%. The reason is that the luminosity classes shown in Figure 19-17 are not thin lines on the H-R diagram but are moderately broad bands. Hence, even if a star's spectral type and luminosity class are known, there is still some uncertainty in the luminosity that we read off an H-R diagram. Nonetheless, spectroscopic parallax is often the only means that an astronomer has to estimate the distance to remote stars.

What has been left out of this discussion is *why* different stars have different spectral types and luminosities. One key factor, as we shall see, turns out to be the mass of the star.

19-9 Binary star systems provide crucial information about stellar masses

We now know something about the sizes, temperatures, and luminosities of stars. To complete our picture of the physical properties of stars, we need to know their masses. In this section, we will see that stars come in a wide range of masses. We will also discover an important relationship between the mass

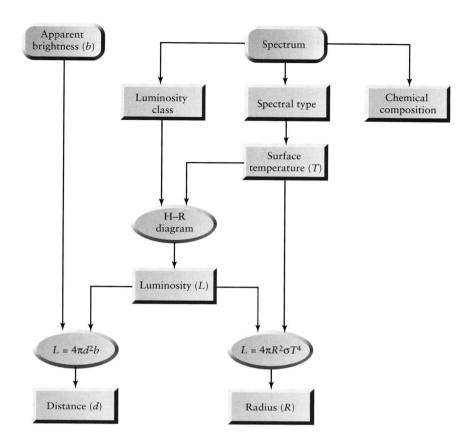

figure 19-18

The Method of Spectroscopic Parallax If a star is too far away, its parallax angle is too small to allow a direct determination of its distance. This flowchart shows how astronomers deduce the properties of these distant stars. The star's spectrum reveals its spectral type and luminosity class. Given these, the surface temperature and luminosity can be inferred using the H–R diagram. And given the star's luminosity and apparent brightness, the star's distance from the Earth can be calculated using the inverse-square law. As for nearby stars (see Figure 19-13), the star's chemical composition is determined from its spectrum, and the star's radius is calculated from the luminosity and surface temperature using the Stefan-Boltzmann law.

and luminosity of main-sequence stars. This relationship is crucial to understanding why some main-sequence stars are hot and luminous, while others are cool and dim. It will also help us understand what happens to a star as it ages and evolves.

Determining the masses of stars is not trivial, however. The problem is that there is no practical, direct way to measure the mass of an isolated star. Fortunately for astronomers, about half of the visible stars in the night sky are not isolated individuals. Instead, they are multiple-star systems, in which two or more stars orbit each other. By carefully observing the motions of these stars, astronomers can glean important information about their masses.

A pair of stars located at nearly the same position in the night sky is called a **double star.** The Anglo-German astronomer William Herschel made the first organized search for such pairs. Between 1782 and 1821, he published three catalogs listing more than 800 double stars. Late in the nineteenth century, his son, John Herschel, discovered 10,000 more doubles. Some of these double stars are **optical double stars,** which are two stars that lie along nearly the same line of sight but are actually at very different distances from us. But many double stars are true **binary stars,** or **binaries**—pairs of stars that actually orbit each other.

When astronomers can actually see the two stars orbiting each other, a binary is called a **visual binary** (Figure 19-19). After many years of patient observation, astronomers can plot the orbits of the stars in a visual binary (Figure 19-20).

Because the two stars in a binary system orbit each other because of their mutual gravitational attraction, their orbital motions obey Kepler's third law as formulated by Isaac Newton (see Section 4-7 and Box 4-4). This law can be written as follows:

Kepler's third law for binary star systems

$$M_1 + M_2 = \frac{a^3}{P^2}$$

M_1, M_2 = masses of two stars in binary system, in solar masses

a = semimajor axis of one star's orbit around the other, in AU

P = orbital period, in years

Here a is the semimajor axis of the elliptical orbit that one star appears to describe around the other, plotted as in Figure 19-20. As this equation indicates, if we can measure this semimajor axis (a) and the orbital period (P), we can learn something about the masses of the two stars.

In principle, the orbital period of a visual binary is easy to determine. All you have to do is see how long it takes for the two stars to revolve once about each other. As Figure 19-20 suggests, however, the period may be so long that more than one astronomer's lifetime is needed to complete the observations.

| 1908 | 1915 | 1920 |

𝒻igure 19-19   R I **V** U X G

The Binary Star Krüger 60 About half of the stars visible in the night sky are actually binary stars. This series of photographs shows the visual binary star Krüger 60 (in the constellation Cepheus) at the upper left of each photograph. Notice how the two stars orbit about each other. This binary system has a period of 44½ years. The maximum angular separation of the two stars is about 2.5 arcsec. (Yerkes Observatory)

Determining the semimajor axis of an orbit is also a challenge. The *angular* separation between the stars can be determined by observation. To convert this angle into a physical distance between the stars, we need to know the distance between the binary and the Earth. This can be found from parallax measurements or by using spectroscopic parallax. The astronomer must also take into account how the orbit is tilted to our line of sight.

Once both P and a have been determined, Kepler's third law can be used to calculate $M_1 + M_2$, the sum of the masses of the two stars in the binary system. But this analysis tells us nothing about the *individual* masses of the two stars. To

obtain these, more information about the motions of the two stars is needed.

Each of the two stars in a binary system actually moves in an elliptical orbit about the **center of mass** of the system. Imagine two children sitting on opposite ends of a seesaw (Figure 19-21*a*). For the seesaw to balance properly, they must position themselves so that their center of mass—an imaginary point that lies along a line connecting their two bodies—is at the fulcrum, or pivot point of the seesaw. If the two children have the same mass, the center of mass lies midway between them, and they should sit equal distances from the fulcrum. If their masses are different, the center of mass is closer to the heavier child.

Just as the seesaw naturally balances at its center of mass, the two stars that make up a binary system naturally orbit around their center of mass (Figure 19-21*b*). The center of mass always lies along the line connecting the two stars and is closer to the more massive star.

The center of mass of a visual binary is located by plotting the separate orbits of the two stars, as in Figure 19-21*b*, using the background stars as reference points. The center of mass lies at the common focus of the two elliptical orbits. Comparing the relative sizes of the two orbits around the center of mass yields the ratio of the two stars' masses, M_1/M_2. The sum $M_1 + M_2$ is already known from Kepler's third law, so the individual masses of the two stars can then be determined.

Years of careful, patient observations of binaries have slowly yielded the masses of many stars. As the data accumulated, an important trend began to emerge: For main-sequence stars, there is a direct correlation between mass and luminosity. The more massive a main-sequence star, the more luminous it is. Figure 19-22 depicts this **mass-luminosity relation** as a graph. The range of stellar masses extends from less than 0.1 of a solar mass to more than 50 solar masses. The Sun's mass lies between these extremes.

The mass-luminosity relation shows that the main sequence on an H-R diagram is a progression in mass as well as in luminosity and surface temperature (Figure 19-23). The hot, bright, bluish stars in the upper left corner of an H-R diagram are

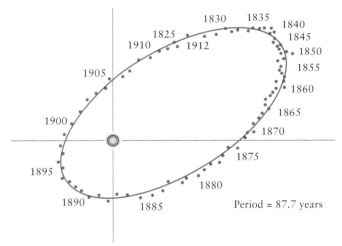

𝒻igure 19-20

The Orbit of 70 Ophiuchi After plotting observations of a visual binary over the years, astronomers can draw the orbit of one star with respect to the other. Once the orbit is known, Kepler's third law can be used to learn about the masses of the two stars. This illustration shows the orbit of a visual binary star in the constellation Ophiuchus. In plotting the orbit, either star may be regarded as the stationary one; the shape of the orbit will be the same in either case.

a

b

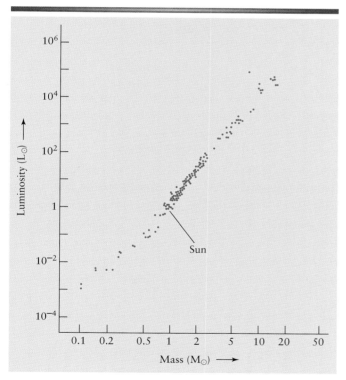

figure 19-21

ANIMATION 19.2 **Center of Mass in a Binary Star System (a)** For a seesaw to balance, the center of mass of the two children must be at the fulcrum. The center of mass is closer to the more massive of the two children. **(b)** In the same way, the center of mass of a binary star system is always nearer the more massive of the two stars (in this case, the reddish star). Both stars move in elliptical orbits about their common center of mass. Although their orbits cross each other, the two stars are always on opposite sides of the center of mass and thus never collide.

figure 19-22

The Mass-Luminosity Relation For main-sequence stars, there is a direct correlation between mass and luminosity—the more massive a star, the more luminous it is. A main-sequence star of mass 10 $M_\odot$ (that is, 10 times the Sun's mass) has roughly 3000 times the Sun's luminosity (3000 $L_\odot$); one with 0.1 $M_\odot$ has a luminosity of only about 0.001 $L_\odot$.

the most massive main-sequence stars. Likewise, the dim, cool, reddish stars in the lower right corner of an H-R diagram are the least massive. Main-sequence stars of intermediate temperature and luminosity also have intermediate masses.

Figure 19-23 strongly suggests that main-sequence stars are objects like the Sun, with essentially the same chemical composition as the Sun but with different masses. In this picture, all main-sequence stars shine because of thermonuclear reactions at their cores that convert hydrogen to helium. The greater the total mass of the star, the greater the pressure and temperature at the core, the more rapidly thermonuclear reactions take place in the core, and the greater the energy output—that is, the luminosity—of the star. In other words, the greater the mass of a main-sequence star, the greater its luminosity. This statement is just the mass-luminosity relation, which we can now recognize as a natural consequence of the nature of main-sequence stars.

If a star's mass is less than about $0.08 M_\odot$, the core pressure and temperature will be too low for thermonuclear reactions to take place. The "star" is then a brown dwarf. Brown dwarfs also obey a mass-luminosity relation: The greater the mass, the faster the brown dwarf contracts because of its own gravity, the more rapidly it radiates energy into space, and, hence, the more luminous the brown dwarf is.

There are no simple mass-luminosity relations for giant, supergiant, or white dwarf stars. Why these stars lie where they do on an H-R diagram will become apparent when we study the evolution of stars in Chapters 21 and 22. We will find that main-sequence stars evolve into giant and supergiant stars, and that some of these eventually end their lives as white dwarfs. As we will see, the mass-luminosity relation helps tell us how rapidly this evolution takes place for stars of different masses.

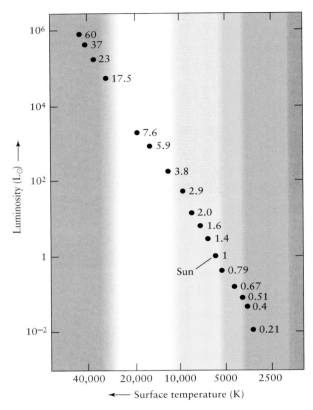

Figure 19-23

The Main Sequence and Masses On this H-R diagram, each dot represents a main-sequence star. The number next to each dot is the mass of that star in solar masses ($M_\odot$). As you move up the main sequence from the lower right to the upper left of the H-R diagram, the mass, luminosity, and surface temperature of main-sequence stars all increase.

19-10 Spectroscopy makes it possible to study binary systems in which the two stars are close together

We have described how the masses of stars can be determined from observations of visual binaries, in which the two stars can be distinguished from each other. But if the two stars in a binary system are too close together, the images of the two stars can blend to produce the semblance of a single star. Happily, in many cases we can use spectroscopy to decide whether a seemingly single star is in fact a binary system. Spectroscopic observations of binaries provide additional useful information about star masses.

Some binaries are discovered when the spectrum of a star shows incongruous spectral lines. For example, the spectrum of what appears to be a single star may include both strong hydrogen lines (characteristic of a type A star) and strong absorption bands of titanium oxide (typical of a type M star). Because a single star cannot have the differing physical properties of these

two spectral types, such a star must actually be a binary system that is too far away for us to resolve its individual stars. A binary system detected in this way is called a **spectrum binary**.

Other binary systems can be detected using the Doppler effect. If a star is moving toward the Earth, its spectral lines are displaced toward the short-wavelength (blue) end of the spectrum. Conversely, the spectral lines of a star moving away from us are shifted toward the long-wavelength (red) end of the spectrum. The upper portion of Figure 19-24 applies these ideas to a hypothetical binary star system with an orbital plane that is edge-on to our line of sight.

As the two stars move around their orbits, they periodically approach and recede from us. Hence, the spectral lines of the two stars are alternately blueshifted and redshifted. The two stars in this hypothetical system are so close together that they appear through a telescope as a single star with a single spectrum. Because one star shows a blueshift while the other is showing a redshift, the spectral lines of the binary system appear to split apart and rejoin periodically. Stars whose binary character is revealed by such shifting spectral lines are called **spectroscopic binaries**.

 To analyze a spectroscopic binary, astronomers measure the wavelength shift of each star's spectral lines and use the Doppler shift formula (introduced in Section 5-9 and Box 5-6) to determine the **radial velocity** of each star—that is, how fast and in what direction it is moving along our line of sight. The lower portion of Figure 19-24 shows a graph of the radial velocity versus time, called a **radial velocity curve**, for the binary system HD 171978. Each of the two stars alternately approaches and recedes as it orbits around the center of mass. The pattern of the curves repeats every 15 days, which is the orbital period of the binary. Also note that the entire wavy pattern is displaced upward from the zero-velocity line by about 12 km/s, which is the overall velocity of the binary system away from the Earth.

Figure 19-25 shows two spectra of the spectroscopic binary κ (kappa) Arietis taken a few days apart. In Figure 19-25a, two sets of spectral lines are visible, offset slightly in opposite directions from the normal positions of these lines. This corresponds to stage 1 or stage 3 in Figure 19-24; one of the orbiting stars is moving toward the Earth and has its spectral lines blueshifted, and the other star is moving away from the Earth and has its lines redshifted. A few days later, the stars have progressed along their orbits so that neither star is moving toward or away from the Earth, corresponding to stage 2 or stage 4 in Figure 19-24. At this time there are no Doppler shifts, and the spectral lines of both stars are at the same positions. That is why only one set of spectral lines appears in Figure 19-25b.

 It is important to emphasize that the Doppler effect applies only to motion along the line of sight. Motion perpendicular to the line of sight does not affect the observed wavelengths of spectral lines. Hence, the ideal orientation for a spectroscopic binary is to have the stars orbit in a plane that is edge-on to our line of sight. (By contrast, a *visual* binary is best observed if the orbital plane is face-on to our line of sight.) For the Doppler shifts to be noticeable, the orbital speeds of the two stars should be at least a few kilometers per second.

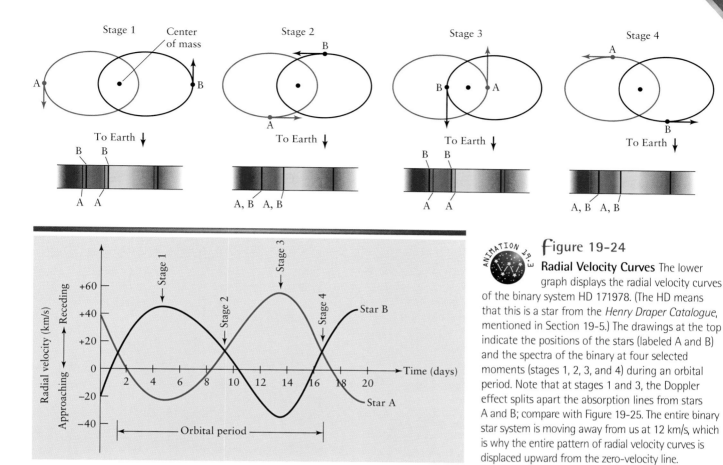

ANIMATION 19.3

figure 19-24

Radial Velocity Curves The lower graph displays the radial velocity curves of the binary system HD 171978. (The HD means that this is a star from the *Henry Draper Catalogue*, mentioned in Section 19-5.) The drawings at the top indicate the positions of the stars (labeled A and B) and the spectra of the binary at four selected moments (stages 1, 2, 3, and 4) during an orbital period. Note that at stages 1 and 3, the Doppler effect splits apart the absorption lines from stars A and B; compare with Figure 19-25. The entire binary star system is moving away from us at 12 km/s, which is why the entire pattern of radial velocity curves is displaced upward from the zero-velocity line.

The binaries depicted in Figure 19-24 and Figure 19-25 are called *double-line* spectroscopic binaries, because the spectral lines of both stars in the binary system can be seen. Most spectroscopic binaries, however, are *single-line* spectroscopic binaries: One of the stars is so dim that its spectral lines cannot be detected. The star is obviously a binary, however, because its spectral lines shift back and forth, thereby revealing the orbital motions of two stars about their center of mass.

As for visual binaries, spectroscopic binaries allow astronomers to learn about stellar masses. From a radial velocity curve, one can find the *ratio* of the masses of the two stars in a binary. The *sum* of the masses is related to the orbital speeds of the two stars by Kepler's laws and Newtonian mechanics. If both the ratio of the masses and their sum are known, the individual masses can be determined using algebra. However, determining the sum of the masses requires that we know how the binary orbits are tilted from our line of sight. This is because the Doppler shifts reveal only the radial velocities of the stars rather than their true orbital speeds. This tilt is often impossible to determine, because we cannot see the individual stars in the binary. Thus, the masses of stars in spectroscopic binaries tend to be uncertain.

There is one important case in which we can determine the orbital tilt of a spectroscopic binary. If the two stars are observed to eclipse each other periodically, then we must be viewing the orbit nearly edge-on. As we will see next, indi-

vidual stellar masses—as well as other useful data—can be determined if a spectroscopic binary also happens to be such an *eclipsing* binary.

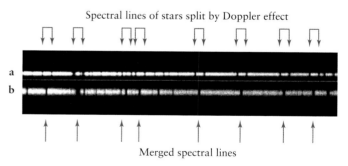

Spectral lines of stars split by Doppler effect

a

b

Merged spectral lines

figure 19-25   R I **V** U X G

A Spectroscopic Binary The visible-light spectrum of the double-line spectroscopic binary κ (kappa) Arietis has spectral lines that shift back and forth as the two stars revolve about each other.
(a) The stars are moving parallel to the line of sight, as in stage 1 or stage 3 in Figure 19-24, producing two sets of spectral lines.
(b) Both stars are moving perpendicular to our line of sight, as in stage 2 or stage 4 in Figure 19-24, and the spectral lines of the two stars merge. (Lick Observatory)

19-11 Light curves of eclipsing binaries provide detailed information about the two stars

Some binary systems are oriented so that the two stars periodically eclipse each other as seen from the Earth. These **eclipsing binaries** can be detected even when the two stars cannot be resolved visually as two distinct images in the telescope. The apparent brightness of the image of the binary dims briefly each time one star blocks the light from the other.

Using a sensitive detector at the focus of a telescope, an astronomer can measure the incoming light intensity quite accurately and create a **light curve**. The shape of the light curve for an eclipsing binary reveals at a glance whether the eclipse is partial or total (compare Figures 19-26*a* and 19-26*b*). Figure 19-26*e* shows an observation of a binary system undergoing a total eclipse.

In fact, the light curve of an eclipsing binary can yield a surprising amount of information. For example, the ratio of the surface temperatures can be determined from how much their combined light is diminished when the stars eclipse each other. Also, the duration of a mutual eclipse tells astronomers about the relative sizes of the stars and their orbits.

a Partial eclipse

b Total eclipse

c Tidal distortion

d Hot-spot reflection

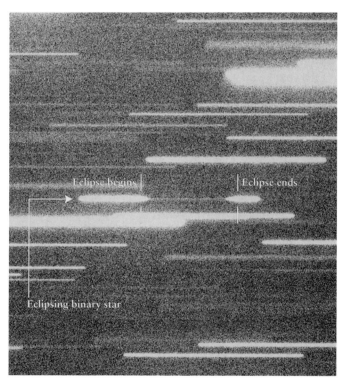

e

R I **V** U X G

figure 19-26

Representative Light Curves of Eclipsing Binaries
The shape of the light curve of an eclipsing binary can reveal many details about the two stars that make up the binary. Illustrated here are **(a)** a partial eclipse, **(b)** a total eclipse, **(c)** tidal distortion, and **(d)** hot-spot reflection. **(e)** This image from the Very Large Telescope shows the binary star NN Serpens (indicated by the arrow) undergoing a total eclipse. The telescope was moved during the exposure so that the sky drifted slowly from left to right across the field of view. During the exposure, the dimmer star of the NN Serpens system (an M6 main-sequence star) passed in front of the other, more luminous star (a white dwarf). During this eclipse, NN Serpens became so dim that it almost disappeared. The overall duration of the eclipse was just 10.5 minutes. (European Southern Observatory)

If the eclipsing binary is also a double-line spectroscopic binary, an astronomer can calculate the mass and radius of each star from the light curves and the velocity curves. Unfortunately, very few binary stars are of this ideal type. Stellar radii determined in this way agree well with the values found using the Stefan-Boltzmann law, as described in Section 19-6.

The shape of a light curve can reveal many additional details about a binary system. In some binaries, for example, the gravitational pull of one star distorts the other, much as the Moon distorts the Earth's oceans in producing tides (see Figure 4-24). Figure 19-26c shows how such tidal distortion affects the light curve. Another example is *hot-spot reflection*, produced when one of the stars in an eclipsing binary is so hot that its radiation creates a hot spot on its cooler companion star. Every time this hot spot is exposed to our view from the Earth, we receive a little extra light energy, which produces a characteristic bump on the binary's light curve (see Figure 19-26d).

Information about stellar atmospheres can also be derived from light curves. Suppose that one star of a binary is a luminous main-sequence star and the other is a bloated red giant. By observing exactly how the light from the bright main-sequence star is gradually cut off as it moves behind the edge of the red giant during the beginning of an eclipse, astronomers can infer the pressure and density in the upper atmosphere of the red giant.

Binary systems are tremendously important because they enable astronomers to measure stellar masses as well as other key properties of stars. In the next several chapters, we will use this information to help us piece together the story of *stellar evolution*—how stars are born, evolve, and eventually die.

KEY WORDS

Terms preceded by an asterisk (*) are discussed in the Boxes.

absolute magnitude, p. 427

apparent brightness (brightness), p. 424

apparent magnitude, p. 427

binary star (binary), p. 441

brown dwarf, p. 434

center of mass, p. 442

color ratio, p. 431

*distance modulus, p. 429

double star, p. 441

eclipsing binary, p. 446

giant, p. 438

Hertzsprung-Russell diagram (H-R diagram), p. 437

inverse-square law, p. 424

light curve, p. 446

luminosity, p. 420

luminosity class, p. 439

luminosity function, p. 426

magnitude scale, p. 426

main sequence, p. 437

main-sequence star, p. 438

mass-luminosity relation, p. 442

metals, p. 434

OBAFGKM, p. 432

optical double star, p. 441

parallax, p. 420

parsec, p. 421

photometry, p. 424

*proper motion, p. 423

radial velocity, pp. 423, 444

radial velocity curve, p. 444

red giant, p. 438

*space velocity, p. 423

spectral classes, p. 432

spectral types, p. 432

spectroscopic binary, p. 444

spectroscopic parallax, p. 440

spectrum binary, p. 444

stellar parallax, p. 420

supergiant, p. 439

*tangential velocity, p. 423

UBV photometry, p. 430

visual binary, p. 441

white dwarf, p. 439

KEY IDEAS

Measuring Distances to Nearby Stars: Distances to the nearer stars can be determined by parallax, the apparent shift of a star against the background stars observed as the Earth moves along its orbit.

• Parallax measurements made from orbit, above the blurring effects of the atmosphere, are much more accurate than those made with Earth-based telescopes.

• Stellar parallaxes can only be measured for stars within a few hundred parsecs.

The Inverse-Square Law: A star's luminosity (total light output), apparent brightness, and distance from the Earth are related by the inverse-square law. If any two of these quantities are known, the third can be calculated.

The Population of Stars: Stars of relatively low luminosity are more common than more luminous stars. Our own Sun is a rather average star of intermediate luminosity.

The Magnitude Scale: The apparent magnitude scale is an alternative way to measure a star's apparent brightness.

• The absolute magnitude of a star is the apparent magnitude it would have if viewed from a distance of 10 parsecs. A version of the inverse-square law relates a star's absolute magnitude, apparent magnitude, and distance.

Photometry and Color Ratios: Photometry measures the apparent brightness of a star. The color ratios of a star are the ratios of brightness values obtained through different standard filters, such as the U, B, and V filters. These ratios are a measure of the star's surface temperature.

Spectral Types: Stars are classified into spectral types (subdivisions of the spectral classes O, B, A, F, G, K, and M), based on the major patterns of spectral lines in their spectra. The spectral class and type of a star is directly related to its surface temperature: O stars are the hottest and M stars are the coolest.

• Most brown dwarfs are in even cooler spectral classes called L and T. Unlike true stars, brown dwarfs are too small to sustain thermonuclear fusion.

Hertzsprung-Russell Diagram: The Hertzsprung-Russell (H-R) diagram is a graph plotting the absolute magnitudes of stars against their spectral types—or, equivalently, their luminosities against surface temperatures.

• The positions on the H-R diagram of most stars are along the main sequence, a band that extends from high luminosity and high surface temperature to low luminosity and low surface temperature.

• On the H-R diagram, giant and supergiant stars lie above the main sequence, while white dwarfs are below the main sequence.

• By carefully examining a star's spectral lines, astronomers can determine whether that star is a main-sequence star, giant, supergiant, or white dwarf. Using the H-R diagram and the inverse-square law, the star's luminosity and distance can be found without measuring its stellar parallax.

Binary Stars: Binary stars, in which two stars are held in orbit around each other by their mutual gravitational attraction, are surprisingly common. Those that can be resolved into two distinct star images by an Earth-based telescope are called visual binaries.

• Each of the two stars in a binary system moves in an elliptical orbit about the center of mass of the system.

• Binary stars are important because they allow astronomers to determine the masses of the two stars in a binary system. The masses can be computed from measurements of the orbital period and orbital dimensions of the system.

Mass-Luminosity Relation for Main-Sequence Stars: The mass-luminosity relation expresses a direct correlation between mass and luminosity for main-sequence stars. The greater the mass of a main-sequence star, the greater its luminosity.

Spectroscopic Observations of Binary Stars: Some binaries can be detected and analyzed, even though the system may be so distant or the two stars so close together that the two star images cannot be resolved.

• A spectrum binary appears to be a single star but has a spectrum with the absorption lines for two distinctly different spectral types.

• A spectroscopic binary has spectral lines that shift back and forth in wavelength. This is caused by the Doppler effect, as the orbits of the stars carry them first toward then away from the Earth.

• An eclipsing binary is a system whose orbits are viewed nearly edge-on from the Earth, so that one star periodically eclipses the other. Detailed information about the stars in an eclipsing binary can be obtained from a study of the binary's radial velocity curve and its light curve.

REVIEW QUESTIONS

1. Describe how the parallax method of finding a star's distance is similar to binocular (two-eye) vision in humans.

2. Why are measurements of stellar parallax difficult to make? What are the advantages of making these measurements from orbit?

3. What is the inverse-square law? Use it to explain why an ordinary lightbulb can appear brighter than a star, even though the lightbulb emits far less light energy per second.

4. Briefly describe how you would determine the luminosity of a nearby star. Of what value is knowing the luminosity of various stars?

5. Why is the magnitude scale called a "backward" scale? What is the difference between apparent magnitude and absolute magnitude?

6. The star Zubenelgenubi (from the Arabic for "scorpion's southern claw") has apparent magnitude +2.75, while the star Sulafat (Arabic for "tortoise") has apparent magnitude +3.25. Which star appears brighter? From this information alone, what can you conclude about the luminosities of these stars? Explain.

7. Would it be possible for a star to appear bright when viewed through a U filter or a V filter, but dim when viewed through a B filter? Explain.

8. Explain why the color ratios of a star are related to the star's surface temperature.

9. Which gives a more accurate measure of a star's surface temperature, its color ratios or its spectral lines? Explain.

10. A fellow student expresses the opinion that since the Sun's spectrum has only weak absorption lines of hydrogen, this element cannot be a major constituent of the Sun. How would you enlighten this person?

11. What are the most prominent absorption lines you would expect to find in the spectrum of a star with a surface temperature of (**a**) 35,000 K, (**b**) 2800 K, and (**c**) 5800 K (like the Sun)? Briefly describe why these stars have such different spectra even though they have essentially the same chemical composition.

12. If a red star and a blue star both have the same radius and both are the same distance from the Earth, which one looks brighter in the night sky? Explain why.

13. Sketch a Hertzsprung-Russell diagram. Indicate the regions on your diagram occupied by (**a**) main-sequence stars, (**b**) red giants, (**c**) supergiants, (**d**) white dwarfs, and (**e**) the Sun.

14. Most of the bright stars in the night sky (see Appendix 5) are giants and supergiants. How can this be, if giants and supergiants make up only 1% of the population of stars?

15. Explain why the dashed lines in Figure 19-15 slope down and to the right.

16. Some giant and supergiant stars are of the same spectral type (G2) as the Sun. What aspects of the spectrum of a G2 star would you concentrate on to determine the star's luminosity class? Explain what you would look for.

17. Suppose that you want to determine the temperature, diameter, and luminosity of an isolated star (not a member of a binary system). Which of these physical quantities require you to know the distance to the star? Explain.

18. What is the mass-luminosity relation? Does it apply to stars of all kinds?

19. Use Figure 19-22 to (**a**) estimate the mass of a main-sequence star that is 1000 times as luminous as the Sun, and (**b**) estimate the luminosity of a main-sequence star whose mass is one-fifth that of the Sun. Explain your answers.

20. How do white dwarfs differ from brown dwarfs? Which are more massive? Which are larger in radius? Which are denser?

21. Sketch the radial velocity curves of a binary consisting of two identical stars moving in circular orbits that are (**a**) perpendicular to and (**b**) parallel to our line of sight.

22. Give two reasons why a visual binary star is unlikely to also be a spectroscopic binary star.

23. Sketch the light curve of an eclipsing binary consisting of two identical stars in highly elongated orbits oriented so that (**a**) their major axes are pointed toward the Earth and (**b**) their major axes are perpendicular to our line of sight.

Questions preceded by an asterisk (*) involve topics discussed in the Boxes.

Problem-solving tips and tools

Look carefully at the worked examples in Boxes 19-1, 19-2, 19-3, and 19-4 before attempting these exercises. For data on the solar system, see Appendices 1 and 2 at the back of this book. Remember that a telescope's light-gathering power is proportional to the area of its objective or primary mirror. The volume of a sphere of radius R is $\frac{4}{3}\pi R^3$. Make use of the H-R diagrams in this chapter to answer questions involving spectroscopic parallax. As Box 19-3 shows, some of the problems concerning magnitudes may require facility with logarithms.

24. Find the average distance from the Sun to Pluto in parsecs. Compared to Pluto, how many times farther away from the Sun is Proxima Centauri?

25. Suppose that a dim star were located 2 million AU from the Sun. Find (**a**) the distance to the star in parsecs and (**b**) the parallax angle of the star. Would this angle be measurable with present-day techniques?

26. Kapteyn's star, named after the Dutch astronomer who found it, has a parallax angle of 0.255 arcsec. How far away is the star?

*27. Kapteyn's star (see the previous question) has a proper motion of 8.67 arcsec per year and a radial velocity of +246 km/s. (**a**) What is the star's tangential velocity? (**b**) What is the star's actual speed relative to the Sun? (**c**) Is Kapteyn's star moving toward the Sun or away from the Sun? Explain.

*28. How far away is a star that has a proper motion of 0.08″/year and a tangential velocity of 40 km/s? For a star at this distance, what would its tangential velocity have to be in order for it to exhibit the same proper motion as Barnard's star (Box 19-1)?

*29. The space velocity of a certain star is 120 km/s and its radial velocity is 48 km/s. Find the star's tangential velocity.

*30. In the spectrum of a particular star, the Balmer line H_α has a wavelength of 656.41 nm. The laboratory value for the wavelength of H_α is 656.28 nm. (**a**) Find the star's radial velocity. (**b**) Is this star approaching us or moving away? Explain. (**c**) Find the wavelength at which you would expect to find H_β in the spectrum of this star, given that the laboratory wavelength of H_β is 486.13 nm. (**d**) Do your answers depend on the distance from the Sun to this star? Why or why not?

*31. Derive the equation given in Box 19-1 relating proper motion and tangential velocity. (*Hint:* See Box 1-1.)

32. How much dimmer does the Sun appear from Neptune than from Earth? (*Hint:* The average distance between a planet and the Sun equals the semimajor axis of the planet's orbit.)

33. Stars *A* and *B* are both equally bright as seen from Earth, but *A* is 60 pc away while *B* is 15 pc away. Which star has the greater luminosity? How many times greater is it?

34. Stars *C* and *D* both have the same luminosity, but *C* is 36 pc from Earth while *D* is 12 pc from Earth. Which star appears brighter as seen from Earth? How many times brighter is it?

35. Suppose two stars have the same apparent brightness, but one star is 12 times farther away than the other. What is the ratio of their luminosities? Which one is more luminous, the closer star or the farther star?

36. The *solar constant*, equal to 1370 W/m², is the amount of light energy from the Sun that falls on 1 square meter of the Earth's surface in 1 second (see Section 19-2). What would the distance between the Earth and the Sun have to be in order for the solar constant to be 1 watt per square meter (1 W/m²)?

37. The star Procyon in Canis Minor (the Small Dog) is a prominent star in the winter sky, with an apparent brightness 1.3×10^{-11} that of the Sun. It is also one of the nearest stars, being only 3.50 parsecs from Earth. What is the luminosity of Procyon? Express your answer as a multiple of the Sun's luminosity.

38. The star HIP 92403 (also called Ross 154) is only 2.97 parsecs from Earth but can be seen only with a telescope, because it is 60 times dimmer than the dimmest star visible to the unaided eye. How close to us would this star have to be in order for it to be visible without a telescope? Give your answer in parsecs and in AU. Compare with the semimajor axis of Pluto's orbit around the Sun.

*39. The star HIP 72509 has an apparent magnitude of +12.1 and a parallax angle of 0.222 arcsecond. (**a**) Determine its absolute magnitude. (**b**) Find the approximate ratio of the luminosity of HIP 72509 to the Sun's luminosity.

*40. Suppose you can just barely see a twelfth-magnitude star through an amateur's 6-inch telescope. What is the magnitude of the dimmest star you could see through a 60-inch telescope?

*41. A certain type of variable star is known to have an average absolute magnitude of 0.0. Such stars are observed in a particular star cluster to have an average apparent magnitude of +14.0. What is the distance to that star cluster?

*42. (**a**) Find the absolute magnitudes of the brightest and dimmest of the labeled stars in Figure 19-6*b*. Assume that all of these stars are 110 pc from Earth. (**b**) If a star in the Pleiades cluster is just bright enough to be seen from Earth

with the naked eye, what is its absolute magnitude? Is such a star more or less luminous than the Sun? Explain.

43. (**a**) On a copy of Figure 19-8, sketch the intensity curve for a blackbody at a temperature of 3000 K. Note that this figure shows a smaller wavelength range than Figure 19-7*a*. (**b**) Repeat part (a) for a blackbody at 12,000 K (see Figure 19-7*c*). (**c**) Use your sketches from parts (a) and (b) to explain why the color ratios b_V/b_B and b_B/b_U are less than 1 for very hot stars but greater than 1 for very cool stars.

*44. Astronomers usually express a star's color using apparent magnitudes. The star's apparent magnitude as viewed through a B filter is called m_B, and its apparent magnitude as viewed through a V filter is m_V. The difference $m_B - m_V$ is called the *B–V color index* ("B minus V"). Is the B–V color index positive or negative for very hot stars? What about very cool stars? Explain your answers.

*45. (See Advanced Question 44.) The B–V color index is related to the color ratio b_V/b_B by the equation

$$ m_B - m_V = 2.5 \log \left(\frac{b_V}{b_B} \right) $$

(**a**) Explain why this equation is correct. (**b**) Use the data in Table 19-1 to calculate the B–V color indices for Bellatrix, the Sun, and Betelgeuse. From your results, describe a simple rule that relates the value of the B–V color index to a star's color.

46. The bright star Rigel in the constellation Orion has a surface temperature about 1.6 times that of the Sun. Its luminosity is about 64,000 $L_\odot$. What is Rigel's radius compared to the radius of the Sun?

47. Suppose a star experiences an outburst in which its surface temperature doubles but its average density (its mass divided by its volume) decreases by a factor of 8. The mass of the star stays the same. By what factors do the star's radius and luminosity change?

48. The Sun experiences solar flares (see Section 18-10). The amount of energy radiated by even the strongest solar flare is not enough to have an appreciable effect on the Sun's luminosity. But when a flare of the same size occurs on a main-sequence star of spectral class M, the star's brightness can increase by as much as a factor of 2. Why should there be an appreciable increase in brightness for a main-sequence M star but not for the Sun?

49. (See Figure 19-12.) What temperature and spectral classification would you give to a star with equal line strengths of hydrogen (H) and neutral helium (He I)? Explain.

50. The bright star Zubeneschmali (β Librae) is of spectral type B8 and has a luminosity of 130 $L_\odot$. What is the star's approximate surface temperature? How does its radius compare to that of the Sun?

51. Castor (α Geminorum) is an A1 V star with an apparent brightness of 4.4×10^{-12} that of the Sun. Determine the approximate distance from the Earth to Castor (in parsecs).

52. A brown dwarf called CoD–33°7795 B has a luminosity of $0.0025L_{\odot}$. It has a relatively high surface temperature of 2550 K, which suggests that it is very young and has not yet had time to cool down by emitting radiation. (**a**) What is this brown dwarf's spectral class? (**b**) Find the radius of CoD–33°7795 B. Express your answer in terms of the Sun's radius and in kilometers. How does this compare to the radius of Jupiter? Is the name "dwarf" justified?

53. The visual binary 70 Ophiuchi (see Figure 19-20) has a period of 87.7 years. The parallax of 70 Ophiuchi is 0.2 arcsec, and the apparent length of the semimajor axis as seen through a telescope is 4.5 arcsec. (**a**) What is the distance to 70 Ophiuchi in parsecs? (**b**) What is the actual length of the semimajor axis in AU? (**c**) What is the sum of the masses of the two stars? Give your answer in solar masses.

54. An astronomer observing a binary star finds that one of the stars orbits the other once every 5 years at a distance of 10 AU. (**a**) Find the sum of the masses of the two stars. (**b**) If the mass ratio of the system (M_1/M_2) is 4.0, find the individual masses of the stars. Give your answers in terms of the mass of the Sun.

DISCUSSION QUESTIONS

*55. As seen from the starship *Enterprise* in the *Star Trek* television series and movies, stars appear to move across the sky due to the starship's motion. How fast would the *Enterprise* have to move in order for a star 1 pc away to appear to move 1° per second? (*Hint:* The speed of the star as seen from the *Enterprise* is the same as the speed of the *Enterprise* relative to the star.) How does this compare with the speed of light? Do you think the stars appear to move as seen from an orbiting space shuttle, which moves at about 8 km/s?

56. From its orbit around the Earth, the Hipparcos satellite could measure stellar parallax angles with acceptable accuracy only if the angles were larger than about 0.002 arcsec. Discuss the advantages or disadvantages of making parallax measurements from a satellite in a large solar orbit, say at the distance of Jupiter from the Sun. If this satellite can also measure parallax angles of 0.002 arcsec, what is the distance of the most remote stars that can be accurately determined? How much bigger a volume of space would be covered compared to the Earth-based observations? How many more stars would you expect to be contained in that volume?

57. It is desirable to be able to measure the radial velocity of stars (using the Doppler effect) to an accuracy of 1 km/s

or better. One complication is that radial velocities refer to the motion of the star relative to the Sun, while the observations are made using a telescope on the Earth. Is it important to take into account the motion of the Earth around the Sun? Is it important to take into account the Earth's rotational motion? To answer this question, you will have to calculate the Earth's orbital speed and the speed of a point on the Earth's equator (the part of the Earth's surface that moves at the greatest speed because of the planet's rotation). If one or both of these effects are of importance, how do you suppose astronomers compensate for them?

 WEB/CD-ROM QUESTIONS

58. Search the World Wide Web for recent discoveries about brown dwarfs. Are all brown dwarfs found orbiting normal stars, as in Figure 7-22, or are they also found in isolation? The Sun experiences flares (see Section 18-10), as do other normal stars; is there any evidence that brown dwarfs also experience flares? If so, is there anything unusual about these flares?

 59. Distances to Stars Using Parallax. Access the Active Integrated Media Module "Using Parallax to Determine Distance" in Chapter 19 of the *Universe* web site or CD-ROM. Use this to determine the distance in parsecs and in light-years to each of the following stars: (**a**) Betelgeuse (parallax $p = 0.00763''$); (**b**) Vega ($p = 0.129''$); (**c**) Antares ($p = 0.00540''$); (**d**) Sirius ($p = 0.379''$).

OBSERVING PROJECTS

Observing tips and tools

Even through a telescope, the colors of stars are sometimes subtle and difficult to see. To give your eye the best chance of seeing color, use the "averted vision" trick: When looking through the telescope eyepiece, direct your vision a little to one side of the star you are most interested in. This places the light from that star on a more sensitive part of your eye's retina.

60. The table at the top of page 452 lists five well-known red stars. It includes their right ascension and declination (celestial coordinates described in Box 2-1), apparent magnitudes, and color ratios. As their apparent magnitudes indicate, all these stars are somewhat variable. Observe at least two of these stars both by eye and through a small telescope. Is the reddish color of the stars readily apparent, especially in contrast to neighboring stars? (The Jesuit priest and astronomer Angelo Secchi named Y Canum

Star	Right ascension	Declination	Apparent magnitude	b_V/b_B
Betelgeuse	5^h 55.2^m	+7° 24′	0.4–1.3	5.5
Y Canum Venaticorum	12 45.1	+45 26	5.5–6.0	10.4
Antares	16 29.4	−26 26	0.9–1.8	5.4
μ Cephei	21 43.5	+58 47	3.6–5.1	8.7
TX Piscium	23 46.4	+3 29	5.3–5.8	11.0

Note: The right ascensions and declinations are given for epoch 2000.

Venaticorum "La Superba," and μ Cephei is often called William Herschel's "Garnet Star.")

61. The table of double stars at the bottom of this page includes vivid examples of contrasting star colors. The table lists the angular separation between the stars of each double. Observe at least four of these double stars through a telescope. Use the spectral types listed to estimate the difference in surface temperature of the stars in each pair you observe. Does the double with the greatest difference in temperature seem to present the greatest color contrast? From what you see through the telescope and on what you know about the H-R diagram, explain why all the cool stars (spectral types K and M) listed are probably giants or supergiants.

62. Observe the eclipsing binary Algol (β Persei), using nearby stars to judge its brightness during the course of an eclipse. Algol has an orbital period of 2.87 days, and, with the onset of primary eclipse, its apparent magnitude drops from 2.1 to 3.4. It remains this faint for about 2 hours. The entire eclipse, from start to finish, takes about 10 hours. Consult the "Celestial Calendar" section of the current issue of *Sky & Telescope* for the predicted dates and times of the minima of Algol. Note that the schedule is given in Universal Time (the same as Greenwich Mean Time), so you will have to convert the time to that of your own time zone. Algol is normally the second brightest star in the constellation of Perseus. Because of its position on the celestial sphere (R.A. = 3^h 08.2^m, Decl. = +40° 57′), Algol is readily visible from northern latitudes during the fall and winter months.

 63. Use the *Starry Night* program to investigate the brightest stars. **(a)** In the Control Panel, click on the "Home" button and set the date to today's date and the time to midnight (12:00:00 A.M.). Display the outlines of the constellations (select **Boundaries, Labels,** and **Astronomical** in the **Constellations** menu). Scroll around the sky and identify at least five of the brighter stars by moving the cursor over them. (Brighter stars are shown as larger dots.) In which constellation does each of the stars you selected lie? Which of the stars you selected are listed in Appendix 5? Of these, which is the most luminous? Which is the most distant? **(b)** In the Control Panel, set the date to six months from today, and again set the time to 12:00:00 A.M. Which of the stars that you selected in part (a) are visible? Which are not? Explain why the passage of six months should make a difference.

Star	Right ascension	Declination	Apparent magnitudes	Angular separation	Spectral types
55 Piscium	0^h 39.9^m	+21° 26′	5.4 and 8.7	6.5″	K0 and F3
γ Andromedae	2 03.9	+42 20	2.3 and 4.8	9.8″	K3 and A0
32 Eridani	3 54.3	− 2 57	4.8 and 6.1	6.8″	G5 and A2
ι Cancri	8 46.7	+28 46	4.2 and 6.6	30.5″	G5 and A5
γ Leonis	10 20.0	+19 51	2.2 and 3.5	4.4″	K0 and G7
24 Comae Berenices	12 35.1	+18 23	5.2 and 6.7	20.3″	K0 and A3
ν Boötis	14 45.0	+27 04	2.5 and 4.9	2.8″	K0 and A0
α Herculis	17 14.6	+14 23	3.5 and 5.4	4.7″	M5 and G5
59 Serpentis	18 27.2	+ 0 12	5.3 and 7.6	3.8″	G0 and A6
β Cygni	19 30.7	+27 58	3.1 and 5.1	34.3″	K3 and B8
δ Cephei	22 29.2	+58 25	4* and 7.5	20.4″	F5 and A0

*The brighter star in the δ Cephei binary system is a variable star of approximately the fourth magnitude.
Note: The right ascensions and declinations are given for epoch 2000.

The Birth of Stars

The stars that illuminate our nights seem eternal and unchanging. But this permanence is an illusion. Each of the stars visible to the naked eye shines due to thermonuclear reactions and has only a finite amount of fuel available for these reactions. Hence, stars cannot last forever: They form from material in interstellar space, evolve over millions or billions of years, and eventually die. In this chapter our concern is with how stars are born and become part of the main sequence.

Stars form within cold, dark clouds of gas and dust that are scattered abundantly throughout our Galaxy. One such cloud appears as a dark area on the right-hand side of the accompanying photograph, which shows a region some 1400 pc (4000 ly) from Earth in the southern constellation Ara (the Altar). Perhaps a dark cloud like this encounters one of the Galaxy's spiral arms, or perhaps a supernova detonates nearby. From the shock of events like these, the cloud begins to contract under the pull of gravity, forming protostars—the fragments that will one day become stars. As a protostar develops, its internal pressure builds and its temperature rises. In time, hydrogen burning can ignite, and a star is born. The hottest, bluest, and brightest young stars, like those in the accompanying image, emit ultraviolet radiation that excites the surrounding interstellar gas. The result is a beautiful glowing nebula, which typically has the red color characteristic of excited hydrogen (as shown here).

(European Southern Observatory)

R I **V** U X G

In Chapters 21 and 22 we will see how stars mature and grow old. Some even blow themselves apart in death throes that enrich interstellar space with the material for future generations of stars. Thus, like the mythical phoenix, new stars arise from the ashes of the old.

As you read the sections of this chapter, look for the answers to the following questions.

20-1 Why do astronomers think that stars evolve?

20-2 What kind of matter exists in the spaces between the stars?

20-3 Where do new stars form?

20-4 What steps are involved in forming a star like the Sun?

20-5 When a star forms, why does it end up with only a fraction of the available matter?

20-6 What do star clusters tell us about the formation of stars?

20-7 Where in the Galaxy does star formation take place?

20-8 How can the death of one star trigger the birth of many other stars?

20-1 Understanding how stars evolve requires both observation and ideas from physics

Over the past several decades, astronomers have labored to develop an understanding of **stellar evolution**: that is, how stars are born, live their lives, and finally die. Our own Sun provides evidence that stars are not permanent. The energy radiated by the Sun comes from thermonuclear reactions in its core, which consume 6×10^{11} kg of hydrogen each second and convert it into helium (Section 18-1). While the amount of hydrogen in the Sun's core is vast, it is not infinite; therefore, the Sun cannot always have been shining, nor can it continue to shine forever. The same is true for all other main-sequence stars, which are fundamentally the same kinds of objects as the Sun but with different masses (Section 19-9). Thus, stars must have a beginning as well as an end.

Stars last very much longer than the lifetime of any astronomer—indeed, far longer than the entire history of human civilization. Thus, it is impossible to watch a single star go through its formation, evolution, and eventual demise. Rather, astronomers have to piece together the evolutionary history of stars by studying different stars at different stages in their life cycles.

ANALOGY To see the magnitude of this task, imagine that you are a biologist from another planet who sets out to understand the life cycles of human beings. You send a spacecraft to fly above the Earth and photograph humans in action. Unfortunately, the spacecraft fails after collecting only 20 seconds of data, but during that time its sophisticated equipment sends back observations of thousands of different humans. From this brief snapshot of life on the Earth—only 10^{-8} (a hundred-millionth) of a typical human lifetime—how would you decide which were the young humans and which were the older ones? Without a look inside our bodies to see the biological processes that shape our lives, could you tell how humans are born and how they die? And how could you

deduce the various biological changes that humans undergo as they age?

Astronomers, too, have data spanning only a tiny fraction of any star's lifetime. A star like the Sun can last for about 10^{10} years, whereas astronomers have been observing stars in detail for only about a century—as in our analogy, roughly 10^{-8} of the life span of a typical star. Astronomers are also frustrated by being unable to see the interiors of stars. For example, we cannot see the thermonuclear reactions that convert hydrogen into helium. But astronomers have an advantage over the biologist in our story. Unlike humans, stars are made of relatively simple substances, primarily hydrogen and helium, that are found almost exclusively in the form of gases. Of the three phases of matter—gas, liquid, and solid—gases are by far the simplest to understand.

Astronomers use our understanding of gases to build theoretical models of the interiors of stars, like the model of the Sun we saw in Section 18-2. Models help to complete the story of stellar evolution. In fact, like all great dramas, the story of stellar evolution can be regarded as a struggle between two opposing and unyielding forces: Gravity continually tries to make a star shrink, while the star's internal pressure tends to make the star expand. When these two opposing forces are in balance, the star is in a state of hydrostatic equilibrium (see Figure 18-3).

But what happens when changes within the star cause either pressure or gravity to predominate? The star must then either expand or contract until it reaches a new equilibrium. In the process, it will change not only in size but also in luminosity and color.

In the following chapters, we will find that giant and supergiant stars are the result of pressure gaining the upper hand over gravity. Both giants and supergiants turn out to be aging stars that have become tremendously luminous and ballooned to hundreds or thousands of times their previous size. White dwarfs, by contrast, are the result of the balance tipping in gravity's favor. These are even older stars that have collapsed to a fraction of the size they had while on the main sequence. In this chapter, however, we will see how the combat between gravity and pressure explains the birth of stars. We start our journey within the diffuse clouds of gas and dust that permeate our galaxy.

20-2 Interstellar gas and dust pervade the Galaxy

Where do stars come from? At first glance, the space between the stars seems to be empty, but there must be some material there from which new stars can form. As we look carefully at interstellar space, we find that it is filled with a thin gas laced with microscopic dust particles. This combination of gas and dust is called the **interstellar medium.** Evidence for this medium includes interstellar clouds of various types, curious lines in the spectra of binary star systems, and an apparent dimming and reddening of distant stars.

You can see evidence for the interstellar medium with the naked eye. Look carefully at the constellation Orion (Figure 20-1a), visible on winter nights in the northern hemisphere and summer nights in the southern hemisphere. While most of the stars in the constellation appear as sharply defined points of light, the middle "star" in Orion's sword

has a fuzzy appearance. This becomes more obvious with binoculars or a telescope. As Figure 20-1b shows, this "star" is actually not a star at all, but the Orion Nebula—a cloud in interstellar space. Any interstellar cloud is called a **nebula** (plural **nebulae**) or **nebulosity.**

The Orion Nebula emits its own light, with the characteristic emission line spectrum of a hot, thin gas. For this reason it is called an **emission nebula.** Many emission nebulae can be seen with a small telescope. Figure 20-2 shows some of these nebulae in a different part of the constellation Orion. Emission nebulae are direct evidence of gas atoms in the interstellar medium.

Typical emission nebulae have masses that range from about 100 to about 10,000 solar masses. Because this mass is spread over a huge volume that is light-years across, the density is quite low by Earth standards, only a few thousand hydrogen atoms per cubic centimeter. (By comparison, the air you are breathing contains more than 10^{19} atoms per cm^3.)

Emission nebulae are found near hot, luminous stars of spectral types O and B. Such stars emit copious amounts of ultraviolet radiation. When atoms in the nearby interstellar gas absorb these energetic ultraviolet photons, the atoms become ionized. Indeed, emission nebulae are composed primarily of ionized hydrogen atoms, that is, free protons

a

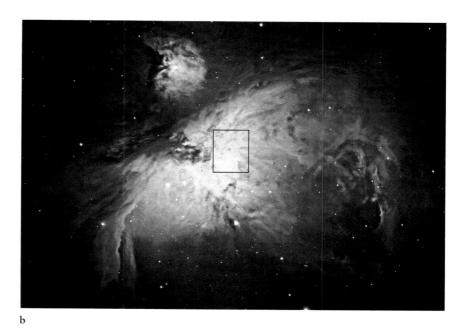

b

figure 20-1 R I **V** U X G

The Orion Nebula (a) This wide-angle telescopic view (angular diameter 20°) shows the constellation Orion. The middle "star" of the three that make up Orion's sword is actually an interstellar cloud called the Orion Nebula. **(b)** This close-up view of the Orion Nebula, which spans about half a degree across the sky, shows its intricate structure. The nebula lies about 450 pc (1500 ly) from Earth and contains about 300 solar masses of material. Within the area shown by the box are four hot, massive stars called the Trapezium. They produce the ultraviolet light that causes the gases to glow. Figure 20-10a shows an even closer view of the area within the box. (a: R. C. Mitchell, Central Washington University; b: Anglo-Australian Observatory)

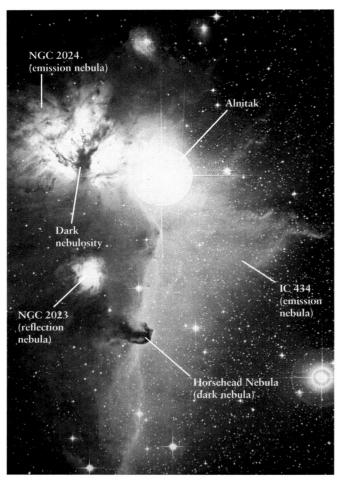

Figure 20-2 R I **V** U X G

Emission, Reflection, and Dark Nebulae in Orion
A variety of different nebulae appear in the sky around Alnitak, the easternmost star in Orion's belt (see Figure 20-1a). To the upper left and lower right of Alnitak are the glowing gases of two red emission nebulae called NGC 2024 and IC 434. Dark streaks of dust obscure part of NGC 2024, while the distinctively shaped dust cloud called the Horsehead Nebula blocks part of the light from IC 434. The Horsehead Nebula is part of a larger complex of dark nebulosity, seen at the lower left corner of the image. In front of the Horsehead Nebula is the reflection nebula NGC 2023, whose dust grains scatter blue light more effectively than red. All the nebulae lie approximately 500 pc (1600 ly) from Earth. They are actually nowhere near Alnitak, which is only 250 pc (820 ly) distant. The area shown in this photograph spans about 1.5° × 2° of the sky. (Royal Observatory, Edinburgh)

(hydrogen nuclei) and electrons. Astronomers use the notation H I for neutral, un-ionized hydrogen atoms and H II for ionized hydrogen atoms, which is why emission nebulae are also called **H II regions.**

H II regions emit visible light when some of the free protons and electrons get back together to form hydrogen atoms, a process called **recombination.** When an atom forms by recombination, the electron is typically captured into a high-energy orbit. As the electron cascades downward through the atom's energy levels toward the ground state, the atom emits visible-light photons. Particularly important is the transition from $n = 3$ to $n = 2$. It produces H_α photons with a wavelength of 656 nm, in the red portion of the visible spectrum (see Section 5-8, especially Figure 5-21). These photons give H II regions their distinctive reddish color.

For each high-energy, ultraviolet photon absorbed by a hydrogen atom to ionize it, several photons of lower energy are emitted when a proton and electron recombine. As Box 20-1 describes, a similar effect takes place in an ordinary fluorescent lightbulb. In this sense, H II regions are immense fluorescent light fixtures!

Further evidence of interstellar gas comes from the spectra of binary star systems. As the two stars that make up a binary system orbit around their common center of mass, their spectral lines shift back and forth (see Figure 19-24). But certain calcium and sodium lines are found to remain at fixed wavelengths. These **stationary absorption lines** are therefore not associated with the binary star. Instead, they must be caused by interstellar gas between us and the binary system.

In addition to the presence of gas atoms in H II regions, Figure 20-2 also shows two kinds of evidence for larger bits of matter, called **dust grains,** in the interstellar medium. These dust grains make their appearance in *dark nebulae* and *reflection nebulae.*

A **dark nebula** is so opaque that it blocks any visible light coming from stars that lie behind it. One such dark nebula, called the Horsehead Nebula for its shape, protrudes in front of one of the H II regions in Figure 20-2. Figure 20-3 shows another dark nebula, called Barnard 86. These nebulae have a relatively dense concentration of microscopic dust grains, which scatter and absorb light much more efficiently than single atoms. Typical dark nebulae have temperatures between 10 K to 100 K, which is low enough for hydrogen atoms to form molecules. Such nebulae contain from 10^4 to 10^9 particles (atoms, molecules, and dust grains) per cubic centimeter. Although tenuous by Earth standards, dark nebulae are large enough—typically many light-years deep—that they block the passage of light. In the same way, a sufficient depth of haze or smoke in our atmosphere can make it impossible to see distant mountains.

The other evidence for dust in Figure 20-2 is the bluish haze surrounding the star immediately above and to the left of the Horsehead Nebula. Figure 20-4 shows a similar haze around a different set of stars. A haze of this kind, called a **reflection nebula,** is caused by fine grains of dust in a lower concentration than that found in dark nebulae. The light we see coming from the nebula is starlight that has been scattered and reflected by these dust grains. The grains are only about 500 nm across, no larger than a typical wavelength of visible light, and they scatter short-wavelength blue light more efficiently than long-wavelength red light. Hence, reflection nebulae have a characteristic blue color. Box 5-4 explains how a similar process—the scattering of sunlight in our atmosphere—gives rise to the blue color of the sky.

Fluorescent Lights

The light that comes from glowing interstellar clouds is, quite literally, otherworldly. But the same principles that explain how such clouds emit light are also at the heart of light phenomena that we see here on the Earth.

A fluorescent lamp produces light in a manner not too different from an emission nebula (H II region). In both cases, the physical effect is called **fluorescence:** High-energy ultraviolet photons are absorbed, and the absorbed energy is re-radiated as lower-energy photons of visible light.

Within the glass tube of a fluorescent lamp is a small amount of the element mercury. When you turn on the lamp, an electric current passes through the tube, vaporizing the mercury and exciting its atoms. This excited mercury vapor emits light with an emission-line spectrum, including lines in the ultraviolet. The white fluorescent coating on the inside of the glass tube absorbs these ultraviolet photons, exciting the coating's molecules to high energy levels.

The molecules then cascade down through a number of lower levels before reaching the ground state. During this cascade, visible-light photons of many different wavelengths are emitted, giving an essentially continuous spectrum and a very white light. (By comparison, the hydrogen atoms in an H II region emit at only certain discrete wavelengths, because the spectrum of hydrogen is much simpler than that of the fluorescent coating's molecules. Another difference is that the molecules in the fluorescent tube never become ionized.)

Many common materials display fluorescence. Among them are teeth, fingernails, and certain minerals; when illuminated with ultraviolet light, these materials glow with a blue or green color. (Most natural history museums and science museums have an exhibit showing fluorescent minerals.) Laundry detergent also contains fluorescent material. After washing, your laundry absorbs ultraviolet light from the Sun and glows faintly, making it appear "whiter than white."

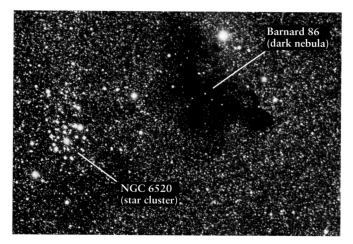

figure 20-3 R I **V** U X G

A Dark Nebula The black area near the top of this image (just below the bright yellow star) is a dark nebula called Barnard 86. When such nebulae were discovered in the late 1700s, they were thought to be "holes in the heavens" where very few stars are present. We now know that dark nebulae are opaque regions that block out light from the stars beyond them. The few stars that appear to be within the dark nebula are between us and the nebula. Barnard 86 is located in the constellation Sagittarius and has an angular diameter of 4 arcminutes, about one-seventh the angular diameter of the full moon. (Anglo-Australian Observatory)

In the 1930s, the American astronomer Robert Trumpler discovered two other convincing pieces of evidence for the existence of interstellar matter—*interstellar extinction* and *interstellar reddening*. While studying the brightness and distances of certain star clusters, Trumpler noticed that remote clusters seem to be dimmer than would be expected from their distance alone. His observations demonstrated that the intensity of light from remote stars is reduced as the light passes through material in interstellar space. This process is called **interstellar extinction.** (In the same way, the headlights of oncoming cars appear dimmer when there is smoke or fog in the air.) The light from remote stars is also reddened as it passes through the interstellar medium, because the blue component of their starlight is scattered and absorbed by interstellar dust. This effect, shown in Figure 20-5, is called **interstellar reddening.** The same effect makes the setting Sun look red (see Box 5-4).

Interstellar reddening is a subtle effect that can be detected only in telescopic images. If a star looks red as seen with the naked eye or through even a moderately large telescope, the cause is not the interstellar medium between that star and Earth. Instead, the red color means that star has a low surface temperature (see Figure 19-7). It is also important to understand the distinction between interstellar reddening and the Doppler effect. If an object is moving away from us, the Doppler effect *shifts* all of that object's light toward longer wavelengths (a redshift). Interstellar reddening,

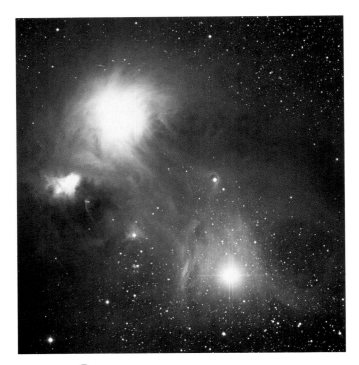

WEB LINK 20.5 **Figure 20-4**   R I **V** U X G

Reflection Nebulae Wispy reflection nebulae called
NGC 6726-27-29 surround several stars in the southern
constellation of Corona Australis (the Southern Crown). Unlike emission
nebulae, reflection nebulae do not emit their own light. Instead, they
scatter and reflect light from the stars that they surround. This scattered
starlight is quite blue in color. The region shown here is about
23 arcminutes across. (David Malin/Anglo-Australian Observatory)

by contrast, makes objects appear red not by shifting wavelengths but by filtering out short wavelengths. The effect is the same as if we looked at an object through red-colored glasses, which let red light pass but block out blue light. The one similarity between Doppler shifts and interstellar reddening is that neither has any discernible effect on what you see with the naked eye.

Measurements of both extinction and reddening in various directions in space indicate that interstellar gas and dust are largely confined to the plane of the Milky Way—that faint, hazy band of stars you can see stretching across the sky on a dark, moonless night. Infrared views like Figure 20-6 clearly show clouds of gas and dust clumped along the Milky Way.

As we will see in Chapter 25, the Milky Way—which appears as a band across the sky—is actually our inside view of our Galaxy, which is a disk-shaped collection of several hundred billion stars about 50,000 pc (160,000 ly) in diameter. The Sun is located about 8000 pc (26,000 ly) from our Galaxy's center. Astronomers know these dimensions because they can map our Galaxy from the locations of bright stars and nebulae and also by using radio telescopes. Radio observations indicate that bright stars, gas, and glowing nebulae are mostly located along arching spiral arms that wind outward from our Galaxy's center. If we could view our Galaxy from a great distance, it would look somewhat like one of the galaxies shown in Figure 20-7.

Interstellar gas and dust are the raw material from which stars are made. The disk of our Galaxy, where most of this matter is concentrated, is therefore the site of ongoing star formation. Our next goal is to examine the steps by which stars are formed.

a

b   R I **V** U X G

WEB LINK 20.6 **Figure 20-5**

Interstellar Reddening (a) As a beam of light from a
distant object passes through interstellar space, tiny dust
grains scatter or absorb blue light but allow red light to pass through
almost undisturbed. Thus, the light beam is deficient in blue light
when it reaches us, and the object appears redder in color than it
really is. This effect is called interstellar reddening. **(b)** The nebula on

the left, NGC 3603, is about 7200 pc (23,000 ly) from Earth, while the
one on the right, NGC 3576, is only about 2400 pc (7800 ly) distant.
Because NGC 3603 is farther away, its light has to pass through more
interstellar dust to reach us, more interstellar reddening occurs, and
NGC 3603 is a ruddier shade of red. The two nebulae are about 1°
apart in the sky. (Anglo-Australian Observatory)

a

R **I** V U X G

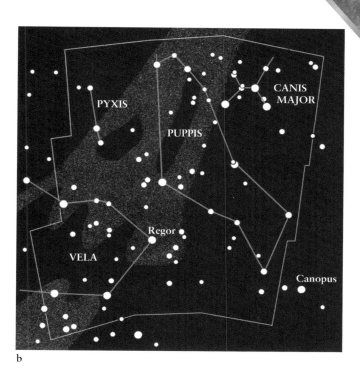

b

figure 20-6

The Infrared Milky Way (a) The dust that occupies the space between the stars is very cold, and so emits only long-wavelength, infrared light (see Section 5-3). This mosaic of infrared images from the Infrared Astronomical Satellite (IRAS) shows light emitted from interstellar dust at a wavelength of 60 μm (colored red in this picture). The blue dots are stars that shine brightly at 12 μm. This mosaic covers a region of the sky approximately 35° × 35°. **(b)** This star chart shows the southern constellations covered by the infrared mosaic. Grayish areas indicate the visible Milky Way. By comparing (a) and (b), you can see that interstellar dust is concentrated within the plane of the Milky Way. Other observations show that the same is true for interstellar gas. (IPAC, NASA)

a

b

figure 20-7   R I **V** U X G

Spiral Galaxies Spiral galaxies, like our own Milky Way Galaxy, consist of stars, gas, and dust that are largely confined to a flattened, rotating disk. **(a)** The galaxy M83 is oriented nearly face-on to us. Spiral arms, delineated by luminous stars and glowing gas, wind outward from the galaxy's center. The reddish knots strung along the spiral arms are H II regions. **(b)** The disk of spiral galaxy NGC 891 is nearly edge-on to our line of sight. The dark band across the midplane is caused by dust in this galaxy's interstellar medium. While M83 and NGC 891 are in different parts of the sky, both are about 7 million pc (23 million ly) from Earth and both have angular diameters of about 13 arcminutes. (a: David Malin/Anglo-Australian Observatory; b: Instituto de Astrofísica de Canarias/Royal Greenwich Observatory/David Malin)

20-3 Protostars form in cold, dark nebulae

How do stars form in the interstellar medium? In this section, we discuss how a contracting cloud gives birth to a clump called a *protostar*—a future main-sequence star.

In order for a star to condense, gravity—which tends to draw interstellar material together—must overwhelm the pressure pushing it apart. This means that stars will most easily form in regions where the interstellar material is relatively dense, so that atoms and dust grains are close together and gravitational attraction is enhanced.

To assist star formation, the pressure of the interstellar medium should be relatively low. This means that the interstellar medium should be as cold as possible, because the pressure of a gas goes down as the gas temperature decreases. (Conversely, increasing the temperature makes the pressure increase. That's why the air pressure inside automobile tires is higher after the auto has been driven a while and the tires have warmed up.)

The only parts of the interstellar medium with high enough density and low enough temperature for stars to form are the dark nebulae. Many of these, such as the Horsehead Nebula (see Figure 20-2) and Barnard 86 (see Figure 20-3), were discovered and catalogued around 1900 by Edward Barnard and are known as **Barnard objects.** (The Horsehead Nebula is also known as Barnard 33.) Other relatively small, nearly spherical dark nebulae are known as **Bok globules,** after the Dutch-American astronomer Bart Bok, who first called attention to them during the 1940s (Figure 20-8). A Bok globule resembles the inner core of a Barnard object with the outer, less dense portions stripped away. The density of the gas and dust within a Barnard object or Bok globule is indeed quite high by cosmic standards, in the range from 100 to 10,000 particles per cm^3; by comparison, most of the interstellar medium contains only 0.1 to 20 particles per cm^3. Radio emissions from molecules within these clouds indicate that their internal temperatures are very low, only about 10 K.

A typical Barnard object contains a few thousand solar masses of gas and dust spread over a volume roughly 10 pc (30 ly) across; a typical Bok globule is about one-tenth as large. The chemical composition of this material is the standard "cosmic abundance" of about 74% (by mass) hydrogen, 25% helium, and 1% heavier elements (review Table 7-3). Within these clouds, the densest portions can contract under their own mutual gravitational attraction and form clumps called **protostars.** Each protostar will eventually evolve into a main-sequence star. Because dark nebulae contain many solar masses of material, it is possible for a large number of protostars to form out of a single such nebula. Thus, we can think of dark nebulae as "stellar nurseries."

In the 1950s, astrophysicists such as Louis Henyey in the United States and C. Hayashi in Japan performed calculations that enabled them to describe the earliest stages of a protostar. At first, a protostar is merely a cool blob of gas several times larger than our solar system. The pressure inside the protostar is too low to support all this cool gas, and so the protostar contracts. As it does, gravitational energy is converted into

thermal energy, making the gases heat up and start glowing. (We discussed this process, called Kelvin-Helmholtz contraction, in Section 7-7 and Section 18-1). Energy from the interior of the protostar is transported outward by convection, warming its surface.

After only a few thousand years of gravitational contraction, the protostar's surface temperature reaches 2000 to 3000 K. At this point the protostar is still quite large, so its glowing gases produce substantial luminosity. (The greater a star's radius and surface temperature, the greater its luminosity. You can review the details in Section 19-6 and Box 19-4.) For example, after only 1000 years of contraction, a protostar of 1 solar mass is 20 times larger in radius than the Sun and has 100 times the Sun's luminosity. Unlike the Sun, however, the luminosity of a young protostar is not the result of thermonuclear fusion, because the protostar's core is not yet hot enough for fusion reactions to begin. Instead, the radiated energy comes exclusively from the heating of the protostar as it contracts.

In order to determine the conditions inside a contracting protostar, astrophysicists use computers to solve equations similar to those used for calculating the structure of the Sun (see Section 18-2). The results tell how the protostar's luminosity and surface temperature change at various stages in its contraction. This information, when plotted on a Hertzsprung-Russell diagram, provides a protostar's **evolutionary track.** The track shows us how the protostar's appearance changes because of changes in its interior. These theoretical tracks agree quite well with actual observations of protostars.

figure 20-8 R I **V** U X G

Bok Globules The black spots on this photograph, which shows a cluster of stars lying in front of a glowing, red H II region, are not imperfections in the image. Rather, they are clouds of dust and gas called Bok globules. A typical Bok globule is a parsec or less in size and contains from one to a thousand solar masses of material. Because of their relatively high density and low temperature, Bok globules are ideal sites for the formation of new stars. The foreground stars are part of a cluster called IC 2944, which lies in the constellation Centaurus (the Centaur). The image shows an area about 8 × 10 arcminutes, about one-third as wide as the apparent size of the full moon. (Anglo-Australian Observatory)

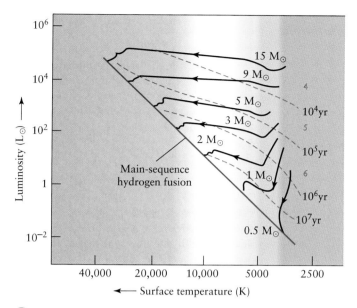

figure 20-9

Pre-Main-Sequence Evolutionary Tracks The evolutionary tracks of seven protostars of different masses are shown on this H-R diagram. The dashed lines indicate the stage reached after the indicated number of years of evolution. All the tracks end on the main sequence, so these protostars all end up as main-sequence stars in whose cores hydrogen is converted into helium. The more massive the protostar, the greater its luminosity and surface temperature when it becomes a main-sequence star. Protostars eject quite a bit of mass into space as they form (as we will see in Section 20-5), so the mass shown for each evolutionary track is that of the final main-sequence star. (Based on stellar model calculations by I. Iben)

 When astronomers refer to a star "following an evolutionary track" or "moving on the H-R diagram," they mean that the star's luminosity, surface temperature, or both change. Hence, the point that represents the star on an H-R diagram changes its position. This is completely unrelated to how the star physically moves through space!

Figure 20-9 shows evolutionary tracks for protostars of seven different masses, ranging from 0.5 $M_\odot$ to 15 $M_\odot$. Because protostars are relatively cool when they begin to shine at visible wavelengths, these tracks all begin near the right (low-temperature) side of the H-R diagram. The subsequent evolution is somewhat different, however, depending on the protostar's mass.

Observing the evolution of a protostar can be quite a challenge. Indeed, while Figure 20-9 shows that young protostars are much more luminous than the Sun, it is quite unlikely that you have ever seen a protostar shining in the night sky. The reason is that protostars form within clouds that contain substantial amounts of interstellar dust. The dust in a protostar's immediate surroundings, called its **cocoon nebula**, absorbs the vast amounts of visible light emitted by the protostar and makes it very hard to detect using visible wavelengths.

WEB LINK 20.10 Protostars can be seen, however, using infrared wavelengths. Because it absorbs so much energy from the protostar that it surrounds, a dusty cocoon nebula becomes heated to a few hundred kelvins. The warmed dust then reradiates its thermal energy at infrared wavelengths, to which the cocoon nebula is relatively transparent. So, by using infrared telescopes, astronomers can see protostars within the "stellar nursery" of a dark nebula.

ANIMATION 20.1 Figure 20-10 shows visible-light and infrared views of a stellar nursery within the Orion Nebula. The visible-light view (Figure 20-10a) shows glowing

figure 20-10

WEB LINK 20.11 **Newborn Stars in the Orion Nebula (a)** This visible-light view shows the inner regions of the Orion Nebula (the boxed area in Figure 20-1b), including the four bright stars called the Trapezium. The area within the irregular outline appears relatively dark and devoid of stars. **(b)** This is an infrared image of the outlined area in (a). The diagonal dimension of the image is about 0.1 parsec (0.4 light-year, or 20,000 AU). Infrared light can penetrate interstellar dust more easily than can visible light, revealing features that cannot be seen in part (a). Emission from young stars and interstellar dust is colored yellow-orange, while emission from excited hydrogen molecules is shown in blue. Several curved arcs are thought to be shock waves caused by material flowing out of protostars faster than the speed of sound waves in the nebula. These shock waves can trigger the formation of even more stars (see Section 20-7). (a: C. R. O'Dell, S. K. Wong, and NASA; b: R. Thompson, M. Rieke, G. Schneider, S. Stolovy, E. Erickson, D. Axon, and NASA) gas clouds that lie in front of a dark, dusty nebula that almost completely blocks visible light. The infrared image (Figure 20-10b) allows us to see through the dust, revealing a number of newly formed protostars within the dark nebula. The properties of these protostars agree well with the theoretical ideas outlined in this section.

20-4 Protostars evolve into main-sequence stars

Evolutionary tracks show how a protostar matures into a star as its gases contract. The details of this evolution depend on the star's mass. To follow these details, remember the basic principle that luminosity is proportional to the square of radius and to the fourth power of surface temperature (see Section 19-6).

For a protostar with the same mass as our Sun (1 M$_\odot$), the outer layers are cool and quite opaque (for the same reasons that the Sun's photosphere is opaque; see Section 18-5). This means that energy released from the shrinking inner layers in the form of radiation cannot reach the surface. Instead, energy flows outward by the slower and less effective method of convection. Hence, the surface temperature of the contracting protostar stays roughly constant, the luminosity decreases as the radius decreases, and the evolutionary track moves downward on the H-R diagram in Figure 20-9.

Although its surface temperature changes relatively little, the internal temperature of the shrinking protostar increases. After a time, the interior becomes ionized, which makes it less opaque. Energy is then conveyed outward by radiation in the interior and by convection in the opaque outer layers, just as in the present-day Sun (see Section 18-2, especially Figure 18-5). This makes it easier for energy to escape from the protostar, so the luminosity increases. As a result, the evolutionary track bends upward (higher luminosity) and to

the left (higher surface temperature, caused by the increased energy flow).

In time, the 1-M$_\odot$ protostar's interior temperature reaches a few million kelvins, hot enough for thermonuclear reactions to begin converting hydrogen into helium. As we saw in Section 18-1 and Box 18-1, these reactions release enormous amounts of energy. Eventually, these reactions provide enough heat and internal pressure to stop the star's gravitational contraction, and hydrostatic equilibrium is reached. The protostar's evolutionary track has now led it to the main sequence, and the protostar has become a full-fledged main-sequence star.

More massive protostars evolve a bit differently. If its mass is more than about 4 M$_\odot$, a protostar contracts and heats more rapidly, and hydrogen burning begins quite early. As a result, the luminosity quickly stabilizes at nearly its final value, while the surface temperature continues to increase as the star shrinks. Thus, the evolutionary tracks of massive protostars traverse the H-R diagram roughly horizontally (signifying approximately constant luminosity) in the direction from right to left (from low to high surface temperature). You can see this most easily for the 9-M$_\odot$ and 15-M$_\odot$ evolutionary tracks in Figure 20-9.

Greater mass means greater pressure and temperature in the interior, which means that a massive star has an even larger temperature difference between its core and its outer layers than the Sun. This causes convection deep in the interior of a massive star. By contrast, a massive star's outer layers are of such low density that energy flows through them more easily by radiation than by convection. Therefore, main-sequence stars with masses more than about 4 M$_\odot$ have convective interiors but radiative outer layers (Figure 20-11a). By contrast, less massive main-sequence stars such as the Sun have radiative interiors and convective outer layers (Figure 20-11b).

The internal structure is also different for main-sequence stars of very low mass. As such a star forms from a protostar,

figure 20-11

Main-Sequence Stars of Different Masses Stellar models show that when a protostar evolves into a main-sequence star, its internal structure depends on its mass. **(a)** If the star's mass is more than about 4 M$_\odot$, energy flows outward from the core by convection in the inner part of the star and by radiation in the outer part. **(b)** If the star's mass is less than about 4 M$_\odot$ but more than about 0.8 M$_\odot$, energy flows by radiation in the interior and by convection in the outer regions. **(c)** Convection takes place throughout the interior of a star less massive than about 0.8 M$_\odot$. Note: The three stars shown here are *not* drawn to scale. Compared with a 1-M$_\odot$ main-sequence star like that shown in (b), a 6-M$_\odot$ main-sequence star like that in (a) has more than 4 times the radius, and a 0.2-M$_\odot$ main-sequence star like that in (c) has only one-third the radius.

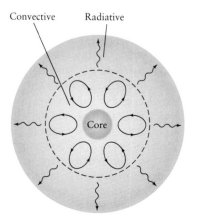

a **Mass more than about 4 M$_\odot$**

b **Mass between about 4 M$_\odot$ and 0.8 M$_\odot$**

c **Mass less than 0.8 M$_\odot$**

the interior temperature is never high enough to ionize the interior. The interior remains too opaque for radiation to flow efficiently, so energy is transported by convection throughout the volume of the star (Figure 20-11*c*).

All the protostar evolutionary tracks in Figure 20-9 end on the main sequence. The main sequence therefore represents stars in which thermonuclear reactions are converting hydrogen into helium. For most stars, this is a stable situation. For example, our Sun will remain on or very near the main sequence, quietly burning hydrogen at its core, for a total of some 10^{10} years. Each evolutionary track ends at a point along the main sequence that agrees with the mass-luminosity relation (recall Figure 19-22). The most massive stars are the most luminous, while the least massive stars are the least luminous.

The theory of how protostars evolve helps explain why the main sequence has both an upper mass limit and a lower mass limit. As we saw in Section 19-5, protostars less massive than about 0.08 $M_\odot$ can never develop the necessary pressure and temperature to start hydrogen burning in their cores. Instead, such "failed stars" end up as *brown dwarfs*, which shine faintly by Kelvin-Helmholtz contraction (see Figure 7-22).

 Protostars with masses greater than about 200 solar masses also do not become main-sequence stars. Such a protostar rapidly becomes very luminous, resulting in tremendous internal pressures. This pressure is so great that it overwhelms the effects of gravity, expelling the outer layers into space and disrupting the star. Main-sequence stars therefore have masses between about 0.08 and 200 $M_\odot$, although the high-mass stars are extremely rare.

Two words of caution are in order here. First, while the evolutionary tracks of protostars begin in the red giant region of the H-R diagram (the upper right), protostars are *not* red giants. As we will see in Chapter 21, red giant stars represent a stage in the evolution of stars that comes *after* being a main-sequence star. Second, it is worth remembering that stars live out most of their lives on the main sequence, after only a relatively brief period as protostars. A 15-$M_\odot$ protostar takes only 20,000 years to become a main-sequence star, and a 1-$M_\odot$ protostar takes about 2×10^7 years. By contrast, the Sun has been a main-sequence star for the past 4.6×10^9 years. By astronomical standards, pre–main-sequence stars are quite transitory.

20-5 During the birth process, stars both gain and lose mass

After reading the previous section, you may think that a main-sequence star forms simply by collapsing inward. In fact, much of the material of a cold, dark nebula is ejected into space and never incorporated into stars. As it is ejected, this material may help sweep away the dust surrounding a young star, making the star observable at visible wavelengths.

Mass ejection into space is a hallmark of **T Tauri stars.** These are protostars with emission lines as well as absorption lines in their spectra and whose luminosity can change irregularly on time scales of a few days. T Tauri stars have masses less than about 3 $M_\odot$ and ages around 10^6 years, so on an H-R diagram such as Figure 20-9 they appear above the right-hand end of the main sequence. The emission lines show that these protostars are surrounded by a thin, hot gas. The Doppler shifts of these emission lines suggest that the protostars eject gas at speeds around 80 km/s (300,000 km/h, or 180,000 mi/h).

On average, T Tauri stars eject about 10^{-8} to 10^{-7} solar masses of material per year. This may seem like a small amount, but by comparison the present-day Sun loses only about 10^{-14} $M_\odot$ per year. The T Tauri phase of a protostar may last 10^7 years or so, during which time the protostar may eject roughly a solar mass of material. Thus, the mass of the final main-sequence star is quite a bit less than that of the cloud of gas and dust from which the star originated. (The stellar masses shown in Figure 20-9 are those of the final, main-sequence stars.)

Young stars that are more massive than about 3 $M_\odot$ do not vary in luminosity like T Tauri stars. They do lose mass, however, because the pressure of radiation at their surfaces is so strong that it blows gas into space. Figure 20-12 shows a young, massive star experiencing mass loss driven by such stellar winds.

 ƒigure 20-12   R **I** V U X G

Mass Loss from a Supermassive Star Astronomers calculate that the star in the middle of this pistol-shaped gas cloud originally had a mass of 100 to 200 $M_\odot$. The structure of the cloud suggests that the gas we see was ejected in two episodes 6000 and 4000 years ago. The gas from previous ejections is now so thinly spread that it cannot be seen. Because this star is about 7500 pc (25,000 ly) from Earth, this image had to be made at infrared wavelengths in order to see through the intervening interstellar medium. (D. F. Figer, NASA)

HH 2

HH 1

Protostar
(hidden by dark, dusty nebula)

figure 20-13 R I **V** U X G

Bipolar Outflow and Herbig-Haro Objects This Hubble Space Telescope image shows two bright knots of glowing, ionized gas. Such Herbig-Haro objects are created when fast-moving gas ejected from a protostar slams into the surrounding interstellar medium, heating the gas to high temperature. These two Herbig-Haro objects, HH 1 and HH 2, are 0.34 parsec (1.1 light-year) apart and lie 470 pc (1500 ly) from Earth in the constellation Orion. (J. Hester, the WFPC-2 Investigation Definition Team, and NASA)

In the early 1980s, it was discovered that many young stars, including T Tauri stars, also lose mass by ejecting gas along two narrow, oppositely directed jets—a phenomenon called **bipolar outflow.** As this material is ejected into space at speeds of several hundred kilometers per second, it collides with the surrounding interstellar medium and produces knots of hot, ionized gas that glow with an emission-line spectrum. These glowing knots are called **Herbig-Haro objects** after the two astronomers, George Herbig in the United States and Guillermo Haro in Mexico, who discovered them independently.

Figure 20-13 is a Hubble Space Telescope image of the Herbig-Haro objects HH 1 and HH 2, which are produced by the two jets from a single young star in the constellation Orion. Herbig-Haro objects like these change noticeably in position, size, shape, and brightness from year to year, indicating the dynamic character of bipolar outflows.

Observations suggest that most protostars eject material in the form of jets at some point during their evolution. These bipolar outflows are very short-lived by astronomical standards, a mere 10^4 to 10^5 years, but they are so energetic that they typically eject into space more mass than ends up in the final protostar.

Protostars slowly add mass to themselves at the same time that they rapidly eject it into space. In fact, the two processes are related. As a protostar's nebula contracts, it spins faster and flattens into a disk with the protostar itself at the center. The same flattening took place in the solar nebula, from which the Sun and planets formed (Section 7-7). Particles orbiting the protostar within this disk collide with each other, causing them to lose energy, spiral inward onto the protostar, and add to the protostar's mass. This is a process of **accretion,**

and the disk of material being added to the protostar in this way is called a **circumstellar accretion disk.** Figure 20-14 is an edge-on view of a circumstellar accretion disk, showing two

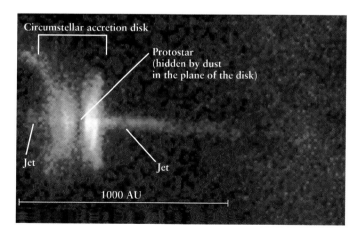

Circumstellar accretion disk

Protostar
(hidden by dust
in the plane of the disk)

Jet

Jet

1000 AU

figure 20-14 R I **V** U X G

A Circumstellar Accretion Disk and Jets This false-color Hubble Space Telescope image shows a star surrounded by an accretion disk, which we see nearly edge-on. Red denotes emission from ionized gas, while green denotes starlight scattered from dust particles in the disk. It appears as though there are two disks around the star, but in fact there is only one—the midplane of this disk is so dusty and opaque that it appears dark. Two oppositely directed jets of ionized gas flow away from the star, perpendicular to the plane of the disk and along the disk's rotation axis. This star lies 140 pc (460 ly) from Earth. (C. Burrows, the WFPC-2 Investigation Definition Team, and NASA)

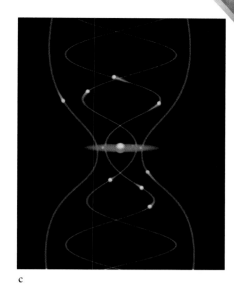

a   b   c

ƒigure 20-15

A Magnetic Model for Bipolar Outflow (a) Observations suggest that circumstellar accretion disks are threaded by magnetic field lines, as shown here. **(b)** As disk material falls onto the central protostar, it pulls the magnetic field lines with it and distorts them into the shape of an hourglass. **(c)** The swirling motions of disk material distort the field lines into helical shapes. Some of the infalling disk material is channeled outward along these helices. (Adapted from Alfred T. Kamajian/Thomas P. Ray, "Fountain of Youth: Early Days in the Life of a Star," *Scientific American*, August 2000)

oppositely directed jets emanating from a point at or near the center of the disk (where the protostar is located).

What causes some of the material in the disk to be blasted outward in a pair of jets? One model involves the magnetic field of the dark nebula in which the star forms (Figure 20-15). As material in the circumstellar accretion disk falls inward, it drags the magnetic field lines along with it. (We saw in Section 18-9 how a similar mechanism may explain the Sun's 22-year cycle.) Parts of the disk at different distances from the central protostar orbit at different speeds, and this can twist the magnetic field lines into two helix shapes, one on each side of the disk. The helices then act as channels that guide infalling material away from the protostar, forming two opposing jets.

Many astronomers suspect that interactions among the protostar, the accretion disk, and the jets help to slow the protostar's rotation. If so, this would explain why main-sequence stars generally spin much more slowly than protostars of the same final mass.

In the 1990s astronomers using the Hubble Space Telescope discovered many examples of disks around newly formed stars in the Orion Nebula (see Figure 20-1), one of the most prominent star-forming regions in the northern sky. Figure 7-13 shows a number of these **protoplanetary disks,** or **proplyds,** that surround young stars within the nebula. As the name suggests, protoplanetary disks are thought to contain the material from which planets form around stars. They are what remains of a circumstellar accretion disk after much of

the material has either fallen onto the star or been ejected by bipolar outflows.

Not all stars are thought to form protoplanetary disks; the exceptions probably include stars with masses in excess of about 3 $M_\odot$, as well as many stars in binary systems. But surveys of the Orion Nebula show that these disks are found with most young, low-mass stars. Thus, disk formation may be a natural stage in the birth of many stars.

20-6 Young star clusters give insight into star formation and evolution

Dark nebulae contain tens or hundreds of solar masses of gas and dust, enough to form many stars. As a consequence, these nebulae tend to form groups or **clusters** of young stars. Such clusters typically include stars with a range of different masses, all of which began to form out of the parent nebula at roughly the same time. Hence, star clusters are valuable laboratories for comparing the evolution of different stars. (In an analogous way, foot races are useful ways to compare the performance of sprinters because all the competitors start the race at the same time.) Star clusters are also objects of great natural beauty. One example is M16, shown in Figure 20-16; another is NGC 6520, depicted in Figure 20-3.

All the stars in a cluster may begin to form nearly simultaneously, but they do not all become main-sequence stars

figure 20-16 R I **V** U X G

A Star Cluster with an H II Region This glowing emission nebula, called the Eagle Nebula because of its shape, surrounds the star cluster M16. The cluster, which is roughly 2200 pc (7000 ly) away, is thought to be no more than 800,000 years old. Star formation is still taking place within dark, dusty globules inside the nebula. The nebula as a whole is about 20 pc (65 ly) across. The inset shows three dense, cold pillars of gas and dust silhouetted against the glowing background of the emission nebula. Hot, luminous stars beyond the upper edge of the image emit copious amounts of ultraviolet radiation, which evaporate away the dark nebula of which the pillars are part. Each of the tiny "fingers" extending from the surface of the pillars has a young star and its cocoon nebula at its tip. Eventually, the cocoon nebula evaporates and the star itself is revealed. The pillar at the upper left extends about 1 light-year from base to tip, and each "finger" is somewhat broader than our entire solar system. (Anglo-Australian Observatory; J. Hester and P. Scowen, Arizona State University; NASA)

Newly revealed stars

Protostars hidden within "fingers" of nebular material

at the same time. As you can see from their evolutionary tracks (see Figure 20-9), high-mass stars evolve more rapidly than low-mass stars. The more massive the protostar, the sooner it develops the central pressures and temperatures needed for steady hydrogen burning to begin.

Upon reaching the main sequence, *high-mass* protostars become hot, ultraluminous stars of spectral types O and B. As we saw in Section 20-2, these are the sorts of stars whose ultraviolet radiation ionizes the surrounding interstellar medium to produce an H II region. Figure 20-16 shows such an H II region, called the Eagle Nebula, surrounding the young star cluster M16. A few hundred thousand years ago, this region of space would have had a far less dramatic appearance. It was then a dark nebula, with protostars just beginning to form. Over the intervening millennia, mass ejection from these evolving protostars swept away the obscuring dust. The exposed young, hot stars heated the relatively thin remnants of the original dark nebula, creating the H II region that we see today.

When the most massive protostars to form out of a dark nebula have reached the main sequence, other *low-mass* protostars are still evolving nearby within their dusty cocoons. The evolution of these low-mass stars can be disturbed by their more massive neighbors. As an example, the inset in Figure 20-16 is a close-up of part of the

Eagle Nebula. Within these opaque pillars of cold gas and dust, protostars are still forming. At the same time, however, the pillars are being eroded by intense ultraviolet light from hot, massive stars that have already shed their cocoons. As each pillar evaporates, the embryonic stars within have their surrounding material stripped away prematurely, limiting the total mass that these stars can accrete.

Star clusters tell us still more about how high-mass and low-mass stars evolve. Figure 20-17 shows the young star cluster NGC 2264 and its associated emission nebula. Astronomers have measured each star's apparent brightness and color ratio. Knowing the distance to the cluster, they have deduced the luminosities and surface temperatures of the stars (see Section 19-2 and Section 19-4). Figure 20-17 shows all these stars on an H-R diagram. Note that the hottest and most massive stars, with surface temperatures around 20,000 K, are on the main sequence. Stars cooler than about 10,000 K, however, have not yet quite arrived at the main sequence. These are less massive stars in the final stages of pre–main-sequence contraction and are just now beginning to ignite thermonuclear reactions at their centers. To find the ages of these stars, we can compare Figure 20-17 with the theoretical calculations of protostar evolution in Figure 20-9. It turns out that this particular cluster is probably about 2 million years old.

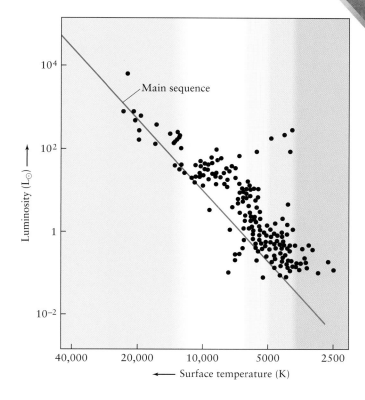

R I **V** U X G

ƒigure 20-17

A Young Star Cluster and Its H-R Diagram The photograph shows an H II region and the young star cluster NGC 2264 in the constellation Monoceros (the Unicorn). It is located about 800 pc (2600 ly) from Earth and contains numerous T Tauri stars. Each dot plotted on the H-R diagram represents a star in this cluster whose luminosity and surface temperature have been determined. Note that most of the cool, low-mass stars (on the right-hand side of the H-R diagram) have not yet arrived at the main sequence. This star cluster probably started forming only 2 million years ago. (Anglo-Australian Observatory)

Figure 20-18 shows another young star cluster called the Pleiades. The photograph shows that the Pleiades must be older than NGC 2264, the cluster in Figure 20-17. The gas that must once have formed an H II region around this cluster has dissipated into interstellar space, leaving only traces of dusty material that forms reflection nebulae around the cluster's stars. The H-R diagram for the Pleiades in Figure 20-18 bears out this idea. In contrast to the H-R diagram for NGC 2264, nearly all the stars in the Pleiades are on the main sequence. The cluster's age is about 50 million years, which is how long it takes for the least massive stars to finally begin hydrogen burning in their cores.

Note that the data points for the most massive stars in the Pleiades (at the upper left of the H-R diagram in Figure 20-18) lie above the main sequence. This is *not* because these stars have yet to arrive at the main sequence. Rather, these stars were the first members of the cluster to arrive at the main sequence some time ago and are now the first members to leave it. They have used up the hydrogen in their cores, so the steady process of core hydrogen burning that characterizes main-sequence stars cannot continue. In

Chapter 21 we will see why massive stars spend a rather short time as main-sequence stars and will study what happens to stars after the main-sequence phase of their lives.

A loose collection of stars such as NGC 2264 or the Pleiades is referred to as an **open cluster** (or *galactic cluster,* since they are usually found in the plane of the Milky Way Galaxy). Such clusters possess barely enough mass to hold themselves together by gravitation. Occasionally, a star moving faster than average will escape, or "evaporate," from a cluster. Indeed, by the time the stars are a few billion years old, they may be so widely separated that a cluster no longer exists.

If a group of stars is gravitationally unbound from the very beginning—that is, if the stars are moving away from one another so rapidly that gravitational forces cannot keep them together—then the group is called a **stellar association.** Because young stellar associations are typically dominated by luminous O and B main-sequence stars, they are also called **OB associations.** The image that opens this chapter shows part of an OB association in the southern constellation Ara (the Altar).

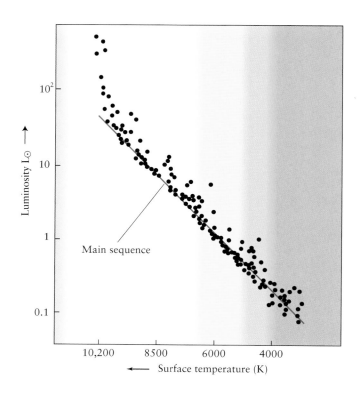

R I **V** U X G

ƒigure 20-18

The Pleiades and Its H-R Diagram The Pleiades star cluster is 117 pc (380 ly) from Earth in the constellation Taurus. It includes some 500 stars, of which a half dozen can be seen with the naked eye. Each dot plotted on the H-R diagram represents a star in the Pleiades whose luminosity and surface temperature have been

measured. (*Note:* The scales on this H-R diagram are different from those in Figure 20-17.) All of the cool, low-mass stars have arrived at the main sequence, indicating that hydrogen burning has begun in their cores. This suggests that the Pleiades is about 50 million (5×10^7) years old. (Anglo-Australian Observatory)

20-7 Star birth can begin in giant molecular clouds

We have seen that star formation takes place within dark nebulae. But where within our Galaxy are these dark nebulae found? Does star formation take place everywhere within the Milky Way, or only in certain special locations? The answers to such questions can enhance our understanding of star formation and of the nature of our home Galaxy.

Dark nebulae are a challenge to locate simply because they *are* dark—they do not emit visible light. Nearby dark nebulae can be seen silhouetted against background stars or H II regions (see Figure 20-2), but sufficiently distant dark nebulae are impossible to see with visible light because of interstellar extinction. They can, however, be detected using longer-wavelength radiation that can pass unaffected through interstellar dust. In fact, dark nebulae actually emit radiation at millimeter wavelengths.

Such emission takes place because in the cold depths of interstellar space, atoms combine to form molecules. The laws of quantum mechanics predict that just as electrons within atoms can occupy only certain specific energy levels (see Section 5-8),

molecules can vibrate and rotate only at certain specific rates. When a molecule goes from one vibrational state or rotational state to another, it either emits or absorbs a photon. (In the same way, an atom emits or absorbs a photon as an electron jumps from one energy level to another.) Most molecules are strong emitters of radiation with wavelengths of around 1 to 10 mm. Consequently, observations with radio telescopes tuned to millimeter wavelengths make it possible to detect interstellar molecules of different types. More than 100 different kinds of molecules have so far been discovered in interstellar space, and the list is constantly growing.

Hydrogen is by far the most abundant element in the universe. Unfortunately, in cold nebulae much of it is in a molecular form (H_2) that is difficult to detect. The reason is that the hydrogen molecule is symmetric, with two atoms of equal mass joined together, and such molecules do not emit many photons at radio frequencies. In contrast, asymmetric molecules that consist of two atoms of unequal mass joined together, such as carbon monoxide (CO), are easily detectable at radio frequencies. When a carbon monoxide molecule makes a transition from one rate of rotation to another, it emits a photon at a wavelength of 2.6 mm or shorter.

The ratio of carbon monoxide to hydrogen in interstellar space is reasonably constant: For every CO molecule, there are about 10,000 H_2 molecules. As a result, carbon monoxide is an excellent "tracer" for molecular hydrogen gas. Wherever astronomers detect strong emission from CO, they know molecular hydrogen gas must be abundant.

The first systematic surveys of our Galaxy looking for 2.6-mm CO radiation were undertaken in 1974 by the American astronomers Philip Solomon and Nicholas Scoville. In mapping the locations of CO emission, they discovered huge clouds, now called **giant molecular clouds,** that must contain enormous amounts of hydrogen. These clouds have masses in the range of 10^5 to 2×10^6 solar masses and diameters that range from about 15 to 100 pc (50 to 300 ly). Inside one of these clouds, there are about 200 hydrogen molecules per cubic centimeter. This is several thousand times greater than the average density of matter in the disk of our Galaxy, but 10^{17} times less dense than the air we breathe. Astronomers now estimate that our Galaxy contains about 5000 of these enormous clouds.

Figure 20-19 is a map of radio emissions from carbon monoxide in the constellations Orion and Monoceros. Note the extensive areas of the sky covered by giant molecular clouds. This part of the sky is of particular interest because it

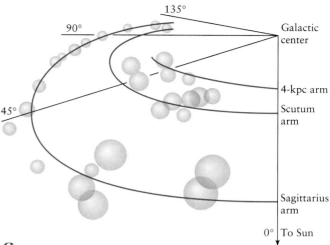

ꜰigure 20-20

Giant Molecular Clouds in the Milky Way This perspective drawing shows the locations of giant molecular clouds in an inner part of our Galaxy as seen from a vantage point above the Sun. These clouds lie primarily along the Galaxy's spiral arms, shown by red arcs. The distance from the Sun to the galactic center is about 8000 pc (26,000 ly). (Adapted from T. M. Dame and colleagues)

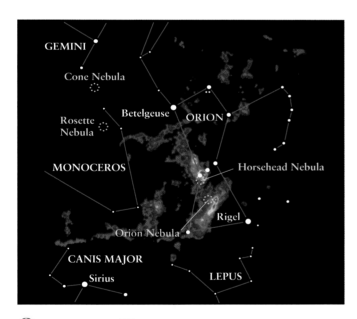

ꜰigure 20-19   R I V U X G

A Carbon Monoxide Map This map of a 35° × 40° section of the sky shows giant molecular clouds in the constellations Orion and Monoceros. To create this map, a radio telescope was tuned to a wavelength of 2.6 mm to detect emissions from carbon monoxide (CO) molecules. Colors in the order of the rainbow show the intensity of emission, from violet for the weakest to red for the strongest. Regions with no detectable emission are shown in black. The star chart overlay shows the locations of four prominent star-forming nebulae. The Orion and Horsehead nebulae are located at sites of intense CO emission, indicating the presence of a particularly dense molecular cloud at these sites of star formation. (Courtesy of R. Maddalena, M. Morris, J. Moscowitz, and P. Thaddeus)

includes several star-forming regions. By comparing the radio map with the star chart overlay, you can see that the areas where CO emission is strongest, and, thus, where giant molecular clouds are densest, are sites of star formation. Therefore, giant molecular clouds are associated with the formation of stars. Particularly dense regions within these clouds form dark nebulae, and within these stars are born.

By using CO emissions to map out giant molecular clouds, astronomers can find the locations in our Galaxy where star formation occurs. These investigations reveal that molecular clouds clearly outline our Galaxy's spiral arms, as Figure 20-20 shows. (Figure 20-7a shows the similar arrangement of spiral arms in another galaxy.) These clouds lie roughly 100 pc (3000 ly) apart, strung along the spiral arms like beads on a string. This arrangement is much like the spacing of H II regions along the arms of other spiral galaxies, such as M83 shown in Figure 20-7a. The presence of both molecular clouds and H II regions shows that spiral arms are sites of ongoing star formation.

In Chapter 25 we will learn that spiral arms are locations where matter "piles up" temporarily as it orbits the center of the Galaxy. You can think of matter in a spiral arm as analogous to a freeway traffic jam. Just as cars are squeezed close together when they enter a traffic jam, a giant molecular cloud is compressed when it passes through a spiral arm. When this happens, vigorous star formation begins in the cloud's densest regions.

As soon as massive O and B stars form, they emit ultraviolet light that ionizes the surrounding hydrogen, and an H II region is born. An H II region is thus a small, bright "hot spot" in a giant molecular cloud. An example is the Orion Nebula, shown in Figure 20-1b. Four hot, luminous O and B

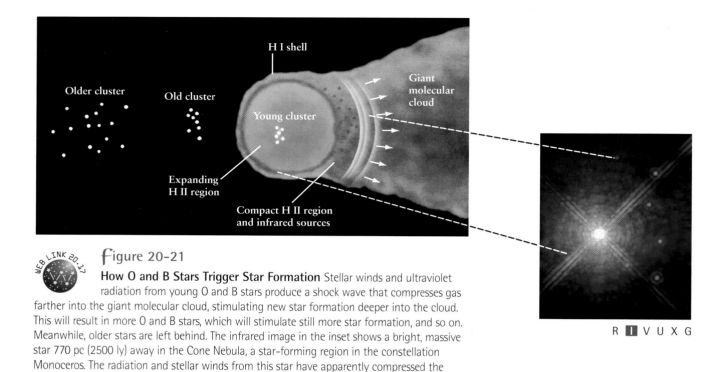

figure 20-21

How O and B Stars Trigger Star Formation Stellar winds and ultraviolet radiation from young O and B stars produce a shock wave that compresses gas farther into the giant molecular cloud, stimulating new star formation deeper into the cloud. This will result in more O and B stars, which will stimulate still more star formation, and so on. Meanwhile, older stars are left behind. The infrared image in the inset shows a bright, massive star 770 pc (2500 ly) away in the Cone Nebula, a star-forming region in the constellation Monoceros. The radiation and stellar winds from this star have apparently compressed the surrounding gas, triggering the formation of several smaller stars (the small dots on the right side of the image). The younger stars are just 0.04 to 0.08 ly from the central star. (Adapted from C. Lada, L. Blitz, and B. Elmegreen; inset: R. Thompson, M. Rieke, G, Schneider, and NASA)

R **I** V U X G

stars at the heart of the nebula produce the ionizing radiation that makes the surrounding gases glow (see Figure 20-10a). The Orion Nebula is embedded in the edge of a giant molecular cloud whose mass is estimated at 500,000 $M_\odot$. The H II regions in Figure 20-2 are located at a different point on the edge of the same molecular cloud, some 25 pc (80 ly) from the Orion Nebula.

Once star formation has begun and an H II region has formed, the massive O and B stars at the core of the H II region induce star formation in the rest of the giant molecular cloud. Ultraviolet radiation and vigorous stellar winds from the O and B stars carve out a cavity in the cloud, and the H II region, heated by the stars, expands into it. These winds travel faster than the speed of sound in the gas—that is, they are **supersonic.** Just as an airplane creates a shock wave (a sonic boom) when it flies faster than sound waves in our atmosphere, a shock wave forms where the expanding H II region pushes at supersonic speed into the rest of the giant molecular cloud. This shock wave compresses the gas through which it passes, stimulating more star birth. Newborn O and B stars further expand the H II region into the giant molecular cloud. Meanwhile, the older O and B stars, which were left behind, begin to disperse (Figure 20-21). In this way, an OB association "eats into" a giant molecular cloud, leaving stars in its wake.

Infrared observations reveal many features that resemble protostars in the swept-up layer immediately behind the shock wave from an OB association. For example, Figure 20-10 shows both optical and infrared views of the core of the Orion Nebula. Four O and B stars, called the Trapezium, and glowing gas and dust dominate the view at visible wavelengths (see Figure 20-10a). Infrared observations penetrate this obscuring material to reveal dozens of infrared objects (see Figure 20-10b). These are thought to be cocoons of warm dust enveloping newly formed stars. Shock waves from the Trapezium may have helped to trigger the formation of these stars. In turn, these newly formed stars generate shock waves of their own, which you can see in Figure 20-10b.

20-8 Supernovae compress the interstellar medium and can trigger star birth

Spiral arms are not the only mechanism for triggering the birth of stars. Presumably, anything that compresses interstellar clouds will do the job. The most dramatic is a *supernova*, caused by the violent death of a massive star after it has left the main sequence. As we will see in Chapter 22, the core of the doomed star collapses suddenly, releasing vast quantities of particles and energy that

WEB LINK 20.18

Figure 20-22

A Supernova Remnant (a) The Cygnus Loop is the remnant of a supernova that occurred about 15,000 years ago. The expanding shock wave from this supernova, located about 800 pc (2,600 ly) from Earth, now has a diameter of about 40 pc (120 ly). The interstellar material that has been compressed by the shock wave is so highly excited that it emits X rays. These were detected by the ROSAT spacecraft to make this false-color image. **(b)** This visible-light image of a portion of the Cygnus Loop is color-coded to show emission from different excited atoms: blue from oxygen, red from sulfur, and green from hydrogen. (a: N. Levenson, MPE, NASA; b: J. Hester, NASA)

a   R I V U **X** G

b   R I **V** U X G

blow the star apart. The star's outer layers are blasted into space at speeds of several thousand kilometers per second.

Astronomers have found many nebulae across the sky that are the shredded funeral shrouds of these dead stars. Such nebulae, like the Cygnus Loop shown in Figure 20-22*a*, are known as **supernova remnants.** Many supernova remnants have a distinctly arched appearance, as would be expected for an expanding shell of gas. This wall of gas is typically moving away from the dead star faster than sound waves can travel through the interstellar medium. As we saw in Section 20-7, such supersonic motion produces a shock wave that abruptly compresses the medium through which it passes. Astronomers have studied the Cygnus Loop with the Hubble Space Telescope to examine in detail a supernova's shock wave plowing through the interstellar medium. Figure 20-22*b* shows the shock wave overrunning dense clumps of gas. This collision heats and compresses the gas, causing it to glow.

When the expanding shell of a supernova remnant slams into an interstellar cloud, it squeezes the cloud, stimulating star birth. This kind of star birth is taking place in the stellar association seen in Figure 20-23. This stellar nursery is located along a luminous arc of gas about 30 pc (100 ly) in length that is presumably the remnant of an ancient supernova explosion. In fact, this arc is part of an almost complete ring of glowing gas with a diameter of about 60 pc (200 ly). Spectroscopic observations of stars along this arc reveal substantial T Tauri activity. This activity results from newborn stars undergoing mass loss in their final stages of contraction before they become main-sequence stars.

By analyzing the composition of meteorites, scientists have found evidence that the Sun was once a member of a loose stellar association created by a supernova. Individual stellar motions soon carry the stars of such an association in various directions away from their birthplaces. Nearly 5 billion years have passed since the birth of our star, so the Sun's brothers and sisters are now widely scattered across the Galaxy.

Many other processes can also trigger star formation. For example, a collision between two interstellar clouds can create new stars. Compression occurs at the interface between the two colliding clouds and vigorous star formation follows. Similarly, stellar winds from a group of O and B stars may exert strong enough pressure on interstellar clouds to cause compression, followed by star formation. This process is taking place in the Rosette Nebula (Figure 20-24).

Our understanding of star birth has improved dramatically in recent years, primarily through infrared- and millimeter-wavelength observations. Nevertheless, many puzzles and mysteries remain. For example, astronomers have generally

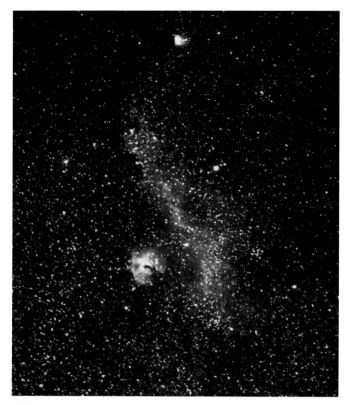

figure 20-23 R I **V** U X G

The Canis Major R1 Association This luminous arc of gas, about 30 pc (100 ly) long, is studded with numerous young stars. Both the luminous arc and the young stars can be traced to the same source, a supernova explosion. The shock wave from the supernova explosion is exciting the gas and making it glow; the same shock wave also compresses the interstellar medium through which it passes, triggering star formation. (Courtesy of H. Vehrenberg)

assumed that there must be a lot of interstellar dust to shield a stellar nursery from the disruptive effects of ultraviolet light from nearby massive, hot stars. However, a neighboring galaxy, called the Large Magellanic Cloud (LMC), contains young OB associations with virtually no dust. Does the process of star birth differ slightly from one galaxy to another?

Another problem is that different modes of star birth tend to produce different percentages of different kinds of stars. For example, the passage of a spiral arm through a giant molecular cloud tends to produce an abundance of massive O and B stars. In contrast, the shock wave from a supernova seems to produce fewer O and B stars but many more of the less massive A, F, G, and K stars. We do not yet know why this is so.

Despite all these unanswered questions, it is now clear that star birth involves mechanisms on a colossal scale, from the deaths of massive stars to the rotation of an entire galaxy. In many respects, we have just begun to appreciate these cosmic processes. The study of cold, dark stellar nurseries will be an active and exciting area of astronomical research for many years to come.

In the chapters that follow, we will learn that the interstellar medium is both the birthplace of new stars and a dumping ground for dying stars. At the end of its life, a star can shed most of its mass in an outburst that enriches interstellar space with new chemical elements. The interstellar medium is therefore both nursery and graveyard. Because of this intimate relationship with stars, the interstellar medium evolves as successive generations of stars live out their lives. Understanding the details of this cosmic symbiosis is one of the challenges of modern astronomy.

figure 20-24 R I **V** U X G

The Core of the Rosette Nebula The Rosette Nebula is a large, circular emission nebula located near one end of a sprawling giant molecular cloud in the constellation Monoceros. Radiation from young, hot stars has blown away the center of this nebula. Some of the remaining gas is clumped in dark globules that appear silhouetted against the glowing background gases. New star formation is taking place within these globules. The entire Rosette Nebula has an angular diameter of 1.3° and lies some 900 pc (3000 ly) from Earth. (Anglo-Australian Observatory)

KEY WORDS

KEY IDEAS

Stellar Evolution: Because stars shine by thermonuclear reactions, they have a finite life span. The theory of stellar evolution describes how stars form and change during that life span.

 The Interstellar Medium: Interstellar gas and dust, which make up the interstellar medium, are concentrated in the disk of the Galaxy. Clouds within the interstellar medium are called nebulae.

• Dark nebulae are so dense that they are opaque. They appear as dark blots against a background of distant stars.

• Emission nebulae, or H II regions, are glowing, ionized clouds of gas. Emission nebulae are powered by ultraviolet light that they absorb from nearby hot stars.

• Reflection nebulae are produced when starlight is reflected from dust grains in the interstellar medium, producing a characteristic bluish glow.

Protostars: Star formation begins in dense, cold nebulae, where gravitational attraction causes a clump of material to condense into a protostar.

• As a protostar grows by the gravitational accretion of gases, Kelvin-Helmholtz contraction causes it to heat and begin glowing. Its relatively low temperature and high luminosity place it in the upper right region on an H-R diagram.

• Further evolution of a protostar causes it to move toward the main sequence on the H-R diagram. When its core temperatures become high enough to ignite steady hydrogen burning, it becomes a main-sequence star.

• The more massive the protostar, the more rapidly it evolves.

Mass Loss by Protostars: In the final stages of pre–main-sequence contraction, when thermonuclear reactions are about to begin in its core, a protostar may eject large amounts of gas into space.

• Low-mass stars that vigorously eject gas are called T Tauri stars.

• A circumstellar accretion disk provides material that a young star ejects as jets. Clumps of glowing gas called Herbig-Haro objects are sometimes found along these jets and at their ends.

Star Clusters: Newborn stars may form an open or galactic cluster. Stars are held together in such a cluster by gravity. Occasionally a star moving more rapidly than average will escape, or "evaporate," from such a cluster.

• A stellar association is a group of newborn stars that are moving apart so rapidly that their gravitational attraction for one another cannot pull them into orbit about one another.

O and B Stars and Their Relation to H II Regions: The most massive protostars to form out of a dark nebula rapidly become main sequence O and B stars. They emit strong ultraviolet radiation that ionizes hydrogen in the surrounding cloud, thus creating the reddish emission nebulae called H II regions.

• Ultraviolet radiation and stellar winds from the O and B stars at the core of an H II region create shock waves that move outward through the gas cloud, compressing the gas and triggering the formation of more protostars.

Giant Molecular Clouds: The spiral arms of our Galaxy are laced with giant molecular clouds, immense nebulae so cold that their constituent atoms can form into molecules.

• Star-forming regions appear when a giant molecular cloud is compressed. This can be caused by the cloud's passage through one of the spiral arms of our Galaxy, by a supernova explosion, or by other mechanisms.

REVIEW QUESTIONS

1. If an interstellar medium fills the space between the stars, how is that we are able to see the stars at all?

2. Summarize the evidence that interstellar space contains (a) gas and (b) dust.

3. What are H II regions? Near what kinds of stars are they found?

4. What are stationary absorption lines? In what sort of spectra are they seen? How do they give evidence for the existence of the interstellar medium?

5. Why is the daytime sky blue? Why are distant mountains purple? Why is the Sun red when seen near the horizon at sunrise or sunset? In what ways are your answers analogous to the explanations for the bluish color of reflection nebulae and the process of interstellar reddening?

6. Why are low temperatures necessary in order for protostars to form inside dark nebulae?

7. Compare and contrast Barnard objects and Bok globules. How many Sun-sized stars could you make out of a Barnard object? Out of a Bok globule?

8. The interior of a dark nebula is billions of times less dense than the air that you breathe. How, then, are dark nebulae able to block out starlight?

9. Describe the energy source that causes a protostar to shine. How does this source differ from the energy source inside a main-sequence star?

10. What is an evolutionary track? How can evolutionary tracks help us interpret the H-R diagram?

11. What happens inside a protostar to slow and eventually halt its gravitational contraction?

12. Why are the evolutionary tracks of high-mass stars different from those of low-mass stars? For which kind of star is the evolution more rapid? Why?

13. In what ways is the internal structure of a 1-$M_\odot$ main-sequence star different from that of a 5-$M_\odot$ main-sequence star? From that of a 0.5-$M_\odot$ main-sequence star? What features are common to all these stars?

14. What sets the limits on the maximum and minimum masses of a main-sequence star?

15. What are Herbig-Haro objects? Why are they often found in pairs?

16. Why do disks form around contracting protostars? What is the connection between disks and bipolar outflows?

17. Young open clusters like those shown in Figures 20-17 and 20-18 are found only in the plane of the Galaxy. Explain why this should be.

18. Why are observations at (a) infrared wavelengths and (b) millimeter wavelengths so much more useful in exploring interstellar clouds than observations at visible wavelengths?

19. What are giant molecular clouds? What role do these clouds play in the birth of stars?

20. Giant molecular clouds are among the largest objects in our Galaxy. Why, then, were they discovered only relatively recently?

21. Consider the following stages in the evolution of a young star cluster: (i) H II region, (ii) dark nebula, (iii) formation of O and B stars, (iv) giant molecular cloud. Put these stages in the correct chronological order and discuss how they are related.

22. Briefly describe four mechanisms that compress the interstellar medium and trigger star formation.

ADVANCED QUESTIONS

Problem-solving tips and tools

You may find it helpful to review Box 19-4, which describes the relationship among a star's luminosity, radius, and surface temperature. Orbital periods are described by Kepler's third law, which we discussed in Boxes 4-2 and 4-4. Remember that the Stefan-Boltzmann law (Box 5-2) relates the temperature of a blackbody to its energy flux. Remember, too, that the volume of a sphere of radius R is $\frac{4}{3}\pi R^3$.

23. The visible-light photograph below shows the Trifid Nebula in the constellation Sagittarius. Label the following features on this photograph: (a) reflection nebulae (and the star or stars whose light is being reflected); (b) dark nebulae; (c) H II regions; (d) regions where star formation may be occurring. Explain how you identified each feature.

(Anglo-Australian Observatory)

R I **V** U X G

24. If you looked at the spectrum of a reflection nebula, would you see absorption lines, emission lines, or no lines? Explain your answer. As part of your explanation, describe how the spectrum demonstrates that the light was reflected from nearby stars.

25. In the direction of a particular star cluster, interstellar extinction allows only 15% of a star's light to pass through each kiloparsec (1000 pc) of the interstellar medium. If the star cluster is 3.0 kiloparsecs away, what percentage of its photons survive the trip to the Earth?

26. Find the density (in atoms per cubic centimeter) of a Bok globule having a radius of 1 light-year and a mass of 100 $M_\odot$. How does your result compare with the density of a typical H II region, between 80 and 600 atoms per cm^3? (Assume that the globule is made purely of hydrogen atoms.)

27. The infrared-bright object at the center of Figure 20-10*b* is called the *Becklin-Neugebauer object* after its discoverers. Like the other bright spots in that image, it is a newly formed star. Assuming that all these stars began the process of formation of the same time, what can you conclude about the mass of the Becklin-Neugebauer object compared with those of the other stars? Does your conclusion depend on whether or not the stars have reached the main sequence? Explain your reasoning.

28. The two false-color images below show a portion of the Trifid Nebula (see Question 23). The upper view is an

R **I** V U X G

(ESA/ISO, ISOCAM, and J. Cernicharo et al.;   R I **V** U X G
IAC, Observatorio del Teide, Tenerife)

infrared image from the Infrared Space Observatory, while the lower picture (shown to the same scale) was made with visible light. Explain why the dark streaks in the visible-light image appear bright in the infrared image.

29. A newly formed protostar and a red giant are both located in the same region on the H-R diagram. Explain how you could distinguish between these two.

30. At one stage during its birth, the protosun had a luminosity of 1000 $L_\odot$ and a surface temperature of about 1000 K. At this time, what was its radius? Express your answer in three ways: as a multiple of the Sun's present-day radius, in kilometers, and in astronomical units.

31. (a) Determine the radius of the circumstellar accretion disk in Figure 20-14. (You will need to measure this image with a ruler. Note the scale bar in this figure.) Give your answer in astronomical units and in kilometers. **(b)** Assume that the young star at the center of this disk has a mass of 1 $M_\odot$. What is the orbital period (in years) of a particle at the outer edge of the disk? **(c)** Using your ruler again, determine the length of the jet that extends to the right of the circumstellar disk in Figure 20-14. At a speed of 200 km/s, how long does it take gas to traverse the entire visible length of the jet?

32. The concentration or abundance of ethyl alcohol in a typical molecular cloud is about 1 molecule per 10^8 cubic meters. What volume of such a cloud would contain enough alcohol to make a martini (about 10 grams of alcohol)? A molecule of ethyl alcohol has 46 times the mass of a hydrogen atom (that is, ethyl alcohol has a molecular weight of 46).

33. (a) Find the angular diameter of the Cygnus Loop, shown in Figure 20-22. **(b)** Assuming that the supernova shock wave that spawned the Cygnus Loop has always traveled outward at the same speed, determine that speed in meters per second and kilometers per hour. **(c)** How long would an Earth astronomer have to wait to see the bright regions of the Cygnus Loop move outward by an angle of 1 arcminute?

DISCUSSION QUESTIONS

34. Some science-fiction movies show stars suddenly becoming dramatically brighter when they are "born" (that is, when thermonuclear fusion reactions begin in their cores). Discuss whether this is a reasonable depiction.

35. Suppose that the electrons in hydrogen atoms were not as strongly attracted to the nuclei of those atoms, so that these atoms were easier to ionize. What consequences might this have for the internal structure of main-sequence stars? Explain your reasoning.

36. What do you think would happen if our solar system were to pass through a giant molecular cloud? Do you think the Earth has ever passed through such clouds?

37. Many of the molecules found in giant molecular clouds are *organic* molecules (that is, they contain carbon). Speculate about the possibility of life-forms and biological processes occurring in giant molecular clouds. In what ways might the conditions existing in giant molecular clouds favor or hinder biological evolution?

38. Speculate on why a shock wave from a supernova seems to produce relatively few high-mass O and B stars, compared to the lower-mass A, F, G, and K stars.

WEB/CD-ROM QUESTIONS

39. In recent years astronomers have been able to learn about the character of the interstellar medium in the vicinity of the Sun. Search the World Wide Web for information about aspects of the nearby interstellar medium, including features called the Local Interstellar Cloud and the Local Bubble. How do astronomers study the nearby interstellar medium? What makes these studies difficult? Is the interstellar medium relatively uniform in our neighborhood, or is it clumpy? If the latter, is our solar system in a relatively thin or thick part of the interstellar medium? How is our solar system moving through the interstellar medium?

40. Search the World Wide Web for recent discoveries about how brown dwarfs form. Do they tend to form in the same locations as "real" stars? Do they form in relatively small or relatively large numbers compared to "real" stars? What techniques are used to make these discoveries?

 41. **Measuring a Stellar Jet.** Access the animation "A Stellar Jet in the Trifid Nebula" in Chapter 20 of the *Universe* web site/CD-ROM. (a) The Trifid Nebula as a whole has an angular diameter of 28 arcmin. By stepping through the animation, estimate the angular size of the stellar jet shown at the end of the animation. (b) The Trifid Nebula is about 2800 pc (9000 ly) from Earth. Estimate the length of the jet in light-years. (c) If gas in the jet travels at 200 km/s, how long does it take to traverse the length of the jet? Give your answer in years. (*Hint:* There are 3.16×10^7 seconds in a year.)

OBSERVING PROJECTS

Observing tips and tools

After looking at the beautiful color photographs of nebulae in this chapter, you may find the view through a telescope a bit disappointing at first, but fear not. You can see a great deal with even a small telescope. To get the best view of a dim nebula using a telescope, use the same "averted vision" trick we described in "Observing tips and tools" in Chapter 19: If you direct your vision a little to one side of the object that you are looking at, the light from that object will go onto a more sensitive part of the retina.

42. Use a telescope to observe at least two of the following H II regions. In each case, can you guess which stars are probably responsible for the ionizing radiation that causes the nebula to glow? Can you see any obscuration or silhouetted features that suggest the presence of interstellar dust? Draw a picture of what you see through the telescope and compare it with a photograph of the object. Take note of which portions of the nebula were not visible through your telescope.

Nebula	Right ascension	Declination
M42 (Orion)	5^h 35.4^m	−5° 27′
M43	5 35.6	−5 16
M20 (Trifid)	18 02.6	−23 02
M8 (Lagoon)	18 03.8	−24 23
M17 (Omega)	18 20.8	−16 11

Note: The right ascensions and declinations are given for epoch 2000.

43. On an exceptionally clear, moonless night, use a telescope to observe at least one of the following dark nebulae. These nebulae are very difficult to find, because they are recognizable only by the absence of stars in an otherwise starry part of the sky. Are you confident that you actually saw the dark nebula? Does the pattern of background stars suggest a particular shape to the nebula?

Nebula	Right ascension	Declination
Barnard 72 (The Snake)	17^h 23.5^m	−23° 38′
Barnard 86	18 02.7	−27 50
Barnard 133	19 06.1	−6 50
Barnard 142 and 143	19 40.7	−10 57

Note: The right ascensions and declinations are given for epoch 2000.

44. A few fine objects cover such large regions of the sky that they are best seen with binoculars. If you have access to a high-quality pair of binoculars, observe the North American Nebula in Cygnus and the Pipe Nebula in Ophiuchus. Both nebulae are quite faint, so you should attempt to observe them only on an exceptionally dark, clear, moonless night. The North America Nebula is a cloud of glowing hydrogen gas located about 3° east of Deneb, the brightest star in Cygnus. While searching for the North America Nebula, you may glimpse another diffuse H II region, the Pelican Nebula, located about 2° southeast of Deneb. The Pipe Nebula is a 7°-long, meandering, dark nebula to the south and to the east of the star θ (theta) Ophiuchi, which is in a section of Ophiuchus that extends southward between the

constellations of Scorpius and Sagittarius. Located about 12° east of the bright red star Antares, you can locate θ Ophiuchi using the *Starry Night* software on the CD-ROM that accompanies this book.

45. Use the *Starry Night* program to investigate a star-forming region. First turn off daylight (select **Daylight** in the **Sky** menu) and show the entire celestial sphere (select **Atlas** in the **Go** menu). Center the field of view on M20, the Trifid Nebula, shown in the figure that accompanies Question 23 (select **Find...** in the **Edit** menu). Using the controls at the right-hand end of the Control Panel, zoom out to the maximum field of view. Then, using the time controls in the Control Panel, step through the months of the year. (**a**) In which month is M20 highest in the sky at noon? Explain how you determined this. (**b**) In which month is M20 highest in the sky at midnight, so that it is best placed for observing with a telescope? Explain how you determined this.

Stellar Evolution: After the Main Sequence

(Anglo-Australian Observatory)

R I V U X G

1 magine a world like Earth, orbiting a star 100 times larger and 2000 times more luminous than our Sun. Bathed in the star's intense light, the surface of this world is hot enough to melt iron. If you could somehow survive on the daytime side of this world, you would see the star filling almost the entire sky. Around the star's edges you would see some of its matter streaming into space at a prodigious rate, like the star in this photograph.

This bizarre planet is not a creation of science fiction—it is our own Earth some 8 billion years from now. The bloated star is our own Sun, which in that remote era will have become a red-giant star.

A main-sequence star evolves into a red giant when all the hydrogen in its core is consumed. The star's core contracts and heats up, but its outer layers expand and cool. In the hot, compressed core, helium burning becomes a new energy source. The more massive a star, the more rapidly it consumes its core's hydrogen and the sooner it evolves into a giant.

The interiors of stars are hidden from our direct view, so much of the story in this chapter is based on theory. We back up those calculations with observations of star clusters, which contain stars of different masses but roughly the same age. Other observations show that some red-giant stars actually pulsate, and that stars can evolve along very different paths if they are part of a binary star system.

As you read the sections of this chapter, look for the answers to the following questions.

21-1   How will our Sun change over the next few billion years?

21-2   Why are red giants larger than main-sequence stars?

21-3   Do all stars evolve into red giants at the same rate?

21-4   How do we know that many stars lived and died before our Sun was born?

21-5   Why do some giant stars pulsate in and out?

21-6   Why do stars in some binary systems evolve in unusual ways?

21-1   When core hydrogen burning ceases, a main-sequence star becomes a red giant

At the core, main-sequence stars are all fundamentally alike. As we saw in Section 20-4, it is in their cores that all such stars convert hydrogen into helium by thermonuclear reactions. This process is called **core hydrogen burning.** The total time that a star will spend burning hydrogen in its core, and thus the total time that it will spend as a main-sequence star, is called its **main-sequence lifetime.** For our Sun, the main-sequence lifetime is about 10 billion (10^{10}) years. Hydrogen burning has been going on in the Sun's core for the past 4.6 billion (4.6×10^9) years, so our Sun is about halfway through its main-sequence lifetime.

What happens to a star after the core hydrogen has been used up, so that it is no longer a main-sequence star? As we will see, it expands dramatically to become a red giant. To understand why this happens, it is useful to first look at how a star evolves *during* its main-sequence lifetime.

A protostar becomes a main-sequence star when steady hydrogen burning begins in its core and it achieves *hydrostatic equilibrium*—a balance between the inward force of gravity and the outward pressure produced by hydrogen burning (see Section 18-2 and Section 20-4). Such a freshly formed main-sequence star is called a **zero-age main-sequence star.**

We make the distinction between "main sequence" and "zero-age main sequence" because a star undergoes noticeable changes in luminosity, surface temperature, and radius during its main-sequence lifetime. These changes are a result of core hydrogen burning, which alters the chemical composition of the core. As an example, when our Sun first formed, its composition was the same at all points throughout its volume: by mass, about 74% hydrogen, 25% helium, and 1% heavy elements. But as Figure 21-1 shows, the Sun's core now contains a greater mass of helium than of hydrogen. (There is still

a

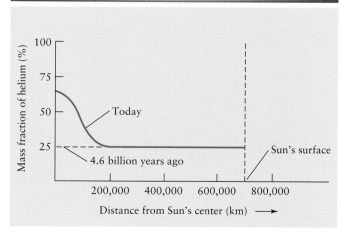

b

ƒigure 21-1

Changes in the Sun's Chemical Composition These graphs show the percentage by mass of hydrogen **(a)** and helium **(b)** at different points within the Sun's interior. In each graph, the solid line shows the situation today and the dashed line shows the situation when our Sun was first formed. Over the past 4.6×10^9 years, thermonuclear reactions at the core have depleted the hydrogen and increased the amount of helium.

box 21-1 | The Heavens on the Earth

Compressing and Expanding Gases

As a star evolves, various parts of the star either contract or expand. When this happens, the gases behave in much the same way as gases here on the Earth when they are forced to compress or allowed to expand.

When a gas is compressed, its temperature rises. You know this by personal experience if you have ever had to inflate a bicycle tire with a hand pump. As you pump, the compressed air gets warm and makes the pump warm to the touch. The same effect happens on a larger scale in southern California during Santa Ana winds or downwind from the Rocky Mountains when there are Chinook winds. Both of these strong winds blow from the mountains down to the lowlands. Even though the mountain air is cold, the winds that reach low elevations can be very hot. (Chinook winds have been known to raise the temperature by as much as 27°C, or 49°F, in only two minutes!) The explanation is compression. Air blown downhill by the winds is compressed by the greater air pressure at lower altitudes, and this compression heats the air.

Expanding gases tend to drop in temperature. When you open a bottle of carbonated beverage, the gases trapped in the bottle expand and cool down. The cooling can be so great that a little cloud forms within the neck of the bottle. Clouds form in the atmosphere in the same way. Rising air cools as it goes to higher altitudes, where the pressure is lower, and the cooling makes water in the air condense into droplets.

Here's an experiment you can do to feel the cooling of expanding gases. Your breath is actually quite warm, as you can feel if you open your mouth wide, hold the back of your hand next to your mouth, and exhale. But if you bring your lips together to form an "o" and again blow on your hand, your breath feels cool. In the second case, your breath has to expand as it passes between your lips to the outside air. That makes its temperature drop.

---

enough hydrogen for another 5×10^9 years or so of core hydrogen burning.)

Thanks to core hydrogen burning, the total number of atomic nuclei in a star's core decreases with time: in each reaction, four hydrogen nuclei are converted to a single helium nucleus (see Figure 18-2 and Box 18-1). With fewer particles bouncing around to provide the core's internal pressure, the core contracts slightly under the weight of the star's outer layers. Compression makes the core denser and increases its temperature. (Box 21-1 gives some everyday examples of how the temperature of a gas changes when it compresses or expands.) As a result of these changes in density and temperature, the pressure in the compressed core is actually higher than before.

As the star's core shrinks, its outer layers expand and shine more brightly. Here's why: As the core's density and temperature increase, hydrogen nuclei in the core collide with one another more frequently, and the rate of core hydrogen burning increases. Hence, the star's luminosity increases. The radius of the star as a whole also increases slightly, because increased core pressure pushes outward on the star's outer layers. The star's surface temperature changes as well, because it is related to the luminosity and radius (see Section 19-6 and Box 19-4). As an example, theoretical calculations indicate that over the past 4.6×10^9 years, our Sun has become 40% more luminous, grown in radius by 6%, and increased in surface temperature by 300 K (Figure 21-2).

As a main-sequence star ages and evolves, the increase in energy outflow from its core also heats the material immediately surrounding the core. As a result, hydrogen burning can begin in this surrounding material. By tapping this fresh supply of hydrogen, a star manages to eke out a few million more years of main-sequence existence.

A star's main-sequence lifetime depends critically on its mass (see Box 21-2 for details). As Table 21-1 shows, massive

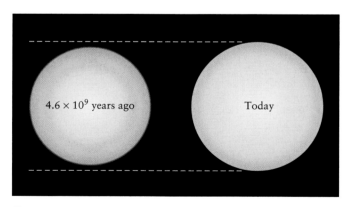

figure 21-2

The Zero-Age Sun and Today's Sun Over the past 4.6×10^9 years, much of the hydrogen in the Sun's core has been converted into helium, the core has contracted a bit, and the Sun's luminosity has gone up by 40%. These changes in the core have made the Sun's outer layers expand in radius by 6% and increased the surface temperature from 5500 K to 5800 K. Figure 21-3a shows how the Sun is expected to change over the next several billion years.

box 21-2 | Tools of the Astronomer's Trade

Main-Sequence Lifetimes

Hydrogen burning converts a portion of a star's mass into energy. We can use Einstein's famous equation relating mass and energy to calculate how long a star will remain on the main sequence.

Suppose that M is the mass of a star and f is the fraction of the star's mass that is converted into energy by hydrogen burning. The total energy E supplied by the hydrogen burning can be expressed as

$$E = fMc^2$$

where c is the speed of light.

This energy is released gradually over many years. If L is the star's luminosity (energy released per unit time) and t is the star's main-sequence lifetime (the total time over which the hydrogen burning occurs), then we can write

$$L = \frac{E}{t}$$

(Actually, this equation is only an approximation. A star's luminosity is not quite constant over its entire main-sequence lifetime. But the variations are not important for our purposes.) We can rewrite this equation as

$$E = Lt$$

From this equation and $E = fMc^2$, we see that

$$Lt = fMc^2$$

which can be rewritten as

$$t = \frac{fMc^2}{L}$$

Thus, a star's lifetime on the main sequence is proportional to its mass divided by its luminosity. Using the symbol $\propto$ to denote "is proportional to," we write

$$t \propto \frac{M}{L}$$

We can carry this analysis further by recalling that main-sequence stars obey the mass-luminosity relation (see Section 19-9, especially Figure 19-22). The distribution of data on the graph in Figure 19-22 tells us that a star's luminosity is roughly proportional to the 3.5 power of its mass:

$$L \propto M^{3.5}$$

Substituting this relationship into the previous proportionality, we find that

$$t \propto \frac{M}{M^{3.5}} = \frac{1}{M^{2.5}} = \frac{1}{M^2\sqrt{M}}$$

This approximate relationship can be used to obtain rough estimates of how long a star will remain on the main sequence. It is often convenient to relate these estimates to the Sun (a typical 1-$M_\odot$ star), which will spend 10^{10} years on the main sequence.

EXAMPLE: Consider a star whose mass is 4 $M_\odot$. This star will be on the main sequence for approximately

$$\frac{1}{4^{2.5}} = \frac{1}{4^2\sqrt{4}} = \frac{1}{32} \text{ solar lifetime}$$

Thus, a 4-$M_\odot$ main-sequence star will burn hydrogen in its core for about $\frac{1}{32} \times 10^{10}$ years, or about 3×10^8 (300 million) years.

table 21-1 | Main-Sequence Lifetimes

Mass ($M_\odot$)	Surface temperature (K)	Spectral class	Luminosity ($L_\odot$)	Main-sequence lifetime (10^6 years)
25	35,000	O	80,000	3
15	30,000	B	10,000	15
3	11,000	A	60	500
1.5	7000	F	5	3000
1.0	6000	G	1	10,000
0.75	5000	K	0.5	15,000
0.50	4000	M	0.03	200,000

stars have short main-sequence lifetimes because they are also very luminous (see Section 19-9, and particularly Figure 19-22). To emit energy so rapidly, these stars must be depleting the hydrogen in their cores at a prodigious rate. Hence, even though a massive O or B main-sequence star contains much more hydrogen fuel in its core than a less-massive M star, it exhausts its hydrogen much sooner. High-mass O and B stars gobble up their hydrogen fuel in only a few million years, but low-mass M stars take hundreds of billions of years to use up their hydrogen. Thus, a main-sequence star's mass determines not only its luminosity and spectral type, but also how long it can remain a main-sequence star.

We saw in Section 20-4 how more massive stars evolve more quickly through the protostar phase to become main-sequence stars (see Figure 20-9). In general, the more massive the star, the more rapidly it goes through *all* the phases of its life. Still, most of the stars we are able to detect are in their main-sequence phase because it lasts so much longer than other luminous phases. (In Chapters 22, 23, and 24 we will explore the final phases of a star's existence, when it ceases to have an appreciable luminosity.)

At the end of a star's main-sequence lifetime, all the hydrogen in its core has been used up and hydrogen burning ceases there. In this new stage, hydrogen burning continues only in the hydrogen-rich material just outside the core, a situation called **shell hydrogen burning**. At first, this process occurs only in the hottest region just outside the core, where the hydrogen fuel has not yet been exhausted.

Strangely enough, the end of core hydrogen burning *increases* the core's temperature. When thermonuclear reac-tions first cease in the core, nothing remains to generate heat there. Hence, the core starts to cool and the pressure there starts to decrease. This pressure decrease allows the star's core to again compress under the weight of the outer layers. As the core contracts, its temperature again increases, and heat begins to flow outward from the core even though no nuclear reactions are taking place there. (Technically, gravitational energy is converted into thermal energy, as in Kelvin-Helmholtz contraction; see Section 7-7 and Section 18-1).

This new flow of heat warms the gases around the core, increasing the rate of shell hydrogen burning and making the shell eat further outward into the surrounding matter. Helium produced by reactions in the shell falls down onto the core, which continues to contract and heat up as it gains mass. Over the course of hundreds of millions of years, the core of a 1-$M_\odot$ star compresses to about one-third of its original radius, while its central temperature increases from about 15 million (1.5×10^7) K to about 100 million (10^8) K.

During this post–main-sequence phase, the star's outer layers expand just as dramatically as the core contracts. As the hydrogen-burning shell burns outward, egged on by heat from the contracting core, the star's luminosity increases substantially. This increases the star's internal pressure and makes the star's outer layers expand to many times their original radius. This tremendous expansion causes those layers to cool down, and the star's surface temperature drops (see Box 21-1). Once the temperature of the star's bloated surface falls to about 3500 K, the gases glow with a reddish hue, in accordance with Wien's law (see Figure 19-7). The star is then appropriately called a **red giant** (Figure 21-3). Thus, we see that red-giant

a

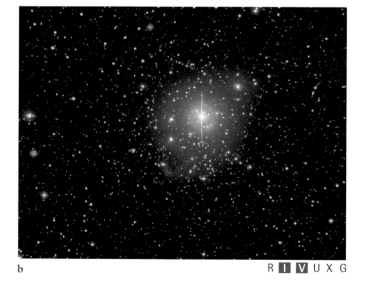

b   R **I** **V** U X G

ƒigure 21-3

Red Giants **(a)** The Sun's energy today is produced in a hydrogen-burning core with a diameter of about 100,000 km. Some 5 billion years from now, when the Sun becomes a red giant, it will draw its energy from a hydrogen-burning shell surrounding a helium-rich core only about 30,000 km in diameter. The contraction of the core converts gravitational energy into thermal energy, heating the hydrogen-burning shell and making the reactions there proceed at a furious rate. As a result, the Sun's luminosity will be about 2000 times greater than today. The increased luminosity will make the Sun's outer layers expand to approximately 100 times their present size. **(b)** This composite of visible and infrared images shows bright red giant stars in the constellation Auriga (the Charioteer). (T. Credner and S. Kohle, Astronomical Institutes of the University of Bonn)

In figure 21-3a the labels read: "The Sun as a main-sequence star (diameter = 1.4×10^6 km ≈ $\frac{1}{100}$ AU)" and "The Sun as a red giant (diameter ≈ 1 AU)".

stars are former main-sequence stars that have evolved into a new stage of existence.

Red-giant stars undergo substantial **mass loss** because of their large diameters and correspondingly weak surface gravity. This makes it relatively easy for gases to escape from the red giant into space. Mass loss can be detected in a star's spectrum, because gas escaping from a red giant toward a telescope on the Earth produces narrow absorption lines that are slightly blueshifted by the Doppler effect (review Figure 5-24). Typical observed blueshifts correspond to a speed of about 10 km/s. A typical red giant loses roughly 10^{-7} $M_\odot$ of matter per year. For comparison, the Sun's present-day mass loss rate is only 10^{-14} $M_\odot$ per year. Hence, an evolving star loses a substantial amount of mass as it becomes a red giant. (The photograph that opens this chapter shows a star losing mass in this way.)

In about 5 billion years, our Sun will finish converting hydrogen into helium at its core. As the Sun's core contracts, its atmosphere will expand to envelop Mercury and reach most of the way to Venus. Three billion years after leaving the main sequence, our red-giant Sun will have swollen to a diameter of about 1 AU—roughly 100 times larger than its present size—and its surface temperature will have dropped to about 3500 K. Shell hydrogen burning will proceed at such a furious rate that our star will shine with the brightness of 2000 present-day Suns. Some of the inner planets will be vaporized, and the thick atmospheres of the outer planets will boil away to reveal tiny, rocky cores. Thus, in its later years, the aging Sun will destroy the planets that have accompanied it since its birth.

21-2 Helium burning begins at the center of a red giant

When a star first becomes a red giant, its hydrogen-burning shell surrounds a small, compact core of almost pure helium. In a moderately low-mass red giant, which the Sun will become

5 billion years from now, the dense helium core is about twice the diameter of the Earth.

Helium, the "ash" of hydrogen burning, is a potential nuclear fuel: **Helium burning,** the thermonuclear fusion of helium nuclei to make heavier nuclei, releases energy. But this reaction does not take place within the core of our present-day Sun because the temperature there is too low. Each helium nucleus contains two protons, so it has twice the positive electric charge of a hydrogen nucleus, and there is a much stronger electric repulsion between two helium nuclei than between two hydrogen nuclei. For helium nuclei to overcome this repulsion and get close enough to fuse together, they must be moving at very high speeds, which means that the temperature of the helium gas must be very high. (For more on the relationship between the temperature of a gas and the speed of atoms in the gas, see Box 7-2.)

When a star first becomes a red giant, the temperature of the contracted helium core is still too low for helium nuclei to fuse. But as the hydrogen-burning shell adds mass to the helium core, the core contracts even more, further increasing the star's central temperature. When the central temperature finally reaches 100 million (10^8) K, **core helium burning**—that is, thermonuclear fusion of helium in the core—begins. As a result, the aging star again has a central energy source for the first time since leaving the main sequence (Figure 21-4).

Helium burning occurs in two steps. First, two helium nuclei combine to form an isotope of beryllium:

$$^4\text{He} + {}^4\text{He} \rightarrow {}^8\text{Be}$$

This particular beryllium isotope is very unstable and breaks into two helium nuclei soon after it forms. However, in the star's dense core a third helium nucleus may strike the ^{8}Be nucleus before it has a chance to fall apart. Such a collision creates a stable isotope of carbon and releases energy in the form of a gamma-ray photon (γ):

$$^8\text{Be} + {}^4\text{He} \rightarrow {}^{12}\text{C} + \gamma$$

This process of fusing three helium nuclei to form a carbon nucleus is called the **triple alpha process,** because helium

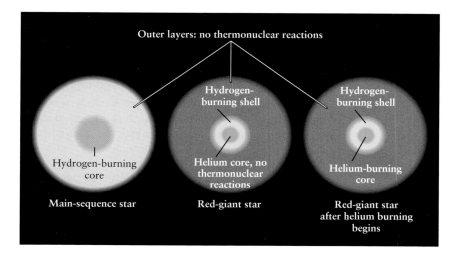

Outer layers: no thermonuclear reactions

Hydrogen-burning shell

Hydrogen-burning shell

Hydrogen-burning core

Helium core, no thermonuclear reactions

Helium-burning core

Main-sequence star

Red-giant star

Red-giant star after helium burning begins

figure 21-4

Stages in Stellar Evolution As a star evolves, thermonuclear fusion occurs at different locations within its interior. In a main-sequence star (left), hydrogen is converted into helium in the core. When the core hydrogen is exhausted, fusion continues in a hydrogen-burning shell, and the star expands to become a red giant (center). When the temperature in the core becomes high enough because of contraction, core helium burning begins (right). (These three pictures are *not* drawn to scale. The star is much larger in its red-giant phase than in its main-sequence phase, then shrinks somewhat when core helium burning begins.)

nuclei are also called **alpha particles** by nuclear physicists. Some of the carbon nuclei created in this process can fuse with an additional helium nucleus to produce a stable isotope of oxygen and release more energy:

$$^{12}C + {}^4He \rightarrow {}^{16}O + \gamma$$

Thus, both carbon and oxygen make up the "ash" of helium burning. It is interesting to note that ^{12}C and ^{16}O are the most common isotopes of carbon and oxygen, respectively; the vast majority of the carbon atoms in your body are ^{12}C, and almost all the oxygen you breathe is ^{16}O. We will discuss the significance of this in the following section.

The second step in the triple alpha process and the process of oxygen formation both release energy. The onset of these reactions reestablishes thermal equilibrium and prevents any further gravitational contraction of the star's core. A mature red giant burns helium in its core for about 20% as long as the time it spent burning hydrogen as a main-sequence star. For example, in the distant future the Sun will consume helium in its core for about 2 billion years.

How helium burning begins at a red giant's center depends on the mass of the star. In high-mass stars (greater than 2 to 3 $M_\odot$), helium burning begins gradually as temperatures in the star's core approach 10^8 K. In low-mass stars (less than 2 to 3 $M_\odot$), helium burning begins explosively and suddenly, in what is called the **helium flash.** Table 21-2 summarizes these differences.

The helium flash occurs because of unusual conditions that develop in the core of a low-mass star as it becomes a red giant. To appreciate these conditions we must first understand how an ordinary gas behaves. Then we can explore how the densely packed electrons at the star's center alter this behavior.

When a gas is compressed, it usually becomes denser and warmer. To describe this process, scientists use the convenient concept of an **ideal gas**, which has a simple relationship between pressure, temperature, and density. Specifically, the pressure exerted by an ideal gas is directly proportional to both the density and the temperature of the gas. Many real gases actually behave like an ideal gas over a wide range of temperatures and densities.

Under most circumstances, the gases inside a star act like an ideal gas. If the gas is compressed, it heats up, and if it expands, it cools down (see Box 21-1). This behavior serves as a safety valve, ensuring that the star does not explode. For example, if energy production overheats the star's core, the

core expands, cooling the gases and slowing the rate of thermonuclear reactions. Conversely, if too little energy is being created to support the star's overlying layers, the core compresses, increasing the temperature and thus speeding up the thermonuclear reactions to increase the energy output.

In a low-mass red giant, however, the core behaves very differently from an ideal gas. The core must be compressed tremendously in order to become hot enough for helium burning to begin. At these extreme pressures and temperatures, the atoms are completely ionized, and most of the core consists of nuclei and detached electrons. Eventually, the free electrons become so closely crowded that a limit to further compression is reached, as predicted by a remarkable law of quantum mechanics called the **Pauli exclusion principle.** Formulated in 1925 by the Austrian physicist Wolfgang Pauli, this principle states that two electrons cannot simultaneously occupy the same quantum state. A quantum state is a particular set of circumstances concerning locations and speeds that are available to a particle. In the submicroscopic world of electrons, the Pauli exclusion principle is analogous to saying that you can't have two things in the same place at the same time.

Just before the onset of helium burning, the electrons in the core of a low-mass star are so closely crowded together that any further compression would violate the Pauli exclusion principle. Because the electrons cannot be squeezed any closer together, they produce a powerful pressure that resists further core contraction.

This phenomenon, in which closely packed particles resist compression as a consequence of the Pauli exclusion principle, is called **degeneracy.** Astronomers say that the electrons in the helium-rich core of a low-mass red giant are "degenerate," and that the core is supported by **degenerate-electron pressure.** This degenerate pressure, unlike the pressure of an ideal gas, does not depend on temperature. Remarkably, you can find degenerate electrons on Earth in an ordinary piece of metal (Figure 21-5).

When the temperature in the core of a low-mass red giant reaches the high level required for the triple alpha process, energy begins to be released. The helium heats up, which makes the triple alpha process happen even faster. However, the pressure provided by the degenerate electrons is independent of the temperature, so the pressure does not change. Without the "safety valve" of increasing pressure, the star's core cannot expand and cool. The rising temperature causes the helium to burn at an ever-increasing rate, producing the helium flash.

Eventually, the temperature becomes so high that the electrons in the core are no longer degenerate. The electrons then behave like an ideal gas and the star's core expands, terminating the helium flash. These events occur so rapidly that the helium flash is over in seconds, after which the star's core settles down to a steady rate of helium burning.

 The term "helium flash" might give you the impression that a star emits a sudden flash of light when the helium flash occurs. If this were true, it would be an incredible sight. During the brief time interval when the helium flash occurs, the helium-burning core is 10^{11} times

Table 21-2	How Helium Burning Begins in Different Red Giants

Mass of star	Onset of helium burning in core
Less than 2–3 solar masses	Explosive (helium flash)
More than 2–3 solar masses	Gradual

figure 21-5   R I **V** U X G

Degenerate Electrons The electrons in a piece of metal, like the chrome grille on this classic car, are so close together that they are affected by the Pauli exclusion principle. The resulting degenerate-electron pressure helps make metals strong and difficult to compress. A more powerful version of this same effect happens inside the cores of low-mass red-giant stars. (Santokh Kochar/PhotoDisc)

more luminous than the present-day Sun, comparable to the total luminosity of all the stars in the Milky Way Galaxy! But in fact the helium flash has no immediately visible consequences—for two reasons. First, much of the energy released during the helium flash goes into heating the core and terminating the degenerate state of the electrons. Second, the energy that does escape the core is largely absorbed by the star's outer layers, which are quite opaque (just like the Sun's present-day interior; see Section 18-2). Therefore, the explosive drama of the helium flash takes place where it cannot be seen directly.

Whether a helium flash occurs or not, the onset of core helium burning actually causes a *decrease* in the luminosity of the star. This is the opposite of what you might expect—after all, turning on a new energy source should make the luminosity greater, not less. What happens is that after the onset of core helium burning, a star's superheated core expands like an ideal gas. (If the star is of sufficiently low mass to have had a degenerate core, the increased temperature after the helium flash makes the core too hot to remain degenerate. Hence, these stars also end up with cores that behave like ideal gases.) Temperatures drop around the expanding core, so the hydrogen-burning shell reduces its energy output and the star's luminosity decreases. This allows the star's outer layers to contract and heat up. Consequently, a post–helium-flash star is less luminous, hotter at the surface, and smaller than a red giant.

Core helium burning lasts for only a relatively short time. Calculations suggest that a 1-$M_\odot$ star like the Sun sustains core hydrogen burning for about 10 billion (10^{10}) years, followed by about 3 billion (3×10^9) years of shell hydrogen burning leading up to the helium flash. After the helium flash,

such a star can burn helium in its core for only 100 million (10^8) years, a mere 1% of its main-sequence lifetime. In Chapter 22 we will take up the story of what happens after a star has consumed all the helium in its core.

Here is the story of post–main-sequence evolution in its briefest form: Before the beginning of core helium burning, the star's core compresses and the outer layers expand, and just after core helium burning begins, the core expands and the outer layers compress. We will see in Chapter 22 that this behavior, in which the inner and outer regions of the star change in opposite ways, occurs again and again in the final stages of a star's evolution.

21-3 H-R diagrams and observations of star clusters reveal how red giants evolve

To see how stars evolve during and after their main-sequence lifetimes, it is helpful to follow them on a Hertzsprung-Russell (H-R) diagram. On such a diagram, zero-age main-sequence stars lie along a line called the **zero-age main sequence,** or **ZAMS** (Figure 21-6*a*). These stars have just emerged from their protostar stage, are burning hydrogen steadily in their cores, and have attained hydrostatic equilibrium. With the passage of time, hydrogen in a main-sequence star's core is converted to helium, the luminosity slowly increases, the star slowly expands, and the star's position on the H-R diagram inches away from the ZAMS. As a result, the main sequence on an H-R diagram is a fairly broad band rather than a narrow line (Figure 21-6*b*).

The dashed line in Figure 21-6*a* denotes stars whose cores have been exhausted of hydrogen and in which core hydrogen burning has ceased. From there, the points representing high-mass stars (3 $M_\odot$, 5 $M_\odot$, and 9 $M_\odot$) move rapidly from left to right across the H-R diagram. This means that, although the star's surface temperature is decreasing, its surface area is increasing at a rate that keeps its overall luminosity roughly constant. During this transition, the star's core contracts and its outer layers expand as energy flows outward from the hydrogen-burning shell.

Just before core helium burning begins, the evolutionary tracks of high-mass stars turn upward in the red-giant region of the H-R diagram (to the upper right of the main sequence). After core helium burning begins, however, the cores of these stars expand, the outer layers contract, and the evolutionary tracks back away from these temporary peak luminosities. The tracks then wander back and forth in the red-giant region while the stars readjust to their new energy sources.

Figure 21-6*a* also shows the evolutionary tracks of two low-mass stars (1 $M_\odot$ and 2 $M_\odot$). The onset of core helium burning in these stars occurs with a helium flash, indicated by the red asterisks in the figure. As we saw in the previous section, after the helium flash, these stars shrink and become less luminous. The decrease in size is proportionally greater than the decrease in luminosity, and so the surface temperatures increase. Hence, after the helium flash, the evolutionary tracks move down and to the left.

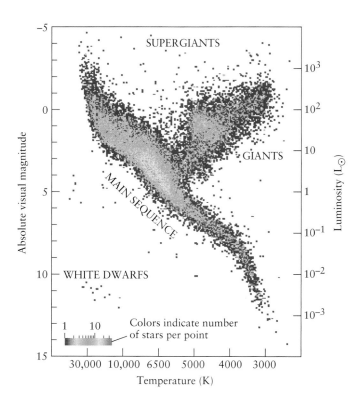

figure 21-6

Stellar Evolution On and Off the Main Sequence **(a)** This H-R diagram shows the post–main-sequence evolutionary tracks of five stars of different mass. In the high-mass stars, core helium burning ignites where the evolutionary tracks make a sharp downward turn in the red-giant region of the diagram. The two lowest-mass stars shown here (1 M$_\odot$ and 2 M$_\odot$) undergo a helium flash at their centers, as shown by the asterisks. **(b)** This H-R diagram shows data collected by the Hipparcos satellite (see Section 19-1) for 20,853 individual stars. The thickness of the main sequence is due in large part to stars following evolutionary tracks like those shown in (a). (a: Adapted from I. Iben; b: Adapted from M. A. C. Perryman)

We can summarize our understanding of stellar evolution from birth through the onset of helium burning by following the evolution of a hypothetical cluster of stars. The eight H-R diagrams in Figure 21-7 are from a computer simulation of the evolution of 100 stars that differ only in initial mass. All the stars, which in this simulation all form at the same moment, start off as cool protostars on the right side of the H-R diagram (see Figure 21-7*a*). The protostars are spread out on the diagram according to their masses, and the greater the mass, the greater the protostar's initial luminosity. As we saw in Section 20-3, the source of a protostar's luminosity is its gravitational energy. As the protostar contracts, this is converted to thermal energy and radiated into space.

The most massive protostars contract and heat up very rapidly, and after only 5000 years, they have already moved across the H-R diagram toward the main sequence (see Figure 21-7*b*). After 100,000 years, these massive stars have ignited hydrogen burning in their cores and have settled down

on the main sequence as O stars (see Figure 21-7*c*). After 3 million years, stars of moderate mass have also ignited core hydrogen burning and become main-sequence stars of spectral classes B and A (see Figure 21-7*d*). Meanwhile, low-mass protostars continue to inch their way toward the main sequence as they leisurely contract and heat up.

After 30 million years (see Figure 21-7*e*), the most massive stars have depleted the hydrogen in their cores and become red giants. These stars have moved from the upper left end of the main sequence to the upper right corner of the H-R diagram. (This simulation follows stars only to the red-giant stage, after which they are simply deleted from the diagram.) Intermediate-mass stars lie on the main sequence, while the lowest-mass stars are still in the protostar stage and lie above the main sequence.

After 66 million years (see Figure 21-7*f*), even the lowest-mass protostars have finally ignited core hydrogen burning and have settled down on the main sequence as cool, dim,

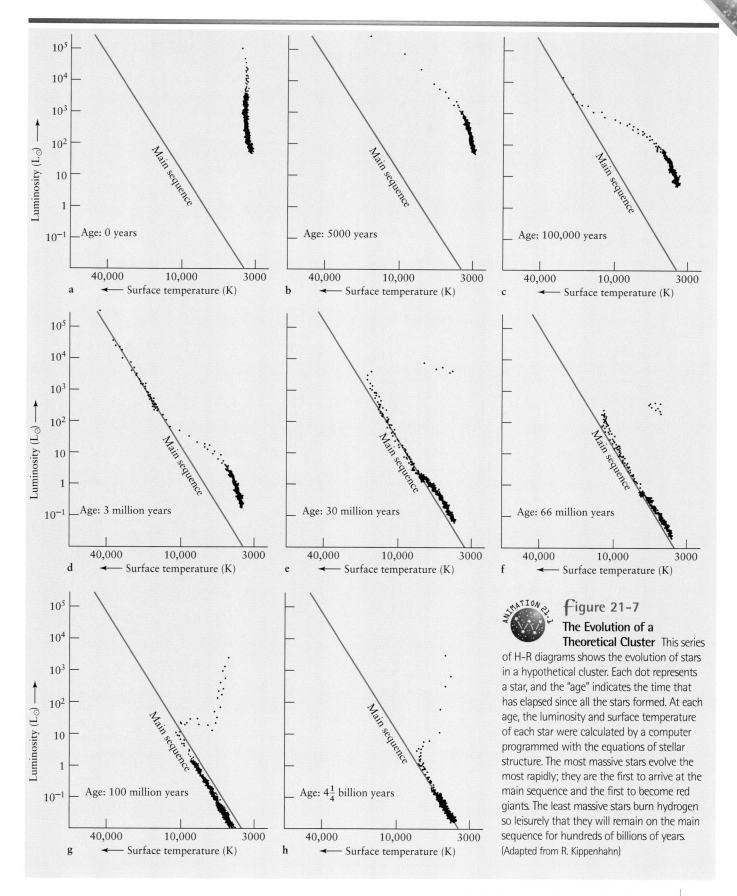

figure 21-7

The Evolution of a Theoretical Cluster This series of H-R diagrams shows the evolution of stars in a hypothetical cluster. Each dot represents a star, and the "age" indicates the time that has elapsed since all the stars formed. At each age, the luminosity and surface temperature of each star were calculated by a computer programmed with the equations of stellar structure. The most massive stars evolve the most rapidly; they are the first to arrive at the main sequence and the first to become red giants. The least massive stars burn hydrogen so leisurely that they will remain on the main sequence for hundreds of billions of years. (Adapted from R. Kippenhahn)

Horizontal-branch stars

Red giants

figure 21-8
R **I** **V** U X G

A Globular Cluster A globular cluster is a spherical grouping of a few hundred thousand stars. This cluster, called M10, lies approximately 5000 pc (16,000 ly) from the Earth in the constellation Ophiuchus. The stars in M10 are all crowded within a diameter of only 20 pc (70 ly). Most of the stars in this image are either red giants or blue, horizontal-branch stars with both core helium burning and shell hydrogen burning. (T. Credner and S. Kohle, Astronomical Institutes of the University of Bonn)

M stars. These lowest-mass stars can continue to burn hydrogen in their cores for hundreds of billions of years.

In the final two H-R diagrams (Figures 21-7*g* and 21-7*h*), the main sequence gets shorter as stars exhaust their core supplies of hydrogen and evolve into red giants. The stars that leave the main sequence between Figure 21-7*g* and Figure 21-7*h* have masses between about 1 M$_\odot$ and 3 M$_\odot$ and undergo the helium flash in their cores.

The evolution of the hypothetical cluster displayed in Figure 21-7 helps us interpret what we see in actual star clusters. We can observe the early stages of stellar evolution in open clusters, many of which are quite young (see Section 20-6, especially Figure 20-17 and Figure 20-18). By contrast, **globular clusters**—so called because of their spherical shape—contain many highly evolved post–main-sequence stars. A typical globular cluster contains up to 1 million stars in a volume less than 100 parsecs across (Figure 21-8).

Globular clusters must be old, because they contain no high-mass main-sequence stars. To show this, you would measure the apparent magnitude (a measure of apparent brightness, which we introduced in Section 19-3) and color ratio of many stars in a globular cluster, then plot the data as shown in Figure 21-9. Such a **color-magnitude diagram** for a cluster is equivalent to an H-R diagram. The color ratio of a star tells you its surface temperature (as described in Section 19-4), and because all the stars in the cluster are at essentially the same distance from us, their relative brightnesses indicate their relative luminosities. What you would discover is that the upper half of the main sequence is missing. All the high-mass main-sequence stars in a globular cluster evolved long ago into red giants. Only low-mass, slowly evolving stars still have core hydrogen burning. (Compare Figure 21-9 with Figure 21-7*h*.)

The color-magnitude diagram of a globular cluster typically shows a horizontal grouping of stars in the left-of-center portion of the diagram (see Figure 21-9). These **horizontal-branch stars** are post–helium-flash, low-mass stars with luminosities of about 50 L$_\odot$ in which there is both core helium

figure 21-9

A Color-Magnitude Diagram of a Globular Cluster
Each dot in this diagram represents the apparent magnitude (measured through a V filter) and surface temperature (as measured by the color ratio b_V/b_B) of a star in the globular cluster M55 in Sagittarius. Because all the stars in M55 are at essentially the same distance from the Earth (about 6000 pc or 20,000 ly), their apparent visual magnitudes (a measure of apparent brightness) are a direct measure of luminosity. Note that the upper half of the main sequence is missing. The horizontal-branch stars are low-mass stars that recently experienced the helium flash and now exhibit core helium burning and shell hydrogen burning. (Adapted from D. Schade, D. VandenBerg, and F. Hartwick)

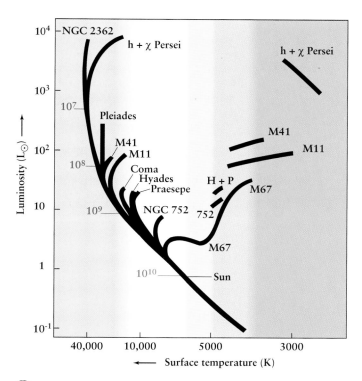

Figure 21-10

An H-R Diagram for Open Star Clusters The black bands indicate where stars from various open clusters fall on the H-R diagram. The age of a cluster can be estimated from the location of the cluster's turnoff point, where the cluster's most massive stars are just now leaving the main sequence. The times for these turnoff points (in years) are listed in red alongside the main sequence. For example, the Pleiades cluster turnoff point is near the 10^8-year point, so this cluster is about 10^8 years old. (Adapted from A. Sandage)

burning and shell hydrogen burning. In years to come, these stars will move back toward the red-giant region as their fuel is devoured. The inset in Figure 21-8 shows some of these stars.

An H-R diagram can be used to determine the age of a cluster. In the H-R diagram for a very young cluster, all the stars are on or near the main sequence. (An example is the open cluster NGC 2264, shown in Figure 20-17). As a cluster gets older, however, stars begin to leave the main sequence. The high-mass, high-luminosity stars are the first to become red giants. Over the years, the main sequence gets shorter and shorter, like a candle burning down.

The age of a cluster can be found from the **turnoff point**, which is the top of the surviving portion of the main sequence on the cluster's H-R diagram (see Figure 21-9). The stars at the turnoff point are just now exhausting the hydrogen in their cores, so their main-sequence lifetime is equal to the age of the cluster. For example, in the case of the globular cluster M55 plotted in Figure 21-9, 0.8-$M_\odot$ stars have just left the main sequence, indicating that the cluster's age is approximately 15 billion (1.5×10^{10}) years (see Table 21-1).

Figure 21-10 shows data for several star clusters plotted on a single H-R diagram. This graph also shows turnoff-point times from which the ages of the clusters can be estimated.

21-4 Stellar evolution has produced two distinct populations of stars

Studies of star clusters reveal a curious difference between the youngest and oldest stars in our Galaxy. Stars in the youngest clusters (those with most of their main sequences still intact) are said to be **metal rich,** because their spectra contain many prominent spectral lines of heavy elements. (Recall from Section 19-5 that astronomers use the term "metal" to denote any element other than hydrogen and helium, which are the two lightest elements.) Such stars are also called **Population I stars.** The Sun is a relatively young, metal-rich, Population I star.

By contrast, the spectra of stars in the oldest clusters show only weak lines of heavy elements. These ancient stars are thus said to be **metal poor,** because heavy elements are only about 3% as abundant in these stars as in the Sun. They are also called **Population II stars.** The stars in globular clusters are metal poor, Population II stars. Figure 21-11 shows the difference in spectra between a metal-poor, Population II star and the Sun (a metal-rich, Population I star).

To explain why there are two distinct populations of stars, we must go back to the Big Bang, the explosive origin of the universe that took place some 10 to 15 billion years ago. As we will discuss in Chapter 29, the early universe consisted almost exclusively of hydrogen and helium, with almost no heavy elements (metals). The first stars to form were likewise metal poor. The least massive of these have survived to the present day and are now the ancient stars of Population II.

The more massive of the original stars evolved more rapidly and no longer shine. But as these stars evolved,

Figure 21-11   R I **V** U X G

Spectra of a Metal-Poor Star and a Metal-Rich Star These spectra compare **(a)** a metal-poor, Population II star and **(b)** a metal-rich, Population I star (the Sun) of the same surface temperature. Many prominent spectral lines in the Sun's spectrum are caused by elements heavier than hydrogen and helium. The corresponding lines in the metal-poor star's spectrum are weak or absent. Both spectra include the strong hydrogen absorption line H_δ (wavelength 410 nm) described in Section 5-8. (Lick Observatory) helium burning in their cores produced metals—carbon and oxygen. In the most massive stars, as we will learn in Chapter 23, further thermonuclear reactions produced even heavier elements. As these massive original stars aged and died, they expelled their metal-enriched gases into space. (The star in the photograph that opens this chapter is going through such a mass-loss phase.). This expelled material joined the interstellar medium and was eventually incorporated into a second generation of stars that have a higher concentration of heavy elements. These metal-rich members of the second stellar generation are the Population I stars, of which our Sun is an example.

Be careful not to let the designations of the two stellar populations confuse you. Population *I* stars are members of a *second* stellar generation, while Population *II* stars belong to an older *first* generation.

The relatively high concentration of heavy elements in the Sun means that the solar nebula, from which both the Sun and planets formed (see Section 7-7), must likewise have been metal rich. The Earth is composed almost entirely of heavy elements, as are our bodies. Thus, our very existence is intimately linked to the Sun's being a Population I star. A planet like the Earth probably could not have formed from the metal-poor gases that went into making Population II stars.

The concept of two stellar populations provides insight into our own origins. Recall from Section 21-2 that helium burning in red-giant stars produces the same isotopes of carbon and oxygen that are found most commonly on the Earth. The reason is that the Earth's carbon and oxygen atoms actually *were* produced by helium burning. These reactions occurred billions of years ago within an earlier generation of stars that died and gave up their atoms to the interstellar medium—the same atoms that later became part of our solar system, our planet, and our bodies. We are literally children of the stars.

21-5 Many mature stars pulsate

We saw in Section 18-3 that the surface of our Sun vibrates in and out, although by only a small amount. But other stars undergo substantial changes in size, alternately swelling and shrinking. As these stars pulsate, they also vary dramatically in brightness. We now understand that these **pulsating variable stars** are actually evolved, post–main-sequence stars.

The first pulsating variable star was discovered in 1595 by the Dutch minister and amateur astronomer David Fabricius. He noticed that the star o (omicron) Ceti is sometimes bright enough to be easily seen with the naked eye but at other times fades to invisibility (Figure 21-12). By 1660, astronomers realized that these brightness variations repeated with a period of 332 days. Seventeenth-century astronomers were so enthralled by this variable that they renamed the star Mira ("wonderful").

a   b

figure 21-12 R I **V** U X G

Mira—A Long-Period Variable Star Mira, or o (omicron) Ceti, is a variable star whose luminosity varies with a 332-day period. At its dimmest, as in **(a)** (photographed in December 1961), Mira is less than 1% as bright as when it is at maximum, as in **(b)** (January 1965). These brightness variations occur because Mira pulsates and thus varies both in luminosity and surface temperature. (Lowell Observatory)

Mira is an example of a class of pulsating stars called **long-period variables**. These stars are cool red giants that vary in brightness by a factor of 100 or more over a period of months or years. With surface temperatures of about 3500 K and average luminosities that range from about 10 to 10,000 $L_\odot$, they occupy the upper right side of the H-R diagram (Figure 21-13). Some, like Mira, are periodic, but others are irregular. Many eject large amounts of gas and dust into space.

Astronomers do not fully understand why some cool red giants become long-period variables. It is difficult to calculate accurate stellar models to describe such huge stars with extended, tenuous atmospheres.

Astronomers have a much better understanding of other pulsating stars, called **Cepheid variables**, or simply Cepheids. A Cepheid variable is recognized by the characteristic way in which its light output varies—rapid brightening followed by gradual dimming. They are named for δ (delta) Cephei, an example of this type of star discovered in 1784 by John Goodricke, a deaf, mute, 19-year-old English amateur astronomer. He found that at its most brilliant, δ Cephei is 2.3 times as bright as at its dimmest. The cycle of brightness variations repeats every 5.4 days. (Sadly, Goodricke paid for his discoveries with his life; he caught pneumonia while making his nightly observations and died before his twenty-second birthday.) The surface temperatures and luminosities of the Cepheid variables place them in the upper middle of the H-R diagram (see Figure 21-13).

After core helium burning begins, mature stars move across the middle of the H-R diagram. Figure 21-6 shows the evolutionary tracks of high-mass stars crisscrossing the

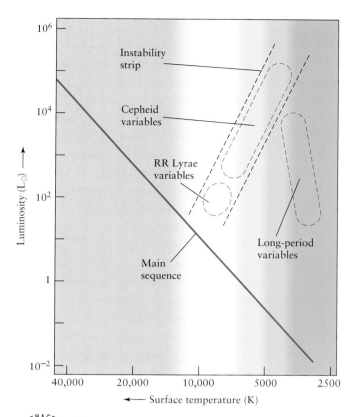

![interactive exercise icon] **figure 21-13**

Variable Stars on the H–R Diagram Pulsating variable stars are found in the upper right of the H-R diagram. Long-period variables like Mira are cool red giant stars that pulsate slowly, changing their brightness in a semiregular fashion over months or years. Cepheid variables and RR Lyrae variables are located in the instability strip, which lies between the main sequence and the red-giant region. A star passing through this strip along its evolutionary track becomes unstable and pulsates.

a

b

c

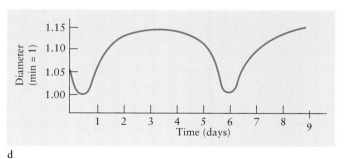

d

![web link icon] **figure 21-14**

δ Cephei—A Pulsating Star **(a)** The light curve of δ Cephei shows that as the star pulsates, it brightens quickly (the curve moves upward sharply) but fades more slowly (the curve declines more gently). **(b)** The radial velocity curve has positive values when the star is shrinking (so the surface is moving away from us) and negative values when it is expanding (so the surface is approaching us). **(c)** The star's surface temperature rises and falls nearly in step with the increases and decreases in brightness (see part a). **(d)** This graph shows the periodic variations in the star's diameter. The star has the smallest diameter slightly before it is at its brightest and hottest (compare with parts a and b).

H-R diagram. Post–helium-flash, low-mass stars also cross the middle of the H-R diagram between the red-giant region and the horizontal branch.

During these transitions across the H-R diagram, a star can become unstable and pulsate. In fact, there is a region on the H-R diagram between the upper main sequence and the red-giant branch called the **instability strip** (see Figure 21-13). When an evolving star passes through this region, the star pulsates and its brightness varies periodically. Figure 21-14*a* shows the brightness variations of δ Cephei, which lies within the instability strip.

A Cepheid variable brightens and fades because the star's outer envelope cyclically expands and contracts. The first to observe this was the Russian astronomer Aristarkh Belopol'skii, who noticed in 1894 that spectral lines in the spectrum of δ Cephei shift back and forth with the same 5.4-day period as that of the magnitude variations. From the

Doppler effect, we can translate these wavelength shifts into radial velocities and draw a velocity curve (Figure 21-14*b*). Negative velocities mean that the star's surface is expanding toward us; positive velocities mean that the star's surface is receding. Note that the light curve and velocity curve are mirror images of each other. The star is brighter than average while it is expanding and dimmer than average while contracting.

When a Cepheid variable pulsates, the star's surface oscillates up and down like a spring. During these cyclical expansions and contractions, the star's gases alternately heat up and cool down. Figure 21-14*c* shows the resulting changes in the star's surface temperature. Figure 21-14*d* graphs the periodic changes in the star's diameter.

Just as a bouncing ball eventually comes to rest, a pulsating star would soon stop pulsating without something to keep its oscillations going. In 1914, the British astronomer Arthur Eddington suggested that a Cepheid pulsates because the star is more opaque when compressed than when expanded. When the star is compressed, trapped heat increases the internal pressure, which pushes the star's surface outward. When the star expands, the heat escapes, the internal pressure drops, and the star's surface falls inward.

In the 1960s, the American astronomer John Cox followed up on Eddington's idea and proved that helium is what keeps Cepheids pulsating. Normally, when a star's helium is compressed, the gas increases in temperature and becomes more transparent. But in certain layers near the star's surface, compression may ionize helium (remove one of its electrons) instead of raising its temperature. Ionized helium gas is quite opaque, so these layers effectively trap heat and make the star expand, as Eddington suggested. This expansion cools the outer layers and makes the helium ions recombine with electrons, which makes the gas more transparent and releases the trapped energy. The star's surface then falls inward, recompressing the helium, and the cycle begins all over again.

Cepheid variables are important because they have two properties that allow astronomers to determine the distances to very remote objects. First, Cepheids can be seen even at distances of millions of parsecs. This is because they are very luminous, ranging from a few hundred times solar luminosity to more than 10^4 L$_\odot$. Second, there is a direct relationship between a Cepheid's period and its average luminosity: The dimmest Cepheid variables pulsate rapidly, with periods of one to two days, while the most luminous Cepheids pulsate with much slower periods of about 100 days.

Figure 21-15 shows this **period-luminosity relation.** By measuring the period of a distant Cepheid's brightness variations and using a graph like Figure 21-15, an astronomer can determine the star's luminosity. By also measuring the star's apparent brightness, the distance to the Cepheid can be found by using the inverse-square law (see Section 19-2). By applying the period-luminosity relation in this way to Cepheids in other galaxies, astronomers have been able to calculate the distances to those galaxies with great accuracy. (Box 19-2 gives an example of such a calculation.) As we will see in Chapters 26 and 28, such measurements play an important role in determining the overall size and structure of the universe.

igure 21-15

Period–Luminosity Relations for Cepheids The period of a Cepheid variable is directly related to its average luminosity: the more luminous the Cepheid, the longer its period and the slower its pulsations. Note that there are actually two distinct period-luminosity relations—one for Type I Cepheids, which are metal-rich, Population I stars, and one for the less luminous Type II Cepheids, which are metal-poor. (Adapted from H. C. Arp)

How a Cepheid pulsates depends on the amount of heavy elements in the star's outer layers, because even trace amounts of these elements can have a large effect on how opaque the stellar gases are. Hence, Cepheids are classified according to their metal content. If the star is a metal-rich, Population I star, it is called a **Type I Cepheid;** if it is a metal-poor, Population II star, it is called a **Type II Cepheid.** As Figure 21-15 shows, these two types of Cepheids exhibit different period-luminosity relations. In order to know which period-luminosity relation to apply to a given Cepheid, an astronomer must determine the star's metal content from its spectrum (see Figure 21-11).

The evolutionary tracks of mature, high-mass stars pass back and forth through the upper end of the instability strip on the H-R diagram. These stars become Cepheids when helium ionization occurs at just the right depth to drive the pulsations. For stars on the high-temperature (left) side of the instability strip, helium ionization occurs too close to the surface and involves only an insignificant fraction of the star's mass. For stars on the cool (right) side of the instability strip, convection in the star's outer layers prevents the storage of the energy needed to drive the pulsations. Thus, Cepheids exist only in a narrow temperature range on the H-R diagram.

Stars of lower mass do not become Cepheids. Instead, after undergoing the helium flash, their evolutionary tracks pass through the lower end of the instability strip as they move along the horizontal branch. Some of these stars become **RR Lyrae variables,** named for their prototype in the constellation Lyra. RR Lyrae variables all have periods shorter than one day and roughly the same average luminosity as horizontal-branch stars, about 100 $L_\odot$. In fact, the RR Lyrae region of the instability strip (see Figure 21-13) is actually a segment of the horizontal branch. RR Lyrae stars are all metal-poor, Population II stars. Many have been found in globular clusters, and they have been used to determine the distances to those clusters in the same way that Cepheids are used to find the distances to other galaxies. In Chapter 25 we will see how RR Lyrae stars helped astronomers determine the size of the Milky Way Galaxy.

In some cases the expansion speed of a pulsating star exceeds the star's escape speed. When this happens, the star's outer layers are ejected completely. We will see in Chapter 22 that dying stars eject significant amounts of mass in this way, renewing and enriching the interstellar medium for future generations of stars.

21-6 Mass transfer can affect the evolution of close binary star systems

We have outlined what happens when a main-sequence star evolves into a red giant. What we have ignored is that more than half of all stars are members of multiple-star systems, including binaries. If the stars in such a system are widely separated, the individual stars follow the same course of evolution as if they were isolated. In a **close binary,** however, when one star expands to become a red giant, its outer layers can be gravitationally captured by the nearby companion star. In other words, a bloated red giant in a close binary system can dump gas onto its companion, a process called **mass transfer.**

Our modern understanding of mass transfer in close binaries is based on the work of French mathematician Edouard Roche. In the mid-1800s, Roche studied how rotation and mutual tidal interaction affect the stars in a binary system. Tidal forces cause the two stars in a close binary to keep the same sides facing each other, just as our Moon keeps its same side facing the Earth (see Section 4-8). But because stars are gaseous, not solid, rotation and tidal forces can have significant effects on their shapes.

In widely separated binaries, the stars are so far apart that tidal effects are small, and, therefore, the stars are nearly perfect spheres. In close binaries, where the separation between the stars is not much greater than their sizes, tidal effects are strong, causing the stars to be somewhat egg-shaped.

Roche discovered a surface that marks the gravitational domain of each star in a close binary. Figure 21-16a shows the outline of this surface as a dashed line. The two halves of this surface, each of which encloses one of the stars, are known as **Roche lobes.** The more massive star is always located inside the larger Roche lobe. If gas from a star leaks over its Roche lobe, it is no longer bound by gravity to that star. This escaped gas is free either to fall onto the companion star or to escape from the binary system.

The point where the two Roche lobes touch, called the **inner Lagrangian point,** is a kind of balance point between the two stars in a binary. It is here that the effects of gravity and rotation cancel each other. When mass transfer occurs in a close binary, gases flow through the inner Lagrangian point from one star to the other.

In many binaries, the stars are so far apart that even during their red-giant stages the surfaces of the stars remain well inside their Roche lobes. As a result, little mass transfer can occur and each star lives out its life as if it were single and isolated. A binary system of this kind is called a **detached binary** (Figure 21-16a).

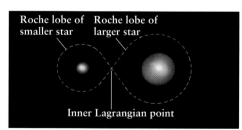
Roche lobe of smaller star   Roche lobe of larger star

Inner Lagrangian point

a Detached binary

b Semi-detached binary

c Contact binary

d Overcontact binary

Figure 21-16
Close Binary Star Systems The gravitational domain of a star in a close binary system is called its Roche lobe. The two Roche lobes meet at the inner Lagrangian point. **(a)** In a detached binary, neither star fills its Roche lobe. Both stars are relatively spherical. **(b)** In a semidetached binary, one star has expanded to fill its Roche lobe. Mass can flow from this star to the other across the inner Lagrangian point. **(c)** In a contact binary, both stars fill their Roche lobes. **(d)** The two stars in an overcontact binary both overfill their Roche lobes. The two stars actually share the same outer atmosphere.

However, if the two stars are close enough, when one star expands to become a red giant, it may fill or overflow its Roche lobe. Such a system is called a **semidetached binary** (Figure 21-16b). If both stars fill their Roche lobes, the two stars actually touch and the system is called a **contact binary** (Figure 21-16c). It is quite unlikely, however, that both stars exactly fill their Roche lobes at the same time. It is more likely that they overflow their Roche lobes, giving rise to a common envelope of gas. Such a system is called an **overcontact binary** (Figure 21-16d).

The binary star system Algol (from an Arabic term for "demon") provided the first clear evidence of mass transfer in close binaries. Also called β (beta) Persei, Algol can easily be seen with the naked eye. Ancient astronomers knew that Algol varies periodically in brightness by a factor of more than 2. In 1782, John Goodricke (the discoverer of δ Cephei's variability) first suggested that these brightness variations take place because Algol is an *eclipsing* binary, discussed in Section 19-11. Because the orbital plane of the two stars that make up the binary system is nearly edge-on to our line of sight, one star periodically eclipses the other. Algol's light curve (Figure 21-17a) and spectrum show that Goodricke's brilliant hypoth-

esis is correct, and that Algol is a semidetached binary. The detached star (on the right in Figure 21-17a) is a luminous blue main-sequence star, while its less massive companion is a dimmer red giant that fills its Roche lobe.

According to stellar evolution theory, the more massive a star, the more rapidly it should evolve. But in Algol and similar binaries, the more massive star (on the right in Figure 21-17a) is still on the main sequence, whereas the less massive star (on the left in Figure 21-17a) has evolved to become a red giant. This seems backward, since the two stars in a binary system form simultaneously and thus are the same age. The explanation for this apparent contradiction is that the red giant in Algol-type binaries was originally the more massive star. As it left the main sequence to become a red giant, this star overflowed its Roche lobe, dumping gas onto the originally less massive companion. Because of the resulting mass transfer, that companion became the more massive star.

Mass transfer is also important in another class of semidetached binaries, called β (beta) Lyrae variables, after their prototype in the constellation Lyra (the Harp). As with Algol, the less massive star in β Lyrae (on the left in Figure 21-17b)

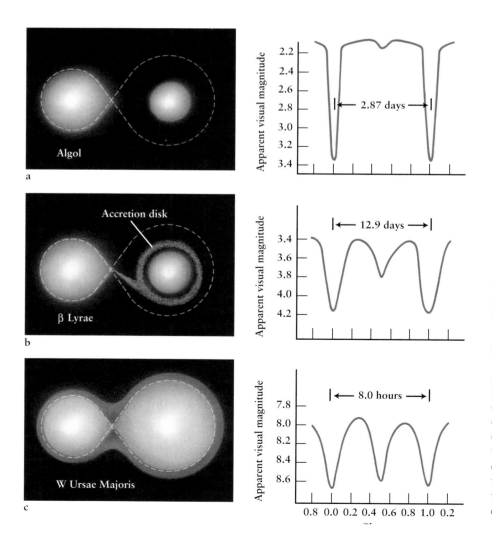

figure 21-17

Three Eclipsing Binaries
Compare the light curves of these three eclipsing binaries to the light curves in Figure 19-26. **(a)** Algol is a semidetached binary. The deep eclipse occurs when the large red-giant star (on the left) blocks the light from the smaller but more luminous main-sequence star. (Compare Figure 19-25b.) **(b)** β Lyrae is a semidetached binary in which mass transfer from the left-hand star has produced an accretion disk around the right-hand star. The light curve is at its lowest when the larger star completely eclipses the smaller one. Half an orbital period later, the smaller star partially eclipses the larger one, making a shallower dip in the light curve. **(c)** W Ursae Majoris is an overcontact binary in which both stars overfill their Roche lobes. The extremely short period of this binary indicates that the two stars are very close to each other.

fills its Roche lobe. Unlike Algol, however, the more massive detached star (on the right in Figure 21-17b) is the dimmer of the two stars. Apparently, this detached star is enveloped in a rotating *accretion disk* of gas captured from its bloated companion. This disk partially blocks the light coming from the detached star, making it appear dimmer.

What is the fate of an Algol or β Lyrae system? If the detached star is massive enough, it will evolve rapidly, expanding to also fill its Roche lobe. The result will be an overcontact binary in which the two stars share the gases of their outer layers. Such binaries are sometimes called W Ursae Majoris stars, after the prototype of this class (Figure 21-17c). Mass transfer can also continue even after nuclear reactions cease in one of the stars in a close binary. In Chapters 22 and 23 we will see how mass transfer onto dead stars produces some of the most unusual objects in the sky.

KEY WORDS

alpha particle, p. 484

Cepheid variable, p. 490

close binary, p. 493

color-magnitude diagram, p. 488

contact binary, p. 494

core helium burning, p. 483

core hydrogen burning, p. 479

degeneracy, p. 484

degenerate-electron pressure, p. 484

detached binary, p. 493

globular cluster, p. 488

helium burning, p. 483

helium flash, p. 484

horizontal-branch star, p. 488

ideal gas, p. 484

inner Lagrangian point, p. 493

instability strip, p. 491

long-period variable, p. 490

main-sequence lifetime, p. 479

mass loss, p. 483

mass transfer, p. 493

metal-poor star, p. 489

metal-rich star, p. 489

overcontact binary, p. 494

Pauli exclusion principle, p. 484

period-luminosity relation, p. 492

Population I and Population II stars, p. 489

pulsating variable star, p. 490

red giant, p. 482

Roche lobe, p. 493

RR Lyrae variable, p. 493

semidetached binary, p. 494

shell hydrogen burning, p. 482

triple alpha process, p. 483

turnoff point, p. 489

Type I and Type II Cepheids, p. 492

zero-age main sequence (ZAMS), p. 485

zero-age main-sequence star, p. 479

KEY IDEAS

The Main-Sequence Lifetime: The duration of a star's main-sequence lifetime depends on the amount of hydrogen in the star's core and the rate at which the hydrogen is consumed.

• The more massive a star, the shorter is its main-sequence lifetime. The Sun has been a main-sequence star for about 4.6 billion years and should remain one for about another 5 billion years.

• During a star's main-sequence lifetime, the star expands somewhat and undergoes a modest increase in luminosity.

Becoming a Red Giant: Core hydrogen burning ceases when the hydrogen has been exhausted in the core of a main-sequence star. This leaves a core of nearly pure helium surrounded by a shell through which hydrogen burning works its way outward in the star. The core shrinks and becomes hotter, while the star's outer layers expand and cool. The result is a red giant star.

• As a star becomes a red giant, its evolutionary track moves rapidly from the main sequence to the red-giant region of the H-R diagram. The more massive the star, the more rapidly this evolution takes place.

Helium Burning: When the central temperature of a red giant reaches about 100 million K, helium burning begins in the core. This process, also called the triple alpha process, converts helium to carbon and oxygen.

• In a more massive red giant, helium burning begins gradually; in a less massive red giant, it begins suddenly, in a process called the helium flash.

• After the helium flash, a low-mass star moves quickly from the red-giant region of the H-R diagram to the horizontal branch.

Star Clusters and Stellar Populations: The age of a star cluster can be estimated by plotting its stars on an H-R diagram.

• The cluster's age is equal to the age of the main-sequence stars at the turnoff point (the upper end of the remaining main sequence).

• As a cluster ages, the main sequence is "eaten away" from the upper left as stars of progressively smaller mass evolve into red giants.

• Relatively young Population I stars are metal rich; ancient Population II stars are metal poor. The metals (heavy elements) in Population I stars were manufactured by thermonuclear reactions in an earlier generation of Population II stars, then ejected into space and incorporated into a later stellar generation.

Pulsating Variable Stars: When a star's evolutionary track carries it through a region in the H-R diagram called the instability strip, the star becomes unstable and begins to pulsate.

• Cepheid variables are high-mass pulsating variables. There is a direct relationship between their periods of pulsation and their luminosities.

• RR Lyrae variables are low-mass, metal-poor pulsating variables with short periods.

• Long-period variable stars also pulsate but in a fashion that is less well understood.

Close Binary Systems: Mass transfer in a close binary system occurs when one star in a close binary overflows its Roche lobe. Gas flowing from one star to the other passes across the inner Lagrangian point. This mass transfer can affect the evolutionary history of the stars that make up the binary system.

REVIEW QUESTIONS

1. Why do high-mass main-sequence stars have shorter lifetimes than those of lower mass?

2. On what grounds are astronomers able to say that the Sun has about 5×10^9 years remaining in its main-sequence stage?

3. What will happen inside the Sun 5 billion years from now, when it begins to mature into a red giant?

4. Explain how it is possible for the core of a red giant to contract at the same time that its outer layers expand.

5. How is a degenerate gas different from ordinary gases?

6. What is the helium flash? Why does it happen in some stars but not in others?

7. Explain why the majority of the stars in the sky are main-sequence stars.

8. What does it mean when an astronomer says that a star "moves" from one place to another on an H-R diagram?

9. On an H-R diagram, main-sequence stars do not lie along a single narrow line but are spread out over a band (see Figure 21-6b). On the basis of how stars evolve during their main-sequence lifetime, explain why this should be so.

10. Explain how and why the turnoff point on the H-R diagram of a cluster is related to the cluster's age.

11. There is a good deal of evidence that our universe is no more than about 15 billion years old (see Chapter 28). Explain why no main-sequence stars of spectral class M have yet evolved into red-giant stars.

12. (a) The main-sequence stars Sirius (spectral type A1), Vega (A0), Spica (B1), Fomalhaut (A3), and Regulus (B7) are among the 20 brightest stars in the sky. Explain how you can tell that all these stars are younger than the Sun. (b) The third-brightest star in the sky, although it can be seen only south of 29° north latitude, is α (alpha) Centauri A. It is a main-sequence star of spectral type G2, the same as the Sun. Can you tell from this whether α Centauri A is younger than the Sun, the same age, or older? Explain your reasoning.

13. Using the same horizontal and vertical scales as in Figure 21-6a, make points on an H-R diagram for each of the stars listed in Table 21-1. Label each point with the star's mass and its main-sequence lifetime. Which of these stars will remain on the main sequence after 10^9 years? After 10^{11} years?

14. How do astronomers know that globular clusters are made of old stars?

15. Red giant stars appear more pronounced in composites of infrared images and visible-light images, like those in Figure 21-3b and Figure 21-8. Explain why.

16. The horizontal-branch stars in Figure 21-8 appear blue. (a) Explain why this is consistent with the color-magnitude diagram shown in Figure 21-9. (b) All horizontal-branch stars were once red giants. Explain what happened to these stars to change their color.

17. What is the difference between Population I and Population II stars? In what sense can the stars of one population be regarded as the "children" of the other population?

18. The star whose spectrum is shown in Figure 21-11a has a lower percentage of heavy elements than the Sun, whose spectrum is shown in Figure 21-11b. Hence, the star in Figure 21-11a has a higher percentage of hydrogen. Why, then, isn't the H_δ absorption line of hydrogen noticeably darker for the star in Figure 21-11a?

19. Why do astronomers attribute the observed Doppler shifts of a Cepheid variable to pulsation, rather than to some other cause, such as orbital motion?

20. Why do Cepheid stars pulsate? Why are these stars important to astronomers who study galaxies beyond the Milky Way?

21. Would you expect the color of a Cepheid variable star (see Figure 21-14) to change during the star's oscillation period? If not, why not? If so, describe why the color should change, and describe the color changes you would expect to see during an oscillation period.

22. What is a Roche lobe? What is the inner Lagrangian point? Why are Roche lobes important in close binary star systems?

23. What is the difference between a detached binary, a semi-detached binary, a contact binary, and an overcontact binary?

24. Massive main-sequence stars turn into red giants before less massive stars. Why, then, is the more massive star in Algol a main-sequence star and the less massive star a red giant?

ADVANCED QUESTIONS

Questions preceded by an asterisk (*) involve topics discussed in the Boxes in this chapter or in Chapter 19.

Problem-solving tips and tools
Recall from Section 18-1 that 6×10^{11} kg of hydrogen is converted into helium each second at the Sun's center. Recall also that you must use absolute (Kelvin) temperatures when using the Stefan-Boltzmann law. You may find it helpful to review the discussion of apparent magnitude, absolute magnitude, and luminosity in Box 19-3. Section 19-4 discusses the connection between the surface temperatures and colors of stars. Newton's form of Kepler's third law (see Section 19-9) describes the orbits of stars in binary systems.

25. The radius of the Sun has increased over the past several billion years. Over the same time period, the size of the Moon's orbit around the Earth has also increased. A few billion years ago, were annular eclipses of the Sun (see Figure 3-12) more or less common than they are today? Explain.

26. The Sun has increased in radius by 6% over the past 4.6 billion years. Its present-day radius is 696,000 km. What was its radius 4.6 billion years ago? (*Hint:* The answer is *not* 654,000 km.)

*27. Calculate the escape speed from (**a**) the surface of the present-day Sun and (**b**) from the surface of the Sun when it becomes a red giant, with essentially the same mass as today but with a radius that is 100 times larger. (**c**) Explain how your results show that a red-giant star can lose mass more easily than a main-sequence star.

*28. Calculate the average speed of a hydrogen atom (mass 1.67×10^{-27} kg) (**a**) in the atmosphere of the present-day Sun, with temperature 5800 K, and (**b**) in the atmosphere of a 1-$M_\odot$ red giant, with temperature 3500 K. (**c**) Compare your results with the escape speeds that you calculated in Question 27. Use this comparison to discuss how well the present-day Sun and a 1-$M_\odot$ red giant can retain hydrogen in their atmospheres.

29. What mass of hydrogen will the Sun convert into helium during its entire main-sequence lifetime of 10^{10} years? What fraction does this represent of the total mass of hydrogen that was originally in the Sun? Assume that

the Sun's luminosity remains nearly constant during the entire 10^{10} years.

*30. Calculate the main-sequence lifetimes of (**a**) a 9-$M_\odot$ star and (**b**) a 0.25-$M_\odot$ star. Compare these lifetimes with that of the Sun.

*31. The earliest fossil records indicate that life appeared on the Earth about a billion years after the formation of the solar system. What is the most mass that a star could have in order that its lifetime on the main sequence is long enough to permit life to form on one or more of its planets? Assume that the evolutionary processes would be similar to those that occurred on the Earth.

32. As a red giant, the Sun's luminosity will be about 2000 times greater than it is now, so the amount of solar energy falling on the Earth will increase to 2000 times its present-day value. Hence, to maintain thermal equilibrium, each square meter of the Earth's surface will have to radiate 2000 times as much energy into space as it does now. Use the Stefan-Boltzmann law to determine what the Earth's surface temperature will be under these conditions. (*Hint:* The present-day Earth has an average surface temperature of 9°C.)

33. When the Sun becomes a red giant, its luminosity will be about 2000 times greater than it is today. Assuming that this luminosity is caused *only* by the burning of the Sun's remaining hydrogen, calculate how long our star will be a red giant. (In fact, only a fraction of the remaining hydrogen will be consumed, so your answer will be an *over*estimate.)

34. What observations would you make of a star to determine whether its primary source of energy is hydrogen or helium burning?

35. The brightness of a certain Cepheid variable star increases and decreases with a period of 10 days. (**a**) What must this star's luminosity be if its spectrum has strong absorption lines of hydrogen and helium, but no strong absorption lines of heavy elements? (**b**) Repeat part (a) for the case in which the star's spectrum also has strong absorption lines of heavy elements.

36. The star X Arietis is an RR Lyrae variable. Its apparent brightness varies between 2.0×10^{-15} and 4.9×10^{-15} that of the Sun with a period of 0.65 day. Interstellar extinction dims the star by 37%. Approximately how far away is the star?

37. The apparent brightness of δ Cephei (a Type I Cepheid variable) varies with a period of 5.4 days. Its average apparent brightness is 5.1×10^{-13} that of the Sun. Approximately how far away is δ Cephei? (Ignore interstellar extinction.)

38. Suppose the detached star in β Lyrae (Figure 21-17*b*) did not have an accretion disk. Would the deeper dips in the light curve be deeper, shallower, or about the same? What about the shallower dip? Explain your answers.

39. The two stars that make up the overcontact binary W Ursae Majoris (Figure 21-17c) have estimated masses of 0.99 $M_\odot$ and 0.62 $M_\odot$. **(a)** Find the average separation between the two stars. Give your answer in kilometers. **(b)** The radii of the two stars are estimated to be 1.14 $R_\odot$ and 0.83 $R_\odot$. Show that these values and your result in part (a) are consistent with the statement that this is an overcontact binary.

 40. Consult recent issues of *Sky & Telescope* and *Astronomy* to find out when Mira will next reach maximum brightness. Look up the star's location in the sky using the *Starry Night* program on the CD-ROM that comes with this book. (Use the **Find...** command in the **Edit** menu to search for Omicron Ceti.) Why is it unlikely that you will be able to observe Mira at maximum brightness?

DISCUSSION QUESTIONS

41. The half-life of the ^{8}Be nucleus is 2.6×10^{-16} second, which is the average time that elapses before this unstable nucleus decays into two alpha particles. How would the universe be different if instead the ^{8}Be half-life were zero? How would the universe be different if the ^{8}Be nucleus were stable?

42. If the universe has a finite age (as a broad array of evidence indicates to be the case), what observational consequences does this have for H-R diagrams of star clusters? Could we use these consequences to establish constraints on the possible age of the universe? Explain.

 ## WEB/CD-ROM QUESTIONS

 43. Suppose that an oxygen nucleus were fused with a helium nucleus. What element would be formed? Look up the relative abundance of this element in, for example, the *Handbook of Chemistry and Physics* or on the World Wide Web. Comment on whether such a process is likely.

44. Although Polaris, the North Star, is a Cepheid variable, it pulsates in a somewhat different way than other Cepheids. Search the World Wide Web for information about this star's pulsations and how they have been measured by astronomers at the U.S. Naval Observatory. How does Polaris pulsate? How does this differ from other Cepheids?

 45. Observing Stellar Evolution. Step through the animation "The Hertzsprung-Russell Diagram and Stellar Evolution" in Chapter 21 of the *Universe* web site or CD-ROM. Use this animation to answer the following questions. **(a)** How does a 1-$M_\odot$ star move on the H-R diagram during its first 4.6 billion (4600 million) years of existence? Compare this with the discussion in Section 21-1 of how the Sun has evolved over the past 4.6 billion years. **(b)** What is the zero-age spectral class of a 2-$M_\odot$ star? At what age does such a star evolve into a red giant of spectral class K? **(c)** What is the approximate zero-age luminosity of a 1.3-$M_\odot$ star? What is its approximate luminosity when it becomes a red giant? **(d)** Suppose a star cluster has no main-sequence stars of spectral classes O or B. What is the approximate age of the cluster? **(e)** Approximately how long do the most massive stars of spectral class B live before leaving the main sequence? What about the most massive stars of spectral class F?

OBSERVING PROJECTS

Observing tips and tools

 An excellent resource for learning how to observe variable stars is the web site of the American Association of Variable Star Observers. A wealth of data about specific variable stars can be found on the *Sky & Telescope* web site and in the three volumes of *Burnham's Celestial Handbook: An Observer's Guide to the Universe Beyond the Solar System* (Dover, 1978). This book also provides useful information about star clusters, as does the web site of Students for the Exploration and Development of Space.

46. Observe several of the red giants and supergiants listed below with the naked eye and through a telescope. (Note that γ Andromedae, α Tauri, and ε Pegasi are all multiple-star systems. The spectral type and luminosity class refer to the brightest star in these systems.) You can locate these stars in the sky using the *Starry Night* program on the CD-ROM that comes with this book. Is the reddish color of these stars apparent when they are compared with neighboring stars?

Star	Spectral type	R.A.	Decl.
Almach (γ Andromedae)	K3 II	2^h 03.9^m	+42° 20″
Aldebaran (α Tauri)	K5 III	4 35.9	+16 31
Betelgeuse (α Orionis)	M2 I	5 55.2	+07 24
Arcturus (α Boötis)	K2 III	14 15.7	+19 11
Antares (α Scorpii)	M1 I	16 29.5	−26 26
Eltanin (γ Draconis)	K5 III	17 56.7	+51 29
Enif (ε Pegasi)	K2 I	21 44.2	+09 52

Note: The right ascensions and declinations are given for epoch 2000.

47. Several of the open clusters referred to in Figure 21-10 can be seen with a good pair of binoculars. Observe as many of these clusters as you can, using both a telescope and binoculars. (Some are actually so large that they will not fit in the field of view of many telescopes.) You can locate these clusters in the sky using the *Starry Night* program on the CD-ROM that comes with this book. Note that in the following table, the M prefix refers to the Messier Catalog, NGC refers to the New General Catalogue, and Mel refers to the Melotte Catalog. In making your own observations, note the overall distribution of stars in each cluster. Which clusters are seen better through binoculars than through a telescope? Which clusters can you see with the naked eye?

Star cluster	Constellation	R.A.	Decl.
h Persei (NGC 869)	Perseus	2^h 19.0^m	+57° 09′
χ Persei (NGC 884)	Perseus	2 22.4	+57 07
Pleiades (M45)	Taurus	3 47.0	+24 07
Hyades (Mel 25)	Taurus	4 27	+16 00
Praesepe (M44)	Cancer	8 40.1	+19 59
Coma (Mel 111)	Coma Berenices	12 25	+26 00
Wild Duck (M11)	Scutum	18 51.1	−06 16

Note: The right ascensions and declinations are given for epoch 2000.

48. There are many beautiful globular clusters scattered around the sky that can be easily seen with a small telescope. Several of the brightest and nearest globulars are listed below. You can locate these clusters in the sky using the *Starry Night* program on the CD-ROM that comes with this book. Observe as many of these globular clusters as you can. Can you distinguish individual stars toward the center of each cluster? Do you notice any differences in the overall distribution of stars between clusters?

Star cluster	Constellation	R.A.	Decl.
M3 (NGC 5272)	Canes Venatici	13^h 42.2^m	+28° 23′
M5 (NGC 5904)	Serpens	15 18.6	+ 2 05
M4 (NGC 6121)	Scorpius	16 23.6	−26 32
M13 (NGC 6205)	Hercules	16 41.7	+36 28
M12 (NGC 6218)	Ophiuchus	16 47.2	− 1 57
M28 (NGC 6626)	Sagittarius	18 24.5	−24 52
M22 (NGC 6656)	Sagittarius	18 36.4	−23 54
M55 (NGC 6809)	Sagittarius	19 40.0	−30 58
M15 (NGC 7078)	Pegasus	21 30.0	+12 10

Note: The right ascensions and declinations are given for epoch 2000.

 49. Use the *Starry Night* program to view the sky as it appears at midnight (12:00:00 A.M.) on today's date from your location. (Use the time controls in the Control Panel and the **Viewing Locations...** command in the **Go Settings** menu.) Locate the giant star Aldebaran, the open cluster M44, and the globular cluster M12. (Use the **Find...** command in the **Edit** menu.) Which of these are visible from your location tonight at midnight? For each object that is visible, in which direction of the compass would you have to look to see it (that is, what is its *azimuth*), and how far above the horizon would you have to look (that is, what is its *altitude*)? (*Hint:* Double-click on each object to bring up an information window.)

Stellar Evolution: The Deaths of Stars

22

During a star's main-sequence lifetime, it swells slightly in radius, its luminosity gradually increases, and its surface temperature changes a little. But these changes pale compared with the transformation that lies ahead. When core hydrogen burning ceases and the aging star becomes a red giant, it comes to have a compressed core and a bloated atmosphere. Finally, it devours its remaining nuclear fuels and begins to die.

A star's death may be relatively gentle or almost inconceivably violent. The character of its death depends crucially on the star's mass.

A star of relatively low mass—such as our own Sun—ends its evolution by gently expelling its outer layers into space. These ejected gases form a glowing cloud called a *planetary nebula*. (The accompanying image shows such a nebula. The different colors indicate gases at different temperatures; blue is hottest and red is coolest.) The burned-out core that remains is called a *white dwarf*.

In contrast, at the end of the short life of a high-mass star, the star's core collapses suddenly. This triggers a powerful supernova explosion that can be as luminous as an entire galaxy of stars. Such supernovae liberate a torrent of neutrinos, and they can leave behind a relic of the star's collapsed core—a neutron star or a black hole. A white dwarf can also become a supernova if it accretes gas from a companion star in a close binary system.

WEB LINK 22.1

(NASA/The Hubble Heritage Team, STScI/AURA)

R I **V** U X G

Thermonuclear reactions in supernovae produce a wide variety of heavy elements, enriching the interstellar medium. In this way, the end of a star's life can provide new material for succeeding generations of stars.

As you read the sections of this chapter, look for the answers to the following questions.

22-1   What kinds of nuclear reactions occur within a star like the Sun as it ages?

22-2   Where did the carbon atoms in our bodies come from?

22-3   What is a planetary nebula, and what does it have to do with planets?

22-4   What is a white dwarf star?

22-5   Why do high-mass stars go through more evolutionary stages than low-mass stars?

22-6   What happens within a high-mass star to turn it into a supernova?

22-7   Why was SN 1987A an unusual supernova?

22-8   What was learned by detecting neutrinos from SN 1987A?

22-9   How do white dwarf stars give rise to certain types of supernovae?

22-10   What vestiges are left after a supernova explosion?

22-1 Low-mass stars go through two distinct red-giant stages

The life story of a low-mass star begins on the main sequence. (By *low mass*, we mean less than about 4 M$_\odot$.) As we saw in Chapter 21, a main-sequence star eventually uses up the hydrogen in its core. The aged star then leaves the main sequence.

We can describe a star's post–main-sequence evolution using an evolutionary track on a H-R diagram. Figure 22-1 shows the track for a 1-M$_\odot$ star like the Sun. Once core hydrogen burning ceases, the core shrinks, heating the surrounding hydrogen and triggering shell hydrogen burning. The new outpouring of energy causes the star's outer layers to expand and cool, and the star becomes a red giant. As the luminosity increases and the surface temperature drops, the post–main-sequence star moves up and to the right along the **red-giant branch** on an H-R diagram (Figure 22-1*a*).

Next, the helium-rich core of the star shrinks and heats until eventually **core helium burning** begins. This second post–main-sequence stage begins gradually in stars more massive

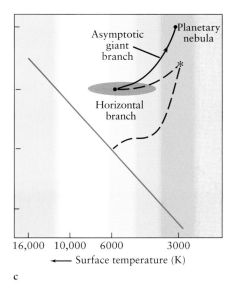

a   b   c

figure 22-1

ANIMATION 22.1

The Post–Main-Sequence Evolution of a 1-M$_\odot$ Star
(a) When a star becomes a red giant, its core shrinks and its outer layers expand, and the star moves up and to the right on the H-R diagram. The asterisk shows the star's position on the H-R diagram when the helium flash occurs. **(b)** After the helium flash, helium burning in the core produces carbon and oxygen. The core

expands and cools while the outer layers contract again and heat up. This continues until all of the helium in the core has been depleted. **(c)** After the core has been converted completely to carbon and oxygen, it again shrinks. Helium is still being burned in a shell around the core, and this makes the outer layers expand once again. The star eventually evolves into a planetary nebula (described in Section 22-3).

than about 2–3 $M_\odot$, but for less massive stars it comes suddenly—in a *helium flash*. During core helium burning, the surrounding hydrogen-burning shell still provides most of the red giant's luminosity.

As we learned in Section 21-2, the core expands when core helium burning begins, which makes the core cool down a bit. (We saw in Box 21-1 that letting a gas expand tends to lower its temperature, while compressing a gas tends to increase its temperature.) The cooling of the core also cools the surrounding hydrogen-burning shell, so that the shell releases energy more slowly. Hence, the luminosity goes down a bit after core helium burning begins.

The slower rate of energy release also lets the star's outer layers contract. As they contract, they heat up, so the star's surface temperature increases and its evolutionary track moves to the left on the H-R diagram in Figure 22-1b. The luminosity changes relatively little during this stage, so the evolutionary track moves almost horizontally, along a path called the **horizontal branch.** Horizontal-branch stars have helium-burning cores surrounded by hydrogen-burning shells. Figure 21-8 shows horizontal-branch stars in a globular cluster.

Helium burning produces nuclei of carbon and oxygen. After about a hundred million (10^8) years, essentially all the helium in the core of a 1-$M_\odot$ star has been converted into carbon and oxygen, and core helium burning ceases. Without thermonuclear reactions to maintain the core's internal pressure, the core again contracts, until it is stopped by degenerate-electron pressure (described in Section 21-2). This contraction releases heat into the surrounding helium-rich gases, and a new stage of helium burning begins in a thin shell around the core. This process is called **shell helium burning.**

History now repeats itself—the star enters a *second* red-giant phase. A star first becomes a red giant at the end of its main-sequence lifetime, when the outpouring of energy from shell hydrogen burning makes the star's outer layers expand and cool. In the same way, the outpouring of energy from shell *helium* burning causes the outer layers to expand again. The low-mass star ascends into the red-giant region of the H-R diagram for a second time (Figure 22-1c), but now with even greater luminosity than during its first red-giant phase.

Stars in this second red-giant phase are commonly called **asymptotic giant branch stars,** or **AGB stars,** and their evolutionary tracks follow what is called the **asymptotic giant branch.** (*Asymptotic* means "approaching," and, indeed, a star on the asymptotic giant branch approaches the red-giant branch from the left on an H-R diagram.)

When a low-mass star first becomes an AGB star, it consists of an inert, degenerate carbon-oxygen core and a helium-burning shell, both inside a hydrogen-burning shell, all within a volume approximately the size of the Earth. This small, dense central region is surrounded by an enormous hydrogen-rich envelope about as big as Earth's orbit. After a while, the expansion of the star's outer layers causes the hydrogen-burning shell to also expand and cool, and thermonuclear reactions in this shell temporarily cease. This leaves the aging star's structure as shown in Figure 22-2.

We saw in Section 21-1 that the more massive a star, the shorter the amount of time it remains on the main sequence. Similarly, the greater the star's mass, the more rapidly it goes through the stages of post–main-sequence evolution. Hence, we can see all of these stages by studying star clusters, which contain stars that are all the same age but that have a range of masses (see Section 21-3). Figure 22-3 shows a color-magnitude diagram for the globular cluster M55, which is about 15 billion years old. The least massive stars in this cluster are still on the main sequence. Progressively more massive stars have evolved to the red giant branch, the horizontal branch, and the asymptotic giant branch.

A 1-$M_\odot$ AGB star can reach a maximum luminosity of nearly $10^4\ L_\odot$, as compared with approximately $10^3\ L_\odot$ when it reached the helium flash and a relatively paltry 1 $L_\odot$ during its main-sequence lifetime. When the Sun becomes an AGB star some 8×10^9 years from now, this tremendous increase in luminosity will cause Mars and the Jovian planets to largely evaporate away. The Sun's bloated outer layers will reach to the Earth's orbit, and Mercury and Venus will simply be swallowed whole.

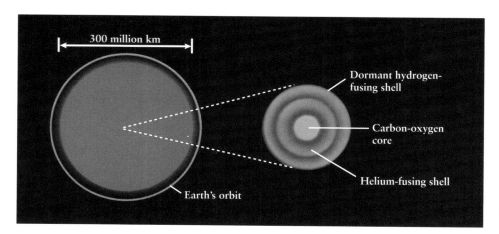

ƒigure 22–2

The Structure of an Old, Low–Mass AGB Star Near the end of its life, a low-mass star such as the Sun becomes a red asymptotic giant branch (AGB) star comparable in size to Earth's orbit around the Sun. The star's inert core, active helium-burning shell, and dormant hydrogen-burning shell are all contained within a volume roughly the size of the Earth. Thermonuclear reactions within the helium-burning shell take place at a furious rate, giving the star a luminosity thousands of times that of the present-day Sun.

WEB LINK 22.2

figure 22-3

Stellar Evolution in a Globular Cluster This color-magnitude diagram for the globular cluster M55 shows several stages in the evolution of stars. The lowest-mass stars in this cluster (less than about 0.8 $M_\odot$) are still on the main sequence, converting hydrogen into helium in their cores. Slightly more massive stars have consumed their core hydrogen and are ascending the red-giant branch. Other, even more massive stars have begun helium core burning; these are found on the horizontal branch. The cluster's most massive stars (which still have less than 4 $M_\odot$) have consumed all the helium in their cores and are ascending the asymptotic giant branch. (Compare with Figure 21-9.) (Adapted from D. Schade, D. VandenBerg, and F. Hartwick)

22-2 Dredge-ups bring the products of nuclear fusion to a giant star's surface

As we saw in Section 18-2, energy is transported outward from a star's core by one of two processes—radiative diffusion or convection. The first is the passage of energy in the form of electromagnetic radiation, and it dominates only when a star's gases are relatively transparent. The second involves up-and-down movement of the star's gases. Convection plays a very important role in giant stars, because it helps supply the cosmos with the elements essential to life.

ANIMATION 22.2
In the Sun, convection dominates only the outer layers, from around 0.71 solar radius (measured from the center of the Sun) up to the photosphere (recall Figure 18-5). During the final stages of a star's life, however, the convective zone can become so broad that it extends down to the star's core. At these times, convection can "dredge up" the heavy elements produced in and around the

core by thermonuclear fusion, transporting them all the way to the star's surface.

LOOKING DEEPER 22.1
The *first* dredge-up takes place after core hydrogen burning stops, when the star becomes a red giant for the first time. Convection dips so deeply into the star that material processed by the CNO cycle of hydrogen burning (see Section 18-1) is carried up to the star's surface, changing the relative abundances of carbon, nitrogen, and oxygen. A *second* dredge-up occurs after core helium burning ceases, further altering the abundances of carbon, nitrogen, and oxygen. Still later, during the AGB stage, a *third* dredge-up can occur if the star has a mass greater than about 2 $M_\odot$. This third dredge-up transports large amounts of freshly synthesized carbon to the star's surface, and the star's spectrum thus exhibits prominent absorption bands of carbon-rich molecules like C_2, CH, and CN. For this reason, an AGB star that has undergone a third dredge-up is called a **carbon star**.

All AGB stars have very strong stellar winds that cause them to lose mass at very high rates, up to 10^{-4} $M_\odot$ per year (a thousand times greater than that of a red giant, and 10^{10} times greater than the rate at which our present-day Sun loses mass). The surface temperature of AGB stars is relatively low, around 3000 K, so any ejected carbon-rich molecules can condense to form tiny grains of soot. Indeed, carbon stars are commonly found to be obscured in sooty cocoons of ejected matter (Figure 22-4).

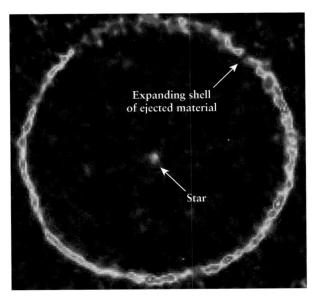

WEB LINK 22.3

figure 22-4 **R** I V U X G

A Carbon Star TT Cygni is an AGB star in the constellation Cygnus that ejects some of its carbon-rich outer layers into space. Some of the ejected carbon combines with oxygen to form molecules of carbon monoxide (CO), whose emissions can be detected with a radio telescope. This radio image shows the CO emissions from a shell of material that TT Cygni ejected some 7000 years ago. Over that time, the shell has expanded to a diameter of about ½ light-year. (H. Olofsson [Stockholm Observatory] et al./NASA)

Carbon stars are important because they enrich the interstellar medium with carbon and some nitrogen and oxygen. The triple alpha process that occurs in helium burning is the *only* way that carbon can be made, and carbon stars are the primary avenue by which this element is dispersed into interstellar space. Indeed, most of the carbon in your body was produced long ago inside a star by the triple alpha process (see Section 21-2), dredged up to the star's surface, ejected into space, and later incorporated into the solar nebula from which our Earth—and all of the life on it—eventually formed.

22-3 Low-mass stars die by gently ejecting their outer layers, creating planetary nebulae

The AGB stage in the evolution of a star of low mass (less than about 4 $M_\odot$) is a dramatic turning point. Before this stage, a star loses mass only gradually through steady stellar winds. But as it evolves during its AGB stage, a star divests itself completely of its outer layers. The aging star undergoes a series of bursts in luminosity, and in each burst it ejects a shell of material into space (see Figure 22-4). Eventually, all that remains of a low-mass star is a fiercely hot, exposed core, surrounded by glowing shells of ejected gas. This late stage in the life of a star is called a **planetary nebula** (see the image that opens this chapter).

To understand how this process works, consider the internal structure of an AGB star as shown in Figure 22-2. As the helium in the helium-burning shell is used up, the pressure that holds up the dormant hydrogen-burning shell decreases. Hence, the dormant hydrogen shell contracts and heats up, and hydrogen burning begins anew. This revitalized hydrogen burning creates helium, which rains downward onto the temporarily dormant helium-burning shell. As the helium shell gains mass, it shrinks and heats up. When the temperature of the helium shell reaches a certain critical value, it re-ignites in a **helium shell flash** that is similar to (but less intense than) the helium flash that occurred earlier in the evolution of a low-mass star (Section 21-2). The released energy pushes the hydrogen-burning shell outward, making it cool off, so that hydrogen burning ceases and this shell again becomes dormant. The process then starts over again.

When a helium shell flash occurs, the luminosity of an AGB star increases substantially in a relatively short-lived burst called a **thermal pulse**. Figure 22-5 shows a theoretical calculation of how the luminosity of a 0.7-$M_\odot$ AGB star varies because of helium shell flashes. The calculations predict that thermal pulses occur at ever-shorter intervals of about 300,000 years.

During these thermal pulses, the dying star's outer layers can separate completely from its carbon-oxygen core. As the ejected material expands into space, dust grains condense out of the cooling gases. Radiation pressure from the star's hot, burned-out core acts on the specks of dust, propelling them further outward, and the star sheds its outer layers altogether. In this way an aging 1-$M_\odot$ star loses as much as 40% of its

Figure 22-5

Thermal Pulses This graph shows how the luminosity of an 0.7-$M_\odot$ AGB star varies with time. The spikes in luminosity, called *thermal pulses*, are caused by brief periods of runaway helium burning. Such a thermal pulse may have caused the AGB star TT Cygni to eject the shell shown in Figure 22-4. These sudden increases in luminosity eventually cause the star to eject its outer layers completely. (Adapted from I. Iben)

mass. More massive stars eject even greater fractions of their original mass.

As a dying star ejects its outer layers, the star's hot core becomes exposed. With a surface temperature of about 100,000 K, this exposed core emits ultraviolet radiation intense enough to ionize and excite the expanding shell of ejected gases. These gases therefore glow and emit visible light through the process of fluorescence (see Box 20-1), producing a planetary nebula like those in Figure 22-6.

 Despite their name, planetary nebulae have nothing to do with planets. This misleading term was introduced in the nineteenth century because these glowing objects looked like distant Jovian planets when viewed through the small telescopes then available. The difference between planets and planetary nebulae became obvious with the advent of spectroscopy: Planets have absorption line spectra (see Section 7-3), but the excited gases of planetary nebulae have emission line spectra.

Planetary nebulae are quite common. Astronomers estimate that there are 20,000 to 50,000 planetary nebulae in our Galaxy alone. Many planetary nebulae, such as those in Figure 22-6, are more or less spherical in shape. This is a result of the symmetrical way in which the gases were ejected. But if the rate of expansion is not the same in all directions, the resulting nebula takes on an hourglass or dumbbell appearance (Figure 22-7).

Spectroscopic observations of planetary nebulae show emission lines of ionized hydrogen, oxygen, and nitrogen.

a   R I **V** U X G

b   R I **V** U X G

c   R **I** V U X G

WEB LINK 22.4

figure 22-6

Planetary Nebulae **(a)** The pinkish blob is a planetary nebula surrounding one of the 30,000 old stars in the globular cluster M15. The cluster is about 10,000 pc (33,000 ly) from Earth in the constellation Pegasus. **(b)** The closest of all known planetary nebulae is the Helix Nebula, which lies 140 pc (450 ly) from Earth in the constellation Aquarius and has an angular diameter of about ½° (the same as the full Moon). The "ring" is actually a spherical shell; it appears as a ring because we see a substantial thickness of the shell only when we look near its rim. **(c)** This infrared image of the planetary nebula NGC 7027 shows wispy shells of cooler gas surrounding a more luminous shell of hot gas, suggesting a complex evolutionary history. NGC 7027 is about 900 pc (3000 ly) from Earth in the constellation Cygnus and is roughly 14,000 AU across. (a: NASA/The Hubble Heritage Team, STScI/AURA; b: Anglo-Australian Observatory; c: William B. Latter, SIRTF Science Center/Caltech; and NASA)

From the Doppler shifts of these lines, astronomers have concluded that the expanding shell of gas moves outward from a dying star at speeds from 10 to 30 km/s. For a shell expanding at such speeds to have attained the typical diameter of a planetary nebula, about 1 light-year, it must have begun expanding about 10,000 years ago. Thus, by astronomical standards, the planetary nebulae we see today were created only very recently.

We do not observe planetary nebulae that are more than about 50,000 years old. After this length of time, the shell has

a

b

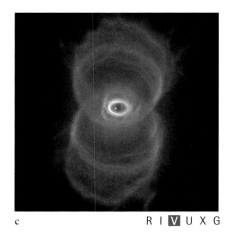

c   R I **V** U X G

WEB LINK 22.5

figure 22-7

Making an Elongated Planetary Nebula Many planetary nebulae have an elongated shape. One proposed explanation is that an aging star **(a)** first ejects a doughnut-shaped cloud of gas and dust (shown in gray) from its equator, then **(b)** ejects gas (shown in orange) from its entire surface. The doughnut blocks the flow of this second ejection, channeling it into two oppositely directed streams perpendicular to the plane of the doughnut. **(c)** The planetary nebula MyCn18 appears to have formed in just this way. It lies some 2500 pc (8000 ly) from Earth in the constellation Musca (the Fly). In this false-color image, the star is the white dot slightly off the center of the greenish "eye." (R. Sahai and J. Trauger, Jet Propulsion Laboratory; the WFPC-2 Science Team; and NASA) spread out so far from the cooling central star that its gases cease to glow and simply fade from view. The nebula's gases then mix with the surrounding interstellar medium. Astronomers estimate that all the planetary nebulae in the Galaxy return a total of about 5 $M_\odot$ to the interstellar medium each year. This amounts to 15% of all the matter expelled by all the various sorts of stars in the Galaxy each year. Because of this significant contribution, planetary nebulae play an important role in the chemical evolution of the Galaxy as a whole.

22-4 The burned-out core of a low-mass star cools and contracts until it becomes a white dwarf

Stars less massive than about 4 $M_\odot$ never develop the necessary central pressures or temperatures to ignite thermonuclear reactions that use carbon or oxygen as fuel. Instead, as we have seen, the process of mass ejection just strips away the star's outer layers and leaves behind the hot carbon-oxygen core. With no thermonuclear reactions taking place, the core simply cools down like a dying ember. These burnt-out relics of a star's former glory are called **white dwarfs.** White dwarfs are quite small by stellar standards—approximately the same size as the Earth (see Section 19-7).

You might think that without thermonuclear reactions to provide internal heat and pressure, a white dwarf should keep on shrinking under the influence of its own gravity as it cools. Actually, however, a cooling white dwarf maintains its size, because the burnt-out stellar core is so dense that most of its electrons are degenerate (see Section 21-2). Thus, degenerate-electron pressure supports the star against further collapse. This pressure does not depend on temperature, so it continues to hold up the star even as the temperature drops.

Many white dwarfs are found in the solar neighborhood, but all are too faint to be seen with the naked eye. One of the first white dwarfs to be discovered is a companion to Sirius, the brightest star in the night sky. In 1844 the German astronomer Friedrich Bessel noticed that Sirius was moving back and forth slightly, as if it was being orbited by an unseen object. This companion, designated Sirius B (Figure 22-8), was first glimpsed in 1862 by the American astronomer Alvan Clark. Recent satellite observations at ultraviolet wavelengths, where hot white dwarfs emit most of their light, show that the surface temperature of Sirius B is about 30,000 K.

Observations of white dwarfs in binary systems like Sirius allow astronomers to determine the mass, radius, and density of these stars (see Sections 19-9, 19-10, and 19-11). Such observations show that the density of the degenerate matter in a white dwarf is typically 10^9 kg/m^3 (a million times denser than water). A teaspoonful of white dwarf matter brought to Earth would weigh nearly 5.5 tons—as much as an elephant!

As we learned in Section 21-2, degenerate matter has a very different relationship between its pressure, density, and temperature than that of ordinary gases. Consequently, white dwarf stars have an unusual **mass-radius relation:** The more massive a white dwarf star, the *smaller* it is.

figure 22-8 R I **V** U X G

Sirius and Its White Dwarf Companion Sirius, the brightest-appearing star in the sky, is actually a binary star: The secondary star, called Sirius B, is a white dwarf. In this photograph Sirius B is almost obscured by the glare of the primary star. The spikes and rays around Sirius are the result of optical effects within the telescope. (Courtesy of R. B. Minton)

 Figure 22-9 displays the mass-radius relation for white dwarfs. Note that the more degenerate matter you pile onto a white dwarf, the smaller it becomes. However, there is a limit to how much pressure degenerate electrons can produce. As a result, there is an upper limit to the mass that a white dwarf can have. This

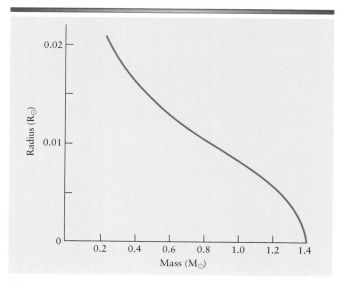

figure 22-9

The Mass-Radius Relationship for White Dwarfs The more massive a white dwarf is, the smaller its radius. This unusual relationship is a result of the degenerate-electron pressure that supports the star. The maximum mass of a white dwarf, called the Chandrasekhar limit, is 1.4 $M_\odot$. Incidentally, 0.01 $R_\odot$ is 6960 km, just slightly larger than the Earth's radius of 6378 km.

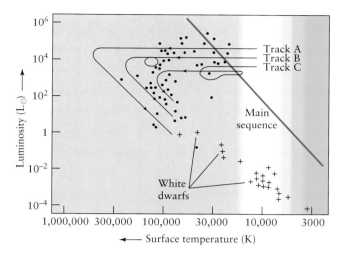

The initial red giants had masses of 0.8, 1.5, and 3.0 M$_\odot$. Mass ejection strips these dying stars of up to 60% of their matter. During their final spasms, the luminosity and surface temperature of these stars change quite rapidly. The points representing these stars on an H-R diagram race along their evolutionary tracks, sometimes executing loops corresponding to thermal pulses. Finally, as the ejected nebulae fade and the stellar cores cool, the evolutionary tracks of these dying stars take a sharp turn toward the white dwarf region of the H-R diagram.

Although a white dwarf maintains the same size as it cools, its luminosity and surface temperature both decrease with time. Consequently, the evolutionary tracks of aging white dwarfs point toward the lower right corner of the H-R diagram. You can see this in Figure 22-10; Figure 22-11 shows it in more detail. The energy that the white dwarf radiates into space comes only from the star's internal heat, which is a relic from the white dwarf's past existence as a stellar core. Over billions of years, white dwarfs grow dimmer and dimmer as their surface temperatures drop toward absolute zero.

After ejecting much of its mass into space, our own Sun will eventually evolve into a white dwarf star about the size of the Earth and with perhaps one-tenth of its present luminosity. It will become even dimmer as it cools. After 5 billion years as a white dwarf, the Sun will radiate with no more than one ten-thousandth of its present brilliance. With the passage of eons, our Sun will simply fade into obscurity.

Evolutionary track	Giant star	Ejected nebula	White dwarf
A	3.0	1.8	1.2
B	1.5	0.7	0.8
C	0.8	0.2	0.6

Mass (M$_\odot$)

ƒigure 22-10

Evolution from Giants to White Dwarfs This H-R diagram shows the evolutionary tracks of three low-mass giant stars as they eject planetary nebulae. The table gives the extent of mass loss in each case. The dots represent the central stars of planetary nebulae whose surface temperatures and luminosities have been determined. The crosses are white dwarfs for which similar data exist. (Adapted from B. Paczynski)

maximum mass is called the **Chandrasekhar limit,** after the Indian-American scientist Subrahmanyan Chandrasekhar, who pioneered theoretical studies of white dwarfs in the 1930s. (The orbiting Chandra X-ray Observatory, described in Section 6-7, is named in his honor.) The Chandrasekhar limit is equal to 1.4 M$_\odot$, meaning that all white dwarfs must have masses less than 1.4 M$_\odot$.

The material inside a white dwarf consists mostly of ionized carbon and oxygen atoms floating in a sea of degenerate electrons. As the dead star cools, the particles in this material slow down, and electric forces between the ions begin to prevail over the random thermal motions. The ions no longer move freely. Instead, they arrange themselves in orderly rows, like an immense crystal lattice. From this time on, you could say that the star is "solid." The degenerate electrons move around freely in this crystal material, just as electrons move freely through an electrically conducting metal like copper or silver. A diamond is also crystallized carbon, so a cool carbon-oxygen white dwarf resembles an immense spherical diamond!

Figure 22-10 shows the evolutionary tracks followed by three burned-out stellar cores as they become white dwarfs.

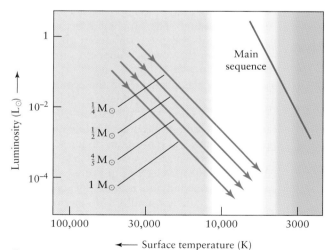

ƒigure 22-11

White Dwarf "Cooling Curves" As white dwarf stars radiate their internal energy into space, they become dimmer and cooler, and their evolutionary tracks move toward the lower right on the H-R diagram. The star's radius remains constant as it cools, however; compare these "cooling curves" with the lines of constant radius in Figure 19-15. The diagram shows tracks for white dwarfs of four different masses: the more massive a white dwarf, the smaller and hence fainter it is.

22-5 High-mass stars create heavy elements in their cores

During the entire lifetime of a low-mass star (mass less than about 4 $M_\odot$), just two kinds of thermonuclear reactions take place—hydrogen burning and helium burning. The heaviest elements manufactured by these reactions are carbon and oxygen.

The life story of a *high-mass* star (with a zero-age mass greater than about 4 $M_\odot$) begins with these same reactions. But theoretical calculations show that high-mass stars can also go through several additional stages of thermonuclear reactions involving the "burning" of carbon, oxygen, and other heavy nuclei. As a result, high-mass stars end their lives quite differently from low-mass stars.

Why is fusion of heavy nuclei possible only in a high-mass star? The reason is that such a star has more gravity trying to pull it together and more pressure at its center. Hence, the star's central temperature is also higher—high enough to fuse heavy elements. Heavy-nucleus fusion reactions require high temperatures, because these nuclei have large electric charges, which means that there are strong electric forces that tend to keep the nuclei apart. Only at the great speeds associated with very high temperatures can the nuclei travel fast enough to overcome their mutual electric repulsion and fuse together.

As we discussed in Section 21-1, when most main-sequence stars use up their core hydrogen, they begin shell hydrogen burning and enter a red-giant phase. These stars then begin core helium burning when the core temperature becomes high enough. (The exceptions are stars with mass less than 0.4 $M_\odot$. Convection in their interiors prevents the ignition of shell hydrogen burning, and they have too little internal pressure to trigger core helium burning.) The differences between low-mass and high-mass stars become pronounced after helium core burning ends, when the core is composed primarily of carbon and oxygen.

Let us consider how the late stages in the evolution of a high-mass star differ from those of a low-mass star. In low-mass stars, as we saw in Section 22-4, the carbon-oxygen core eventually becomes exposed and becomes a white dwarf. But in stars whose overall mass is more than about 4 $M_\odot$, the carbon-oxygen core is more massive than the Chandrasekhar limit of 1.4 $M_\odot$, so degenerate-electron pressure cannot prevent the core from contracting and heating. Hence, a high-mass star is able to enter a new round of core thermonuclear reactions. When the central temperature of such a high-mass star reaches 600 million kelvins (6×10^8 K), the first of the new thermonuclear reactions, **carbon burning**, begins. Carbon burning consumes carbon and produces oxygen, neon, sodium, and magnesium.

If a star has an even larger main-sequence mass of about 8 $M_\odot$ or so (before mass ejection), even more thermonuclear reactions can take place. After the cessation of carbon burning, the core will again contract, and the star's central temperature can rise to 1 billion kelvins (10^9 K). At this temperature **neon burning** begins. This uses up the neon accumulated

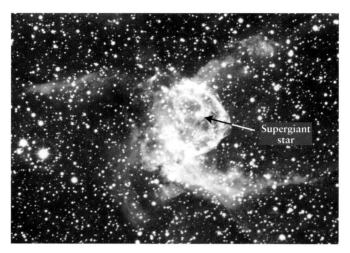

Supergiant star

f̄igure 22-12 R I **V** U X G

Mass Loss from a Supergiant Star At the heart of this nebulosity, called NGC 2359, lies a supergiant star with a mass in excess of 40 $M_\odot$. This star is losing mass at a rapid rate in a strong stellar wind. As this wind collides with the surrounding interstellar gas and dust, it creates the "bubble" shown here. NGC 2359, which has an angular size of 10 by 5 arcmin, lies in the constellation Canis Major. (Anglo-Australian Observatory)

from carbon burning and further increases the concentrations of oxygen and magnesium in the star's core.

After neon burning ends, the core will again contract, and **oxygen burning** will begin when the central temperature of the star reaches about 1.5 billion kelvins (1.5×10^9 K). The principal product of oxygen burning is sulfur. Once oxygen burning is over, the core will contract yet again. If the central temperature reaches 2.7 billion kelvins (2.7×10^9 K), **silicon burning** begins, producing a variety of nuclei from sulfur to iron.

Between each new stage of core thermonuclear reactions comes a period of shell burning and a new red-giant phase. This means that the evolutionary tracks of high-mass stars go through a number of back-and-forth gyrations on an H-R diagram. When each new stage of core burning begins, the luminosity and radius decrease and the surface temperature increases; when that stage ceases and only shell burning is taking place, the luminosity and radius increase, the surface temperature decreases, and the star again becomes a giant. While all of this is going on, the star is also losing mass at a rapid rate (Figure 22-12).

As a high-mass star consumes increasingly heavier nuclei, the thermonuclear reactions produce a wider variety of products. For example, oxygen burning also produces silicon, phosphorus, and magnesium. Some thermonuclear reactions that create heavy elements also release neutrons. A neutron is like a proton except that it carries no electric charge. Therefore, neutrons are not repelled by positively charged nuclei, and so can easily collide and combine with them. This absorption of neutrons by nuclei, called **neutron capture**, creates

Stage	Core temperature (K)	Core density (kg/m³)	Duration of stage
Hydrogen burning	4×10^7	5×10^3	7×10^6 years
Helium burning	2×10^8	7×10^5	7×10^5 years
Carbon burning	6×10^8	2×10^8	600 years
Neon burning	1.2×10^9	4×10^9	1 year
Oxygen burning	1.5×10^9	10^{10}	6 months
Silicon burning	2.7×10^9	3×10^{10}	1 day
Core collapse	5.4×10^9	3×10^{12}	¼ second
Core bounce	2.3×10^{10}	4×10^{15}	milliseconds
Explosive (supernova)	about 10^9	varies	10 seconds

table 22-1   Evolutionary Stages of a 25-M$_\odot$ Star

many elements and isotopes that are not produced directly in fusion reactions.

Each stage of thermonuclear reactions in a high-mass star helps to trigger the succeeding stage. In each stage, when the star exhausts a given variety of nuclear fuel in its core, gravitational contraction takes the core to ever-higher densities and temperatures, thereby igniting the "ash" of the previous fusion stage—and possibly the outlying shell of unburned fuel as well.

The increasing density and temperature of the core make each successive thermonuclear reaction more rapid than the one that preceded it. As an example, Stanford Woosley at the University of California, Santa Cruz, and Thomas Weaver at Lawrence Livermore National Laboratory have made detailed calculations for a star with a zero-age mass of 25 M$_\odot$. They find that carbon burning in such a star lasts for 600 years, neon burning for 1 year, and oxygen burning for only 6 months. The last, and briefest, stage of nuclear reactions is silicon burning.

The entire core supply of silicon in a 25-M$_\odot$ star is used up in only one day! Table 22-1 summarizes the evolutionary stages in the life of such a 25-M$_\odot$ star.

Each stage of core burning in a high-mass star generates a new shell of material around the core. After several such stages, the internal structure of a truly massive star—say, 25 M$_\odot$ or greater—resembles that of an onion, as Figure 22-13 shows. Because thermonuclear reactions can take place simultaneously in several shells, energy is released at such a rapid rate that the star's outer layers expand tremendously. The result is a **supergiant** star, whose luminosity and radius are much larger than those of a giant (see Section 19-7).

Several of the brightest stars in the sky are supergiants, including Betelgeuse and Rigel in the constellation Orion and Antares in the constellation Scorpius. (Figure 19-14 shows the locations of these stars on an H-R diagram.) They appear bright not because they are particularly close, but because they are extraordinarily luminous.

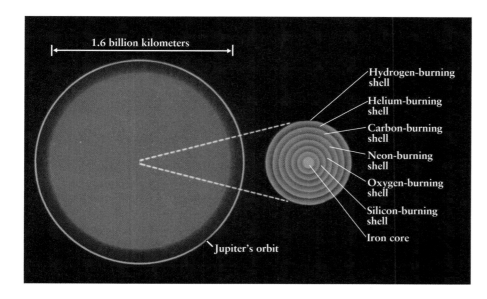

figure 22-13

The Structure of an Old High-Mass Star
Near the end of its life, a high-mass star becomes a supergiant whose overall size can be as large as Jupiter's orbit around the Sun. The star's energy comes from six concentric burning shells, all contained within a volume roughly the same size as the Earth. Thermonuclear reactions cannot occur within the iron core, because fusion reactions that involve iron absorb energy rather than releasing energy.

A supergiant star cannot keep adding shells to its "onion" structure forever, because the sequence of thermonuclear reactions cannot go on indefinitely. In order for an element to serve as a thermonuclear fuel, energy must be given off when its nuclei collide and fuse. This released energy is a result of the strong nuclear force of attraction that draws nucleons (neutrons and protons) together. However, protons also repel one another by the weaker electric force. As a result of this electric repulsion, adding extra protons to nuclei larger than iron, which has 26 protons, requires an *input* of energy rather than causing energy to be released. Nuclei of this size or larger cannot act as fuel for thermonuclear reactions. Hence, the sequence of burning stages ends with silicon burning. One of the products of silicon burning is iron, and the result is a star with an iron-rich core in which no thermonuclear reactions take place (see Figure 22-13).

Shell burning in the layers surrounding the iron-rich core consumes the star's remaining reserves of fuel. At this stage the entire energy-producing region of the star is contained in a volume no bigger than the Earth, some 10^6 times smaller in radius than the overall size of the star. This state of affairs will soon come to an end, because the buildup of an inert, iron-rich core signals the impending violent death of a massive supergiant star.

22-6 High-mass stars violently blow apart in supernova explosions

Our present understanding is that all stars of about 8 M$_\odot$ or less divest most of their mass in the form of planetary nebulae. The burned-out core that remains settles down to become a white dwarf star. But the truly massive stars—stellar heavyweights that begin their lives with more than 8 solar masses of material—do not pass through a planetary nebula phase. Instead, they die in spectacular explosions.

To understand what happens in such a supernova explosion, we must look deep inside a massive star at the end of its life. Of course, we cannot do this in actuality, because the interiors of stars are opaque. But astronomers have developed theoretical models based on what we know about the behavior of gases and atomic nuclei. The story that follows, while largely theoretical, describes our observations of supernovae fairly well. And as we will see in Section 22-8, a special kind of "telescope" has allowed us to glimpse the interior of at least one famous supernova.

The core of an aging, massive star gets progressively hotter as it contracts to ignite successive stages of thermonuclear burning. Wien's law (Section 5-4) and Planck's law (Section 5-5) together tell us that as the temperature of an object like a star increases, so does the energy of the photons it emits. When temperature in the core of a massive star reaches a few hundred million kelvins, the photons are energetic enough to initiate a host of nuclear reactions that create neutrinos. These neutrinos, which carry off energy, escape from the star's core, just as solar neutrinos flow freely out of the Sun (see Section 18-4 and the essay "Searching for Neutrinos Beyond the Textbooks" at the end of Chapter 18).

To compensate for the energy drained by the neutrinos, the star must provide energy either by burning more thermonuclear fuel, by contracting, or both. But when the star's core is converted into iron (as shown in Figure 22-13), no more energy-producing thermonuclear reactions are possible, and the only source of energy is contraction and rapid heating.

Once a star with an original mass of about 8 M$_\odot$ or more develops an iron-rich core, the core contracts very rapidly, so that the core temperature skyrockets to 5×10^9 K within a tenth of a second. The gamma-ray photons emitted by the intensely hot core have so much energy that when they collide with iron nuclei, they begin to break the nuclei down into alpha particles (^{4}He nuclei). This process is called **photodisintegration**. As Table 22-1 shows, it takes a high-mass star millions of years and several stages of thermonuclear reactions to build up an iron core; within a fraction of a second, photodisintegration undoes much of those millions of years of effort.

Within another tenth of a second, the core becomes so dense that the negatively charged electrons within it are forced to combine with the positively charged protons to produce electrically neutral neutrons. This process also releases a flood of neutrinos, denoted by the Greek letter ν (nu):

$$e^- + p \rightarrow n + \nu$$

Although neutrinos interact only very weakly with matter (see Section 18-4), the core is now so dense that even neutrinos cannot escape from it immediately. But because these neutrinos carry away a substantial amount of energy, as they escape from the core they cause the core to cool down and to condense even further.

At about 0.25 second after its rapid contraction begins, the core is less than 20 km in diameter and its density is in excess of 4×10^{17} kg/m^3. This is **nuclear density**, the density with which neutrons and protons are packed together inside nuclei. (If the Earth were compressed to this density, it would be only 300 meters, or 1000 feet, in diameter.)

Matter at nuclear density or higher is extraordinarily difficult to compress. Thus, when the density of the neutron-rich core begins to exceed nuclear density, the core suddenly becomes very stiff and rigid. The core's contraction comes to a sudden halt, and the innermost part of the core actually bounces back and expands somewhat. This *core bounce* sends a powerful wave of pressure, like an unimaginably intense sound wave, outward into the outer core.

During this critical stage, the cooling of the core has caused the pressure to decrease profoundly in the regions surrounding the core. Without pressure to hold it up against gravity, the material from these regions plunges inward at speeds up to 15% of the speed of light. When this inward-moving material crashes down onto the rigid core, it encounters the outward-moving pressure wave. In just a fraction of a second, the material that fell onto the core begins to move back out toward the star's surface, propelled in part by the flood of neutrinos trying to escape from the star's core.

5 milliseconds

10 milliseconds

20 milliseconds

ANIMATION 22.3

ƒigure 22-14

The Core of a Supernova These images from a supercomputer simulation show the core of a massive star during the first 20 milliseconds of a supernova explosion. The colors indicate temperatures that range from 2×10^{10} K (red) to 6×10^8 K (blue). Matter that bounces off the stiffened core of the star (colored black) produces a shock wave. Instabilities generate turbulent eddies that grow as the shock wave moves outward. (Courtesy of Adam Burrows, University of Arizona; and Bruce Fryxell, NASA/GSFC)

The outward-moving wave accelerates as it encounters less and less resistance and soon reaches a speed greater than the speed of sound waves in the star's outer layers. When this happens, the wave becomes a shock wave, like the sonic boom produced by a supersonic airplane.

After a few hours, the shock wave reaches the star's surface, by which time the star's outer layers have begun to lift away from the core. The energy released in this titanic event is an incomprehensibly large 10^{46} joules—a hundred times more energy than the Sun has emitted over the past 4.6 billion years. When the star's outer layers thin out sufficiently, a portion of this energy escapes in a torrent of light. The star has become a **supernova** (plural **supernovae**).

Supercomputer simulations provide many insights into the violent, complex, and rapidly changing conditions deep inside a star as it is torn apart by a supernova explosion. For example, Figure 22-14 shows the first 20 milliseconds after the stiffening of the inner core. Note the turbulent swirls and eddies that grow behind the shock wave that moves outward from the star's core.

Recent research suggests that turbulence is an important feature of a supernova explosion. Without turbulence, the shock wave might not be able to plow through the bulk of the doomed star, which continues to fall inward. Turbulence, which involves plumes of matter moving at up to half the speed of light, can revitalize the explosion and ensure that the star blows up.

Figure 22-14 depicts the death of a 25-$M_\odot$ star. (The final stages in the evolution of other massive stars probably follow similar scenarios, although details may differ.) Detailed computer calculations suggest that a 25-$M_\odot$ star ejects about 96% of its material to the interstellar medium for use in producing future generations of stars. Less massive stars eject proportionally less of their mass into space when they become supernovae.

Before this material is ejected into space, it is compressed so much by the passage of the shock wave through the star's outer layers that a new wave of thermonuclear reactions set in. These reactions produce many more chemical elements, including *all* the elements heavier than iron. These reactions require a tremendous input of energy, and thus cannot take place during the star's pre-supernova lifetime.

The energy-rich environment of a supernova shock wave is the *only* place where such heavy elements as zinc, silver, tin, gold, mercury, lead, and uranium can be produced. Remarkably, all of these elements are found on the Earth. This means that our solar system, our Earth, and our bodies include material that long ago was part of a star that lived, evolved, and died as a supernova.

22-7 In 1987 a nearby supernova gave us a close-up look at the death of a massive star

Our discussion of supernovae in the previous section was based on theoretical models and supercomputer calculations. But supernovae are not just theory; several dozen of them are seen each year. Supernovae have peak luminosities as great as 10^9 $L_\odot$, rivaling the light output of an entire galaxy. This makes it possible to see supernovae in galaxies far beyond our own Milky Way Galaxy. These observations provide a crucial test of the theoretical ideas in Section 22-6.

On February 23, 1987, a supernova was discovered in the Large Magellanic Cloud (LMC), a companion galaxy to our Milky Way some 50,000 pc (160,000 ly) from Earth. The supernova, designated SN 1987A because it was the first discovered that year, occurred near an enormous H II region in the LMC called the Tarantula Nebula (Figure 22-15). The supernova was so bright that observers in the southern hemisphere could see it without a telescope.

Such a bright supernova is a rare event. The last supernova bright enough to be seen with the naked eye (and the last one

figure 22-15   R I **V** U X G

Supernova 1987A A supernova appeared in a nearby galaxy called the Large Magellanic Cloud (LMC) in 1987. This photograph, taken soon after the discovery, shows a portion of the LMC that includes the supernova (SN 1987A) and a huge H II region called the Tarantula Nebula. SN 1987A was the first supernova in almost 400 years that was bright enough to be seen without a telescope. (European Southern Observatory)

placed into service. These enabled astronomers to study the supernova's evolution with unprecedented resolution and in wavelength ranges not accessible from the Earth's surface (see Section 6-7). As a result, SN 1987A has given astronomers an unprecedented view of the violent death of a massive star.

The light from a supernova such as SN 1987A does not all come in a single brief flash; the outer layers continue to glow as they expand into space. For the first 20 days after the detonation of SN 1987A, this glow was powered primarily by the tremendous heat that the shock wave deposited in the star's outer layers. As the expanding gases cooled, the light energy began to be provided by a different source—the decay of radioactive isotopes of cobalt, nickel, and titanium produced in the supernova explosion.

Astronomers have been able to pinpoint the specific isotopes involved because different radioactive nuclei emit gamma rays of different wavelengths when they decay. These emissions have been detected by orbiting gamma-ray telescopes (see Figure 6-33). Thanks to these radioactive decays, the brightness of SN 1987A actually *increased* for the first 85 days after the detonation, then settled into a slow decline as the radioactive isotopes were used up. The supernova remained visible to the naked eye for several months after the detonation.

Ideally, SN 1987A would have confirmed the theories of astronomers about typical supernovae. But SN 1987A was *not* typical. Its luminosity peaked at roughly 10^8 $L_\odot$, only a tenth of the maximum luminosity observed for other, more distant supernovae. Fortunately, the doomed star had been observed prior to becoming a supernova, and these observations helped explain why SN 1987A was an exceptional case.

The star that exploded and became SN 1987A, called the **progenitor star,** was identified as a blue B3 I supergiant (Figure 22-16). When this star was on the main sequence its mass was about 20 $M_\odot$, although by the time it exploded—some 10^7 years after it first formed—it probably had shed a few solar

to have been observed in the Milky Way Galaxy) was in 1604. Although outside our galaxy, SN 1987A occurred relatively close to us in a part of the heavens that is obscured only slightly by the Milky Way's interstellar dust. Furthermore, not long after SN 1987A appeared, several new orbiting telescopes were

a

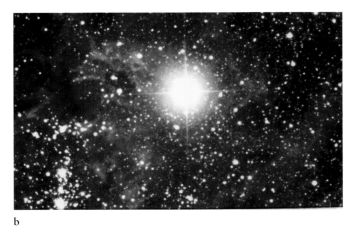

b

figure 22-16   R I **V** U X G

SN 1987A—Before and After (a) This photograph shows a small section of the Large Magellanic Cloud as it appeared before the explosion of SN 1987A. The arrow points to the supernova's progenitor star, a B3 blue supergiant. (b) This image shows a somewhat larger region of the sky a few days after the supernova exploded into brightness. (Anglo-Australian Observatory)

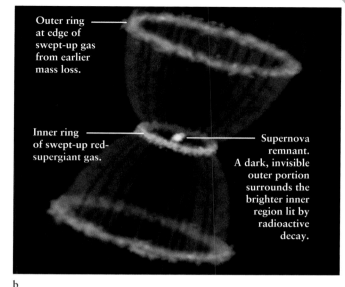

Outer ring
at edge of
swept-up gas
from earlier
mass loss.

Inner ring
of swept-up red-
supergiant gas.

Supernova
remnant.
A dark, invisible
outer portion
surrounds the
brighter inner
region lit by
radioactive
decay.

a   R I **V** U X G   b

ƒigure 22-17

SN 1987A and Its "Three-Ring Circus" **(a)** This true-color view from the Hubble Space Telescope shows three bright rings around SN 1987A (the bright dot at the center of the rings). The inner ring is about 1.3 light-years (0.4 parsec) in diameter. The two stars near the outer rings are not part of the supernova. **(b)** This drawing shows the probable origin of the rings. A wind from the progenitor star of SN 1987A formed an hourglass-shaped shell surrounding the star. (Compare with Figure 22-7.) Ultraviolet light from the supernova detonation ionized ring-shaped regions in the shell, causing them to glow. (Robert Kirshner and Peter Challis, Harvard-Smithsonian Center for Astrophysics; STScI)

masses. The evolutionary track for an aging 20-M$_\odot$ star wanders back and forth across the top of the H-R diagram as the star alternates between being a hot (blue) supergiant and a cool (red) supergiant. The star's size changes significantly as the surface temperature changes. A blue supergiant is 10 times larger in diameter than the Sun, but a red supergiant of the same luminosity has a diameter 1000 times that of the Sun.

Because the progenitor star was relatively small when it exploded, the star's outer layers were close to the core and thus held more strongly by the core's gravitational attraction. When the detonation occurred, a relatively large fraction of the shock wave's energy had to be used against this gravitational attraction to push the outer layers into space. Hence, the amount of shock wave energy available to be converted into light was smaller than for most supernovae. The result was that SN 1987A was only a tenth as bright as an exploding red supergiant.

Three and a half years after SN 1987A exploded, astronomers used the newly launched Hubble Space Telescope to obtain a picture of the supernova. To their surprise, the image showed a ring of glowing gas around the exploded star. In 1994, soon after the optics of the Hubble Space Telescope were repaired (see Section 6-7), SN 1987A was observed again and a set of *three* glowing rings was revealed (Figure 22-17*a*).

These rings are relics of a hydrogen-rich outer atmosphere that was ejected by gentle stellar winds from the progenitor

star when it was a red supergiant, about 20,000 years ago. This diffuse gas expanded in a hourglass shape (Figure 22-17*b*), because it was blocked from expanding around the star's equator either by a preexisting ring of gas or by the orbit of an as yet unseen companion star. (Figure 22-7 shows a similar model used to explain the shapes of certain planetary nebulae.) The outer rings in Figure 22-17*a* are parts of the hourglass that were ionized by the initial flash of ultraviolet radiation from the supernova; as electrons recombine with the ions, the rings emit visible-light photons. (We described this process of *recombination* in Section 20-2.)

 At this writing (early 2001), the shock wave from the supernova is beginning to collide with the "waist" of the hourglass shown in Figure 22-17*b*. This collision is making the hourglass glow more brightly in visible wavelengths—not enough, unfortunately, to make the supernova again visible to the naked eye—and emit copious radiation at X-ray and ultraviolet wavelengths.

By studying this cosmic collision, astronomers hope to learn more about the shock wave and the supernova explosion that spawned it. They hope as well to learn about the matter in the inner ring, which will give us insight into the stellar winds that blew from the progenitor star thousands of years ago. Because it provides a unique laboratory for studying the evolution of a supernova, SN 1987A will be carefully monitored by astronomers for decades to come.

22-8 Neutrinos emanate from supernovae like SN 1987A

In addition to electromagnetic radiation, supernovae also emit a brief but intense burst of neutrinos from their collapsing cores. In fact, theory suggests that *most* of the energy released by the exploding star is in the form of neutrinos. If it were possible to detect the flood of neutrinos from an exploding star, astronomers would have direct evidence of the nuclear processes that occur within the star during its final seconds before becoming a supernova.

Unfortunately, detecting neutrinos is a difficult, tricky business. The problem is that under most conditions matter is transparent to neutrinos (see Section 18-4). Consequently, when supernova neutrinos encounter the Earth, almost all of them pass completely through the planet as if it were not there. The challenge to scientists is to detect the tiny fraction of neutrinos that *do* interact with the matter through which they pass.

 During the 1980s, two "neutrino telescopes" uniquely suited for detecting supernova neutrinos went into operation—the Kamiokande detector in Japan (a joint project of the University of Tokyo and the University of Pennsylvania), and the IMB detector (a collaboration of the University of California at Irvine, the University of Michigan, and Brookhaven National Laboratory). Both detectors consisted of large tanks containing thousands of tons of water. On the rare occasion when a neutrino collided with one of the water molecules, it produced a brief flash of light. Any light flash was recorded by photomultiplier tubes lining the walls of the tank (Figure 22-18). Because other types of subatomic particles besides neutrinos could also produce similar light flashes, the detectors were placed deep underground so that hundreds of meters of earth would screen out almost all particles except neutrinos.

What causes the light flashes? And how do we know whether a neutrino comes from a supernova? The key is that a single supernova neutrino carries a relatively large amount of energy, typically 20 MeV or more. (We introduced the unit of energy called the electron volt, or eV, in Section 5-5. One MeV is equal to 10^6 electron volts. Only the most energetic nuclear reactions produce particles with energies of more than 1 MeV.) If such a high-energy neutrino hits a proton in the water-filled tank of a neutrino telescope, the collision produces a positron (see Box 18-1). The positron then recoils at a speed greater than the speed of light in water, which is 2.3×10^5 km/s. (Such motion does not violate the ultimate speed limit in the universe, 3×10^5 km/s, which is the speed of light in a *vacuum*.)

Just as an airplane that flies faster than sound produces a shock wave (a sonic boom), a positron that moves through a substance such as water faster than the speed of light in that substance produces a shock wave of light. This shock wave is called **Cerenkov radiation** after the Russian physicist Pavel A. Cerenkov, who first observed it in 1934. It is this radiation that is detected by the photomultiplier tubes that line the detector walls.

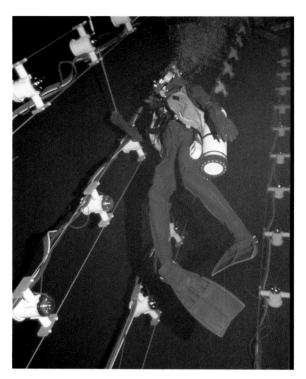

Figure 22-18 R I **V** U X G

Inside a Neutrino Detector This diver is servicing one of the light-sensitive photomultiplier tubes lining the walls of the Irvine-Michigan-Brookhaven (IMB) neutrino detector. Eight neutrinos from SN 1987A struck protons in the water, producing brief flashes of light that were picked up by the photomultipliers. (Courtesy of K. S. Luttrell)

By measuring the properties of the Cerenkov radiation from a recoiling positron, scientists can determine the positron's energy and, therefore, the energy of the neutrino that created the positron. This allows them to tell the difference between the high-energy neutrinos from supernovae and neutrinos from the Sun, which typically have energies of 1 MeV or less.

Both the Kamiokande and IMB detectors were operational on February 23, 1987, when SN 1987A was first observed in the Large Magellanic Cloud. Soon afterward, the physicists working with these detectors excitedly reported that they had detected Cerenkov flashes from a 12-second burst of neutrinos that reached the Earth 3 hours before astronomers saw the light from the exploding star. Only a few neutrinos were seen: The Kamiokande detector saw flashes from 12 neutrinos at about the same time that 8 were recorded by the IMB detector. But when the physicists factored in the sensitivity of their detectors, they calculated that Kamiokande and IMB had actually been exposed to a torrent of more than 10^{16} neutrinos.

Given the flux of neutrinos measured by the detectors and the 160,000–light-year distance from SN 1987A to the Earth, physicists used the inverse-square law to determine the total number of neutrinos that had been emitted from the supernova. (This law applies to neutrinos just as it does to electromagnetic radiation; see Section 19-2.) They found that over a 10-second period, SN 1987A emitted 10^{58} neutrinos with a total energy of 10^{46} joules. This is more than 100 times as much energy as the Sun has emitted in its entire history and more than 100 times the amount of energy that the supernova emitted in the form of electromagnetic radiation. Indeed, for a few seconds the supernova's neutrino luminosity—that is, the *rate* at which it emitted energy in the form of neutrinos—was 10 times greater than the total luminosity in electromagnetic radiation of all of the stars in the observable universe! Such comparisons give a hint of the incomprehensible violence with which a supernova explodes.

Why did the neutrinos from SN 1987A arrive 3 hours *before* the first light was seen? As we saw in Section 22-6, neutrinos are produced when thermonuclear reactions cease in the core of a massive star and the core collapses. These neutrinos encounter little delay as they pass through the volume of the star. The tremendous increase in the star's light output, by contrast, occurs only when the shock wave reaches the star's outermost layers (which are thin enough to allow light to pass through them). It took 3 hours for this shock wave to travel outward from the star's core to its surface, by which time the neutrino burst was already billions of kilometers beyond the dying star. For the next 160,000 years, the neutrinos that would eventually produce light flashes in Kamiokande and IMB remained in front of the photons emitted from the star's surface, and so the neutrinos were detected before the supernova's light. Thus, the neutrino data from SN 1987A gave astronomers direct confirmation of theoretical ideas about how supernova explosions take place.

Kamiokande and IMB have both been replaced by a new generation of neutrino telescopes (see Section 18-4). While these new detectors are intended primarily to observe neutrinos from the Sun, they are also fully capable of measuring neutrino bursts from nearby supernovae. Astronomers have identified a number of supergiant stars in our Galaxy that are likely to explode into supernovae, among them the bright red supergiant Betelgeuse in the constellation Orion (see Figure 2-2 and Figure 6-29).

Unfortunately, astronomers do not yet know how to predict precisely when such stars will explode into supernovae. It may be many thousands of years before Betelgeuse explodes. Then again, it could happen tomorrow. If it does, neutrino telescopes will be ready to record the collapse of its massive core.

22-9 White dwarfs in close binary systems can also become supernovae

Astronomers discover dozens of supernovae in distant galaxies every year (Figure 22-19), but not all of these are the result of massive stars dying violently. A totally different type of supernova occurs when a white dwarf star in a binary system blows itself completely apart. The first clue that two entirely distinct chains of events could produce supernovae was rather subtle: Some supernovae have prominent hydrogen emission lines in their spectra but others do not.

Supernovae with hydrogen emission lines, called **Type II supernovae**, are the sort we described in Section 22-6. They are caused by the death of highly evolved massive stars that still have ample hydrogen in their atmospheres when they explode. When the star explodes, the hydrogen atoms are excited and glow prominently, producing hydrogen emission

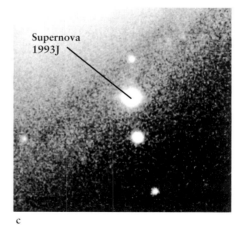

a   b   c

WEB LINK 22.12

ƒigure 22-19   R I V U X G

A Supernova in a Distant Galaxy   On the night of March 28, 1993, Francisco Garcia Diaz, a Spanish amateur astronomer, discovered supernova SN 1993J in the galaxy M81 in Ursa Major. **(a)** M81 lies some 3.6 million pc (12 million ly) from Earth. Its angular size is about half that of the full Moon. The supernova occurred within the region outlined by the yellow box; this region is shown in close-up in parts (b) and (c). **(b)** This close-up image made before the supernova explosion shows the progenitor star of SN 1993J, a K0 I red supergiant. **(c)** This image shows the same part of the sky as (b), but after the explosion of SN 1993J. This supernova, like SN 1987A, resulted from the explosion of a massive star. (a: Palomar Observatory; b, c: D. Jones and E. Telles, Isaac Newton Telescope)

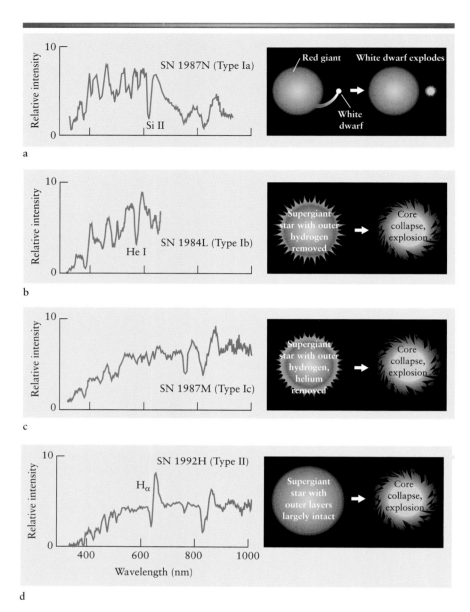

figure 22-20

Supernova Types **(a)** The spectrum of a Type Ia supernova has no hydrogen or helium lines. These supernovae are thought to be produced by runaway carbon burning in a white dwarf star in a close binary system. The strong absorption line is due to ionized silicon (Si II), a by-product of carbon burning. **(b)** The spectrum of a Type Ib supernova has no hydrogen lines, but does have a strong absorption line of unionized helium (He I). Such a supernova may result from core collapse in a massive star that lost the hydrogen from its outer layers. **(c)** Type Ic supernovae lack both hydrogen and helium lines in their spectra. The progenitors are probably massive stars that lose both hydrogen and helium from their outer layers prior to core collapse. **(d)** Type II supernovae, like their cousins Type Ib and Type Ic, result from core collapse within a massive star. In this case, however, enough hydrogen remained with the star before exploding to produce strong hydrogen lines such as H_α. (Spectra courtesy of Alexei V. Filippenko, University of California, Berkeley)

lines. SN 1987A, the topic of Sections 22-7 and 22-8, was a Type II supernova.

Hydrogen lines are missing in the spectrum of a **Type I supernova**, which tells us that little or no hydrogen is left in the debris from the explosion. Type I supernovae are further divided into three important subclasses. **Type Ia** supernovae have spectra that include a strong absorption line of ionized silicon. **Type Ib** and **Type Ic** supernovae both lack the ionized silicon line. The difference between them is that the spectra of Type Ib supernovae have a strong helium absorption line, while those of Type Ic supernovae do not. Figure 22-20 shows these different supernova spectra.

Astronomers suspect that Type Ib and Ic supernovae are caused by dying massive stars, just like Type II supernovae. The difference is that the progenitor stars of Type Ib and Ic supernovae have been stripped of their outer layers before they explode. A star can lose its outer layers to a strong stellar wind

(see Figure 22-12) or, if it is part of a close binary system, by transferring mass to its companion star (see Figure 21-17b). If enough mass remains for the star's core to collapse, the star dies as a Type Ib supernova. Because the outer layers of hydrogen are absent, the supernova's spectrum exhibits no hydrogen lines but many helium lines (Figure 22-20b). Type Ic supernovae have apparently undergone even more mass loss prior to their explosion; their spectra show that they have lost much of their helium as well as their hydrogen (Figure 22-20c).

The evidence that Type II, Type Ib, and Type Ic supernovae all begin as massive stars is that all three types are found only near sites of recent star formation. The life span of a massive star from its formation to its explosive death as a supernova is only about 10^7 years, less time than it took our own Sun to condense from a protostar to a main-sequence star (see Figure 20-9) and a mere blink of an eye on the time scale of stellar evolution. Because massive stars live such a short time, it makes sense that they should meet their demise very close to where they were formed.

Type Ia supernovae, by contrast, are found even in galaxies where there is no ongoing star formation. Hence, they are probably *not* the death throes of massive supergiant stars. Instead, Type Ia supernovae may result from the thermonuclear explosion of a white dwarf star. This may sound contradictory, because we saw in Section 22-4 that white dwarf stars have no thermonuclear reactions going on in their interiors. But these reactions *can* occur if a carbon-oxygen–rich white dwarf is in a close, semidetached binary with a red giant star (see Figure 21-16b).

As the red giant evolves and its outer layers expand, it overflows its Roche lobe and dumps gas from its outer layers onto the white dwarf (Figure 22-20a). When the total mass of the white dwarf approaches the Chandrasekhar limit, the increased pressure applied to the white dwarf's interior causes carbon burning to begin there. Hence, the interior temperature of the white dwarf increases.

 If the white dwarf were made of ordinary matter, the temperature increase would cause a further increase in pressure, the white dwarf would expand and cool, and the carbon-burning reactions would abate. But because the white dwarf is composed of degenerate matter, this "safety valve" between temperature and pressure does not operate. Instead, the increased temperature just makes the reactions proceed at an ever-increasing rate, in a catastrophic runaway process reminiscent of the helium flash in low-mass stars. Soon the temperature gets so high that the electrons are no longer degenerate and the white dwarf blows apart, dispersing all of its mass into space.

Before exploding, the white dwarf contained primarily carbon and oxygen and almost no hydrogen or helium, which explains the absence of hydrogen and helium lines in the spectrum of the resulting supernova. Silicon is a by-product of the carbon-burning reaction and gives rise to the silicon absorption line characteristic of Type Ia supernovae.

 Different types of supernovae have fundamentally different energy sources. Type II, Ib, and Ic supernovae are powered by *gravitational* energy. The star's iron-rich core collapses inward under the influence of its own gravity, releasing a flood of energy in the form of neutrinos; the outer layers of the star also fall inward under gravity, providing the energy to power the nuclear reactions that generate the supernova's electromagnetic radiation. Type Ia supernovae, by contrast, are powered by *nuclear* energy released in the explosive thermonuclear burning of a white dwarf star. While Type Ia supernovae typically emit more energy in the form of electromagnetic radiation than supernovae of other types, they do not emit copious numbers of neutrinos because there is no core collapse. If we include the energy emitted in the form of neutrinos, the most luminous supernovae by far are those of Type II.

In addition to the differences in their spectra, different types of supernovae can be distinguished by their light curves (Figure 22-21). All supernovae begin with a sudden rise in

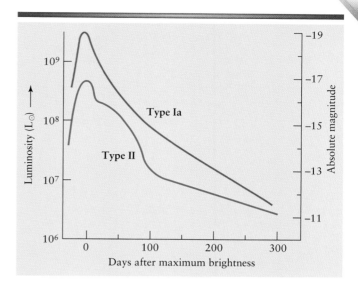

figure 22-21

Supernova Light Curves A Type Ia supernova reaches maximum brightness in about a day, followed by a gradual decline in brightness. A Type II supernova reaches a maximum brightness only about one-fourth that of a Type Ia supernova and usually has alternating intervals of steep and gradual declines.

brightness that occurs in less than a day. After reaching peak luminosity, Type Ia, Ib, and Ic supernovae settle into a steady, gradual decline in luminosity. By contrast, the Type II light curve has a steplike appearance caused by alternating periods of steep and gradual declines in brightness.

For all supernova types, the energy source during the period of declining brightness is the decay of radioactive isotopes produced during the supernova explosion. Because a different set of thermonuclear reactions occur for each type of supernova, each type produces a unique set of isotopes that decay at different rates. This helps explains the distinctive light curves for different supernova types.

For the same reason, each type of supernovae ejects a somewhat different mix of elements into the interstellar medium. As an example, Type Ia supernovae are primarily responsible for the elements near iron in the periodic table, because they generate these elements in more copious quantities than Type II supernovae.

A number of astronomers are now measuring the distances to remote galaxies by looking for Type Ia supernovae in those galaxies. Because all such supernovae have about the same maximum luminosity, a measurement of a Type Ia supernova's peak apparent brightness tells us (through the inverse-square law) the distance to the supernova, and, therefore, the distance to the supernova's host galaxy. The tremendous luminosity of Type Ia supernovae allows this method to be used for galaxies more than 10^9 light-years distant. In Chapter 28 we will learn what such studies tell us about the size and evolution of the universe as a whole.

22-10 A supernova remnant can be detected at many wavelengths for centuries after the explosion

Astronomers find the debris of supernova explosions, called **supernova remnants,** scattered across the sky. A beautiful example is the Veil Nebula, shown in Figure 22-22. The doomed star's outer layers were blasted into space with such violence that they are still traveling through the interstellar medium at supersonic speeds 15,000 years later. As this expanding shell of gas plows through space, it collides with atoms in the interstellar medium, exciting the gas and making it glow.

A few nearby supernova remnants cover sizable areas of the sky. The largest is the Gum Nebula, named after the astronomer Colin Gum, who first noticed its faint glowing wisps on photographs of the southern sky (Figure 22-23). Its 40° angular diameter is centered on the constellation Vela.

The Gum Nebula looks big simply because it is quite close to us. Its center is only about 460 pc (1300 ly) from Earth, and its near side is just 100 pc (330 ly) away. Studies of the nebula's expansion rate suggest that this supernova exploded around 9000 B.C. At maximum brilliance, the exploding star probably was as bright as the Moon at first quarter. Like the first quarter Moon, it would have been visible in the daytime!

Many supernova remnants are virtually invisible at optical wavelengths. However, when the expanding gases collide with the interstellar medium, they radiate energy at a wide range of wavelengths, from X rays through radio waves. For example, Figure 22-24 shows both X-ray and radio images of the super-

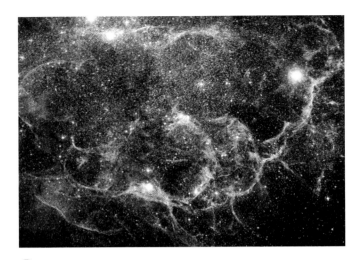

ʄigure 22-23   R I **V** U X G

The Gum Nebula—A Supernova Remnant The Gum Nebula, which spans 40° of the sky, has the largest angular size of any known supernova remnant. Only the central regions of the nebula are shown here. The supernova explosion occurred about 11,000 years ago, and the remnant now has a diameter of about 700 pc (2300 ly). (Royal Observatory, Edinburgh)

nova remnant Cassiopeia A. Optical photographs of this part of the sky reveal only a few small, faint wisps. In fact, radio searches for supernova remnants are more fruitful than optical searches. Only two dozen supernova remnants have been found in visible-light images, but more than 100 remnants have been discovered by radio astronomers.

From the expansion rate of Cassiopeia A, astronomers conclude that this supernova explosion occurred about 300 years ago. Although telescopes were in wide use by the late 1600s, no one saw the outburst (and no one today knows why). The last supernova seen in our Galaxy, which occurred in 1604, was observed by Johannes Kepler. In 1572, Tycho Brahe also recorded the sudden appearance of an exceptionally bright star in the sky. To find any other accounts of nearby bright supernovae, we must delve into astronomical records that are almost 1000 years old.

At first glance, this apparent lack of nearby supernovae may seem puzzling. From the frequency with which supernovae occur in distant galaxies, it is reasonable to suppose that a galaxy such as our own should have as many as five supernovae per century. Where have they been?

As we will learn from the study of galaxies in Chapters 25 and 26, the plane of our Galaxy is where massive stars are born and supernovae explode. This region is so rich in interstellar dust, however, that we simply cannot see very far into space in the directions occupied by the Milky Way (see Section 20-2). In other words, supernovae probably do in fact erupt every few decades in remote parts of our Galaxy, but their detonations are hidden from our view by intervening interstellar matter.

A supernova remnant may be all that is left after some supernovae explode. But for supernovae of Types II, Ib, and

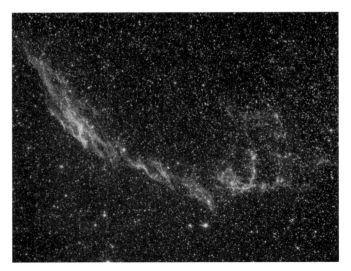

ʄigure 22-22   R I **V** U X G

The Veil Nebula—A Supernova Remnant This nebulosity is a portion of the Cygnus Loop (see Figure 20-22), which is the remnant of a supernova that exploded about 15,000 years ago. The distance to the nebula is about 800 pc (2600 ly), and the overall diameter of the loop is about 35 pc (120 ly). (Palomar Observatory)

a   R I V U **X** G

b   **R** I V U X G

WEB LINK 22.14

figure 22-24

Cassiopeia A—A Supernova Remnant **(a)** This false-color X-ray image of Cassiopeia A was taken by the Chandra X-ray Observatory (Figure 6-32a). The colors represent emission at different wavelengths by different types of atoms. **(b)** This false-color radio image of Cassiopeia A was produced by the Very Large Array (Figure 6-25). In this image, color indicates the intensity of radio emission (red is strongest). The impact of supernova material on the interstellar medium causes ionization, and the liberated electrons generate radio waves as they move. Cassiopeia A is 3000 pc (10,000 ly) from Earth. (a: NASA/CXC/SAO/Rutgers/J. Hughes; b: Very Large Array)

Ic, which are caused by core collapse in a massive star, the core itself may also remain. If there is a relic of the core, it may be either a *neutron star* or a *black hole*, depending on the mass of the core and the conditions within it during the collapse. Neutron stars, as the name suggests, are made primarily of neutrons. Wholly unlike anything we have studied so far, these exotic objects are the subject of Chapter 23. We will study

black holes, which are far stranger even than neutron stars, in Chapter 24.

CAUTION! Although neutron stars and black holes can be part of the debris from a supernova explosion, they are *not* called "supernova remnants." That term is applied exclusively to the gas that spreads away from the site of the supernova explosion.

We have seen in this chapter that only the most massive stars end their lives as supernovae of Types II, Ib, or Ic. We learned earlier that mass plays a central role in determining the speed with which a star forms and joins the main sequence (see Figure 20-9), the star's luminosity and surface temperature while on the main sequence (see Figures 19-22 and 19-23), and how long a star can remain on the main sequence (see Table 21-1). Now we see that mass also determines the eventual fate of a star—whether it ends in the whimper of a planetary nebula, as do low-mass stars like the Sun, or follows the evolutionary path of a high-mass star and ends with an almost inconceivably intense bang (Figure 22-25).

INTERACTIVE EXERCISE 22.3

figure 22-25

Pathways of Stellar Evolution This flowchart shows how the evolution of an isolated star (one that is not part of a close multiple-star system) depends on the star's mass. The scale at the top gives the mass of the star when it is on the main sequence, and the scale at the bottom shows the mass of the resulting stellar corpse. Stars whose original masses are less than about 4 $M_\odot$ eject enough mass over their lifetimes to end up as white dwarfs (which requires that the corpse be less massive than the Chandrasekhar limit of 1.4 $M_\odot$). High-mass stars can evolve into supernovae of Type Ib, Ic, or II and become neutron stars or black holes. It also possible that a high-mass star can leave no corpse at all.

KEY WORDS

asymptotic giant branch, p. 502

asymptotic giant branch star (AGB star), p. 502

carbon burning, p. 508

carbon star, p. 503

Cerenkov radiation, p. 514

Chandrasekhar limit, p. 507

core helium burning, p. 501

dredge-up, p. 503

helium shell flash, p. 504

horizontal branch, p. 502

mass-radius relation, p. 506

neon burning, p. 508

neutron capture, p. 508

nuclear density, p. 510

oxygen burning, p. 508

photodisintegration, p. 510

planetary nebula, p. 504

progenitor star, p. 512

red-giant branch, p. 501

shell helium burning, p. 502

silicon burning, p. 508

supergiant, p. 509

supernova (*plural* supernovae), p. 511

supernova remnant, p. 518

thermal pulses, p. 504

Type Ia supernova, p. 516

Type Ib supernova, p. 516

Type Ic supernova, p. 516

Type II supernova, p. 515

white dwarf, p. 506

KEY IDEAS

 Late Evolution of Low-Mass Stars: A low-mass star becomes a red giant when shell hydrogen burning begins, a horizontal-branch star when core helium burning begins, and an asymptotic giant branch (AGB) star when the helium in the core is exhausted and shell helium burning begins.

• As a low-mass star ages, convection occurs over a larger portion of its volume. This takes heavy elements formed in the star's interior and distributes them throughout the star.

Planetary Nebulae and White Dwarfs: Helium shell flashes in an old, low-mass star produce thermal pulses during which more than half the star's mass may be ejected into space. This exposes the hot carbon-oxygen core of the star.

• Ultraviolet radiation from the exposed core ionizes and excites the ejected gases, producing a planetary nebula.

• No further nuclear reactions take place within the exposed core. Instead, it becomes a degenerate, dense sphere about the size of the Earth and is called a white dwarf. It glows from thermal radiation; as the sphere cools, it becomes dimmer.

Late Evolution of High-Mass Stars: Unlike a low-mass star, a high-mass star undergoes an extended sequence of thermonuclear reactions in its core and shells. These include carbon burning, neon burning, oxygen burning, and silicon burning.

• In the last stages of its life, a high-mass star has an iron-rich core surrounded by concentric shells hosting the various thermonuclear reactions. The sequence of thermonuclear reactions stops here, because the formation of elements heavier than iron requires an input of energy rather than causing energy to be released.

The Deaths of Massive Stars: A high-mass star dies in a violent cataclysm in which its core collapses and most of its matter is ejected into space at high speeds. The luminosity of the star increases suddenly by a factor of around 10^8 during this explosion, producing a supernova.

• More than 99% of the energy from such a supernova is emitted in the form of neutrinos from the collapsing core.

• The matter ejected from the supernova, moving at supersonic speeds through interstellar gases and dust, glows as a nebula called a supernova remnant.

Other Types of Supernovae: An accreting white dwarf in a close binary system can also become a supernova when carbon burning ignites explosively throughout such a degenerate star.

• Type Ia supernovae are those produced by accreting white dwarfs in close binaries. Type II supernovae are created by the deaths of massive stars, as are supernovae of Type Ib and Type Ic; these latter types occur when the star has lost a substantial part of its outer layers before exploding.

• Most supernovae occurring in our Galaxy are hidden from our view by interstellar dust and gases.

REVIEW QUESTIONS

1. What is the horizontal branch? Where is it located on an H-R diagram? How do stars on the horizontal branch differ from red giants or main-sequence stars?

2. What is the asymptotic giant branch? Where is it located on an H-R diagram? How do asymptotic giant branch stars differ from red giants or main-sequence stars?

3. What is the connection between dredge-ups in old stars and life on Earth?

4. How is a planetary nebula formed?

5. What is a white dwarf? Does it produce light in the same way as a star like the Sun?

6. How does the radius of a white dwarf depend on its mass? How is this different from other types of stars?

7. What is the significance of the Chandrasekhar limit?

8. On an H-R diagram, sketch the evolutionary track that the Sun will follow from when it leaves the main sequence to when it becomes a white dwarf. Approximately how much mass will the Sun have when it becomes a white dwarf? Where will the rest of the mass go?

9. Why do you suppose that all the white dwarfs known to astronomers are relatively close to the Sun?

10. What prevents thermonuclear reactions from occurring at the center of a white dwarf? If no thermonuclear reactions are occurring in its core, why doesn't the star collapse?

11. Why does the mass of a star play such an important role in determining the star's evolution?

12. Why is the temperature in a star's core so important in determining which nuclear reactions can occur there?

13. What is the difference between a red giant and a red supergiant?

14. What is nuclear density? Why is it significant when a star's core reaches this density?

15. Why is SN 1987A so interesting to astronomers? In what ways was it not a typical supernova?

16. What are the differences among Type Ia, Type Ib, Type Ic, and Type II supernovae? Which type is most unlike the other three, and why?

17. How can a supernova continue to shine for many years after it explodes?

18. Why have radio searches for supernova remnants been more fruitful than optical searches?

19. Is our own Sun likely to become a supernova? Why or why not?

ADVANCED QUESTIONS

Questions preceded by an asterisk (*) involve topics discussed in the Boxes in [Chapter 7](#) or [Chapter 19](#).

Problem-solving tips and tools

You may find it useful to review [Box 19-4](#), which discusses stellar radii and their relationship to temperature and luminosity. [Section 4-7](#) explains the formula for gravitational force, and [Box 7-2](#) explains the concept of escape speed. [Sections 5-2 and 5-4](#) describe some key properties of light, especially blackbody radiation. We discussed the relationship among luminosity, apparent brightness, and distance in [Box 19-2](#). The relationship among absolute magnitude, apparent magnitude, and distance was the topic of [Box 19-3](#).

20. The globular cluster M15 depicted in Figure 22-6*a* contains 30,000 old stars, but only one of these stars is presently in the planetary nebula stage of its evolution. Explain why planetary nebulae are not more prevalent in M15.

21. Explain why infrared light was used to detect the faint wisps in planetary nebula NGC 7027 (Figure 22-6*c*).

22. The central star in a newly formed planetary nebula has a luminosity of 1000 $L_\odot$ and a surface temperature of 100,000 K. What is the star's radius? Give your answer as a multiple of the Sun's radius.

23. You want to determine the age of a planetary nebula. What observations should you make, and how would you use the resulting data?

24. The Ring Nebula in the constellation Lyra has an angular size of 1.4×1.0 arcmin and is expanding at the rate of about 20 km/s. Approximately how long ago did the central star shed its outer layers? Assume that the nebula is 2,700 ly from Earth.

25. (a) Calculate the wavelength of maximum emission of the white dwarf Sirius B. In what part of the electromagnetic spectrum does this wavelength lie? (b) In a visible-light photograph such as Figure 22-8, Sirius B appears much fainter than its primary star. But in an image made with an X-ray telescope, Sirius B is the brighter star. Explain the difference.

26. (a) Find the average density of a 1-$M_\odot$ white dwarf having the same diameter as the Earth. (b) What speed is required to eject gas from the white dwarf's surface? (This is also the speed with which interstellar gas falling from a great distance would strike the star's surface.)

27. (a) What kinds of stars would you monitor if you wished to observe a supernova explosion from its very beginning? (b) Examine [Appendices 4 and 5](#), which list the nearest and brightest stars, respectively. Which, if any, of these stars are possible supernova candidates? Explain.

*28. Consider a high-mass star just prior to a supernova explosion, with a core of diameter 20 km and density 4×10^{17} kg/m^3. (a) Calculate the mass of the core. Give your answer in kilograms and in solar masses. (b) Calculate the force of gravity on a 1-kg object at the surface of the core. How many times larger is this than the gravitational force on such an object at the surface of the Earth, about 10 newtons? (c) Calculate the escape speed from the surface of the star's core. Give your answer in m/s and as a fraction of the speed of light. What does this tell you about how powerful a supernova explosion must be in order to blow material away from the star's core?

29. The shock wave that traveled through the progenitor star of SN 1987A took 3 hours to reach the star's surface. (a) Given the size of a blue supergiant star (see Section 22-7), estimate the speed with which the shock wave traveled through the star's outer layers. (The core of the progenitor star was very small, so you may consider the shock wave

to have started at the very center of the star.) Give your answer in meters per second. (b) Compare your answer with the speed of sound waves in our atmosphere, about 340 m/s, and with the speed of light. (c) A shock wave traveling through a gas is a special case of a sound wave. In general, sound waves travel faster through denser, less easily compressed materials. Thus, sound travels faster through water (about 1500 m/s) than through our atmosphere, and faster still through steel (about 5900 m/s). Use this idea to compare the gases within the progenitor star of SN 1987A with the gases in our atmosphere in terms of their average density and how easily they are compressed.

30. The neutrinos from SN 1987A arrived 3 hours before the visible light. While they were en route to the Earth, what was the distance between the neutrinos and the first photons from SN 1987A? Assume that neutrinos are massless and thus travel at the speed of light. Give your answers in kilometers and in AU.

*31. Compared to SN 1987A (Figure 22-15), the supernova SN 1993J (see Figure 22-19) had a maximum apparent brightness only 9.1×10^{-4} as great. Using the distances from Earth to each of these supernovae, determine the ratio of the maximum luminosity of SN 1993J to that of SN 1987A. Which of the two supernovae had the greater maximum luminosity?

*32. Suppose that the brightness of a star becoming a supernova increases by 20 magnitudes. Show that this corresponds to an increase of 10^8 in luminosity.

*33. Suppose that the red-supergiant star Betelgeuse, which lies some 1400 light-years from the Earth, becomes a Type II supernova. (a) At the height of the outburst, how bright would it appear in the sky? Give your answer as a fraction of the brightness of the Sun ($b_\odot$). (b) How would it compare with the brightness of Venus (about 10^{-9} $b_\odot$)?

*34. In July 1997, a supernova named SN 1997cw exploded in the galaxy NGC 105 in the constellation Cetus (the Whale). It reached an apparent magnitude of +16.5 at maximum brilliance, and its spectrum showed an absorption line of ionized silicon. Use this information to find the distance to NGC 105. (*Hint:* Inspect the light curves in Figure 22-21 to find the *absolute* magnitudes of typical supernovae at peak brightness.)

35. Figure 22-22 shows a portion of the Veil Nebula in Cygnus. Use the information given in the caption to find the average speed at which material has been moving away from the site of the supernova explosion over the past 15,000 years. Express your answer in km/s and as a fraction of the speed of light.

DISCUSSION QUESTIONS

36. Suppose that you discover a small, glowing disk of light while searching the sky with a telescope. How would you decide if this object is a planetary nebula? Could your object be something else? Explain.

37. Suppose the convective zone in AGB stars did *not* reach all the way down into their carbon-rich cores. How might this have affected the origin and evolution of life on Earth?

38. In the classic 1960s science-fiction comic book *The Atom*, a physicist discovers a basketball-sized meteorite (about 10 cm in radius) that is actually a fragment of a white dwarf star. With some difficulty, he manages to hand-carry the meteorite back to his laboratory. Estimate the mass of such a fragment, and discuss the plausibility of this scenario.

39. SN 1987A did not agree with the theoretical picture outlined in Section 22-6. Does this mean that the theory was wrong? Discuss.

40. The major final product of silicon burning is ^{56}Fe, an isotope of iron with 26 protons and 30 neutrons. This is also the most common isotope of iron found on Earth. Discuss what this tells you about the origin of the solar system.

 WEB/CD-ROM QUESTIONS

41. It has been claimed that the Dogon tribe in western Africa has known for thousands of years that Sirius is a binary star. Search the World Wide Web for information about these claims. What is the basis of these claims? Why are scientists skeptical, and how do they refute these claims?

42. Search the World Wide Web for recent information about SN 1993J. Has the shape of the supernova's light curve been adequately explained? Has the supernova produced any surprises?

43. Search the World Wide Web for information about SN 1994I, a supernova that occurred in the galaxy M51 (NGC 5194). Why was this supernova unusual? Was it bright enough to have been seen by amateur astronomers?

 44. **Convection Inside a Giant Star.** Access and view the animation "Convection Inside a Giant Star" in Chapter 22 of the *Universe* web site or CD-ROM. Describe the motion of material in the interior of the star. In what ways is it similar to convection within the present-day Sun (see Section 18-2)? In what ways is it different? Is a dredge-up taking place in this animation? How can you tell?

  45. **Types of Supernovae.** Access and view the animations "In the Heart of a Type II Supernova" and "A Type Ia Supernova" in Chapter 22 of the *Universe* web site or CD-ROM. Describe how these two types are fundamentally different in their origin.

OBSERVING PROJECTS

Observing tips and tools

While planetary nebulae are rather bright objects, their brightness is spread over a relatively large angular size, which can make seeing them a challenge for the beginning observer. For example, the Helix Nebula (see Figure 22-6) has the largest angular size of any planetary nebula but is also one of the most difficult to see. To improve your view, make your observations on a dark, moonless night from a location well shielded from city lights. Another useful trick, mentioned in Chapters 19 and 20, is to use "averted vision." Once you have the nebula centered in the telescope, you will get a brighter and clearer image if you look at the nebula out of the corner of your eye. The so-called Blinking Planetary in Cygnus affords an excellent demonstration of this effect; the nebula seems to disappear when you look straight at it, but it reappears as soon as you look toward the side of your field of view.

Another useful tip is to view the nebula through a green filter (a #58, or O III, filter available from telescope supply houses). Green light is emitted by excited, doubly ionized oxygen atoms, which are common in planetary nebulae but not in most other celestial objects. Using such a filter can make a planetary nebula stand out more distinctly against the sky. As a side benefit, it also helps to block out stray light from street lamps. The same tips also apply to observing supernova remnants.

46. Although they represent a fleeting stage at the end of a star's life, planetary nebulae are found all across the sky. Some of the brightest are listed in Chapter 22 of the *Universe* web site and CD-ROM. Observe as many of these planetary nebulae as you can on a clear, moonless night using the largest telescope at your disposal. Note and compare the various shapes of the different nebulae. In how many cases can you see the central star? The central star in the Eskimo Nebula is supposed to be the "nose" of an Eskimo wearing a parka (see the cover of this book). Can you see this pattern?

47. Northern hemisphere observers with modest telescopes can see two supernova remnants, one in the winter sky and the other in the summer sky. Both are quite faint, however, so you should schedule your observations for a moonless night. The winter sky contains the Crab Nebula, which is discussed in detail in Chapter 23. The coordinates are R.A. = 5^h 34.5^m and Decl. = $+22°$ $00'$, which places the object near the star marking the eastern horn of Taurus (the Bull). Whereas the entire Crab Nebula easily fits in the field of view of an eyepiece, the Veil or Cirrus Nebula in the summer sky is so vast that you can see only a small fraction of it at a time. The easiest way to find the Veil Nebula is to aim the telescope at the star 52 Cygni (R.A. = 20^h 45.7^m and Decl. = $+30°$ $43'$), which lies on one of the brightest portions of the nebula. If you then move the telescope slightly north or south until 52 Cygni is just out of the field of view, you should see faint wisps of glowing gas. (52 Cygni is the bright star toward the lower right corner of Figure 20-22, which shows the entire Cygnus Loop.)

48. Use a telescope to observe the remarkable triple star 40 Eridani, whose coordinates are R.A. = 4^h 15.3^m and Decl. = $-7°$ $39'$. The primary, a 4.4-magnitude yellowish star like the Sun, has a 9.6-magnitude white dwarf companion, the most easily seen white dwarf in the sky. On a clear, dark night with a moderately large telescope, you should also see that the white dwarf has an 11th-magnitude companion, which completes this most interesting trio.

49. The red supergiant Betelgeuse in the constellation Orion will someday explode as a supernova. Use the *Starry Night* program to investigate how the supernova might appear if this was to happen tonight. Click on the "Now" button in the Control Panel to make the program display the sky as it is right now. Then use the **Find...** command in the **Edit** menu to center the field of view on Betelgeuse. (If Betelgeuse is below the horizon, allow the program to reset the time to when it rises.) Then determine the times when Betelgeuse rises and sets on today's date at your location. (You can use the controls in the Control Panel to move forward or backward in time as needed.) If Betelgeuse became a supernova today, would it be visible in the daytime? How would it appear at night? Do you think it would cast shadows? Are Betelgeuse and the Moon both in the night sky tonight? If they are, what kinds of shadows might they both cast if Betelgeuse became a supernova?

Neutron Stars

Almost all the light we see from objects in the night sky is due to thermonuclear fusion reactions. Such reactions are what make stars shine, and in turn make any surrounding nebulae glow. Even the light we see from the Moon and planets—which is simply reflected sunlight—can be traced back to thermonuclear reactions in the Sun's core.

By contrast, the exotic object shown here—the Crab Nebula in Taurus—is undergoing no thermonuclear fusion at all. Despite this, the object is so energetic that it is emitting copious amounts of X rays. This image from the orbiting Chandra X-ray Observatory shows tilted, glowing rings more than a light-year across, with oppositely directed jets flowing outward perpendicular to the plane of the rings. At the object's very center, too small to pick out in this image, is the source of all this dynamic activity: a neutron star, the leftover core of what was once a supergiant star.

Forty years ago neutron stars were considered pure speculation. But today we know of more than a thousand of these exotic stellar corpses. As we will discover in this chapter, neutron stars have powerful magnetic fields that sweep beams of radiation across the sky. We detect these beams as pulsating radio signals. We will find that neutron stars in close binary systems display even more outrageous behavior, including bouts of explosive helium burning that yield a vast outpouring of X rays. And we

1 arcmin

WEB LINK 23.1   (NASA/CXC/SAO)   R I V U **X** G

will see that a similar process involving a white dwarf in a close binary system produces a short-lived intense burst of visible light called a nova.

QUESTIONS TO GUIDE YOUR READING

As you read the sections of this chapter, look for the answers to the following questions.

23-1   What led scientists to the idea of a neutron star?

23-2   What are pulsars, and how were they discovered?

23-3   How did astronomers determine the connection between pulsars and neutron stars?

23-4   How can a neutron star supply energy to a surrounding nebula?

23-5   What are conditions like inside a neutron star?

23-6   How are some neutron stars able to spin several hundred times per second?

23-7   Why do some pulsars emit fantastic amounts of X rays?

23-8   Are X-ray bursters and novae similar to supernovae?

23-9   How massive can a neutron star be?

23-1   Scientists first proposed the existence of neutron stars in the 1930s

On the morning of July 4, 1054, Yang Wei-T'e, the imperial astronomer to the Chinese court, made a startling discovery. Just a few minutes before sunrise, a new and dazzling object ascended above the eastern horizon. This "guest star," as Yang called it, was far brighter than Venus and more resplendent than any star he had ever seen.

Yang's records show that the "guest star" was so brilliant that it could easily be seen during broad daylight for the rest of July. Works of art made by the Anasazi tribe in the American southwest suggest that they, too, saw and recorded this object (Figure 23-1). Over the next year, however, the "guest star" faded to invisibility.

The "guest star" of 1054 was actually a remarkable stellar transformation: A massive star some 6500 light-years away perished in a supernova explosion, leaving behind both a supernova remnant (see Figure 1-6) and a bizarre object called a *neutron star*. But it was not until 1932 that the English physicist James Chadwick discovered the subatomic particle called the neutron. The neutron has about the same mass as the positively charged proton, but has no electric charge. Within a year of Chadwick's discovery, two astronomers were inspired to predict the existence of neutron stars.

As we saw in Section 22-6, a proton and an electron can combine under high pressure to form a neutron. Could such pressures be found within the core of a dying star? If so, they could transform the core into a sphere of neutrons—a **neutron star.** This is just what was proposed by Fritz Zwicky at the California Institute of Technology (Caltech) and his colleague Walter Baade at Mount Wilson Observatory. "With all reserve," Zwicky and Baade theorized, "we advance the view that supernovae represent the transition from ordinary stars into neutron stars, which in their final stages consist of extremely closely packed neutrons." In other words, there could be at least two types of stellar corpses—white dwarfs and neutron stars.

Zwicky and Baade realized that once a neutron star had formed in response to tremendous pressures, it would be able to resist any further compression. The neutrons themselves would provide a counterbalancing outward pressure. Motivated by the idea that white dwarfs are supported by degenerate electron pressure, Zwicky and Baade proposed that a highly compact ball of neutrons would similarly produce a **degenerate neutron pressure.** (Like electrons, neutrons obey the Pauli exclusion principle that we described in Section 21-2.) This powerful pressure could support a stellar corpse—perhaps one even more massive than a white dwarf, because degenerate neutron pressure can be much stronger than degenerate electron pressure. So, while a white dwarf collapses if its mass is above the Chandrasekhar limit (Section 22-4), a neutron star might not.

Although Zwicky and Baade's proposal would prove to be very close to the mark, most scientists politely ignored it

figure 23-1   R I **V** U X G

A Supernova Pictograph? This drawing on the wall of an eleventh-century structure in New Mexico shows a ten-pointed star next to a crescent. It may depict the scene on the morning of July 5, 1054, when a "guest star" appeared next to the waning crescent Moon. (Courtesy of National Parks Service) for years. After all, a neutron star must be a rather weird object. A single thimbleful of neutron star matter would weigh 100 million tons on the Earth's surface!

A star compacted to nuclear density must be very small. A 1-$M_\odot$ neutron star would have a diameter of only 30 km (19 mi), about the size of a large city on Earth. Its surface gravity would be so strong that an object would have to travel at one-half the speed of light to escape into space. These conditions seemed outrageous until the late 1960s, when astronomers discovered pulsating radio sources.

23-2 The discovery of pulsars in the 1960s stimulated interest in neutron stars

As a young graduate student at Cambridge University, Jocelyn Bell spent many months helping construct an array of radio antennas covering 4½ acres of the English countryside. The instrument was completed by the summer of 1967, and Bell and her colleagues in Anthony Hewish's research group began using it to scrutinize radio emissions from the sky. They were looking for radio sources that "twinkle" like stars; that is, they looked for random small fluctuations in brightness caused by the motion of gas between the source and the observer. While searching for such random flickering, Bell noticed that the antennas had detected regular pulses of radio noise from one particular location in the sky. These radio pulses were arriving with a regular period of 1.3373011 seconds.

The pulsations were much more rapid than those of any other astronomical object known at that time. Indeed, they were so rapid and regular that the Cambridge team at first suspected that they might not be of natural origin. Instead, they might be signals from an advanced alien civilization. That possibility had to be discarded within a few months after several more of these pulsating radio sources, which came to be called **pulsars,** were discovered across the sky. In all cases, the periods were extremely regular, ranging from about 0.25 second for the fastest to about 1.5 seconds for the slowest (Figure 23-2).

What could these objects possibly be? As is often done in science, astronomers first decided what pulsars could *not* be. It was clear that pulsars were not ordinary stars or nebulae, because while these objects emit some radio waves, their emissions do not pulsate (that is, their light output does not increase and decrease). One type of star that does appear to pulsate is an eclipsing binary star system, which appears dimmer when one star in the binary system passes in front of the other (see Section 19-11). However, in order to vary in brightness with a typical pulsar period of 1 second or less, an eclipsing binary would have to have an equally short orbital period. Newton's form of Kepler's third law (Section 4-7) implies that the two stars would have to orbit only 1000 kilometers or so from each other. This is less than the diameter of any ordinary star, even a white dwarf star. Hence, the eclipsing binary hypothesis can be ruled out, because it leads to the highly implausible picture of two stars that overlap!

figure 23-2

A Recording of a Pulsar This chart recording shows the intensity of radio emission from one of the first pulsars to be discovered, PSR 0329+54. (The designation means "pulsar at a right ascension of 03^h 29^m and a declination of +54°"). Note that some pulses are weak and others are strong. Nevertheless, the spacing between pulses is exactly 0.714 seconds. (Adapted from R. N. Manchester and J. H. Taylor)

Astronomers also ruled out the idea that pulsars might be variable stars, which vary in luminosity as they increase and decrease in size (see Section 21-5). However, pulsating variables have periods of days or weeks, while pulsars have periods of seconds or less. Just as the broad wings of an eagle could never flap as fast as a hummingbird's wings without breaking, an ordinary star or even a white dwarf is too large to pulsate in and out in less than a second.

Another possibility, however, was given serious consideration. Perhaps pulsars were rapidly rotating white dwarf stars with some sort of radio-emitting "hot spots" on their surfaces. But even this model had its limits, because a white dwarf that rotated on its axis in just 1 second would be on the verge of flying apart. What finally ruled out the spinning white dwarf model was the discovery of a pulsar in the middle of the Crab Nebula, the supernova remnant left behind by the supernova that Yang Wei-T'e saw in 1054 (see Figure 1-6). That discovery clarified the true nature of pulsars.

At the time of its discovery, the Crab pulsar was the fastest pulsar known to astronomers. Its period is 0.033 second, which means that it flashes 30 times each second. It was immediately apparent to astrophysicists that white dwarfs are too big and bulky to generate 30 signals per second, because calculations demonstrated that a white dwarf could not rotate that fast without tearing itself apart. Hence, the existence of the Crab pulsar means that the stellar corpse at the center of the Crab Nebula has to be much smaller and more compact than a white dwarf.

This was a truly unsettling realization, because most astronomers in the mid-1960s thought that *all* stellar corpses are white dwarfs. The number of white dwarfs in the sky seemed to account for all the stars that must have died since our Galaxy was formed. It was generally assumed that all dying stars—even the most massive ones, which produce supernovae—somehow manage to eject enough matter so that their corpses do not exceed the Chandrasekhar limit. But the discovery of the Crab pulsar showed these conservative opinions to be in error.

It was now clear that pulsars had to be something far more exotic than rotating white dwarfs. As Thomas Gold of Cornell University emphasized, astronomers would now have to seriously consider the existence of neutron stars.

23-3 Pulsars are rapidly rotating neutron stars with intense magnetic fields

Science is, by its very nature, based on skepticism. If a scientist proposes a new and exotic explanation for an unexplained phenomenon, that explanation will be accepted only if it satisfies two demanding criteria. First, there must be excellent reasons why more conventional explanations cannot work. Second, the new explanation must help scientists understand a range of other phenomena.

A good example is the Copernican idea that the Earth orbits the Sun, which gained general acceptance only after Galileo's observations (see Section 4-3). First, Galileo showed the older geocentric explanation of celestial motions to be untenable, and, second, it became clear that the Copernican model could also explain the motions of Jupiter's moons.

We have seen how the discovery of the Crab pulsar ruled out the idea that pulsars were white dwarfs. But astronomers soon accepted the more exotic idea that pulsars were actually neutron stars. To understand why, we must assume the skeptical outlook of the scientist and ask some probing questions about this idea. Why should neutron stars emit radiation at all? In particular, why should they emit radio waves? Why should the emissions be pulsed? And how could the pulses occur as rapidly as 30 times per second, as in the Crab pulsar? The answers to all of these questions are intimately related to the small size of neutron stars.

Because neutron stars are very small, they should also rotate rapidly. A typical star, such as our Sun, takes nearly a full month to rotate once about its axis. But just as an ice skater doing a pirouette speeds up when she pulls in her arms, a collapsing star also speeds up as its size shrinks. If our Sun were compressed to the size of a neutron star, it would spin about 1000 times per second! Because neutron stars are so small and dense, they can spin this rapidly without flying apart.

The small size of neutron stars also implies that they have intense magnetic fields. It seems safe to say that every star possesses some magnetic field, but the strength is typically quite low. The magnetic field of a main-sequence star is spread out over billions of square kilometers of the star's surface. However, if such a star collapses down to a neutron star, its surface area (which is proportional to the square of its radius) shrinks by a factor of about 10^{10}. The magnetic field, which is bonded to the star's gases, becomes concentrated onto an area 10^{10} times smaller than before the collapse, and thus the field strength increases by a factor of 10^{10}.

A star's magnetic field splits its spectral lines into two or more lines (see Figure 18-23), whose spacing reveals the field's

strength. The strength of a magnetic field is usually measured in the unit called the *gauss* (G), named after the famous German mathematician and astronomer Karl Friedrich Gauss. For example, the magnetic splitting of lines in the spectra of main-sequence stars reveal field strengths in the range from about 1 G (the value for the Sun) to several thousand gauss. The splitting of lines in the spectra of certain white dwarfs corresponds to field strengths in excess of 10^6 G, comparable to the strongest fields ever produced in a laboratory. But the magnetic fields surrounding typical neutron stars are a million times stronger still, on the order of 1 trillion gauss (10^{12} G).

The magnetic field of a neutron star makes it possible for the star to radiate pulses of energy toward our telescopes. A number of different models have been proposed for how rotating neutron stars generate radiation; we'll discuss just one leading model. The idea behind this model is that the magnetic axis of a neutron star, the line connecting the north and south magnetic poles, is likely to be inclined at an angle to the rotation axis (Figure 23-3). After all, there is no

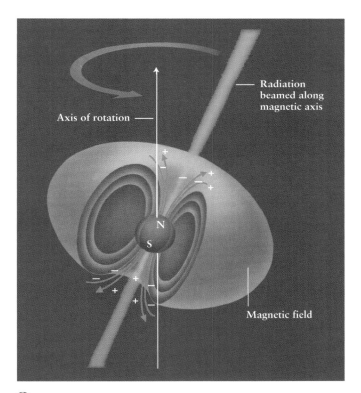

Figure 23-3

A Rotating, Magnetized Neutron Star When a neutron star forms, it is rotating rapidly and has a powerful magnetic field. Charged particles are accelerated near the star's magnetic poles, producing two oppositely directed beams of radiation that emerge from the magnetic poles. In general, a star's magnetic axis (a line that connects the north and south magnetic poles) is tilted at an angle from the axis of rotation, as shown here. Therefore, as the star rotates, the beams sweep around the sky. If the Earth happens to lie in the path of the beams, we detect radiation that appears to pulse on and off—in other words, a pulsar.

fundamental reason for these two axes to coincide. (Indeed, these two axes do not coincide for any of the planets of our solar system.)

In 1969, Peter Goldreich at Caltech pointed out that the combination of such a powerful magnetic field and rapid rotation would act like a giant electric generator, creating very strong electric fields near the neutron star's surface. Indeed, these fields are so intense that part of their energy is used to create electrons and positrons out of nothingness in a process called **pair production.** The powerful electric fields push these charged particles out from the neutron star's surface and into its curved magnetic field, as sketched in Figure 23-3. As the particles spiral along the curved field, they are accelerated and emit energy in the form of electromagnetic radiation. (A radio transmission antenna works in a similar way. Electrons are accelerated back and forth along the length of the antenna, producing radio waves.) The result is that two narrow beams of radiation pour out of the neutron star's north and south magnetic polar regions.

 A rotating, magnetized neutron star is somewhat like a lighthouse beacon. As the star rotates, the beams of radiation sweep around the sky. If at some point during the rotation one of those beams happens to point toward the Earth, we will detect a brief flash as the beam sweeps over us. Hence, a radio telescope will detect regular pulses of radiation, with one pulse being received for each rotation of the neutron star.

The name *pulsar* may have led you to the idea that the source of radio waves is actually pulsing. But in the model just described, this is not the case at all. Instead, beams of radiation are emitted continuously from the magnetic poles of the neutron star. The pulsing that astronomers detect here on the Earth is simply a result of the rapid rotation of the neutron star, which brings one of the beams periodically into our line of sight. In this sense, the analogy between a pulsar and a lighthouse beacon is a very close one.

Soon after the rotating neutron star model of pulsars was proposed in the late 1960s, a team of astronomers at the University of Arizona began wondering if pulsars might also emit pulses at wavelengths other than the radio part of the spectrum. To investigate, they began a search for *visible-light* pulses from the Crab pulsar. They aimed a telescope at the center of the Crab Nebula, then used a spinning disk with a slit in it to "chop," or interrupt, incoming light at rapid intervals. To their surprise and delight, they found that one of the stars at the center of the nebula appears to be flashing on and off 30 times each second (Figure 23-4)—exactly the same pulse rate as observed with radio waves.

This observation was strong evidence in favor of pulsars being rotating, magnetized neutron stars, because charged particles accelerating in a magnetic field can radiate strongly over a very wide range of wavelengths. By contrast, the black-body radiation from an ordinary star, which has a surface

a
b

figure 23-4   R I **V** U X G

The Crab Pulsar in Visible Light (a) A pulsar is located at the center of the Crab Nebula, which is about 2000 pc (6500 ly) from Earth and about 3 pc (10 ly) across. (b) These two close-up images show the Crab pulsar when its beam is directed toward us (upper image) and when it is not (lower image). These are called the "on" and "off" states, respectively. Like the radio pulses, these visual flashes have a period of 0.033 s. (a: The FORS Team, VLT, European Southern Observatory; b: Lick Observatory)

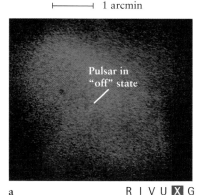

Pulsar in "on" state

⊢————⊣ 1 arcmin

Pulsar in "off" state

a   R I V U X G

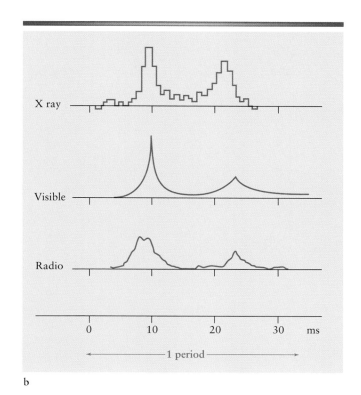

X ray

Visible

Radio

0   10   20   30   ms

←————— 1 period —————→

b

ƒigure 23-5

X-Ray Views of the Crab Pulsar **(a)** These views of the Crab pulsar were obtained with an X-ray telescope on the Earth-orbiting Einstein Observatory. They show the pulsar in its "on" state and "off "state. Compare these images to the visible-light images in Figure 23-4b. **(b)** These three graphs show the intensity of radiation emitted by the Crab pulsar at X-ray, visible, and radio wavelengths. At all wavelengths, the pulses repeat with a period of 33 milliseconds (ms), or 0.033 s. (One millisecond equals 10^{-3} s.) Note that there is a second, weaker pulse about halfway through the pulsar's period. One interpretation of this effect is that we are seeing emissions from both of the rotating neutron star's magnetic poles. If one of the magnetic poles is more directly aligned toward the Earth than the other, we detect one strong pulse and one weak one during each rotation of the neutron star. (a: Harvard-Smithsonian Center for Astrophysics)

temperature in thousands of degrees, is very weak at radio wavelengths. (The Sun's weak radio emissions come not from its glowing surface but rather from charged particles spiraling in the solar magnetic field.)

Since the pioneering days of the 1960s, periodic flashes of radiation have been detected from the Crab pulsar at still other wavelengths, including X rays (Figure 23-5a). The pulsar period is the same at all wavelengths (Figure 23-5b), just as we would expect if the emissions are coming from a portion of the neutron star, which comes into view periodically as the star rotates.

The many successes of the neutron star model of pulsars led to the rapid acceptance of the model. Since 1968 radio astronomers have discovered more than 1000 pulsars scattered across the sky, and it is estimated that some 10^5 more are strewn around the disk of the Milky Way Galaxy. Each one is presumed to be the neutron star corpse of an extinct, massive star. Radio telescopes have detected pulsars with a wide variety of pulse periods, from a

relatively sluggish 8.51 s to a blindingly fast 0.00156 s. In each case the pulse period is thought to be the same as the rotation period of the neutron star.

Supernovae have been seen throughout history, but they did not necessarily produce pulsars. For example, those observed by Tycho Brahe in 1572 and by Johannes Kepler in 1604 were probably Type Ia supernovae (see Section 22-9), and it is difficult to imagine how an exploding white dwarf might become a neutron star. Indeed, no pulsar has been found at the locations of either of those supernovae. The primary sources of pulsars are thought to be Type II supernovae, in which the cores of massive stars collapse (as described in Section 22-6). The supernova that left behind the Crab Nebula and the Crab pulsar is thought to have been of Type II.

The Crab pulsar is not far from the center of the Crab Nebula. This is what we might expect: When the progenitor star of the supernova of 1054 exploded, its core remained at the center of the explosion and became a rotating neutron star, while the star's outer layers were blasted into space to form the

figure 23-6

R I V U X G

A Fast-Moving Pulsar
These false-color radio images show the supernova remnant G5.4-1.2, which lies about 5000 pc (16,000 ly) from Earth in the constellation Sagittarius. Red denotes areas of strongest radio emission. When the progenitor star exploded into a supernova, it ejected the star's core at such a high speed that the core overtook the blast wave and moved beyond the boundary of the supernova remnant. The ejected core became a pulsar, denoted B1757-24. Tens of thousands of years after the explosion, the pulsar is still moving at about 600 km/s (350 mi/s). (National Radio Astronomy Observatory)

nebula. However, Figure 23-6 shows a pulsar that lies *outside* its supernova remnant and is moving away from the remnant at hundreds of kilometers per second. It appears that when the progenitor star exploded, the explosion was much more powerful on one side of the core than the other. As a result, the core was sent flying at immense speed. Many such fast-moving pulsars have been observed, which suggests that supernova explosions are often asymmetrical. In this way studies of pulsars have given us new insights into the nature of supernovae.

The most intensely studied of all recent Type II supernovae is the great supernova of 1987 (SN 1987A, described in Section 22-7). Astronomers have been carefully observing its remains for signs of a pulsar. To date, however, no confirmed observations of pulses have been made, and the search for a pulsar in SN 1987A goes on.

23-4 Pulsars gradually slow down as they radiate energy into space

 As remarkable as the Crab pulsar is, the Crab Nebula that surrounds it is even more so. Although the nebula is 2000 pc (6500 ly) away, it can be spotted in dark skies with even a small telescope. It shines with a luminosity 75,000 times that of the Sun (and some 10^8 times the luminosity of the pulses from the Crab pulsar).

But why is the Crab Nebula so luminous? Certain other nebulae—the H II regions described in Section 20-2—can also be seen at great distances. But H II regions contain young, hot stars that give off ultraviolet radiation. This radiation ionizes the surrounding gas and provides the energy that the H II region emits into space. The Crab Nebula, however, is differ-

ent from H II regions—it contains *no* young stars. What, then, is the source of this nebula's prodigious energy output? As we will see, the Crab pulsar provides the energy not only for radio pulses but for the entire Crab Nebula.

In 1966, two years before the discovery of pulsars, John A. Wheeler at Princeton University and Franco Pacini in Italy had already speculated that the ultimate source of the Crab Nebula's immense energy output might be a spinning neutron star. This speculation was quickly accepted as the correct explanation of the Crab Nebula's luminosity. What confirmed their idea was the discovery that the Crab pulsar is slowing down.

Although pulsars were first noted for their very regular pulses, careful measurements by radio telescopes soon revealed that the rotation period of a typical pulsar—that is, the time for it to spin once on its axis—increases by a few billionths of a second each day. The Crab pulsar, which is one of the most rapidly rotating pulsars, is also slowing more quickly than most—its period increases by 3×10^{-8} second each day. Thirty billionths of a second may sound trivial, but there is so much energy in the rotation of a rapidly spinning neutron star that even the slightest slowdown corresponds to a tremendous loss of energy. In fact, the observed rate of slowing for the Crab pulsar corresponds to a rate of energy loss equal to the entire luminosity of the Crab Nebula.

The interpretation is that the energy lost by the Crab pulsar as it slows down is transferred to the surrounding nebula. But *how* does this transfer of energy take place, and how is the energy radiated into space? The answer lies in the diffuse part of the Crab Nebula, which shines with an eerie bluish light (see Figure 23-4a). This same type of light was first produced on the Earth in 1947 in a particle accelerator in Schenectady, New York. This machine, called a *synchrotron*, was built by the General Electric Company to accelerate electrons to nearly the speed of light for experiments in nuclear physics. (The electrons are said to be *relativistic*, because their velocities are near the speed of light and Einstein's theory of relativity must be applied to understand their motions.)

As a beam of electrons whirled along a circular path within the cyclotron, held in orbit by powerful magnets, scientists noticed that it emitted a strange light. It is now known that this light, called **synchrotron radiation,** is emitted whenever high-speed electrons move along curved paths through a magnetic field.

The total energy output of the Crab Nebula in synchrotron radiation is 3×10^{31} watts, compared to the relatively paltry 4×10^{26} watts emitted by the Sun. Thus, the Crab Nebula must contain quite a large number of relativistic electrons spiraling in an extensive magnetic field. The X-ray image that opens this chapter shows where these electrons come from: They are ejected from the neutron star's vicinity, forming immense jets that emanate from the star's poles and a disk that spreads outward from its equator. As the Crab pulsar slows, the energy of its rotation is transferred to these electrons through the magnetic field and is then emitted by them in the form of synchrotron radiation. So much energy is transferred in this way that the Crab Nebula shines very brightly.

High-resolution visible-light images of the Crab pulsar's surroundings show that the gases around the pulsar are in a state of great agitation. This indicates the tremendous amount of energy transferred from the star to the nebula every second (Figure 23-7).

ƒigure 23-7 R I **V** U X G

The Dynamic Heart of the Crab Nebula The pulsar at the center of the Crab Nebula not only powers the synchrotron radiation by which the nebula shines but also stirs up the gases surrounding the pulsar. (The bright star that appears near the pulsar is located much closer to us than the Crab Nebula. It just happens to be along the line of sight to the nebula.) The bright clumps of gas have been excited by the pulsar, perhaps by jets of material being thrown off the neutron star. These clumps move outward from the pulsar at half the speed of light. (Jeff Hester and Paul Scowen, Arizona State University; and NASA)

Because a spinning neutron star slows down as it radiates away its rotational energy, it follows that an old neutron star should be spinning more slowly than a young one. Slower pulsars are thought to have been formed in supernova explosions that occurred hundreds of thousands of years ago. Over the ages, the supernova remnants have dispersed into the interstellar medium. The pulsars left behind from the exploded stars have slowed to periods of a second or more.

Like all general rules, the one relating a pulsar's age and its period has exceptions. In Section 23-6 we will see how some old pulsars can be re-accelerated to truly dizzying rotation speeds.

23-5 Superfluidity and superconductivity are among the strange properties of neutron stars

The radiation from pulsars gives us information about the surroundings of neutron stars, as well as how they rotate. To deduce what conditions are like *inside* a neutron star, however, astrophysicists must construct theoretical models. We saw examples of such models for the Sun in Section 18-2 and for post–main-sequence stars in Sections 22-1 and 22-5.

The models for neutron stars are very different. For one thing, neutron stars are thought to have a solid crust on their surface. Furthermore, the interior of a neutron star is a sea of densely packed, degenerate neutrons, with properties quite unlike those of ordinary gases or even degenerate electrons. Detailed calculations strongly suggest that degenerate neutron matter can flow without any friction whatsoever, a phenomenon called **superfluidity.**

Superfluidity can be observed in the laboratory by cooling liquid helium to temperatures near absolute zero. Because it is frictionless, superfluid liquid helium exhibits strange properties such as being able to creep up the walls of a container in apparent defiance of gravity. Within a neutron star, elongated, friction-free whirlpools of superfluid neutrons may form. Their interaction with the star's crust may be the cause of sudden changes in a pulsar's rotation.

In addition to the general slowing down of pulsars, astronomers have found that they sometimes exhibit a sudden speedup, prosaically called a **glitch.** For example, Figure 23-8 shows measurements of the Vela pulsar's period during 1975 and 1976. On this graph, the pulsar's gradual slowdown is shown as a steady increase in its period. In September 1975, however, there was an abrupt speedup, after which the pulsar continued to slow down at its usual rate. Similar glitches are seen for other pulsars.

Current opinion among pulsar theorists is that glitches are caused by the superfluid neutrons within the neutron star. As a rotating neutron star radiates energy into space, the rotation of its crust slows down, but the neutron whirlpools in the star's interior continue to rotate with the same speed. Some of these whirlpools cling to the crust, as though they were bungee cords with one end attached to the crust and the other to the star's interior. As the crust slows down relative to the interior, these superfluid "bungee cords" stretch out. When

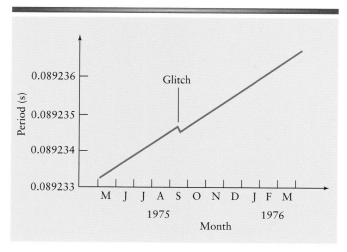

figure 23-8

A Pulsar Glitch This graph shows how the period of the Vela pulsar varied during 1975 and 1976. As the neutron star's rotational energy was converted into radiation, its rotation slowed and the period increased slightly. During September 1975, however, the period *decreased* suddenly: The neutron star's rotation underwent a sudden speedup, called a glitch. Glitches are thought to be caused by interaction between the neutron star's solid crust and its superfluid interior.

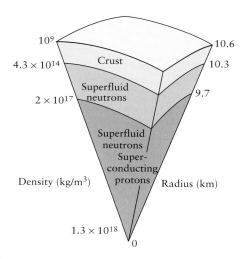

figure 23-9

A Model of a Neutron Star This theoretical model for a 1.4-$M_\odot$ neutron star has a superconducting, superfluid core 9.7 km in radius. The core is surrounded by a 0.6-km thick mantle containing nuclei, electrons, and superfluid neutrons. The star's crust is only 0.3 km thick (the length of three football fields) and is composed of heavy nuclei (such as iron) and free electrons.

the tension in them gets too great, they deliver a sharp jolt that makes the crust speed up suddenly. (An older model, in which pulsar glitches were caused by a "starquake" in the neutron star's settling crust, does not appear to agree with the accumulated observational data on glitches.)

Superfluidity is not the only exotic property of neutron star interiors. Models of the internal structure of a neutron star strongly suggest that the protons in the core can move around without experiencing any electrical resistance whatsoever. This phenomenon, called **superconductivity,** also occurs on the Earth with certain substances at low temperatures. Once you start an electric current moving in a superconducting material, it keeps moving forever.

You may have been surprised to read in the preceding paragraph about *protons* in a *neutron* star. While a neutron star is made up predominantly of neutrons, some protons and electrons must be scattered throughout the star's interior. Indeed, a pulsar's magnetic field must be anchored to the neutron star by charged particles. Neutrons are electrically neutral, so without the protons and electrons in its interior a neutron star would rapidly lose its magnetic field.

As Figure 23-9 shows, the structure of a neutron star probably consists of a core with superfluid neutrons and superconducting protons, a mantle of superfluid neutrons in which some of the neutrons combine with protons to form nuclei, and a brittle crust less than a kilometer thick. At the center of a neutron star, the density is on the order of 10^{18} kg/m³. This is some 10 trillion (10^{13}) times greater than

the central density of our present-day Sun and approximately twice the density of an atomic nucleus.

23-6 The fastest pulsars were probably created by mass transfer in close binary systems

In 1982 radio astronomers discovered a pulsar whose period is only 1.558 ms. (One millisecond, or 1 ms, equals a thousandth of a second). This remarkable pulsar, called PSR 1937+21 after its coordinates in the sky, is a neutron star spinning 642 times per second—about 3 times faster than the blades of a kitchen blender! Such incredibly rapid rotation is a clue to this pulsar's unusual history.

As we have seen, pulsars should slow down as they age. The fact that PSR 1937+21 is spinning so rapidly suggests that it has hardly aged at all, so it must be very young. A young pulsar should also be slowing down very rapidly. The faster it rotates, the more rapidly it can transfer rotational energy to its surroundings and the faster the slowdown. But PSR 1937+21 is actually slowing down at a very gradual rate, millions of times more gradually than the Crab pulsar. Such a gradual slowdown is characteristic of a pulsar hundreds of millions of years old. But if PSR 1937+21 is so old, how can it be spinning so rapidly?

PSR 1937+21 is not the only pulsar with unusually rapid rotation. Since 1982, astronomers have discovered more than 50 very fast pulsars, which are now called **millisecond pulsars.**

a

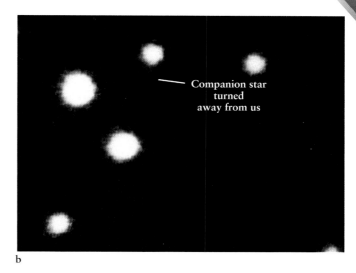

b

figure 23-10   R I **V** U X G

The Black Widow Pulsar The millisecond pulsar PSR 1957+20, called the Black Widow Pulsar, is a rapidly spinning neutron star in a close binary system. The plane of the orbit is aligned so that the two stars periodically eclipse each other. Most of the visible light from this system comes from the side of the companion star facing the neutron star, because this side is heated by emissions from the pulsar. That hemisphere is exposed to our view in **(a)** but turned away from us in **(b)**. An unrelated background star less than 1 arcsec from the pulsar shines constantly. Just as the female black widow spider kills and eats its mate, the Black Widow Pulsar is slowly eating away its companion star by blasting it with energetic photons and particles. (Courtesy of Jan van Paradijs)

All have periods between 1 and 10 ms, which means that these neutron stars are spinning at rates of 100 to 1000 rotations per second.

An important clue to understanding millisecond pulsars is that the majority are in close binary systems, with only a small separation between the spinning neutron star and its companion. (We know the separation must be small because the orbital periods of these binary systems are short, between 10 and 100 days. Kepler's third law for binary star systems, described in Section 19-9, tells us that the shorter the period, the smaller the distance between the two stars.) This fact suggests a scenario for how millisecond pulsars acquired their very rapid rotation.

Imagine a binary system consisting of a high-mass star and a low-mass star. The high-mass star evolves more rapidly than the low-mass star, and within a few million years becomes a Type II supernova that creates a neutron star. Like most newborn neutron stars, it spins several times per second, and, thus, initially we see a pulsar rather like the Crab or Vela pulsar.

Over the next few billion years, the pulsar slows down as it radiates energy into space. Meanwhile, the slowly evolving low-mass star begins to expand as it evolves away from the main sequence to become a red giant. When the red giant gets big enough to fill its Roche lobe, it starts to spill gas over the inner Lagrangian point onto the neutron star. (See Section 21-6 for a review of close binary systems.) The infalling gas strikes the neutron star's surface at high speed and at an angle that causes the star to spin faster. In this way, a slow, aging pulsar is "spun up" by mass transfer from its bloated companion.

What about the few millisecond pulsars, like PSR 1937+21, that are *not* members of close binaries? It may be that solitary millisecond pulsars were once part of close binary systems, but the companion stars have been eroded away by the high-energy particles emitted by the pulsar after it was spun up. The Black Widow Pulsar may be caught in the act of destroying its companion in just such a process (Figure 23-10).

The Black Widow Pulsar (a neutron star spinning 622 times per second) and its companion orbit each other with a period of 9.16 hours. The orbital plane of the system is nearly edge-on to our line of sight, so this system is an eclipsing binary (described in Section 19-11): The companion blocks out the pulsar's radio signals for 45 minutes during each eclipse. Just before and after an eclipse, however, the pulsar's signals are delayed substantially, as if the radio waves were being slowed as they pass through a cloud of ionized gas surrounding the companion star. This circumstellar material is probably the star's outer layers, dislodged by radiation and fast-moving particles emitted from the pulsar. In a few hundred million years or so, the companion of the pulsar will have completely disintegrated, leaving behind only a spun-up, solitary millisecond pulsar.

23-7   Pulsating X-ray sources are also neutron stars in close binary systems

Millisecond pulsars are not the only result of having a neutron star in a close binary system. Even more exotic are **pulsating X-ray sources**, in which material from the companion star is drawn onto the magnetic poles of the neutron star, producing *f*igure 23-11

X-Ray Pulses from Centaurus X-3 This graph shows the intensity of X rays detected by the *Uhuru* spacecraft as Centaurus X-3 moved across the satellite's field of view. Because the pulse period is only 4.84 s, Centaurus X-3 is in all probability a rotating neutron star. The gradual variation in the height of the pulses from left to right is the result of the changing orientation of the X-ray detectors toward the source as *Uhuru* rotated. (Adapted from R. Giacconi and colleagues)

intense hot spots that emit breathtaking amounts of X rays. As the neutron star spins and the X-ray beams are directed alternately toward us and away from us, we see the X rays flash on and off. Such pulsating X-ray sources, while similar in many ways to ordinary pulsars, are far more luminous.

Pulsating X-ray sources were first discovered in 1971 by the *Uhuru* spacecraft, the first to give astronomers a comprehensive look at the X-ray sky. (As we saw in Section 6-7, the Earth's atmosphere is opaque to X rays.) The first X-ray source found to emit pulses was Centaurus X-3, the third X-ray source found in the southern constellation Centaurus (Figure 23-11). The X-ray pulses have a regular period of 4.84 s. A few months later, similar pulses were discovered coming from a source designated Hercules X-1, which had a period of 1.24 s. Because the periods of these two X-ray sources are so short, astronomers began to suspect that they were actually observing rapidly rotating neutron stars.

It soon became clear that Centaurus X-3 and Hercules X-1 are members of binary systems. One piece of evidence is that every 2.087 days, Centaurus X-3 appears to turn off for

almost 12 hours. Apparently, Centaurus X-3 is a neutron star in an eclipsing binary system, and it takes nearly 12 hours for the X-ray source to pass behind its companion star. Hercules X-1, too, appears to turn off periodically, corresponding to a 6-hour eclipse every 1.7 days by its visible companion, the star HZ Herculis. Moreover, careful timing of its X-ray pulses shows a periodic Doppler shifting every 1.7 days, which is direct evidence of orbital motion about a companion star. When the X-ray source is approaching us, the pulses from Hercules X-1 are separated by slightly less than 1.24 seconds. When the source is receding from us, slightly more than 1.24 seconds elapse between the pulses.

Putting all the pieces together, astronomers now realize that pulsating X-ray sources like Centaurus X-3 and Hercules X-1 are binary systems in which one of the stars is a neutron star. All these binary systems have very short orbital periods, which means that the distance between the two stars is quite small. The neutron star can therefore capture gases escaping from its ordinary companion (Figure 23-12). More than 20 such exotic binary systems, also called *X-ray binary pulsars*,

*f*igure 23-12

A Model of a Pulsating X-Ray Source Pulsating X-ray sources are close binary systems in which one member is a rotating neutron star. Gas from the ordinary star flows across the star's critical surface and is captured by the neutron star (compare with Figure 21-16b). The infalling gas is funneled down onto the neutron star's magnetic poles and strikes the star with enough energy to create two X-ray–emitting hot spots. As the neutron star spins, beams of X rays from the hot spots sweep around the sky. If one of the beams is oriented so that it sweeps over the Earth, we see a pulse of X rays on each sweep.

are known. A typical rate of mass transfer from the ordinary star to the neutron star is roughly 10^{-9} solar masses per year.

Because of its strong gravity, a neutron star in a pulsating X-ray source easily captures much of the gas escaping from its companion. But like an ordinary pulsar, the neutron star is rotating rapidly and has a powerful magnetic field inclined to the axis of rotation (recall Figure 23-3). As the gas falls toward the neutron star, its magnetic field funnels the incoming matter down onto the star's north and south magnetic polar regions. The neutron star's gravity is so strong that the gas is traveling at nearly half the speed of light by the time it crashes onto the star's surface. This violent impact creates spots at both poles with searing temperatures of about 10^8 K. These temperatures are so high that the hot spots emit abundant X rays. In fact, their X-ray luminosity is roughly 10^{31} watts, nearly 100,000 times greater than the combined luminosity of the Sun at all wavelengths.

As the neutron star rotates, the beams of X rays from its magnetic poles sweep around the sky. If the Earth happens to be in the path of one of the two beams, we can observe a pulsating X-ray source. We detect a pulse once per rotation period, which means that the neutron star in Hercules X-1 spins at the rate of once every 1.24 seconds. If the orbital plane of the binary system is nearly edge-on to our line of sight, as for Centaurus X-3 and Hercules X-1, the pulsating X rays appear to turn off when the neutron star is eclipsed by the companion star. Other pulsating X-ray sources, such as Scorpius X-1, do not undergo this sort of turning off. The orbital planes of these systems are oriented more nearly face-on to our line of sight, so no eclipses occur.

 As time passes, the neutron star in a pulsating X-ray source will accrete more and more mass from its companion. In the process, it will spin up and may eventually become a millisecond pulsar (Section 23-6). In 1998 the Rossi X-ray Timing Explorer spacecraft first observed an evolutionary "missing link" between pulsating X-ray sources and millisecond pulsars. This object, designated SAX J1808.4-3658, is an X-ray-emitting millisecond pulsar in a close binary system with a pulse period of just 2.5 ms. Apparently the pulsar has accreted a substantial amount of mass from its companion star. Like the Black Widow Pulsar (see Figure 23-10), it has also blown much of the companion's mass into space. Within a billion years, the companion star may disappear altogether.

23-8 Explosive thermonuclear processes on white dwarfs and neutron stars produce novae and bursters

From time to time astronomers observe a faint star suddenly brighten by a factor of 10^4 to 10^8 over a few days or hours, reaching a peak luminosity of about 10^5 $L_\odot$. This phenomenon is called a **nova** (plural **novae**). By contrast, a supernova has a peak luminosity of about 10^9 $L_\odot$.

A nova's abrupt rise in brightness is followed by a gradual decline that may stretch out over several months or more (Figures 23-13 and 23-14). Every year two or three novae are seen in our Galaxy, and several dozen more are thought to take place in remote regions of the Galaxy that are obscured from our view by interstellar dust.

In the 1950s, painstaking observations of numerous novae by Robert Kraft, Merle Walker, and their colleagues at the University of California's Lick Observatory led to the conclusion that all novae are members of close binary systems containing

a

b

f **igure 23-13** R I **V** U X G

Nova Herculis 1934 These two pictures show a nova **(a)** shortly after peak brightness as a star of apparent magnitude +3, bright enough to be seen easily with the naked eye, and **(b)** two months later, when it had faded by a factor of 4000 in brightness to apparent magnitude +12. (See Figure 19-6 for a description of the apparent magnitude scale.) Novae are named after the constellation and year in which they appear. (Lick Observatory)

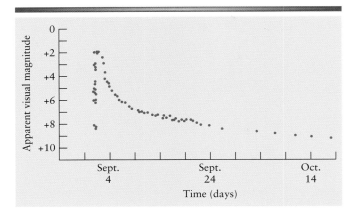

figure 23-14

The Light Curve of Nova Cygni 1975 This graph shows the history of a nova that blazed forth in the constellation Cygnus in September 1975. Its rapid rise and gradual decline in apparent brightness are characteristic of all novae. This nova, also designated V1500 Cyg, was easily visible to the naked eye (that is, was brighter than an apparent magnitude of +6) for nearly a week.

a white dwarf. Gradual mass transfer from the ordinary companion star, which presumably fills its Roche lobe, deposits fresh hydrogen onto the white dwarf (see Figure 21-17b for a schematic diagram of this sort of mass transfer).

Because of the white dwarf's strong gravity, this hydrogen is compressed into a dense layer covering the hot surface of the white dwarf. As more gas is deposited and compressed, the temperature in the hydrogen layer increases. When the temperature reaches about 10^7 K, hydrogen burning ignites throughout the gas layer, embroiling the white dwarf's surface in a thermonuclear holocaust that we see as a nova.

It is important to understand the similarities and differences between novae and Type Ia supernovae. Both kinds of celestial explosions are thought to occur in close binary systems where one of the stars is a white dwarf. But, as befits their name, supernovae are much more energetic. A Type Ia supernova explosion radiates 10^{44} joules of energy into space, while the corresponding figure for a typical nova is 10^{37} joules. (To be fair to novae, this relatively paltry figure is as much energy as our Sun emits in 1000 years.) The difference is thought to be that in a Type Ia supernova, the white dwarf accretes much more mass from its companion. This added mass causes so much compression that nuclear reactions can take place *inside* the white dwarf. Eventually, these reactions blow the white dwarf completely apart. In a nova, by contrast, nuclear reactions occur only within the accreted material. The reaction is more sedate because it takes place only on the white dwarf's surface (perhaps because the accretion rate is less or because the white dwarf had less mass in the first place).

Because the white dwarf itself survives a nova explosion, it is possible for the same star to undergo more than one nova. As an example, the star T Coronae Borealis erupted as a nova in 1866, then put in a repeat performance in 1946. By contrast, a given star can only be a supernova once.

A surface explosion similar to a nova also occurs with neutron stars. In 1975 it was discovered that some objects in the sky emit sudden, powerful bursts of X rays. Figure 23-15 shows the record of a typical burst. The source emits X rays at a constant low level until suddenly, without warning, there is an abrupt increase in X rays, followed by a gradual decline. An entire burst typically lasts for only 20 s. Unlike pulsating X-ray sources, there is a fairly long interval of hours or days between bursts. Sources that behave in this fashion are known as **X-ray bursters.** Several dozen X-ray bursters have been discovered in our Galaxy.

X-ray bursters, like novae, are believed to involve close binaries whose stars are engaged in mass transfer. With a burster, however, the stellar corpse is a neutron star rather than a white dwarf. Gases escaping from the ordinary companion star fall onto the neutron star. The X-ray burster's magnetic field is probably not strong enough to funnel the falling material toward the magnetic poles, so the gases are distributed more evenly over the surface of the neutron star. The energy released as these gases crash down onto the neutron star's surface produces the low-level X rays that are continuously emitted by the burster.

Most of the gas falling onto the neutron star is hydrogen, which the star's powerful gravity compresses against its hot surface. In fact, temperatures and pressures in this accreting

figure 23-15

The Light Curve of an X-Ray Burster A burster emits a constant low intensity of X rays interspersed with occasional powerful X-ray bursts. X-ray bursters are thought to occur when gas accretes onto a neutron star from a normal companion star. When the gas layer becomes a few meters thick, it undergoes explosive thermonuclear reactions. These heat the star's surface to such a high temperature that copious X rays are emitted. This typical burst was recorded on September 28, 1975, by an Earth-orbiting X-ray telescope pointed toward the globular cluster NGC 6624. About one-third of all known bursters are located in globular clusters (see Figure 21-8). (Adapted from W. Lewin) layer become so high that the arriving hydrogen is converted into helium by hydrogen burning. As a result, the accreted gases develop a layered structure that covers the entire neutron star, with a few tens of centimeters of hydrogen lying atop a similar thickness of helium. The structure is reminiscent of the layers within an evolved giant star (see Figure 22-2), although the layers atop a neutron star are much more compressed, thanks to the star's tremendous surface gravity.

When the helium layer is about 1 m thick, helium burning ignites explosively and heats the neutron star's surface to about 3×10^7 K. At this temperature the surface predominantly emits X rays, but the emission ceases within a few seconds as the surface cools. Hence, we observe a sudden burst of X rays only a few seconds in duration. New hydrogen then flows onto the neutron star, and the whole process starts over. Indeed, X-ray bursters typically emit a burst every few hours or days.

Whereas explosive *hydrogen* burning on a white dwarf produces a nova, explosive *helium* burning on a neutron star produces an X-ray burster. In both cases, the process is explosive, because the fuel is compressed so tightly against the star's surface that it becomes degenerate, like the star itself. As with the helium flash inside red giants (described in Section 21-2), the ignition of a degenerate thermonuclear fuel involves a sudden thermal runaway. This is because an increase in temperature does not produce a corresponding increase in pressure that would otherwise relieve compression of the gases and slow the nuclear reactions.

One of the great puzzles in modern astronomy has been the nature of an even more remarkable class of events called *gamma-ray bursters*. As the name suggests, these objects emit sudden, intense bursts of high-energy gamma rays. A number of explanations for gamma-ray bursters have been proposed, but their true nature remains unknown. We will discuss these mysterious objects further in Chapter 27.

23-9 Like a white dwarf, a neutron star has an upper limit on its mass

A white dwarf will collapse if its mass is greater than the Chandrasekhar limit of 1.4 $M_\odot$. At that point, degenerate electron pressure cannot support the overpowering weight of the star's matter, which presses inward from all sides (see Section 22-4). The mass of a neutron star also has an upper limit. However, the pressure within a neutron star is harder to analyze, because it comes from *two* sources. One is the degenerate nature of the neutrons, and the other is the strong nuclear force that acts between the neutrons themselves.

The strong nuclear force is what holds protons and neutrons together in atomic nuclei. Neutrons exert strong nuclear forces on one another only when they are almost touching. This force behaves somewhat like the force that billiard balls exert on one another when they touch: It strongly resists further compression. (Try squeezing two billiard balls together and see how much success you have.) Hence the strong nuclear force is a major contributor to the star's internal pressure. Unfortunately, there is a good deal of uncertainty about the details of this force. This translates into uncertainties about how much weight the neutron star's internal pressure can support—that is, the neutron star's maximum mass. Theoretical estimates of this maximum mass range from 2 to 3 $M_\odot$.

Before pulsars were discovered, most astronomers believed all dead stars to be white dwarfs. Dying stars were thought to somehow eject enough material so that their corpses could be below the Chandrasekhar limit. The discovery of neutron stars proved this idea incorrect. Inspired by this lesson, astronomers soon began wondering what might happen if a dying massive star failed to eject enough matter to get below the upper limit for a neutron star. For example, what might a 5-$M_\odot$ stellar corpse be like?

The gravity associated with a neutron star is so strong that the escape speed from it is roughly one-half the speed of light. But if a stellar corpse has a mass greater than 3 $M_\odot$, so much matter is crushed into such a small volume that the escape speed actually *exceeds* the speed of light. Because nothing can travel faster than light, nothing—not even light—can leave this dead star. Its gravity is so powerful that it leaves a hole in the fabric of space and time. Thus the discovery of neutron stars inspired astrophysicists to examine seriously one of the most bizarre and fantastic objects ever predicted by modern science, the black hole. We take up its story in the next chapter.

KEY WORDS

degenerate neutron pressure, p. 525

glitch, p. 531

millisecond pulsar, p. 532

neutron star, p. 525

nova (*plural* novae), p. 535

pair production, p. 528

pulsar, p. 526

pulsating X-ray source, p. 533

superconductivity, p. 532

superfluidity, p. 531

synchrotron radiation, p. 531

X-ray burster, p. 536

KEY IDEAS

Neutron Stars: A neutron star is a dense stellar corpse consisting primarily of closely packed degenerate neutrons.

• A neutron star typically has a diameter of about 30 km, a mass less than 3 $M_\odot$, a magnetic field 10^{12} times stronger than that of the Sun, and a rotation period of roughly 1 second.

• A neutron star consists of a superfluid, superconducting core surrounded by a superfluid mantle and a thin, brittle crust.

• Intense beams of radiation emanate from regions near the north and south magnetic poles of a neutron star. These beams are produced by streams of charged particles moving in the star's intense magnetic field.

Pulsars: A pulsar is a source of periodic pulses of radio radiation. These pulses are produced as beams of radio waves from a neutron star's magnetic poles sweep past the Earth.

• The pulse rate of many pulsars is slowing steadily. This reflects the gradual slowing of the neutron star's rotation as it radiates energy into space. Sudden speedups of the pulse rate, called glitches, may be caused by interactions between the neutron star's crust and its superfluid interior.

Neutron Stars in Close Binary Systems: If a neutron star is in a close binary system with an ordinary star, tidal forces will draw gas from the ordinary star onto the neutron star.

• The transfer of material onto the neutron star can make it rotate extremely rapidly, giving rise to a millisecond pulsar.

• Magnetic forces can funnel the gas onto the neutron star's magnetic poles, producing hot spots. These hot spots then radiate intense beams of X rays. As the neutron star rotates, the X-ray beams appear to flash on and off. Such a system is called a pulsating X-ray variable.

Novae and Bursters: Material from an ordinary star in a close binary can fall onto the surface of the companion white dwarf or neutron star to produce a surface layer in which thermonuclear reactions can explosively ignite.

• Explosive hydrogen burning may occur in the surface layer of a companion white dwarf, producing the sudden increase in luminosity that we call a nova. The peak luminosity of a nova is only 10^{-4} of that observed in a supernova.

• Explosive helium burning may occur in the surface layer of a companion neutron star. This produces a sudden increase in X-ray radiation, which we call a burster.

REVIEW QUESTIONS

1. What are neutron stars? What led scientists to propose their existence?

2. What prevents a neutron star from collapsing under its own gravity?

3. Why do astronomers think that pulsars are rapidly rotating neutron stars?

4. During the weeks immediately following the discovery of the first pulsar, one suggested explanation was that the pulses might be signals from an extraterrestrial civilization. Why did astronomers soon discard this idea?

5. Why do neutron stars rotate so much more rapidly than ordinary stars? Why do they have such strong magnetic fields?

6. Do all supernova remnants contain pulsars? Are all pulsars found within supernova remnants? For each question, explain why or why not.

7. Is our Sun likely to end up as a neutron star? Why or why not?

8. Compare the internal structure of a white dwarf to that of a neutron star. What are the similarities? What are the differences?

9. What is the difference between superconductivity and superfluidity?

10. How does a glitch affect a pulsar's period? What could be the cause of glitches?

11. Astronomers have deduced that the Vela pulsar is about 11,000 years old. How do you suppose they did this?

12. Why do astronomers think that millisecond pulsars are very old?

13. Describe a pulsating X-ray source like Hercules X-1 or Centaurus X-3. What produces the pulsation?

14. What is the connection between pulsating X-ray sources and millisecond pulsars?

15. What is the difference between a nova and a Type Ia supernova?

16. What are the similarities between novae and X-ray bursters? What are the differences?

17. What are the similarities between pulsating X-ray sources and X-ray bursters? What are the differences?

18. Why is the maximum mass of neutron stars not known as accurately as the Chandrasekhar limit for white dwarfs?

ADVANCED QUESTIONS

Problem-solving tips and tools

You will need the formulas for the Doppler effect introduced in Section 5-9 and Box 5-6 (see Figure 5-24). The volume of a sphere of radius r is $^4/_3\pi r^3$. The small-angle formula is given in Box 1-1. Section 5-4 describes how to relate the temperature of a blackbody to its wavelength of maximum emission. Section 19-6 gives the formula relating a star's luminosity, surface temperature, and radius (see Box 19-4 for worked examples). Section 3-5 gives the angular diameter of the Moon.

19. Using a diagram like Figure 23-3, explain why the number of pulsars that we observe in nearby space is probably quite a bit less than the number of rotating, magnetized neutron stars in nearby space.

20. There are many more main-sequence stars of low mass (less than 8 $M_\odot$) than of high mass (8 $M_\odot$ or more). Use this fact to explain why white dwarf stars are far more common than neutron stars.

21. The distance to the Crab Nebula is about 2000 parsecs. In what year did the star actually explode? Explain your answer.

22. How do we know that the Crab pulsar is really embedded in the Crab Nebula and not simply located at a different distance along the same line of sight?

23. The apparent expansion rate of the Crab Nebula is 0.23 arcsec per year. The apparent size of the Crab Nebula is about 4 by 6 arcmin. If the expansion rate remains constant, calculate how long it will be until the long axis of the Crab Nebula has the same apparent size as the full moon.

24. Emission lines in the spectrum of the Crab Nebula exhibit a Doppler shift, which indicates an expansion velocity of about 1200 km/s. (a) From this and the data given in the preceding question, calculate the distance to the Crab Nebula and estimate the date on which it exploded. (b) Do your answers agree with figures quoted in this chapter? If not, can you point to assumptions you made in your computations that led to the discrepancies? Or do you think your calculations suggest additional physical effects are at work in the Crab Nebula, over and above a constant rate of expansion?

25. To determine accurately the period of a pulsar, astronomers must take into account the Earth's orbital motion about the Sun. (a) Explain why. (b) Knowing that the Earth's orbital velocity is 30 km/s, calculate the maximum correction to a pulsar's period because of the Earth's motion. Explain why the size of the correction is greatest for pulsars located near the ecliptic.

26. A neutron has a mass of about 1.7×10^{-27} kg and a radius of about 10^{-15} m. (a) Compare the density of matter in a neutron with the average density of a neutron star. (b) If the neutron star's density is more than that of a neutron, the neutrons within the star are overlapping; if it is less, the neutrons are not overlapping. Which of these seems to be the case? What do you think is happening at the center of the neutron star, where densities are higher than average?

27. X-ray pulsars are speeding up but ordinary (radio) pulsars are slowing down. Propose an explanation for this difference.

28. If the model for Hercules X-1 discussed in the text is correct, at what orientation of the binary system do we see its maximum optical brightness? Explain your answer.

29. In an X-ray burster, the surface of a neutron star 10 km in radius is heated to a temperature of 3×10^7 K. (a) Determine the wavelength of maximum emission of the heated surface (which you may treat as a blackbody). In what part of the electromagnetic spectrum does this lie? (See Figure 5-7.) (b) Find the luminosity of the heated neutron star. Give your answer in watts and in terms of the luminosity of the Sun, given in Table 18-1. How does this compare with the peak luminosity of a nova? Of a Type Ia supernova?

30. The nearest neutron star, called RX J185635-3754, is just 60 pc (200 ly) from Earth. It is thought to be the relic of a star that underwent a supernova explosion about 1 million years ago. The explosion ejected the neutron star at high speed, so it is now moving through nearly empty space. (a) RX J185635-3754 is *not* a pulsar, that is, it does not emit pulses of radiation. Suggest why this might be so. (b) The neutron star has a surface temperature of 600,000 K. Find the wavelength at which it emits most strongly, and explain why the neutron star appears as a steady, nonpulsating object in an X-ray telescope. (c) RX J185635-3754 has a total luminosity at all wavelengths of about $0.046L_\odot$. Calculate its radius, and explain why astronomers conclude that it is a neutron star.

DISCUSSION QUESTIONS

31. How might astronomers be able to detect the presence of an accretion disk in a close binary system?

32. When neutrons are very close to one another, they repel one another through the strong nuclear force. If this repulsion were made even stronger, what effect might this have on the maximum mass of a neutron star? Explain your answer.

WEB/CD-ROM QUESTIONS

33. Search the World Wide Web for information about the latest observations of the stellar remnant at the center of SN 1987A. Has a pulsar been detected? If so, how fast is it spinning? Has the supernova's debris thinned out enough to give a clear view of the neutron star?

34. Search the World Wide Web for information about a class of neutron stars called *magnetars*. How do they differ from ordinary neutron stars? What is their connection to objects called soft gamma-ray repeaters?

35. Monitoring the Crab Pulsar. Access the video "The Crab Pulsar" in Chapter 23 of the *Universe* web site or CD-ROM. View the video and use it to answer the following questions. For each part, explain how you determined your answer. (**a**) How many rotations does the neutron star complete during the duration of the video? (**b**) Is the neutron star visible at the beginning of the video? If not, explain why not. (**c**) How does the peak brightness of the Crab pulsar compare to the steady brightness of the nearby star? (**d**) What total amount of time is depicted in the video?

OBSERVING PROJECTS

36. If you did not take the opportunity to observe the Crab Nebula as part of the exercises in Chapter 22, do so now. The Crab Nebula is visible in the night sky from October through March. Its epoch 2000 coordinates are R.A. = 5^h 34.5^m and Decl. = $22° 00'$, which is near the star marking the eastern horn of Taurus (the Bull). Be sure to schedule your observations for a moonless night. The larger the telescope you use, the better, because the Crab Nebula is quite dim.

37. Consult the World Wide Web to see if any novae have been sighted recently. If by good fortune one has been sighted, what is its apparent magnitude? Is it within reach of a telescope at your disposal? If so, arrange to observe it. Draw what you see through the eyepiece, noting the object's brightness in comparison with other stars in the field of view. If possible, observe the same object a few weeks or months later to see how its brightness has changed.

38. Use the *Starry Night* program to observe the sky in July 1054, when the supernova that spawned the Crab Nebula was visible from the American Southwest. Select **Viewing Location…** in the **Go** menu, then set the latitude to 36° N and the longitude to 109° W. Set the time to 5:00 A.M. on July 5, 1054. Use **Find…** in the **Edit** menu to find and center on the Crab Nebula. Zoom in or out until you can see both the position of the nebula and the Moon. You may find it helpful to turn off daylight (select **Daylight** in the **Sky** menu). What is the phase of the Moon? Investigate how the relative positions of the Moon and the Crab Nebula change when you set the date to July 4, 1054, or July 6, 1054. On which date do the relative positions of the Moon and the Crab Nebula give the best match to the pictograph in Figure 23-1?

Black Holes

(Chandra X-ray Observatory/Harvard-Smithsonian Center for Astrophysics)

WEB LINK 24.1

Imagine a swirling disk of gas and dust, orbiting around an object that has more mass than the Sun but is so dark that it cannot be seen. Imagine the material in this disk being compressed and heated as it spirals into the unseen object, reaching temperatures so high that it emits X rays. And imagine that the unseen object has such powerful gravity that any material that falls into it simply disappears, never to be seen again.

Such hellish maelstroms, like the one shown in the illustration, really do exist. The disks are called *accretion disks*, and the unseen objects with immensely strong gravity are called *black holes.*

The matter that makes up a black hole has been so greatly compressed that it violently warps space and time. If you get too close to a black hole, the speed you would need to escape exceeds the speed of light. Because nothing can travel faster than light, nothing—not even light—can escape from a black hole.

Black holes, whose existence was predicted by Einstein's general theory of relativity (our best description of what gravity is and how it behaves), are both strange and simple. The structure of a black hole is completely specified by only three quantities—its mass, electric charge, and angular momentum. Perhaps strangest of all, there is compelling evidence that black holes really exist. In recent years astronomers have found that certain binary star systems contain black holes. Even more remarkable is the discovery that black holes of more than a million solar masses may lie at the centers of many galaxies, including our own Milky Way.

As you read the sections of this chapter, look for the answers to the following questions.

24-1   What are the two central ideas behind Einstein's special theory of relativity?

24-2   How did Einstein describe gravity, and how is his description different from Newton's?

24-3   How do astronomers search for black holes?

24-4   What are supermassive black holes, and where are they found?

24-5   In what sense is a black hole "black"?

24-6   In what way are black holes actually simpler than any other objects in astronomy?

24-7   What happens to an object that falls into a black hole?

24-8   Do black holes last forever?

24-1   The special theory of relativity changes our conceptions of space and time

The idea of a black hole is a truly strange one. To truly appreciate and understand black holes, we must discard some "commonsense" notions about the nature of space and time.

According to the classical physics of Newton, space is perfectly uniform and fills the universe like a rigid framework. Similarly, time passes at a monotonous, unchanging rate. It is always possible to know exactly how fast you are moving through this rigid fabric of space and time.

Those ideas were upset forever in 1905 when Albert Einstein proposed his **special theory of relativity**. This theory describes how motion affects our measurements of distance and time. It is "special" in the sense of being specialized. In particular, it does not include the effects of gravity. The word "relativity" implies the opposite of "absolutism," and hints at one of the key results of the theory: The distance between two points is not an absolute, nor is the time interval between two events. Instead, the values you measure for these quantities depend on how you are moving.

Remarkably, Einstein's theory is based on just two basic principles. The first is quite simple:

Your description of physical reality is the same regardless of the constant velocity at which you move.

In other words, as long as you are moving in a straight line at a constant speed, you experience the same laws of physics as anyone else moving at any other constant speed and in any other direction. As an example, suppose you were inside a closed railroad car moving due north in a straight line at 100 km/h. Any measurements you make inside the car—for example, how long your thumb is or how much time elapses between ticks of your watch—will give exactly the same results as if the car were moving in any other direction or at any other speed, or was not moving at all.

Einstein's second principle is much more bizarre:

Regardless of your speed or direction of motion, you always measure the speed of light to be the same.

To see what this implies, imagine that you are in a spaceship moving toward a flashlight. Even if you are moving at 99% of the speed of light, you will measure the photons from the flashlight to be moving at the same speed ($c = 3 \times 10^8$ m/s $= 3 \times 10^5$ km/s) as if your spaceship were motionless. This statement is in direct conflict with the Newtonian view that a stationary person and a moving person should measure different speeds (Figure 24-1).

Speed involves both distance and time. Since speed has a very different behavior in the special theory of relativity than in Newton's physics, it follows that both space and time behave differently as well. Indeed, in relativity time proves to be so intimately intertwined with the three dimensions of space that we regard them as a single *four*-dimensional entity called **spacetime**.

 Einstein expressed his ideas about spacetime in a mathematical form and used this to make a number of predictions about nature. All these predictions have been verified in innumerable experiments. One prediction is that the length you measure an object to have depends on how that object is moving; the faster it moves, the shorter its length along its direction of motion. This is called **length contraction**. In other words, if a railroad car moves past you, from your perspective on the ground you will measure it to be shorter than if it were at rest (Figure 24-2a). However, if you are on board the railroad car and moving with it, you will measure its length to be the same as measured on the ground when it was at rest. The word "relativity" emphasizes the importance of the relative speed between the observer and the object being measured.

If moving objects are shortened, why don't we see this effect in fast-moving cars or airplanes? The reason is that length contraction is noticeable only at very high speeds, comparable to that of light. But even the fastest spacecraft ever built by humans travels at a mere 1/25,000 of the speed of

As seen by outfielder, ball is approaching her at
(30 m/s) + (10 m/s) = 40 m/s

a

Incorrect Newtonian description:
As seen by astronaut in spaceship, light is approaching her at
$(3 \times 10^8$ m/s$) + (1 \times 10^8$ m/s$) = 4 \times 10^8$ m/s

Correct Einsteinian description:
As seen by astronaut in spaceship, light is approaching her at
3×10^8 m/s

b

figure 24-1

The Speed of Light Is the Same to All Observers (a) In
Newtonian physics, the speed you measure for an object depends on
how you are moving. Thus, the batter sees the ball moving at 30 m/s,
but the outfielder sees it moving at 40 m/s relative to her. (b) Einstein
showed that this commonsense principle does not apply to light. No
matter how fast or in what direction the astronaut in the spaceship is
moving, she and the astronaut holding the flashlight will always
measure light to have the same speed.

light. As Box 24-1 describes, however, we can easily see the
effects of length contraction for subatomic particles that do
indeed travel at nearly the speed of light.

A second result of relativity is no less strange: A clock
moving past you runs more slowly than a clock that is at rest.
Like length contraction, this **time dilation** is a very small effect
unless the clock is moving at extremely high speeds (Figure
24-2*b*). Nonetheless, physicists have confirmed the existence
of time dilation by using an extremely accurate atomic clock
carried on a jet airliner. When the airliner landed, they found
that the on-board clock had actually ticked off slightly less
time than an identical, stationary clock on the ground. For the
passengers on board the airliner during this experiment, however, time flowed at a normal rate: the on-board clock ticked
off the seconds as usual, their hearts beat at a normal rate, and
so on. You only measure a clock (or a beating heart) to be running slow if it is *moving* relative to you. Box 24-1 discusses
time dilation in more detail.

The special theory of relativity predicts that the astronaut
with the flashlight in Figure 24-1*b* will see the astronaut flying in the spaceship as shortened along the direction of
motion and as having slowly ticking clocks. But the theory
also says that these effects depend only on *relative* motion.
Hence, the *flying* astronaut will see the astronaut with the
flashlight (who is moving relative to her) as being shortened
and as having slowly ticking clocks. There is no contradiction: Each astronaut's measurements are correct relative to
his or her frame of reference.

CAUTION! You may have the idea that the special theory of relativity implies that there are *no* absolutes in nature
and that everything is relative. But, in fact, the special theory is based on the principles that the laws of physics and the speed of light in a vacuum *are* absolutes. Only
certain quantities, such as distance and time, depend on your

a **Length contraction**

At rest

In motion

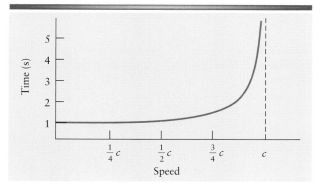

b **Time dilation**

figure 24-2

Length Contraction and Time Dilation (a) The faster an object
moves, the shorter it becomes along its direction of motion. The object
becomes infinitesimally small as its speed approaches the speed of light.
Speed has no effect on the object's dimensions perpendicular to the
direction of motion. (b) The faster a clock moves, the slower it runs. This
graph shows how many seconds (as measured on your clock) it takes a
clock that moves relative to you to tick off one second. The effect is
pronounced only for speeds near the speed of light *c*.

Box 24-1   Tools of the Astronomer's Trade

Time Dilation and Length Contraction

The special theory of relativity describes how motion affects measurements of time and distance. By using the two basic principles of his theory—that no matter how fast you move, the laws of physics and the speed of light are the same—Einstein concluded that these measurements must depend on how the person making the measurements is moving.

To describe how measurements depend on motion, Einstein derived a series of equations named the **Lorentz transformations,** after the famous Dutch physicist Hendrik Antoon Lorentz. (Lorentz was a contemporary of Einstein who developed these equations independently but did not grasp their true meaning.) These equations tell us exactly how moving clocks slow down and how moving objects shrink.

To appreciate the Lorentz transformations, imagine two observers named Sam and Maria. Sam is at rest on Earth, while Maria is flying past in her spaceship at a speed v. Sam and Maria both observe the same phenomenon on Earth—say, the beating of Sam's heart or the ticking of Sam's watch—which appears to occur over an interval of time. According to Sam's clock (which is not moving relative to the phenomenon), the phenomenon lasts for T_0 seconds. This is called the **proper time** of the phenomenon. But according to Maria's clock (which *is* moving relative to the phenomenon), the same phenomenon lasts for a different length of time, T seconds. These two time intervals are related as follows:

Lorentz transformation for time

$$T = \frac{T_0}{\sqrt{1 - (v/c)^2}}$$

T = time interval measured by an observer moving relative to the phenomenon

T_0 = time interval measured by a observer *not* moving relative to the phenomenon

v = speed of the moving observer
c = speed of light

EXAMPLE: Suppose that Maria is moving at 98% of the speed of light. Then,

$$\frac{v}{c} = 0.98$$

so that

$$T = \frac{T_0}{\sqrt{1 - (0.98)^2}} = 5T_0$$

Thus, a phenomenon that lasts for $T_0 = 1$ second on Sam's clock is stretched out to $T = 5T_0 = 5$ seconds as measured on Maria's clock moving at 98% of the speed of light. As measured by our fast-moving friend Maria, a 60-second commercial on Sam's TV will last for 5 minutes and the minute hand on Sam's clock will take 5 hours to make a complete sweep. The converse is also true. As measured by *Sam*, the minute hand on *Maria*'s clock will take 5 hours to make a complete sweep. This phenomenon, in which events moving relative to an observer happen at a slower pace, is called *time dilation.*

The Lorentz transformation for time is plotted in Figure 24-2b, which shows how 1 second measured on a stationary clock (say, Sam's) is stretched out when measured using a clock carried by a moving observer (such as Maria). For speeds less than about half the speed of light, the mathematical factor of $\sqrt{1 - (v/c)^2}$ is not too different from 1, and there is little difference between the recordings of the stationary and moving clocks. At the fastest speeds that humans have ever traveled (en route from the Earth to the Moon), the difference between 1 and the factor $\sqrt{1 - (v/c)^2}$ is less than 10^{-9}. So, for any speed associated with human activities, stationary and slowly moving clocks tick at essentially the

state of motion. You may also have the idea that length contraction and time dilation are just optical illusions caused by high speeds. But these effects are *real*. A moving railroad car really is shorter, and a moving clock really does tick more slowly. Einstein's theory is supported by every experiment designed to test it. Relativity is very strange, but it is also very real!

Einstein's special theory of relativity predicts another famous relationship: Any object with mass (m) has energy (E)

embodied in its mass. This is expressed by the well-known formula $E = mc^2$, where c is the speed of light. In thermonuclear reactions, part of the energy embodied in the mass of atomic nuclei is converted into other forms of energy. This released energy is what makes the Sun and the stars shine (see Section 18-1). The sunlight we see by day and the starlight that graces our nights are a resounding verification of the special theory of relativity.

The special theory of relativity also explains why it is impossible for a spaceship to move at the speed of light. If it same rate. As the next example shows, however, time dilation is important for subatomic particles that travel at speeds comparable to c.

EXAMPLE: Fast-moving protons from interstellar space frequently collide with atoms in the Earth's upper atmosphere. When they do, they can create unstable particles called *muons* (pronounced "mew-ons") that decay in an average time of 2.2×10^{-6} s. Such muons typically move at 99.9% of the speed of light and are formed at an altitude of 10 km. As measured by an observer on the Earth, the time that a muon would take to reach the Earth's surface is

$$\frac{\text{distance}}{\text{speed}} = \frac{10{,}000 \text{ m}}{0.999 \times 3.00 \times 10^8 \text{ m/s}} = 3.3 \times 10^{-5} \text{ s}$$

This is 15 times longer than the life expectancy of a muon, so it would seem that muons could never reach the Earth's surface before decaying. In fact, these muons *are* detected by experiments on the surface! The reason is that as seen by an Earth observer, the muon is a "moving clock," and hence its decay is slowed down by time dilation. To an Earth observer, the actual lifetime of a muon is

$$T = \frac{2.2 \times 10^{-6} \text{ s}}{\sqrt{1 - (0.999)^2}} = 4.9 \times 10^{-5} \text{ s}$$

Thus, as measured by an Earth observer, muons live more than long enough for them to reach the surface. The detection at the Earth's surface of muons from the upper atmosphere is compelling evidence for the reality of time dilation.

In the same terminology as "proper time," we say that a ruler at rest measures **proper length** or **proper distance** (L_0). According to the Lorentz transformations, distances perpendicular to the direction of motion are unaffected. However, a ruler of proper length L_0 held parallel to the direction of motion shrinks to a length L, given by

Lorentz transformation for length

$$L = L_0 \sqrt{1 - (v/c)^2}$$

L = length of a moving object along the direction of motion
L_0 = length of the same object at rest (proper length)
v = speed of the moving object
c = speed of light

EXAMPLE: If Maria is traveling at 98% of the speed of light relative to Sam, he concludes that her clocks are ticking only one-fifth as fast as his. He also concludes that Maria's 1-meter ruler is only one-fifth as long (20 cm) when held parallel to the direction of motion:

$$L = (1 \text{ m})\sqrt{1 - (0.98)^2} = 0.20 \text{ m} = 20 \text{ cm}$$

This shrinkage of length is called *length contraction*.

EXAMPLE: Length contraction gives us another way to explain why unstable muons created high in the atmosphere are able to reach the Earth's surface. The proper distance from where the muon is created to the Earth's surface is 10 km. But as measured by a muon moving at 99.9% of the speed of light, the distance is much shorter:

$$L = (10 \text{ km})\sqrt{1 - (0.999)^2} = 0.45 \text{ km} = 450 \text{ m}$$

As measured by the muon, the time required to travel this contracted distance is

$$\frac{\text{distance}}{\text{speed}} = \frac{450 \text{ m}}{0.999 \times 3.00 \times 10^8 \text{ m/s}} = 1.5 \times 10^{-6} \text{ s}$$

This is less time than the 2.2×10^{-6} s that an average muon takes to decay. Hence, muons can successfully reach the Earth's surface.

could, then a light beam traveling in the same direction as the spaceship would appear to the ship's crew to be stationary. But this would contradict the second of Einstein's principles, which says that all observers, including those on a spaceship, must see light traveling at the speed of light. Therefore, it cannot be possible for a spaceship, or indeed any material object, to travel at the speed of light. This, too, has been verified by experiments. Subatomic particles can be made to travel at 99.99999999% of the speed of light c, but they can never make it all the way to c.

24-2 The general theory of relativity predicts black holes

Einstein's special theory of relativity is a comprehensive description of the behavior of light and, by extension, of electricity and magnetism. (Recall from Section 5-2 that light is both electric and magnetic in nature.) Einstein's next goal was to develop an even more comprehensive theory that also

a   b

figure 24-3
The Equivalence Principle The equivalence principle asserts that you cannot distinguish between **(a)** being at rest in a gravitational field and **(b)** being accelerated upward in a gravity-free environment. This idea was an important step in Einstein's quest to develop the general theory of relativity.

explained gravity. This was the **general theory of relativity,** which he published in 1915.

Einstein found Newton's description of gravity disappointing, because it assumes that space and time are absolute. However, gravity causes a falling object to accelerate and, according to the special theory of relativity, the resulting motion also affects rulers and clocks that are moving *with* the falling object. Thus, Einstein argued that gravity must affect the shape of space and the flow of time.

Einstein began by demonstrating that it is not necessary to think of gravity as a force. According to Newton's theory, an apple falls to the floor because the force of gravity pulls the apple down. Einstein pointed out that the apple would appear to behave in exactly the same way in space, far from the Earth's gravity, if the floor were to accelerate upward (in other words, if the floor came up to meet the apple).

In Figure 24-3, two famous gentlemen are watching an apple fall toward the floor of their closed compartments. They have no way of telling who is at rest on the Earth and who is in the hypothetical elevator moving upward at a constantly increasing speed. This is an example of Einstein's **equivalence principle**, which states that in a small volume of space, the downward pull of gravity can be accurately and completely duplicated by an upward acceleration of the observer.

The equivalence principle allowed Einstein to focus entirely on motion, rather than force, in discussing gravity. A hallmark of gravity is that it causes the same acceleration no matter the mass of the object. For example, a baseball and a cannon ball have very different masses, but if you drop them side by side in a vacuum, they accelerate downward at exactly the same rate. To explain this, Einstein envisioned gravity as being caused by a curvature of space. In fact, his general theory of relativity describes gravity entirely in terms of the geometry of both space *and* time, that is, of spacetime. Far from a source of gravity, like a planet or a star, spacetime is "flat" and clocks tick at their normal rate. Closer to a source of gravity, however, clocks slow down and space is curved.

A useful analogy is to picture the spacetime near a massive object such as the Sun as being curved like the surface in Figure 24-4. Imagine a ball rolling along this surface. Far from the "well" that represents the Sun, the surface is fairly flat and the ball moves in a straight line. If the ball passes near the well, however, it curves in toward it. If the ball is moving at an appropriate speed, it might move in an orbit around the sides of the well. The curvature of the well has the same effect on a ball of any size, which explains why gravity produces the same acceleration on objects of different mass.

Einstein's general theory of relativity and its picture of curved spacetime have been tested in a variety of different ways. In what follows we discuss some key experimental tests of the theory.

Experimental Test 1: The gravitational bending of light. Light rays naturally travel in straight lines. But if the space through which the rays travel is curved, as happens when light passes near the surface of a massive object like the Sun, the paths of the rays will likewise be curved (Figure 24-5). In other words, gravity should bend light rays, an effect not predicted by Newtonian mechanics because light has no mass. This prediction was first tested in 1919 during a total solar eclipse. During totality, when the Moon blocked out the Sun's disk, astronomers photographed the stars around the Sun. Careful measurements afterward revealed that the stars were shifted from their usual positions by an amount consistent with Einstein's theory.

Experimental Test 2: The precession of Mercury's orbit. As the planet Mercury moves along its elliptical orbit, the orbit itself slowly changes orientation or *precesses* (Figure 24-6). Most of Mercury's precession is caused by the gravitational pull of the other planets, as explained by Newtonian mechanics. But once the effects of all the other planets had been accounted for, there remained an unexplained excess rotation of Mercury's major axis of 43 arcsec per century. Although this

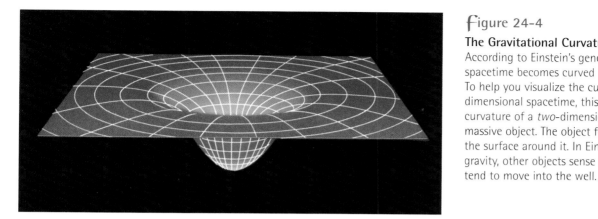

figure 24-4

The Gravitational Curvature of Spacetime
According to Einstein's general theory of relativity, spacetime becomes curved near a massive object. To help you visualize the curvature of four-dimensional spacetime, this figure shows the curvature of a *two*-dimensional space around a massive object. The object forms a curved "well" in the surface around it. In Einstein's picture of gravity, other objects sense this curvature and tend to move into the well.

discrepancy may seem very small, it frustrated astronomers for half a century. Some astronomers searched for a missing planet even closer to the Sun that might be tugging on Mercury; none has ever been found. Einstein showed that at Mercury's position close to the Sun, the general theory of relativity predicts a small correction to Newton's description of gravity. This correction is just enough to account for the excess precession.

Experimental Test 3: The gravitational slowing of time and the gravitational red shift. In the general theory of relativity, a massive object such as the Earth warps time as well as space. Einstein predicted that clocks on the ground floor of a building should tick slightly more slowly than clocks on the top floor, which are farther from the Earth (Figure 24-7*a*).

A light wave can be thought of as a clock; just as a clock makes a steady number of ticks per minute, an observer sees a steady number of complete cycles of a light wave passing by each second. If a light beam is aimed straight up from the

ground floor of a building, an observer at the top floor will measure a "slow-ticking" light wave with a lower frequency, and thus a longer wavelength, than will an observer on the ground floor (Figure 24-7*b*). The increase in wavelength means that a photon reaching the top floor has less energy than when it left the ground floor. (You may want to review the discussion of frequency and wavelength in Section 5-2 as well as the description of photons in Section 5-5.) These effects, which have no counterpart in Newton's theory of gravity, are called the **gravitational redshift**.

CAUTION! Be careful not to confuse the gravitational redshift with a Doppler shift. In the Doppler effect, redshifts are caused by a light source moving away from an observer. Gravitational redshifts, by contrast, are caused by time flowing at different rates at different locations. No motion is involved.

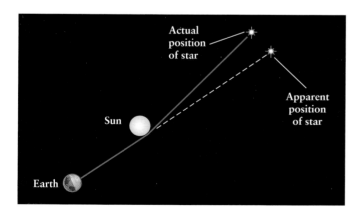

figure 24-5

The Gravitational Deflection of Light Light rays are deflected by the curved spacetime around a massive object like the Sun. The maximum deflection is very small, only 1.75 arcsec for a light ray grazing the Sun's surface. By contrast, Newton's theory of gravity predicts *no* deflection at all. The deflection of starlight by the Sun was confirmed during a solar eclipse in 1919.

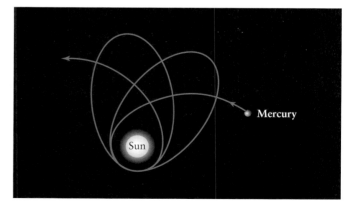

figure 24-6

The Precession of Mercury's Orbit The long axis of Mercury's orbit changes orientation at a rate of 574 arcsec, or about one-sixth of a degree, per century. (This is greatly exaggerated in the figure.) Newtonian mechanics predicts that the gravitational influences of the other planets should cause Mercury's perihelion to advance by only 531 arcsec per century. Einstein was able to explain the excess 43 arcsec per century using his general theory of relativity.

a
b

Light
source

ƒigure 24-7

The Gravitational Slowing of Time and the Gravitational Redshift **(a)** A clock on the ground floor of a building is closer to the Earth than one on the top floor. Consequently, according to the general theory of relativity, the clock on the ground floor ticks more slowly than the one on the top floor. **(b)** The oscillations of a light wave constitute a type of clock. As a light wave climbs from the ground floor toward the top floor, its oscillation frequency becomes lower and its wavelength becomes longer. This effect, which has been measured in experiments, is called the gravitational redshift.

The American physicists Robert Pound and Glen Rebka first measured this gravitational redshift in 1960 using gamma rays fired between the top and bottom of a shaft 20 meters tall. Because the Earth's gravity is relatively weak, they measured only a very small redshift ($\Delta\lambda/\lambda = 2.5 \times 10^{-15}$), but this value was in complete agreement with Einstein's prediction.

Much larger shifts are seen in the spectra of white dwarfs, whose spectral lines are redshifted as light climbs out of the white dwarf's intense surface gravity. As an example, the gravitational redshift of the spectral lines of the white dwarf Sirius B (see Figure 22-8) is $\Delta\lambda/\lambda = 3.0 \times 10^{-4}$, which also agrees with the general theory of relativity.

Experimental Test 4: Gravitational waves. Electric charges oscillating up and down in a radio transmitter's antenna produce electromagnetic radiation. In a similar way, the general theory of relativity predicts that oscillating massive objects should produce **gravitational radiation,** or **gravitational waves.** (Newton's theory of gravity makes no such prediction.)

Gravitational radiation is exceedingly difficult to detect, because it is by nature much weaker than electromagnetic radiation. Although physicists have built a number of sensitive "antennas" for gravitational radiation, no confirmed detections have been made as of this writing (2001). But compelling indirect evidence for the existence of gravitational radiation has come from a binary system of two neutron stars. Russell Hulse and Joseph Taylor at the University of Massachusetts discovered that these stars are slowly spiraling toward each other and losing energy in the process. The rate at which they lose energy is just what would be expected if the two neutron stars are emitting gravitational radiation as predicted by Einstein.

The general theory of relativity has never made an incorrect prediction. It now stands as our most accurate and complete description of gravity. Einstein demonstrated that

Newtonian mechanics is accurate only when applied to low speeds and weak gravity. If extremely high speeds or powerful gravity are involved, only a calculation using relativity will give correct answers.

Perhaps the most dramatic prediction of the general theory of relativity concerns what happens when a large amount of matter is concentrated in a small volume. We have seen that if a dying star is not too massive, it ends up as a white dwarf star. If the dying star is more massive than the Chandrasekhar limit of about 1.4 $M_\odot$, it cannot exist as a stable white dwarf star and, instead, shrinks down to form a neutron star. But if the dying star has more mass than the maximum permissible for a neutron star, about 3 $M_\odot$, not even the internal pressure of neutrons can hold the star up against its own gravity, and the star contracts rapidly.

As the star's matter becomes compressed to enormous densities, the strength of gravity at the surface of this rapidly shrinking sphere also increases dramatically. According to the general theory of relativity, the space immediately surrounding the star becomes so highly curved that it closes on itself. Photons flying outward at an angle from the star's surface arc back inward, while photons that fly straight outward undergo such a strong gravitational redshift that they lose all their energy and cease to exist (Figure 24-8).

Ordinary matter can never travel as fast as light. Hence, if light cannot escape from the collapsing star, neither can anything else. An object from which neither matter nor electromagnetic radiation can escape is called a **black hole.** In a sense, a hole is punched in the fabric of the universe, and the dying star disappears into this cavity (Figure 24-9).

None of the star's mass is lost when it collapses to form a black hole, however. This mass gives the spacetime around the black hole its strong curvature. Thanks to this curvature, the black hole's gravitational influence can still be felt by other objects.

figure 24-8

The Formation of a Black Hole **(a)** Photons emitted from a main-sequence, giant, or supergiant star are affected very little by the star's gravity, and travel in essentially straight lines. **(b)** At the surface of a collapsing neutron star, gravity is very strong and photons emitted from the surface follow curved paths. **(c)** As the star continues to collapse, the curvature of the surrounding spacetime increases and the trajectories of photons bend more sharply. **(d)** When the neutron star shrinks to a certain critical size, it becomes a black hole. The spacetime around the black hole is so highly curved that none of the emitted photons can escape. Not even photons emitted directly upward from the surface can escape; they undergo an infinite gravitational redshift and disappear.

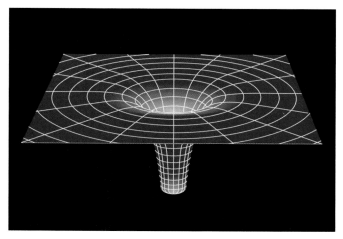

figure 24-9

Curved Spacetime around a Black Hole This diagram suggests how spacetime is distorted by a black hole's mass. As in Figure 24-4, spacetime is represented as a two-dimensional surface. Far from the hole, gravity is weak and spacetime is nearly "flat." Near the hole, gravity is strong and spacetime is highly curved. Unlike the situation in Figure 24-4, a black hole's gravitational "well" is infinitely deep.

CAUTION! Some low-quality science-fiction movies and books suggest that black holes are evil things that go around gobbling up everything in the universe. Not so! The bizarre effects created by highly warped spacetime are limited to a region quite near the hole. For example, the effects of the general theory of relativity predominate only within 1000 km of a 10-$M_\odot$ black hole. Beyond 1000 km, gravity is weak enough that Newtonian physics can adequately describe everything. If our own Sun somehow turned into a black hole (an event that, happily, seems to be quite impossible), the orbits of the planets would hardly be affected at all.

Perhaps the most remarkable aspect of black holes is that they really exist. As we will see, astronomers have located a number of black holes with masses a few times that of the Sun. What is truly amazing is that they have also discovered many truly immense black holes containing millions or billions of solar masses. These discoveries are a resounding confirmation of the ideas of the general theory of relativity. In the next two sections we will see how astronomers hunt down black holes in space.

24-3 Certain binary star systems probably contain black holes

Finding black holes is a difficult business. Because light cannot escape from inside the black hole, you cannot observe one directly in the same way that you can observe a star or a planet. The best you can hope for is to detect the effects of a black hole's powerful gravity.

Close binary star systems offer one way to find black holes. Imagine that one of the stars in a binary system evolves into a black hole. If the black hole orbits close enough to the other, ordinary star in the system, tidal forces can draw matter from the ordinary star onto the black hole. If we can detect radiation coming from this "stolen" matter, we can infer the presence of the black hole.

The first sign of such emissions from a binary system with a black hole came shortly after the launch of the *Uhuru* X-ray-detecting satellite in 1971. Astronomers became intrigued with an X-ray source designated Cygnus X-1. This source is quite unlike pulsating X-ray sources, which emit regular bursts of X rays every few seconds (see Section 23-7). Instead, the X-ray emissions from Cygnus X-1 are highly variable and irregular; they flicker on time scales as short as one-hundredth of a second. One of the fundamental concepts in physics is that nothing can travel faster than the speed of light (recall Section 24-1). Because of this limitation, an object cannot flicker faster than the time required for light to travel across the object. Because light travels 3000 km in a hundredth of a second, Cygnus X-1 can be no more than 3000 km across, or about a quarter the size of the Earth.

Cygnus X-1 occasionally emits bursts of radio radiation, and in 1971 radio astronomers used these bursts to show that the source was at the same location in the sky as the star

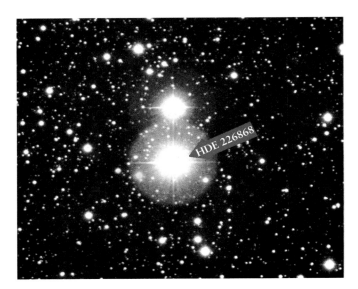

figure 24-10 R I **V** U X G

HDE 226868 This photograph from the 5-m telescope on Palomar Mountain shows the star at the location of the X-ray source Cygnus X-1. HDE 226868, a B0 supergiant located 2500 pc (8000 ly) from Earth in the constellation Cygnus, is part of a binary system. The other, unseen member of the system has a mass of at least 7 M$_\odot$ and is probably a black hole. The bright star directly above HDE 226868 is not part of the binary system. (Courtesy of J. Kristian, Carnegie Observatories)

HDE 226868 (Figure 24-10). Spectroscopic observations revealed that HDE 226868 is a B0 supergiant with a surface temperature of about 31,000 K. Such stars do not emit significant amounts of X rays, so HDE 226868 alone cannot be the

X-ray source Cygnus X-1. Because binary stars are very common, astronomers began to suspect that the visible star and the X-ray source are in orbit about each other.

Further spectroscopic observations soon showed that the lines in the spectrum of HDE 226868 shift back and forth with a period of 5.6 days. This behavior is characteristic of a single-line spectroscopic binary (see Section 19-10); the companion of HDE 226868 is too dim to produce its own set of spectral lines. The clear implication is that HDE 226868 and Cygnus X-1 are the two components of a binary star system.

From what we know about the masses of other supergiant stars, HDE 226868 is estimated to have a mass of roughly 30 M$_\odot$. As a result, the unseen member of the binary system must have a mass of about 7 M$_\odot$ or more. Otherwise, it would not exert enough gravitational pull to make the B0 star wobble by the amount deduced from the periodic Doppler shifting of its spectral lines. Because the unseen companion does not emit visible light, it cannot be an ordinary star. Furthermore, because 7 solar masses is too large for either a white dwarf or a neutron star, Cygnus X-1 is likely to be a black hole.

The case for Cygnus X-1 being a black hole is not airtight. HDE 226868 might have a low mass for its spectral type, which would imply a somewhat lower mass for Cygnus X-1. In addition, uncertainties in the distance to the binary system could further reduce estimates of the mass of Cygnus X-1. If all these uncertainties combined in just the right way, the estimated mass of Cygnus X-1 could be pushed down to about 3 M$_\odot$. Thus, there is a slim chance that Cygnus X-1 might contain the most massive possible neutron star rather than a black hole.

If Cygnus X-1 does contain a black hole, the X rays do not come from the black hole itself. Gas captured from HDE 226868 goes into orbit about the hole, forming an accretion disk about 4 million kilometers in diameter (Figure 24-11). As material in the disk gradually spirals in toward

a

b

figure 24-11

The Cygnus X-1 System (a) Gas from the blue supergiant HDE 226868 is captured into orbit about the black hole, forming an accretion disk. As gases spiral in toward the black hole, friction heats them to high temperatures. At the inner edge of the disk, just above the black hole, the gases become so hot that they emit vast quantities of X rays. (b) This artist's rendering shows how the Cygnus X-1 system might look at close range. At even closer range, the accretion disk would appear as shown in the illustration that opens this chapter. (Courtesy of D. Norton, Science Graphics) the hole, friction heats the gas to temperatures approaching 2×10^6 K. In the final 200 kilometers above the hole, these extremely hot gases emit the X rays that our satellites detect. Presumably the X-ray flickering is caused by small hot spots on the rapidly rotating inner edge of the accretion disk. In this way, the black hole's existence is announced by doomed gases just before they plunge to oblivion.

Astronomers have found several other black hole candidates. Like Cygnus X-1, all are compact X-ray sources orbiting ordinary stars in a spectroscopic binary system. Perhaps the best candidate is V404 Cygni in the constellation Cygnus. Doppler shift measurements reveal that as the visible star moves around its 6.47-day orbit, its radial velocity (see Section 19-10) varies by more than 400 km/s. These data give a firm lower limit of 6.26 $M_\odot$ for the mass of the unseen companion. It is probably impossible for a neutron star to be more massive than 3 $M_\odot$, so V404 Cygni must almost certainly be a black hole.

Another particularly convincing case is the flickering X-ray source A0620-00 in the constellation Monoceros. (The A refers to the British satellite Ariel 5, which discovered the source; the numbers denote its position in the sky). The visible companion to A0620-00 is an orange K5 main-sequence star called V616 Monocerotis, which orbits the X-ray source every 7.75 hours. Because this star is relatively faint, it is possible to observe the shifting spectral lines from *both* the visible companion and the X-ray source as they orbit each other. With this more complete information about the orbits, astronomers have determined the mass of A0620-00 to be more than 3.2 $M_\odot$, and more probably about 9 $M_\odot$.

If there really are black holes in close binary systems, how did they get there? One possibility is that one member of the binary explodes as a Type II supernova, leaving a burned-out core whose mass exceeds 3 $M_\odot$. This core undergoes gravitational collapse to form a black hole.

Another possibility is that a white dwarf or a neutron star in a binary system can become a black hole if it accretes enough matter from its companion star. This transformation can occur when the companion star becomes a red giant and dumps a significant part of its mass over its Roche lobe (see Figure 21-17b).

Another possibility is two dead stars coalescing to form a black hole. For example, imagine two neutron stars orbiting each other, like the binary system discussed in Section 24-2. Because the two stars emit gravitational radiation, they gradually spiral in toward each other and eventually merge. If their total mass exceeds 3 $M_\odot$, the entire system may become a black hole.

24-4 Supermassive black holes exist at the centers of most galaxies

Stellar evolution makes it possible to form black holes with masses several times that of the Sun. But calculations suggest other ways to form black holes that are either extremely large or extraordinarily small.

As we will discuss in Chapter 26, galaxies formed in the early universe by the coalescence of gas clouds. During this formation process, some of a galaxy's gas could have plunged straight toward the center of the galaxy, where it collected and compressed together under its own gravity. If this central clump became sufficiently dense, it could have formed a black hole. Typical galaxies have masses of 10^{11} $M_\odot$, so even a tiny percentage of the galaxy's gas collected at its center would give rise to a **supermassive black hole** with a truly stupendous mass.

Since the 1970s astronomers have found circumstantial evidence suggesting the existence of supermassive black holes in other galaxies. With the advent of a new generation of optical telescopes in the 1990s, it became possible to see the environments of these suspected black holes with unprecedented detail. One example is Figure 24-12, a Hubble Space Telescope image of the galaxy M87 and a tiny, bright source of light at its center. By measuring the Doppler shifts of material in the vicinity of the central source, astronomers found that this material was orbiting this source at speeds of hundreds of kilometers per second. (By comparison, the Earth orbits the Sun at 30 km/s.) Given the size of the material's orbit, and using Newton's form of Kepler's third law (see Section 4-7 and Box 4-4), they were able to calculate the mass of the object within the central source. The answer is an amazing three *billion* solar masses! What is more, the images show that this object can be no larger than our solar system. The only

figure 24-12 R I **V** U X G

A Supermassive Black Hole The galaxy M87 lies some 15 million pc (50 million ly) from Earth in the constellation Virgo. The image in the inset shows stars and gas orbiting around this galaxy's bright nucleus, which is about the size of our solar system. From the speed of the orbiting material, astronomers conclude that the object in the nucleus is a black hole of some three billion (3×10^9) solar masses. In the large image, the bright band that extends up and to the right is a jet of material that failed to fall into the black hole. A jet of this kind is shown in the illustration that opens this chapter. (H. Ford, STScI/Johns Hopkins; and collaborators) possible explanation is that the object at the center of M87 is a black hole.

Dozens of other black holes in the centers of galaxies have been identified by their gravitational effect on surrounding gas and dust. Surveys of galaxies have shown that supermassive black holes are not at all unusual; most large galaxies appear to have them at their centers. As we will see in the next chapter, a black hole with several million solar masses lies at the center of our own Milky Way Galaxy, some 8000 pc (26,000 ly) from Earth.

Black holes seem to fall into two very different groups depending on their size. Supermassive black holes have masses from 10^6 to 10^9 $M_\odot$, while stellar-mass black holes like those discussed in Section 24-3 have masses around 10 $M_\odot$. In 2000 a team of astronomers from Japan, the United States, and Great Britain reported finding a "missing link" between these two groups.

The new discovery was a fluctuating X-ray source located some 200 pc (600 ly) from the center of the galaxy M82 (Figure 24-13). Images do not reveal material orbiting around this source, so we cannot determine the source's mass using Newton's form of Kepler's third law. But by relating the source's X-ray luminosity to the rate at which matter would have to fall onto the source to produce that luminosity, it is possible to show that the source must

have a mass of at least 500 $M_\odot$ or so. The small size of the source suggests that it is a black hole. If these conclusions are confirmed, the observations of M82 will be the first detection of a "mid-mass" black hole. Theorists have suggested that such black holes might form by the coalescence of many normal stars or by the direct merger of stellar-mass black holes.

During the early 1970s Stephen Hawking of Cambridge University proposed the existence of yet another type of black hole. He suggested that during the Big Bang some 15 billion years ago, local pockets of the universe could have been dense enough and under sufficient pressure to form small black holes. According to his theory, these **primordial black holes** could have had masses as large as the Earth or as small as 5×10^{-8} kg, about $\frac{1}{40}$ the mass of a single raindrop. To date, however, no evidence has been found for the existence of primordial black holes.

24-5 A nonrotating black hole has only a "center" and a "surface"

Understanding in detail how black holes form is a challenging task. But once a black hole forms, it has a remarkably simple structure.

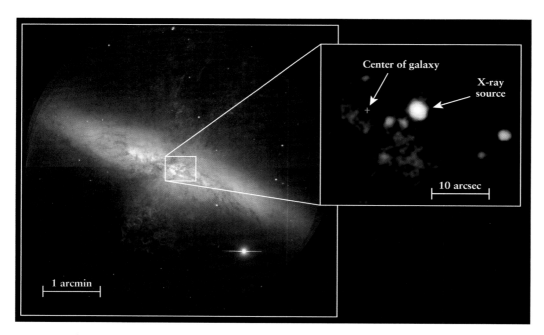

Figure 24-13 R I **V** U X G   R I V U **X** G

A "Mid-Mass" Black Hole? M82 is an unusual galaxy in the constellation Ursa Major. The inset shows an image of the central region of M82 from the orbiting Chandra X-ray Observatory (see Section 6-7). The bright, compact X-ray source shown by the arrow varies in its light output over a period of months. The properties of this source suggest that it may be a black hole of roughly 500 solar masses or more. (Subaru Telescope, National Astronomical Observatory of Japan; inset: NASA/SAO/CXC)

box 24-2 | Tools of the Astronomer's Trade

The Schwarzschild Radius

The Schwarzschild radius (R_{Sch}) is the distance from the center of a nonrotating black hole to its event horizon. You can think of it as the "size" of the black hole. The Schwarzschild radius is related to the mass M of the black hole by

Schwarzschild radius

$$R_{Sch} = \frac{2GM}{c^2}$$

R_{Sch} = Schwarzschild radius of a black hole
G = universal constant of gravitation
M = mass of black hole
c = speed of light

When using this formula, be careful to express the mass in kilograms, not solar masses. That is because of the units

in which G and c are commonly expressed: $G = 6.67 \times 10^{-11}$ N m^2/kg^2 and $c = 3.00 \times 10^8$ m/s. If you use M in kilograms, the answer that you get for R_{Sch} will be in meters.

EXAMPLE: Consider a black hole of mass 10 M$_\odot$. One solar mass is 1.99×10^{30} kg, so in this case $M = 10 \times 1.99 \times 10^{30}$ kg $= 1.99 \times 10^{31}$ kg. The Schwarzschild radius of this black hole is then

$$R_{Sch} = \frac{2(6.67 \times 10^{11})(1.99 \times 10^{30})}{(3.00 \times 10^8)^2} = 3.0 \times 10^4 \text{ m}$$

One kilometer is 10^3 meters, so the Schwarzschild radius of this 10-M$_\odot$ black hole is 30 km (19 mi).

Surrounding a black hole, where the escape speed from the hole just equals the speed of light, is the **event horizon.** You can think of this sphere as the "surface" of the black hole, although the black hole's mass all lies well inside this "surface." Once a massive dying star collapses to within its event horizon, it disappears permanently from the universe. The term "event horizon" is quite appropriate, because this surface is like a horizon beyond which we cannot see any events.

If the dying star is not rotating before it collapses, the black hole will likewise not be rotating. The distance from the center of a nonrotating black hole to its event horizon is called the **Schwarzschild radius** (denoted R_{Sch}), after the German physicist Karl Schwarzschild who first determined its properties. Box 24-2 describes how to calculate the Schwarzschild radius, which depends only on the black hole's mass. The more massive the black hole, the larger its event horizon.

Once a nonrotating star has contracted inside its event horizon, nothing can prevent its complete collapse. The star's entire mass is crushed to zero volume—and hence infinite density—at a single point, known as the **singularity**, at the center of the black hole. We now can see that the structure of a nonrotating black hole is quite simple. As drawn in Figure 24-14, it has only two parts: a singularity at the center surrounded by a spherical event horizon.

To understand why the complete collapse of such a doomed star is inevitable, think about your own life on the Earth, far from any black holes. You have the freedom to move as you wish through the three dimensions of space: up and down, left and right, or forward and back. But you do not have the free-

dom to move at will through the dimension of time. Whether we like it or not, we are all carried inexorably from the cradle to the grave.

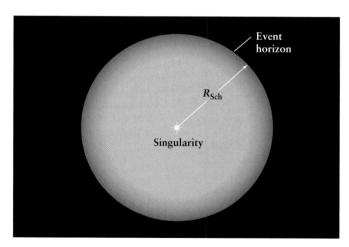

figure 24-14

The Structure of a Nonrotating (Schwarzschild) Black Hole
A nonrotating black hole has only two parts: a singularity, where all of the mass is located, and a surrounding event horizon. The distance from the singularity to the event horizon is called the Schwarzschild radius (R_{Sch}). Inside the event horizon, the escape speed exceeds the speed of light, so the event horizon is a one-way surface. Things can fall in, but nothing can get out.

Inside a black hole, gravity distorts the structure of space-time so severely that the directions of space and time become interchanged. In a sense, inside a black hole you acquire a limited ability to affect the passage of time. This seeming gain does you no good, however, because you lose a corresponding amount of freedom to move through space. Whether you like it or not, you will be dragged inexorably from the event horizon toward the singularity. Just as no force in the universe can prevent the forward march of time from past to future outside a black hole, no force in the universe can prevent the inward march of space from event horizon to singularity inside a black hole. Once an object dropped into a black hole crosses the event horizon, it is gone forever, like an object dropped into a bottomless pit.

The same is true for a light beam aimed at the black hole. Because all this light will be absorbed and none reflected back, a black hole is totally and utterly black.

At a black hole's singularity, the strength of gravity is infinite, so the curvature of spacetime there is infinite. Space and time at the singularity are thus all jumbled up. They do not exist as separate, distinctive entities.

This confusion of space and time has profound implications for what goes on inside a black hole. All the laws of physics require a clear, distinct background of space and time. Without this identifiable background, we cannot speak rationally about the arrangement of objects in space or the ordering of events in time. Because space and time are all jumbled up at the center of a black hole, the singularity there does not obey the laws of physics. The singularity behaves in a random, capricious fashion, totally devoid of rhyme or reason.

Fortunately, we are shielded from the singularity by the event horizon. In other words, although random things do happen at the singularity, none of their effects manages to escape beyond the event horizon. Consequently, the outside universe remains understandable and predictable.

The chaotic, random behavior of the singularity has been so disturbing to physicists that in 1969 the British mathematician Roger Penrose and his colleagues proposed the **law of cosmic censorship:** "Thou shalt not have naked singularities." In other words, every singularity must be completely surrounded by an event horizon, because an exposed singularity could affect the universe in an unpredictable and random way.

24-6 Just three numbers completely describe the structure of a black hole

In addition to shielding us from singularities, the event horizon prevents us from ever knowing much about anything that falls into a black hole. A black hole is, in fact, an "information sink." Many properties of matter falling into a black hole, such as its chemical composition, texture, color, shape, and size, simply vanish as soon as the matter crosses the event horizon.

Because this information has completely vanished, it cannot affect the structure or properties of the hole. For example,

consider two hypothetical black holes, one made from the gravitational collapse of 10 $M_\odot$ of iron and the other made from the gravitational collapse of 10 $M_\odot$ of peanut butter. Obviously, quite different substances went into the creation of the two holes. Once the event horizons of these two black holes have formed, however, both the iron and the peanut butter will have permanently disappeared from the universe. As seen from the outside, the two holes are absolutely identical, making it impossible for us to tell which ate the peanut butter and which ate the iron. A black hole is thus unaffected by the information it destroys.

Because a black hole is indeed an information sink, it is reasonable to wonder whether we can determine anything at all about a black hole. What properties *do* characterize a black hole?

First, we can measure the *mass* of a black hole. One way to do this would be by placing a satellite into orbit around the hole. After measuring the size and period of the satellite's orbit, we could use Newton's form of Kepler's third law (recall Section 4-7 and Box 4-4) to determine the mass of the black hole. This is equal to the total mass of all the material that has gone into the hole.

Second, we can also measure the total *electric charge* possessed by a black hole. Like gravity, the electric force acts over long distances—it is a long-range interaction that is felt in the space around the hole. Appropriate equipment on a space probe passing near the hole could measure the strength of the electric force, and the electric charge could thus be determined.

In actuality, we would not expect a black hole to possess much, if any, electric charge. For example, if a hole did happen to start off with a sizable positive charge, it would vigorously attract vast numbers of negatively charged electrons from the interstellar medium, which would soon neutralize the hole's charge. For this reason, astronomers usually neglect electric charge when discussing real black holes.

Although a black hole might theoretically have a tiny electric charge, it can have no magnetic field of its own whatsoever. When a black hole is created, however, the collapsing star from which it forms may possess an appreciable magnetic field. The star must therefore radiate this magnetic field away before it can settle down inside its event horizon. Theory predicts that the star does this by emitting electromagnetic and gravitational waves. As we saw in Section 24-2, gravitational waves are ripples in the overall geometry of space. Various experiments soon to be in operation may detect bursts of gravitational radiation emitted by massive stars as they collapse to form black holes.

Third, we can detect the effects of a black hole's rotation, that is, measure its *angular momentum*. An object's angular momentum is related to how fast it rotates and how the object's mass is distributed over its volume. As a dead star collapses into a black hole, its rotation naturally speeds up as its mass moves toward the center, just as a figure skater rotates faster when she pulls her arms and legs in. This same effect explains the rotation of the solar nebula (see Section 7-7), as well as why neutron stars spin so fast (see Section 23-3). We expect a black hole that forms from a rotating star to be spinning very rapidly.

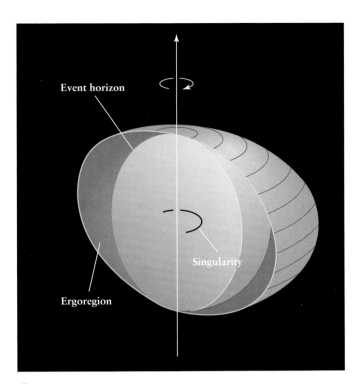

Figure 24-15

The Structure of a Rotating (Kerr) Black Hole The singularity of a rotating black hole is an infinitely thin ring centered on the geometrical center of the hole. (It appears as an arc in this cutaway diagram.) Outside the spherical event horizon is the doughnut-shaped ergoregion, where the dragging of spacetime around the hole is so severe that nothing can remain at rest. The tan-colored surface marks the outer boundary of the ergoregion.

When the matter that collapses to form a black hole is rotating, that matter does not compress to a point. Instead, it collapses into a ring-shaped singularity located between the center of the hole and the event horizon (Figure 24-15). The structure of such rotating black holes was first worked out in 1963 by the New Zealand mathematician Roy Kerr. Most rotating black holes should be spinning thousands of times per second, even faster than the most rapid pulsars (see Section 23-6).

Einstein's theory makes the startling prediction that a rotating object drags spacetime around it. (Spacecraft measurements indicate that the rotating Earth drags spacetime in just this way.) Surrounding the event horizon of every rotating black hole is a region where this dragging of space and time is so severe that it is impossible to stay in the same place. No matter what you do, you get pulled around the hole, along with the rotating geometry of space and time. This region, where it is impossible to be at rest, is called the **ergoregion** (see Figure 24-15).

To measure a black hole's angular momentum, we could hypothetically place two satellites in orbit about the hole. Suppose that one satellite circles the hole in the same direction the hole rotates and the other in the opposite direction. One satellite is thus carried along by the geometry of space and time, but the other is constantly fighting its way "upstream." The two satellites will thus have different orbital periods. By comparing these two periods, astronomers can deduce the total angular momentum of the hole.

Because the ergoregion is outside the event horizon, this bizarre region is accessible to us, and spacecraft could travel through it without disappearing into the black hole. According to detailed calculations, objects grazing the ergoregion could be catapulted back out into space at tremendous speeds. In other words, the ejected object could leave the ergoregion with more energy than it had initially, having extracted added energy from the hole's rotation. This is called the *Penrose process*, after Roger Penrose, the British mathematician who proposed it.

Mass, charge, and angular momentum are the only three properties that a black hole possesses. This simplicity is the essence of the **no-hair theorem**, first formulated in the early 1970s: "Black holes have no hair." Once matter has fallen into the hole, any and all additional properties ("hair") carried by the matter have disappeared from the universe. Hence, these properties can have no effect on the structure of the hole.

24-7 Falling into a black hole is an infinite voyage

Imagine that you are on board a spacecraft a safe distance from a 5-$M_\odot$ black hole. (A distance of 1000 Schwarzschild radii, or 1500 km, would suffice.) You now release a space probe and let it fall into the black hole. To make the probe easier to track, it is coated with a phosphorescent paint that emits a blue glow. What will you see as the probe falls?

You might expect that as the probe falls, its speed should continue to increase. This is true up to a point. But as the probe approaches the event horizon, where the black hole's gravity is extremely strong, the gravitational slowing of time becomes so pronounced that the probe will appear to slow down! From your point of view, the spacecraft takes an infinite amount of time to reach the event horizon, where it will appear to remain suspended for all eternity.

To watch these effects, however, you will need special equipment. The reason is the gravitational redshift: As the probe falls, the light reaching you from the probe is shifted to longer and longer wavelengths. The probe's blue glow will turn green, then yellow, then red, and eventually fade into infrared wavelengths that your unaided eye cannot detect.

If you could view the falling probe with infrared goggles, you would see it come to an unpleasant end. Near the event horizon, the strength of the black hole's gravity increases dramatically as the probe moves just a short distance closer to the hole. In fact, the side of the probe nearest the black hole feels a much stronger gravitational pull than the side opposite the hole. These *tidal forces* are like those that the Moon exerts on the Earth (see Section 4-8), but tremendously stronger.

Probe far from
black hole

Probe approaching black hole

Black
hole

Event
horizon

a   b   c   d

ƒigure 24-16

Falling into a Black Hole (a) A cube-shaped probe is dropped from a distance of 1000 Schwarzschild radii from a 5-$M_\odot$ black hole. (b), (c), (d) As the probe approaches the event horizon, it is distorted into a long, thin shape by the black hole's extreme gravity. A distant observer sees the probe change color as photons from the probe undergo a strong gravitational redshift.

Furthermore, the sides of the probe are pulled together, since the hole's gravity makes them fall in straight lines aimed at the center of the hole. The net effect is that the probe will be stretched out along the line pointing toward the hole, and squeezed together along the perpendicular directions (Figure 24-16). The stretching is so great that it will rip the probe into atoms, and even rip the atoms apart.

Someone foolhardy enough to ride along with the probe would have a very different view of the fall than yours. From this astronaut's point of view, there is no slowing of time, and the probe continues its fall through the event horizon into the singularity. The astronaut is unlikely to enjoy the ride, however, since he will be disintegrated by tidal forces as he falls inward toward the singularity.

Could a black hole somehow be connected to another part of spacetime, or even some other universe? General relativity predicts that such connections, called **wormholes**, can exist for rotating black holes. Box 24-3 discusses the possibility that such wormholes really exist.

24-8 Black holes evaporate

With all the mass of a black hole hidden behind its event horizon and collapsed into a singularity, it may seem that there is no way of getting mass from the black hole back out into the universe. But in fact there is, as Stephen Hawking pointed out in the 1970s. To understand how this is possible, we must look at the behavior of matter at the submicroscopic scale of nuclei, and electrons.

The **Heisenberg uncertainty principle** is a basic tenet of submicroscopic physics. This principle states that you cannot determine precisely both the position and the speed of a subatomic particle. Over extremely short distances or times, a certain amount of "fuzziness" is built into the nature of reality.

The Heisenberg uncertainty principle leads to the concept of **virtual pairs**: At every point in space, pairs of particles and antiparticles are constantly being created and destroyed. An antiparticle is quite like an ordinary particle except that it has an opposite electric charge and can annihilate an ordinary particle so that both disappear, usually leaving a pair of photons in their place. In the case of virtual pairs, the process of creation and annihilation occurs over such incredibly brief time intervals that these virtual particles and antiparticles cannot be observed directly.

Think about a tiny black hole. Furthermore, think about the momentary creation of a virtual pair of an electron and a positron (the antiparticle of an electron) just outside the hole's event horizon. It may happen that one of the particles falls into the black hole. Its partner is then deprived of a counterpart with which to annihilate and must therefore become a real particle. To accomplish this conversion, some of the energy of the black hole's gravity must be converted into matter, according to $E = mc^2$. This decreases the mass of the black hole by a corresponding amount, and the particle is free to escape from the hole. In this way, particles can quantum-mechanically "leak" out of a black hole, carrying some of the hole's mass with them (Figure 24-17).

The less mass a black hole has, the more easily particles can leak out through its event horizon. Jacob Bekenstein and Stephen Hawking proved mathematically that you can speak of the *temperature* of a black hole as a way of describing the amounts of energy carried away by particles leaking out of it. For example, a 1-trillion-ton (10^{15} kg) black hole emits particles and energy as if were a blackbody with a temperature of nearly 10^9 K.

For stellar-mass black holes, such as Cygnus X-1, this effect is negligible over time spans of billions of years. The reason is that the temperature of a black hole is inversely proportional to the black hole's mass. Compared to a trillion-ton black hole, a 5-$M_\odot$ (10^{31} kg) hole has a mass that is 10^{16} times greater and a temperature that is only 10^{-16} as much:

$$10^{-16} \times 10^9 \text{ K} = 10^{-7} \text{ K}$$

box 24-3   **T**he Heavens on the Earth

Wormholes and Time Machines

Wormholes have become a staple of science fiction. They are certainly a convenient plot device: The gallant crew flies their spaceship into one end of a wormhole, and a short time later they emerge many light-years away. Time machines, too, have been part of science fiction ever since H. G. Wells published his classic novel *The Time Machine* at the end of the nineteenth century. What would it be like to go back in time and watch famous historical events—or even change history?

Wormholes and time machines challenge our normal ideas about space and time. So, too, does the general theory of relativity. Could it be that this theory really makes it possible to travel through wormholes and to travel in time?

In the 1930s, Einstein and his colleague Nathan Rosen discovered that a black hole could possibly connect our universe with a second domain of space and time that is separate from ours. The first diagram shows this connection, called an *Einstein-Rosen bridge*. You could think of the upper surface as our universe and the lower surface as a "parallel universe." Alternatively, the upper and lower surfaces could be different regions of our own universe. In that case, an Einstein-Rosen bridge would connect our universe with itself, forming a wormhole, shown in the second diagram.

A wormhole may seem like a shortcut to distant places in our universe, but detailed calculations reveal a major obstacle. The powerful gravity of a black hole causes the wormhole to collapse almost as soon as it forms. As a result, to get from one side of a wormhole to the other, you would have to travel faster than the speed of light, which is not possible.

Caltech physicists Kip Thorne, Michael Morris, and Ulvi Yurtsever have proposed a scheme that might get around the difficulty of a collapsing wormhole. According to general relativity, they point out, pressure as well as mass can be a source of gravity. Normally we do not see the gravitational effects of pressure because they are so small. Thorne and his colleagues speculate that a technologically advanced civilization might someday be able to use pressure to produce *antigravity* strong enough to keep the wormhole open.

If a wormhole could be held open, it could also be a time machine. To see how, imagine you take one end of a wormhole and move it around for a while at speeds very near the speed of light. As we saw in Box 24-1, such motion causes clocks to slow down. Thus, when you stop moving that end of the wormhole, you find that it has not aged as much as the stationary end. In other words, one side of the wormhole has a different time from the other. As a result, you could go into one end of the wormhole at a late time and come out at an early time. For example, you might go in at 10 A.M. and come out at 9 A.M.

Time machines challenges ordinary logic. If you could get back from a trip an hour before you left, you could meet yourself and tell yourself what a nice journey you had. Then both of you could take the trip. If you and your twin return just before you both left, there would be four of you. And all four of you could take the trip again. And then all eight of you. Then all sixteen of you.

Making copies of yourself is an example of how time machines violate *causality*, the notion that effects must follow their causes. To date, we have never seen a violation of causality. Many scientists would therefore like to show that time machines cannot exist. The British astrophysicist Stephen Hawking points to one strong bit of observational evidence against time machines: We are not being visited by hordes of tourists from the future. If we could discover why nature seems to preclude time machines, we would have a much deeper understanding of the nature of space and time.

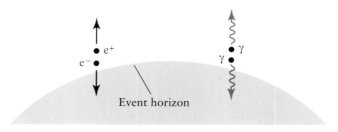

figure 24-17

Evaporation of a Black Hole Throughout the universe, pairs of virtual particles appear and disappear so rapidly that they do not violate any law of physics. But if such a pair appears just outside the event horizon of a black hole, the hole's tidal forces can pull the pair apart before they annihilate each other. If one member of the pair crosses the event horizon, the other member can escape into space. The net result is that energy is taken from the black hole and escapes into space. The illustration shows two sets of pairs: an electron (e⁻) and a positron (e⁺), and a pair of photons (γ).

This temperature is barely above absolute zero. In other words, ordinary black holes have such low temperatures that particles hardly ever manage to escape from them.

The story is different for low-mass black holes, such as the primordial black holes we discussed in Section 24-4. As particles escape from a small black hole, the mass of the black hole decreases, making its temperature go up. As its temperature rises, still more particles escape, further decreasing the hole's mass and forcing the temperature still higher. This runaway process of **black hole evaporation** finally causes the hole to vanish altogether. During its final seconds of evaporation, the hole gives up the last of its mass in a violent burst of energy equal to the detonation of a billion megaton hydrogen bombs.

A 10^{10}-kg primordial black hole (comparable to the mass of Mount Everest) would take about 15 billion years to evaporate. This is close to the present age of the universe. If primordial black holes were formed in the Big Bang, we would expect to see some of them going through the explosive final stages of evaporation. To date, however, there is no compelling evidence that we have seen such evaporations taking place, and many astronomers suspect that there are no primordial black holes.

By contrast, a 5-$M_\odot$ black hole would take more than 10^{62} years to evaporate, and a supermassive black hole of 5 million solar masses would take more than 10^{80} years. These time spans are far longer than the age of the universe. We can safely predict that the black holes we have observed to date will remain as black holes for the foreseeable future.

KEY WORDS

Terms preceded by an asterisk (*) are discussed in the Boxes.

black hole, p. 548

black hole evaporation, p. 558

equivalence principle, p. 546

ergoregion, p. 555

event horizon, p. 553

general theory of relativity, p. 546

gravitational radiation (gravitational waves), p. 548

gravitational redshift, p. 547

Heisenberg uncertainty principle, p. 556

law of cosmic censorship, p. 554

length contraction, p. 542

*Lorentz transformations, p. 544

no-hair theorem, p. 555

primordial black hole, p. 552

*proper length (proper distance), p. 545

*proper time, p. 544

Schwarzschild radius (R_{Sch}), p. 553

singularity, p. 553

spacetime, p. 542

special theory of relativity, p. 542

supermassive black hole, p. 551

time dilation, p. 543

virtual pairs, p. 556

wormhole, p. 556

KEY IDEAS

The Special Theory of Relativity: This theory, published by Einstein in 1905, is based on the notion that there is no such thing as absolute space or time.

• The speed of light is the same to all observers, no matter how fast they are moving.

• An observer will note a slowing of clocks and a shortening of rulers that are moving with respect to the observer. This

effect becomes significant only if the clock or ruler is moving at a substantial fraction of the speed of light.

• Space and time are not wholly independent of each other, but are aspects of a single entity called spacetime.

The General Theory of Relativity: Published by Einstein in 1915, this is a theory of gravity. Any massive object causes space to curve and time to slow down, and these effects manifest themselves as a gravitational force. These distortions of space and time are most noticeable in the vicinity of large masses or compact objects.

• The general theory of relativity is our most accurate description of gravitation. It predicts a number of phenomena, including the bending of light by gravity and the gravitational redshift, whose existence has been confirmed by observation and experiment.

• The general theory of relativity also predicts the existence of gravitational waves, which are ripples in the overall geometry of space and time produced by moving masses. Gravitational waves have been detected indirectly, and specialized antennas are under construction to make direct measurement of the gravitational waves from cosmic cataclysms.

Black Holes: If a stellar corpse has a mass greater than about 3 $M_\odot$, gravitational compression will overwhelm any and all forms of internal pressure. The stellar corpse will collapse to such a high density that its escape speed exceeds the speed of light.

Observing Black Holes: Black holes have been detected using indirect methods.

• Some binary star systems contain a black hole. In such a system, gases captured from the companion star by the black hole emit detectable X rays.

• Many galaxies have supermassive black holes at their centers. These are detected by observing the motions of material around the black hole.

 Properties of Black Holes: The entire mass of a black hole is concentrated in an infinitely dense singularity.

• The singularity is surrounded by a surface called the event horizon, where the escape speed equals the speed of light. Nothing—not even light—can escape from inside the event horizon.

• A black hole has only three physical properties: mass, electric charge, and angular momentum.

• A rotating black hole (one with angular momentum) has an ergoregion around the outside of the event horizon. In the ergoregion, space and time themselves are dragged along with the rotation of the black hole.

• Black holes can evaporate, but in most cases at an extremely slow rate.

REVIEW QUESTIONS

1. In Einstein's special theory of relativity, two different observers moving at different speeds will measure the same value of the speed of light. Will these same observers measure the same value of, say, the speed of an airplane? Explain.

2. Sally flies past Martin in her spaceship at nearly the speed of light. According to Martin, Sally's clock runs slow. According to Sally, does Martin's clock run slow, fast, or at the normal rate? Explain.

3. Why does the speed of light represent an ultimate speed limit?

4. Why is Einstein's general theory of relativity a better description of gravity than Newton's universal law of gravitation?

5. Describe two different predictions of the general theory of relativity and how these predictions were tested experimentally. Do the results of the experiments agree with the theory?

6. In what circumstances are degenerate electron pressure and degenerate neutron pressure incapable of preventing the complete gravitational collapse of a dead star?

7. Should we worry about the Earth's being pulled into a black hole? Why or why not?

8. All the stellar-mass black hole candidates mentioned in the text are members of very short-period binary systems. Explain how this makes it possible to detect the presence of the black hole.

9. Astronomers cannot actually see the black hole candidates in close binary systems. How, then, do they know that these candidates are not white dwarfs or neutron stars?

10. How do astronomers locate supermassive black holes in galaxies?

11. In what way is a black hole blacker than black ink or a black piece of paper?

12. If the Sun suddenly became a black hole, how would the Earth's orbit be affected? Explain.

13. According to the general theory of relativity, why can't some sort of yet-undiscovered degenerate pressure prevent the matter inside a black hole from collapsing all the way down to a singularity?

14. What is the law of cosmic censorship?

15. What is the no-hair theorem?

16. What kind of black hole is surrounded by an ergoregion? What happens inside the ergoregion?

17. As seen by a distant observer, how long does it take an object dropped from a great distance to fall through the event horizon of a black hole? Explain.

18. If even light cannot escape from a black hole, how is it possible for black holes to evaporate?

ADVANCED QUESTIONS

Questions preceded by an asterisk (*) involve topics discussed in the Boxes.

> ### Problem-solving tips and tools
>
> Remember that the time to travel a certain distance is equal to the distance divided by the speed, and that the density of an object is its mass divided by its volume. The volume of a sphere of radius r is $\frac{4}{3}\pi r^3$. Section 4-7 describes Newton's law of universal gravitation. Box 4-4 shows how to use Newton's formulation of Kepler's third law, which explicitly includes masses; when using this formula, note that the period P must be expressed in seconds, the semimajor axis a in meters, and the masses in kilograms. For another version of this formula, in which period is in years, semimajor axis in AU, and masses in solar masses, see Section 19-9.

*19. A clock on board a moving starship shows a total elapsed time of 2 minutes, while an identical clock on the Earth shows a total elapsed time of 10 minutes. What is the speed of the starship relative to the Earth?

*20. How fast should a meter stick be moving in order to appear to be a "centimeter stick"?

*21. An astronaut flies from the Earth to a distant star at 80% of the speed of light. As measured by the astronaut, the one-way trip takes 15 years. (a) How long does the trip take as measured by an observer on the Earth? (b) What is the distance from the Earth to the star (in light-years) as measured by an Earth observer? As measured by the astronaut?

22. In the binary system of two neutron stars discovered by Hulse and Taylor (Section 24-2), one of the neutron stars is a pulsar. The distance between the two stars varies between 1.1 and 4.8 times the radius of the Sun. The time interval between pulses from the pulsar is not constant: It is greatest when the two stars are closest to each other and least when the two stars are farthest apart. Explain why this is consistent with the gravitational slowing of time (Figure 24-7a).

23. Find the total mass of the neutron star binary system discovered by Hulse and Taylor (Section 24-2), for which the orbital period is 7.75 hours and the average distance between the neutron stars is 2.8 solar radii. Is your result reasonable for a pair of neutron stars? Explain.

24. Estimate how long it will be until the two neutron stars that make up the binary system discovered by Hulse and Taylor collide with each other. Assume that the distance between the two stars will continue to decrease at its present rate of 3 mm every 7.75 hours, and use the data given in Question 23. (You can assume that the two stars are very small, so they will collide when the distance between them is equal to zero.)

25. The orbital period of the binary system containing A0620-00 is 0.32 day, and Doppler shift measurements reveal that the radial velocity of the X-ray source peaks at 457 km/s (about 1 million miles per hour). (a) Assuming that the orbit of the X-ray source is a circle, find the radius of its orbit in kilometers. (This is actually an estimate of the semimajor axis of the orbit.) (b) By using Newton's form of Kepler's third law, prove that the mass of the X-ray source must be at least 3.1 times the mass of the Sun. (*Hint:* Assume that the mass of the K5V visible star—about 0.5 $M_\odot$ from the mass-luminosity relationship—is negligible compared to that of the invisible companion.)

26. Find the orbital period of a star moving in a circular orbit of radius 500 AU around the supermassive black hole in M87.

*27. Find the Schwarzschild radius for an object having a mass equal to that of the planet Jupiter.

*28. What is the Schwarzschild radius of a black hole whose mass is that of (a) the Earth, (b) the Sun, (c) the supermassive black hole in M87? In each case, also calculate what the density would be if the matter were spread uniformly throughout the volume of the event horizon.

*29. To what density must the matter of a dead 10-$M_\odot$ star be compressed in order for the star to disappear inside its event horizon? How does this compare with the density at the center of a neutron star, about 3×10^{18} kg/m^3?

*30. Prove that the density of matter needed to produce a black hole is inversely proportional to the square of the mass of the hole. If you wanted to make a black hole from matter compressed to the density of water (1000 kg/m^3), how much mass would you need?

DISCUSSION QUESTIONS

31. The speed of light is the same for all observers, regardless of their motion. Discuss why this requires us to abandon the Newtonian view of space and time.

32. Describe the kinds of observations you might make in order to locate and identify black holes.

33. Speculate on the effects you might encounter on a trip to the center of a black hole (assuming that you could survive the journey).

WEB/CD-ROM QUESTIONS

34. Search the World Wide Web for information about a stellar-mass black hole candidate named V4641 Sgr. In

what ways does it resemble other black hole candidates such as Cygnus X-1 and V404 Cygni? In what ways is it different and more dramatic? How do astronomers explain why V4641 Sgr is different?

35. Search the World Wide Web for information about the "mid-mass" black hole candidate in M82. Is this still thought to be a black hole? What new evidence has been used to either support or oppose the idea that this object is a black hole?

 36. The Equivalence Principle. Access the animation "The Equivalence Principle" in Chapter 24 of the *Universe* web site or CD-ROM. View the animation and answer the following questions. (**a**) Describe what is happening as viewed from the frame of reference of the elevator. What causes the apple to fall to the floor of each elevator? (**b**) Describe what is happening as viewed from the frame of reference of the stars. What causes the apple to fall to the floor of each elevator? (**c**) Think of another experiment you could perform with the apple. Describe what would happen during this experiment as seen by Newton (in the left-hand box) and by Einstein (in the right-hand box).

OBSERVING PROJECT

37. You cannot see a black hole with a telescope. Nevertheless, you might want to observe the visible companion of Cygnus X-1. The epoch 2000 coordinates of this ninth-magnitude star are R.A. = 19^h 58.4^m and Decl. = $+35°$ $12'$, which is quite near the bright star η (eta) Cygni. Compare what you see with the photograph in Figure 24-10.

38. Use the *Starry Night* program to plan observations of Cygnus X-1. First turn off daylight (select **Daylight** in the **Sky** menu). Center the field of view on Cygnus X-1 (select **Find...** in the **Edit** menu). If Cygnus X-1 is below the horizon, allow the program to reset the time to when it rises. Using the controls at the right-hand end of the Control Panel, zoom out to the maximum field of view. (**a**) Using the time controls in the Control Panel, step through time and determine when Cygnus X-1 rises and sets on today's date from your location. (**b**) At approximately what time on today's date is Cygnus X-1 highest in the sky? Is tonight a good night for observing this star with a visible-light telescope? Would it be better placed in the sky for observation six months from now? Explain how you determined this.

Our Galaxy

(Dirk Hoppe)

R I **V** U X G

O n a clear, moonless night, away from the glare of city lights, you can often see a hazy, luminous band stretching across the sky. This band, called the Milky Way, extends all the way around the celestial sphere. The accompanying photograph shows the brightest part of the Milky Way, in the constellation Sagittarius.

Galileo, the first person to view the Milky Way with a telescope, discovered that it is composed of countless dim stars. Today we realize that the Milky Way is actually a disk tens of thousands of parsecs across containing several hundred billion stars, one of which is our own Sun, as well as vast quantities of gas and dust.

Recently, astronomers have discovered that most of the Milky Way's mass is not in its stars, but in a halo of *dark matter* that emits no measurable radiation. The character of this dark matter remains mysterious. Even more mysterious is the very center of the Milky Way, where radio and infrared observations reveal the presence of a black hole with a mass of nearly 3 million Suns.

The Milky Way is just one of myriad *galaxies*, or systems of stars and interstellar matter, that are spread across the observable universe. By studying our home Milky Way Galaxy, we begin to explore the universe on a grand scale. Instead of focusing on individual stars, we look at the overall arrangement and history of a huge stellar community of which the Sun is a member. In this way, we gain insights into galaxies in general and prepare ourselves to ask fundamental questions about the cosmos.

As you read the sections of this chapter, look for the answers to the following questions.

25-1 What is our Galaxy? How do astronomers know where we are located within it?

25-2 What is the shape and size of our Galaxy?

25-3 How do we know that our Galaxy has spiral arms?

25-4 What is most of the Galaxy made of? Is it stars, gas, dust, or something else?

25-5 What is the nature of the spiral arms?

25-6 What lies at the very center of our Galaxy?

25-1 The Sun is located in the disk of our Galaxy, about 8,000 parsecs from the galactic center

Eighteenth-century astronomers were the first to suspect that because the Milky Way completely encircles us, all the stars in the sky are part of an enormous disk of stars—the **Milky Way Galaxy**. (In astronomy, the term **galaxy** refers to an immense collection of stars and interstellar matter, far larger than a star cluster.) We have an edge-on view from inside this disk, which is why the Milky Way appears as a band around the sky (Figure 25-1).

But where within this disk is our own Sun? For well over a century, the prevailing opinion was that the Sun and planets lie at the Galaxy's center. We now know that the Milky Way contains not only stars but also copious interstellar dust. This dust hides most of the Galaxy's stars from view and thus obscures its true dimensions. As we will see, we are actually nowhere near its center.

An early attempt to deduce the Sun's location in the Galaxy was made by the English astronomer William Herschel, who discovered the planet Uranus and was a pioneering cataloger of binary star systems (see Section 19-9). Herschel's approach was to count the number of stars in each of 683 regions of the sky. He reasoned that he should see the greatest number of stars toward the Galaxy's center and a lesser number toward the Galaxy's edge.

Herschel found approximately the same density of stars all along the Milky Way. Therefore, he concluded that we are at the center of our Galaxy (Figure 25-2). In the early 1900s, the

a

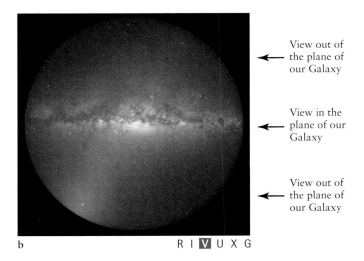

b   R I V U X G

Figure 25-1

Our View of the Milky Way (a) The Milky Way Galaxy is a disk-shaped collection of hundreds of billions of stars (of which our Sun is one). These stars are so numerous that when we look out at the night sky in the plane of the disk, they appear as a continuous band of light that stretches all the way around the sky. When we look in a direction perpendicular to the plane of the Galaxy, we see only those relatively few stars that lie between us and the "top" or "bottom" of the Galaxy's disk. **(b)** This wide-angle photograph shows a 180° view of the Milky Way centered on the constellation Sagittarius (compare with the photograph that opens this chapter). The dark streaks across the Milky Way are due to interstellar dust in the plane of our Galaxy. (Courtesy of Dennis di Cicco)

figure 25-2

Herschel's Map of Our Galaxy In the late eighteenth century, the English astronomer William Herschel attempted to map the Milky Way Galaxy by counting the numbers of stars in various parts of the sky. Because of interstellar dust that blocked his view of distant stars, Herschel erroneously concluded that the Sun is at the center of the Galaxy. (Yerkes Observatory)

Dutch astronomer Jacobus Kapteyn came to essentially the same conclusion by analyzing the brightness and proper motion of a large number of stars. According to Kapteyn, the Milky Way is about 17 kpc (17 kiloparsecs = 17,000 parsecs, or 55,000 light-years) in diameter, with the Sun near its center.

Both Herschel and Kapteyn were wrong about the Sun being at the center of our Galaxy. The reason for their mistake was finally discerned in 1930 by Robert J. Trumpler of Lick Observatory. While studying star clusters, Trumpler discovered that the more remote clusters appear unusually dim—more so than would be expected from their distances alone. As a result, Trumpler concluded that interstellar space must not be a perfect vacuum: It must contain dust that absorbs or scatters light from distant stars.

We first encountered interstellar dust in Section 20-2. Like the stars themselves, this material is concentrated in the plane of the Galaxy. As a result, it obscures our view within the plane and makes distant objects appear dim, an effect called **interstellar extinction**. Great patches of interstellar dust are clearly visible in such wide-angle photographs as the one that opens this chapter. Thanks to interstellar extinction, Herschel and Kapteyn were actually seeing only the nearest stars in the Galaxy. Hence, they had no idea of either the enormous size of the Galaxy or of the vast number of stars concentrated around the galactic center.

 Herschel and Kapteyn faced much the same dilemma as a lost motorist on a foggy night. Unable to see more than a city block in any direction, he would have a hard time deciding what part of town he was in. If the fog layer were relatively shallow, however, our hapless motorist would be able to see the lights from tall buildings that extend above the fog, and in that way he could determine his location (Figure 25-3a).

a

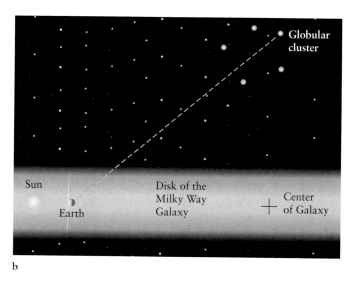

b

figure 25-3

Finding the Center of the Galaxy (a) A motorist lost on a foggy night can determine his location by looking for tall buildings that extend above the fog. (b) In the same way, astronomers determine our location in the Galaxy by observing globular clusters that are part of the Galaxy but lie outside the obscuring material in the galactic disk. Observations by Harlow Shapley showed that the globular clusters form a spherical halo centered several thousand parsecs from the Sun in the direction of the constellation Sagittarius. The center of the halo is also the center of the Galaxy.

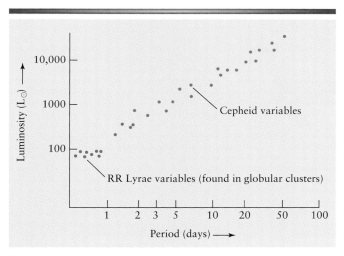

figure 25-4

The Period-Luminosity Relation for Cepheids This graph shows the relationship between the periods and luminosities of Cepheid variables. The brighter the Cepheid, the longer is its pulsation period. Data for RR Lyrae variables, which are similar variable stars found in globular clusters, are also shown. RR Lyrae variables are horizontal branch stars with average luminosities of about 100 $L_\odot$ and pulsation periods of less than a day.

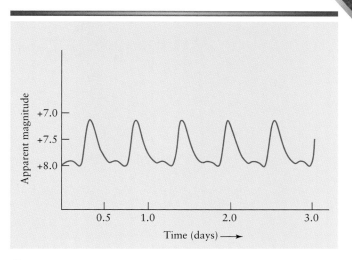

figure 25-5

The Light Curve of RR Lyrae An RR Lyrae variable star varies in brightness in a characteristic way. All RR Lyrae variables have periods of less than a day and brightnesses that vary by about 1 magnitude (that is, by a factor of about 2.5; see Section 19-3). The example shown here is the light curve of RR Lyrae, the prototype of this class of stars.

The same principle applies to our Galaxy. While interstellar dust in the plane of our Galaxy hides the sky covered by the Milky Way, we have an almost unobscured view out of the plane (that is, to either side of the Milky Way). To find our location in the Galaxy, we need to locate bright objects that are part of the Galaxy but lie outside its plane in unobscured regions of the sky.

 Fortunately, bright objects like this do exist. They are the **globular clusters,** a class of star clusters associated with the Galaxy but which lie outside its plane (Figure 25-3b). As we saw in Section 21-3, a typical globular cluster is a spherical distribution of roughly 10^6 stars packed in a volume only a few hundred light-years across (see Figure 21-8).

However, to use globular clusters to determine our location in the Galaxy, we must first determine the distances from the Earth to these clusters. (Think again of our lost motorist—glimpsing the lights of a skyscraper through the fog is useful to the motorist only if he can tell how far away the skyscraper is.) Pulsating variable stars in globular clusters provide the distances, giving astronomers the key to the dimensions of our Galaxy.

In 1912, the American astronomer Henrietta Leavitt reported her important discovery of the period-luminosity relation for Cepheid variables. As we saw in Section 21-5, Cepheid variables are pulsating stars that vary periodically in brightness (see Figure 21-14). Leavitt studied numerous Cepheids in the Small Magellanic Cloud (a small galaxy near the Milky Way) and found their periods to be directly related

to their average luminosities. As Figure 25-4 shows, the longer a Cepheid's period, the greater its luminosity.

The period-luminosity law is an important tool in astronomy because it can be used to determine distances. For example, suppose you find a Cepheid variable in the sky. By measuring its period and using a graph like Figure 25-4, you can determine the star's average luminosity. Knowing the star's average luminosity, you can find out how far away the star must be in order to give the observed brightness. (Box 19-2 explains how this is done.)

Shortly after Leavitt's discovery of the period-luminosity law, Harlow Shapley, a young astronomer at the Mount Wilson Observatory in California, began studying a family of pulsating stars closely related to Cepheid variables called **RR Lyrae variables.** The light curve of an RR Lyrae variable (Figure 25-5) is similar to that of a Cepheid, and their period-luminosity relationship is similar to that of short-period Cepheids like those Leavitt had studied.

The tremendous importance of RR Lyrae variables is that they are commonly found in globular clusters (Figure 25-6). By using the period-luminosity relationship for these stars, Shapley was able to determine the distances to the 93 globular clusters then known. He found that some of them were more than 100,000 light-years from Earth. The large values of these distances immediately suggested that the Galaxy was much larger than Herschel or Kapteyn had thought.

Another striking property of globular clusters is how they are distributed across the sky. Ordinary stars and open clusters of stars (see Section 20-6, especially Figures 20-17 and

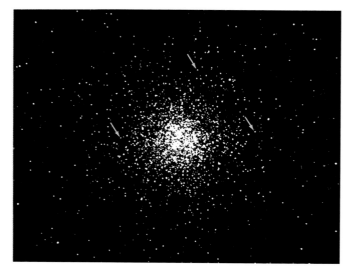

figure 25-6   R I **V** U X G

RR Lyrae Variables in a Globular Cluster The arrows point to three RR Lyrae variables in the globular cluster M55, located in the constellation Sagittarius. From the average apparent brightness (as seen in this photograph) and average luminosity (known to be roughly 100 $L_{\odot}$) of these variable stars, astronomers have deduced that the distance to M55 is 6500 pc (20,000 ly). (Harvard Observatory)

20-18) are rather uniformly spread along the Milky Way. However, the majority of the 93 globular clusters that Shapley studied are located in one-half of the sky, widely scattered around the portion of the Milky Way that is in the constellation Sagittarius.

From the directions to the globular clusters and their distances from us, Shapley mapped out the three-dimensional distribution of these clusters in space. In 1920 he concluded that the globular clusters form a huge spherical distribution centered not on the Earth but rather about a point in the Milky Way several kiloparsecs away in the direction of Sagittarius (see Figure 25-3b). This point, reasoned Shapley, must coincide with the center of our Galaxy, because of gravitational forces between the disk of the Galaxy and the "halo" of globular clusters. Therefore, by locating the center of the distribution of globular clusters, Shapley was in effect measuring the location of the galactic center.

Since Shapley's pioneering observations, many astronomers have measured the distance from the Sun to the **galactic nucleus,** the center of our Galaxy. Shapley's estimate of this distance was too large by about a factor of 2, because he did not take into account the effects of interstellar extinction (which were not well understood at the time). Today, the generally accepted distance to the galactic nucleus is about 8 kpc (26,000 ly), with an uncertainty of about 1 kpc (3300 ly).

Just as Copernicus and Galileo showed that the Earth was not at the center of the solar system, Shapley and his successors showed that the solar system lies nowhere near the center of the Galaxy. We see that the Earth indeed occupies no special position in the universe.

25-2 Observations at nonvisible wavelengths reveal the shape of the Galaxy

At visible wavelengths, light suffers so much interstellar extinction that the galactic nucleus is totally obscured from view. But the amount of interstellar extinction is roughly inversely proportional to wavelength. In other words, the longer the wavelength, the farther radiation can travel through interstellar dust without being scattered or absorbed. As a result, we can see farther into the plane of the Milky Way at infrared wavelengths than at visible wavelengths, and radio waves can traverse the Galaxy freely. For this reason, telescopes sensitive to these nonvisible wavelengths are important tools for studying the structure of our Galaxy.

Infrared light is particularly useful for tracing the location of interstellar dust in the Galaxy. Starlight warms the dust grains to temperatures in the range of 10 to 90 K; thus, in accordance with Wien's law (see Section 5-4), the dust emits radiation predominately at wavelengths from about 30 to 300 μm. These are called **far-infrared** wavelengths, because they lie in the part of the infrared spectrum most different in wavelength from visible light (see Figure 5-7). At these wavelengths, interstellar dust radiates more strongly than stars, so a far-infrared view of the sky is principally a view of where the dust is. In 1983 the Infrared Astronomical Satellite (IRAS) scanned the sky with a 60-cm reflecting telescope at far-infrared wavelengths, giving the panoramic view of the Milky Way's dust shown in Figure 25-7a.

In 1990 an instrument on the Cosmic Background Explorer (COBE) satellite scanned the sky at **near-infrared** wavelengths, that is, relatively short wavelengths closer to the visible spectrum. Figure 25-7b shows the resulting near-infrared view of the plane of the Milky Way. At near-infrared wavelengths, interstellar dust does not emit very much light. Hence, the light sources in Figure 25-7b are stars, which do emit strongly in the near-infrared (especially the cool stars, such as red giants and supergiants). Because interstellar dust causes little interstellar extinction in the near-infrared, many of the stars whose light is recorded in Figure 25-7b lie deep within the Milky Way.

Observations such as those shown in Figure 25-7, along with the known distance to the center of the Galaxy, have helped astronomers to establish the dimensions of the Galaxy. The **disk** of our Galaxy is about 50 kpc (160,000 ly) in diameter and about 0.6 kpc (2000 ly) thick, as shown in Figure 25-8. The center of the Galaxy is surrounded by a distribution of stars, called the **central bulge,** that is about 2 kpc (6500 ly) in diameter. This central bulge is clearly visible in Figure 25-7b.

a

b

figure 25-7   R **I** V U X G

The Infrared Milky Way **(a)** This far-infrared view was constructed from observations made by the IRAS spacecraft at wavelengths of 25, 60, and 100 μm. Interstellar dust is the principal source of radiation in this wavelength range, so the image shows how this dust is distributed along the Milky Way. Note that the dust is very narrowly confined to the plane of the Galaxy, which runs horizontally across the image. **(b)** In this near-infrared view, constructed from COBE observations at 1.2, 2.2, and 3.4 μm, we see much farther through the dust in the Milky Way than we can at visible wavelengths. The light recorded in this image comes principally from stars in the plane of the Galaxy. Note the bulge at the center of the Galaxy, which is in the middle of this image. (NASA)

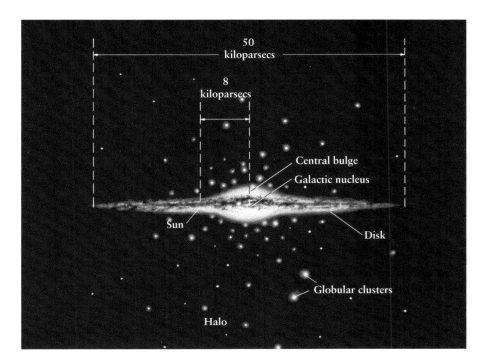

figure 25-8

Our Galaxy (Schematic Edge-on View)
There are three major components of our Galaxy: a thin disk, a central bulge, and a halo. The disk contains gas and dust along with metal-rich (Population I) stars. The halo is composed almost exclusively of old, metal-poor (Population II) stars. The central bulge is a mixture of Population I and Population II stars.

Figure 25-9 R I **V** U X G

Edge-on View of the Galaxy NGC 4565 If we could view our Galaxy edge-on from a great distance, it would probably look like this galaxy in the constellation Coma Berenices. A layer of dust and gas is clearly visible in the plane of the galaxy. Also note the reddish color of the bulge that surrounds the galaxy's nucleus. NGC 4565 is about 15 million pc (50 million ly) from Earth. (U.S. Naval Observatory)

Observations suggest that this bulge is not shaped like a flattened sphere but may be elongated somewhat into a bar or peanut shape. (We discuss the potential significance of this elongated shape in Section 25-5.) The spherical distribution of globular clusters traces the **halo** of the Galaxy. Our Galaxy, seen edge-on from a great distance, would probably look somewhat like the galaxy shown in Figure 25-9.

Different kinds of stars are found in the various components of our Galaxy. The globular clusters in the halo are composed of old, metal-poor, Population II stars. (We introduced the different populations of stars in Section 21-4.) Although these clusters are conspicuous, they contain only about 1% of the total number of stars in the halo. Most halo stars are single Population II stars in isolation, called **high-velocity stars** because of their high speed relative to the Sun. These ancient stars orbit the Galaxy along paths tilted at random angles to the plane of the Milky Way.

The stars in the disk are mostly young, metal-rich, Population I stars like the Sun. The disk of a galaxy appears bluish, as in Figure 25-9, because its light is dominated by radiation from hot O and B main-sequence stars. Such stars have very short main-sequence lifetimes (see Section 21-1, especially Table 21-1), so they must be quite young by astronomical standards. Hence, their presence shows that there must be active star formation in the galactic disk. By contrast, no O or B stars are present in the halo, which implies that star formation ceased there long ago.

The central bulge contains both Population I and Population II stars, suggesting that some stars there are quite ancient whereas others were created more recently. The central bulge looks reddish because it contains many red giants and red supergiants.

Why are there such different populations of stars in the halo, disk, and central bulge? Why has star formation stopped in some regions of the Galaxy but continues in other regions? The answers to these questions are related to the way that stars, as well as the gas and dust from which stars form, move within the Galaxy.

25-3 Observations of star-forming regions reveal that our Galaxy has spiral arms

Because interstellar dust obscures our visible-light view in the plane of our Galaxy, a detailed understanding of the structure of the galactic disk had to wait until the development of radio astronomy. Thanks to their long wavelengths, radio waves can penetrate the interstellar medium even more easily than infrared light and can travel without being scattered or absorbed. As we shall see in this section, both radio and optical observations reveal that our Galaxy has **spiral arms,** spiral-shaped concentrations of gas and dust that extend outward from the center in a shape reminiscent of a pinwheel.

Hydrogen is by far the most abundant element in the universe. Hence, by looking for concentrations of hydrogen gas, we should be able to detect important clues about the structure of the disk of our Galaxy. Unfortunately, the major electron transitions in the hydrogen atom produce photons at ultraviolet and visible wavelengths (review Section 5-8), which do not penetrate the interstellar medium. Furthermore, to emit even these photons hydrogen atoms must be excited to at least the second Bohr orbit (see Figure 5-22), which is very unlikely in the cold depths of interstellar space.

Fortunately, hydrogen atoms also emit radio waves, which makes it possible to detect even remote, cold hydrogen clouds. The hydrogen in such clouds is neutral—that is, not ionized—and is called **H I.** (This distinguishes it from ionized hydrogen, which is designated H II.) To understand how these radio waves are emitted, we must probe a bit more deeply into the structure of protons and electrons, the particles of which hydrogen atoms are made.

In addition to having mass and charge, particles such as protons and electrons possess a tiny amount of angular momentum (that is, rotational motion) commonly called **spin.** Very roughly, you can visualize a proton or electron as a tiny, electrically charged sphere that spins on its axis. Because electric charges in motion generate magnetic fields, a proton or electron behaves like a tiny magnet with a north pole and a south pole.

If you have ever played with magnets, you know that two magnets attract when the north pole of one magnet is next to the south pole of the other and repel when two like poles (both north or both south) are next to each other (Figure 25-10*a*). In other words, the energy of the two magnets is least when opposite poles are together and highest when like poles are

a

figure 25-10

Magnetic Interactions in the Hydrogen Atom **(a)** Two magnets are in a state of high energy when they have like poles (two north poles or two south poles) near each other. They are in a state of low energy when they have opposite poles (one north pole and one south pole) near each other. **(b)** Thanks to their spin, electrons and protons are both tiny magnets. When the electron and the proton spin in the same direction, their energy is higher than when they spin in opposite directions. When the electron flips from the higher-energy to the lower-energy configuration, the atom loses a tiny amount of energy. This energy is emitted as a radio photon with a wavelength of 21 cm.

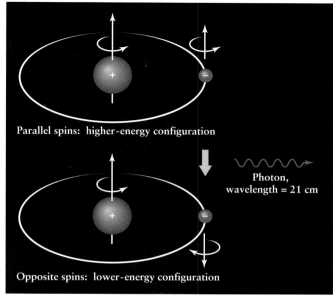

b

together. Hence, as shown in Figure 25-10*b*, the energy of a hydrogen atom is slightly different depending on whether the spins of the proton and electron are in the same direction or opposite directions. (According to the laws of quantum mechanics, these are the only two possibilities; the spins cannot be at random angles.)

If the spin of the electron changes its orientation from the higher-energy configuration to the lower-energy one—called a **spin-flip transition**—a photon is emitted. The energy difference between the two spin configurations is very small, only about 10^{-6} as great as those between different electron orbits (see Figure 5-22). Therefore, the photon emitted in a spin-flip transition between these configurations has only a small energy, and thus its wavelength is a relatively long 21 cm—a radio wavelength.

The spin-flip transition in neutral hydrogen was first predicted in 1944 by Dutch astronomer Hendrik van de Hulst. His calculations suggested that it should be possible to detect the **21-cm radio emission** from interstellar hydrogen, although a very sensitive radio telescope would be required. In 1951, Harold Ewen and Edward Purcell at Harvard University first succeeded in detecting this faint emission from hydrogen between the stars.

Figure 25-11 shows the results of a more recent 21-cm survey of the entire sky. Neutral hydrogen gas (H I) in the plane of the Milky Way stands out prominently as a bright band across the middle of this image.

 The distribution of gas in the Milky Way is not uniform but is actually quite frothy. In fact, our Sun lies near the edge of an irregularly shaped

figure 25-11   R I V U X G

The Entire Sky at 21 Centimeters This false-color image shows the results of an all-sky survey of 21-cm emission from interstellar hydrogen. The entire sky has been mapped onto an oval, with the plane of the Galaxy extending horizontally across the image. Colors indicate the density of hydrogen gas, ranging from black and blue for the lowest to red and white for the highest. Note how hydrogen gas is concentrated along the plane of the Galaxy. (Courtesy of C. Jones and W. Forman, Harvard-Smithsonian Center for Astrophysics)

box 25-1   The Heavens on the Earth

Spin-Flip Transitions in Medicine

Thanks to their spin, protons and electrons act like microscopic bar magnets. In a hydrogen atom, the interaction between the magnetism of the electron and that of the proton gives rise to the 21-cm radio emission. But these particles can also interact with outside magnetic fields, such as that produced by a large electromagnet. This is the physical principle behind **magnetic resonance imaging** (**MRI**), an important diagnostic tool of modern medicine.

Much of the human body is made of water. Every water molecule has two hydrogen atoms, each of which has a nucleus made of a single proton. If a person's body is placed in a strong magnetic field, the spins of the protons in the hydrogen atoms of their body can either be in the same direction as the field ("aligned") or in the direction opposite to the field ("opposed"). The aligned orientation has lower energy, and therefore the majority of protons end up with their spins in this orientation. But if a radio wave of just the right wavelength is now sent through the person's body, an aligned proton can absorb a radio photon and flip its spin into the higher-energy, opposed orientation. How much of the radio wave is absorbed depends on the number of protons in the body, which in turns depends on how much water (and, thus, how much water-containing tissue) is in the body.

In magnetic resonance imaging, a magnetic field is used whose strength varies from place to place. The difference in energy between the opposed and aligned orientations of a proton depends on the strength of the magnetic field, so radio waves will only be absorbed at places where this energy difference is equal to the energy of a radio photon. (This equality is called *resonance*, which is how magnetic resonance imaging gets its name.) By varying the magnetic field strength over the body and the wavelength of the radio waves, and by measuring how much of the radio wave is absorbed by different parts of the body, it is possible to map out the body's tissues. The accompanying figure shows such a map of a patient's head.

Unlike X-ray images, which show only the densest parts of the body, such as bones and teeth, magnetic resonance imaging can be used to view less dense (but water-containing) soft tissue. Just as the 21-cm radio emission has given astronomers a clear view of what were hidden regions of our Galaxy, magnetic resonance imaging allows modern medicine to see otherwise invisible parts of the human body.

R I V U X G

---

region within which the interstellar medium is very thin but at very high temperatures (about 10^6 K). This region, called the **Local Bubble,** is several hundred parsecs across. The Local Bubble may have been carved out by a supernova that exploded nearby some 300,000 years ago.

Remarkably, spin-flip transitions can be used not only to map our Galaxy but also to map out the internal structure of the human body. Box 25-1 discusses this application, called magnetic resonance imaging.

The detection of 21-cm radio radiation was a major breakthrough that permitted astronomers to probe the galactic disk. Figure 25-12 shows how this was done. Suppose that you aim a radio telescope along a particular line of sight across the Galaxy. Your radio receiver, located at *S* (the position of the solar system), picks up 21-cm emission from H I clouds at points 1, 2, 3, and 4. However, the radio waves from these various clouds are Doppler shifted by slightly different amounts, because the clouds are moving at different speeds as they travel with the rotating Galaxy.

It is important to remember that the Doppler shift reveals only motion along the line of sight (review <u>Figure 5-24</u>). In Figure 25-12, cloud 2 has the highest speed along our line of sight, because it is moving directly toward us. Consequently, the radio waves from cloud 2 exhibit a larger Doppler shift than those from the other three clouds along our line of sight. Because clouds 1 and 3 are at the same distance from the

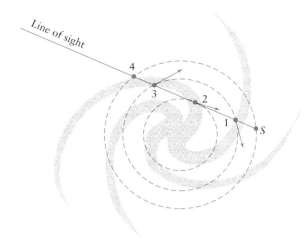

figure 25-12

A Technique for Mapping Our Galaxy If we look within the plane of our Galaxy from our position at *S*, hydrogen clouds at different locations (shown as 1, 2, 3, and 4) along our line of sight are moving at slightly different speeds. As a result, radio waves from these various gas clouds are subjected to slightly different Doppler shifts. This permits radio astronomers to sort out the gas clouds and thus map the Galaxy.

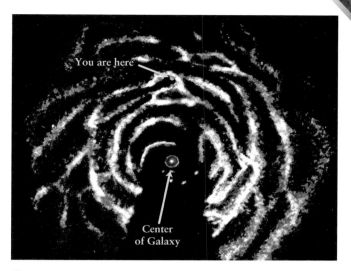

figure 25-13   **R**   I V U X G

A Map of Neutral Hydrogen in Our Galaxy This map, constructed from radio-telescope surveys of 21-cm radiation, shows the distribution of hydrogen gas in a face-on view of our Galaxy. The map suggests a spiral structure. Details in the blank, wedge-shaped region at the bottom of the map are unknown. Gas in this part of the Galaxy is moving perpendicular to our line of sight and thus does not exhibit a detectable Doppler shift. (Courtesy of G. Westerhout)

galactic center, they have the same orbital speed. The fraction of their velocity parallel to our line of sight is also the same, so their radio waves exhibit the same Doppler shift, which is less than the Doppler shift of cloud 2. Finally, cloud 4 is the same distance from the galactic center as the Sun. This cloud is thus orbiting the Galaxy at the same speed as the Sun, resulting in no net motion along the line of sight. Radio waves from cloud 4, as well as from hydrogen gas near the Sun, are not Doppler shifted at all.

These various Doppler shifts cause the 21-cm radiation to be smeared out over a range of wavelengths, that is, radio waves from gases in different parts of the Galaxy arrive at our radio telescopes with slightly different wavelengths. It is therefore possible to sort out the various gas clouds and thus produce a map of the Galaxy like that shown in Figure 25-13.

You can see from Figure 25-13 that neutral hydrogen gas is not spread uniformly around the disk of the Galaxy but is concentrated into numerous arched lanes. Similar features are seen in other galaxies beyond the Milky Way. As an example, the galaxy in Figure 25-14*a* has prominent spiral arms outlined by hot, luminous, blue main-sequence stars and the red emission nebulae (H II regions) found near many such stars. Stars of this sort are very short-lived, so these features indicate that spiral arms are sites of active, ongoing star formation. The 21-cm radio image of this same galaxy, shown in Figure 25-14*b*, shows that spiral arms are also regions where neutral hydrogen gas is concentrated, similar to the structures in our own Galaxy visible in Figure 25-13. This similarity is a strong indication that our Galaxy also has spiral arms.

CAUTION! Photographs such as Figure 25-14*a* can lead to the impression that there are very few stars between the spiral arms of a galaxy. Nothing could be further from the truth! In fact, stars are distributed rather uniformly throughout the disk of a galaxy like the one in Figure 25-14*a*; the density of stars in the spiral arms is only about 5% higher than in the rest of the disk. The spiral arms stand out nonetheless because that is where hot, blue O and B stars are found. One such star is about 10^4 times more luminous than an average star in the disk, so the light from O and B stars completely dominates the visible appearance of a spiral galaxy. An infrared image such as Figure 25-14*c* gives a better impression of how stars of all kinds are distributed through a spiral galaxy's disk.

Figure 25-14*a* suggests that we can confirm the presence of spiral structure in our own Galaxy by mapping the locations of star-forming regions. Such regions are marked by OB associations, H II regions, and molecular clouds (see Section 20-7). Unfortunately, the first two of these are best observed using visible light, and interstellar extinction limits the range of visual observations in the plane of the Galaxy to less than 3 kpc (10,000 ly) from the Earth. But there are enough OB associations and H II regions within this range to plot the spiral arms in the vicinity of the Sun.

Molecular clouds are easier to observe at great distances, because molecules of carbon monoxide (CO) in these clouds emit radio waves that are relatively unaffected by interstellar extinction. Hence, the positions of molecular clouds have been plotted even in remote regions of the Galaxy (see Figure 20-20).

a   R I **V** U X G

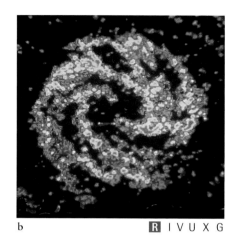

b   **R** I V U X G

c   R **I** V U X G

figure 25-14

A Spiral Galaxy The galaxy M83 lies in the southern constellation Hydra about 5 million pc (15 million ly) from Earth. **(a)** This visible-light image clearly shows the spiral arms traced out by the light from hot, young, blue stars and the red glow of H II regions. The presence of these stars and H II regions indicates that star formation takes place in spiral arms. **(b)** This radio view at a wavelength of 21 cm shows the emission from neutral hydrogen gas (H I). Note that essentially the same pattern of spiral arms is traced out in this image as in the visible-light photograph. Hence, spiral arms are also regions where interstellar hydrogen is most densely concentrated. **(c)** M83 has a much smoother appearance in this near-infrared view. This shows that cooler stars, which emit strongly in the infrared, are spread more uniformly across the galaxy's disk. Note the elongated bar shape of the central bulge. (a: Anglo-Australian Observatory; b: VLA, NRAO; c: S. Van Dyk/IPAC)

(We saw in <u>Section 20-7</u> that the CO in molecular clouds emits more strongly than the hydrogen, even though hydrogen is the principal constituent of these clouds.)

Taken together, all these observations demonstrate that our Galaxy has four major spiral arms and several short arm segments (Figure 25-15). The Sun is located on a relatively short arm segment called the Orion arm, which includes the Orion Nebula and neighboring sites of vigorous star formation in that constellation.

Two major spiral arms border either side of the Sun's position. The Sagittarius arm is on the side toward the galactic center. This is the arm you see on June and July nights when you look at the portion of the Milky Way stretching across Scorpius and Sagittarius (see the photograph that opens this chapter). In December and January, when our nighttime view is directed away from the galactic center, we see the Perseus arm. The remaining two major spiral arms are usually referred to as the Centaurus arm and the Cygnus arm.

Why are the young stars, star-forming regions, and clouds of neutral hydrogen in our Galaxy all found predominantly in the spiral arms? To answer this question, we must understand why spiral arms exist at all. Spiral arms are essentially cosmic "traffic jams," places where matter piles up as it orbits around the center of the Galaxy. This orbital motion, which is essential to grasping the significance of spiral arms, is our next topic as we continue the exploration of the Galaxy.

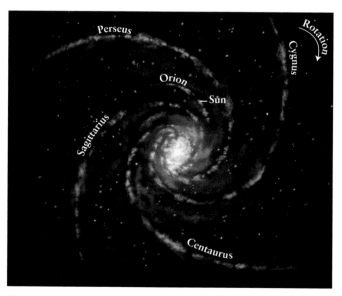

figure 25-15

Our Galaxy (Face-on View) Our Galaxy has four major spiral arms and several shorter arm segments. The Sun is located on the Orion arm, between two major spiral arms. The Galaxy's diameter is about 50,000 pc (160,000 ly), and the Sun is about 8000 pc (26,000 ly) from the galactic center. Note that the central bulge is somewhat elongated, as is also the case for M83 (see Figure 25-14).

25-4 The rotation of our Galaxy reveals the presence of dark matter

The spiral arms in the disk of our Galaxy suggest that the disk rotates. This means that the stars, gas, and dust in our Galaxy are all orbiting the galactic center. Indeed, if this were not the case, mutual gravitational attraction would cause the entire Galaxy to collapse into the galactic center. In the same way, the Moon is kept from crashing into the Earth and the planets from crashing into the Sun because of their motion around their orbits (see Section 4-7).

Measuring the rotation of our Galaxy accurately is a difficult business. But such challenging measurements have been made, as we shall see, and the results lead to a remarkable conclusion: Most of the mass of the Galaxy is in the form of *dark matter,* a mysterious sort of material that emits no light at all.

Radio observations of 21-cm radiation from hydrogen gas provide important clues about our Galaxy's rotation. Doppler shift measurements of this radiation indicate that stars and gas all orbit in the same direction around the galactic center, just as the planets all orbit in the same direction around the Sun. Measurements also show that the orbital speed of stars and gas about the galactic center is fairly uniform throughout much of the Galaxy's disk. As a result, stars inside the Sun's orbit complete a trip around the galactic center more quickly than the Sun, because the stars have a shorter distance to travel. Conversely, stars outside the Sun's orbit take longer to go once around the galactic center because they have farther to travel. As seen by Earth-based astronomers moving along with the Sun, stars inside the Sun's orbit overtake and pass us, while we overtake and pass stars outside the Sun's orbit (Figure 25-16a).

Figure 25-16 shows that the rotation of the Galaxy is very different from that of a solid disk. All parts of a rotating solid disk—a CD or DVD, for example—take the same time to complete one rotation. Because the outer part of the disk has to travel around a larger circle than the inner part, the speed (distance per time) is greater in the outer part (Figure 25-16b). By contrast, the orbital speed of material in our Galaxy is roughly the same at all distances from the galactic center. This

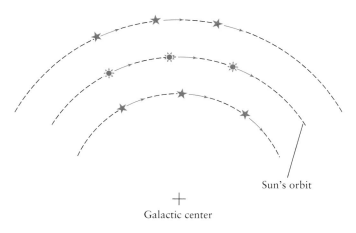

a   Actual rotation of our Galaxy

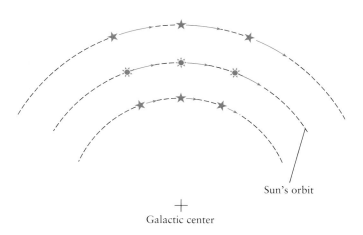

b   If our Galaxy rotated like a solid disk

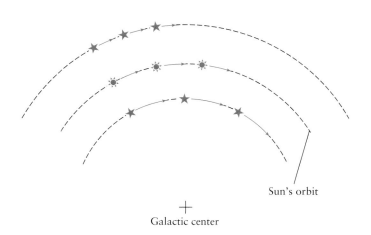

c   If the Sun and the stars obeyed Kepler's third law

ƒigure 25-16

The Rotation of Our Galaxy (a) The orbital speed of stars and gas around the galactic center is fairly uniform throughout most of our Galaxy. This schematic diagram shows the Sun and two other stars traveling at the same speed along different orbits. Although they start off lined up, the Sun and the stars become increasingly separated as they move along their orbits. Stars inside the Sun's orbit overtake and move ahead of the Sun, while stars far from the galactic center lag behind the Sun. **(b)** If the Galaxy rotated like a solid disk, the Sun and stars would always keep the same relative positions. This is not what is observed. **(c)** If the Sun and stars orbited the galactic center in the same way that planets orbit the Sun, the speed would be less for stars farther from the galactic center. This model cannot be correct because it predicts that stars inside the Sun's orbit would overtake us faster than they are observed to do.

is also quite unlike the orbits of the planets. The larger a planet's orbit, the slower the speed at which it moves around the orbit (Figure 25-16c), as we saw in Section 4-2.

The 21-cm observations reveal only how fast things are moving relative to the Sun. But the Sun itself is also moving. To get a complete picture of our Galaxy's rotation, we must therefore find out how fast the Sun is traveling along its own orbit. In other words, we need to find a kind of background that is not rotating along with the rest of the Galaxy. Such a background is provided by distant galaxies beyond the Milky Way and by the globular clusters. (Since globular clusters lie outside the plane of the Galaxy, they do not take part in the rotation of the disk.) By measuring the Doppler shifts of these objects and averaging their velocities, astronomers deduce that the Sun is moving along its orbit around the galactic center is about 220 km/s—about 790,000 kilometers per hour or 490,000 miles per hour!

We know that we are about 8000 parsecs from the galactic center, so we can use the Sun's speed v to calculate the time required for one complete trip around the Galaxy, a time called the Sun's orbital period P. For the sake of simplicity, we assume that the Sun travels along a circular orbit about the center of the Galaxy. (The orbit is in fact nearly circular.) The distance traveled by the Sun is the circumference of this circle, which is given by $2\pi r$ (with r being the radius of the circle). The time required for one orbit is equal to the distance traveled divided by the Sun's speed:

Period of the Sun's orbit around the galactic center

$$P = \frac{2\pi r}{v}$$

P = orbital period of the Sun

r = distance from the Sun to the galactic center

v = orbital speed of the Sun

or

$$P = \frac{2\pi \times 8000 \text{ pc}}{220 \text{ km/s}} \times \frac{3.09 \times 10^{13} \text{ km}}{1 \text{ pc}}$$

$$= 7.1 \times 10^{15} \text{ s} = 2.2 \times 10^8 \text{ years}$$

Traveling at 790,000 kilometers per hour, it takes the Sun about 220 million years to complete one trip around the Galaxy. The Galaxy is a very large place!

The Sun's orbit around the center of the Galaxy is roughly circular, just as is the Earth's orbit around the Sun. There is, however, a key difference between these two orbits. While the most important gravitational force holding the Earth in orbit comes from just one object (the Sun), no single object at the center of the Galaxy is primarily responsible for holding the Sun in its 220-million-year orbit. Instead, what keeps the Sun in this orbit is the combined gravitational force exerted on it by *all* of the matter (including stars, gas, and dust) that lies inside the Sun's orbit. (It turns out that the gravitational forces from matter *outside* the Sun's orbit have little or no net effect on the Sun's motion.)

We can use Newton's form of Kepler's third law to determine the total mass of all the matter that lies inside the Sun's orbit. As Box 25-2 shows, such calculations yield 9.0×10^{10} M$_\odot$. Because the Galaxy extends well beyond the Sun's orbit (see Figure 25-8), the mass of the entire Galaxy must be larger than this.

In recent years, astronomers have been astonished to discover how much matter may lie outside the Sun's orbit. The clues come from 21-cm radiation emitted by hydrogen in spiral arms that extend to the outer reaches of the Galaxy. Because we know the true speed of the Sun, we can convert the Doppler shifts of this radiation into actual speeds for the spiral arms. This calculation gives us a **rotation curve,** a graph of the speed of galactic rotation measured outward from the galactic center (Figure 25-17). Note that the rotation curve keeps rising slowly, even out to a distance of more than 18,000 pc from the galactic nucleus.

According to Kepler's third law, the orbital speeds of stars or gas clouds beyond the confines of most of the Galaxy's mass should decrease with increasing distance from the Galaxy's center, just as the orbital speeds of the planets decrease with increasing distance from the Sun (see Figure 25-16c). But the Galaxy's rotation curve is quite flat, indicating roughly uniform orbital speeds well beyond the visible edge of the galactic disk.

To explain these uniform orbital speeds in the outer parts of the Galaxy, astronomers conclude that a large amount of matter must lie outside the Sun's orbit. When this matter is included, the total mass of our Galaxy could exceed 10^{12} M$_\odot$ or more, of which about one-tenth is in the form of stars. This implies that our Galaxy contains roughly 200 billion stars.

 These observations lead to a profound mystery. Stars, gas, and dust account for only about 10% of the Galaxy's total mass. What, then, makes up the remaining 90% of the matter in our Galaxy? Whatever it is, it is dark. It does not show up on photographs, nor indeed in images made in any part of the electromagnetic spectrum. This unseen material, which is by far the predominant constituent of our Galaxy, is called **dark matter.** We sense its presence only through its gravitational influence on the orbits of stars and gas clouds.

 Be careful not to confuse dark *matter* with dark *nebulae*. A dark nebula like the one in Figure 20-3 emits no visible light, but does radiate at longer wavelengths. By contrast, no electromagnetic radiation of any kind has yet been discovered coming from dark matter.

Observations of star groupings outside the Milky Way suggest that our Galaxy's dark matter forms a spherical halo centered on the galactic nucleus, like the halo of globular clusters and high-velocity stars shown in Figure 25-8. However, the dark matter halo is much larger; it may extend to a distance of 100–200 kpc from the center of our Galaxy, some 2 to 4 times the extent of the visible halo. Analysis of the rotation curve in Figure 25-17 shows that the density of the dark matter halo decreases with increasing distance from the center of the Galaxy.

box 25-2 | Tools of the Astronomer's Trade

Estimating the Mass Inside the Sun's Orbit

The force that keeps the Sun in orbit around the center of the Galaxy is the gravitational pull of all the matter interior to the Sun's orbit. We can estimate the total mass of all of this matter using Newton's form of Kepler's third law (see Section 4-7 and Box 4-4):

$$P^2 = \frac{4\pi^2 a^3}{G(M + M_\odot)}$$

In this equation P is the orbital period of the Sun, a is the semimajor axis of the Sun's orbit around the galactic center, G is the universal constant of gravitation, M is the amount of mass inside the Sun's orbit, and $M_\odot$ is the mass of the Sun.

Because the Sun is only one of more than 10^{11} stars in the Galaxy, the Sun's mass is minuscule compared to M. Hence, we can safely replace the sum $M + M_\odot$ in the above equation by simply M. If we now assume that the Sun's orbit is a circle, the semimajor axis a of the orbit is just the radius of this circle, which we call r. From Section 25-4, the period P of the orbit is equal to $2\pi r/v$, where v is the Sun's orbital speed. You can then show that

$$M = \frac{rv^2}{G}$$

(The derivation of this equation is left as an exercise at the end of this chapter.)

Now we can insert known values to obtain the mass inside the Sun's orbit. Being careful to express distance in meters and speed in meters per second, we have $v = 220$ km/s $= 2.2 \times 10^5$ m/s, $G = 6.67 \times 10^{-11}$ newton m²/kg², and

$$r = 26{,}000 \text{ light-years} \times \frac{9.46 \times 10^{12} \text{ km}}{1 \text{ light-year}} \times \frac{10^3 \text{ m}}{1 \text{ km}}$$

$$= 2.5 \times 10^{20} \text{ m}$$

Hence, we find that

$$M = \frac{2.5 \times 10^{20} \times (2.2 \times 10^5)}{6.67 \times 10^{-11}} = 1.8 \times 10^{41} \text{ kg}$$

or

$$M = 1.8 \times 10^{41} \text{ kg} \times \frac{1 \text{ M}_\odot}{1.99 \times 10^{30} \text{ kg}} = 9.0 \times 10^{10} \text{ M}_\odot$$

This estimate involves only mass that is interior to the Sun's orbit. Matter outside the Sun's orbit has no net gravitational effect on the Sun's motion and thus does not enter into Kepler's third law.

 What is the nature of this mysterious dark matter? One proposal is that the dark matter halo is composed, at least in part, of dim objects with masses less than 1 M$_\odot$. These objects, which could include brown dwarfs, white dwarfs, or black holes, are called **massive compact halo objects**, or **MACHOs**. Astronomers have searched for MACHOs by monitoring the light from distant stars. If a MACHO passes between us and the star, its gravity will bend

figure 25-17

The Galaxy's Rotation Curve The blue curve shows the orbital speeds of stars and gas in the disk of the Galaxy out to a distance of 18,000 parsecs from the galactic center. (Very few stars are found beyond this distance.) The dashed red curve indicates how this orbital speed should decline beyond the confines of most of the Galaxy's visible mass. Because the data (blue curve) do not show any such decline, there must be an abundance of invisible dark matter that extends to great distances from the galactic center.

a

b

ƒigure 25-18

Microlensing by Dark Matter in the Galactic Halo (a) If a dense object such as a brown dwarf or black hole passes between the Earth and a distant star, the gravitational curvature of space around the dense object deflects the starlight and focuses it in our direction. This effect is called microlensing. (b) This light curve shows the gravitational microlensing of light from a star in the Galaxy's central bulge. Astronomers do not know the nature of the object that passed between the Earth and this star to cause the microlensing. (Courtesy of the Macho and Gman Collaborations)

the light coming from the star. (In Section 24-2 we described how gravity can bend starlight.) As Figure 25-18 shows, the MACHO's gravity acts like a lens that focuses the light from the star. This effect, called **microlensing,** makes the star appear to brighten substantially for a few days.

Astronomers have indeed detected MACHOs in this way, but not enough to completely solve the dark matter mystery. MACHOs with very low mass (10^{-6} to 0.1 $M_\odot$ each) do not appear to be a significant part of the dark matter halo. MACHOs of roughly 0.5 $M_\odot$ are more prevalent, but account for only about half of the dark matter halo.

The remainder of the dark matter is thought to be much more exotic. One candidate is a neutrino with a small amount of mass. If these neutrinos are sufficiently massive, and if enough of them are present in the halo of the Galaxy, they might constitute a reasonable fraction of the dark matter. As we saw in Section 18-4, recent data show that one type of neutrino can transform into another. These transformations can take place only if neutrinos have a nonzero amount of mass. Thus, neutrinos must comprise at least part of the dark matter, though it is not known how much.

Another speculative possibility is a new class of subatomic particle called **weakly interacting massive particles,** or **WIMPs.** These particles, whose existence is suggested by certain theories but has not yet been

confirmed experimentally, would not emit or absorb electromagnetic radiation. Physicists are attempting to detect these curious particles, which would have masses 10 to 10,000 times greater than a proton or neutron, by using a large crystal cooled almost to absolute zero. If a WIMP should enter this crystal and collide with one of its atoms, the collision will deposit a tiny but measurable amount of heat in the crystal.

As yet, the true nature of dark matter remains a mystery. Furthermore, this mystery is not confined to our own Galaxy. In Chapter 26 we will find that other galaxies have the same sort of rotation curve as in Figure 25-17, indicating that they also contain vast amounts of dark matter. Indeed, dark matter appears to make up most of the mass in our universe. Hence, the quest to understand dark matter is one of the most important in modern astronomy.

25-5 Spiral arms are caused by density waves that sweep around the Galaxy

The disk shape of our Galaxy is not difficult to understand. In Section 7-7 we described what happens when a large number of objects are put into orbit around a common center: Over time the objects tend naturally to orbit in the same plane. This is what happened when our solar system formed from the solar nebula. There a giant cloud of material eventually organized itself into planets, all of which orbit in nearly the same plane. In like fashion, the disk of our Galaxy, which is also made up of a large number of individual objects orbiting a common center, is very flat, as Figure 25-8 shows.

Understanding why our Galaxy has spiral arms presents more of a challenge. One early idea was that the material in the Galaxy somehow condensed into a spiral pattern from the very start. In this view, once stars, gas, and dust had become concentrated within the spiral arms, the pattern would remain fixed. This would be possible only if the Galaxy rotated like a solid disk (see Figure 25-16b); the fixed pattern would be like the spokes on a rotating bicycle wheel. But the reality is that the Galaxy is not a solid disk. As we have seen, stars, gas, and dust all orbit the galactic center with approximately the same speed, as shown in Figure 25-16a. Let us see why this makes it impossible for a rigid spiral pattern to persist.

Imagine four stars A, B, C, and D that originally lie on a line extending outward from the galactic center (Figure 25-19a). In a given amount of time, each of the stars travels the same distance around its orbit. But because the innermost star has a smaller orbit than the others, it takes less time to complete one orbit. As a result, a line connecting the four stars is soon bent into a spiral (Figure 25-19b). Moreover, the spiral becomes tighter and tighter with the passage of time (Figure 25-19c and Figure 25-19d). This "winding up" of the spiral arms causes the spiral structure to disappear completely after a few hundred million years—a very brief time compared to the age of our Galaxy, thought to be roughly 10 billion (10^{10}) years.

Figure 25-19 suggests that the Milky Way's spiral arms ought to have disappeared by now. The fact that they have not is called the **winding dilemma.** It shows that the spiral arms cannot simply be assemblages of stars and interstellar matter that travel around the Galaxy together, like a troop of soldiers marching in formation around a flagpole. In other words, the spiral arms cannot be made of anything *material.* What, then, can the spiral arms be?

In the 1940s, the Swedish astronomer Bertil Lindblad proposed that the spiral arms of a galaxy are actually a pattern that moves through the Galaxy like ripples on water. This idea was greatly enhanced and embellished in the 1960s by the American astronomers Chia Chiao Lin and Frank Shu. In this picture, spiral arms are a kind of wave, like the waves that move across the surface of a pond when you toss a stone into the water. Water molecules pile up at a crest of the wave but spread out again when the crest passes. By analogy, Lindblad, Lin, and Shu pictured a pattern of **density waves** sweeping around the Galaxy. These waves make matter pile up in the spiral arms, which are the crests of the waves. Individual parts of the Galaxy's material are compressed only temporarily when they pass through a spiral arm. The pattern of spiral arms persists, however, just as the waves made by a stone dropped in the water can persist for quite awhile after the stone has sunk.

To understand better how a density wave operates in a galaxy, think again about a water wave in a pond. If one part of the pond is disturbed by dropping a stone into it, the mol-

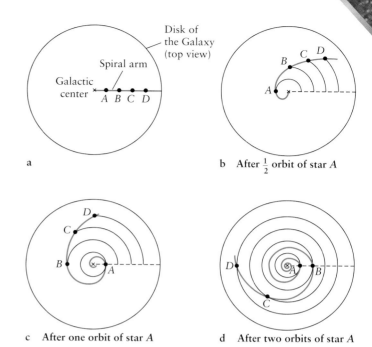

a

b After $\frac{1}{2}$ orbit of star A

c After one orbit of star A

d After two orbits of star A

ꞙigure 25-19

The Winding Dilemma **(a)** Imagine four stars, A, B, C, and D, that lie along a line that extends out from the galactic center. **(b)** All four stars have roughly the same orbital speed but travel in orbits of different sizes. Hence, when star A has completed one-half of an orbit, star B has completed less than one-half, star C has completed less than star B, and star D has completed the least of all. The line connecting the four stars has become a spiral. **(c)** By the time star A has completed one orbit, the spiral has wound tighter. **(d)** The spiral continues to tighten with each successive orbit. This shows that the spiral structure of galaxies (including our own) cannot be caused by matter moving together like soldiers in formation. Such formations would "wind up" and erase the spiral structure in just a few hundred million years.

ecules in that part will be displaced a bit. They will nudge the molecules next to them, causing those molecules to be displaced and to nudge the molecules beyond them. In this way the wave disturbance spreads throughout the pond.

In a galaxy, stars play the role of water molecules. Although stars and interstellar clouds of gas and dust are separated by vast distances, they can nonetheless exert forces on each other because they are affected by each other's gravity. If a region of above-average density should form, its gravitational attraction will draw nearby material into it. The displacement of this material will change the gravitational force that it exerts on other parts of the galaxy, causing additional displacements. In this way a spiral-shaped density wave can travel around the disk of a galaxy.

 A key feature of density waves is that they move more slowly around a galaxy than do stars or interstellar matter. To visualize this, imagine workers painting a

figure 25-20

A Density Wave on the Highway A slow-moving crew of painters can cause a traffic jam on the highway. As cars slow down in the traffic jam, they form a region where the cars are more densely packed, a region that moves along with the painters. Individual cars, by contrast, enter this region on the left and exit it on the right. In a similar way, a density wave in a spiral galaxy is a slow-moving region where stars, gas, and dust are more densely packed than in the rest of the galaxy. As the material of the galaxy passes through the density wave, it is compressed. This triggers star formation, as Figure 25-21 shows.

line down a busy freeway (Figure 25-20). The cars normally cruise along the freeway at high speed, but the crew of painters is moving much more slowly. When the cars come up on the painters, they must slow down temporarily to avoid hitting anyone. As seen from the air, cars are jammed together around the painters. An individual car spends only a few moments in the traffic jam before resuming its usual speed, but the traffic jam itself lasts all day, inching its way along the road as the painters advance.

A similar crowding takes place when interstellar matter enters a spiral arm. This crowding plays a key role in the formation of stars and the recycling of the interstellar medium. As interstellar gas and dust moves into a spiral arm, it is compressed into new nebulae (Figure 25-21). This compression begins the process by which new stars form, which we described in Section 20-3.

These freshly formed stars continue to orbit around the center of their galaxy, just like the matter from which they formed. The most luminous among these are the hot, massive, blue O and B stars, which may have emission nebulae (H II regions) associated with them. These stars have main-sequence lifetimes of only 3 to 15 million years (see Table 21-1), which is very short compared to the 220 million years required for the Sun to make a complete orbit around the Galaxy.

 As a result, these luminous O and B stars can travel only a relatively short distance before dying off. Therefore, these stars, and their associated H II regions, are only seen in or close to the spiral arm in which they formed. (For an example of this in another galaxy, see Figure 25-14a.) Less massive stars have much longer main-sequence lifetimes, and thus their orbits are able to carry them all around the galactic disk. These less-luminous stars are found throughout the disk, including between the spiral arms.

The density-wave model of spiral arms explains why the disk of our Galaxy is dominated by metal-rich Population I stars. Because the material left over from the death of ancient

 figure 25-21

Star Formation in the Density-Wave Model In the density-wave model, a spiral arm is a region where the density of material is higher than in the surrounding parts of a galaxy. Interstellar matter moves around the galactic center more rapidly than the spiral arms and is compressed as it passes through the spiral arms. This compression triggers the process of star formation, and new stars appear immediately "downstream" of the densest part of the spiral arms.

a   Grand-design spiral galaxy

b   Flocculent spiral galaxy

ƒigure 25-22   R I **V** U X G

Variety in Spiral Arms   The differences from one spiral galaxy to another suggest that more than one process can create spiral arms. **(a)** NGC 628 is a grand-design spiral galaxy with thin, well-defined spiral arms. **(b)** NGC 7793 is a flocculent spiral galaxy with fuzzy, poorly defined spiral arms. (Courtesy of P. Seiden, D. Elmegreen, B. Elmegreen, and A. Mobarak; IBM)

stars is enriched in heavy elements, new generations of stars formed in spiral arms are likely to be more metal-rich than their ancestors.

The density-wave model is still under development. One problem is finding a driving mechanism that keeps density waves going in spiral galaxies. After all, density waves expend an enormous amount of energy to compress the interstellar gas and dust. Hence, we would expect that density waves should eventually die away, just as do ripples on a pond. The American astronomers Debra and Bruce Elmegreen have suggested that gravity can supply that needed energy. As mentioned in Section 25-2, the central bulge of our Galaxy may be elongated somewhat into a bar shape, much like the central bulge of the galaxy M83 (see Figure 25-14*a*). The asymmetric gravitational field of such a bar pulls on the stars and interstellar matter of a galaxy to generate density waves. Another factor that may help to generate and sustain spiral arms are gravitational interactions *between* galaxies. We will discuss this in Chapter 26.

Spiral density waves may not be the whole story behind spiral arms in our Galaxy and other galaxies. The reason is that spiral density waves should produce very well defined spiral arms. We do indeed see many so-called **grand-design spiral galaxies** (Figure 25-22*a*), with thin, graceful, and well-defined spiral arms. But in some galaxies, called **flocculent spiral galaxies** (Figure 25-22*b*), the spiral arms are broad, fuzzy, chaotic, and poorly defined. ("Flocculent" means "resembling wool.")

To explain such flocculent spirals, M. W. Mueller and W. David Arnett in 1976 proposed a theory of **self-propagating star formation.** Imagine that star formation begins in a dense interstellar cloud within the disk of a galaxy that does not yet have spiral arms. As soon as hot, massive stars form, their radiation and stellar winds compress nearby matter, triggering the formation of additional stars in that gas. When massive stars become supernovae, they produce shock waves that further compress the surrounding interstellar medium, thus encouraging still more star formation.

Although all parts of this broad, star-forming region have approximately the same orbital speed about the galaxy's center, the inner regions have a shorter distance to travel to complete one orbit than the outer regions. As a result, the inner edges of the star-forming region move ahead of the outer edges as the Galaxy rotates. The bright O and B stars and their nearby glowing nebulae soon become stretched out in the form of a spiral arm. These spiral arms come and go essentially at random across a galaxy. Bits and pieces of spiral arms appear where star formation has recently begun but fade and disappear at other locations where all the massive stars have died off. Self-propagating star formation therefore tends to produce flocculent spiral galaxies that have a chaotic appearance with poorly defined spiral arms, like the galaxy in Figure 25-22*b*.

The two theories presented here are very different in character. In the density-wave model, star formation is caused by the spiral arms; in the self-propagating star formation model, by contrast, the spiral arms are caused by star formation. The correct description of spiral arms in our Galaxy may involve a combination of these models and remains a topic of active research.

25-6 Infrared and radio observations are used to probe the galactic nucleus

The innermost region of our Galaxy is an active, crowded place. If you lived on a planet near the galactic center, you could see a million stars as bright as Sirius, the brightest single star in our own night sky. The total intensity of starlight from all those nearby stars would be equivalent to 200 of our full moons. In effect, night would never really fall on a planet near the center of the Milky Way. At the center of this empire of light, however, may lie the darkest of all objects in the universe—a black hole millions of times more massive than the Sun.

Because of the severe interstellar absorption at visual wavelengths, most of our information about the galactic center comes from infrared and radio observations. Figure 25-23 shows three infrared views of the center of our Galaxy. Figure 25-23a is a wide-angle view covering a 50° segment of the Milky Way from Sagittarius through Scorpius. (The photograph that opens this chapter shows this same region at visible wavelengths, viewed at a different angle.) The prominent reddish band through the center of this false-color infrared image is a layer of dust in the plane of the Galaxy. Figure 25-23b is an IRAS view of the galactic center. It is surrounded by numerous streamers of dust (shown in blue). The strongest infrared emission (shown in white) comes from **Sagittarius A**, which is a grouping of several powerful sources of radio waves. One of these sources, Sagittarius A* (say "A star"), is thought to be the galactic nucleus. The infrared view in Figure 25-23c shows hundreds of stars crowded within 1 ly (0.3 pc) of Sagittarius A*. Compare this to our region of the Galaxy, where the average distance between stars is more than a light-year.

Radio observations give a different and complementary view of the galactic center. In addition to 21-cm radiation from neutral hydrogen atoms, astronomers have also detected synchrotron radiation coming from the Sagittarius A region at the galactic center. (We saw in Section 23-4 that synchrotron

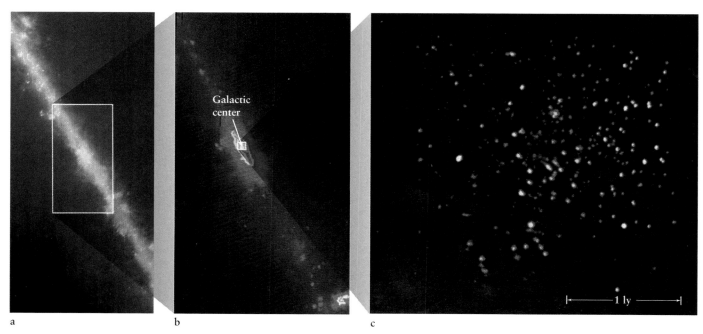

a   b   c

Figure 25-23 R ▮ V U X G

The Galactic Center **(a)** This infrared image shows a 50° view of the Milky Way centered on the galactic nucleus. Colors indicate the intensity of infrared emission: white is brightest, followed by red, yellow, and blue; regions shown in black emit essentially no infrared radiation. The reddish band extending from upper left to lower right is dust in the plane of the Galaxy. The bluish knots and blobs along the dust layer are interstellar clouds heated by young O and B stars. **(b)** This close-up infrared view covers the area outlined by the white rectangle in (a). **(c)** Even higher magnification shows about 300 bright stars within 1 ly (0.3 pc) from Sagittarius A*, a radio source which lies at the center of the picture and at the very center of the Galaxy. The distribution and motions of these stars indicate that this innermost part of the Galaxy contains hundreds of thousands of other, dimmer stars. (a, b: NASA; c: A. Eckart, R. Genzel, R. Hofman, B. J. Sams, and L. E. Tacconi-Garman, ESO) a   R I V U X G   b   R I V U X G

Figure 25-24

Three Views of the Galactic Nucleus Images (a) and (b) show the center of our Galaxy at radio wavelengths. The strength of radio emission is shown by the colors of the rainbow: Red is brightest and violet is dimmest. Black shows regions that emit negligible amounts of radio waves. **(a)** This view covers an area of the sky with about the same angular size as the full Moon, corresponding to a diameter of about 80 pc (250 ly). The galactic nucleus is within the rectangle in the red-colored region of strong emission. **(b)** This high-resolution radio view shows an area about 10 pc (30 ly) across. The material that makes up the bright pinwheel is orbiting around Sagittarius A*. **(c)** The cross at the center of this infrared image marks the location of the black hole at the Galaxy's center. (a, b: VLA, NRAO; c: K. M. Menten and M. J. Reid [Harvard-Smithsonian Center for Astrophysics], A. Eckart and R. Genzel [Max-Planck-Institut für extraterrestrische Physik])

radiation is produced by high-speed electrons spiraling around a magnetic field.) Despite its compact size, Sagittarius A is one of the brightest radio sources in the entire sky.

Some of the most detailed views of the galactic center have been provided by the Very Large Array (see Figure 6-25). Figure 25-24a is a wide-angle radio image of Sagittarius A covering an area 60 pc (200 ly) across. Huge filaments of gas extend perpendicular to the plane of the Galaxy. They stretch for 20 pc (65 ly) northward of the galactic center (to the right and upward in Figure 25-24a), then abruptly arch southward (straight down in the figure) toward Sagittarius A, at the core of the strongest emission. The orderly arrangement of these filaments is reminiscent of prominences on the Sun (see Section 18-10, especially Figure 18-30). This suggests that, as on the Sun, a magnetic field may be controlling the flow of ionized gas at the galactic center.

The inner core of Sagittarius A, shown in Figure 25-24b, covers an area about 7 pc (20 ly) across. Here, a pinwheel-like feature is centered on the radio source called Sagittarius A*. One of the arms of this pinwheel is part of a ring of gas and dust (shown in a faint blue color in Figure 25-24b) that orbits the galactic center. The total amount of gas orbiting the galactic nucleus along with this ringlike structure is estimated at 3×10^4 $M_\odot$. All the material less than about 2 pc from the galactic nucleus seems to have been ionized by a source that coincides with Sagittarius A*.

Sagittarius A* is thought to be the galactic nucleus, the very center of the Milky Way Galaxy. Astronomers think this because Sagittarius A* is the brightest radio source within several degrees of the galactic center. Furthermore, its position, pinpointed with simultaneous observations by radio telescopes scattered around the world, seems to be very near the dynamical and gravitational center of the Galaxy.

But what *is* Sagittarius A*? It is neither a star nor a pulsar, from its high luminosity and radio spectrum. It also cannot be a supernova remnant because it is not expanding rapidly. Instead, Sagittarius A* is almost certainly an extremely massive black hole.

The strongest evidence for this comes from recent infrared observations of the motions of stars in the vicinity of Sagittarius A* (Figure 25-24c). Two research groups—Andreas Eckart and Reinhard Genzel of the Max Planck Institute for Extraterrestrial Physics in Garching, Germany, and Andrea Ghez, Mark Morris, and Eric Becklin of UCLA—independently observed stars orbiting around Sagittarius A* at speeds in excess of 1500 km/s (Figure 25-25). A source of gravity must be keeping these stars in orbit about the galactic center.

 Using Newton's form of Kepler's third law, astronomers calculate that this source of gravity must have a mass of roughly 3×10^6 $M_\odot$. Yet all this mass is concentrated in a volume no larger than our solar

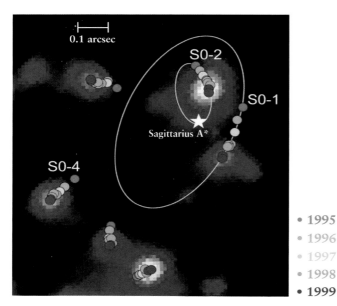

- 1995
- 1996
- 1997
- 1998
- **1999**

figure 25-25   R ■ V U X G

Stars Orbiting the Center of the Galaxy The colored dots superimposed on this infrared image show the motions of six stars at the galactic center. Each star is orbiting around an unseen massive object at the position of the radio source Sagittarius A*. (Sagittarius A* is actually at one focus of each orbit. It appears otherwise because the orbits are not oriented face-on to us.) By measuring how much each star accelerated between 1995 and 1999, astronomers can deduce the gravitational pull and hence the mass of this object. The conclusion is that Sagittarius A* contains about 3 million solar masses within a very small volume less than 1 AU across. The only plausible explanation is that Sagittarius A* is a supermassive black hole. (Andrea Ghez, Department of Astronomy, UCLA)

system. It appears conclusive that an object this massive and this compact could only be a supermassive black hole (see Section 24-4). Observations show that such supermassive black holes are a feature of most large galaxies. In Chapter 27 we will see how black holes of this kind power *quasars*, the most luminous sustained light sources in the cosmos.

Astronomers are still groping for a better understanding of the galactic center. With future developments in very-long-baseline interferometry (described in Section 6-6), it may be possible to actually obtain a picture of the supermassive black hole lurking there. During the coming years, observations from Earth-orbiting satellites as well as from radio and infrared telescopes on the ground will certainly add to our knowledge of the core of the Milky Way.

KEY WORDS

The term preceded by an asterisk (*) is discussed in Box 25-1.

central bulge (of a galaxy), p. 566

dark matter, p. 574

density wave, p. 577

disk (of a galaxy), p. 566

far-infrared, p. 566

flocculent spiral galaxy, p. 579

galactic nucleus, p. 566

galaxy, p. 563

globular cluster, p. 565

grand-design spiral galaxy, p. 579

H I, p. 568

halo (of a galaxy), p. 568

high-velocity star, p. 568

interstellar extinction, p. 564

Local Bubble, p. 570

*magnetic resonance imaging (MRI), p. 570

massive compact halo object (MACHO), p. 575

microlensing, p. 576

Milky Way Galaxy, p. 563

near-infrared, p. 566

rotation curve, p. 574

RR Lyrae variable, p. 565

Sagittarius A, p. 580

self-propagating star formation, p. 579

spin (of a particle), p. 568

spin-flip transition, p. 569

spiral arm, p. 568

21-cm radio emission, p. 569

weakly interacting massive particle (WIMP), p. 576

winding dilemma, p. 577

KEY IDEAS

The Shape and Size of the Galaxy: Our Galaxy has a disk about 50 kpc (160,000 ly) in diameter and about 600 pc (2000 ly) thick, with a high concentration of interstellar dust and gas in the disk.

• The galactic center is surrounded by a large distribution of stars called the central bulge. This bulge is not perfectly symmetrical, but may have a bar or peanut shape.

• The disk of the Galaxy is surrounded by a spherical distribution of globular clusters and old stars, called the galactic halo.

• There are about 200 billion (2×10^{11}) stars in the Galaxy's disk, central bulge, and halo.

The Sun's Location in the Galaxy: Our Sun lies within the galactic disk, some 8000 pc (26,000 ly) from the center of the Galaxy.

• Interstellar dust obscures our view at visible wavelengths along lines of sight that lie in the plane of the galactic disk. As a result, the Sun's location in the Galaxy was unknown for many years. This dilemma was resolved by observing parts of the Galaxy outside the disk.

• The Sun orbits around the center of the Galaxy at a speed of about 790,000 km/h. It takes about 220 million years to complete one orbit.

The Rotation of the Galaxy and Dark Matter: From studies of the rotation of the Galaxy, astronomers estimate that the total mass of the Galaxy is about 10^{12} M$_\odot$. Only about 10% of this mass is in the form of visible stars, gas, and dust. The remaining 90% is in some nonvisible form, called dark matter, that extends beyond the edge of the luminous material in the Galaxy.

• Our Galaxy's dark matter may be a combination of MACHOs (dim, star-sized objects), massive neutrinos, and WIMPs (relatively massive subatomic particles).

The Galaxy's Spiral Structure: OB associations, H II regions, and molecular clouds in the galactic disk outline huge spiral arms.

• Spiral arms can be traced from the positions of clouds of atomic hydrogen. These can be detected throughout the galactic disk by the 21-cm radio waves emitted by the spin-flip transition in hydrogen. These emissions easily penetrate the intervening interstellar dust.

Theories of Spiral Structure: There are two leading theories of spiral structure in galaxies.

• According to the density-wave theory, spiral arms are created by density waves that sweep around the Galaxy. The gravitational field of this spiral pattern compresses the interstellar clouds through which it passes, thereby triggering the formation of the OB associations and H II regions that illuminate the spiral arms.

• According to the theory of self-propagating star formation, spiral arms are caused by the birth of stars over an extended region in a galaxy. Differential rotation of the galaxy stretches the star-forming region into an elongated arch of stars and nebulae.

The Galactic Nucleus: The innermost part of the Galaxy, or galactic nucleus, has been studied through its radio and infrared emissions (which are able to pass through interstellar dust).

• A strong radio source called Sagittarius A* is located at the galactic center. This marks the position of a supermassive black hole with a mass of about 3×10^6 M$_\odot$.

REVIEW QUESTIONS

1. How did interstellar extinction mislead astronomers into believing that we are at the center of our Galaxy?

2. How did observations of globular clusters help astronomers determine our location in the Galaxy?

3. Why are infrared telescopes useful for exploring the structure of the Galaxy? Why is it important to make observations at both near-infrared and far-infrared wavelengths?

4. The galactic halo is dominated by Population II stars, whereas the galactic disk contains predominantly Population I stars. In which of these parts of the Galaxy has star formation taken place recently? Explain.

5. Most interstellar hydrogen atoms emit only radio waves at a wavelength of 21 cm, but some hydrogen clouds emit profuse amounts of visible light (see, for example, Figures 20-1 and 20-2). What causes this difference?

6. How do astronomers determine the distances to H I (neutral hydrogen) clouds?

7. In a spiral galaxy, are stars in general concentrated in the spiral arms? Why are spiral arms so prominent in photographs of spiral galaxies?

8. What kinds of objects (other than H I clouds) do astronomers observe to map out the Galaxy's spiral structure? What is special about these objects? Which of these can be observed at great distances?

9. Why don't astronomers detect 21-cm radiation from the hydrogen in giant molecular clouds?

10. How do astronomers determine how fast the Sun moves in its orbit around the Galaxy?

11. In what way are the orbits of stars in the galactic disk different from the orbits of planets in our solar system?

12. How do astronomers conclude that vast quantities of dark matter surround our Galaxy? How is this dark matter distributed in space?

13. What is the difference between dark matter and dark nebulae?

14. What proposals have been made to explain the nature of dark matter? What experiments or observations have been made to investigate these proposals? What are the results of this research?

15. What is the winding dilemma? What does it tell us about the nature of spiral arms?

16. Do density waves form a stationary pattern in a galaxy? If not, do they move more rapidly, less rapidly, or at the same speed as stars in the disk?

17. In our Galaxy, why are stars of spectral classes O and B found in or near the spiral arms? Is the same true for stars of other spectral classes? Explain why or why not.

18. Compare the kinds of spiral arms produced by density waves with those produced by self- propagating star formation. By examining Figure 25-15, cite evidence that both processes probably occur in our Galaxy.

19. Describe the evidence for a supermassive black hole at the center of our Galaxy.

ADVANCED QUESTIONS

Questions preceded by an asterisk (*) involve topics discussed in the Boxes.

Problem-solving tips and tools

Several of the following questions make extensive use of Newton's form of Kepler's third law, and you might find it helpful to review Box 4-4. Another useful version of Kepler's third law is given in Section 19-9. We discussed the relationship between the energy and wavelength of a photon in Section 5-5. According to the Pythagorean theorem, an isosceles right triangle has a hypotenuse that is longer than its sides by a factor of $\sqrt{2} = 1.414$. The formula for the Schwarzschild radius of a black hole is given in Box 24-2. You will find the small-angle formula in Box 1-1 useful. It is also helpful to remember that an object 1 AU across viewed at a distance of 1 parsec has an angular size of 1 arcsecond. Remember, too, that the volume of a cylinder is equal to its height multiplied by the area of its base, and that the area of a circle of radius r is πr^2. You can find other geometrical formulae in Appendix 8.

20. Discuss how the Milky Way would appear to us if the Sun were relocated to (a) the edge of the Galaxy; (b) the galactic halo; (c) the galactic bulge.

21. Explain why globular clusters spend most of their time in the galactic halo, even though their eccentric orbits take them close to the galactic center.

22. The disk of the Galaxy is about 50 kpc in diameter and 600 pc thick. (a) Find the volume of the disk in cubic parsecs. (b) Find the volume (in cubic parsecs) of a sphere 300 pc in radius centered on the Sun. (c) If supernovae occur randomly throughout the volume of the Galaxy, what is the probability that a given supernova will occur within 300 pc of the Sun? If there are about three supernovae each century in our Galaxy, how often, on average, should we expect to see one within 300 pc of the Sun?

23. A typical hydrogen atom in interstellar space undergoes a spin-flip transition only once every 10^7 years. How, then, is it at all possible to detect the 21-cm radio emission from interstellar hydrogen?

24. Calculate the energy of the photon emitted when a hydrogen atom undergoes a spin-flip transition. How many such photons would it take to equal the energy of a single H_α photon of wavelength 656.3 nm?

25. Suppose you were to use a radio telescope to measure the Doppler shift of 21-cm radiation in the plane of the Galaxy. (a) If you observe 21-cm radiation from clouds of atomic hydrogen at an angle of 45° from the galactic center, you will see the highest Doppler shift from a cloud that is as far from the galactic center as it is from the Sun. Explain this statement using a diagram. (b) Find the distance from the Sun to the particular cloud mentioned in (a).

26. Calculate approximately how many times our solar system has orbited the center of our Galaxy since the Sun and planets were formed some 5×10^9 years ago.

27. Sketch the rotation curve you would obtain if the Galaxy were rotating like a rigid body.

28. A gas cloud located in the spiral arm of a distant galaxy is observed to have an orbital velocity of 400 km/s. If the cloud is 20,000 pc from the center of the galaxy and is moving in a circular orbit, find (a) the orbital period of the cloud and (b) the mass of the galaxy contained within the cloud's orbit.

29. The mass of our Galaxy interior to the Sun's orbit is calculated from the radius of the Sun's orbit and its orbital speed. By how much would this estimate be in error if the calculated distance to the galactic center were off by 10%? By how much would this estimate be in error if the calculated orbital velocity were off by 10%?

30. Speculate on the reasons for the rapid rise in the Galaxy's rotation curve (see Figure 25-17) at distances close to the galactic center.

31. According to the Galaxy's rotation curve in Figure 25-17, a star 16 kpc from the galactic center has an orbital speed of about 270 km/s. Calculate the mass within that star's orbit.

*32. Show that the form of Kepler's third law stated in Box 25-2, $P^2 = 4\pi^2 a^3/G(M + M_\odot)$, is equivalent to $M = rv^2/G$, provided the orbit is a circle. [*Hint:* The mass of the Sun ($M_\odot$) is much less than the mass of the Galaxy inside the Sun's orbit (M).]

33. The accompanying image shows the spiral galaxy M74, located about 55 million light-years from the Earth in the constellation Pisces (the Fish). It is actually a superposition of two false-color images: The red portion is an optical image taken at visible wavelengths, while the blue portion is an ultraviolet image made by NASA's Ultraviolet Imaging Telescope, which was carried into orbit by the space shuttle *Columbia* during the *Astro-1* mission in 1990. Compare the visible and ultraviolet images and, from what you know about stellar evolution and spiral structure, explain the differences you see.

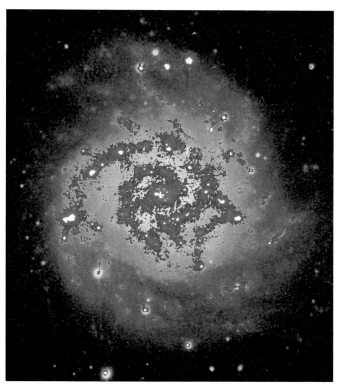

(Astronomical Society of the Pacific)　　　　R I **V** **U** X G

*34. (a) Calculate the Schwarzschild radius of a supermassive black hole of mass $3 \times 10^6 \ M_\odot$, the estimated mass of the black hole at the galactic center. Give your answer in both kilometers and in astronomical units. (b) What is the angular diameter of such a black hole as seen at a distance of 8 kpc, the distance from the Earth to the galactic center? Give your answer in arcseconds. Observing an object with such a small angular size will be a challenge indeed!

35. The orbital periods of the stars S0-1 and S0-2 shown in Figure 25-25 are estimated to be 63 and 17 years, respectively. (a) Assuming that the supermassive black hole in Sagittarius A* has a mass of $3 \times 10^6 \ M_\odot$, determine the semimajor axes of the orbits of these two stars. Give your answers in AU. (b) Calculate the angular size of each

orbit's semimajor axis as seen from Earth, which is 8000 pc from the center of the Galaxy. Explain why extremely high-resolution infrared images are required to observe the motions of these stars.

36. Consider a star that orbits around Sagittarius A* in a circular orbit of radius 1000 AU. (a) If the star's orbital speed is 1500 km/s, what is its orbital period? Give your answer in years. (b) Determine the sum of the masses of Sagittarius A* and the star. Give your answer in solar masses. (Your answer is an estimate of the mass of Sagittarius A*, because the mass of a single star is negligibly small by comparison.)

DISCUSSION QUESTIONS

37. From what you know about stellar evolution, the interstellar medium, and the density-wave theory, explain the appearance and structure of the spiral arms of grand-design spiral galaxies.

38. What observations would you make to determine the nature of the dark matter in our Galaxy's halo?

39. Describe how the appearance of the night sky might change if dark matter were visible to our eyes.

WEB/CD-ROM QUESTIONS

40. Clouds of positrons have been discovered recently near the center of the Galaxy. These are particles with the same properties as electrons except that their charge is positive rather than negative (see Box 18-1). Search the World Wide Web for information about this discovery. How were the positrons discovered? What is thought to be the origin of these particles? What happens when these positrons encounter ordinary electrons?

41. Fast-Moving Stars at the Galactic Center. Access and view the video "Fast-Moving Stars at the Galactic Center" in Chapter 25 of the *Universe* web site or CD-ROM. Explain how you can tell which of the stars in the video are actually close to Sagittarius A* and which just happen to lie along our line of sight to the center of the Galaxy.

OBSERVING PROJECTS

42. Use the *Starry Night* program to observe the Milky Way. First turn off daylight (select **Daylight** in the **Sky** menu) and show the entire celestial sphere (select **Atlas** in the **Go** menu). To make the Milky Way visible, make sure that the selection **Milky Way** in the **Sky** menu has a check mark next to it. If you find the Milky Way difficult to see, select **Sky Settings...** in the **Sky** menu and click on **Milky Way Colour**. Choose a color that will stand out against the dark night sky, then click on

OK. (a) Using the "hand" cursor, look at different parts of the Milky Way. Can you identify the direction toward the galactic nucleus? In this direction the Milky Way appears broadest. Check your identification by centering the field of view on the constellation Sagittarius (use the **Find...** command in the **Edit** menu). (b) Click on the "Home" button in the Control Panel to turn the horizon back on, then set the local time to midnight (12:00:00 A.M.). Use the controls in the Time palette to step through the months of the year. In which month is the galactic nucleus highest in the sky at midnight, so that it is most easily seen?

43. Use the *Starry Night* program to determine how the plane of the Galaxy is oriented on the celestial sphere. As in the preceding Observing Project, turn off daylight, show the entire celestial sphere, and make the Milky Way visible. Next, adjust your field of view so that the celestial equator, shown as a red line, runs across the middle of the screen. (Select **Celestial Grid** in the **Guide** menu if you can't see this.) Finally, move your field of view until you can see where the Milky Way crosses the celestial equator. (a) Measure the angle between the Milky Way and the celestial equator either on the screen or on a printout of the screen. (Some estimation is needed here.) How well aligned with each other are the plane of the Milky Way and the plane of the Earth's equator?
(b) A third plane of interest is the plane of the ecliptic. The ecliptic is shown as a green line (select **The Ecliptic** in the **Guide** menu if you can't see this.) Adjust your field of view so that the ecliptic appears as a straight line rather than as a curve (which means you are viewing in a direction that lies in the ecliptic plane) and so that you can see where the ecliptic crosses the Milky Way. Measure the angle between the Milky Way and the ecliptic either on the screen or on a printout of the screen. (Again, some estimation will be needed.) How well aligned with each other are the plane of the Milky Way and the plane of the Earth's orbit around the Sun?

Galaxies

26

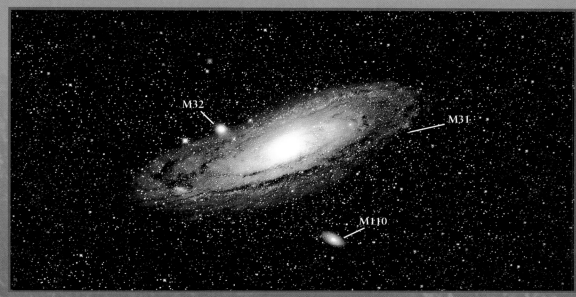

M32

M31

M110

(Bill and Sally Fletcher/Tom Stack & Associates)

R I **V** U X G

A century ago, most astronomers thought that the entire universe was only a few thousand light-years across and that nothing lay beyond our Milky Way Galaxy. One of the most important discoveries of the twentieth century was that this picture was utterly wrong. We now understand that the Milky Way is just one of billions of galaxies strewn across billions of light-years. The accompanying image shows three of them—the great spiral galaxy M31 in Andromeda and the two smaller galaxies (M32 and M110) that orbit it.

Some galaxies are spirals like M31 or the Milky Way, with arching spiral arms that are active sites of star formation. Others, like M32 and M110, are featureless, ellipse-shaped agglomerations of stars, virtually devoid of interstellar gas and dust. Some galaxies are only one-hundredth the size and one

ten-thousandth the mass of the Milky Way. Others are giants, with 5 times the size and 50 times the mass of the Milky Way. Only about 10% of a typical galaxy's mass emits radiation of any kind; the remainder is made up of the mysterious dark matter.

Just as most stars are found within galaxies, most galaxies are located in groups and clusters. These clusters of galaxies stretch across the universe, forming huge, lacy patterns. Remarkably, remote clusters of galaxies are receding from us; the greater their distance, the more rapidly they are moving away. This relationship between distance and recessional velocity, called the Hubble law, reveals that our immense universe is expanding. In Chapters 28 and 29 we will learn what this implies about the distant past and remote future of the universe.

As you read the sections of this chapter, look for the answers to the following questions.

26-1 How did astronomers first discover other galaxies?

26-2 How did astronomers first determine the distances to galaxies?

26-3 Do all galaxies have spiral arms, like the Milky Way?

26-4 How do modern astronomers tell how far away galaxies are?

26-5 How do the spectra of galaxies tell astronomers that the universe is expanding?

26-6 Are galaxies isolated in space, or are they found near other galaxies?

26-7 What happens when galaxies collide with each other?

26-8 Is dark matter found in galaxies beyond the Milky Way?

26-9 How do astronomers think galaxies formed?

26-1 When galaxies were first discovered, it was not clear that they lie far beyond the Milky Way

As early as 1755, the German philosopher Immanuel Kant suggested that vast collections of stars lie outside the confines of the Milky Way. Less than a century later, an Irish astronomer observed the structure of some of the "island universes" that Kant proposed.

William Parsons was the third Earl of Rosse in Ireland. He was rich, he liked machines, and he was fascinated with astronomy. Accordingly, he set about building gigantic telescopes. In February 1845 his pièce de résistance was finished. The telescope's massive mirror measured 6 feet in diameter and was mounted at one end of a 60-foot tube controlled by cables, straps, pulleys, and cranes (Figure 26-1a). For many years, this triumph of nineteenth-century engineering was the largest telescope in the world.

With this new telescope, Lord Rosse examined many of the nebulae that had been discovered and cataloged by William Herschel. He observed that some of these nebulae have a distinct spiral structure. One of the best examples is M51, also called NGC 5194.

Lacking photographic equipment, Lord Rosse had to make drawings of what he saw. Figure 26-1b shows a drawing he made of M51, which compares favorably with modern photographs (Figure 26-2). Views such as this inspired Lord Rosse to echo Kant's proposal that such nebulae are actually "island universes."

Many astronomers of the nineteenth century disagreed with this notion of island universes. A considerable number of nebulae are in fact scattered throughout the Milky Way. (Figures 20-2 and 20-3 show some examples.) It therefore seemed reasonable that "spiral nebulae," even though they

a   R I **V** U X G   b

Figure 26-1

A Pioneering View of Another Galaxy **(a)** Built in 1845, this structure housed a 1.8-m-diameter telescope called the "Leviathan," which was the largest of its day. The telescope itself (the black cylinder pointing at a 45° angle above the horizontal) was restored to its original state during 1996–1998. **(b)** Using his telescope, Lord

Rosse made this sketch of spiral structure in M51. This galaxy, whose angular size is 8 × 11 arcminutes (about a third the angular size of the full moon), is today called the Whirlpool Galaxy because of its distinctive appearance. (a: Birr Castle Demesne; b: Courtesy of Lund Humphries)

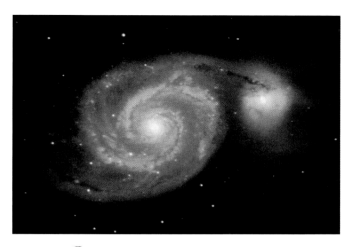

figure 26-2　R I **V** U X G

A Modern View of the Spiral Galaxy M51 This spiral galaxy, also called NGC 5194, lies in the constellation Canes Venatici (the Hunting Dogs). It is some 8.5 Mpc (8.5 million parsecs, or 28 million ly) from Earth and is about 20 kpc (20,000 parsecs, or 65,000 ly) in diameter. The "blob" at the end of one spiral arm is the companion galaxy NGC 5195. (NOAO)

are very different in shape from other sorts of nebulae, could also be components of our Galaxy.

The astronomical community became increasingly divided over the nature of the spiral nebulae. In April 1920, two opposing ideas were presented before the National Academy of Sciences in Washington, D.C. On one side was Harlow Shapley, a young, brilliant astronomer renowned for his recent determination of the size of the Milky Way Galaxy (see Section 25-1). Shapley thought the spiral nebulae were relatively small, nearby objects scattered around our Galaxy like the globular clusters he had studied. Opposing Shapley was Heber D. Curtis of the University of California's Lick Observatory. Curtis championed the island universe theory, arguing that each of these spiral nebulae is a rotating system of stars much like our own Galaxy.

The Shapley-Curtis "debate" generated much heat but little light. Nothing was decided, because no one could present conclusive evidence to demonstrate exactly how far away the spiral nebulae are. Astronomy desperately needed a definitive determination of the distance to a spiral nebula. Such a measurement became the first great achievement of a young man who studied astronomy at the Yerkes Observatory, near Chicago. His name was Edwin Hubble.

26-2　Hubble proved that the spiral nebulae are far beyond the Milky Way

After completing his studies, Edwin Hubble joined the staff of the Mount Wilson Observatory in Pasadena, California. On October 6, 1923, he took a historic photograph of the Andromeda "Nebula," one of the spiral nebulae around which controversy raged. (The image that opens this chapter is a modern photograph of this object.)

Hubble carefully examined his photographic plate and discovered what he at first thought to be a nova. Referring to previous plates of that region, he soon realized that the object was actually a Cepheid variable star. Further scrutiny of additional plates over the next several months revealed several more Cepheids. Figure 26-3 shows modern observations of a Cepheid in another "spiral nebula."

As we saw in Section 21-5, Cepheid variables help astronomers determine distances. An astronomer begins by carefully measuring the variations in apparent brightness of a Cepheid variable, then recording the results in the form of a plot of brightness versus time, or light curve (see Figure 21-14a). This graph gives the variable star's period and average brightness. Given the star's period, the astronomer then uses the period-luminosity relation shown in Figure 21-15 to find the Cepheid's average luminosity. (The astronomer must also examine the spectrum of the star to determine whether it is a metal-rich Type I Cepheid or a metal-poor Type II Cepheid. As Figure 21-15 shows, these have somewhat different period-luminosity relations.) Knowing both the apparent brightness and luminosity of the Cepheid, the astronomer can then use the inverse-square law to calculate the distance to the star (see Box 19-2). Box 26-1 presents an example of how this is done.

figure 26-3　R I **V** U X G

A Cepheid Variable in a Distant Galaxy One of the most reliable ways to determine the distance to a galaxy is to observe Cepheid variable stars in it. Using this technique, astronomers have found that M100, the galaxy shown here, is about 17 Mpc (56 million ly) from Earth. The insets show one of the Cepheids in M100 at different stages in its brightness cycle, which lasts several weeks. (Wendy L. Freedman, Carnegie Institution of Washington, and NASA)

Box 26-1 Tools of the Astronomer's Trade

Cepheids and Supernovae as Indicators of Distance

Because their periods are directly linked to their luminosities, Cepheid variables are one of the most reliable tools astronomers have for determining the distances to galaxies. To this day, astronomers use this link—much as Hubble did back in the 1920s—to measure intergalactic distances. More recently, they have begun to use Type Ia supernovae, which are far more luminous and thus can be seen much farther away, to determine the distances to very remote galaxies.

EXAMPLE: In 1992, a team of astronomers used Cepheid variables in a galaxy called IC 4182 to deduce this galaxy's distance from the Earth. The team of astronomers used the Hubble Space Telescope on 20 separate occasions to record images of the stars in IC 4182. By comparing these images, the astronomers could pick out which stars vary in brightness. In this way, they discovered 27 Cepheids in IC 4182. Using their observations, the astronomers estimated the brightnesses of the Cepheid variables and plotted their light curves. The Hubble Space Telescope is particularly well suited for studies of this kind, because its extraordinary angular resolution makes it possible to pick out individual stars at great distances.

One such Cepheid has a period is 42.0 days and an average apparent magnitude (m) of +22.0. (See Box 19-3 for an explanation of the apparent magnitude scale.) By comparison, the dimmest star you can see with the naked eye has $m = +6$; this Cepheid in IC 4182 appears less than one one-millionth as bright. The star's spectrum shows that it is a metal-rich Type I Cepheid variable.

According to the period-luminosity relation shown in Figure 21-15, such a Type I Cepheid with a period of 42.0 days has an average luminosity of 33,000 $L_\odot$. This can expressed by saying that this Cepheid has an average absolute magnitude (M) of −6.5. (This compares to $M = +4.8$ for the Sun). Hence, the difference between the Cepheid's apparent and absolute magnitudes, called its distance modulus, is

$$m - M = (+22.0) - (-6.5) = 22.0 + 6.5 = 28.5$$

In Box 19-3, we saw that the distance modulus of a star is related to its distance in parsecs (d) by

$$m - M = 5 \log d - 5$$

This can be rewritten as

$$d = 10^{(m - M + 5)/5} \text{ parsecs}$$

Inserting the value for the distance modulus in this equation, we get the distance to the Cepheid variable and, hence, the distance to the galaxy of which the star is part:

$$d = 10^{(28.5 + 5)/5} \text{ parsecs} = 10^{6.7} \text{ parsecs} = 5 \times 10^6 \text{ parsecs}$$

The galaxy is 5 Mpc (1 Mpc = 1 megaparsec = 10^6 parsecs), or 16 million (1.6×10^7) light-years, from Earth.

Astronomers are interested in IC 4182 because a Type Ia supernova was observed there in 1937. All Type Ia supernovae are exploding white dwarfs that reach nearly the same maximum brightness at the peak of their outburst (see Section 22-9). Once astronomers know the peak absolute magnitude of Type Ia supernovae, they can use these supernovae as distance indicators. Because the distance to IC 4182 is known from its Cepheids, the 1937 observations of the supernova in that galaxy allow us to calibrate Type Ia supernovae as distance indicators.

At maximum brightness, the 1937 supernova reached an apparent magnitude of $m = +8.6$. Since the distance modulus of the galaxy ($m - M$) is 28.5, we see that when a Type Ia supernova is at maximum brightness, its absolute magnitude is

$$M = m - (m - M) = 8.6 - 28.5 = -19.9$$

Whenever astronomers find a Type Ia supernova in a remote galaxy, they can combine this absolute magnitude with the observed maximum apparent magnitude to get the galaxy's distance modulus, from which the galaxy's distance can be easily calculated (just as we did above for the Cepheids in IC 4182). This technique has been used to determine the distances to galaxies hundreds of millions of parsecs away (see Section 26-4).

Cepheid variables are intrinsically quite luminous, with average luminosities that can exceed 10^4 $L_\odot$. Hubble realized that for these luminous stars to appear as dim as they were on his photographs of the Andromeda "Nebula," they must be extremely far away. Straightforward calculations using modern data reveal that M31 is some 750 kiloparsecs (2.5 million light-years) from Earth. Based on its angular size, M31 has a diameter of 70 kiloparsecs—larger than the diameter of our own Milky Way Galaxy!

These results prove that the Andromeda "Nebula" is not a nearby nebula but an enormous stellar system, far beyond the confines of the Milky Way. Today, this system is properly called the Andromeda Galaxy. (Under good observing conditions, you can actually see this galaxy's central bulge with the naked eye. If you could see the entire Andromeda Galaxy, it would cover an area of the sky roughly 5 times as large as the full moon.)

Hubble's results, which were presented at a meeting of the American Astronomical Society on December 30, 1924, settled the Shapley-Curtis "debate" once and for all. The universe was recognized to be far larger and populated with far bigger objects than anyone had seriously imagined. Hubble had discovered the realm of the galaxies.

 In everyday language, many people use the words "galaxy" and "universe" interchangeably. It is true that before Hubble's discoveries our Milky Way Galaxy was thought to constitute essentially the entire universe. But we now know that the universe contains literally billions of galaxies. A single galaxy, vast though it may be, is just a tiny part of the entire observable universe.

26-3 Hubble devised a system for classifying galaxies according to their appearance

Millions of galaxies are visible across every unobscured part of the sky. Although all galaxies are made up of large numbers of stars, they come in a variety of shapes and sizes.

 Hubble classified galaxies into four broad categories based on their appearance. These categories form the basis for the **Hubble classification,** a scheme that is still used today. The four classes of galaxies are the spirals, classified S; barred spirals, or SB; ellipticals, E; and irregulars, Irr. Table 26-1 summarizes some key properties of each class. These various types of galaxies differ not only in their shapes but also in the kinds of processes taking place within them.

 M31 (shown in the image that opens this chapter), M51 (see Figure 26-2), and M100 (see Figure 26-3) are examples of **spiral galaxies.** Figure 26-4 shows that spiral galaxies are characterized by arched lanes of stars, just as is our own Milky Way Galaxy (see Section 25-3). The spiral arms contain young, hot, blue stars and their associated H II regions, indicating ongoing star formation.

Thermonuclear reactions within stars create heavy elements, which are dispersed into space as the stars evolve and die. So, if new stars are being formed from the interstellar matter in spiral galaxies, they will incorporate these heavy elements and be Population I stars (see Section 21-4 and Section 25-2). Indeed, the visible-light spectrum of the disk of a spiral galaxy has strong absorption lines of heavy elements. Such a spectrum is a composite of the spectra of many stars and shows that the stars in the disk are principally of Population I. By contrast, there is relatively little star formation in the central bulges of spiral galaxies, and these regions are dominated by old Population II stars that contain few heavy elements. This also explains why the central bulges of spiral galaxies have a yellowish or reddish color; as a population of stars ages, the massive, luminous blue stars die off first, leaving only the longer-lived, low-mass red stars.

Hubble further classified spiral galaxies according to the texture of the spiral arms and the relative size of the central bulge. Spirals with smooth, broad spiral arms and a fat central bulge are called Sa galaxies, for spiral type *a* (Figure 26-4*a*); those with moderately well-defined spiral arms and a moderate-sized central bulge, like M31 and M51, are Sb galaxies (Figure 26-4*b*); and galaxies with narrow, well-defined spiral arms and a tiny central bulge are Sc galaxies (Figure 26-4*c*).

The differences between Sa, Sb, and Sc galaxies may be related to the relative amounts of gas and dust they contain. Observations with infrared telescopes (which detect the emission from interstellar dust) and radio telescopes (which detect radiation from interstellar gases such as hydrogen and carbon monoxide) show that about 4% of the mass of an Sa galaxy is in the form of gas and dust. This percentage is 8% for Sb galaxies and 25% for Sc galaxies.

Interstellar gas and dust is the material from which new stars are formed, so an Sc galaxy has a greater proportion of its

Table 26-1	Some Properties of Galaxies		
	Spiral (S) and barred spiral (SB) galaxies	**Elliptical galaxies (E)**	**Irregular galaxies (Irr)**
Mass ($M_\odot$)	10^9 to 4×10^{11}	10^5 to 10^{13}	10^8 to 3×10^{10}
Luminosity ($L_\odot$)	10^8 to 2×10^{10}	3×10^5 to 10^{11}	10^7 to 10^9
Diameter (kpc)	5 to 250	1 to 200	1 to 10
Stellar populations	disk: young Population I central bulge and halo: Population II and old Population I	Population II and old Population I	mostly Population I
Percentage of observed galaxies	77%	20%*	3%

This percentage does not include dwarf elliptical galaxies that are as yet too dim and distant to detect. Hence, the actual percentage of galaxies that are ellipticals may be higher than shown here.

a   Sa (NGC 1357)

b   Sb (M81)

c   Sc (NGC 4321)

ƒigure 26-4   R I **V** U X G

Various Spiral Galaxies Edwin Hubble classified spiral galaxies according to the texture of their spiral arms and the relative size of their central bulges. Sa galaxies have smooth, broad spiral arms and the largest central bulges, while Sc galaxies have narrow, well-defined arms and the smallest central bulges. (a: STScI Digital Sky Survey; b: N. A. Sharp/AURA/NOAO/NSF; c: Anglo-Australian Observatory)

mass involved in star formation than an Sb or Sa galaxy. Hence, an Sc galaxy has a large disk (where star formation occurs) and a small central bulge (where there is little or no star formation). By comparison, an Sa galaxy, which has relatively little gas and dust, and thus less material from which to form stars, has a large central bulge and only a small star-forming disk.

In **barred spiral galaxies,** such as those in Figure 26-5, the spiral arms originate at the ends of a bar-shaped region running through the galaxy's nucleus rather than from the nucleus itself. As with ordinary spirals, Hubble subdivided barred spirals according to the relative size of their central bulge and the character of their spiral arms. An SBa galaxy has a large central bulge and thin, tightly wound spiral arms (Figure 26-5a). Likewise, an SBb galaxy is a barred spiral with a moderate central bulge and moderately wound spiral arms (Figure 26-5b), while an SBc galaxy has lumpy, loosely wound spiral arms and a tiny central bulge (Figure 26-5c). As for ordinary spiral galaxies, the difference between SBa, SBb, and SBc galaxies may be related to the amount of gas and dust in the galaxy.

Bars appear to form naturally in many spiral galaxies. This conclusion comes from computer simulations of galaxies, which set hundreds of thousands of simulated "stars" into orbit around a common center. As the "stars" orbit and exert gravitational forces on one another, a bar structure forms in most cases. Indeed, barred spiral galaxies outnumber ordinary spirals by about two to one. (As we saw in Section 25-2, there is evidence that the Milky Way

a   SBa (NGC 4650)

b   SBb (M83)

c   SBc (NGC 1365)

ƒigure 26-5   R I **V** U X G

Various Barred Spiral Galaxies As with spiral galaxies, Edwin Hubble classified barred spirals according to the texture of their spiral arms (which correlates to the sizes of their central bulges). SBa galaxies have the smoothest spiral arms and the largest central bulges, while SBc galaxies have narrow, well-defined arms and the smallest central bulges. (a: AURA/NOAO/NSF; b, c: Anglo-Australian Observatory)

a E0 (M105)

b E3 (NGC 4365)

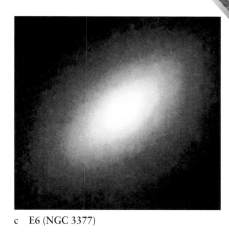

c E6 (NGC 3377)

ƒigure 26-6 R I **V** U X G

Various Elliptical Galaxies Edwin Hubble classified elliptical galaxies according to how round or flattened they look. A galaxy that appears round is labeled E0, and the flattest-appearing elliptical galaxies are designated E7. Three examples are shown here. (Courtesy of J. D. Wray; McDonald Observatory)

Galaxy is a barred spiral. The bar points roughly toward the Earth so that we see it almost end-on.)

Why don't all spiral galaxies have bars? According to calculations by Jeremiah Ostriker and P. J. E. Peebles of Princeton University, a bar will not develop if a galaxy is surrounded by a sufficiently massive halo of nonluminous *dark matter*. (In Section 25-4 we saw evidence that our Milky Way Galaxy is surrounded by such a dark matter halo.) The difference between barred spirals and ordinary spirals may thus lie in the amount of dark matter the galaxy possesses. In Section 26-8 we will see compelling evidence for the existence of dark matter in spiral galaxies.

Elliptical galaxies, so named because of their distinctly elliptical shapes, have no spiral arms. Hubble subdivided these galaxies according to how round or flattened they look. The roundest elliptical galaxies are called E0 galaxies and the flattest, E7 galaxies. Elliptical galaxies with intermediate amounts of flattening are given designations between these extremes (Figure 26-6).

CAUTION! Unlike the designations for spirals and barred spirals, the classifications E0 through E7 may not reflect the true shape of elliptical galaxies. An E1 or E2 galaxy might actually be a very flattened disk of stars that we just happen to view face-on, and a cigar-shaped E7 galaxy might look spherical if seen end-on. The Hubble scheme classifies galaxies entirely by how they *appear* to us on the Earth.

Elliptical galaxies look far less dramatic than their spiral and barred spiral cousins. The reason, as shown by radio and infrared observations, is that ellipticals are virtually devoid of interstellar gas and dust. Consequently, there is little material from which stars could have recently formed, and indeed there is no evidence of young stars in most elliptical galaxies. For the most part, star formation in elliptical galaxies ended long ago. Hence, these galaxies are composed

of old, red, Population II stars with only small amounts of heavy elements.

Elliptical galaxies come in a wide range of sizes and masses. Both the largest and the smallest galaxies in the known universe are elliptical. Figure 26-7 shows two **giant elliptical galaxies** that are about 20 times larger than an average galaxy. These giant ellipticals are located near the middle

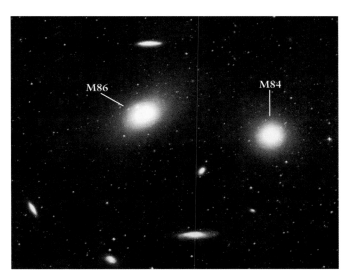

ƒigure 26-7 R I **V** U X G

Giant Elliptical Galaxies The Virgo cluster is a rich, sprawling collection of more than 2000 galaxies about 15 Mpc (50 million light-years) from the Earth. Only the center of this huge cluster appears in this photograph. The two largest members of this cluster are the giant elliptical galaxies M84 and M86. These galaxies have angular sizes of 5 to 7 arcmin. (Royal Observatory, Edinburgh) of a large cluster of galaxies in the constellation Virgo. (We will discuss this cluster in Section 26-6.)

Giant ellipticals are rather rare, but **dwarf elliptical galaxies** are quite common. Dwarf ellipticals are only a fraction the size of their normal counterparts and contain so few stars—only a few million, compared to more than 100 billion (10^{11}) stars in our Milky Way Galaxy—that these galaxies are completely transparent. You can actually see straight through the center of a dwarf galaxy and out the other side, as Figure 26-8 shows.

The visible light from a galaxy is emitted by its stars, so the visible-light spectrum of a galaxy has absorption lines. But because the galaxy's stars are in motion, with some approaching us and others moving away, the Doppler effect smears out and broadens the absorption lines. The average motions of stars in a galaxy can be deduced from the details of this broadening.

For elliptical galaxies, studies of this kind show that star motions are quite random. In a very round (E0) elliptical galaxy, this randomness is **isotropic,** meaning "equal in all directions." Because the stars are whizzing around equally in all directions, the galaxy is genuinely spherical. In a flattened (E7) elliptical galaxy, the randomness of the stellar motions is **anisotropic,** which means that the range of star speeds is different in different directions.

Edwin Hubble summarized his classification scheme for spiral, barred spiral, and elliptical galaxies in a diagram, now called the **tuning fork diagram** for its shape (Figure 26-9). According to this scheme, galaxies midway in appearance between ellipticals and the two kinds of spirals are denoted as S0 and SB0 galaxies, also called **lenticular galaxies.** Although they look somewhat elliptical, lenticular ("lens-shaped") galaxies have both a central bulge and a disk like spiral galaxies, but no discernible spiral arms. They are therefore sometimes referred to as "armless spirals." The giant elliptical galaxies M84 and M86, shown in Figure 26-7, may both actually be lenticular S0 galaxies.

figure 26-8 R I **V** U X G

A Dwarf Elliptical Galaxy This diffuse cloud of stars is a nearby E4 dwarf elliptical called Leo I. It actually orbits the Milky Way at a distance of about 180 kpc (600,000 ly). Leo I is about 1 kpc (3000 ly) in diameter but contains so few stars that you can see through the galaxy's center. (Anglo-Australian Observatory)

CAUTION! When Hubble first drew his tuning fork diagram, he had the idea that it represented an evolutionary sequence. He thought that galaxies evolved over time from the left to the right of the diagram, beginning as ellipticals and eventually becoming either spiral or barred spiral galaxies. We now understand that this is not the case at all! For one thing, elliptical galaxies have little or no overall rotation, while spiral and barred spiral galaxies have a substantial

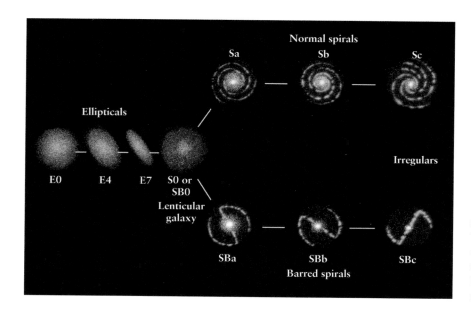

figure 26-9

Hubble's Tuning Fork Diagram Hubble's classification of regular galaxies is shown in his tuning fork diagram. An elliptical galaxy is classified according to how flattened it appears. A spiral or barred spiral galaxy is classified according to the texture of its spiral arms as well as the size of its central bulge. An S0 or SB0 galaxy, also called a lenticular galaxy, is midway in appearance between ellipticals and spirals. Irregular galaxies do not fit into this simple classification scheme.

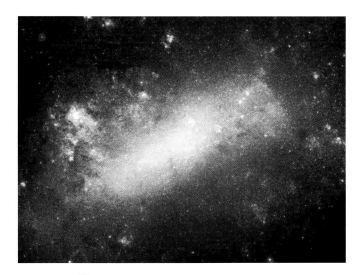

Figure 26-10 R I **V** U X G

The Large Magellanic Cloud (LMC) At a distance of only 50 kpc (160,000 ly), this galaxy is the second closest companion of our Milky Way Galaxy. (The closest is the Sagittarius Dwarf, shown in Figure 26-19*a*.) About 19 kpc (62,000 ly) across, the LMC spans 22° across the sky, or about 50 times the size of the full moon. Note the huge H II region (called the Tarantula Nebula or 30 Doradus) toward the left side of the photograph. Its diameter of 250 pc (800 ly) and mass of 5×10^6 M$_\odot$ make it the largest known H II region. (Anglo-Australian Observatory)

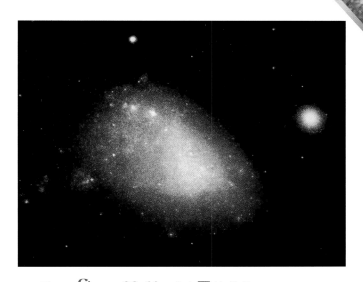

Figure 26-11 R I **V** U X G

The Small Magellanic Cloud (SMC) The SMC is 63 kpc (200,000 ly) away from us and just 8 kpc (26,000 ly) across. Because of its sprawling, asymmetrical shape, the SMC is classified as an irregular galaxy. Note that the SMC is rich in young, blue stars. (Anglo-Australian Observatory)

amount of overall rotation. There is no way that an elliptical galaxy could suddenly start rotating, which means that it could not evolve into a spiral galaxy.

A more modern interpretation is that the Hubble tuning fork diagram is an arrangement of galaxies according to their overall rotation. A rapidly rotating collection of matter in space tends to form a disk, while a slowly rotating collection does not. Thus, the elliptical galaxies at the far left of the tuning fork diagram have little internal rotation and hence no disk. Sa and SBa galaxies have enough overall rotation to form a disk, though their central bulges are still dominant. The galaxies with the greatest amount of overall rotation are Sc and SBc galaxies, in which the central bulges are small and most of the gas, dust, and stars are in the disk.

Galaxies that do not fit into the scheme of spirals, barred spirals, and ellipticals are usually referred to as **irregular galaxies.** Hubble defined two types of irregulars. Irr I galaxies, like underdeveloped spiral galaxies, contain many OB associations and H II regions. Irr II galaxies have asymmetrical, distorted shapes that seem to have been caused by collisions with other galaxies or by violent activity in their nuclei.

The best-known examples of Irr I galaxies are the Large Magellanic Cloud (LMC) and the Small Magellanic Cloud (SMC). Both are nearby companions of our Milky Way and can be seen with the naked eye from southern latitudes. Telescopic views of the Magellanic clouds can easily resolve indi-

vidual stars (Figures 26-10 and 26-11). The SMC does not exhibit any of the geometric symmetry characteristic of spirals or ellipticals, but the LMC does have a faint barlike structure. Both these galaxies contain substantial amounts of interstellar gas. Tidal forces exerted on these irregular galaxies by the Milky Way help to compress the gas, which is why both the LMC and SMC are sites of active star formation.

26-4 Astronomers use various techniques to determine the distances to remote galaxies

A key question that astronomers ask about galaxies is "How far away are they?" Unfortunately, many of the techniques that are used to measure distances within our Milky Way Galaxy cannot be used for the far greater distances to other galaxies. The extremely accurate parallax method (see Section 19-1) can be used only for stars within about 500 pc. Beyond that distance, parallax angles become too small to measure. Spectroscopic parallax, in which the distance to a star or star cluster is found with the help of the H-R diagram (see Section 19-8), is accurate only out to roughly 10 kpc from Earth; more distant stars or clusters are too dim to give reliable results.

To determine the distance to a remote galaxy, astronomers look instead for a **standard candle**—an object, such as a star, that lies within that galaxy and for which we know the luminosity (or, equivalently, the absolute magnitude, described in Section 19-3). By measuring how bright the standard candle appears, astronomers can calculate its distance—and hence the distance to the galaxy of which it is part—using the inverse-square law.

The challenge is to find standard candles that are luminous enough to be seen across the tremendous distances to galaxies. To be useful, standard candles should have four properties:

1. They should be luminous, so we can see them out to great distances.

2. We should be fairly certain about their luminosities, so we can be equally certain of any distance calculated from a standard candle's apparent brightness and luminosity.

3. They should be easily identifiable—for example, by the shape of the light curve of a variable star.

4. They should be relatively common, so that astronomers can use them to determine the distances to many different galaxies.

For nearby galaxies, Cepheid variable stars make reliable standard candles. These variables can be seen out to about 30 Mpc (100 million ly) using the Hubble Space Telescope, and their luminosity can be determined from their period through the period-luminosity relation (Figure 21-15). Box 26-1 gives an example of using Cepheid variables to determine distances. RR Lyrae stars, which are Population II variable stars often found in globular clusters, can be used as standard candles in a similar way. (We saw in Section 25-1 how RR Lyrae variables helped determine the size of our Galaxy.) Because they are less luminous than Cepheids, RR Lyrae variables can be seen only out to 100 kpc (300,000 ly).

Beyond about 30 Mpc even the brightest Cepheid variables, which have luminosities of about $2 \times 10^4\ L_\odot$, fade from view. Astronomers have tried to use even more luminous stars such as blue supergiants to serve as standard candles. However, this idea is based on the assumption that there is a fixed upper limit on the luminosities of stars, which may not be the case. Hence, these standard candles are not very "standard," and distances measured in this way are somewhat uncertain.

One class of standard candles that astronomers have used beyond 30 Mpc are Type Ia supernovae (Section 22-9). These occur when a white dwarf in a close binary system accretes enough matter from its companion to blow itself apart in a thermonuclear conflagration. A Type Ia supernova can reach a maximum luminosity of about $3 \times 10^9\ L_\odot$ (Figure 26-12). If a Type Ia supernova is seen in a distant galaxy and its maximum apparent brightness measured, the inverse-square law can be used to find the galaxy's distance (see Box 26-1).

One complication is that not all Type Ia supernovae are equally luminous. Fortunately, there is a simple relationship between the peak luminosity of a Type Ia supernova and the rate at which the luminosity decreases after the peak: The more slowly the brightness decreases, the more luminous the super-

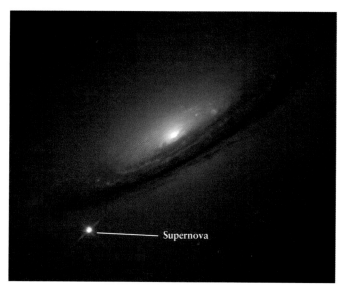

figure 26-12 R I **V** U X G

A Supernova in a Spiral Galaxy This image from the Hubble Space Telescope shows a Type Ia supernova observed in 1994 in the outskirts of spiral galaxy NGC 4526. Note that the supernova is comparable in brightness to the entire galaxy. Such supernovae, which can be seen at extreme distances, are important standard candles used to determine the distances to faraway galaxies. (Harvard-Smithsonian Center for Astrophysics)

nova. Using this relationship, astronomers have measured distances to supernovae more than 1000 Mpc (3 billion ly) from Earth.

Unfortunately, this technique can be used only for galaxies in which we happen to observe a Type Ia supernova. But telescopic surveys now identify many dozens of these supernovae every year, so the number of galaxies whose distances can be measured in this way is continually increasing.

Other methods for determining the distances to galaxies do not make use of standard candles. One was discovered in the 1970s by the astronomers Brent Tully and Richard Fisher. They found that the width of the hydrogen 21-cm emission line of a spiral galaxy (see Section 25-3) is related to the galaxy's luminosity. This correlation is the **Tully-Fisher relation**—the broader the line, the more luminous the galaxy.

Such a relationship exists because radiation from the approaching side of a rotating galaxy is blueshifted while that from the galaxy's receding side is redshifted. Thus, the 21-cm line is Doppler broadened by an amount directly related to how fast a galaxy is rotating. Rotation speed is related to the galaxy's mass by Newton's form of Kepler's third law. The more massive a galaxy, the more stars it contains and thus the brighter it is. Thus, the width of a galaxy's 21-cm line is directly related to its luminosity.

Because line widths can be measured quite accurately, astronomers can use the Tully-Fisher relation to determine the luminosity of a distant spiral galaxy. By combining this information with measurements of apparent brightness, they can calculate the distance to the galaxy. This technique can be used to measure distances of 100 Mpc or more.

Elliptical galaxies do not rotate, so the Tully-Fisher relation cannot be used to determine their distances. But in 1987 the American astronomers Marc Davis and George Djorgovski pointed out a correlation between the size of an elliptical galaxy, the average motions of its stars, and how the galaxy's brightness appears distributed over its surface.

In geometry, three points define a plane, so the relationship among size, motion, and brightness is called the **fundamental plane**. By measuring the last two quantities, an astronomer can use the fundamental plane relationship to determine a galaxy's actual size. And by comparing this to the galaxy's apparent size, the astronomer can calculate the distance to the galaxy using the small-angle formula (Box 1-1). Ellipticals can be substantially larger and more luminous than spirals, so the fundamental plane can be used at somewhat greater distances than the Tully-Fisher relation.

Figure 26-13 shows the ranges of applicability of several important means of determining astronomical distances. Because these ranges overlap, one technique can be used to calibrate another. As an example, astronomers have studied Cepheids in nearby galaxies that have been host to Type Ia supernovae. The Cepheids provide the distances to these nearby galaxies, making it possible to determine the peak luminosity of each supernova using its maximum apparent brightness and the inverse-square law. Once the peak luminosity is known, it can be used to determine the distance to Type Ia supernovae in more distant galaxies. Because one measuring technique leads us to the next one like rungs on a ladder, the techniques shown in Figure 26-13 (along with others) are referred to collectively as the **distance ladder.**

ANALOGY If you give a slight shake to the bottom of a tall ladder, the top can wobble back and forth alarmingly. A change in distance-measuring techniques used for nearby objects can also have substantial effects on the distances to remote galaxies. For instance, if astronomers discovered that the distances to nearby Cepheids were in error, distance measurements using any technique that is calibrated by Cepheids (for example, Type Ia supernovae) would be affected as well. For this reason, astronomers go to great lengths to check the accuracy and reliability of their standard candles.

WEB LINK 26-12 One distance-measuring technique has broken free of the distance ladder by using observations of molecular clouds called **masers.** ("Maser" is an acronym for "microwave amplification by stimulated emission of radiation.") Just as an electric current stimulates a laser to emit an intense beam of visible light, nearby luminous stars can stimulate water molecules in a maser to emit intensely at microwave wavelengths.

During the 1990s, Jim Herrnstein and his collaborators used the Very Long Baseline Array (see Section 6-6) to observe a number of masers orbiting in a disk around the center of the spiral galaxy M106. They determined the orbital speed of the masers from the Doppler shift of masers near the edges of the disk, where they are moving most directly toward or away from Earth. They also measured the apparent change in postion of masers moving across the face of M106. By relating this apparent speed to the true speed determined from the Doppler

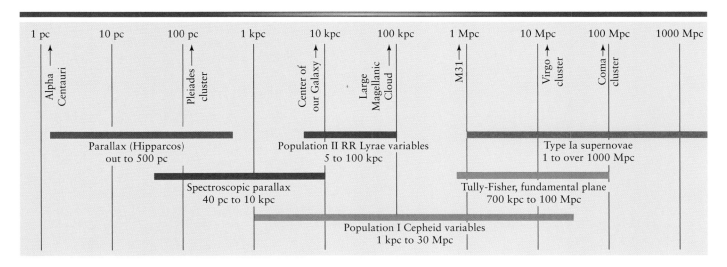

Figure 26-13

The Distance Ladder Astronomers employ a variety of techniques for determining the distances to objects beyond the solar system. Because their ranges of applicability overlap, one technique can be used to calibrate another. The arrows indicate distances to several important objects. Note that each division on the scale indicates a tenfold increase in distance, such as from 1 to 10 Mpc.

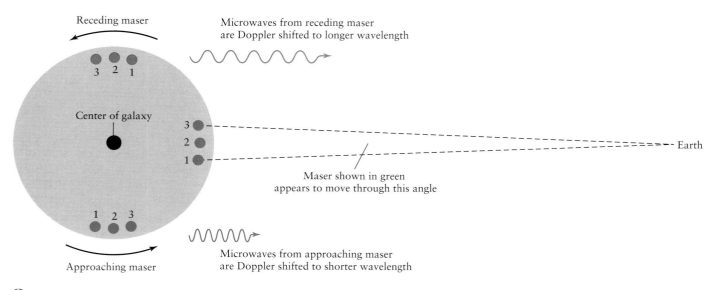

Receding maser

Microwaves from receding maser
are Doppler shifted to longer wavelength

3 2 1

Center of galaxy

3
2
1
- Earth

Maser shown in green
appears to move through this angle

1 2 3

Approaching maser

Microwaves from approaching maser
are Doppler shifted to shorter wavelength

Figure 26-14

Measuring the Distance to a Galaxy Using Masers Masers are interstellar clouds that emit such intense microwave radiation that they can be observed millions of parsecs away. This drawing shows masers (the colored dots) moving from position 1 to 2 to 3 as they orbit the center of a galaxy. The redshift and blueshift of microwaves from the masers shown in red and blue tell us the speed at which the masers orbit the center of their galaxy. This speed tells us how far the masers shown in green move in a certain amount of time. By comparing this with the angle through which these masers appear to move, we can calculate the distance to the galaxy using the small-angle formula.

shift, they calculated that the masers and hence the galaxy are 7.2 Mpc (23 million ly) from Earth (Figure 26-14).

The maser technique is still in its infancy. But because it is independent of all other distance measuring methods, it is likely to play an important role in calibrating the rungs of the distance ladder.

26-5 The Hubble law states that the redshifts of remote galaxies are proportional to their distances from the Earth

Whenever an astronomer finds an object in the sky that can be seen or photographed, the natural inclination is to attach a spectrograph to a telescope and record the spectrum. As long ago as 1914, Vesto M. Slipher, working at the Lowell Observatory in Arizona, began taking spectra of "spiral nebulae." He was surprised to discover that of the 15 spiral nebulae he studied, the spectral lines of 11 were shifted toward the red end of the spectrum, indicating that they were moving away from the Earth.

This marked dominance of redshifts was presented by Curtis in the 1920 Shapley-Curtis "debate" as evidence that

these spiral nebulae could not be ordinary nebulae in our Milky Way Galaxy. It was only later that astronomers realized that the redshifts of spiral nebulae—that is, galaxies—reveal a basic law of our expanding universe.

 During the 1920s, Edwin Hubble and Milton Humason photographed the spectra of many galaxies with the 100-inch (2.5-meter) telescope on Mount Wilson in California. By observing the apparent brightnesses and pulsation periods of Cepheid variables in these galaxies, they were also able to measure the distance to each galaxy (see Section 26-2). Hubble and Humason found that most galaxies show a redshift in their spectrum. They also found a direct correlation between the distance to a galaxy and its redshift: The more distant a galaxy, the greater its redshift and the more rapidly it is receding from us. In other words, nearby galaxies are moving away from us slowly, and more distant galaxies are rushing away from us much more rapidly. Figure 26-15 shows this relationship for five representative elliptical galaxies. This universal recessional movement is referred to as the **Hubble flow.**

Hubble estimated the distances to a number of galaxies and the redshifts of those galaxies. The **redshift**, denoted by the symbol z, is found by taking the wavelength (λ) observed for a given spectral line, subtracting from it the ordinary, unshifted wavelength of that line (λ_0) to get the wavelength difference ($\Delta\lambda$), and then dividing that difference by λ_0:

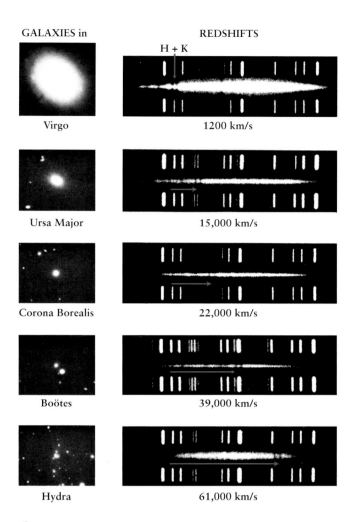

GALAXIES in   REDSHIFTS

H + K

Virgo   1200 km/s

Ursa Major   15,000 km/s

Corona Borealis   22,000 km/s

Boötes   39,000 km/s

Hydra   61,000 km/s

figure 26-15   R I **V** U X G

Five Galaxies and Their Spectra These five elliptical galaxies are all shown at the same magnification. They are arranged, from top to bottom, in order of increasing distance from the Earth. The spectrum of each galaxy is the hazy band between the comparison spectra. (The comparison spectra were produced by a light source at the observatory on Earth.) In each case, the galaxy's spectrum shows the so-called H and K lines of singly ionized calcium. Below each spectrum is the recessional velocity calculated from the Doppler shifts of these lines. Note that the more distant a galaxy is, the greater its redshift. (Carnegie Observatories)

Redshift of a receding object

$$z = \frac{\lambda - \lambda_0}{\lambda_0} = \frac{\Delta\lambda}{\lambda_0}$$

z = redshift of an object

λ_0 = ordinary, unshifted wavelength of a spectral line

λ = wavelength of that spectral line that is actually observed from the object

From the redshifts, Hubble used the Doppler formula to calculate the speed at which these galaxies are receding from us. (See Box 5-6 for an example of how this is done.) Plotting the data on a graph of distance versus speed, Hubble found that the points lie near a straight line. Figure 26-16 is a modern version of Hubble's graph based on recent data.

This relationship between the distances to galaxies and their redshifts is one of the most important astronomical discoveries of the twentieth century. As we will see in Chapter 28, this relationship tells us that we are living in an expanding universe. In 1929, Hubble published this discovery, which is now known as the **Hubble law**. The Hubble law is most easily stated as a formula:

The Hubble law

$$v = H_0 d$$

v = recessional velocity of a galaxy

H_0 = Hubble constant

d = distance to the galaxy

This formula is the equation for the straight line displayed in Figure 26-16, and the **Hubble constant** H_0 is the slope of this straight line. From the data plotted on this graph we find that $H_0 = 65$ km/s/Mpc (say "65 kilometers per second per megaparsec"). In other words, for each million parsecs to a galaxy, the galaxy's speed away from us increases by 65 km/s. For example, a galaxy located 100 million parsecs from the Earth should be rushing away from us with a speed of

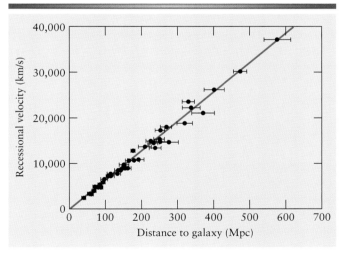

figure 26-16

The Hubble Law This graph plots the distances and recessional velocities of a sample of galaxies. The short horizontal lines from each data point, called error bars, represent uncertainties in the distances (which were determined using Type Ia supernovae). The straight line is the best fit to the data. This linear relationship between distance and recessional velocity is called the Hubble law. (Adapted from Adam Riess)

6500 km/s. (In other books you may see the units of the Hubble constant written with exponents: 65 km s^{-1} Mpc^{-1}.)

 A common misconception about the Hubble law is that *all* galaxies are moving away from the Milky Way. The reality is that galaxies have their own motions relative to one another, thanks to their mutual gravitational attraction. For nearby galaxies, the speed of the Hubble flow is small compared to these intrinsic velocities. Hence, some of the nearest galaxies, including M31 (see Figure 26-3), are actually approaching us and have blueshifts rather than redshifts. But for distant galaxies, the Hubble speed $v = H_0 d$ is much greater than any intrinsic motion that the galaxies might have. Even if the intrinsic velocity of such a distant galaxy is toward the Milky Way, the fast-moving Hubble flow sweeps that galaxy away from us.

The exact value of the Hubble constant has been a topic of heated debate among astronomers for several decades. The problem is that while redshifts are relatively easy to measure in a reliable way, distances to galaxies (especially remote galaxies) are not, as we saw in Section 26-4. Hence, astronomers who use different methods of determining galactic distances have obtained different values of H_0. To see why this is so, it is helpful to rewrite the Hubble law as

$$H_0 = \frac{v}{d}$$

This shows that if galaxies of a given recessional velocity (v) are far away (so d is large), the Hubble constant H_0 must be relatively small. But if these galaxies are relatively close (so d is small), then H_0 must have a larger value.

In the past, astronomers who used Type Ia supernovae for determining galactic distances found galaxies to be farther away than their colleagues who employed the Tully-Fisher relation. Therefore, the supernova adherents found values of H_0 in the range from about 40 to 65 km/s/Mpc, while the values from the Tully-Fisher relation ranged from about 80 to 100 km/s/Mpc. (The data shown in Figure 26-16, which indicate that $H_0 = 65$ km/s/Mpc, are from observations of Type Ia supernovae.)

In the past few years, the Hubble Space Telescope (HST) has been used to observe Cepheid variables with unprecedented precision and in more distant galaxies than ever before. These observations suggest a value of the Hubble constant of about 72 km/s/Mpc, with an uncertainty of about 10%. At the same time, reanalysis of the supernova and Tully-Fisher results have brought the values of H_0 from these techniques closer to the HST Cepheid value. Although a definite value of the Hubble constant has yet to be agreed upon by all researchers, we will use a compromise value of $H_0 = 70$ km/s/Mpc in this book.

The value of H_0 is a subject of heated debate for a very simple reason: The Hubble constant is one of the most important numbers in all astronomy. It expresses the rate at which the universe is expanding and, as we will see in Chapter 28, even gives the age of the universe. Furthermore, the Hubble law can be used to determine the distances to extremely remote galaxies and quasars. If the redshift of a galaxy is known, the Hubble law can be used to convert that redshift into a distance from the Earth, as Box 26-2 shows. Thus, the value of the Hubble constant helps determine the distances of the most remote objects in the universe that astronomers can observe.

26-6 Galaxies are grouped into clusters and superclusters

Galaxies are not scattered randomly throughout the universe but are found in **clusters.** Figure 26-17 shows a typical cluster. Like stars within a star cluster, the members of a cluster of galaxies are in continual motion around one another. They appear to be at rest only because they are so distant from us.

A cluster is said to be either **poor** or **rich,** depending on how many galaxies it contains. Poor clusters, which far outnumber rich ones, are often called **groups.** For example, the Milky Way Galaxy, the Andromeda Galaxy, and the Large and Small Magellanic Clouds belong to a poor cluster familiarly

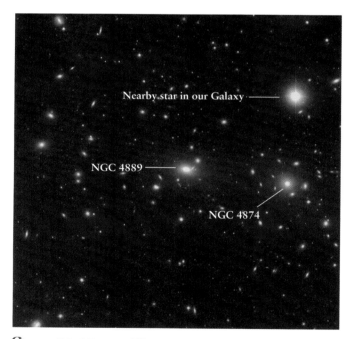

Figure 26-17   R I **V** U X G

The Coma Cluster of Galaxies This rich, regular cluster is about 90 Mpc (300 million light-years) from the Earth. Almost all of the spots of light in this image are individual galaxies of the cluster, which may total 10,000 or more. Regular clusters such as the Coma cluster are composed mostly of elliptical and lenticular galaxies. Two giant elliptical galaxies, NGC 4889 and NGC 4874, dominate the center of the cluster (see also Figure 26-7). The bright star at the upper right is within our own Milky Way Galaxy, a million times closer than any of the galaxies shown here. (O. Lopez-Cruz and I. K. Shelton, University of Toronto; KPNO) box 26-2   Tools of the Astronomer's Trade

The Hubble Law as a Distance Indicator

Suppose that you aim a telescope at an extremely distant galaxy. You take a spectrum of the galaxy and find that the spectral lines are shifted toward the red end of the spectrum. For example, you find that a spectral line whose normal wavelength is λ_0 appears in the galaxy's spectrum at a longer wavelength λ. The spectral line has thus been shifted by an amount $\Delta\lambda = \lambda - \lambda_0$.

The redshift of the galaxy, z, is given by

$$z = \frac{\lambda - \lambda_0}{\lambda_0} = \frac{\Delta\lambda}{\lambda_0}$$

According to the Doppler effect (discussed in Section 5-9 and Box 5-6), the ratio $\Delta\lambda/\lambda_0$ is equal to v/c, the ratio of the recessional velocity (v) to the speed of light (c). Hence, the redshift is also equal to v/c:

$$z = \frac{v}{c}$$

This equation is valid only if the velocity v is much less than the speed of light.

According to the Hubble law, the recessional velocity v of a galaxy is related to its distance d from the Earth by

$$v = H_0 d$$

where H_0 is the Hubble constant. Combining the equation for the Doppler shift with the equation for the Hubble law, the distance to a galaxy is related to its redshift by

$$d = \frac{zc}{H_0}$$

EXAMPLE: When measured in a laboratory on Earth, the so-called K line of singly ionized calcium has a wavelength $\lambda_0 = 393.3$ nm. But when you observe the spectrum of the giant elliptical galaxy NGC 4889, you find this spectral line at $\lambda = 401.8$ nm. Thus, the redshift of this galaxy is

$$z = \frac{401.8 \text{ nm} - 393.3 \text{ nm}}{393.3 \text{ nm}} = 0.0216$$

NGC 4889 is therefore moving away from us with a speed of

$$v = zc = (0.0216)(3 \times 10^5 \text{ km/s}) = 6500 \text{ km/s}$$

With $H_0 = 70$ km/s/Mpc, the Hubble law gives the distance to the galaxy:

$$d = \frac{zc}{H_0} = \frac{6500 \text{ km/s}}{70 \text{ km/s/Mpc}} = 93 \text{ Mpc} = 300 \text{ million ly}$$

Note that if you had used $H_0 = 65$ km/s/Mpc, you would have concluded that the distance is

$$d = \frac{zc}{H_0} = \frac{6500 \text{ km/s}}{65 \text{ km/s/Mpc}} = 100 \text{ Mpc} = 330 \text{ million ly}$$

Unfortunately, astronomers must deal with this kind of uncertainty. At present, the value of H_0 is thought to lie in the range between 60 and 90 km/s/Mpc. Thus, the distance to NGC 4889 is somewhere between 72 and 108 megaparsecs.

---

known as the **Local Group.** The Local Group contains about 30 galaxies, most of which are dwarf ellipticals. Figure 26-18 is a map of a portion of the Local Group.

In recent years astronomers have discovered several previously unknown dwarf elliptical galaxies in the Local Group. One of these is the Sagittarius Dwarf, first observed in 1994, which is actually the galaxy nearest to the Milky Way (Figure 26-19*a*). Another is the galaxy Antlia, named for the southern constellation in which it resides (Antlia, the Air Pump) and discovered in 1997 (Figure 26-19*b*).

We may never know the total number of galaxies in the Local Group, because dust in the plane of the Milky Way obscures our view over a considerable region of the sky. Nevertheless, we can be certain that no additional large spiral galaxies are hidden by the Milky Way, because radio astronomers would have detected 21-cm radiation from them, even though their visible light is completely absorbed by interstellar dust.

The nearest fairly rich cluster is the Virgo cluster, a collection of more than 2000 galaxies covering a 10° × 12° area of the sky. Measurements of Cepheid variables in the spiral galaxy M100, a member of the Virgo cluster, give a distance of about 15 Mpc (50 million ly); the Tully-Fisher relation and observations of Type Ia supernovae give similar distances to this cluster. The overall diameter of the cluster is about 3 Mpc (9 million ly).

The center of the Virgo cluster is dominated by three giant elliptical galaxies, two of which can be seen in Figure 26-7. The diameter of each of these enormous galaxies is comparable to the 750-kpc distance between the Milky Way and M31.

figure 26-18
The Local Group Our galaxy is a member of a poor, irregularly shaped cluster of galaxies called the Local Group. This map shows how the 21 nearest members of the Local Group are distributed in the space around our own Galaxy. (The galaxies are actually distributed in three-dimensional space. In this figure they have been drawn in a plane to represent their distances from the Milky Way.) The Andromeda Galaxy (M31, shown in the photograph that opens this chapter) is the largest and most massive galaxy in the Local Group; the Milky Way takes second place. Most of the members of the Local Group are dwarf ellipticals scattered around these two large spiral galaxies. One of these, the Sagittarius Dwarf (see Figure 26-19a), is the Milky Way's nearest neighbor.

In other words, one giant elliptical is approximately the same size as the entire Local Group!

Astronomers also categorize clusters of galaxies as regular or irregular, depending on the overall shape of the cluster. The Virgo cluster, for example, is called an **irregular cluster,** because its galaxies are scattered throughout a sprawling region of the sky. Our own Local Group is also an irregular cluster. In contrast, a **regular cluster** has a distinctly spherical appearance, with a marked concentration of galaxies at its center.

The nearest example of a rich, regular cluster is the Coma cluster, located about 90 Mpc (300 million light-years) from us toward the constellation of Coma Berenices (Berenice's Hair).

a

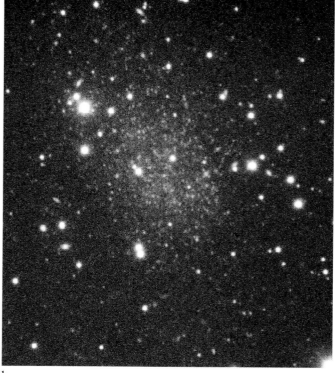

b

figure 26-19 R I **V** U X G
The Nearest and Newest Members of the Local Group (a) The Sagittarius Dwarf is a dwarf elliptical galaxy that lies some 25 kpc (80,000 ly) from the Milky Way. (The curved lines in this visible-light image outline areas of equal brightness and emphasize the structure of the galaxy.) The Sagittarius Dwarf is the closest galaxy to the Milky Way. (b) Antlia is the most recently discovered member of the Local Group. First detected in 1997, it lies about 1 Mpc (3 million ly) away, outside the region depicted in Figure 26-18. Antlia contains only about 10^6 stars within a diameter of 1 to 2 kpc. (a: R. A. Ibata, University of British Columbia, and G. Gilmore and M. J. Irwin, Royal Greenwich Observatory; b: M. J. Irwin, Royal Greenwich Observatory, and A. B. Whiting and G. K. T. Hau, Institute of Astronomy, Cambridge University)

Despite its great distance, telescopic images of this cluster show more than 1000 galaxies (see Figure 26-17). The Coma cluster almost certainly contains many thousands of dwarf ellipticals, so the total membership of the cluster may be as many as 10,000 galaxies. The core of the Coma cluster is dominated by two giant ellipticals surrounded by many normal-sized galaxies.

The overall shape of a cluster is related to the dominant types of galaxies it contains. Rich, regular clusters contain mostly elliptical and lenticular galaxies. For example, about 80% of the brightest galaxies in the Coma cluster are ellipticals; only a few spiral galaxies are scattered around the cluster's outer regions. Irregular clusters, such as the Virgo cluster and the Hercules cluster (Figure 26-20), have a more even mixture of galaxy types.

Clusters of galaxies are themselves grouped together in huge associations called **superclusters.** A typical supercluster contains dozens of individual clusters spread over a region of space up to 30 Mpc (100 million ly) across. In maps of the sky such as the one in Figure 26-21, which covers many hundreds of square degrees as seen from Earth, you can see the delicate, filamentary pattern of superclusters spread across the sky.

The arrangement of superclusters in space was clarified in the early 1980s, when several teams of astronomers began the monumental task of mapping the locations of galaxies in space. They used redshifts of galaxies in unobscured regions of the sky to construct maps that display the actual distribution of the galaxies in three dimensions. Figure 26-22 shows an example of such a map. This particular map encompasses two wedge-shaped slices of the universe, one in the northern half and one in the southern half of the celestial sphere. The Earth is at the apex of the wedge-shaped map; each dot represents a galaxy. The measurements used to create this map included galaxies with redshifts as large as $z = 0.04$, corresponding to a recessional velocity of 12,000 km/s. If the Hubble constant is 70 km/s/Mpc, this means that the map in Figure 26-22 extends out to a distance of 170 Mpc (550 million light-years) from the Earth.

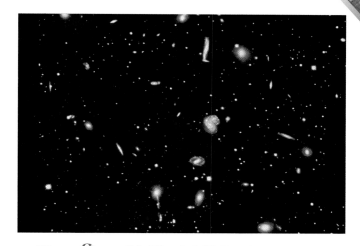

figure 26-20   R I **V** U X G

The Hercules Cluster   This irregular cluster, which is about 200 Mpc (650 million ly) from Earth, is much less dense than the Coma cluster. The Hercules cluster contains many spiral galaxies, often associated in pairs and small groups. (NOAO)

Maps such as Figure 26-22 reveal enormous **voids** where exceptionally few galaxies are found. These voids are roughly spherical and measure 30 to 120 Mpc (100 million to 400 million ly) in diameter. They may not all be entirely empty, however. There is evidence for hydrogen clouds in some voids, while others may be subdivided by strings of dim galaxies.

Recent observations show that most galaxies are concentrated in sheets on the surfaces between voids. This pattern is similar to that of soapsuds in a kitchen sink, with sheets of soap film surrounding air bubbles (analogous to voids). As we will see in Chapter 29, many astronomers suspect that this

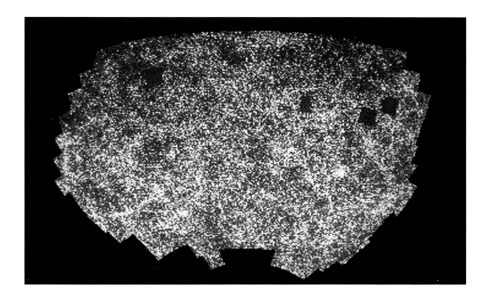

figure 26-21

Two Million Galaxies   This map shows the distribution of roughly 2 million galaxies over about 10% of the sky. Each dot on the map is shaded according to the number of galaxies it contains. White dots indicate more than 20 galaxies and blue dots between 1 and 19 galaxies. Black dots contain no galaxies. The small, bright patches are individual clusters of galaxies. Note that the clusters of galaxies are not smoothly distributed across the sky; instead, they form a lacy, filamentary structure. The larger, elongated bright areas are superclusters and filaments, which generally surround darker voids containing few galaxies. (S. J. Maddox, W. J. Sutherland, G. P. Efstathiou, and J. Loveday, Oxford University)

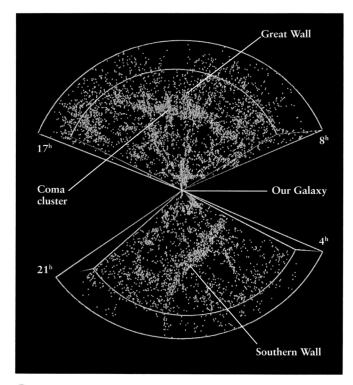

ƒigure 26-22

The Large-Scale Distribution of Galaxies This map shows the distribution of 9325 galaxies in two wedges extending out to a redshift of $z = 0.04$. Our Milky Way is located at the apex of the wedge-shaped slices. If you squint while looking at this map, you can see the Great Wall in the upper wedge and the Southern Wall in the bottom wedge. The Coma cluster (see Figure 26-17) forms part of the Great Wall. Note the prominent voids along the inner edge of the Great Wall and on either side of the Southern Wall. (Courtesy of M. J. Geller, Harvard-Smithsonian Center for Astrophysics)

pattern contains important clues about how clusters of galaxies formed in the early universe.

One notable sheet of galaxies, which appears as a ribbon in the northern part of the map in Figure 26-22, has been dubbed the Great Wall. The arc of this "wall of galaxies" runs for some 150 Mpc (500 million ly) across the map. The Great Wall also extends at least 5 Mpc (15 million ly) perpendicular to the plane of Figure 26-22. A similar structure called the Southern Wall, visible in the southern part of the map, contains more than a thousand galaxies and extends across approximately 100 Mpc.

The Great Wall and the Southern Wall are among the largest structures known in the universe. On scales much larger than 100 Mpc, the distribution of galaxies in the universe appears to be roughly uniform.

 Astrophysicists are working to explain how these enormous structures came into existence. Ongoing redshift surveys, which are exploring ever larger volumes of the universe, may well uncover even more surprises.

26-7 Colliding galaxies produce starbursts, spiral arms, and other spectacular phenomena

Occasionally, two galaxies within a cluster or from adjacent clusters can collide with each other. Past collisions have hurled vast numbers of stars into intergalactic space. In some cases, we can even observe a collision in progress, a cosmic catastrophe that gives birth to new stars. And astronomers can predict collisions that will not take place for billions of years, such as the collision that is fated to occur between the galaxy M31 and our own Milky Way Galaxy.

When two galaxies collide at high speed, the huge clouds of interstellar gas and dust in the galaxies slam into each other and can be completely stopped in their tracks. In this way, two colliding galaxies can be stripped of their interstellar gas and dust.

The best evidence that such collisions take place is that many rich clusters of galaxies are strong sources of X rays (Figure 26-23). This emission reveals the presence of substantial amounts of hot **intracluster gas** (that is, gas within the cluster) at temperatures between 10^7 and 10^8 K. The only way that such large amounts of gas could be heated to such extremely high temperatures is in violent collisions between galaxies.

 ƒigure 26-23   R I V U Ⓧ G

X-Ray Emission from a Cluster of Galaxies This false-color view from the orbiting Chandra X-ray Observatory shows the rich cluster Abell 2142, one of many such clusters cataloged by the UCLA astronomer George O. Abell. The X-ray emission shown here comes from gas between the galaxies with temperatures from 5×10^7 K to 10^8 K. The gas may have been raised to such high temperatures when two smaller clusters collided to form Abell 2142. The galaxies themselves are too dim at X-ray wavelengths to appear in this image. (NASA/CXC/SAO)

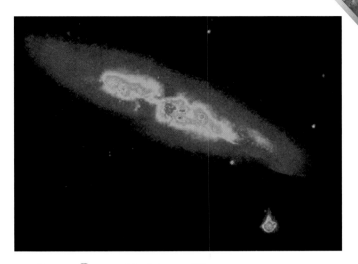

Although galaxies can and do collide, it is highly unlikely that the *stars* from two colliding galaxies actually run into each other. The reason is that the stars within a galaxy are very widely separated from one another, with a tremendous amount of space between them.

In a less violent collision or a near-miss between two galaxies, the compressed interstellar gas may have more time to cool, allowing many protostars to form. Such collisions may account for **starburst galaxies,** which blaze with the light of numerous newborn stars. These galaxies have bright centers surrounded by clouds of warm interstellar dust, indicating recent, vigorous star birth. Because the warm dust is so abundant, starburst galaxies are among the most luminous objects in the universe at infrared wavelengths (Figure 26-24).

The starburst galaxy M82 shown in Figure 26-24 is one member of a nearby cluster that includes the beautiful spiral galaxy M81 and a fainter elliptical companion called NGC 3077 (Figure 26-25a). Radio surveys of that region of the sky reveal enormous streams of hydrogen gas connecting the three galaxies (Figure 26-25b). The loops and twists in these streamers suggest that the three galaxies have had several close encounters over the ages. A similar stream of hydrogen gas connects our Galaxy with its second nearest neighbor, the Large Magellanic Cloud (LMC),

figure 26-24   R **I** V U X G

A Starburst Galaxy Prolific star formation is occurring at the center of this irregular galaxy, called M82, causing it to glow brightly in this false-color infrared image. This activity was probably triggered by gravitational interactions with neighboring galaxies. Figure 24-13 shows this same galaxy at visible wavelengths. (SAO/Harvard Smithsonian Center for Astrophysics)

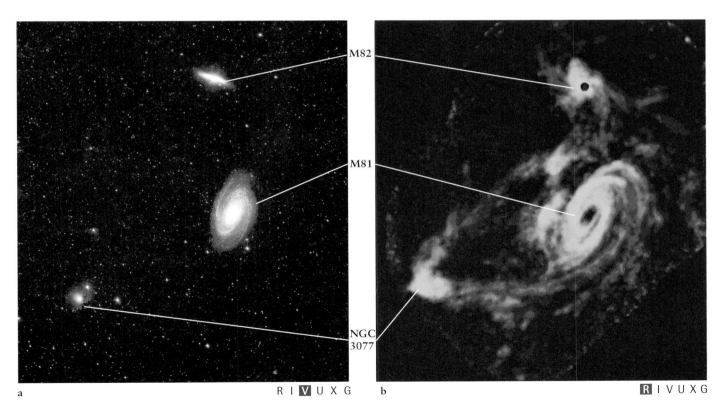

a   R I **V** U X G

b   **R** I V U X G

figure 26-25

The M81 Group The starburst galaxy M82 (see Figure 26-24) is in a cluster of about a dozen galaxies some 3.5 Mpc (12 million ly) from Earth. **(a)** This wide-angle visible-light photograph shows the three brightest galaxies of the cluster. The area shown is about 1° across.

(b) This false-color radio image of the same region, crated from data taken by the Very Large Array, shows streamers of hydrogen gas that connect the three bright galaxies as well as several dim ones. (Palomar Sky Survey; M. S. Yun, VLA and Harvard)

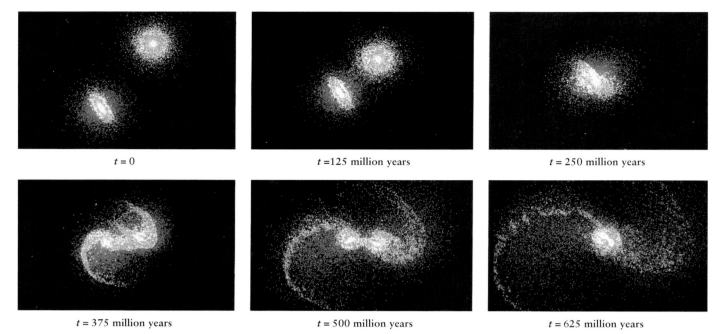

$t = 0$ $t = 125$ million years $t = 250$ million years

$t = 375$ million years $t = 500$ million years $t = 625$ million years

figure 26-26

A Simulated Collision Between Two Galaxies These frames from a supercomputer simulation show the collision and merger of two galaxies accompanied by an ejection of stars into intergalactic space. Stars in the disk of each galaxy are colored blue, while stars in their central bulges are yellow. Red indicates dark matter that surrounds each galaxy. The frames progress at 125-million-year intervals. Compare the bottom frames with Figure 26-27. (Courtesy of J. Barnes, University of Hawaii)

suggesting a history of close encounters between our Galaxy and the LMC.

Tidal forces between colliding galaxies can deform the galaxies from their original shapes, just as the tidal forces of the Moon on the Earth deform the oceans and help give rise to the tides (see Section 4-8, especially Figure 4-26). The galactic deformation is so great that thousands of stars can be hurled into intergalactic space along huge, arching streams. Supercomputer simulations of such collisions show that while some of the stars are flung far and wide, other stars slow down and the galaxies may merge.

Figure 26-26 shows one such simulation, by Joshua Barnes, now at the Institute for Astronomy at the University of Hawaii. As the two galaxies pass through each other, they are severely distorted by gravitational interactions and throw out a pair of extended tails. The interaction also prevents the galaxies from continuing on their original paths. Instead, they fall back together for a second encounter (at 625 million years), when they look remarkably similar to the actual colliding galaxies shown in Figure 26-27. The simulated galaxies merge soon thereafter, leaving a single object.

Our own Milky Way Galaxy is expected to undergo a galactic collision like that shown in Figure 26-27. The Milky Way and the Andromeda Galaxy, shown in the image that opens this chapter, are actu-

ally approaching each other and should collide in another 6 billion years or so. (Recall that our solar system is only 4.6 billion years old.) When this happens, the sky will light up with a plethora of newly formed stars, followed in rapid succession by a string of supernovae, as the most massive of these stars complete their life spans. Any inhabitants of either galaxy will see a night sky far more dramatic and tempestuous than ours.

When two galaxies merge, the result is a bigger galaxy. If this new galaxy is located in a rich cluster, it may capture and devour additional galaxies, growing to enormous dimensions by **galactic cannibalism.** Cannibalism differs from mergers in that the galaxy that does the devouring is bigger than its "meal," whereas merging galaxies are about the same size.

Many astronomers suspect that galactic cannibalism is the reason that giant ellipticals are so huge. As we have seen, giant galaxies typically occupy the centers of rich clusters. In many cases, smaller galaxies are located around these giants. As they pass through the extended halo of a giant elliptical, these smaller galaxies slow down and are eventually devoured by the larger galaxy.

Close encounters between galaxies provide a third way of forming spiral arms (in addition to density waves and self-propagating star formation, discussed in Section 25-5). Supercomputer simulations clearly

Figure 26-27   R I **V** U X G

Colliding Galaxies with "Antennae" Pairs of colliding galaxies often exhibit long tendrils of stars ejected by the collision. The left-hand image shows these for the colliding spiral galaxies NGC 4038 and NGC 4039, which are located about 20 Mpc (63 million ly) from Earth in the southern constellation Corvus (the Crow). **Inset:** The collision of the two galaxies has compressed vast amounts of interstellar gas and dust, triggering a firestorm of star formation. This is shown by the numerous red H II regions and blue clusters of hot, young stars. (Brad Whitmore, Space Telescope Science Institute; NASA)

demonstrate that spiral arms can be created during a collision, either by drawing out long streamers of stars or by compressing clouds of interstellar gas. For example, the spiral arms of M51 (examine Figure 26-2) may have been produced by a close encounter with a second galaxy. The disruptive galaxy is now located at the end of one of the spiral arms created by the collision.

Some astronomers argue that the spiral arms of our Milky Way Galaxy were similarly produced by a close encounter with the Large Magellanic Cloud. If this is so, then the chain of events that led to the formation of our Sun, our solar system, and ourselves was initiated by this long-ago interaction between two galaxies.

26-8   Most of the matter in the universe has yet to be discovered

A cluster of galaxies must be held together by gravity. In other words, there must be enough matter in the cluster to prevent the galaxies from wandering away. Nevertheless, careful examination of a rich cluster, like the Coma cluster, reveals that the mass of the visually luminous matter (principally the stars in the galaxies) is not at all sufficient to bind the cluster gravitationally. The observed line-of-sight speeds of the galax-

ies, measured by Doppler shifts, are so large that the cluster should have broken apart long ago. Considerably more mass than is visible is needed to keep the galaxies bound in orbit about the center of the cluster.

We encountered a similar situation in studying our own Milky Way Galaxy in Section 25-4: The total mass of our Galaxy is more than the amount of visible mass. As for our Galaxy, we conclude that clusters of galaxies must contain significant amounts of nonluminous *dark matter*. If this dark matter were not there, the galaxies would have long ago dispersed in random directions and the cluster would no longer exist today. Analyses demonstrate that the total mass needed to bind a typical rich cluster is about 10 times greater than the mass of material that shows up on visible-light images.

As for our Galaxy, the problem is to determine what form the invisible mass takes. A partial solution to this **dark-matter problem,** which dates from the 1930s, was provided by the discovery in the late 1970s of hot, X-ray-emitting gas within clusters of galaxies (see Figure 26-23). By measuring the amount of X-ray emission, astronomers find that the total mass of intracluster gas in a typical rich cluster can be greater than the combined mass of all the stars in all the cluster's galaxies. This is sufficient to account for only about 10% of the invisible mass, however. The remainder is dark matter of unknown composition.

Although we do not know what dark matter is made of, it is possible to investigate how dark matter is distributed in galaxies and clusters of galaxies. It appears that dark matter lies within and immediately surrounding galaxies, not in the vast spaces between galaxies. The evidence for this comes principally from observations of the rotation curves of galaxies and of the gravitational bending of light by clusters of galaxies.

As we saw in Section 25-4, a rotation curve is a graph that shows how fast stars in a galaxy are moving at different distances from that galaxy's center. For example, Figure 25-17 is the rotation curve for our Galaxy. Many other galaxies have similar rotation curves, as Figure 26-28 shows. These curves for four spiral galaxies all remain remarkably flat out to surprisingly great distances from each galaxy's center. In other words, the orbital speed of the stars remains roughly constant out to the visible edges of these galaxies.

This observation tells us that we still have not detected the *true* edges of these galaxies (and many similar ones). Near the true edge of a galaxy we should see a decline in orbital speed, in accordance with Kepler's third law (see Figure 25-17). Because this decline has not been observed, astronomers conclude that there must be a considerable amount of dark material that extends well beyond the visible portion of the disk.

Further evidence about how dark matter is distributed comes from the gravitational bending of light rays, which we described in Section 24-2. As Figure 24-5 shows, the gravity of a single star like the Sun can deflect light by only a few arcseconds. But a more massive object such as a galaxy can produce much greater deflections, and the amount of this deflection can be used to determine the galaxy's mass. For example, suppose that the Earth, a massive galaxy, and a background light source (such as a more distant galaxy) are in

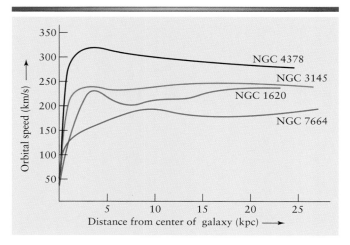

The Rotation Curves of Four Spiral Galaxies This graph shows how the orbital speed of material in the disks of four spiral galaxies varies with the distance from the center of each galaxy. If most of each galaxy's mass were concentrated near its center, these curves would fall off at large distances. But these and many other galaxies have flat rotation curves that do not fall off. This indicates the presence of extended halos of dark matter. (Adapted from V. Rubin and K. Ford)

nearly perfect alignment, as sketched in Figure 26-29. Because of the warped space around the massive galaxy, light from the background source curves around the galaxy as it heads toward us. As a result, light rays can travel along two paths from the background source to us here on the Earth. Thus, we should see two images of the background source.

A powerful source of gravity that distorts background images is called a **gravitational lens.** For gravitational lensing to work, the alignment between the Earth, the massive galaxy, and a remote background light source must be almost perfect. Without nearly perfect alignment, the second image of the background star is too faint to be noticeable.

A number of gravitational lenses have been discovered, involving images of remote, luminous objects called *quasars.* As we will see in Chapter 27, typical quasars shine as brightly as hundreds of galaxies and can be detected billions of parsecs from Earth. Several thousand quasars have been discovered across the sky.

In 1979, Dennis Walsh, Robert Carswell, and Ray Weymann discovered two faint quasars separated by only 6 arcsec. They took spectra of both quasars and discovered that they were nearly identical. They thus concluded that they were looking at two images of the same quasar. Further observations revealed a galaxy between the quasar images (Figure 26-30). The gravitational field of this galaxy bends the light from the remote quasar and thus acts like a gravitational lens.

The Hubble Space Telescope (HST) has found many cases in which galaxies and clusters of galaxies act as gravitational lenses. Figure 26-31 is an HST image of an ordinary-looking rich cluster of yellowish elliptical and spiral galaxies—but with a number of curious blue arcs. Reconstruction of the light paths through the cluster show that all these blue arcs are actually distorted images of a single galaxy that lies billions of light-years beyond the cluster.

By measuring the distortion of the images of such background galaxies, J. Anthony Tyson of Bell Laboratories and his colleagues have determined that dark matter, which constitutes about 90% of the cluster's mass, is distributed much like the visible matter in the cluster. In other words, the overall arrangement of visible galaxies seems to trace the location of dark matter.

Many proposals have been made to explain the nature of dark matter. One reasonable suggestion was that clusters might contain a large number of faint, red, low-mass (0.2 $M_\odot$ or less) stars. These faint stars could be located in extended halos surrounding individual galaxies or scattered throughout the spaces between the galaxies of a cluster. They would have escaped detection because their luminosity and hence apparent brightness would be very low. (Recall from Section 19-9 the mass-luminosity relation for main-sequence stars.) Searches for these stars around other galaxies as well as around the Milky Way have been carried out using the Hubble Space Telescope. None have yet been detected, so it is

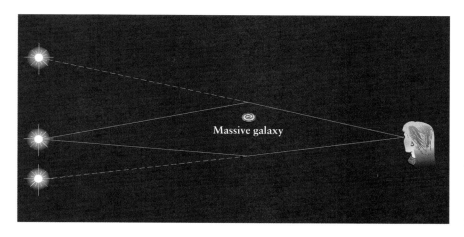

figure 26-29

A Gravitational Lens A massive object such as a galaxy can deflect light rays from a distant light source so that an observer sees more than one image of the distant source. (Compare with Figure 25-18, which shows the same effect on a much smaller scale.) Depending on the alignment of observer, massive galaxy, and distant source, the observer can see several images of the light source, as in Figure 26-31.

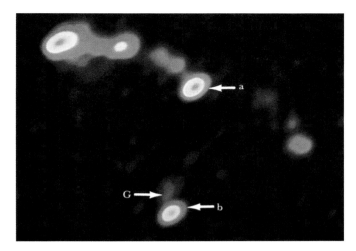

figure 26-30   R I V U X G

A "Double" Quasar This radio image from the Very Large Array shows two images of the same quasar, labeled **a** and **b.** Light from the distant quasar is deflected to either side of a massive galaxy located between us and the quasar. A faint image of the deflecting galaxy **G** is seen directly above image **b.** The jetlike feature protruding to the left and right of the upper image **a** does not appear alongside the lower image because the jet is too far away from the required quasar-galaxy-Earth alignment. (VLA, NRAO)

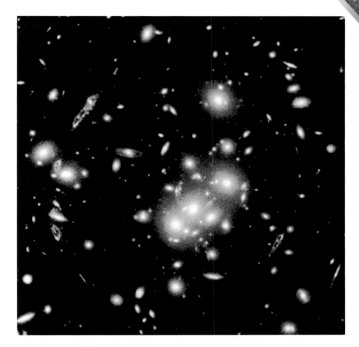

figure 26-31   R I V U X G

Gravitational Lensing by a Cluster of Galaxies
The Hubble Space Telescope image of the rich cluster 0024 + 1654 records not only the cluster's galaxies, which appear as yellowish blobs, but also a number of blue arcs. These blue arcs are distorted multiple images of a single more distant galaxy, the result of gravitational lensing by the matter in 0024 + 1654. The cluster is about 1600 Mpc (5 billion ly) from Earth; the blue galaxy is about twice as distant. The blue color of the remote galaxy suggests that it is very young and is actively forming stars. (W. N. Colley and E. Turner, Princeton University; J. A. Tyson, Bell Labs, Lucent Technologies; NASA)

thought that faint stars are unlikely to constitute the majority of the dark matter in the universe.

As we described in Section 25-4, more exotic dark matter candidates include massive neutrinos, subatomic particles called WIMPs (weakly interacting massive particles), and MACHOs (massive compact halo objects, such as small black holes or brown dwarfs). To date, however, the true nature of dark matter remains unknown.

26-9   Galaxies may have formed from the merger of smaller objects

How do galaxies form and how do they evolve? Why are some spiral, others elliptical, and still others irregular? Astronomers can gain important clues about galactic evolution simply by looking deep into space. The more distant a galaxy is, the longer its light takes to reach us. As we examine galaxies that are at increasing distances from the Earth, we are actually looking further and further back in time. By looking into the past, we can see galaxies in their earliest stages.

By observing remote galaxies, astronomers have discovered that galaxies were bluer in the past than they are today. This trait was noticed in 1978 by the astronomers Harvey Butcher and Augustus Oemler, who drew attention to large numbers of blue galaxies in the remote, rich clusters they were studying. Spectroscopic studies of such galaxies in the 1980s by James Gunn and Alan Dressler demonstrated that most owe their

blue color to vigorous star formation, often occurring in intense, episodic bursts. The hot, luminous, and short-lived O and B stars produced in these bursts of star formation give blue galaxies their characteristic color. (The arcs in Figure 26-31 are gravitational-lens images of such a young, distant, blue galaxy.)

In 1992, Dressler, Oemler, Gunn, and Butcher teamed up to use the Hubble Space Telescope to examine two remote, rich clusters of galaxies with redshifts $z = 0.4$. This redshift corresponds to looking about 4 billion years into the past. Careful examination of the HST images revealed a surprisingly large percentage of spiral galaxies, about 30%. (Only 5% of the galaxies in nearby rich clusters are spirals.) The astronomers also found that most of the remote, blue, star-forming galaxies are spirals. Many show signs of collisions or mergers.

If spiral galaxies were more common in the distant past, where have they gone? Galactic collisions and mergers are probably responsible for depleting rich clusters of their spiral galaxies. During a collision, interstellar gas in the colliding galaxies is vigorously compressed, triggering a burst of star formation. A succession of collisions produces a series of star-forming episodes that create numerous bright, hot O and

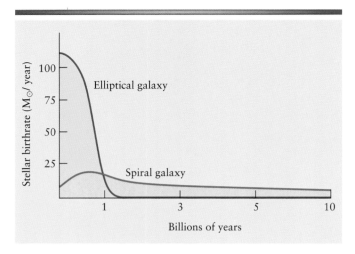

figure 26-32

The Stellar Birthrate in Galaxies Most of the stars in an elliptical galaxy are created in a brief burst of star formation when the galaxy is very young. In spiral galaxies, star formation occurs at a more leisurely pace and may extend over billions of years. (Adapted from J. Silk)

An alternative theory, proposed in 1977, suggests that galaxies formed by the merging of several gas clouds, rather than from the gravitational contraction of huge, isolated clouds. A third theory also contends that galaxies formed from mergers of gas clouds, except that the ancestral fragments were rather small and quite numerous. Figure 26-33 illustrates these three competing theories of galaxy formation.

Observations from the Hubble Space Telescope favor the idea that galaxies formed from the merging of gas clouds. Figure 26-34 shows a number of galaxylike objects some 11 billion light-years (3400 Mpc) away and are thus seen as they were 11 billion years ago. These objects are smaller than even the smallest galaxies we see in the present-day universe and have unusual, irregular shapes. Furthermore, these objects are scattered over an area only 600 kpc (2 million light-years) across—less than the distance between the Milky Way Galaxy and M31—making it quite probable that they would collide and merge with each other. These collisions would be aided by the dark matter associated with each subgalactic object, which increases the object's mass and, hence, the gravitational forces pulling the objects together. Such mergers would eventually give rise to a normal-sized galaxy.

B stars that become dispersed along arching spiral arms by the galaxy's rotation. Eventually, however, the gas is used up; star formation then ceases and the spiral arms become less visible. Furthermore, tidal forces tend to disrupt colliding galaxies, strewing their stars across intergalactic space until the galaxies are completely disrupted (see Figure 26-26).

Although collisions and mergers deplete the population of spiral galaxies, these events do not produce elliptical or lenticular galaxies. The HST images show that ellipticals and lenticulars were already well developed 4 billion years ago. Indeed, the preponderance of old Population II stars in elliptical galaxies, as well as the lack of gas and dust in these galaxies, demonstrates that their stars formed in a burst of activity 10 to 15 billion years ago. In contrast, spiral galaxies have been forming stars continually over the past 10 to 15 billion years, although at a gradually decreasing rate as their interstellar gas gets used up. Figure 26-32 compares the rates at which spiral and elliptical galaxies form stars.

One of the challenges that faces astronomers today is to understand how galaxies were created. One theory, popularized in the 1960s, argues that galaxies formed from the gravitational contraction of huge clouds of primordial gas. The rate of star formation in a contracting gas cloud determines whether it becomes a spiral or an elliptical galaxy. If the rate of star formation is low, then the gas has plenty of time to settle in a flattened disk. A flattened disk is the natural consequence of the overall rotation of the original gas cloud. Star formation continues in the disk because it contains an abundance of hydrogen, and thus a spiral galaxy is created. But if the stellar birthrate is high, then virtually all of the gas is used up in the formation of stars before a disk has time to form. In this case, an elliptical galaxy is created.

figure 26-33

Theories of Galaxy Formation Theories of galaxy formation fall into three categories. **(a)** A single huge primordial gas cloud contracts to form a galaxy. **(b)** Several moderate-sized gas clouds merge to form a galaxy. **(c)** Numerous small gas clouds coalesce to form a galaxy. (Adapted from van den Bergh and Hesser)

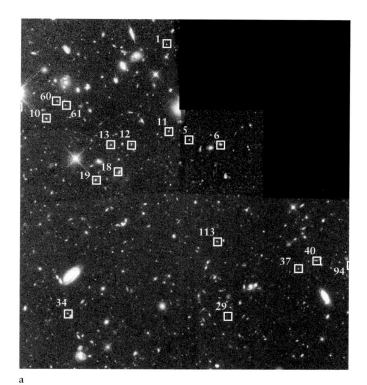

a

b

WEB LINK 26-24

ℱigure 26-34   R I **V** U X G

The Building Blocks of Galaxies **(a)** This Hubble Space Telescope image shows a portion of the sky 2 arcmin across in the constellation Hercules. The objects outlined by boxes are about 3400 Mpc (11 billion ly) from Earth and are only 600 to 900 pc (2000 to 3000 ly) across—larger than a star cluster but smaller than even dwarf elliptical galaxies like those in Figure 26-19. **(b)** Close-up images of these objects show their irregular shapes and blue color, indicating the presence of young stars. If these objects were to coalesce, the result would be a full-sized galaxy, such as we see in the nearby universe today. (Rogier Windhorst and Sam Pascarelle, Arizona State University; NASA)

Galactic evolution is a difficult and controversial subject full of unresolved questions. Where did the subgalactic clouds of gas come from? And what happened in the early universe to cause the primordial hydrogen and helium to clump up in clouds destined to evolve into galaxies, instead of becoming objects a million times bigger or smaller? Even more troublesome is the issue of the dark matter. We know that the observable stars, gas, and dust in a galaxy constitute only 10% of the mass associated with the galaxy. We have no idea what the remaining 90% looks like or what it is made of. These questions are among the most challenging that face astronomers today.

KEY WORDS

KEY IDEAS

 The Hubble Classification: Galaxies can be grouped into four major categories: spirals, barred spirals, ellipticals, and irregulars.

• The disks of spiral and barred spiral galaxies are sites of active star formation.

• Elliptical galaxies are nearly devoid of interstellar gas and dust, and so star formation is severely inhibited.

• Lenticular galaxies are intermediate between spiral and elliptical galaxies.

• Irregular galaxies have ill-defined, asymmetrical shapes. They are often found associated with other galaxies.

Distance to Galaxies: Standard candles, such as Cepheid variables and the most luminous supergiants, globular clusters, H II regions, and supernovae in a galaxy, are used in estimating intergalactic distances.

• The Tully-Fisher relation, which correlates the width of the 21-cm line of hydrogen in a spiral galaxy with its luminosity, can also be used for determining distance. A method that can be used for elliptical galaxies is the fundamental plane, which relates the galaxy's size to its surface brightness distribution and to the motions of its stars.

The Hubble Law: There is a simple linear relationship between the distance from the Earth to a remote galaxy and the redshift of that galaxy (which is a measure of the speed with which it is receding from us). This relationship is the Hubble law, $v = H_0 d$.

• Because of difficulties in measuring the distances to galaxies, the value of the Hubble constant, H_0, is not known with certainty but is probably close to 70 km/s/Mpc. As a result of this uncertainty, distances to remote galaxies calculated from the Hubble law are uncertain by about 10%.

 Clusters and Superclusters: Galaxies are grouped into clusters rather than being scattered randomly throughout the universe.

• A rich cluster contains hundreds or even thousands of galaxies; a poor cluster, often called a group, may contain only a few dozen.

• A regular cluster has a nearly spherical shape with a central concentration of galaxies; in an irregular cluster, galaxies are distributed asymmetrically.

• Our Galaxy is a member of a poor, irregular cluster called the Local Group.

• Rich, regular clusters contain mostly elliptical and lenticular galaxies; irregular clusters contain spiral, barred spiral, and irregular galaxies along with ellipticals.

• Giant elliptical galaxies are often found near the centers of rich clusters.

Galactic Collisions and Mergers: When two galaxies collide, their stars pass each other, but their interstellar media collide violently, either stripping the gas and dust from the galaxies or triggering prolific star formation.

• The gravitational effects during a galactic collision can throw stars out of their galaxies into intergalactic space.

• Galactic mergers may occur; a large galaxy in a rich cluster may tend to grow steadily through galactic cannibalism, perhaps producing in the process a giant elliptical galaxy.

The Dark-Matter Problem: The luminous mass of a cluster of galaxies is not large enough to account for the observed motions of the galaxies; a large amount of unobserved mass must also be present. This situation is called the dark-matter problem.

• Hot intergalactic gases in rich clusters account for a small part of the unobserved mass. These gases are detected by their X-ray emission. The remaining unobserved mass is probably in the form of dark-matter halos that surround all the galaxies in these clusters.

• Gravitational lensing of remote galaxies by a foreground cluster enables astronomers to glean information about the distribution of dark matter in the foreground cluster.

Formation and Evolution of Galaxies: Observations of remote clusters of galaxies reveal that spiral galaxies were more common billions of years ago than they are today. This excess of spiral galaxies probably resulted from collisions and mergers when galaxies had an abundance of interstellar gas with which to form stars.

• One theory states that galaxies formed by the gravitational contraction of huge, individual clouds of gas; others explain that galaxies arose from mergers of several smaller gas clouds. Hubble Space Telescope observations suggest that the latter model explains at least some of the galaxies we see today.

REVIEW QUESTIONS

1. What was the Shapley-Curtis "debate" all about? Was a winner declared at the end of the "debate?" Whose ideas turned out to be correct?

2. How did Edwin Hubble prove that the Andromeda "Nebula" is not a nebula within our Milky Way Galaxy?

3. An educational publication for children included the following statement: "The Sun is in fact the only star in our galaxy. All of the other stars in the sky are located in other galaxies." How would you correct this statement?

4. What is the Hubble classification scheme? Which category includes the largest galaxies? Which includes the smallest? Which category of galaxy is the most common?

5. Explain why the apparent shape of an elliptical galaxy may be quite different from its real shape.

6. Which types of galaxies are most likely to have new stars forming? Describe the observational evidence that supports your answer.

7. Are any galaxies besides our own visible with the naked eye? If so, which one(s)?

8. Why are Cepheid variable stars useful for finding the distances to galaxies? Are there any limitations on their use for this purpose?

9. What is the Tully-Fisher relation? How is it used for measuring distances? Can it be used for galaxies of all kinds? Why or why not?

10. What is the Hubble law? How can it be used to determine distances?

11. Why do you suppose there are many discordant determinations of H_0?

12. Some galaxies in the Local Group exhibit blueshifted spectral lines. Why aren't these blueshifts violations of the Hubble law?

13. Describe what voids are and what they tell us about the large-scale structure of the universe.

14. Why is the intracluster gas in galaxy clusters at such high temperatures?

15. Why do giant elliptical galaxies dominate rich clusters but not poor clusters?

16. What evidence is there for the existence of dark matter in clusters of galaxies?

17. On what grounds do astronomers believe that spiral galaxies were more numerous in the past than they are today? What could account for this excess of spiral galaxies in the distant past?

18. What observations suggest that present-day galaxies formed from smaller assemblages of matter?

ADVANCED QUESTIONS

Questions preceded by an asterisk (*) involve topics discussed in the Boxes.

Problem-solving tips and tools

Box 1-1 explains the small-angle formula, and Box 19-3 discusses the relationship among apparent magnitude, absolute magnitude, and distance. As Box 25-2 explains, a useful form of Kepler's third law is $M = rv^2/G$, where M is the mass within an orbit of radius r, v is the orbital speed, and G is the gravitational constant. Another form of Kepler's third law, particularly useful for two stars or two galaxies orbiting each other, is given in Section 19-9. The volume of a sphere of radius r is $\frac{4}{3}\pi r^3$. The mass of a hydrogen atom (^{1}H) is given in Appendix 7.

19. When the results from the Hipparcos mission were released, with new and improved measurements of the parallaxes of *nearby* stars within 500 pc, astronomers had to revise the distances to many *remote* galaxies millions of parsecs away. Explain why.

20. As Figure 21-15 shows, there are two types of Cepheid variables. Type I Cepheids are metal-rich stars of Population I, while Type II Cepheids are metal-poor stars of Population II. (a) Which type of Cepheid variables would you expect to be found in globular clusters? Which type would you expect to be found in the disk of a spiral galaxy? Explain your reasoning. (b) When Hubble discovered Cepheid variables in M31, the distinction between Type I and Type II Cepheids was not yet known. Hence, Hubble thought that the Cepheids seen in the disk of M31 were identical to those seen in globular clusters in our own Galaxy. As a result, his calculations of the distance to M31 were in error. Using Figure 21-15, explain whether Hubble's calculated distance was too small or too large.

*21. Astronomers often state the distance to a remote galaxy in terms of its distance modulus, which is the difference between the apparent magnitude m and the absolute magnitude M (see Box 19-3). (a) By measuring the brightness of supernova 1994I in the galaxy M51 (see Figure 26-2), the distance modulus for this galaxy was determined to be $m - M = 29.2$. Find the distance to M51 in megaparsecs (Mpc). (b) A separate distance determination, which involved measuring the brightnesses of planetary nebulae in M51, found $m - M = 29.6$. What is the distance to M51 that you calculate from this information? (c) What is the difference between your answers to parts (a) and (b)? Compare this difference with the 750-kpc distance from the Earth to M31, the Andromeda Galaxy. The difference between your answers illustrates the uncertainties involved in determining the distances to galaxies!

*22. Suppose you discover a Type Ia supernova in a distant galaxy. At maximum brilliance, the supernova reaches an apparent magnitude of +10. How far away is the galaxy? (*Hint:* See Box 26-1.)

23. The masers that orbit the center of the spiral galaxy M106 travel at an orbital speed of about 1000 km/s. Astronomers observed these masers at intervals of 4 months. (a) What distance does a single maser move during a 4-month period? Give your answer in kilometers and in AU. (b) During this period, a maser moving across the line of sight (like the maser shown in green in Figure 26-14) appeared to move through an angle of only 10^{-5} arcsec. Calculate the distance to the galaxy.

24. The average radial velocity of galaxies in the Hercules cluster pictured in Figure 26-20 is 10,800 km/s. (a) Using $H_0 = 70$ km/s/Mpc, find the distance to this cluster. Give your answer in megaparsecs and in

light-years. (**b**) How would your answer to (a) differ if the Hubble constant had a smaller value? A larger value? Explain.

25. A certain galaxy is observed to be receding from the Sun at a rate of 7500 km/s. The distance to this galaxy is measured independently and found to be 1.4×10^8 pc. From these data, what is the value of the Hubble constant?

26. It is estimated that the Coma cluster (see Figure 26-17) contains about 10^{13} M$_\odot$ of intracluster gas. (**a**) Assuming that this gas is made of hydrogen atoms, calculate the total number of intracluster gas atoms in the Coma cluster. (**b**) The Coma cluster is roughly spherical in shape, with a radius of about 3 Mpc. Calculate the number of intracluster gas atoms per cubic centimeter in the Coma cluster. Assume that the gas fills the cluster uniformly. (**c**) Compare the intracluster gas in the Coma cluster with the gas in our atmosphere (3×10^{19} molecules per cubic centimeter, temperature 300 K); a typical gas cloud within our own Galaxy (a few hundred molecules per cubic centimeter, temperature 50 K or less); and the corona of the Sun (10^5 atoms per cubic centimeter, temperature 10^6 K).

*27. Two galaxies separated by 600 kpc are orbiting each other with a period of 40 billion years. What is the total mass of the two galaxies?

*28. Figure 26-28 shows the rotation curve of the Sa galaxy NGC 4378. Using data from that graph, calculate the orbital period of stars 20 kpc from the galaxy's center. How much mass lies within 20 kpc from the center of NGC 4378?

29. How might you determine what part of a galaxy's redshift is caused by the galaxy's orbital motion about the center of mass of its cluster?

DISCUSSION QUESTIONS

30. The Earth is composed principally of heavy elements, such as silicon, nickel, and iron. Would you be likely to find such planets orbiting stars in the disk of a spiral galaxy? In the nucleus of a spiral galaxy? In an elliptical galaxy? In an irregular galaxy? Explain your answers.

31. Discuss whether the various Hubble types of galaxies actually represent some sort of evolutionary sequence.

32. Discuss the advantages and disadvantages of using the various standard candle distance indicators to obtain extragalactic distances.

33. How would you distinguish star images from unresolved images of remote galaxies on a CCD?

34. Describe what sorts of observations you might make to search for as yet undiscovered galaxies in our Local Group. How is it possible that such galaxies might still remain to be discovered? In what part of the sky would these galaxies be located? What sorts of observations might reveal these galaxies?

35. Explain why the dark matter in galaxy clusters could not be neutral hydrogen.

 ## WEB/CD-ROM QUESTIONS

36. When galaxies pass close to one another, as should happen frequently in a rich cluster, tidal forces between the galaxies can strip away their outlying stars. The result should be a loosely dispersed sea of "intergalactic stars" populating the space between galaxies in a cluster. Search the World Wide Web for information about intergalactic stars. Have they been observed? Where are they found? What would our nighttime sky look like if our Sun were an intergalactic star?

 37. The Hubble Space Telescope (HST) has made extensive observations of very distant galaxies. Visit the HST web site to learn about these investigations. How far back in time has HST been able to look? What sorts of early galaxies are observed? What is the current thinking about how galaxies formed and evolved?

 38. **Radiation from a Rotating Galaxy.** Access and view the animation "Radiation from a Rotating Galaxy" in Chapter 26 of the *Universe* web site or CD-ROM. Describe how the animation would have to be changed for a spiral galaxy of the same size but (**a**) greater mass and (**b**) smaller mass.

OBSERVING PROJECTS

39. Using a telescope with an aperture of at least 30 cm (12 in.), observe as many of the spiral galaxies listed in the following table as you can. Use the *Starry Night* program to determine when they can best be viewed. Many of these galaxies are members of the Virgo cluster, which can best be seen from March through June. Because all galaxies are quite faint, be sure to schedule your observations for a moonless night. The best view is obtained when a galaxy is near the meridian. While at the eyepiece, make a sketch of what you see. Can you distinguish any spiral structure? After completing your observations, compare your sketches with photographs found in such popular books as *Galaxies* by Timothy Ferris (Sierra Club Books, 1980) and *The Color Atlas of the Galaxies* by James Wray (Cambridge University Press, 1988).

Spiral galaxy	Right ascension	Declination	Hubble type
M31 (NGC 224)	0^h 42.7^m	+41° 16′	Sb
M58 (NGC 4579)	12 37.7	+11 49	Sb
M61 (NGC 4303)	12 21.9	+4 28	Sc
M63 (NGC 5055)	13 15.8	+42 02	Sb
M64 (NGC 4826)	12 56.7	+21 41	Sb
M74 (NGC 628)	1 36.7	+15 47	Sc
M83 (NGC 5236)	13 37.0	−29 52	Sc
M88 (NGC 4501)	12 32.0	+14 25	Sb
M90 (NGC 4569)	12 36.8	+13 10	Sb
M91 (NGC 4548)	12 35.4	+14 30	SBb
M94 (NGC 4736)	12 50.9	+41 07	Sb
M98 (NGC 4192)	12 13.8	+14 54	Sb
M99 (NGC 4254)	12 18.8	+14 25	Sc
M100 (NGC 4321)	12 22.9	+15 49	Sc
M101 (NGC 5457)	14 03.2	+54 21	Sc
M104 (NGC 4594)	12 40.0	−11 37	Sa
M108 (NGC 3556)	11 11.5	+55 40	Sc

Note: The right ascensions and declinations are given for epoch 2000.

40. Using a telescope with an aperture of at least 30 cm (12 in.), observe as many of the following elliptical galaxies as you can. Six of these galaxies are in the Virgo cluster, which is conveniently located in the evening sky from March through June. Use the *Starry Night* program to help you determine when they can best be viewed. As in the previous exercise, be sure to schedule your observations for a moonless night, when the galaxies you wish to observe will be near the meridian. Do these elliptical galaxies differ in appearance from spiral galaxies?

Elliptical galaxy	Right ascension	Declination	Hubble type
M49 (NGC 4472)	12^h 29.8^m	+8° 00′	E4
M59 (NGC 4621)	12 42.0	+11 39	E3
M60 (NGC 4649)	12 43.7	+11 33	E1
M84 (NGC 4374)	12 25.1	+12 53	E1
M86 (NGC 4406)	12 26.2	+12 57	E3
M89 (NGC 4552)	12 35.7	+12 33	E0
M110 (NGC 205)	00 40.4	+41 41	E6

Note: The right ascensions and declinations are given for epoch 2000.

41. Using a telescope with an aperture of at least 30 cm (12 in.), observe as many of the following interacting galaxies as you can. As in the previous exercises, be sure to schedule your observations for a moonless night, when the galaxies you wish to observe will be near the meridian. While at the eyepiece, make a sketch of what you see. Can you distinguish hints of interplay among the galaxies? After completing your observations, compare your sketches with photographs found in such popular books as *Galaxies* by Timothy Ferris (Sierra Club Books, 1980).

Interacting galaxies	Right ascension	Declination
M51 (NGC 5194)	13^h 29.9^m	+47° 12′
NGC 5195	13 30.0	+47 16
M65 (NGC 3623)	11 18.9	+13 05
M66 (NGC 3627)	11 20.2	+12 59
NGC 3628	11 20.3	+13 36
M81 (NGC 3031)	9 55.6	+69 04
M82 (NGC 3034)	9 55.8	+69 41
M95 (NGC 3351)	10 44.0	+11 42
M96 (NGC 3368)	10 46.8	+11 49
M105 (NGC 3379)	10 47.8	+12 35

Note: The right ascensions and declinations are given for epoch 2000.

42. Use the *Starry Night* program to observe other galaxies. First turn off daylight (select **Daylight** in the **Sky** menu) and show the entire celestial sphere (select **Atlas** in the **Go** menu). Center the field of view on each of the following galaxies in turn (select **Find…** in the **Edit** menu), and zoom in as needed until you can see the shape of the galaxy. In each case, use the image of the galaxy to determine its Hubble classification as best you can, and explain your reasoning. (**a**) M87; (**b**) M101; (**c**) M104; (**d**) M102.

Quasars, Active Galaxies, and Gamma-Ray Bursters

27

An ordinary star emits radiation primarily at ultraviolet, visible, and infrared wavelengths, in proportions that reflect the star's surface temperature. Ordinary galaxies, too, emit most strongly in these wavelength regions. But the object shown here, called 3C 273, is outrageously different: It emits strongly over an immense range of wavelengths from radio to X-ray. 3C 273 is also intensely luminous, with thousands of times the radiation output of an ordinary galaxy. And, as this X-ray image shows, some source of energy within 3C 273 causes it to eject a glowing, high-speed jet that extends outward for hundreds of thousands of light-years.

3C 273 is a *quasar*—one of many thousands of distant objects whose luminosity is far too great to come from starlight alone. They are intimately related to *active galaxies*, whose tremendous luminosity can fluctuate substantially over a period of months, weeks, or even days.

What makes quasars and active galaxies so tremendously energetic? In this chapter, we will see evidence that a supermassive black hole lies at the center of each of these objects. As matter accretes onto the black hole, it releases fantastic amounts of energy and can produce powerful jets.

Quasars and active galaxies emit their intense radiation continuously. But from time to time astronomers detect flashes of high-energy gamma

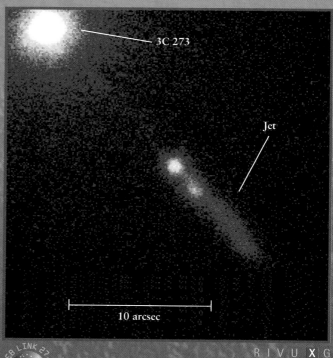

3C 273

Jet

10 arcsec

WEB LINK 27.1

(NASA/CXC/SAO/H. Marshall et al.)

R I V U X G

rays that for a brief time outshine even the most luminous quasar. For many years the nature of these *gamma-ray bursters* was a complete mystery to astronomers. As we will see, however, dramatic recent evidence suggests that these intense bursts are caused by stellar cataclysms taking place within distant galaxies.

As you read the sections of this chapter, look for the answers to the following questions.

27-1 Why are quasars unusual? How did astronomers discover that they are extraordinarily distant and luminous?

27-2 How does the light from quasars differ from starlight?

27-3 How are Seyfert galaxies and radio galaxies related to quasars?

27-4 How can material ejected from quasars appear to travel faster than light?

27-5 What could power the incredible energy output from active galaxies?

27-6 Why do many active galaxies emit ultrafast jets of material?

27-7 What are gamma-ray bursters? How did astronomers discover how far away they are?

27-1 Quasars look like stars but have huge redshifts

Many of the most revolutionary discoveries in astronomy have been the result of advances in technology. So it was in 1610 when Galileo used a new device called the telescope to scan the sky, and in the process found evidence that shattered the ancient geocentric model of the universe (see Section 4-3). And so it was during the middle years of the twentieth century, when the new technique of radio astronomy revealed the existence of remote, dazzlingly luminous objects called quasi-stellar radio sources or quasars.

Grote Reber, a radio engineer and ham radio enthusiast, built the first true radio telescope in 1936 in his backyard in Illinois (see Section 6-6). By 1944 he had detected strong radio emissions from sources in the constellations Sagittarius, Cassiopeia, and Cygnus.

Two of these sources, named Sagittarius A and Cassiopeia A, happen to lie in our own Galaxy; they are the galactic nucleus (see Section 25-6) and a supernova remnant (see Figure 22-24). The nature of the third source, called Cygnus A, proved more elusive. The mystery only deepened in 1951, when Walter Baade and Rudolph Minkowski used the 200-inch (5-meter) optical telescope on Palomar Mountain to discover a dim, strange-looking galaxy at the position of Cygnus A (see the inset in Figure 27-1).

R **V** U X G

Figure 27-1

Cygnus A (3C 405) This radio image from the Very Large Array shows that most of the emission from Cygnus A comes from the luminous radio lobes located on either side of a peculiar galaxy. The two lobes each extend about 70 kpc (230,000 ly) from the galaxy. **Inset:** The galaxy at the heart of Cygnus A has a substantial redshift, which means that it must be extremely far from Earth (about 230 Mpc, or 750 million ly). Cygnus A is one of the brightest radio sources in the sky, so this remote galaxy and its radio lobes must have an enormous energy output. (R. A. Perley, J. W. Dreher, and J. J. Cowan, NRAO; inset: Palomar Observatory)

R I V U X G

When Baade and Minkowski photographed the spectrum of Cygnus A, they were surprised to find a number of bright emission lines. By contrast, a normal galaxy has absorption lines in its spectrum (see Section 26-3 and Figure 26-15). This absorption takes place in the atmospheres of the stars that make up the galaxy, as described in Section 19-5. In order for Cygnus A to have emission lines, something must be exciting and ionizing its atoms. Furthermore, the wavelengths of Cygnus A's emission lines are all shifted by 5.6% toward the red end of the spectrum. Astronomers use the letter z to denote redshift, and thus this object has $z = 0.056$, corresponding to a recessional velocity of 16,000 km/s.

If Cygnus A participates in the same Hubble flow as clusters of galaxies (described in Section 26-5), then this recessional velocity corresponds to a tremendous distance from the Earth—about 230 Mpc (750 million ly) if the Hubble constant equals 70 km/s/Mpc. Yet despite its tremendous distance from the Earth, radio waves from Cygnus A can be picked up by amateur astronomers with backyard equipment. This means that Cygnus A must be one of the most luminous radio sources in the sky. In fact, its radio luminosity is 10^7 times as great as that of an ordinary galaxy like the Milky Way. The object that creates the Cygnus A radio emission must be something quite extraordinary.

Cygnus A was not the only unusual radio source to draw the attention of astronomers. In 1960, Allan Sandage used the 200-inch telescope to discover a "star" at the location of a radio source designated 3C 48 (Figure 27-2). Because ordinary stars are not strong sources of radio emission, astronomers knew 3C 48 must be something unusual. Like Cygnus A, its spectrum showed a series of emission lines, but astronomers were unable to identify the chemical elements that produced these lines.

Although 3C 48 was clearly an oddball, many astronomers thought it was just a strange star in our own Galaxy. This "star" proved not to be unique. Two years later, astronomers discovered a similar starlike optical counterpart to the radio source 3C 273. This "star" is even more unusual, with a luminous jet protruding from one side. (The image that opens this chapter shows this jet at X-ray wavelengths.) Like 3C 48, its visible spectrum contains a series of emission lines that no one could explain.

A breakthrough occurred in 1963, when Maarten Schmidt at Caltech took another look at the spectrum of 3C 273. He realized that four of its brightest emission lines are positioned relative to one another in precisely the same way as the first four Balmer lines of hydrogen. However, these emission lines from 3C 273 were all shifted to much longer wavelengths than the usual wavelengths of the Balmer lines. Schmidt determined that 3C 273 has a redshift of $z = 0.158$, corresponding to a recessional velocity of 44,000 km/s (15% of the speed of light). No star could be moving this fast and remain within our Galaxy, so Schmidt concluded that 3C 273 could not be a nearby star but must lie outside the Milky Way.

According to the Hubble law, the recessional velocity of 3C 273 implies that it is about 620 Mpc (2 billion ly) away. To be detected at such distances, 3C 273 must be an extraordinarily powerful source of both visible light and radio radiation.

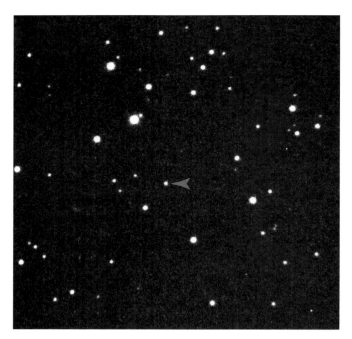

Figure 27-2 R I **V** U X G

The Quasar 3C 48 For several years astronomers erroneously believed that this object (shown by the arrow) in the constellation Triangulum was simply a peculiar nearby star that happened to emit radio waves. Actually, the redshift of this starlike object is so great ($z = 0.367$) that, according to the Hubble law, it must be approximately 1300 Mpc (4 billion ly) away. The "3C" in this object's designation refers to the *Third Cambridge Catalogue*, a compendium of radio sources. (Alex G. Smith, Rosemary Hill Observatory, University of Florida)

Figure 27-3 shows the visible and near-infrared spectrum of 3C 273. As we saw in Section 6-5 (review Figure 6-23), such a spectrum is a graph of intensity versus wavelength on which emission lines appear as peaks and absorption lines appear as valleys. Emission lines are caused by excited atoms that emit radiation at specific wavelengths.

Upon learning how Schmidt deciphered the spectrum of 3C 273, two other Caltech astronomers, Jesse Greenstein and Thomas Matthews, found they could identify the spectral lines of 3C 48 as having suffered an even larger redshift of $z = 0.367$. That shift corresponds to a speed three-tenths that of light, which places 3C 48 twice as far away as 3C 273, or approximately 1300 Mpc (4 billion ly) from Earth.

The only way that we can determine the distances to an object such as 3C 273 or 3C 48 is through its redshift. This tells us the recessional velocity v of the object, and the Hubble law $v = H_0 d$ then tells us the distance d to the object. But because the value of the Hubble constant H_0 is somewhat uncertain, distances determined in this way are equally uncertain. Most astronomers prefer simply to state the redshift of a distant object and use the simple rule that the greater the redshift, the more distant the object.

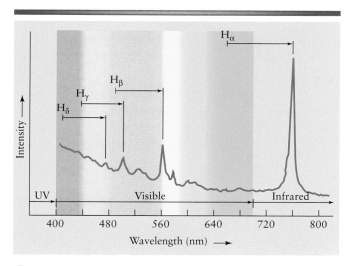

Figure 27-3

The Spectrum of 3C 273 This graph shows the spectrum of the quasar 3C 273 in visible light and in the near infrared. The spectrum is dominated by four bright emission lines of hydrogen (see Section 5-8). The arrows indicate how far these spectral lines are redshifted from their normal wavelengths. For example, the wavelength of H_β is shifted from 486 nm (blue-green) to 563 nm (yellow).

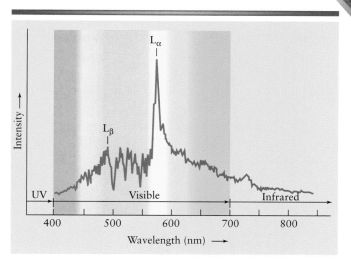

Figure 27-4

The Spectrum of a High-Redshift Quasar This quasar, known as PKS 2000-030, has a redshift $z = 3.773$. This is so large that the Lyman spectral lines L_α and L_β of hydrogen, which are normally at ultraviolet wavelengths, have been shifted into the visible part of the spectrum. Note that there are many deep absorption lines to the left of the L_α emission line (that is, at shorter wavelengths). These lines, collectively called the "Lyman-alpha forest," are due to remote gas clouds along our line of sight to the quasar. Hydrogen atoms in these clouds absorb L_α photons from the quasar; because these clouds are nearer to us than the quasar and hence have smaller redshifts, their absorption lines are less redshifted than the quasar's L_α emission line.

WEB LINK 27.3 Because of their strong radio emission and starlike appearance, 3C 48 and 3C 273 were dubbed *quasi-stellar radio sources*, a term soon shortened to **quasars.** After the first quasars were discovered by their radio emission, many similar, high-redshift, starlike objects were found that emit little or no radio radiation. These "radio-quiet" quasars were originally called *quasi-stellar objects,* or *QSOs,* to distinguish them from radio emitters. Today, however, the term "quasar" is often used to include both types. Only about 10% of quasars are "radio-loud" like 3C 48 and 3C 273.

More than 10,000 quasars are now known. They all look rather like stars (see Figure 1-8), and all have large redshifts, ranging from 0.06 to at least 5.8 (Figure 27-4). Most quasars have redshifts of 0.3 or more, which implies that they are more than 1000 Mpc (3 billion light-years) from Earth. For example, the quasar PC 1247+3406 has a redshift of $z = 4.897$, which corresponds to a recessional velocity of more than 94% of the speed of light and a distance of roughly 3800 Mpc (12 billion ly) from Earth.

CAUTION! A value of z greater than 1 does *not* mean that a quasar is receding from us faster than the speed of light. At high speeds, the relationship between redshift and recessional velocity must be modified by the special theory of relativity, as explained in Box 27-1. As the speed of a receding source approaches the speed of light, its redshift can become very much greater than 1. An infinite redshift ($z = \infty$) corresponds to a recessional velocity equal to the speed of light.

Light takes time to travel across space, so when we observe a very remote object, we are seeing it in the remote past. This means that for very remote objects, the relationship between redshift and distance from the Earth depends on how the universe has evolved over time. As we will see in Chapter 28, the Hubble law reveals that the universe is expanding. In other words, if you could watch the motions of widely separated clusters of galaxies over millions of years, you would see them gradually moving away from one another. Furthermore, in Chapter 28 we will see evidence that the universe has expanded at different rates at different times in the past. Table 27-1 relates the redshift to the recessional velocity and distance in a model consistent with our present understanding of the expansion of the universe.

Because there are no quasars with small redshifts, it follows that there are no nearby quasars. The nearest one is some 250 Mpc (800 million ly) from the Earth. It is useful to remember that the distance to a remote object expressed in light-years tells you the "lookback time"—that is, how many years into the past you are looking when you view that object. Hence, the absence of nearby quasars means that there have been no quasars for nearly a billion years. Indeed, the number of quasars began to decline precipitously roughly 10 billion

box 27-1 | Tools of the Astronomer's Trade

The Relativistic Redshift

As we saw in Section 26-5, the redshift (z) of an object moving toward or away from us is defined as follows:

The redshift formula

$$z = \frac{\lambda - \lambda_0}{\lambda_0}$$

$\lambda =$ observed wavelength of a spectral line in the spectrum of an object (for example, a star, galaxy, or quasar)

$\lambda_0 =$ normal (unshifted) wavelength of that same spectral line, as deduced from laboratory experiments

$z =$ redshift of the object

If the object is moving away from us, the observed wavelength is greater than the normal wavelength (a redshift) and z is a positive number.

EXAMPLE: The strongest emission line in the spectrum of the quasar PKS 2000-330 (see Figure 27-4) is the Lyman-alpha (L_α) line of hydrogen, observed at a wavelength of 582.5 nm.

From laboratory measurements, we know that this spectral line is normally seen in the ultraviolet at a wavelength of 121.6 nm. Thus, the redshift of this spectral line is

$$z = \frac{582.5 \text{ nm} - 121.6 \text{ nm}}{121.6 \text{ nm}} = 3.79$$

The average redshift for this quasar, determined by measuring a number of spectral lines, is $z = 3.773$.

For low recessional velocities v, we can ignore the effects of relativity and use the following equation:

$$z = \frac{v}{c} \text{ (low speeds only)}$$

where c is the speed of light. For example, a 5% shift in wavelength ($z = 0.05$) corresponds to a recessional velocity of 5% of the speed of light ($v = 0.05c$). For redshifts greater than about 0.1, however, we must use the relativistic equation for the Doppler shift:

$$z = \sqrt{\frac{c + v}{c - v}} - 1$$

A useful way to rewrite this relationship is as follows:

Relativistic relationship between redshift and recessional velocity

$$\frac{v}{c} = \frac{(z + 1)^2 - 1}{(z - 1)^2 - 1} = 0.916$$

$z =$ redshift of an object

$v =$ recessional velocity of that object

$c =$ speed of light

The accompanying graph displays this relationship. Note that z approaches infinity as v approaches the speed of light.

EXAMPLE: Quasar PKS 2000-330 has a redshift of $z = 3.773$. Using this value in the above relativistic equation for the quasar's recessional velocity, we obtain

$$\frac{v}{c} = \frac{(4.773)^2 - 1}{(4.773)^2 + 1} = \frac{21.78}{23.78} = 0.916$$

In other words, this quasar appears to be receding from us at 91.6% of the speed of light.

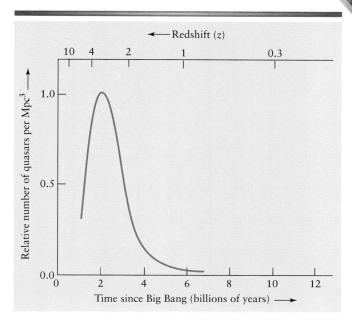

Table 27-1	Redshift and Distance		
		Distance	
z	v/c	(Mpc)	(10^9 ly)
0.000	0.000	0	0.00
0.100	0.095	400	1.30
0.200	0.180	750	2.43
0.300	0.257	1050	3.42
0.400	0.324	1310	4.28
0.500	0.385	1550	5.04
0.750	0.508	2020	6.57
1.00	0.600	2370	7.71
1.50	0.724	2840	9.27
2.00	0.800	3140	10.2
3.00	0.882	3480	11.4
4.00	0.923	3670	12.0
5.00	0.946	3780	12.3
∞	1.00	4130	13.5

Note: This table assumes a Hubble constant $H_0 = 70$ km/s/Mpc, a matter density parameter $\Omega_m = 0.3$, and a dark energy density parameter $\Omega_\Lambda = 0.7$ (see Chapter 28).

figure 27-5

Quasars Are Extinct The greater the redshift of a quasar, the farther it is from Earth and the farther back in time we are seeing it. By observing the number of quasars found at different redshifts, astronomers can calculate how the density of quasars in the universe has changed over the history of the universe. As this graph shows, the number of quasars per volume of space increased during the first 2 billion years after the Big Bang but has since decreased to near zero. This is reminiscent of the history of the dinosaurs, which once populated the entire Earth but today are extinct. (Peter Shaver, European Southern Observatory)

years ago (Figure 27-5). Quasars were a common feature of the universe in the distant past, but there are none in the present-day universe.

27-2 Quasars are the ultraluminous centers of distant galaxies

Quasars must be extraordinarily luminous if they can be seen from the Earth despite their enormous distances. Indeed, quasars are among the most distant objects ever seen. Only a few ancient galaxies have been seen at greater distances. It turns out that quasars are themselves the ultraluminous nuclei of remote galaxies.

A quasar's luminosity can be calculated from its apparent brightness and distance using the inverse-square law (see Section 19-2). For example, 3C 273 has a luminosity of about 10^{40} watts, which is equivalent to 2.5×10^{13} (25 trillion) Suns. Generally, quasar luminosities range from about 10^{38} watts up to nearly 10^{42} watts. For comparison, a typical large galaxy, like our own Milky Way, shines with a luminosity of 10^{37} watts, which equals 2.5×10^{10} (25 billion) Suns. Thus, a bright quasar is a thousand times more luminous than the entire Milky Way Galaxy.

The radiation from quasars is not like the radiation from ordinary galaxies. The light from an ordinary galaxy, be it spiral (see Figures 26-4 and 26-5), elliptical (Figure 26-6), or irregular (Figures 26-10 and 26-11), is dominated by the blackbody-like radiation from the galaxy's stars. Stars have surface temperatures from about 2500 to 50,000 K, so the spectrum of light from an ordinary galaxy is a combination of blackbody curves that span this range of temperatures. By contrast, quasars like 3C 273 emit far more X-ray and gamma-ray radiation than do ordinary galaxies. This radiation can be interpreted as being due to material at temperatures of 100,000 K or more, hotter than even the highest-temperature stars. Hence, whatever the origin of a quasar's radiation, it is not simply starlight.

Further clues about quasars come from examining the details of their spectra. Figure 27-6 shows the spectrum of a typical quasar over a broad range of frequencies. The "spikes" in this spectrum are emission lines from hydrogen gas as well as other elements. The shapes of these emission lines indicate that the light-emitting gas clouds are moving within the quasar at very high speeds (10,000 km/s or so). Some of the clouds are moving toward us, causing the emitted light to have a shorter wavelength and higher frequency, while other clouds

figure 27-6

The Spectrum of a Quasar In this graph the intensity of radiation from the quasar PKS 0405-123 is plotted against frequency. It shows a much larger segment of the electromagnetic spectrum than do Figures 27-3 and 27-4. The two prominent emission lines are the (redshifted) Balmer lines H_α and H_β of hydrogen. These are due to excited gases within the quasar.

are moving away from us and emit light with longer wavelength and lower frequency. This explains why the emission lines in Figure 27-6 are quite broad.

When quasars were first discovered, their energy output seemed so absurdly huge that a few astronomers began to question long-held beliefs,

such as the Hubble law. Traditionally, astronomers had used the Hubble law to calculate the distance to a galaxy from its redshift: the higher the redshift, the greater the distance. In the 1960s the American astronomer Halton C. Arp suggested that the Hubble law might not apply to quasars. If part of a quasar's redshift was caused by some yet-undiscovered phenomenon, then the quasar could be much closer to Earth than the Hubble law would have us believe. If so, a quasar's luminosity would not be so incredibly large.

In support of this theory, Arp and his collaborators drew attention to certain high-redshift quasars that seem to be located in or associated with low-redshift galaxies. These examples of "discordant redshifts" fueled a heated debate during the 1970s reminiscent of the Shapley-Curtis "debate" 50 years earlier (see Section 26-1). By the 1980s, however, the preponderance of evidence clearly favored the standard interpretation of redshifts according to the Hubble law. Most astronomers today regard Arp's discordant redshifts as simply a *projection effect*, whereby a distant quasar just happens to be in the same part of the sky as a nearby galaxy.

An analogy to the projection effect sometimes happens in photography on Earth. If a photographer poses a person with a telephone pole in the background, the photo can give the misleading impression that the unlucky person has a telephone pole growing out of his head.

What largely ended the redshift debate of the 1970s were observations showing that quasars are associated with remote galaxies. Long-exposure images of quasars showed that they are often found in groups or clusters of galaxies (Figure 27-7). In all these cases, the redshift of the quasar is essentially the same as the redshifts of the galaxies that surround it. Thus, the quasar's redshift indicates its distance from Earth, just as do the redshifts of galaxies.

A second important link between quasars and galaxies was established in the 1980s by astronomers who examined

a

b

c

figure 27-7 R I [V] U X G

Quasars in Remote Clusters of Galaxies A quasar is located at the center of each of these images. Very distant galaxies, identified by the numbers, are faintly visible near each of the quasars. In each case, the redshift of the quasar is virtually the same as the redshifts of the galaxies surrounding it. These observations reinforce the idea that quasars are extremely distant objects. (Courtesy of A. Stockton)

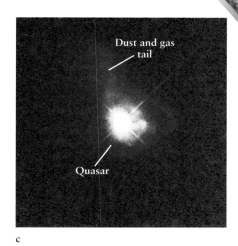

a   b   c

figure 27-8 R I 🗹 U X G

Quasars and Their "Host Galaxies" **(a)** Quasar PG 0052+251 is located at the center of an apparently normal spiral galaxy at redshift $z=0.155$. Other quasars are found at the centers of ordinary-looking elliptical galaxies. **(b)** The galaxy that hosts quasar PG 1012+008 (redshift $z=0.185$) is in the process of merging with a second luminous galaxy. The wispy material surrounding the quasar may have been pulled out of the galaxies by

tidal forces (see Figures 26-26 and 26-27). The two merging galaxies are just 9500 pc (31,000 ly) apart. Another small galaxy to the left of the quasar may also be merging with the others. **(c)** A long tail of gas and dust extends upward in this image from quasar PG 0316-346 (redshift $z=0.260$). This tail was presumably formed when the quasar's host galaxy collided with a second galaxy that lies beyond this image. (J. Bahcall, Institute for Advanced Study; M. Disney, University of Wales; NASA)

the "fuzz" seen around the images of some quasars. The spectrum of this fuzz shows stellar absorption lines, indicating that each of these quasars is embedded in a galaxy. In each case, the absorption lines of the stars have the same redshift as the quasar's emission lines, further supporting the belief that quasars are at distances indicated by their redshifts and the Hubble law.

It is very difficult to observe the "host galaxy" in which a quasar is located because the quasar's light overwhelms light from the galaxy's stars. Nevertheless, painstaking observations have revealed some basic properties of these host galaxies (Figure 27-8a). Relatively nearby radio-quiet quasars (with redshifts of less than 0.2, corresponding to distances of less than about 750 Mpc) tend to be located in spiral galaxies, whereas radio-loud quasars as well as more distant radio-quiet quasars tend to be located in ellipticals. However, a large percentage of these host galaxies have distorted shapes or are otherwise peculiar. Many have nearby companion galaxies, suggesting a link between collisions or mergers and the quasar itself (Figure 27-8b and Figure 27-8c). As we will see in Section 27-6, this link can help to explain why quasars are found only at great distances from Earth—and hence only in the distant past of the universe.

27-3 Active galaxies bridge the gap between normal galaxies and quasars

We have seen that some scientists in the 1970s preferred to challenge the Hubble law rather than accept the existence of

highly luminous objects. One reason for their skepticism was the huge gap in energy output between normal galaxies and quasars. The gap was bridged, however, when astronomers realized that the centers of certain galaxies look like low-luminosity quasars.

The first "missing links" between quasars and ordinary galaxies were actually discovered before quasars themselves. In 1943, Carl Seyfert at the Mount Wilson Observatory made a systematic study of spiral galaxies with bright, compact nuclei that seem to show signs of intense and violent activity (Figure 27-9). Like quasars, the nuclei of these galaxies have strong emission lines in their spectra. These galaxies are now referred to as **Seyfert galaxies.**

A few percent of the most luminous spiral galaxies are Seyfert galaxies. More than 700 Seyferts are known and more continue to be found. Seyfert galaxies range in luminosity from about 10^{36} to 10^{38} watts, which makes the brightest Seyferts as luminous as faint quasars. Indeed, there is no sharp dividing line between the properties of Seyferts and those of quasars. Like radio-quiet quasars, Seyferts tend to have only weak radio emissions.

Some Seyferts are members of interacting pairs or exhibit the vestiges of mergers and collisions. For example, the Seyfert galaxy NGC 1275 (Figure 27-10) is actually two colliding galaxies, a spiral and an elliptical. Some violent process in the nucleus of NGC 1275 has ejected gas outward in all directions at 3000 km/s.

While Seyfert galaxies resemble dim, radio-quiet quasars, certain elliptical galaxies, called **radio galaxies** because of their strong radio emission, are like dim, radio-loud quasars. The first of these peculiar galaxies was discovered in 1918 by

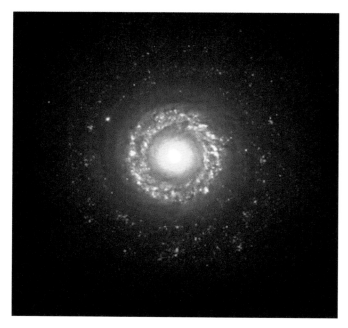

figure 27-9   R I **V** U X G

WEB LINK 27.6
A Seyfert Galaxy This spiral galaxy, called NGC 7742, is a Seyfert galaxy that lies some 22 Mpc (72 million ly) from Earth in the constellation Pegasus. The spectrum of this galaxy's intensely luminous nucleus shows emission lines of highly ionized atoms. (Hubble Heritage Team, AURA/STScI/NASA)

Heber D. Curtis. His short-exposure photograph of the giant elliptical galaxy M87 revealed a bright, starlike nucleus with a protruding jet. Figure 27-11*a* shows an overall view of the entire galaxy. Figure 27-11*b*, from the Hubble Space Telescope, shows the jet extending outward from the nucleus of M87.

Light from the central regions of M87 is **thermal radiation,** with a spectrum like that of a blackbody. This is radiation that depends only on the temperature of the object that emits it. For a given temperature, an object emits the maximum amount of thermal radiation at one particular frequency and emits less at frequencies above or below that value. The thermal radiation from the center of M87 is presumably due to a profusion of stars crowded around the galaxy's nucleus.

By contrast, the light from M87's jet has an intensity that decreases as frequency increases. There is no one single frequency for which the intensity is maximum. This behavior is typical of **nonthermal radiation,** which is *not* due simply to the temperature of the emitting object.

The particular type of nonthermal radiation emitted from the jet is *synchrotron radiation* (Figure 27-12). As we saw in the discussion of the Crab Nebula in Section 23-4, synchrotron radiation is produced by relativistic electrons traveling in a strong magnetic field. (Recall that "relativistic" means "traveling near the speed of light.") As the electrons spiral around the magnetic field, they emit electromagnetic radiation. The presence of synchrotron radiation coming from M87's jet indicates that relativistic particles are being ejected from the galaxy's nucleus and encountering a magnetic field.

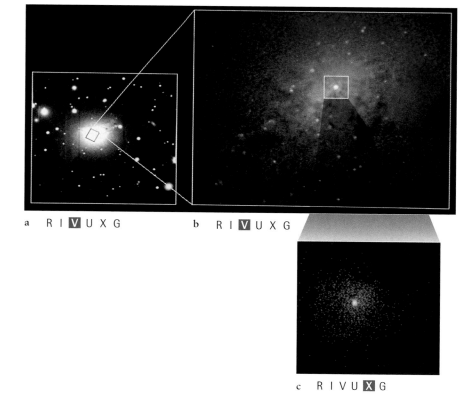

a   R I **V** U X G   b   R I **V** U X G

c   R I V U **X** G

WEB LINK 27.7
figure 27-10
A Seyfert Galaxy at Different Wavelengths

(a) This ground-based image shows the Seyfert galaxy NGC 1275, the largest and most luminous member of the Perseus cluster of galaxies located about 75 Mpc (250 million ly) from the Earth. Spectroscopic studies indicate that NGC 1275 is actually two galaxies in collision. **(b)** This Hubble Space Telescope image at visible wavelengths shows about 50 small, bright, blue objects surrounding NGC 1275. These are thought to be young globular clusters formed by the collision. The small white spot is the nucleus of the galaxy. **(c)** NGC 1275 is also a strong source of X rays and radio radiation. This X-ray image from the Einstein Observatory shows that most of the galaxy's X-ray emission comes from its nucleus. (a: NOAO; b: J. Holtzman, Lowell Observatory; S. Faber, University of California, Santa Cruz; and NASA; c: Harvard-Smithsonian Center for Astrophysics)

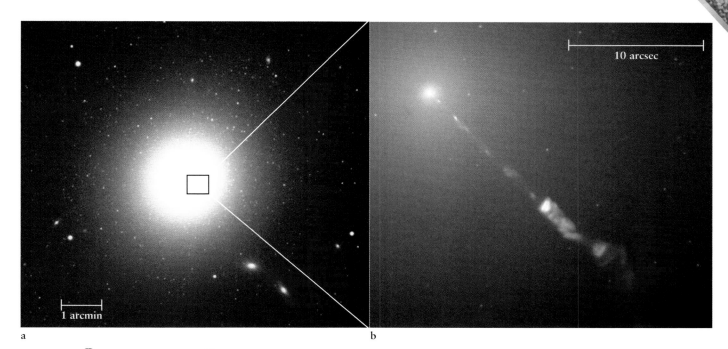

a   b

WEB LINK 27.8

ƒigure 27-11   R I **V** U X G

The Radio Galaxy M87 This giant elliptical galaxy is in the Virgo cluster, about 15 Mpc (50 million ly) from the Earth. **(a)** This photograph shows the full extent of the galaxy, which measures about 100 kpc (300,000 ly) across. The numerous fuzzy dots that surround the galaxy are globular clusters. **(b)** This

Hubble Space Telescope image shows the inner portion of the galaxy. From the galaxy's tiny, bright nucleus—less than 2 pc (6.5 ly) in diameter—a jet extends outward some 1500 pc (5000 ly). (a: David Malin/Anglo-Australian Observatory; b: NASA and The Hubble Heritage Team, STScI/AURA)

One piece of evidence that the radiation from M87's jet is nonthermal is that some of it is **polarized radiation,** which means that the electric fields of the waves are oriented in a specific direction. By contrast, thermal radiation—like light from the Sun, stars, or lightbulbs—is *unpolarized* radiation. It consists of waves whose electric fields are oriented at random angles. (Most radio stations transmit signals that are polarized in a vertical plane. In order to detect these signals, the radio antennas on cars are oriented vertically rather than horizontally.)

During the 1960s and 1970s, astronomers found that a radio galaxy generally consists of two **radio lobes,** one on each side of a parent galaxy. The radio lobes generally span a distance that is 5 to 10 times the size of the parent galaxy, although smaller and much larger examples are known. The parent galaxy—almost always a giant elliptical—sits midway between the radio lobes. Because of their overall shape, radio galaxies (and many similar quasars) are sometimes called **double radio sources.** Cygnus A, shown in Figure 27-1, is a fine example.

The spectrum and polarization of the radio-frequency emission from a radio galaxy bear all the characteristics of synchrotron radiation. This suggests that radio galaxies should have jets of relativistic particles. In fact, some radio galaxies appear to have a "head" of concentrated radio emission, with a weaker "tail" trailing behind it.

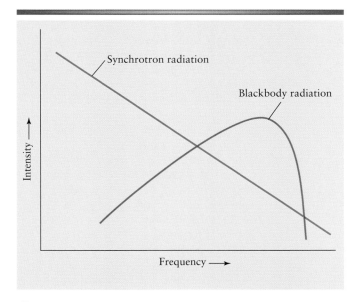

ƒigure 27-12

Thermal and Nonthermal Spectra This schematic graph compares the spectra of synchrotron radiation and blackbody radiation. A blackbody spectrum, like the spectrum of a star or ordinary galaxy, always has a hump. By contrast, the nonthermal spectrum of synchrotron radiation shows a steady decline with frequency.

figure 27-13   R I V U X G

The Head-Tail Source NGC 1265 The elliptical galaxy NGC 1265 would probably be an ordinary double radio source except that this galaxy is moving at a high speed through the intergalactic medium. Because of this motion, its two jets trail behind the galaxy, giving this radio source its distinctly windswept appearance. Like NGC 1275 (Figure 27-10), this galaxy is a member of the Perseus cluster of galaxies. (NRAO)

A good example of these **head-tail sources** is the elliptical galaxy NGC 1265 in the Perseus cluster of galaxies. This galaxy is known to be moving at a high speed (2,500 km/s) relative to the cluster as a whole. Figure 27-13 is a radio image of NGC 1265. Note that its radio emission has a distinctly windswept appearance. Just as smoke pouring from a steam locomotive trails behind a rapidly moving train, particles ejected along two jets from this galaxy are deflected by the galaxy's passage through the sparse intergalactic medium.

Like most giant ellipticals, many radio galaxies are found near the centers of rich clusters of galaxies and thus are probably subjected to collisions and mergers (see Section 26-7). A spectacular example is Centaurus A, one of the brightest radio sources in the southern sky. The peculiar parent galaxy of Centaurus A, shown in Figure 27-14, has a broad dust lane studded with young, hot, massive stars. This galaxy, called NGC 5128, is thought to be an elliptical galaxy that has collided with a spiral galaxy. Radio waves pour from two lobes on opposite sides of the galaxy. A second set of radio lobes farther from the galaxy (not shown in Figure 27-14) spans a volume 2 million light-years across. The inset in Figure 27-14 shows an X-ray-emitting jet emanating from the galaxy's nucleus. This jet, which is perpendicular to the galaxy's dust lane, is aimed toward one of the radio lobes.

The energy output of radio galaxies covers approximately the same range as that of Seyferts. As Table 27-2 shows, these galaxies bridge the gap between quasars and normal galaxies. Because they share so many properties of their remote, ultra-

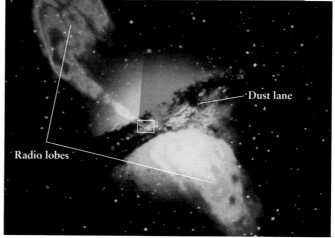

R I V U X G

figure 27-14

The Radio Galaxy NGC 5128 This extraordinary galaxy, also known as the radio source Centaurus A, is roughly 4 Mpc (13 million ly) from Earth. NGC 5128 is so large that despite its great distance, its angular diameter is half that of the full moon. The visible-light image shows a broad dust lane across the face of the galaxy. A bluish glow along the edges of the dust lane discloses the presence of numerous young stars. Superimposed on the visible-light image is a false-color radio image. Vast quantities of radiation pour from radio lobes that emanate from the galaxy nearly perpendicular to the dust lane. **Inset:** This image from the Einstein Observatory shows a jet glowing brightly at X-ray wavelengths. (Jack Burns, University of Missouri; inset: Harvard-Smithsonian Center for Astrophysics)

table 27-2	Galaxy and Quasar Luminosities
Object	**Luminosity (watts)**
Sun	4×10^{26}
Milky Way Galaxy	10^{37}
Seyfert galaxies	$10^{36} - 10^{38}$
Radio galaxies	$10^{36} - 10^{38}$
Quasars	$10^{38} - 10^{42}$

luminous cousins, at least some Seyfert galaxies may be the nearby relics or analogs of radio-quiet quasars, whereas radio galaxies may be the remnants of radio-loud quasars. (The connection between radio galaxies and radio-loud quasars is reinforced by the discovery that some high-redshift double radio sources have quasars midway between their radio lobes.) Thus, as quasars became extinct, they may have evolved into Seyfert and radio galaxies.

27-4 Quasars, blazars, Seyferts, and radio galaxies are active galaxies

Inspired by the discovery of quasars and luminous peculiar galaxies, astronomers during the 1960s and 1970s searched for clues to these powerful energy sources. One important clue turned out to be a new class of unusual starlike objects called *blazars*. Like quasars, blazars are extraordinarily luminous objects that proved to be the nuclei of distant galaxies. The difference is that unlike a quasar, the spectrum of a blazar is totally featureless, with *no* absorption lines and *no* emission lines at all!

The first of this new class of objects to be discovered was BL Lacertae (Figure 27-15), or BL Lac for short, in the constellation Lacerta (the Lizard). When discovered in 1929, it was mistaken for a variable star, largely because its brightness varies by a factor of 15 within only a few months. Careful examination, however, revealed some fuzz (visible in Figure 27-15) around its bright, starlike core. In the early 1970s, Joseph Miller at the University of California's Lick Observatory blocked out the light from the bright center of BL Lac and managed to obtain a spectrum of the fuzz. This spectrum, which contains stellar absorption lines, strongly resembles the spectrum of an elliptical galaxy. In other words, BL Lacertae is an elliptical galaxy with an intensely luminous center.

Objects such as BL Lacertae, also called **blazars** or **BL Lac objects,** emit polarized light with a featureless, nonthermal spectrum typical of synchrotron radiation (see Figure 27-12).

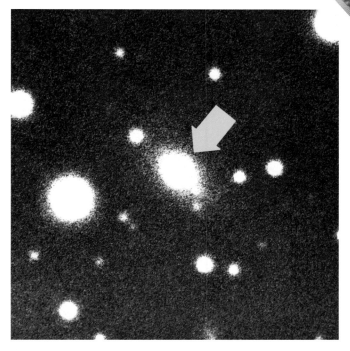

Figure 27-15   R I **V** U X G

BL Lacertae This photograph shows fuzz around the blazar BL Lacertae. The redshift of this fuzz indicates that BL Lacertae is about 280 Mpc (900 million ly) from Earth. Blazars are giant elliptical galaxies with bright, starlike nuclei that have many quasarlike properties. They contain much less gas and dust than Seyfert galaxies. (Courtesy of T. D. Kinman; Kitt Peak National Observatory)

Another important characteristic of blazars is their variability. For example, Figure 27-16 shows brightness fluctuations of the blazar 3C 279, determined by carefully examining old photographic plates on which it had been photographed inadvertently. Note the prominent outbursts that occurred

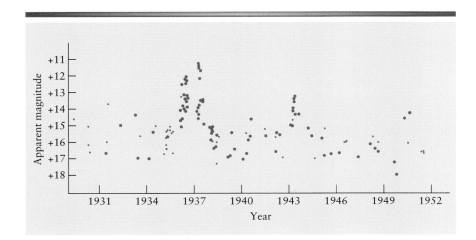

Figure 27-16

Brightness Variations of 3C 279 This graph shows variations in apparent magnitude (a measure of apparent brightness) for the blazar 3C 279, located in the constellation Coma Berenices. The data were obtained by carefully examining old photographic plates at Harvard College Observatory. Note the large outburst in 1937 and the somewhat smaller one in 1943. (Adapted from L. Eachus and W. Liller)

| July 1977 | March 1978 | July 1979 | July 1980 |

figure 27-17 **R** I V U X G

Superluminal Motion in 3C 273 These four images are false-color, high-resolution radio maps of the quasar 3C 273 (shown in the X-ray image that opens this chapter). They show blobs that seem to be moving apart at 10 times the speed of light. In fact, a beam of relativistic particles from 3C 273 is aimed almost directly at the Earth, giving the illusion of faster-than-light motion. (NRAO)

around 1937 and 1943. During these outbursts, the luminosity of 3C 279 increased by a factor of at least 25.

How do blazars relate to the other types of exotic objects we have encountered in this chapter? Detailed radio observations show that blazars are probably double radio sources—but oriented so that we see their jets end-on. For example, high-resolution studies of blazars with the Very Large Array have revealed diffuse radio emission around a bright core in almost all cases. A faint radio halo around a bright radio core is exactly what you would expect if you were looking straight down the jet from a radio galaxy. As seen from this angle, the galaxy's bright nucleus is surrounded by weaker emission from its radio lobes.

The idea that blazars are radio galaxies with their jets aimed toward the Earth was supported by the surprising discovery of movement that appeared to be faster than the speed of light. Such **superluminal motion** is also observed in some quasars, where very-long-baseline interferometry (described in Section 6-6) reveals a lumpy structure that changes with time. For example, Figure 27-17 shows four high-resolution images of the quasar 3C 273 spanning three years. During this interval, the two "blobs" that compose 3C 273 moved away from each other at a rate of almost 0.001 arcsec per year. Taking into account the distance to the source, this rate of angular separation corresponds to a speed 10 times that of light!

Superluminal motion was puzzling when it was first discovered, because it seemed to violate one of the basic tenets of the special theory of relativity: Nothing can move faster than light. Astronomers soon realized, however, that superluminal motion can be explained as movement slower than light—once we take into account the angle at which we view the radio source. If a relativistic beam of material is aimed close to your line of sight, it can appear to be moving faster than the speed of light. Figure 27-18 shows an example. Quasars generally exhibit lower superluminal speeds (1 to 5 c) than blazars (5 to 10 c), probably because the relativistic jets

from quasars are not aimed as close to our line of sight as are blazar jets.

Because of the many properties they share, quasars, blazars, Seyfert galaxies, and radio galaxies are now collectively called **active galaxies**. The activity of such a galaxy comes from an energy source at its center. Hence astronomers say that these galaxies possess **active galactic nuclei.**

An important clue about the energy sources that power active galactic nuclei comes from their variability. Some quasars vary in brightness over a few weeks or months. Some blazars actually fluctuate from night to night. X-ray observations reveal that some blazars vary in brightness over as few as 3 hours.

These fluctuations in brightness allow astronomers to place strict limits on the maximum size of a light source, because an object cannot vary in brightness faster than light can travel across that object. For example, an object that is one light-year in diameter cannot vary significantly in brightness over a period of less than one year.

To understand this limitation, imagine an object that measures one light-year across, as in Figure 27-19. Suppose the entire object emits a brief flash of light. Photons from that part of the object nearest the Earth arrive at our telescopes first. Photons from the middle of the object arrive at Earth six months later. Finally, light from the far side of the object arrives a year after the first photons. Although the object emitted a sudden flash of light, we observe a gradual increase in brightness that lasts a full year. In other words, the flash is stretched out over an interval equal to the difference in the light travel time between the nearest and most remote observable regions of the object.

The rapid flickering exhibited by active galactic nuclei means that they emit their energy from a small volume, possibly less than one light-day across. In other words, a region no larger than our solar system emits as much energy per second as a thousand galaxies! Astrophysicists therefore face the challenge of explaining how so much energy can be produced in such a very small volume.

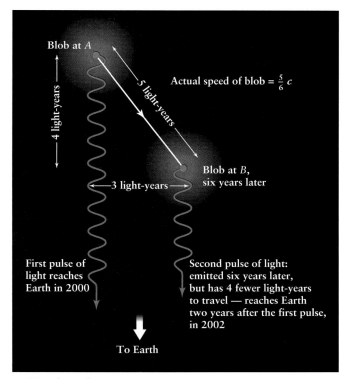

Blob at *A*

4 light-years

5 light-years

Actual speed of blob = $\frac{5}{6}c$

3 light-years

Blob at *B*, six years later

First pulse of light reaches Earth in 2000

Second pulse of light: emitted six years later, but has 4 fewer light-years to travel — reaches Earth two years after the first pulse, in 2002

To Earth

a   View from above

2000 ●————→———● 2002

←— 3 light-years —→

Apparent speed of blob = 1.5*c*

b   View from Earth

INTERACTIVE EXERCISE 27.2

ƒigure 27-18

An Explanation of Superluminal Motion An object moving almost directly toward the Earth at a high speed (but slower than light) can appear to be going faster than light. **(a)** If a blob of material ejected from a quasar moves at five-sixths of the speed of light, it covers the 5 light-years from point *A* to point *B* in six years. In the case shown here, it moves 4 light-years toward the Earth and 3 light-years in a transverse direction. The light emitted by the blob at *A* reaches us in 2000. The light emitted by the blob at *B* reaches us in 2002. The light left the blob at *B* six years later than the light from *A* but had four fewer light-years to travel to reach us. **(b)** From Earth we can see only the blob's transverse motion across the sky, as in Figure 27-18. It appears that the blob has traveled 3 light-years in just two years, so its apparent speed is ³/₂ of the speed of light, or 1.5 times *c*.

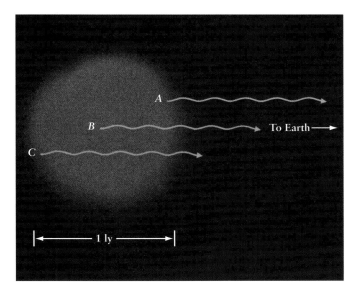

ƒigure 27-19

A Limit on the Speed of Variations in Brightness The rapidity with which the brightness of an object can vary significantly is limited by the time it takes light to travel across the object. Even if the object, shown here to be 1 light-year in size, emits a sudden flash of light, photons from point *A* arrive at Earth one year before photons from point *C*. Thus, as seen from Earth, the sudden flash is observed over a full year.

27-5   Supermassive black holes may be the "central engines" that power active galaxies

As long ago as 1968, the British astronomer Donald Lynden-Bell pointed out that a black hole lurking at the center of a galaxy could be the "central engine" that powers an active galactic nucleus. Lynden-Bell theorized that as gases fall onto a black hole, their gravitational energy would be converted into radiation. (As we saw in Section 24-3, a similar process produces radiation from black holes in close binary star systems.) To produce as much radiation as is seen from active galactic nuclei, the black hole would have to be very massive. But even a gigantic black hole would occupy a volume much smaller than our solar system—exactly what is needed to explain how active galactic nuclei can vary so rapidly in brightness.

How large a black hole would be needed to power an active galactic nucleus? You might think that what really matters is not the size of the black hole, but rather the amount of gas that falls onto it and releases energy. However, there is a natural limit to the luminosity that can be radiated by accretion onto a compact object like a black hole. This is called the **Eddington limit**, after the British astrophysicist Sir Arthur Eddington.

If the luminosity exceeds the Eddington limit, there is so much *radiation pressure*—the pressure produced by photons streaming outward from the infalling material—that the surrounding gas is pushed outward rather than falling inward onto the black hole. Without a source of gas to provide energy, the luminosity naturally decreases to below the Eddington limit, at which point gas can again fall inward. This limit allows us to calculate the minimum mass of an active galactic nucleus.

Numerically, the Eddington limit is:

Eddington limit

$$L_{\text{Edd}} = 30,000 \left(\frac{M}{\text{M}_\odot} \right) \text{L}_\odot$$

L_{Edd} = maximum luminosity that can be radiated by accretion onto a compact object

M = mass of the compact object

$\text{M}_\odot$ = mass of the Sun

$\text{L}_\odot$ = luminosity of the Sun

The luminosity of an active galactic nucleus must be less than its Eddington limit, so this limit must be very high indeed. Hence, the mass of the black hole must also be quite large. For example, consider the quasar 3C 273 (shown in the image that opens this chapter), which has a luminosity of about 3×10^{13} L$_\odot$. To calculate the minimum mass of a black hole that could continue to attract gas to power the quasar, assume that the quasar's luminosity equals the Eddington limit. Inserting $L_{\text{Edd}} = 3 \times 10^{13}$ L$_\odot$ into the above equation, we find that $M = 10^9$ M$_\odot$. Therefore, if a black hole is responsible for the energy output of 3C 273, its mass must be greater than a billion Suns.

Astronomers have indeed found evidence for such **supermassive black holes** at the centers of many nearby normal galaxies (see Section 24-4). As we saw in Section 25-6, at the center of our own Milky Way Galaxy lies what is almost certainly a black hole of about 3×10^6 solar masses—supermassive in comparison to a star but less than 1% the mass of the behemoth black hole at the center of 3C 273.

 Theory suggests that unlike stellar-mass black holes, which require a supernova to produce them (see Figure 22-25), supermassive black holes can be produced without extreme pressures or densities. This may help to explain why they appear to be a feature of so many galaxies.

One galaxy that probably has a black hole at its center is the Andromeda Galaxy (M31), shown in the image that opens Chapter 26. M31 is only 750 kpc (2.5 million ly) from Earth, close enough that details in its core as small as 1 parsec across can be resolved under the best seeing conditions.

In the mid-1980s, astronomers made high-resolution spectroscopic observations of M31's core. By measuring the Doppler shifts of spectral lines at various locations in the core, we can determine the orbital speeds of the stars about the galaxy's nucleus.

Figure 27-20 plots the results for the innermost 80 arcseconds of M31. (At M31's distance, this corresponds to a linear distance of 290 pc or 950 ly.) Note that the rotation

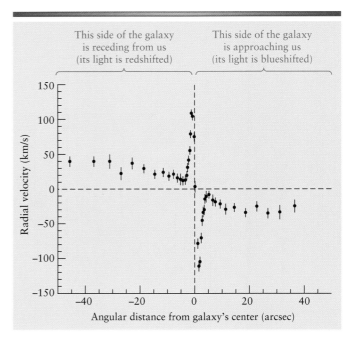

Figure 27-20

The Rotation Curve of the Core of M31 This graph plots radial velocity of matter in the core of M31 versus the angular distance from the galaxy's center. Note the sharp peaks, one blueshifted and one redshifted, within 5 arcsec of the galaxy's center. This indicates the presence of a compact, very massive object at the center of the galaxy. At the distance of M31, 1 arcsec corresponds to 3.6 pc (12 ly). (Adapted from J. Kormendy)

curve in the galaxy's nucleus does not follow the trend set in the outer core. Rather, there are sharp peaks—one on the approaching side of the galaxy and the other on the receding side—within 5 arcsec of the galaxy's center.

The most straightforward interpretation is that the peaks are caused by the rotation of a disk of stars orbiting M31's center. One side of the disk is approaching us while the other side is receding from us. The highest observed radial velocity (at 1.1 arcsec from the galaxy's center; see Figure 27-20) is 110 km/s. This is surely an underestimate, because unsteadiness of the Earth's atmosphere prohibits the detection of features smaller than about 0.5 arcsec across.

 The high-speed stars orbiting close to M31's center indicate the presence of a massive central object. Calculations using Newton's form of Kepler's third law (described in Section 4-7 and Box 4-4) show that there must be about 30 million (3×10^7) solar masses within 5 pc (16 ly) of the galaxy's center. That much matter confined to such a small volume strongly suggests the presence of a supermassive black hole. Observations of M31 with the Chandra X-ray Observatory are consistent with this picture.

By applying high-resolution spectroscopy to the cores of other nearby galaxies, astronomers have discovered a number of supermassive black holes like the one in M31. Unfortunately, this technique for identifying black holes is difficult to apply to quasars, which are very distant and have small angular sizes. Hence, the evidence for supermassive black holes in quasars is circumstantial: No other known energy source could provide enough power to sustain a quasar's intense light output.

If most galaxies have supermassive black holes at their centers, and supermassive black holes are the "central engines" of active galactic nuclei, why aren't all galaxies active? Why are there no nearby quasars? Why do quasars and radio galaxies have jets? As we will see in the next section, the answers to all these questions may be provided by a recent model of what happens in an active galactic nucleus.

27-6 A unified model may explain active galaxies of several different types

Accretion onto a supermassive black hole is the most likely explanation of the immense energy output of active galactic nuclei. The challenge to astrophysicists is to understand how that accretion takes place. A successful model of the accretion process must also explain other properties of active galactic nuclei, including their unusual spectra, variable light output, and energetic jets.

One scenario that has emerged to describe the accretion process is called the *unified model*, because it attempts to explain different types of active galaxies as just being different views of one type of object. In this model, at the heart of an active galaxy is a supermassive black hole surrounded by an **accretion disk**, an immense disk of matter captured by the hole's gravity and spiraling into it.

We saw in Section 20-5 that accretion disks are found around stars in the process of formation. The accretion disks that astrophysicists envision around supermassive black holes are similar but far larger and far more dynamic.

Imagine a billion-solar-mass black hole sitting at the center of a galaxy, surrounded by a rotating accretion disk. According to Kepler's third law, the inner regions of this accretion disk would orbit the hole more rapidly than would the outer parts. Thus, the rapidly spinning inner regions would constantly rub against the slower moving gases in the outer regions. This friction would cause the gases to lose energy and spiral inward toward the black hole.

As the gases move inward within the accretion disk, they are compressed and heated to very high temperatures. This causes the accretion disk to glow, thus producing the brilliant luminosity of an active galactic nucleus. Any variations in the density of the gas will cause the luminosity to fluctuate, giving rise to the brightness variations observed by astronomers (see Figure 27-16).

Supercomputer simulations can help us visualize the flow of matter in such an accretion disk. These simulations combine general relativity with equations describing gas flow. At first, matter accelerates to supersonic speeds as it spirals inward toward a black hole. But because the matter is rotat-

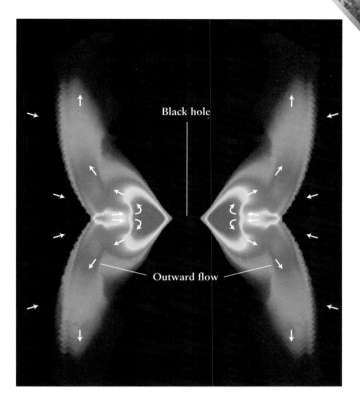

figure 27-21

Flow Patterns along the Inner Edge of an Accretion Disk This computer-generated picture shows the flow pattern of gas in an accretion disk surrounding a black hole. The black hole itself is located at the center of this cross-sectional diagram, where the broad-angled inner edges of the accretion disk point. Colors in the order of the rainbow represent gas pressure, from red for high pressure to blue for low pressure. The flow pattern is shown by white arrows. (Courtesy of J. F. Hawley and L. L. Smarr)

ing around the hole as well as moving toward it, this inward rush stops abruptly near the hole. The reason is the law of conservation of angular momentum (described in Section 7-7). Thanks to this law, rotating objects of all kinds tend to expand (like a spinning lump of dough that expands to make a pizza) rather than contract. The inward motion of gases stops where this tendency to expand outward balances the pull of the black hole's gravity. As a result, not all of the infalling gas reaches the black hole. Instead, part of the gas can become concentrated in high-speed orbits quite close to the hole. The sudden halt in the supersonic inflow creates a shock wave, which defines the inner edge of the accretion disk (Figure 27-21).

Because of the constant inward crowding of hot gases, pressures climb rapidly in the inner accretion disk. To relieve this congestion, matter is expelled from the hole along the inner edges of the empty funnel-shaped cavity. (The white arrows in Figure 27-21 show this outward flow.) What results are two broad, oppositely directed beams of relativistic particles that flow at right angles to the accretion disk.

figure 27-22

Jets from a Supermassive Black Hole This artist's rendering shows the accretion disk (red-yellow) surrounding a supermassive black hole. The rotation of the disk twists the disk's magnetic field lines into a helix, which channels the flow of subatomic particles pouring out of the disk. Over a distance of about a light-year, this channeling changes a broad flow into a tightly focused jet. The same process occurs on the other side of the disk (not shown). Figure 20-15 shows a similar process that takes place on a much smaller scale in protostars. (NASA and Ann Feild, Space Telescope Science Institute)

Magnetic forces play a crucial role in steering these fast-moving particles. These forces arise because the hot gases in the accretion disk are ionized, forming a plasma, and the motions of this plasma generate a magnetic field (see Section 18-9). As the plasma in the disk rotates around the black hole, it pulls the magnetic field along with it. But because the disk rotates faster in its inner regions than at its outer rim, the magnetic field becomes severely twisted. This twisted field forms two helix shapes, one on either side of the plane of the disk. Relativistic particles flowing outward from the accretion disk tend to follow these magnetic field lines. The result is that the outflowing beams of particles are focused into two jets oriented perpendicular to the plane of the accretion disk, as Figure 27-22 shows. Figure 27-23 is an artist's conception of the accretion disk and jets as seen from a nearby planet.

Recent observations of active galaxies provide evidence in support of this model. Figure 27-24a shows two radio lobes that extend some 15 kpc (50,000 ly) from the nucleus of the galaxy NGC 4261. The highly magnified Hubble Space Telescope image in Figure 27-24b shows a disk some 250 pc (800 ly) in diameter at the very center of the galactic nucleus. The speed of gas and dust in the disk

(measured from Doppler shifts) indicates that a black hole of 1.2×10^9 solar masses lies at its center. Presumably a (much smaller) accretion disk lies at the center of this large disk. The plane of the accretion disk is perpendicular to the radio lobes, just as we would expect if the lobes are the endpoints of relativistic jets like those in Figures 27-22 and 27-23.

The model we have discussed—a supermassive black hole surrounded by an accretion disk with oppositely directed high-speed jets—also offers a single explanation for several types of active galaxies. Many astronomers suspect that the main difference between double radio sources, radio-loud quasars, and blazars is only the angle at which the black hole "central engine" is viewed. As Figure 27-25 shows, an observer sees a double radio source when the accretion disk is viewed nearly edge-on, so that the jets are nearly in the plane of the sky. This is the case for the galaxy in Figure 27-24.

figure 27-23

At the Core of an Active Galaxy This artist's rendering shows a supermassive black hole surrounded by an accretion disk. In the inner regions of the accretion disk, matter crowding toward the hole is diverted outward along two oppositely directed jets. As the relativistic particles move in spiral paths within the jets, they emit synchrotron radiation. (*Astronomy*)

a   R I V̅ U X G   b   R I V̅ U X G

figure 27-24

A Disk and Jets in NGC 4261
The giant elliptical galaxy NGC 4261 is a double radio source located in the Virgo cluster, about 30 Mpc (100 million ly) from Earth. **(a)** This image combines a visible-light image of the galaxy (white) with a radio image of the jets (orange). **(b)** This Hubble Space Telescope image of the nucleus of NGC 4261 shows a disk of gas and dust about 250 pc (800 ly) in diameter. The radio jets are perpendicular to the plane of this disk. Observations indicate that a supermassive black hole is at the center of the disk. (NASA, ESA)

At a steeper angle, the observer sees a radio-loud quasar if the luminosity is high and a radio galaxy if the luminosity is low. At this angle the observer is looking at the accretion disk nearly face-on. Excited gases in the disk produce the emission lines that are characteristic of the spectra of quasar and radio galaxies (see Figure 27-6).

If one of the jets is aimed almost directly at the Earth, a blazar is observed. The radiation we see is dominated by synchrotron radiation from the jet, which has a continuous spectrum with no emission or absorption lines (see Figure 27-12). This helps explain why the spectra of blazars are so featureless.

The unified model can also explain the two types of spectra seen in Seyfert galaxies. *Type 1 Seyfert galaxies* have both broad and narrow emission lines, usually with a narrow, weak emission peak superimposed on each of the broad lines. The spectra of *Type 2 Seyfert galaxies* (such as NGC 7742, shown in Figure 27-9) have only narrow emission lines. These spectral differences may simply be the result of viewing the "central engine" from different angles. A Type 2 Seyfert galaxy may be a mostly edge-on view of matter spiraling in toward a supermassive black hole, whereas a Type 1 Seyfert provides a more pole-on view, allowing us to see the turbulent region around the black hole. Because gases move at many different velocities in this region, we see spectral lines that have been broadened by the Doppler effect.

The accretion-disk idea also helps to explain why there are no nearby quasars (see Figure 27-5). Over time, most of the available gas and dust surrounding a quasar's "central engine" is accreted onto the black hole. The "central engine" becomes starved of new infalling matter to act as "fuel," and the quasar becomes much less active. Hence, with the passage of time, a quasar might turn off altogether. As a result, there are essentially no quasars in the present-day universe. Perhaps galaxies such as our own Milky Way, which have supermassive black holes at their centers but are not presently active, once led much more dramatic lives as quasars or radio galaxies.

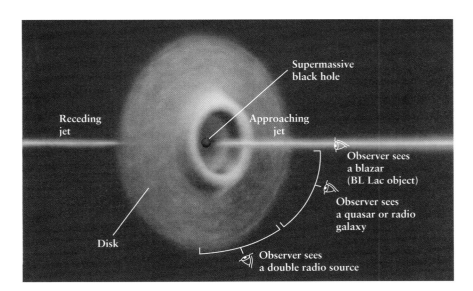

figure 27-25

A Unified Model of Active Galaxies Double radio sources, quasars, and blazars may be the same type of object—a supermassive black hole, its accretion disk, and its relativistic jets—viewed at different angles. If the jets are nearly perpendicular to our line of sight, we see a double radio source. If the jet is somewhat tilted to our line of sight, we see a radio-loud quasar. If one of the jets is aimed almost exactly at the Earth, we see a blazar.

Quasars, Active Galaxies, and Gamma-Ray Bursters | **633**

Finally, the accretion-disk idea explains what happens when active galaxies collide or merge with other galaxies. Such collisions and mergers can transfer gas and dust from one galaxy to another, providing more "fuel" for a supermassive black hole. Thus, it is not surprising that many of the most luminous active galaxies are also those that are undergoing collisions, such as the Seyfert galaxy NGC 1275 in Figure 27-10.

The probable existence of supermassive black holes in active galaxies is one of the most exciting topics in modern astronomy. In the coming years, instruments like the Hubble Space Telescope and the twin Keck telescopes will be used to further probe the nuclei of galaxies with unprecedented resolution. These observations may give us a much better understanding of the powerful engines at the cores of active galaxies and quasars.

27-7 Gamma–ray bursters are among the most powerful explosions in the known universe

 Astronomers in the 1960s used the large redshifts of quasars to show that these exotic objects are very distant. Astronomers in the 1990s used the same logic when they discovered large redshifts for **gamma-ray bursters,** mysterious objects that emit intense bursts of high-energy radiation. They concluded that gamma-ray bursters must, like quasars, be at tremendous distances from the Earth. (Except for their great distance, gamma-ray bursters seem to be quite unrelated to quasars.)

Gamma-ray bursters were discovered in the late 1960s by the orbiting Vela satellites, whose detectors noticed flashes of gamma rays coming from random parts of the sky at random intervals. This discovery was an unexpected consequence of the Cold War. The Vela satellites were originally placed in orbit by the United States to look for high-energy photons coming from Soviet nuclear weapon tests, which had been banned by treaty since 1963.

A typical gamma-ray burst can last from a hundredth of a second to the better part of an hour, after which the source fades to invisibility. Unlike X-ray bursters (see Section 23-8), gamma-ray bursters appear to emit only one burst in their entire history.

What are gamma-ray bursters, and how far away are they? These questions plagued astronomers for almost 30 years after the discovery of gamma-ray bursters. One clue is that gamma-ray bursters are seen with roughly equal probability in all parts of the sky, as Figure 27-26 shows. This suggests that they are not in the disk of our Galaxy, because then most gamma-ray bursters would be in the plane of the Milky Way. One idea was that gamma-ray bursters are relatively close to us and lie in a spherical halo surrounding the Milky Way, rather like globular clusters (see Figure 25-8). Alternatively, gamma-ray bursters could be strewn across space like galaxies, with some of them billions of light-years away.

 Figure 27-26 R I V U X **G**

Gamma-Ray Bursters This map shows the locations of 1776 gamma-ray bursters detected by the Compton Gamma Ray Observatory (see Section 6-7). The entire celestial sphere is mapped onto an oval, with the Milky Way stretching horizontally across the center. Unlike X-ray bursters, which originate in the disk of the Milky Way Galaxy, gamma-ray bursters are seen in all parts of the sky. The colors in the order of the rainbow indicate the total amount of energy detected from each burst; bright bursts appear in red and weak bursts appear in purple. (Compare with Figure 6-34.)

For many years there was no convincing way to decide between these competing models. The problem was that unlike the situation for quasars, astronomers were unable to find optical counterparts for gamma-ray bursters. With no spectral lines to examine, they could not tell whether gamma-ray bursters orbit our Galaxy (in which case their spectra should show little or no redshift) or are at galactic distances (in which case many of them should have substantial redshifts).

This state of affairs changed dramatically in 1997, thanks to the Italian-Dutch BeppoSAX satellite. Unlike previous gamma-ray satellites, which could determine the position of a burst only to within a few degrees, BeppoSAX could pin down its position to within an arcminute. This made it far easier for astronomers using optical telescopes to search for a counterpart to the burster.

Just 21 hours after BeppoSAX observed a gamma-ray burster on February 28, 1997, astronomers led by Jan van Paradis of the University of Amsterdam located a visible-light "afterglow" of the burst. And when BeppoSAX detected another gamma-ray burster on May 8, Mark Metzger and his colleagues from Caltech were able to record the visible-light spectrum of its "afterglow." In the spectrum they found absorption lines of iron and magnesium, and with a substantial redshift: $z = 0.835$. The Hubble law immediately told the astronomers that the gamma-ray burster was billions of light-years away. (The absorption lines could actually have been formed when light from the burster passed through a galaxy that happened to lie between the burster and the Earth. If so, the gamma-ray burster itself could have an even greater redshift and be even farther away.)

Since 1997, astronomers using large visible-light and infrared telescopes have detected the "afterglow" of many

a December 1997

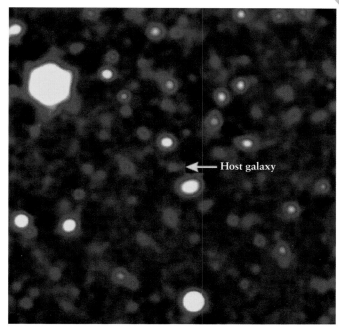

b February 1998

Figure 27-27   R I **V** U X G

The Fading Light of a Gamma-Ray Burster

(a) This false-color image from the 10-meter W. M. Keck telescope shows the visible-light "afterglow" of a gamma-ray burster detected on December 14, 1997. This image was made two days later. This burster had a redshift $z = 3.418$, which implies that the burst of radiation was emitted about 12 billion years ago. **(b)** By the following February the "afterglow" had faded, revealing a faint galaxy at the position of the burster. (S. G. Djorgovski and S. R. Kulkarni, Caltech; the Caltech GRB Team; and W. M. Keck Observatory)

gamma-ray bursters. While the gamma-ray intensity of a typical burster fades away within 100 seconds or so, the intensity at visible-light and infrared wavelengths can persist for days (Figure 27-27). This enables astronomers to make detailed observations of the burster's spectrum and redshift. By early 2001, the most distant gamma-ray burster seen had a redshift $z = 4.51$. From Table 27-1, this implies that the light from this gamma-ray burster was emitted more than 12 billion years ago, when the universe was only about a tenth of its present age.

If gamma-ray bursters are so very distant, they must be incredibly luminous for our gamma-ray satellites to detect them. If we assume that a gamma-ray burster emits light equally in all directions, the inverse-square law (Section 19-2) tells us that the most energetic bursters emit about 3×10^{47} joules of energy in less than a minute. From Einstein's equation $E = mc^2$, this is equivalent to converting more than a solar mass of matter completely into energy! This is about 10 times more electromagnetic radiation than is emitted by a supernova (although, as we saw in Section 22-8, Type II supernovae emit even more energy in the form of neutrinos).

What could be the source of these intense bursts? An important clue has come from visible-light and infrared observations, which in a few cases have

discerned a galaxy at the same location as a gamma-ray burster (see Figure 27-27). As Figure 27-28 shows, the bursters do not appear at the centers of their host galaxies. Hence, they are unlikely to be a flare-up of an active galactic nucleus. Instead, gamma-ray bursters may be the result of some cataclysmic event involving one of a galaxy's stars.

Astrophysicists have proposed a variety of models to explain the energy output of gamma-ray bursters. One idea is the merger of two neutron stars; another is a *hypernova*, a postulated, unusually energetic supernova. In these models, the energy is released through the formation of a black hole.

An alternative proposal is that gamma-ray bursters are produced by relativistic "cannonballs" of plasma ejected from an ordinary supernova. In this model, the "cannonballs" are heated to extraordinarily high temperatures as they fly through the slower-moving gas that expands away from the supernova. Such ultra-hot, fast-moving fireballs would emit high-energy photons in a narrow beam pointed along the direction of their motion. If the beam happens to be aimed toward Earth, we would detect an intense burst of gamma rays. Because the radiation is beamed rather than emitted in all directions, this model does not require the immense release of energy that the inverse-square law would suggest.

figure 27-28   R I **V** U X G

The Host Galaxy of a Gamma-Ray Burster The Hubble Space Telescope recorded this visible-light image on February 8, 1999, 16 days after a gamma-ray burster was observed at this location. The image shows a faint galaxy that is presumably the home of whatever object produced the gamma-ray burst. The galaxy has a very blue color, indicating the presence of many recently formed stars. The gamma-ray burst may have been produced when one of the most massive of these stars became a supernova. (Andrew Fruchter, STScI; and NASA)

 An ordinary flashlight is an example of beamed radiation. The lightbulb in a flashlight is very small and would produce only a weak light if you removed it from the flashlight housing. But the curved mirror in a flashlight channels the bulb's light into a narrow beam. Thus, a flashlight produces an intense beam of light with only a small input of energy from the batteries.

 Although we have learned a great deal about gamma-ray bursters in recent years, the question of their fundamental nature remains unanswered. Astronomers hope that answers will come from a new generation of gamma-ray telescopes to be placed in orbit by NASA and the European Space Agency. Perhaps the next few years will bring us a deeper understanding of the mysterious gamma-ray bursters.

KEY WORDS

accretion disk, p. 631

active galactic nucleus, p. 628

active galaxy, p. 628

blazar, p. 627

BL Lac object, p. 627

double radio source, p. 625

Eddington limit, p. 629

gamma-ray burster, p. 634

head-tail source, p. 626

nonthermal radiation, p. 624

polarized radiation, p. 625

quasar, p. 619

radio galaxy, p. 623

radio lobes, p. 625

Seyfert galaxy, p. 623

superluminal motion, p. 628

supermassive black hole, p. 630

thermal radiation, p. 624

KEY IDEAS

Quasars: A quasar looks like a star but has a huge redshift. This corresponds to an extreme distance of several hundred megaparsecs or more from the Earth, according to the Hubble law.

• To be seen at such large distances, quasars must be very luminous, typically about 1000 times brighter than an ordinary galaxy.

• About 10% of all quasars are strong sources of radio emission and are therefore called "radio-loud"; the remaining 90% are "radio-quiet," because they are weak radio sources at best.

• Some of the energy emitted by quasars is synchrotron radiation produced by high-speed particles traveling in a strong magnetic field.

Seyfert Galaxies: Seyfert galaxies are spiral galaxies with bright nuclei that are strong sources of radiation. Seyfert galaxies seem to be nearby, low-luminosity, radio-quiet quasars.

Radio Galaxies: Radio galaxies are elliptical galaxies located midway between the lobes of a double radio source. Radio galaxies seem to be nearby, low-luminosity, radio-loud quasars.

- Relativistic particles are ejected from the nucleus of a radio galaxy along two oppositely directed beams.

Blazars: Blazars (BL Lac objects) are bright, starlike objects that can vary rapidly in their luminosity. They are probably radio galaxies seen end-on, with a jet of relativistic particles aimed toward the Earth.

- Rapid fluctuations in the brightness of blazars indicate that the energy-producing region is quite small.

Active Galaxies: Quasars, blazars, and Seyfert and radio galaxies are examples of active galaxies. The energy source at the center of an active galaxy is called an active galactic nucleus.

Black Holes and Active Galactic Nuclei: The preponderance of evidence suggests that an active galactic nucleus consists of a supermassive black hole onto which matter accretes.

- As gases spiral in toward the supermassive black hole, some of the gas may be redirected to become two jets of high-speed particles that are aligned perpendicularly to the accretion disk.

- An observer sees a double radio source when the accretion disk is viewed nearly edge-on, so that the jets are nearly in the plane of the sky. At a steeper angle, the observer sees a quasar. If one of the jets is aimed almost directly at the Earth, a blazar is observed.

Gamma-Ray Bursters: Short, intense bursts of gamma rays are observed at random times coming from random parts of the sky.

- By observing the afterglow of gamma-ray bursters, astronomers find that these objects have very large redshifts and appear to be located within distant galaxies. The origin of the bursts is unknown.

REVIEW QUESTIONS

1. How were quasars first discovered? How was it discovered that they are very distant objects?

2. Suppose you saw an object in the sky that you suspected might be a quasar. What sort of observations might you perform to test your suspicion?

3. Explain why astronomers cannot use any of the standard candles described in Section 26-4 to determine the distances to quasars.

4. How does the spectrum of a quasar differ from that of an ordinary galaxy? How do spectral lines help astronomers determine the distances to quasars?

5. It was suggested in the 1960s that quasars might be compact objects ejected at high speeds from the centers of nearby ordinary galaxies. Explain why the absence of blueshifted quasars disproves this hypothesis.

6. How would you distinguish between thermal and nonthermal radiation?

7. What do astronomers learn from the shapes of the spectral lines of quasars?

8. How do astronomers know that quasars are located in galaxies? In what sorts of galaxies are they found?

9. What is a Seyfert galaxy? Why do astronomers think that Seyfert galaxies may be related to radio-quiet quasars?

10. What is a radio galaxy? What is a double radio source? Why do astronomers think these objects may be related to radio-loud quasars?

11. What is a blazar? What is unique about its spectrum? How is it related to other active galaxies?

12. What do the brightness fluctuations of a particular active galaxy tell us about the size of the energy-emitting region within that galaxy?

13. How could a supermassive black hole, from which nothing—not even light—can escape, be responsible for the extraordinary luminosity of a quasar?

14. What is the Eddington limit? Explain how it can be used to set a limit on the mass of a supermassive black hole, and explain why this limit represents a minimum mass for the black hole.

15. Explain how the rotation curve of a galactic nucleus can help determine whether a supermassive black hole is present.

16. How does matter falling inward toward a central black hole find itself being ejected outward in a high-speed jet?

17. Explain how the unified model of active galaxies accounts for the basic properties of quasars, blazars, and double radio sources.

18. Why do you suppose there are no quasars relatively near our Galaxy?

19. What is a gamma-ray burster? What is the evidence that gammy-ray bursters are not located in the disk of our Galaxy or in a halo surrounding our Galaxy?

20. Summarize the evidence that gamma-ray bursters result from some process involving a star in a distant galaxy.

ADVANCED QUESTIONS

Questions preceded by an asterisk (*) involve topics discussed in the Boxes.

Problem-solving tips and tools
Table 27-1 gives the connection between redshift and distance, Box 27-1 discusses relativistic redshift, and Box 24-1 describes time dilation in the special theory of relativity. You may find it useful to recall from Section 1-7 that 1 light-year = 63,240 AU. A speed of 1 km/s is the same as 0.211 AU/yr. You can find Newton's form of Kepler's third law in Box 4-4, and the formula for the Schwarzschild radius in Box 24-2.

21. When we observe a quasar with redshift $z = 0.75$, how far into its past are we looking? If we could see that quasar as it really is right now (that is, if the light from the quasar could somehow reach us instantaneously), would it still look like a quasar? Explain why or why not.

*22. What redshift would be observed for an object receding from the Sun at 99% of the speed of light?

*23. The quasar SDSS 1044–0125 has a redshift $z = 5.80$. At what speed does this quasar seem to be receding from us? Give your answer in km/s and as a fraction of the speed of light c.

*24. Imagine driving down a street toward a traffic light. How fast would you have to go so that the red light ($\lambda = 700$ nm) would appear green ($\lambda = 500$ nm)?

*25. Suppose that an astronomer discovers a quasar with a redshift of 8.0. With what speed would this quasar seem to be receding from us? Give your answer in km/s and as a fraction of the speed of light.

26. Explain how the existence of gravitational lenses involving quasars (review Section 26-8) constitutes evidence that quasars are located at the great distances inferred from the Hubble law.

27. Explain why the blue color of the globular clusters in NGC 1275 (Figure 27-10b) indicates that these clusters formed relatively recently, perhaps as a result of the collision with another galaxy.

*28. Suppose a blazar at $z = 1.00$ goes through a fluctuation in brightness that lasts one week (168 hours) as seen from Earth. (a) At what speed does the blazar seem to be moving away from us? (b) Using the idea of time dilation, determine how long this fluctuation lasted as measured by an astronomer within the blazar's host galaxy. (c) What is the maximum size (in AU) of the region from which this blazar emits energy?

29. Suppose the distance from point A to point B in Figure 27-18 is 26 light-years and the blob moves at 13/15 of the speed of light. As the blob moves from A to B, it moves 24 light-years toward the Earth and 10 light-years in a transverse direction. (a) How long does it take the blob to travel from A to B? (b) If the light from the blob at A reached Earth in 2000, in what year does the light from B reach Earth? (c) As seen from Earth, at what speed does the blob appear to move across the sky?

30. (a) Calculate the maximum luminosity that could be generated by accretion onto a black hole of 3×10^6 solar masses. (This is the size of the black hole found at the center of the Milky Way, as described in Section 25-6.) Compare this to the total luminosity of the Milky Way (see Table 27-2). (b) Speculate on what we might see if the center of our Galaxy became an active galactic nucleus with the luminosity you calculated in (a).

31. Observations of a certain galaxy show that stars at a distance of 16 pc from the center of the galaxy orbit the center at a speed of 200 km/s. Use Newton's form of Kepler's third law to determine the mass of the central black hole.

*32. Calculate the Schwarzschild radius of a 10^9-solar-mass black hole. How does your answer compare with the size of our solar system (given by the diameter of Pluto's orbit)?

33. Figure 27-24 shows the double radio source NGC 4261. Is it possible that somewhere in the universe there is an alien astronomer who observes this same object as a blazar? Explain your answer with a drawing showing the relative positions of the Earth, the alien astronomer, and NGC 4261.

34. Contrast gamma-ray bursters with X-ray bursters (discussed in Section 23-8). From our models of what causes these energetic phenomena, explain why X-ray bursters emit repeated pulses but gamma-ray bursters apparently emit just once.

DISCUSSION QUESTIONS

35. The accompanying image from the Very Large Array (VLA) shows the radio galaxy 3C 75 in the constellation Cetus. This galaxy has several radio-emitting jets. High-resolution optical photographs reveal that the galaxy has two nuclei, which are the two red spots near the center of the VLA image. Propose a scenario that might explain the appearance of 3C 75.

(NRAO)   R I V U X G

36. Some quasars show several sets of absorption lines whose redshifts are less than the redshifts of their emission lines. For example, the quasar PKS 0237-23 has five sets of absorption lines with redshifts in the range from 1.364 to 2.202, whereas the quasar's emission lines have a redshift of 2.223. Propose an explanation for these sets of absorption lines.

WEB/CD-ROM QUESTIONS

37. Search the World Wide Web for information about "microquasars." These are objects that are found *within* the Milky Way Galaxy. How are they detected? What are the similarities and differences between these objects and true quasars? Are they long-lasting or short-lived?

38. Search the World Wide Web for information about the supernova SN 1998bw. What was unusual about this supernova? What did SN 1998bw suggest about the origin of gamma-ray bursters?

39. The *Lockman hole* is a region in the constellation Ursa Minor where the Milky Way's interstellar hydrogen is the thinnest. By observing in this part of the sky, astronomers get the clearest possible view of distant galaxies and quasars. Search the World Wide Web for information about observations of the Lockman hole. What has been learned through X-ray observations made by spacecraft such as ROSAT and XMM-Newton?

 40. Relativistic Redshift. Access the Active Integrated Media Module "Relativistic Redshift" in Chapter 27 of the *Universe* web site or CD-ROM. Use this to calculate the redshift and recessional velocity of a quasar in whose spectrum the H_α emission line of hydrogen (unshifted wavelength 656 nm) appears at a wavelength of (**a**) 937 nm and (**b**) 5000 nm.

OBSERVING PROJECTS

41. Use a telescope with an aperture of at least 20 cm (8 in.) to observe the Seyfert galaxy NGC 1068 (also known as M77). Located in the constellation Cetus (the Whale), this galaxy is most easily seen from September through January. The epoch 2000 coordinates are R.A. = 2^h 2.7^m and Decl. = −0° 01′. Sketch what you see. Is the galaxy's nucleus diffuse or starlike? How does this compare with other galaxies you have observed?

42. Use a telescope with an aperture of at least 20 cm (8 in.) to observe the two companions of the Andromeda Galaxy, M32 and M110. Both are small elliptical galaxies, but only M32 is suspected of harboring a supermassive black hole. They are located on opposite sides of the Andromeda Galaxy and are most easily seen from September through January. The epoch 2000 coordinates are:

Galaxy	Right ascension	Declination
M32 (NGC 221)	0^h 42.7^m	+40° 52′
M110 (NGC 205)	0 40.4	+41 41

Make a sketch of each galaxy. Can you see any obvious difference in the appearance of these galaxies? How does this difference correlate with what you know about these galaxies?

43. If you have access to a telescope with an aperture of at least 30 cm (12 in.), you should definitely observe the Sombrero Galaxy, M104, which is suspected to have a supermassive black hole at its center. It is located in Virgo and can most easily be seen from March through July. The epoch 2000 coordinates are R.A. = 12^h 40.0^m and Decl. = −11° 37′. Make a sketch of the galaxy. Can you see the dust lane? How does the nucleus of M104 compare with the centers of other galaxies you have observed?

44. If you have access to a telescope with an aperture of at least 30 cm (12 in.), observe M87 and compare it with the other two giant elliptical galaxies, M84 and M86, that dominate the central regions of the Virgo cluster. Use the *Starry Night* program to help you plan your observations. The epoch 2000 coordinates are:

Galaxy	Right ascension	Declination
M84 (NGC 4374)	12^h 25.1^m	+12° 53′
M86 (NGC 4406)	12 26.2	+12 57
M87 (NGC 4486)	12 30.8	+12 24

45. If you have access to a telescope with an aperture of at least 40 cm (16 in.), you might try to observe the brightest-appearing quasar, 3C 273, which has an apparent magnitude of nearly +13. It is located in Virgo at coordinates R.A. = 12^h 29^m 07^s and Decl. = +2° 03′ 07″.

 46. If the Milky Way had an active galactic nucleus, with an accretion disk around its central black hole, there might be a pair of relativistic jets emanating from its center. Use the *Starry Night* program to investigate how these jets might appear from Earth. In the Control Panel, set the date and time to June 15 of this year at 12:00:00 A.M. (midnight), when the center of the Milky Way is prominent in the sky. Use the **Find...** command in the **Edit** menu to center the field of view on the star TYC6840-25-1, whose position on the celestial sphere is less than 1° from the black hole at the center of the Milky Way. (To make the Milky Way visible, make sure that the selection **Milky Way** in the **Sky** menu has a check mark next to it. If you find the Milky Way difficult to see, select **Sky Settings...** in the **Sky** menu and click on **Milky Way Colour.** Choose a color that will stand out against the dark night sky, then click on **OK.**) From what you see, make a drawing showing how the night sky would appear on June 15 at 12:00:00 A.M. if our Galaxy had an active galactic nucleus. Label the Milky Way, the jets, the central black hole, and the accretion disk. Assume that the plane of the accretion disk is aligned with the plane of the Milky Way.

Cosmology: The Creation and Fate of the Universe

SN 1997cj
z = 0.55

SN 1997ce
z = 0.44

SN 1997ck
z = 0.97

(Peter Garnavich, Harvard-Smithsonian Center for Astrophysics, the High-Z
Supernova Search Team, and NASA)

R I **V** U X G

So far in this book we have cataloged the contents of the universe. Our scope has ranged from submicroscopic objects, such as atomic nuclei, to superclusters of galaxies hundreds of millions of light-years across. In between, we have studied planets, moons, and stars.

But now we turn our focus beyond what we find in the universe to the nature of the universe itself—the subject of the science called *cosmology*. How large is the universe? What is its structure? How long has it existed, how has it changed over time, and how will it continue to change in the distant future?

In this chapter we will see that the universe is expanding. This expansion began with an explosion at the beginning of time called the Big Bang. We will see direct evidence of the Big Bang in the form of microwave radiation from space. This radiation is the faint afterglow of a primordial fireball that filled all space shortly after the beginning of the universe.

Will the universe continue to expand forever, or will it eventually collapse back on itself? We will find that to predict the future of the universe, we must first understand what happened in the remote past. To this end, astronomers study luminous supernovae like the examples shown in the accompanying image. These can be seen across billions of light-years, and so can tell us about conditions in the universe billions of years ago. We will see how recent results from such supernovae have revolutionized our understanding of cosmology and given us new insights into our place in the cosmos.

As you read the sections of this chapter, look for the answers to the following questions.

28-1 What does the darkness of the night sky tell us about the nature of the universe?

28-2 As the universe expands, what, if anything, is it expanding into?

28-3 Where did the Big Bang take place?

28-4 How do we know that the Big Bang was hot?

28-5 What was the universe like during its first 300,000 years?

28-6 How is it possible to measure how the universe is curved?

28-7 What is "dark energy"?

28-8 Has the universe always expanded at the same rate?

28-9 What will happen if the universe keeps expanding forever?

28-1 The darkness of the night sky tells us about the nature of the universe

Cosmology is the science concerned with the structure and evolution of the universe as a whole. One of the most profound and basic questions in cosmology may at first seem foolish: Why is the sky dark at night? This question, which haunted Johannes Kepler as long ago as 1610, was brought to public attention in the early 1800s by the German amateur astronomer Heinrich Olbers.

Olbers and his contemporaries pictured a universe of stars scattered more or less randomly throughout infinite space. Isaac Newton himself thought that no other model made sense. The gravitational forces between any *finite* number of stars, he argued, would in time cause them all to fall together, and the universe would soon be a compact blob.

Obviously, this has not happened. Newton concluded that we must be living amid a static, infinite expanse of stars. In this model, the universe is infinitely old, and it will exist forever without major changes in its structure. Olbers noticed, however, that a static, infinite universe presents a major puzzle.

If space goes on forever, with stars scattered throughout it, then any line of sight must eventually hit a star. In this case, no matter where you look in the night sky, you should ultimately see a star. The entire sky should be as bright as an average star, so, even at night, the sky should blaze like the surface of the Sun. **Olbers's paradox** is that the night sky is actually *dark* (Figure 28-1).

Olbers's paradox suggests that something is wrong with Newton's infinite, static universe. According to the classical, Newtonian picture of reality, space is like a gigantic flat sheet of inflexible, rectangular graph paper. (Space is actually three-dimensional, but it is easier to visualize just two of its three dimensions. In a similar way, an ordinary map represents the three-dimensional surface of the Earth, with its hills and valleys, as a flat, two-dimensional surface.)

This rigid, flat, Newtonian space stretches on and on, totally independent of stars or galaxies or anything else. The same is true of time in Newton's view of the universe; a New-

tonian clock ticks steadily and monotonously forever, never slowing down or speeding up. Furthermore, Newtonian space and time are unrelated, in that a clock runs at the same rate no matter where in the universe it is located.

Albert Einstein overturned this view of space and time. His special theory of relativity (recall Section 24-1 and Box 24-1) shows that measurements with clocks and rulers depend on the motion of the observer. What is more, Einstein's general theory of relativity (Section 24-2) tells us that gravity curves the fabric of space.

figure 28-1   R I **V** U X G

The Dark Night Sky If the universe were infinitely old and filled uniformly with stars, the night sky would be ablaze with light. The observation that the night sky is dark, punctuated only by the light from isolated stars, tells us that the universe is quite different from this simple picture. (Okiro Fujii, *L'Astronomia*)

As a result, the matter that occupies the universe influences the overall shape of space throughout the universe.

If we represent the universe as a sheet of graph paper, the sheet is not perfectly flat but has a dip wherever there is a concentration of mass, such as a person, a planet, or a star (see Figure 24-4). Because of gravitational effects, clocks run at different rates depending on whether they are close to or far from a massive object, as Figure 24-7a shows.

What does the general theory of relativity, with its many differences from the Newtonian picture, have to say about the structure of the universe as a whole? Einstein attacked this problem shortly after formulating his general theory in 1915. At that time, the prevailing view was that the universe was static, just as Newton had thought.

Einstein was therefore dismayed to find that his calculations could not produce a truly static universe. According to general relativity, the universe must be either expanding or contracting. In a desperate move to force his theory to predict a static universe, he added to the equations of general relativity a term called the **cosmological constant** (denoted by Λ, the capital Greek letter lambda). The cosmological constant was intended to represent a pressure that tends to make the universe as a whole expand. Einstein's idea was that this pressure would just exactly balance gravitational attraction, so that the universe would be static and not collapse.

 Einstein's cosmological constant is analogous to the pressure of gas inside a bicycle tire. This pressure exactly balances the inward force exerted by the stretched rubber of the tire itself, so the tire maintains the same size.

Unlike other aspects of Einstein's theories, the cosmological constant did not have a firm basis in physics. He just added it to make the general theory of relativity agree with the prejudice that the universe is static. (The essay by Professor Stephen Hawking that follows this chapter describes more of the history of cosmological ideas.)

Because Einstein doubted his original equations, he missed an incredible opportunity: He could have postulated that we live in an expanding universe. Einstein has been quoted as saying in his later years that the cosmological constant was "the greatest blunder of my life." (We will see in Section 28-7 that the cosmological constant plays an important role in modern cosmology, although a very different one from what Einstein proposed.)

Instead, the first hint that we live in an expanding universe came more than a decade later from the observations of Edwin Hubble. As we will see in Section 28-3, Hubble's discovery provides the resolution of Olbers's paradox.

28-2 The universe is expanding

Hubble is usually credited with discovering that our universe is expanding. He found a simple linear relationship between the distances to remote

figure 28-2

The Expanding Balloon Analogy The expanding universe can be compared to the expanding surface of an inflating balloon. All the coins on the balloon recede from one another as the balloon expands, just as all the galaxies recede from one another as the universe expands. (Adapted from C. Misner, K. Thorne, and J. Wheeler)

galaxies and the redshifts of the spectral lines of those galaxies (review Section 26-5, especially Figures 26-15 and 26-16). This relationship, now called the *Hubble law*, states that the greater the distance to a galaxy, the greater is the galaxy's redshift. Thus, remote galaxies are moving away from us with speeds proportional to their distances. Specifically, the recessional velocity v of a galaxy is related to its distance d from the Earth by the equation

$$v = H_0 d$$

where H_0 is the Hubble constant. Because clusters of galaxies are getting farther and farther apart as time goes on, astronomers say that the universe is expanding.

 What does it actually mean to say that the universe is expanding? According to general relativity, space itself is not rigid. The amount of space between widely separated locations changes over time. A good analogy is that of blowing up a balloon, as in Figure 28-2. Small coins, each representing a galaxy, are glued onto the surface of the balloon. As the balloon expands, the amount of space between the coins gets larger and larger. In the same way, as the universe expands, the amount of space between widely separated galaxies increases. The expansion of the universe *is* the expansion of space.

It is important to realize that the expansion of the universe occurs primarily in the space that separates clusters of galaxies. Just as the coins in Figure 28-2 do not expand as the balloon inflates, galaxies themselves do not expand. Einstein and others have established that an object that is held together by its own gravity, such as a galaxy, is always contained within a patch of nonexpanding space. A galaxy's gravitational field produces this nonexpanding region, which is indistinguishable from the rigid space described by Newton. Thus, the Earth and your body, for example, are not getting any bigger. Only the distance between widely separated galaxies increases with time.

The Hubble law is a direct proportionality—that is, a galaxy twice as far away is receding from us twice as fast. This is just what we would expect in an expanding universe. To see why this is so, imagine a grid of parallel lines (as on a piece of graph paper) crisscrossing the universe. Figure 28-3a shows a series of such gridlines 100 Mpc apart, along with five galaxies labeled A, B, C, D, and E that happen to lie where gridlines cross. As the universe expands in all directions, the gridlines and the attached galaxies spread apart. (This is just what would happen if the universe were a two-dimensional rubber sheet that was being pulled equally on all sides. Alter-

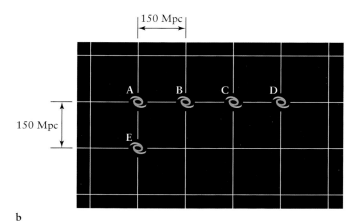

	Original distance (Mpc)	Later distance (Mpc)	Change in distance (Mpc)
A–B	100	150	50
A–C	200	300	100
A–D	300	450	150
A–E	100	150	50

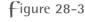 figure 28-3

The Expanding Universe and the Hubble Law
(a) Imagine five galaxies labeled A, B, C, D, and E. At the time shown here, adjacent galaxies are 100 Mpc apart. **(b)** Due to the expansion of the universe, at some later time the spacing between adjacent galaxies has increased to 150 Mpc. The table shows that the greater the original distance between galaxies, the greater the amount that distance has increased. This agrees with the Hubble law.

natively, you can imagine that Figure 28-3a depicts a very small portion of the surface of the balloon in Figure 28-2).

Figure 28-3b shows the universe at a later time, when the gridlines are 50% farther apart (150 Mpc) and all the distances between galaxies are 50% greater than in Figure 28-2a. Imagine that A represents our galaxy, the Milky Way. The table accompanying Figure 28-3 shows how far each of the other galaxies have moved away from us during the expansion: Galaxies A and B and galaxies A and E were originally 100 Mpc apart, and have moved away from each other by an additional 50 Mpc; A and C, which were originally 200 Mpc apart, have increased their separation by an additional 100 Mpc; and the distance between A and D, originally 300 Mpc, has increased by an additional 150 Mpc. In other words, the increase in distance between any pair of galaxies is in direct proportion to the original distance; if the original distance is twice as great, the increase in distance is also twice as great.

To see what this tells us about the recessional velocities of galaxies, remember that velocity is equal to the distance moved divided by the elapsed time. (For example, if you traveled in a straight line for 360 kilometers in 4 hours, your velocity was (360 km)/(4 hours) = 90 kilometers per hour.) Because the distance that each galaxy moves away from A during the expansion is directly proportional to its original distance from A, it follows that the velocity v at which each galaxy moves away from A is also directly proportional to the original distance d. This is just the Hubble law, $v = H_0 d$.

 It may seem that if the universe is expanding, and if we see all the distant galaxies rushing away from us, then we must be in a special position at the very center of the universe. In fact, the expansion of the universe looks the same from the vantage point of *any* galaxy. For example, as seen from galaxy D in Figure 28-3, the initial distances to galaxies A, B, and C are 300 Mpc, 200 Mpc, and 100 Mpc, respectively. Between parts *a* and *b* of the figure, these distances increase by 150 Mpc, 100 Mpc, and 50 Mpc, respectively. So, as seen from D as well, the recessional velocity increases in direct proportion with the distance, and in the same proportion as seen from A. In other words, no matter which galaxy you call home, you will see all the other galaxies receding from you in accordance with the same Hubble law (and the same Hubble constant) that we observe from Earth.

 Figure 28-2 also shows that the expansion of the universe looks the same from one galaxy as from any other. An insect sitting on any one of the coins would see all the other coins moving away. No one coin lies at the center of the balloon; indeed, the surface of the balloon does not actually have a center. Likewise, because every point in the universe appears to be at the center of the expansion, it must be that our universe has no center at all.

 If the universe is expanding, what is it expanding into? This commonly asked question arises only if we take our balloon analogy too literally. In Figure 28-2, the two-dimensional surface of the balloon (representing the universe) expands in three-dimensional space into the surrounding air. But the real universe is three-dimensional, and it is expanding in three dimensions. There is nothing "beyond" it, because there is no "beyond." Asking "What lies beyond the universe?" is as meaningless as asking "Where on the Earth is north of the North Pole?"

The ongoing expansion of space explains why the light from remote galaxies is redshifted. Imagine a photon coming toward us from a distant galaxy. As the photon travels through space, the space is expanding, so the photon's wavelength becomes stretched. When the photon reaches our eyes, we see a longer wavelength than usual, which is the redshift. The longer the photon's journey, the more its wavelength will have been stretched. Thus, photons from distant galaxies have larger redshifts than those of photons from nearby galaxies, as expressed by the Hubble law.

A redshift caused by the expansion of the universe is properly called a **cosmological redshift**. It is *not* the same as a Doppler shift. Doppler shifts are caused by an object's *motion through space*, whereas a cosmological redshift is caused by the *expansion of space*.

We can calculate the factor by which the universe has expanded since some ancient time from the redshift of light emitted by objects at that time. As we saw in Section 26-5, redshift (z) is defined as

$$z = \frac{\lambda - \lambda_0}{\lambda_0}$$

where λ_0 is the unshifted wavelength of a photon and λ is the wavelength we observe. For example, λ_0 could be the wavelength of a particular emission line in the spectrum of light leaving a remote quasar. As the quasar's light travels through space, its wavelength is stretched by the expansion of the universe. Thus, at our telescopes we observe the spectral line to have a wavelength λ. The ratio λ/λ_0 is a measure of the amount of stretching. By rearranging terms in the preceding equation, we can solve for this ratio to obtain

$$\frac{\lambda}{\lambda_0} = 1 + z$$

For example, consider a quasar with a redshift $z = 3$. Since the time that light left that quasar, the universe has expanded by a factor of $1 + z = 1 + 3 = 4$. In other words, when the light left that quasar, representative distances between widely separated galaxies were only one-quarter as large as they are today. A representative volume of space, which is proportional to the cube of its dimensions, was only $(\frac{1}{4})^3 = \frac{1}{64}$ as large as it is today. Thus, the density of matter in such a volume was 64 times greater than it is today.

If you know the redshift z of a distant object such as a quasar, you can calculate that object's recessional velocity v. (See Box 27-1 for how this is done.) Then, using the Hubble law, you can determine the distance d to that object if you know the value of the Hubble constant H_0. The distance in light-years tells you the **lookback time**, that is, how far into the past we are looking when we see that object. For example, if a quasar is a billion light-years away, the light from that quasar took about a billion years to reach us, and we are seeing it as it was about a billion years ago.

Unfortunately, as we saw in Section 26-5, the value of H_0 is uncertain by about 10%. As a result, lookback times determined using the Hubble law are uncertain by that same percentage. Furthermore, as we will see in Section 28-8, the universe has not always expanded at the same rate. This means that the value of the Hubble constant H_0 was different in the distant past. (In other words, the Hubble "constant" is not actually constant in time.)

To avoid dealing with these uncertainties and complications, astronomers commonly refer to times in the distant past in terms of redshift rather than years. For example, instead of asking "How common were quasars 5 billion years ago?", an astronomer might ask "How common were quasars at $z = 1.0$?" In this question, "at $z = 1.0$" is a shorthand way of saying "at the lookback time that corresponds to objects at $z = 1.0$." We will use this terminology later in this chapter. Remember that the greater the redshift, the greater the lookback time and hence the further back into time we are peering.

These ideas about the expanding universe demonstrate the central philosophy of cosmology. In cosmology, unlike other sciences, we cannot carry out controlled experiments or even make comparisons: There is only one observable universe. To make progress in cosmology, we must accept certain philosophical assumptions or abandon hope of understanding the nature of the universe. The Hubble law provides a classic example. It could be interpreted to mean that we are at the center of the universe. We reject this interpretation, however, because it violates a cosmological extension of Copernicus's belief that we do not occupy a special location in space.

When Einstein began applying his general theory of relativity to cosmology, he made a daring assumption: Over very large distances the universe is **homogeneous** (that is, every region is the same as every other region) and **isotropic** (that is, the universe looks the same in every direction). In other words, if you could stand back and look at a very large region of space, any one part of the universe would look basically the same as any other part, with the same kinds of galaxies distributed through space in the same way. The assumption that the universe is homogeneous and isotropic constitutes the **cosmological principle**. It gives precise meaning to the idea that we do not occupy a special location in space.

Models of the universe based on the cosmological principle have proven remarkably successful in describing the structure and evolution of the universe and in interpreting observational data. All the discussion about the universe in this chapter and the next assumes that the universe is homogeneous and isotropic on the largest scale.

28-3 The expanding universe emerged from a cataclysmic event called the Big Bang

The Hubble flow shows that the universe has been expanding for billions of years. This means that in the past the matter in the universe must have been closer together and therefore denser than it is today. If we look far enough into the very distant past, there must have been a time when the density of matter was almost inconceivably high. This leads us to conclude that some sort of colossal explosion of this ultradense matter initiated the expansion of the universe. This explosion, commonly called the **Big Bang,** marks the creation of the universe.

CAUTION! Although we can think of the Big Bang as an explosion, it was not at all like an exploding bomb. When a bomb explodes, pieces of debris fly off *into space* from a central location. If you could trace all the pieces back to their origin, you could find out exactly where the bomb had been. This process is not possible with the universe, however, because the universe itself always has and always will consist of all space. As we have seen, the universe logically cannot have an edge (see the discussion of the expanding balloon in Section 28-2).

How long ago did the Big Bang take place? To estimate an answer, imagine watching a movie of any two galaxies that today are separated by a distance d and receding from each other with a velocity v. Now run the movie backward, and observe the two galaxies approaching each other as time runs in reverse. We can calculate the time T_0 it will take for the galaxies to collide by using the simple equation

$$T_0 = \frac{d}{v}$$

This says that the time to travel a distance d at velocity v is equal to the ratio d/v. (As an example, to travel a distance of 360 km at a velocity of 90 km/h takes (360 km)/(90 km/h) = 4 hours.) If we use the Hubble law, $v = H_0 d$, to replace the velocity v in this equation, we get

$$T_0 = \frac{d}{H_0 d} = \frac{1}{H_0}$$

Note that the distance of separation, d, has canceled out and does not appear in the final expression. This means that T_0 is the same for *all* galaxies. This is the time in the past when all galaxies were crushed together, the time back to the Big Bang. In other words, the reciprocal of the Hubble constant H_0 gives us an estimate of the age of the universe, which is one reason why H_0 is such an important quantity in cosmology.

If we use our standard value of $H_0 = 70$ km/s/Mpc, our estimate for the age of the universe is

$$T_0 = \frac{1}{70 \text{ km/s/Mpc}}$$

To convert this into units of time, we simply need to remember that 1 Mpc equals 3.09×10^{19} km and 1 year equals 3.156×10^7 s. Using the technique we discussed in Box 1-3 for converting units, we get

$$T_0 = \frac{1}{70} \frac{\text{Mpc/s}}{\text{km}} \times \frac{3.09 \times 10^{19} \text{ km}}{1 \text{ Mpc}} \times \frac{1 \text{ year}}{3.156 \times 10^7 \text{ s}}$$

$$= 1.4 \times 10^{10} \text{ years} = 14 \text{ billion years}$$

By comparison, the age of our solar system is only about 4.6 billion years, or about one-third the age of the universe. Thus, the formation of our home planet is a relatively recent event in the history of the cosmos.

As we saw in Section 26-5, the value of the Hubble constant is uncertain by approximately 10%. This means that T_0 is likewise somewhat uncertain: A slightly larger value of H_0 implies a slightly smaller value of T_0, and vice versa. The 10% uncertainty in H_0 corresponds to a range in T_0 from about 13 to 16 billion years.

Of course, the formula $T_0 = 1/H_0$ is only an estimate of the age of the universe. In deriving it, we assumed that the universe expands at a constant rate, and that may not be so. In Section 28-8 we will discuss how the expansion rate of the universe has changed over its history.

 Whatever the true age of the universe, it must be at least as old as the oldest stars. The oldest stars that we can observe readily lie in the Milky Way's globular clusters (see Section 21-3 and Section 25-1). Most calculations based on the theory of stellar evolution indicate that these stars are 12 to 16 billion years old. Encouragingly enough, this is not too different from our rough estimate of the age of the universe obtained from $T_0 = 1/H_0$. As a compromise, in this chapter we use a figure of 14 billion years for the age of the universe.

The Big Bang helps resolve Olbers's paradox. We know that the universe had a definite beginning, and thus its age is finite (as opposed to infinite). If the universe is about 14 billion years old, then light from stars more than about 14 billion light-years away has just not had enough time to get here. As a result, we cannot see any objects that are more than about 14 billion light-years away. This is true even if the universe is infinite, with galaxies scattered throughout its limitless expanse.

You can think of the Earth as being at the center of an enormous sphere having a radius of approximately 14 billion light-years (Figure 28-4). The surface of this sphere is called the **cosmic particle horizon.** Our entire **observable universe** is located inside this sphere. We cannot see anything beyond the cosmic particle horizon, because the time required for light to reach us from these incredibly remote distances is greater than the age of the universe. (As Figure 28-2 shows, the size of this sphere increases as the universe ages. When the universe was 5 billion years old, for example, our observable universe was only 5 billion light-years in radius.) Throughout the observable universe, galaxies are distributed sparsely enough that there are no stars along most of our lines of sight, which helps explain why the night sky is dark.

Besides the finite age of the universe, a second effect also contributes significantly to the darkness of the night sky—the redshift. According to the Hubble law, the greater the distance to a galaxy, the greater the redshift. When a photon is redshifted, its wavelength becomes longer, and its energy—which is inversely proportional to its wavelength (see Section 5-5)—decreases. Consequently, even though there are many galaxies far from the Earth, they have large redshifts and their light does not carry much energy. A galaxy nearly at the cosmic particle horizon has a nearly infinite redshift, meaning that the

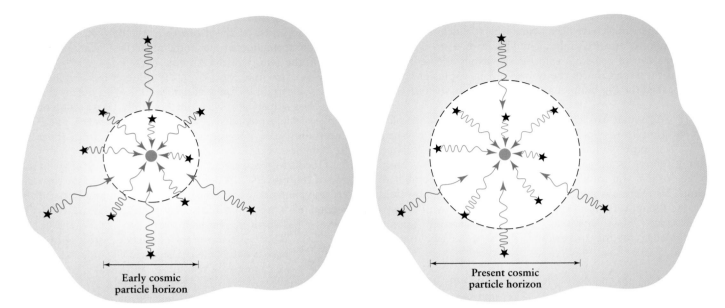

Figure 28-4

The Observable Universe The part of the universe that we can observe lies within a sphere centered on the Earth called the cosmic particle horizon. Its radius is equal to the distance that light has traveled since the Big Bang. This occurred about 14 billion years ago, so at present the cosmic particle horizon is about 14 billion light-years in radius. We cannot see objects beyond the cosmic particle horizon, because their light has not had enough time to reach us. As the universe ages, the radius of the cosmic particle horizon increases and more of the universe becomes observable. In the illustration, light waves have been drawn to show how their wavelength increases (that is, is redshifted) as they travel through the expanding universe.

light we receive from that galaxy carries practically no energy at all. This decrease in photon energy because of the expansion of the universe decreases the brilliance of remote galaxies, helping to make the night sky dark.

The concept of a Big Bang origin for the universe is a straightforward, logical consequence of having an expanding universe. If you can just imagine far enough back into the past, you can arrive at a time approximately 14 billion years ago, when the density throughout the universe was infinite. The entire universe was in effect like the center of a black hole.

Comparing the Big Bang to the center of a black hole can help us appreciate certain aspects of the creation of the universe. As we saw in Section 24-5, matter at the center of a black hole is crushed to infinite density. At this location, called the singularity, the curvature of space and time is infinite, and the very distinction between space and time becomes muddled. Without a clear background of space and time, such concepts as "past," "future," "here," and "now" cease to have meaning. For this reason, a better name for the Big Bang is the **cosmic singularity.**

At the moment of the Big Bang, a state of infinite density filled the universe. Throughout the universe, space and time were completely jumbled up in a condition of infinite curvature like that at the center of a black hole. Thus, we cannot use the known laws of physics to tell us exactly what happened at the moment of the Big Bang. And we certainly can-

not use these laws to tell us what existed before the Big Bang. These things may be fundamentally unknowable. The phrases "before the Big Bang" or "at the moment of the Big Bang" are meaningless, because time did not really exist until after that moment.

A very short time after the Big Bang, space and time began to behave in the way we think of them today. This short time interval, called the **Planck time** (t_{P}), is given by the following expression:

The Planck time

$$t_{\mathrm{P}} = \sqrt{\frac{Gh}{c^5}} = 1.35 \times 10^{-43} \text{ s}$$

t_{P} = Planck time

G = universal constant of gravitation

h = Planck's constant

c = speed of light

 From the beginning of the Big Bang to the Planck time, about 10^{-43} second later, all known science fails us. We do not know how space, time, and matter behaved in that brief but important interval. Hence, the Planck time represents a limit to our knowledge of conditions at the very beginning of the universe. Nevertheless, some physicists have speculated that space and time as we know them today burst forth from a seething, foamlike, space-time mishmash during the Planck time. If we think of the Big Bang as an explosion of space at the beginning of time, then the universe we see today is the debris from that cosmic explosion.

28-4 The microwave radiation that fills all space is evidence of a hot Big Bang

One of the major advances in twentieth-century astronomy was the discovery of the origin of the heavy elements. We know today that essentially all the heavy elements are created by thermonuclear reactions at the centers of stars and in supernovae (see Chapter 22). The starting point of all these reactions is the fusion of hydrogen into helium, which we described in Section 18-1. But as astronomers began to understand the details of thermonuclear synthesis in the 1960s, they were faced with a dilemma: There is far more helium in the universe than could have been created by hydrogen fusion in stars.

For example, the Sun consists of about 74% hydrogen and 25% helium by mass, leaving only 1% for all the remaining heavier elements combined. This 1% can be understood as material produced inside earlier generations of massive stars that long ago cast these heavy elements out into space when they became supernovae. Some freshly made helium, produced by the thermonuclear fusion of hydrogen within the stars, certainly accompanied these heavy elements. But calculations showed that the amount of helium produced in this way was not nearly enough to account for one-quarter of the Sun's mass. Because it was thought that the universe originally contained only hydrogen—the simplest of all the chemical elements—the presence of so much helium posed a major dilemma.

Shortly after World War II, Ralph Alpher and Robert Hermann proposed that the universe immediately following the Big Bang must have been so incredibly hot that thermonuclear reactions occurred everywhere throughout space. Following up this idea in 1960, Princeton University physicists Robert Dicke and P. J. E. Peebles discovered that they could indeed account for today's high abundance of helium by assuming that the early universe had been at least as hot as the Sun's center, where helium is currently being produced. The hot early universe must therefore have been filled with many high-energy, short-wavelength photons. The properties of this radiation field depended on its temperature, as described by Planck's blackbody law (review Figure 5-10).

The universe has expanded so much since those ancient times that all those short-wavelength photons have had their wavelengths stretched by a tremendous factor. As a result, they have become low-energy, long-wavelength photons. The temperature of this cosmic radiation field is now only a few degrees above absolute zero. By Wien's law, radiation at such a low temperature should have its peak intensity at microwave wavelengths of approximately 1 millimeter. Hence, this radia-

 Figure 28-5   R I V U X G

The Bell Labs Horn Antenna Using this microwave horn antenna, originally built for communications purposes, Arno Penzias and Robert Wilson detected a signal that seemed to come from all parts of the sky. After carefully removing all potential sources of electronic "noise" (including bird droppings inside the antenna) that could create a false signal, Penzias and Wilson realized that they were actually detecting radiation from space. This radiation is the afterglow of the Big Bang. (Bell Labs)

tion field, which fills all of space, is called the **cosmic microwave background** or **cosmic background radiation.** In the early 1960s, Dicke and his colleagues began designing an antenna to detect this microwave radiation.

Meanwhile, just a few miles from Princeton University, Arno Penzias and Robert Wilson of Bell Telephone Laboratories in New Jersey were working on a new microwave horn antenna designed to relay telephone calls to Earth-orbiting communications satellites (Figure 28-5). Penzias and Wilson were deeply puzzled when, no matter where in the sky they pointed their antenna, they detected faint background noise. Thanks to a colleague, they happened to learn about the work of Dicke and Peebles and came to realize that they had discovered the cooled-down cosmic background radiation left over from the hot Big Bang.

You can actually detect cosmic background radiation background using an ordinary television set. This radiation is responsible for about 1% of the random noise or "hash" that appears on the screen when you tune a television to a station that is off the air. Using far more sophisticated detectors than TV sets, scientists have made many measurements of the intensity of the background radiation at a variety of wavelengths. Unfortunately, the Earth's atmosphere is almost totally opaque to wavelengths between about 10 μm and 1 cm (see Figure 6-27), which is just the wavelength range in which the background radiation is most intense. As a result, scientists have had to place detectors either on high-altitude balloons (which can fly above the majority of the obscuring atmosphere) or, even better, on board orbiting spacecraft.

ƒigure 28-6

The Cosmic Background Explorer (COBE) This satellite, launched in 1989, measured the spectrum and angular distribution of the cosmic microwave background over a wavelength range from 1 μm to 1 cm. COBE was designed to detect deviations from a perfect blackbody spectrum and from perfect isotropy. (Courtesy of J. Mather; NASA)

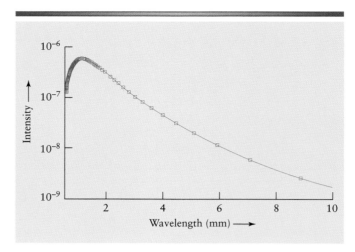

ƒigure 28-7

The Spectrum of the Cosmic Microwave Background The little squares on this graph are COBE's measurements of the brightness of the cosmic microwave background plotted against wavelength. To a remarkably high degree of accuracy, the data fall along a blackbody curve for 2.725 K. The peak of the curve is at a wavelength of 1.1 mm, in accordance with Wien's law (Section 5-4). (Courtesy of E. Cheng; NASA COBE Science Team)

The first high-precision measurements of the cosmic microwave background came from the Cosmic Background Explorer (COBE, pronounced "coe-bee") satellite, which was placed in Earth orbit in 1989 (Figure 28-6). Data from COBE's spectrometer, shown in Figure 28-7, demonstrate that this ancient radiation has the spectrum of a blackbody with a temperature of 2.725 K.

An important feature of the microwave background is that its intensity is almost perfectly isotropic, that is, the same in all directions. In other words, we detect nearly the same background intensity from all parts of the sky. This is a striking confirmation of Einstein's assumption that the universe is isotropic (Section 28-2). However, extremely accurate measurements first made from high-flying airplanes, and later from high-altitude balloons and from COBE, reveal a very slight variation in temperature across the sky. The microwave background appears slightly warmer than average toward the constellation of Leo and slightly cooler than average in the opposite direction toward Aquarius. Between the warm spot in Leo and the cool spot in Aquarius, the background temperature declines smoothly across the sky. Figure 28-8 is a map of the microwave sky showing this variation.

This apparent variation in temperature is caused by the Earth's motion through the cosmos. If we were at rest with respect to the microwave background, the radiation would be even more nearly isotropic. Because we are moving through this radiation field, however, we see a Doppler shift. Specifically, we see shorter-than-average wavelengths in the direction toward which we are moving, as drawn in Figure 28-9. A decrease in wavelength corresponds to an increase in pho-

ton energy and thus an increase in the temperature of the radiation. The slight temperature excess observed, about 0.00337 K, corresponds to a speed of 371 km/s. Conversely, we see longer-than-average wavelengths in that part of the

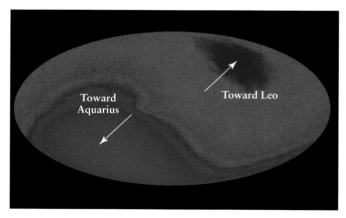

ƒigure 28-8   R I V U X G

The Microwave Sky Data from COBE were used to produce this map of microwave radiation from the entire sky. The plane of the Milky Way runs horizontally across the map, with the galactic center in the middle. Color indicates temperature—red is warm and blue is cool. The temperature variation across the sky is caused by the Earth's motion through the microwave background. The variation is quite small, only 0.0033 K above or below the average radiation temperature of 2.725 K. (NASA)

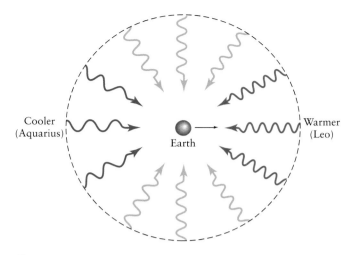

Figure 28-9

Our Motion Through the Microwave Background Because of the Doppler effect, we detect shorter wavelengths in the microwave background and a higher temperature of radiation in that part of the sky toward which we are moving. This part of the sky is the area shown in red in Figure 28-8. In the opposite part of the sky, shown in blue in Figure 28-8, the microwave radiation has longer wavelengths and a cooler temperature.

sky from which we are receding. An increase in wavelength corresponds to a decline in photon energy and, hence, a decline in radiation temperature.

Our solar system is thus traveling away from Aquarius and toward Leo at a speed of 371 km/s. Taking into account the known velocity of the Sun around the center of our Galaxy, we find that the entire Local Group of galaxies, including our Milky Way Galaxy, is moving at about 620 km/s toward the Hydra-Centaurus supercluster. Observations show that thousands of other galaxies are being carried in this direction, as is the Hydra-Centaurus supercluster itself. This tremendous flow of matter is thought to be due to the gravitational pull of an enormous collection of visible galaxies and dark matter lying in that direction. This immense object, dubbed the **Great Attractor,** lies about 50 Mpc (150 million ly) from Earth.

The existence of such concentrations of mass, as well as the existence of superclusters of galaxies (see Section 26-6), shows that the universe is rather "lumpy" on scales of 100 Mpc or smaller. It is only on larger scales that the universe is homogeneous and isotropic.

28-5 The universe was a hot, opaque plasma during its first 300,000 years

Everything in the universe falls into one of two categories—energy or matter. One form of energy is radiation, that is, photons. (We will encounter another type of energy in

Section 28-7.) There are many photons of starlight traveling across space, but the vast majority of photons in the universe belong to the cosmic microwave background. The matter in the universe is contained in such luminous objects as stars, planets, and galaxies, as well as in nonluminous dark matter. A natural question to ask is this: Which plays a more important role in the universe, radiation or matter? As we will see, the answer to this question is different for the early universe from the answer for our universe today.

To make a comparison between radiation and matter, recall Einstein's famous equation $E = mc^2$ (see Section 18-1). We can think of the photon energy in the universe (E) as being equal to a quantity of mass (m) multiplied by the square of the speed of light (c). The amount of this equivalent mass in a volume V, divided by that volume, is the **mass density of radiation** (ρ_{rad}; say "rho sub rad").

We can combine $E = mc^2$ with the Stefan-Boltzmann law (Section 5-4) to give the following formula:

Mass density of radiation

$$\rho_{rad} = \frac{4\sigma T^4}{c^3}$$

ρ_{rad} = mass density of radiation

T = temperature of radiation

σ = Stefan-Boltzmann constant

= 5.67×10^{-8} W m^{-2} K^{-4}

c = speed of light = 3×10^8 m/s

For the present-day temperature of the cosmic background radiation, $T = 2.725$ K, this equation yields

$$\rho_{rad} = 4.6 \times 10^{-31} \text{ kg/m}^3$$

The **average density of matter** (ρ_m; say "rho sub em") in the universe is harder to determine. To find this density, we look at a large volume (V) of space, determine the total mass (M) of all the stars, galaxies, and dark matter in that volume, and divide the volume into the mass: $\rho_m = M/V$. (We emphasize that this is the *average* density of matter. It would be the actual density if all the matter in the universe were spread out uniformly rather than being clumped into galaxies and clusters of galaxies.) Determining how much mass is in a large volume of space is a challenging task. A major part of the challenge involves dark matter, which emits no electromagnetic radiation and can be detected only by its gravitational influence (see Section 25-4 and Section 26-8).

One method that appears to deal successfully with this challenge is to observe clusters of galaxies, within which most of the mass in the universe is concentrated. Rich clusters are surrounded by halos of hot, X-ray-emitting gas, typically at temperatures of 10^7 to 10^8 K (see Figure 26-23). Such a halo should be in hydrostatic equilibrium, so that it neither expands nor contracts but remains the same size (see Section 18-2, especially Figure 18-3). The outward gas pressure associated with the halo's high temperature would then be balanced by the inward gravitational pull due to the total mass of the cluster. Thus, by measuring the temperature of the halo—which can be determined from the properties of the halo's X-ray emission—astronomers can infer the cluster's mass, including the contribution from dark matter.

From galaxy clusters and other measurements, the present-day average density of matter in the universe is thought to be in the range

$$\rho_m = 2 - 4 \times 10^{-27} \text{ kg/m}^3$$

The mass of a single hydrogen atom is 1.7×10^{-27} kg/m^3. Hence, if the mass of the universe were spread uniformly over space, there would be the equivalent of only one or two hydrogen atoms per cubic meter of space. By contrast, there are 5×10^{25} atoms in a cubic meter of the air you breathe! The very small value of ρ_m shows that our universe has very little matter in it.

Furthermore, by counting galaxies, astronomers determine that the average density of *luminous* matter (that is, the stars and gas within clusters of galaxies) is about 3.6×10^{-28} kg/m^3. This is only about 10% of the average density of matter of all forms. Thus, nonluminous dark matter is actually the predominant form of matter in our universe. The "ordinary" matter of which the stars, the planets, and ourselves are made is actually somewhat exceptional.

Although the average density of matter in the universe is tiny by Earth standards, it is thousands of times larger than ρ_{rad}, the mass density of radiation. However, this was not always the case. Matter prevails over radiation today only because the energy now carried by microwave photons is so small. Nevertheless, the number of photons in the microwave background is astounding. From the physics of blackbody radiation it can be demonstrated that there are today 550 million photons in every cubic meter of space. In other words, the photons in space outnumber atoms by roughly a billion to one. In terms of total number of particles, the universe thus consists almost entirely of microwave photons. This radiation field no longer has much "clout," however, because its photons have been redshifted to long wavelengths and low energies after approximately 14 billion years of being stretched by the expansion of the universe.

In contrast, think back toward the Big Bang. The universe becomes increasingly compressed, and thus the density of matter increases as we go back in time. The photons in the background radiation also become more crowded together as we go back in time. But, in addition, the photons become less redshifted and thus have shorter wavelengths and higher energy than they do today. Because of this added energy, the mass density of radiation (ρ_{rad}) increases more quickly as we go back in time than does the average density of matter (ρ_m). In fact, as Figure 28-10 shows, there was a time in the ancient past when ρ_{rad} equaled ρ_m. Before this time, ρ_{rad} was greater than ρ_m, and radiation thus held sway over matter. Astronomers call this state a **radiation-dominated universe**. After ρ_m became greater than ρ_{rad}, so that matter prevailed over radiation, our universe became a **matter-dominated universe**.

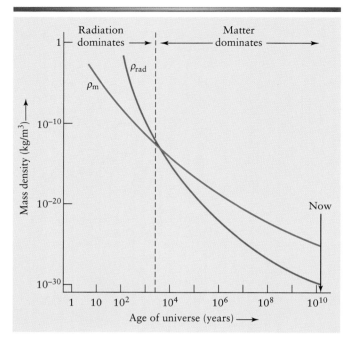

Figure 28-10

The Evolution of Density For approximately 2500 years after the Big Bang, the mass density of radiation (ρ_{rad}, shown in red) exceeded the matter density (ρ_m, shown in blue), and the universe was radiation-dominated. Later, however, continued expansion of the universe caused ρ_{rad} to become less than ρ_m, at which point the universe became matter-dominated.

This transition from a radiation-dominated universe to a matter-dominated universe occurred about 2500 years after the Big Bang, at a time that corresponds to a redshift of about $z = 25{,}000$. In other words, since that time the wavelengths of photons have been stretched by a factor of 25,000. Today, these microwave photons typically have wavelengths of about 1 mm. But when the universe was about 2500 years old, they had wavelengths of about 40 nm, in the ultraviolet part of the spectrum.

To calculate the temperature of the cosmic background radiation at the time of this transition from a radiation-dominated universe to a matter-dominated one, we use Wien's law (Section 5-4). This law says that the wavelength of maximum emission (λ_{max}) of a blackbody is inversely proportional to its temperature (T): A *decrease* of λ_{max} by a factor of 2 corresponds to an *increase* of T by a factor of 2.

The present-day peak wavelength of the cosmic background radiation is roughly 1 mm, corresponding to a blackbody temperature of about 3 K; hence, a peak wavelength 25,000 times smaller (40 nm) corresponds to $T = 25{,}000 \times 3 \text{ K} = 75{,}000 \text{ K}$. In other words, the temperature of the radiation background is in direct proportion to the factor $1 + z$ and has been declining over the ages, as Figure 28-11 shows.

figure 28-11

The Evolution of Radiation Temperature As the universe expanded, the photons in the radiation background became increasingly redshifted and the temperature of the radiation fell. Approximately 300,000 years after the Big Bang, when the temperature fell below 3000 K, hydrogen atoms formed and the radiation field "decoupled" from the matter in the universe. After that point, the temperature of matter in the universe was not the same as the temperature of radiation.

The nature of the universe changed again in a fundamental way some 300,000 years after the Big Bang, when z was roughly 1000 and the temperature of the radiation background was about $1000 \times 3 \text{ K} = 3000 \text{ K}$. To see the significance of this moment in cosmic history, recall that hydrogen is by far the most abundant element in the universe—hydrogen atoms outnumber helium atoms by about 12 to 1. A hydrogen atom consists of a single proton orbited by a single electron, and it takes relatively little energy to knock the proton and electron apart. In fact, ultraviolet radiation warmer than about 3000 K easily ionizes hydrogen. Thus, hydrogen atoms could not survive in the universe that existed before $z = 1000$. That is, in the first 300,000 years after the Big Bang (Figure 28-12a), the background photons had energies great enough to prevent electrons and protons from binding to form hydrogen atoms. Only since $z = 1000$ (that is, since $t = 300,000$ years) have the energies of these photons been low enough to permit hydrogen atoms to exist (Figure 28-12b).

The epoch when atoms first formed at $t = 300,000$ years is called the **era of recombination**. This refers to electrons "recombining" to form atoms. (The name is a bit misleading, because the electrons and protons had never before combined into atoms.)

Prior to $t = 300,000$ years, the universe was completely filled with a shimmering expanse of high-energy photons colliding vigorously with protons and electrons. This state of matter, called a **plasma,** is opaque, just like the glowing gases inside a discharge tube (like a neon advertising sign). The surface and interior of the Sun are also a hot, glowing, opaque plasma (Section 18-9). P. J. E. Peebles coined the term **primordial fireball** to describe the universe during its first 300,000 years of existence.

a   Before recombination

b   After recombination

figure 28-12

The Era of Recombination **(a)** Before recombination, the energy of photons in the cosmic background was high enough to prevent protons and electrons from forming hydrogen atoms. The universe was opaque, like the interior of the Sun, and the temperature of matter was the same as the temperature of radiation. **(b)** As soon as the energy of the cosmic background radiation became low enough that hydrogen atoms could survive, the universe became transparent. This transition from opaque to transparent occurred about 1 million years after the Big Bang. The photons could now pass through the universe with little or no interaction with matter. As a result, the temperatures of matter and radiation were no longer the same.

ƒigure 28-13 R I **V** U X G

In Search of Ancient Photons The silvery object at the right in this photograph is BOOMERANG, a sensitive telescope designed to study variations in the temperature of the cosmic background radiation. Lifting off from Antarctica, the large balloon kept BOOMERANG aloft at an altitude of 35 km (120,000 ft) for 10 days of continuous observation. In the background is Mount Erebus, the southernmost active volcano on Earth. (The BOOMERANG Group, University of California, Santa Barbara)

After $t = 300,000$ years, the photons no longer had enough energy to keep the protons and electrons apart. As soon as the temperature of the radiation field fell below about 3000 K, protons and electrons began combining to form hydrogen atoms. These atoms do not absorb low-energy photons, and so space became transparent! All the photons that heretofore had been vigorously colliding with charged particles could now stream unimpeded across space. Today these same photons constitute the microwave background.

Before recombination, matter and the radiation field had the same temperature, because photons, electrons, and protons were all in continuous interaction with one another. After recombination, photons and atoms hardly interacted at all, and thus the temperature of matter in the universe was no longer the same as the temperature of the background radiation. Thus, $T = 2.725$ K is the (very uniform) temperature of the present-day background radiation field, *not* the temperature of the matter in the universe (which is *not* uniform, ranging from hundreds of millions of kelvins in the interiors of giant stars to a few tens of kelvins in the interstellar medium).

ANALOGY A good analogy is the behavior of a glass of cold water. If you hold the glass in your hand, the water will get warmer and your hand will get colder until both the water and your hand are at the same temperature. But if you set the glass down and do not touch it, so that the glass and your hand do not interact, the water will stay cold and your hand will stay warm for much longer.

Because the universe was opaque prior to $t = 300,000$ years, we cannot see any further into the past than the era of recombination. In particular, we cannot see back to the era when the universe was radiation-dominated. The microwave background, whose photons have suffered a redshift of $z = 1000$, contains the most ancient photons we will ever be able to observe.

Careful analysis of COBE data shows that the cosmic background radiation is not completely isotropic. Even when the effects of the Earth's motion are accounted for, there remain variations in the temperature of the radiation field of about 100 μK (100 microkelvins, or 10^{-4} K) above or below the average 2.725 K temperature. These tiny temperature variations indicate that the matter and radiation in the universe were not totally uniform at the moment of recombination. Regions that were slightly denser than average were also slightly warmer than average; less dense regions were slightly cooler. When radiation decoupled from matter at the time of recombination, the radiation preserved a record of these variations in temperature and density.

Astronomers place great importance on studying temperature variations in the cosmic background radiation. The reason is that concentrations of mass in our present-day universe, such as superclusters of galaxies, are thought to have formed from these tiny nonuniformities in the early universe. Within these immense concentrations formed the galaxies, stars, and planets. Thus, by studying these nonuniformities, we are really studying our origins.

WEB LINK 28.11 Unfortunately, the detectors on board COBE had a relatively coarse angular resolution of 7° and thus could not give a detailed picture of these temperature variations. In 1998 two balloon-borne experiments, BOOMERANG and MAXIMA, carried new, high-resolution telescopes aloft to study the cosmic background radiation with unprecedented precision (Figure 28-13). Figure 28-14 shows a map of the results from the BOOMERANG experiment, which observed the southern part of the celestial sphere; MAXIMA, which observed the northern celestial sphere, gave similar results. Such maps show us the state of the universe when it was only about 0.002% of its present age.

Temperature variations in the cosmic background radiation do more than show us the origins of large-scale structure in the universe. As we will see in the next two sections, they

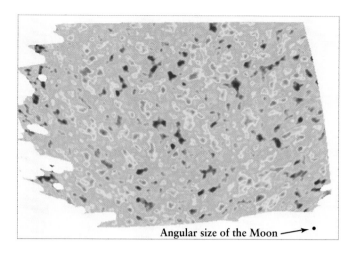

Angular size of the Moon ⟶ •

−300 −200 −100   0   +100 +200 +300

Temperature difference from average (μK)

ƒigure 28-14   **R** I V U X G

Temperature Variations in the Cosmic Microwave Background This map from BOOMERANG data shows small variations in the temperature of the cosmic background radiation. Higher-temperature regions (shown in red) show where the early universe was slightly denser than average. Note that the variations in temperature are no more than 300 μK, or 3×10^{-4} K. The total area shown is many times larger in angular size than the Moon. (The BOOMERANG Group, University of California, Santa Barbara)

actually show us the shape of the universe as a whole, and afford us a glimpse of how our universe may evolve in the future.

28-6   The shape of the universe indicates its matter and energy content

We have seen that by following the mass densities of radiation (ρ_{rad}) and of matter (ρ_m), we can learn about the evolution of the universe. But it is equally important to know the combined mass density of *all* forms of matter and energy. (In an analogous way, an accountant needs to know the overall financial status of a company, not just individual profits or losses.) Remarkably, we can do this by investigating the overall shape of the universe.

Einstein's general theory of relativity explains that gravity curves the fabric of space. Furthermore, the equivalence between matter and energy, expressed by Einstein's equation $E = mc^2$, tells that either matter or energy produces gravity. Thus, the matter and energy scattered across space should give the universe an overall curvature. The degree of curvature

depends on the **combined average mass density** of all forms of matter and energy. This quantity, which we call ρ_0 (say "rho sub zero"), is the sum of the average mass densities of matter, radiation, and any other form of energy. Thus, by measuring the curvature of space, we should be able to determine the value of ρ_0 and, hence, learn about the content of the universe as a whole.

To see what astronomers mean by the curvature of the universe, imagine shining two powerful laser beams out into space so that they are perfectly parallel as they leave the Earth. Furthermore, suppose that nothing gets in the way of these two beams, so that we can follow them for billions of light-years as they travel across the universe and across the space whose curvature we wish to detect. There are only three possibilities:

1. We might find that our two beams of light remain perfectly parallel, even after traversing billions of light-years. In this case, space would not be curved: The universe would have **zero curvature,** and space would be **flat.**

2. Alternatively, we might find that our two beams of light gradually converge. In such a case, space would not be flat. Recall that lines of longitude on the Earth's surface are parallel at the equator but intersect at the poles. Thus, in this case the three-dimensional geometry of the universe would be analogous to the two-dimensional geometry of a spherical surface. We would then say that space is **spherical** and that the universe has **positive curvature.** Such a universe is also called **closed,** because if you travel in a straight line in any direction in such a universe, you will eventually return to your starting point.

3. Finally, we might find that the two initially parallel beams of light would gradually diverge, becoming farther and farther apart as they moved across the universe. In this case, the universe would still have to be curved, but in the opposite sense from the spherical model. We would then say that the universe has **negative curvature.** In the same way that a sphere is a positively curved two-dimensional surface, a saddle is a good example of a negatively curved two-dimensional surface. Parallel lines drawn on a sphere always converge, but parallel lines drawn on a saddle always diverge. Mathematicians say that saddle-shaped surfaces are hyperbolic. Thus, in a negatively curved universe, we would describe space as **hyperbolic.** Such a universe is also called **open** because if you were to travel in a straight line in any direction, you would never return to your starting point.

Figure 28-15 illustrates the three cases of positive curvature, zero curvature, and negative curvature. Real space is three-dimensional, but we have drawn the three cases as analogous, more easily visualized two-dimensional surfaces. Therefore, as you examine the drawings in Figure 28-15, remember that the real universe has one more dimension. For example, if the universe is in fact hyperbolic, then the geometry of space must be the (difficult to visualize) three-dimensional analog of the two-dimensional surface of a saddle.

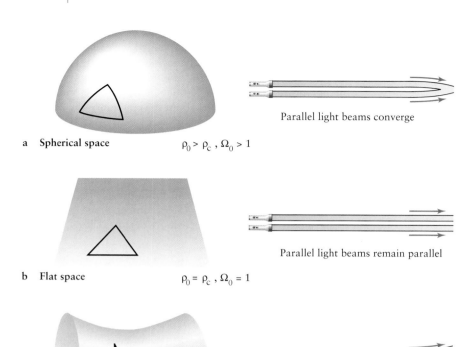

a Spherical space $\rho_0 > \rho_c$, $\Omega_0 > 1$

Parallel light beams converge

b Flat space $\rho_0 = \rho_c$, $\Omega_0 = 1$

Parallel light beams remain parallel

c Hyperbolic space $\rho_0 < \rho_c$, $\Omega_0 < 1$

Parallel light beams diverge

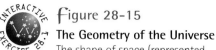

figure 28-15

The Geometry of the Universe

The shape of space (represented here as two-dimensional surfaces for ease in visualization) is determined by the combined density of matter and energy contained in the universe. The curvature of space is either (a) positive, (b) zero, or (c) negative. In theory, the curvature of the universe could be determined by seeing whether two laser beams initially parallel to each other would converge, remain parallel, or spread apart. Alternatively, one could measure the angles of a huge triangle in space. The curvature is positive, zero, or negative, depending on whether the combined mass density is greater than, equal to, or less than the critical density, or, equivalently, on whether the density parameter Ω_0 is greater than, equal to, or less than 1.

Note that in accordance with the cosmological principle, none of these models of the universe has an "edge" or a "center." This is clearly the case for both the flat and hyperbolic universes, because they are infinite and extend forever in all directions. A spherical universe is finite, but it also lacks a center and an edge. You could walk forever around the surface of a sphere (like the surface of the Earth) without finding a center or an edge.

The curvature of the universe is determined by the value of the combined mass density ρ_0. If ρ_0 is greater than a certain value ρ_c (say "rho sub cee"), the universe has positive curvature and is closed. If ρ_0 is less than ρ_c, the universe has negative curvature and is open. In the special case that ρ_0 is exactly equal to ρ_c, the universe is flat. Because of its crucial role in determining the geometry of the universe, ρ_c is called the **critical density**. It is given by the expression

Critical density of the universe

$$\rho_c = \frac{3H_0{}^2}{8\pi G}$$

ρ_c = critical density of the universe

H_0 = Hubble constant

G = universal constant of gravitation

Using a Hubble constant $H_0 = 70$ km/s/Mpc, we get

$$\rho_c = 9.2 \times 10^{-27} \text{ kg/m}^3$$

This density is equivalent to 5.5 hydrogen atoms per cubic meter of space.

Many astronomers prefer to characterize the combined average mass density of the universe in terms of the **density parameter** Ω_0 (say "omega sub zero"). This is just the ratio of the combined average mass density to the critical density:

Density parameter

$$\Omega_0 = \frac{\rho_0}{\rho_c}$$

Ω_0 = density parameter

ρ_0 = combined average mass density

ρ_c = critical density

An open universe has a density parameter Ω_0 between 0 and 1, and a closed universe has Ω_0 greater than 1. In a flat universe, Ω_0 is equal to 1. Thus, we can use the value of Ω_0 as a measure of the curvature of the universe (Table 28-1).

There is no easy way to measure the curvature of space across the universe. In theory, if you drew an enormous tri-

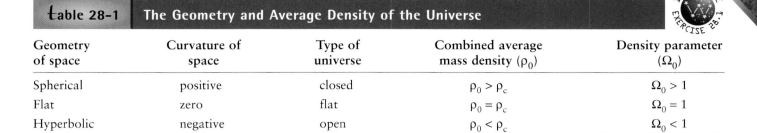

Geometry of space	Curvature of space	Type of universe	Combined average mass density (ρ_0)	Density parameter (Ω_0)
Spherical	positive	closed	$\rho_0 > \rho_c$	$\Omega_0 > 1$
Flat	zero	flat	$\rho_0 = \rho_c$	$\Omega_0 = 1$
Hyperbolic	negative	open	$\rho_0 < \rho_c$	$\Omega_0 < 1$

table 28-1   **The Geometry and Average Density of the Universe**

angle whose sides were each a billion light-years long (see Figure 28-15), you could determine the curvature of space by measuring the three angles of the triangle. If their sum equaled 180°, space would be flat. If the sum was greater than 180°, space would be spherical. And if the sum of the three angles was less than 180°, space would be hyperbolic. Unfortunately, this direct method for measuring the curvature of space is not practical.

A way to determine the curvature of the universe that is both practical and precise is to see if light rays bend toward or away from each other, as shown in Figure 28-15. The greater the distance a pair of light rays has traveled, and, hence, the longer the time the light has been in flight, the more pronounced any such bending should be. Therefore, astronomers test for the presence of such bending by examining the oldest radiation in the universe: the cosmic microwave background.

If the cosmic microwave background were truly isotropic, so that equal amounts of radiation reached us from all directions in the sky, it would be impossible to tell whether individual light rays have been bent. However, as we saw in Section 28-5, there are localized "hot spots" in the cosmic microwave background due to density variations in the early universe. The apparent size of these hot spots depends on the curvature of the universe (Figure 28-16). If the universe is closed, the bending of light rays from a hot spot will make the spot appear larger; if the universe is open, the light rays will bend the other way and the hot spots will appear smaller. Only in a flat universe will the light rays travel along straight lines, so that the hot spots appear with their true size.

By calculating what conditions were like in the primordial fireball, astrophysicists find that in a flat universe, the dominant "hot spots" in the cosmic background radiation should

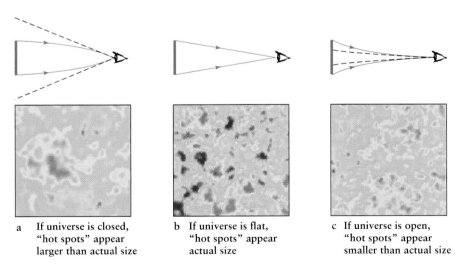

a   If universe is closed, "hot spots" appear larger than actual size

b   If universe is flat, "hot spots" appear actual size

c   If universe is open, "hot spots" appear smaller than actual size

figure 28-16   R I V U X G

The Cosmic Microwave Background and the Curvature of Space Temperature variations in the early universe appear as "hot spots" in the cosmic microwave background. The apparent size of these spots depends on the curvature of space. **(a)** In a closed universe with positive curvature, light rays from opposite sides of a hot spot bend toward each other. Hence, the hot spot appears to us to be larger than it actually is, as shown by the dashed lines. **(b)** The light rays do not bend in a flat universe. **(c)** In an open universe light rays bend apart. The dashed lines show that a hot spot would appear smaller than its actual size. (The BOOMERANG Group, University of California, Santa Barbara) have an angular size of about 1°. (In Chapter 29 we will learn how this is deduced.) This is just what the BOOMERANG and MAXIMA experiments observed (see Figure 28-14). Hence, the curvature of the universe must be very close to zero, and the universe must either be flat or very nearly so.

As Table 28-1 shows, once we know the curvature of the universe, we can determine the density parameter Ω_0 and hence the combined average mass density ρ_0. By analyzing the data shown in Figure 28-14, astrophysicists find that $\Omega_0 = 1.0$ with an uncertainty of about 10%. In other words, ρ_0 is within 10% of the critical density ρ_c. Other, earlier observations had suggested the same result, but with far greater uncertainties.

28-7 The universe appears to be filled with dark energy

The flatness of the universe poses a major dilemma. We saw in Section 28-5 that the average mass density of matter in the universe, ρ_m, is in the range $2 - 4 \times 10^{-27}$ kg/m³. This is about 0.2 to 0.4 of the critical density ρ_c. We can express this in terms of the **matter density parameter** Ω_m (say "omega sub em"), equal to the ratio of ρ_m to the critical density:

$$\Omega_m = \frac{\rho_m}{\rho_c} = 0.2 - 0.4$$

If matter and radiation were all there is in the universe, the combined average mass density ρ_0 would be equal to ρ_m (plus a tiny contribution from radiation, which we can neglect because the average mass density of radiation is only about 1/6000 that of matter). Then the density parameter Ω_0 would be equal to Ω_m—that is, in the range from 0.2 to 0.4—and the universe would be open. But the temperature variations in the cosmic microwave background clearly show that the universe is either flat or very nearly so. These variations also show that the density parameter Ω_0, which includes the effects of *all* kinds of matter and energy, is equal to 1.0. In other words, radiation and matter, including dark matter, together account for only 20% to 40% of the total density of the universe! The dilemma is this: What could account for the rest?

The source of the missing density must be some form of energy that we cannot detect from its gravitational effects (the technique astronomers use to detect dark matter). It must also not emit detectable radiation of any kind. We refer to this mysterious energy as **dark energy.**

Just as we express the average density of matter and radiation by the matter density parameter Ω_m, we can express the average density of dark energy in terms of the **dark energy density parameter** Ω_Λ (say "omega sub lambda"). This is equal to the average mass density of dark energy, ρ_Λ, divided by the critical density ρ_c:

$$\Omega_\Lambda = \frac{\rho_\Lambda}{\rho_c}$$

We can determine the value of Ω_Λ by noting that the combined average mass density ρ_0 must be the sum of the average mass densities of matter, radiation, and dark energy. As we have seen, the contribution of radiation is so small that we can ignore it, so we have

$$\rho_0 = \rho_m + \rho_\Lambda$$

If we divide this through by the critical density ρ_c, we obtain

$$\Omega_0 = \Omega_m + \Omega_\Lambda$$

That is, the density parameter Ω_0 is the sum of the matter density parameter Ω_m and the dark energy density parameter Ω_Λ. Solving for Ω_Λ, we find

$$\Omega_\Lambda = \Omega_0 - \Omega_m$$

Since Ω_0 is close to 1.0 and Ω_m is in the range 0.2 to 0.4, we conclude that Ω_Λ must be in the range from 0.6 to 0.8. Thus, dark energy accounts for between 60% and 80% of the contents of the universe.

The concept of dark energy is actually due to Einstein. When he proposed the existence of a cosmological constant, he was suggesting that the universe is filled with a form of energy that by itself tends to make the universe expand (see Section 28-1). Unlike gravity, which tends to make objects attract, the energy associated with a cosmological constant would provide a form of "antigravity." Hence, it would not be detected in the same way as matter. (The subscript Λ in the symbol for the dark energy density parameter pays homage to the symbol that Einstein chose for the cosmological constant.)

If dark energy is in fact due to a cosmological constant, the value of this constant must be far larger than Einstein suggested. This is needed if we are to explain why Ω_Λ has a large value of 0.6 to 0.8. Einstein felt he erred by introducing the idea of a cosmological constant; it may be that his error was giving it too small a value!

These ideas concerning dark energy are controversial, and require additional observations to verify them. As we will see in the next section, a crucial test is to examine how the rate of expansion of the universe has evolved over the eons.

28-8 Observations of distant supernovae indicate that we live in an accelerating universe

We have seen that the universe is expanding. But does the rate of expansion stay the same? Because there is matter in the universe, and because gravity tends to pull the bits of matter in the universe toward one another, we would expect that the expansion should slow down with time. (In the same way, a cannonball shot upward from the surface of the Earth will slow down as it ascends because of the Earth's gravitational pull.) If there is a cosmological constant, however, its associated dark energy will exert an outward pressure that tends to accelerate the expansion.

To determine which of these effects is the more important, astronomers study the relationship between redshift and distance for extremely remote galaxies. We see these galaxies as they were billions of years ago. If the rate of expansion was the same in the distant past as it is now, the same Hubble law should apply to distant galaxies as to nearby ones. But if the rate of expansion has either increased or decreased, we will find important deviations from the Hubble law.

To see how astronomers approach this problem, first imagine two different parallel universes. Both Universe #1 and Universe #2 are expanding at a constant rate, so that there is a direct proportion between recessional velocity v and distance d as expressed by the Hubble law $v = H_0d$. Hence, a graph of distance versus recessional velocity for either universe is a straight line, as Figure 28-17a shows. The only difference is that Universe #1 is expanding at a slower rate than Universe #2. Hence, a galaxy at a certain distance from Earth in Universe #1 will have a slower recessional velocity than a galaxy at the same distance from Earth in Universe #2. As a result, the graph of distance versus recessional velocity for slowly expanding Universe #1 (shown in blue) has a steeper slope than the graph for rapidly expanding Universe #2 (shown in green). Keep this observation in mind: A slower expansion means a steeper slope on a graph of distance versus recessional velocity.

Now consider our universe and allow for the possibility that the expansion rate may change over time. If we observe very remote galaxies, we are seeing them as they were in the remote past. If the expansion of the universe in the remote past was slower or faster than it is now, the slope of the graph of distance versus recessional velocity will be different for those remote galaxies. If the expansion was slower then, the slope will be steeper for distant galaxies (shown in blue in Figure 28-17b); if the expansion was faster, the slope will be shallower for distant galaxies (shown in green in Figure 28-17b). In either case, there will be a deviation from the straight-line Hubble law (shown in red in Figure 28-17b).

Figure 26-16 shows the observed relationship between distance and recessional velocity for galaxies. The data points appear to lie along a straight line, suggesting that the rate of cosmological expansion has not changed. (Figure 26-16 is actually a graph of recessional velocity versus distance, not the other way around. But a straight line on one kind of graph will be a straight line on the other, because in either case there is a direct proportion between the two quantities being graphed.) However, this graph was based on measurements of galaxies no farther than 600 Mpc (2 billion ly) from the Earth, which means we are looking only 2 billion years into the past. The straightness of the line in Figure 26-16 means only that the expansion of the universe has been relatively constant over the past 2 billion years—a relatively brief interval on the cosmic scale.

Now suppose that you were to measure the redshifts and distances of galaxies *several* billion light-years from the Earth. The light from these galaxies has taken billions of years to

a

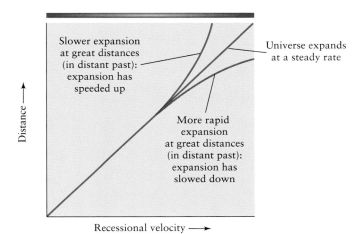

b

ƒigure 28-17

Varying Rates of Cosmic Expansion (a) Imagine two universes, #1 and #2, that each expands at its own constant rate. For a galaxy at a given distance, the recessional velocity will be greater in the more rapidly expanding universe. Hence, the graph of distance d versus recessional velocity v will have a shallower slope for the rapidly expanding universe and a steeper slope for the slowly expanding one. **(b)** If the rate of expansion of our universe was more rapid in the distant past, corresponding to remote distances, the graph of d versus v will have a shallower slope for large distances. If the expansion rate was slower in the distant past, the graph will have a steeper slope for large distances.

arrive at your telescope, so your measurements will reveal how fast the universe was expanding billions of years ago. To do this, we need a technique that will allow us to find the distances to these very remote galaxies. We saw in Section 26-4 that one way to do this is to identify Type Ia supernovae in such galaxies. These supernovae are the most luminous objects in the universe, and hence can be detected even at extremely large distances (see Figure 26-13 and the image that opens this chapter). The maximum brightness of a supernova tells astronomers its distance through the inverse-square law for light, and the redshift of the supernova's spectrum tells them its recessional velocity.

In 1998, two large research groups—the Supernova Cosmology Project, led by Saul Perlmutter of Lawrence Berkeley National Laboratory, and the High-Z Supernova Search Team, led by Brian Schmidt of the Mount Stromlo and Siding Springs Observatories in Australia—reported their results from a survey of Type Ia supernovae in galaxies at redshifts of 0.2 or greater, corresponding to distances beyond 750 Mpc (2.4 billion ly). Figure 28-18 shows some of these data on a graph of apparent magnitude versus redshift. A greater apparent magnitude corresponds to a dimmer supernova (see Section 19-3), which means that the supernova is more distant. A greater redshift implies a greater recessional velocity. Hence, this graph is basically the same as those in Figure 28-17.

To interpret these results, we need guidance from **relativistic cosmology**. This is a theoretical description of the universe and its expansion, based on Einstein's general theory of relativity and developed in the 1920s by Alexander Friedmann in Russia, Georges Lemaître in France, and Willem de Sitter in the Netherlands. Given values of the mass density parameter Ω_m and the dark energy density parameter Ω_Λ, cosmologists can use the equations of relativistic cosmology to predict how the expansion rate of the universe should change over time. Such predictions can be expressed as curves on a graph of distance versus redshift such as Figure 28-18.

The two green curves in Figure 28-18 show what would be expected if there were no cosmological constant, so that $\Omega_\Lambda = 0$ and $\Omega_0 = \Omega_m$ (that is, the density parameter is due to matter and radiation alone). The dashed green curve assumes an open universe with $\Omega_0 = 0.20$, while the solid green curve assumes a flat universe with $\Omega_0 = 1.00$. In either case, the gravitational attraction between galaxies would cause the expansion of the universe to slow down with time, so that the expansion rate would have been greater in the past (compare with the green curve in Figure 28-17b). However, neither one of these models agrees very well with the data: The data points for large redshifts (toward the right of the graph) lie above both these curves.

The data agree very well with the blue curve in Figure 28-18. This curve assumes a flat universe with a cosmological constant ($\Omega_m = 0.24$, $\Omega_\Lambda = 0.76$, $\Omega_0 = \Omega_m + \Omega_\Lambda = 1.00$). In this model, the cosmological constant has made the expansion of the universe speed up over time, so that the expansion was slower in the distant past. Just as in the blue curve in Figure 28-17b, the data show that supernovae of a certain brightness (and hence a given distance) have smaller redshifts (and hence smaller recessional velocities) than would be the case if the expansion rate had always been the same.

The supernova data in Figure 28-18 provide compelling evidence of the existence of dark energy. By themselves, however, they do not determine the value of the dark energy density parameter Ω_Λ. Roughly speaking, the data in Figure 28-18 indicate the relative importance of dark energy (which tends to

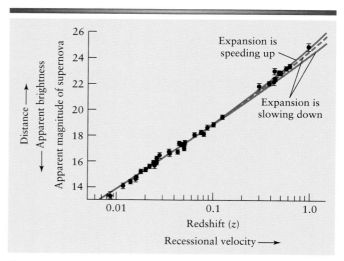

Figure 28-18

The Hubble Diagram for Distant Supernovae This graph shows apparent magnitude versus redshift for supernovae in distant galaxies. The greater the apparent magnitude, the dimmer the supernova and the greater the distance to it and its host galaxy. The three curves are predictions for a flat universe with no cosmological constant (solid green curve), an open universe with no cosmological constant (dashed green curve), and a flat universe with a cosmological constant (blue curve). The last of these gives the best fit to the data, indicating that the expansion of the universe is speeding up. (After Adam Riess, Alex Filippenko, and the High-Z Supernova Search Team)

make the expansion speed up) and gravitational attraction between galaxies (which tends to make the attraction slow down). Thus, these data tell us about the *difference* between the values of Ω_Λ and the matter density parameter Ω_m. By contrast, measurements of the cosmic microwave background (Section 28-6) give information about Ω_0, equal to the *sum* of Ω_Λ and Ω_m. Observations of galaxy clusters (Section 28-5) set limits on the value of Ω_m by itself. By combining these three very different kinds of observations as shown in Figure 28-19, we can set more stringent limits on both Ω_Λ and Ω_m.

Taken together and combined with other observations, all these data suggest the following values:

$$\Omega_m = 0.2 \text{ to } 0.4$$
$$\Omega_\Lambda = 0.6 \text{ to } 0.8$$
$$\Omega_0 = \Omega_m + \Omega_\Lambda = 0.9 \text{ to } 1.1$$

This collection of numbers points to a radically different model of the universe from what was suspected just a few years ago. In the 1980s there was no compelling evidence for an accelerating expansion of the universe, so it was widely assumed that $\Omega_\Lambda = 0$. Evidence from distant galaxies suggested a flat universe, so it was presumed that $\Omega_m = 1$. Figure 28-19 shows that this model is probably excluded by current data.

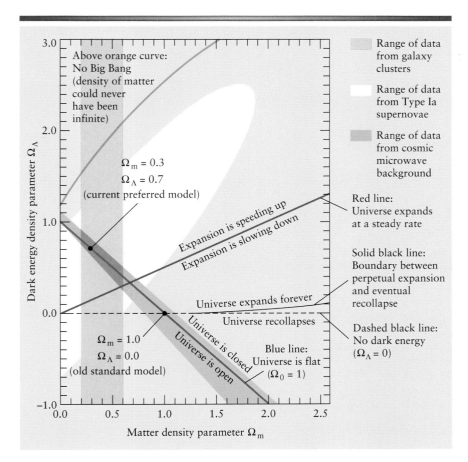

figure 28-19

Limits on the Nature of the Universe The three regions on this graph show values of the mass density parameter Ω_m and the dark energy density parameter Ω_Λ that are consistent with various types of observations. Galaxy cluster measurements set limits on Ω_m but not on Ω_Λ. Observations of the cosmic microwave background set limits on the sum of Ω_m and Ω_Λ: A larger value of Ω_m (to the right in the graph) implies a smaller value of Ω_Λ (downward in the graph) to keep the sum the same, which is why this band slopes downward. Observations of Type Ia supernovae set limits on the difference between Ω_m and Ω_Λ; this band slopes upward since a larger value of Ω_m implies a larger value of Ω_Λ to keep the difference the same. The best agreement to all these observations is where all three regions overlap, in the vicinity of $\Omega_m = 0.3$, $\Omega_\Lambda = 0.7$. This indicates that we live in a flat universe with an accelerating rate of expansion. During the 1980s the preferred model of the universe was $\Omega_m = 1.0$, $\Omega_\Lambda = 0.0$; the current data appear to exclude this possibility. (After Lawrence Krauss)

The model we are left with is one in which the universe is suffused with a curious dark energy due to a cosmological constant. Unlike matter or radiation, whose average densities decrease as the universe expands and thins out, the average density of this dark energy remains constant throughout the history of the universe (Figure 28-20). The dark energy was relatively unimportant over most of the early history of the universe. Today, however, the density of dark energy is greater than that of matter (Ω_Λ is greater than Ω_m). In other words, we live in a **dark-energy-dominated universe.**

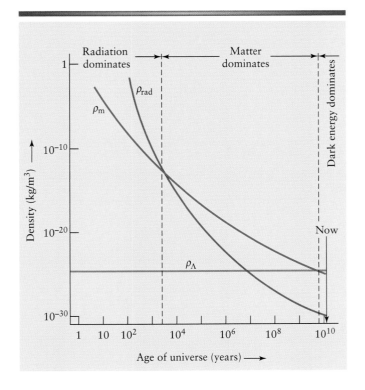

figure 28-20

The Evolution of Density, Revisited The average mass density of matter, ρ_m, and the average mass density of radiation, ρ_{rad}, both decrease as the universe expands and becomes more tenuous. But if the dark energy is due to a cosmological constant, its average mass density ρ_Λ remains constant. In this model, our universe became dominated by dark energy about 2 billion years ago.

There may be other explanations for dark energy besides a cosmological constant. Several physicists have proposed a type of dark energy whose density decreases slowly as the universe expands. Others have proposed that the dark energy is actually concentrated in a network of structures called cosmic strings, which we discuss in Chapter 29. Future observations, including space-based, high-precision measurements of the cosmic background radiation and of Type Ia supernovae, may help resolve the nature of the dark energy. They may also help pin down the values of Ω_m, Ω_Λ, and the Hubble constant, thus giving us a better global understanding of our curious universe.

28-9 The matter and dark energy in the universe determine its future evolution

We have seen that the average densities of matter and dark energy in the universe determine whether the cosmological expansion is speeding up, slowing down, or maintaining the same rate. These average densities also determine what will happen to the universe in the eons to come.

The future evolution of the universe depends on how the rate of cosmological expansion changes with time. Astronomers describe this by the **deceleration parameter**, denoted q_0. Appropriately, $q_0 = 0$ corresponds to no deceleration or acceleration at all. This is possible if the universe is completely empty, so that there is no gravity to slow down the expansion and no dark energy to speed it up. As shown in Figure 28-21, a universe with $q_0 = 0$ expands forever at a constant rate. In this case, the age of the universe would be exactly $1/H_0$, as we saw in Section 28-3.

The case $q_0 = \frac{1}{2}$ corresponds to a **marginally bounded universe**. In this case the rate of expansion continually decreases but never stops, and the universe just barely manages to continue expanding forever. The red line in Figure 28-19 labeled "Universe expands at a steady rate" corresponds to $q_0 = \frac{1}{2}$.

Values of q_0 less than $\frac{1}{2}$ correspond to an **unbounded universe** that continues to expand forever. If q_0 is between 0 and $\frac{1}{2}$, the rate of expansion decreases but never goes to zero, even after an infinitely long time. If q_0 is less than zero, there is "negative deceleration" and the rate of expansion increases.

A **bounded universe** has q_0 greater than $\frac{1}{2}$. The bits of matter in the universe exert so much gravitational attraction on one another that the expansion slows and eventually comes to a halt. The universe then contracts inward, so that the redshifts of receding galaxies change to blueshifts as the galaxies approach. This cosmological contraction ends in a cosmic singularity whimsically called the **Big Crunch.** Just as we cannot use physics to tell us what existed before the Big Bang, science cannot reveal what might happen after the Big Crunch. The infinite warping of space and time in a cosmic singularity prevents us from extrapolating any further into the future.

What is the future of the universe we live in? The answer depends on the values of the matter density parameter Ω_m and

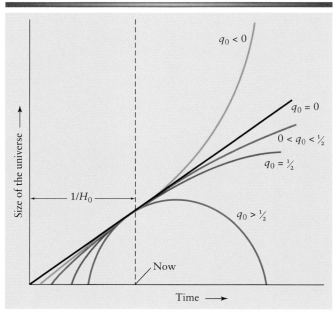

figure 28-21

The Growth of the Universe This diagram shows how the size of the universe—represented by the separation between two widely-separated galaxies—evolves with time in different models. The case $q_0 = 0$ is an empty universe that expands at a steady rate. The case $q_0 = \frac{1}{2}$ is a marginally bounded universe. If $q_0 > \frac{1}{2}$ the universe is bounded and will eventually recollapse. If $q_0 < \frac{1}{2}$, the universe is unbounded and will expand forever. If $q_0 < 0$, as may be the case for our universe, this unbounded expansion gains speed due to the presence of dark energy.

the dark energy density parameter Ω_Λ (see Figure 28-19). If we assume that the dark energy is due to a cosmological constant, the deceleration parameter can be calculated using the formula

$$q_0 = \frac{1}{2}\Omega_0 - \frac{3}{2}\Omega_\Lambda$$

Recall that the density parameter Ω_0 is equal to $\Omega_m + \Omega_\Lambda$.

If there is no cosmological constant so that $\Omega_\Lambda = 0$, then q_0 is just equal to $\Omega_0/2$. In this situation the geometry of the universe determines its destiny. A flat universe with $\Omega_0 = 1$ has $q_0 = \frac{1}{2}$ and is marginally bounded, an open universe ($\Omega_0 < 1$) has $q_0 < \frac{1}{2}$ and expands forever, and a closed universe ($\Omega_0 > 1$) has $q_0 > \frac{1}{2}$ and eventually recollapses.

If there is a cosmological constant, there is no simple connection between the geometry of the universe and its eventual fate. For example, our universe appears to be flat or nearly so ($\Omega_0 = 1$) and appears to have a substantial cosmological constant with Ω_Λ of about 0.7. Using these numbers, we find

$$q_0 = \frac{1}{2}(1.0) - \frac{3}{2}(0.7) = -0.55$$

Since q_0 is negative, we predict that our flat universe will expand forever with an ever-increasing rate of expansion.

Although the ordinary matter in the universe today consists mostly of hydrogen and helium, these gases will eventually be used up in the thermonuclear fires of generation after generation of stars. Astronomers estimate that this depletion will occur in about 10^{12} years. No more stars will form, because the hydrogen and helium required will no longer exist.

Galaxies will grow dim as the final generation of stars dies off. Eventually, roughly a trillion years after the Big Bang, all the matter in the universe will consist of either dead stars (white dwarfs, neutron stars, and black holes) or cold lumps (planets, meteorites, used spaceships, rocks, and so forth).

Still further into the future, extremely rare events will become important. For example, the probability of two stars colliding somewhere in our Galaxy (or any other galaxy) is extremely small. Indeed, the probability of two stars even passing near each other is almost infinitesimal. Over trillions upon trillions of years, however, such near collisions will occur sooner or later.

A close encounter between two stars in a galaxy would likely be significant because one of the stars could easily exert enough gravitational force on the other to kick it out of its galaxy. The remaining star would then fall into a lower-energy orbit closer to the galaxy's nucleus. Eventually, because of the emission of gravitational radiation (recall Section 24-2), these remaining stars would gradually spiral into the galaxy's center, coalescing into an enormous black hole. The mass of the resulting black hole would be roughly 10^{11} $M_\odot$. The time needed for all this to occur is about 10^{27} years. Thus, when the universe is a billion billion billion years old, galaxies will consist of enormous black holes surrounded by dead stars.

The orbital motions of galaxies in clusters also emit gravitational radiation. As a result of this energy loss, entire galaxies spiral toward each other, eventually coalescing into even more gigantic black holes. It will take roughly 10^{31} years for these colossal objects to form, each of which might contain 10^{15} $M_\odot$.

Even further into the future, black hole evaporation (see Section 24-8) will release the matter within black holes back into space. Calculations show that stellar-mass black holes evaporate after roughly 10^{67} years. The gigantic black holes discussed above will evaporate after 10^{97} to 10^{106} years.

Professor Stephen Hawking first showed in the early 1970s that the laws of quantum mechanics permit black holes to evaporate. Hawking has proved that, during the final moments of evaporation, a black hole is actually a **white hole** that randomly pumps new matter and information into the universe in an unpredictable fashion. More of Professor Hawking's thoughts on cosmology can be found in the essay that follows this chapter.

KEY WORDS

average density of matter, p. 649

Big Bang, p. 645

Big Crunch, p. 660

bounded universe, p. 660

closed universe, p. 653

combined average mass density, p. 653

cosmic background radiation, p. 647

cosmic microwave background, p. 647

cosmic particle horizon, p. 645

cosmic singularity, p. 646

cosmological constant, p. 642

cosmological principle, p. 644

cosmological redshift, p. 644

cosmology, p. 641

critical density, p. 654

dark energy, p. 656

dark energy density parameter, p. 656

dark-energy-dominated universe, p. 659

deceleration parameter, p. 660

density parameter, p. 654

era of recombination, p. 651

flat space, p. 653

Great Attractor, p. 649

homogeneous, p. 644

hyperbolic space, p. 653

isotropic, p. 644

lookback time, p. 644

marginally bounded universe, p. 660

mass density of radiation, p. 649

matter density parameter, p. 656

matter-dominated universe, p. 650

negative curvature, p. 653

observable universe, p. 645

Olbers's paradox, p. 641

open universe, p. 653

Planck time, p. 646

plasma, p. 651

positive curvature, p. 653

primordial fireball, p. 651

radiation-dominated universe, p. 650

relativistic cosmology, p. 658

spherical space, p. 653

unbounded universe, p. 660

white hole, p. 661

zero curvature, p. 653

KEY IDEAS

The Expansion of the Universe: The Hubble law describes the continuing expansion of space. The redshifts that we see from distant galaxies are caused by this expansion, not by the motions of galaxies through space.

• The redshift of a distant galaxy is a measure of the scale of the universe at the time the galaxy emitted its light.

• It is meaningless to speak of an edge or center to the universe or of what lies beyond the universe.

The Cosmological Principle: Cosmological theories are based on the idea that on large scales, the universe looks the same at all locations and in every direction.

The Big Bang: The universe began as an infinitely dense cosmic singularity that expanded explosively in the event called the Big Bang, which can be described as an explosion of space at the beginning of time.

• The observable universe extends about 14 billion light-years in every direction from the Earth. We cannot see objects beyond this distance because light from these objects has not had enough time to reach us.

• During the first 10^{-43} second after the Big Bang, the universe was too dense to be described by the known laws of physics.

Cosmic Background Radiation and the Evolution of the Universe: The cosmic microwave background radiation, corresponding to radiation from a blackbody at a temperature of nearly 3 K, is the greatly redshifted remnant of the hot universe that existed about 300,000 years after the Big Bang.

• The background radiation was hotter and more intense in the past. During the first 300,000 years of the universe, radiation and matter formed an opaque plasma called the primordial fireball. When the temperature of the radiation fell below 3000 K, protons and electrons could combine to form hydrogen atoms and the universe became transparent.

• The abundance of helium in the universe is explained by the high temperatures in its early history.

The Geometry of the Universe: The curvature of the universe as a whole depends on how the combined average mass density ρ_0 compares to a critical density ρ_c.

• If ρ_0 is greater than ρ_c, the density parameter Ω_0 has a value greater than 1, the universe is closed and space is spherical (with positive curvature).

• If ρ_0 is less than ρ_c, the density parameter Ω_0 has a value less than 1, the universe is open and space is hyperbolic (with negative curvature).

• If ρ_0 is equal to ρ_c, the density parameter Ω_0 is equal to 1 and space is flat (with zero curvature).

Cosmological Parameters and Dark Energy: Observations of temperature variations in the cosmic microwave background indicate that the universe is flat or nearly so, with a combined average mass density equal to the critical density. Observations of galaxy clusters suggest that the average density of matter in the universe is about 0.2 to 0.4 of the critical density. The remaining contribution to the average density is called dark energy.

• Measurements of Type Ia supernovae in distant galaxies suggest that the expansion of the universe is speeding up. This may be due to the presence of dark energy in the form of a cosmological constant, which provides a pressure that pushes the universe outward.

The Fate of the Universe: If the universe is bounded, it will collapse in a Big Crunch. If the universe is unbounded or marginally bounded, black holes will eventually evaporate, pouring matter and energy randomly back into the universe.

REVIEW QUESTIONS

1. What is Olbers's paradox? How can it be resolved?

2. What is a cosmological constant? Why did Einstein introduce it into cosmology?

3. What does it mean when astronomers say that we live in an expanding universe? What is actually expanding?

4. Using a diagram, explain why the expansion of the universe as seen from a distant galaxy would look the same as seen from our Galaxy.

5. How does modern cosmology preclude the possibility of either a center or an edge to the universe?

6. Explain the difference between a Doppler shift and a cosmological redshift.

7. Explain how redshift can be used as a measure of lookback time. In what ways is it superior to time measured in years?

8. What is the cosmological principle? How is it justified?

9. How was the Big Bang different from an ordinary explosion? Where in the universe did it occur?

10. What is meant by "the observable universe"?

11. Can you see the cosmic background radiation with the naked eye? With a visible-light telescope? Explain why or why not.

12. If the universe continues to expand forever, what will eventually become of the microwave background radiation?

13. How can astronomers measure the average mass density of the universe?

14. What does it mean to say that the universe was once radiation-dominated? What happened when the universe changed from being radiation-dominated to being matter-dominated? When did this happen?

15. What was the era of recombination? What significant events occurred in the universe during this era? Was the universe matter-dominated or radiation-dominated during this era?

16. Describe two different ways in which the cosmic microwave background is not isotropic.

17. What is meant by the critical density of the universe? Why is this quantity important to cosmologists?

18. Explain why the universe does not have a center.

19. Describe how astronomers use the cosmic background radiation to determine the geometry of the universe.

20. Explain why it is important to measure how the expansion rate of the universe has changed over time. How is this rate measured?

21. What is dark energy? Describe two ways that we can infer its presence.

22. What does it mean to say that the universe is dark-energy-dominated? What happened when the universe changed from being matter-dominated to being dark-energy-dominated?

23. What does the geometry of the universe tell you about its eventual fate? Is your answer the same whether or not there is a cosmological constant?

24. In what circumstances can a black hole behave like a white hole?

ADVANCED QUESTIONS

Problem-solving tips and tools

You may find it useful to recall that 1 parsec equals 3.26 light-years, that 1 Mpc equals 3.09×10^{19} km, and that a year contains 3.16×10^7 seconds.

25. The galaxy RD1 has a redshift $z = 5.34$. The light from this galaxy includes the Lyman-alpha (L_α) spectral line of hydrogen, with an unshifted wavelength of 121.6 nm. Calculate the wavelength at which we detect the Lyman-alpha photons from RD1. In what part of the electromagnetic spectrum does this wavelength lie?

26. Estimate the age of the universe for a Hubble constant of (a) 50 km/s/Mpc, (b) 75 km/s/Mpc, and (c) 100 km/s/Mpc. On the basis of your answers, explain how the ages of globular clusters could be used to place a limit on the maximum value of the Hubble constant.

27. Some so-called "creation scientists" claim that the universe came into being about 6000 years ago. Find the value of the Hubble constant for such a universe. Is this a reasonable value for H_0? Explain.

28. The quasar HS 1946+7658 has a redshift $z = 3.02$. At the time when the light we see from HS 1946+7658 left the quasar, how many times more dense was the matter in the universe than it is today?

29. If the mass density of radiation in the universe were 625 times larger than it is now, what would the background temperature be?

30. Suppose that the present-day temperature of the cosmic background radiation were somehow increased by a factor of 100, from 2.725 K to 272.5 K. (a) Calculate ρ_{rad} in this situation. (b) If the average density of matter (ρ_m) remained unchanged, would it be more accurate to describe our universe as matter-dominated or radiation-dominated? Explain.

31. Calculate the mass density of radiation (ρ_{rad}) in each of the following situations, and explain whether each situation is matter-dominated or radiation-dominated: (a) the photosphere of the Sun ($T = 5800$ K, $\rho_m = 3 \times 10^{-4}$ kg/m^3); (b) the center of the Sun ($T = 1.55 \times 10^7$ K, $\rho_m = 1.6 \times 10^5$ kg/m^3); (c) the solar corona ($T = 2 \times 10^6$ K, $\rho_m = 5 \times 10^{-13}$ kg/m^3).

32. What would be the critical density of matter in the universe (ρ_c) if the value of the Hubble constant were (a) 50 km/s/Mpc? (b) 100 km/s/Mpc?

33. Consider the galaxy RD1 (see Advanced Question 25), which has $z = 5.34$. (a) Suppose that in the present-day universe, two clusters of galaxies are 500 Mpc apart. At the time that the light was emitted from RD1 to produce an image tonight, how far apart were those two clusters? b) What was the average density of matter (ρ_m) at that time? Assume that in today's universe, $\rho_m = 2 \times 10^{-27}$ kg/m^3. (c) What were the temperatures of the cosmic background radiation and the mass density of radiation (ρ_{rad}) at that time? (d) At this time in the remote past, was the universe matter-dominated, radiation-dominated, or dark-energy-dominated? Explain.

34. Suppose we lived in a flat universe that had $q_0 = 0$. What would be the value of the dark energy density parameter in such a universe? Which would be dominant in such a universe, matter or dark energy? Explain.

35. In general, the deceleration parameter is not constant but varies with time. For a flat universe, the deceleration parameter at a redshift z is given by the formula

$$q_z = \frac{1}{2} - \frac{3}{2}\left[\frac{\Omega_\Lambda}{\Omega_\Lambda + (1 - \Omega_\Lambda)(1 + z)^3}\right]$$

where Ω_Λ is the dark energy density parameter. Using $\Omega_\Lambda = 0.7$, find the value of q_z for (a) $z = 0.5$ and (b) $z = 1.0$. (c) Explain how your results show that the expansion of the universe was actually decelerating at $z = 1.0$, but changed from deceleration to acceleration between $z = 1.0$ and $z = 0.5$.

36. The dark energy density parameter Ω_Λ is related to the value of the cosmological constant Λ by the formula

$$\Omega_\Lambda = \frac{2\Lambda c^2}{9H_0^2}$$

where $c = 3 \times 10^8$ m/s is the speed of light. Determine the value of Λ if $\Omega_\Lambda = 0.7$ and the Hubble constant is $H_0 = 70$ km/s/Mpc. (*Hint:* You will need to convert units to eliminate kilometers and megaparsecs.)

DISCUSSION QUESTIONS

37. Suppose we were living in a radiation-dominated universe. Discuss how such a universe would be different from what we now observe.

38. How can astronomers be certain that the cosmic microwave background fills the entire cosmos, not just the vicinity of the Earth?

39. Discuss the implications of the idea that we cannot use science to tell us what existed before the Big Bang.

40. Do you think there can be "other universes," regions of space and time that are not connected to our universe? Should astronomers be concerned with such possibilities? Why or why not?

WEB/CD ROM QUESTIONS

41. Before the discovery of the cosmic microwave background, it seemed possible that we might be living in a "steady-state universe" with overall properties that do not change with time. The steady-state model, like the Big Bang model, assumes an expanding universe, but does not assume a "creation event." Instead, matter is assumed to be created continuously everywhere in space to ensure that the average density of the universe remains constant. Search the World Wide Web for information about the "steady-state" theory. Explain why the existence of the cosmic microwave background was a fatal blow to the steady-state theory.

42. Search the World Wide Web for information about the Microwave Anistropy Probe mission. What is the purpose of this mission? Has the spacecraft been launched? What information may this mission provide that the BOOMERANG and MAXIMA experiments could not?

 43. Temperatures in the Early Universe. Access the Active Integrated Media Module "Blackbody Curves" in Chapter 5 of the *Universe* web site or CD-ROM. (**a**) Use the module to determine by trial and error the temperature at which a blackbody spectrum has its peak at a wavelength of 1 μm. (**b**) At the time when the temperature of the cosmic background radiation was equal to the value you found in (a), was the universe matter-dominated or radiation-dominated? Explain your answer.

OBSERVING PROJECT

 44. Use the *Starry Night* program to determine how the solar system moves through the cosmic microwave background. First turn off daylight (select **Daylight** in the **Sky** menu) and show the entire celestial sphere (select **Atlas** in the **Go** menu). You should see a green line representing the ecliptic (select **The Ecliptic** in the **Guide** menu if you can't see this). Use **Find…** in the **Edit** menu to center the field of view on Leo. (**a**) Draw a sketch showing the Sun, the plane in which the Earth orbits the Sun, and the direction in which the solar system moves through the cosmic microwave background. (**b**) Use the time controls in the Control Panel to step through the months of the year. In which month is the Sun placed most nearly in front of the Earth as the solar system travels through the cosmic background radiation?

STEPHEN W. HAWKING

The Edge of Spacetime

Stephen Hawking is a theoretical physicist at the University of Cambridge. Born in Oxford in 1942, he obtained his B.A. at Oxford in 1962 and went on to earn his doctorate working on gravity and cosmology under Dennis W. Sciama at Cambridge. He continued his work on these subjects at Cambridge and, in 1974 and 1975, at the California Institute of Technology. A fellow of the Royal Society since 1974, he has received many honors and awards, including the Eddington Medal of the Royal Astronomical Society, the Dannie Heinemann Prize of the American Institute of Physics and the American Physical Society, the Maxwell Medal and Prize of the Institute of Physics, and the

Einstein Medal. In 1979 he was elected to Newton's old chair, Lucasian Professor at Cambridge.

His contributions have been in the areas of general relativity, gravitation, and quantum theory as it relates to black holes and their thermodynamics. Although afflicted by a progressive disease of the nervous system since 1961, which has confined him to a wheelchair, he continues to be phenomenally productive as both a writer and a researcher.

From the dawn of civilization, people have asked questions like: Did the universe have a beginning in time? Will it have an end? Is the universe bounded or infinite in spatial extent? In this essay I shall outline some answers to these questions that are suggested by modern developments in science. Most of what I describe is now generally accepted, though some of it was controversial not long ago. My final conclusion, however, is based on some very recent work on which there has not yet been time to reach a consensus.

In most early mythological and religious accounts, the universe, or at least its human inhabitants, was created by a Divine Being at some date in the fairly recent past like 4004 B.C. Indeed, the necessity of a "first cause" to account for the creation of the universe was used as an argument to prove the existence of God. Conversely, Greek philosophers such as Plato and Aristotle did not like the thought of such direct divine intervention in the affairs of the world and preferred to believe that the universe had existed and would exist forever. Most people in the ancient world believed that the universe is spatially bounded. In the earliest cosmologies the world was a flat plate with the sky as a pudding basin overhead. The Greeks, however, realized that the world is round. They constructed an elaborate model in which the Earth was a sphere surrounded by a number of spheres that carried the Sun, Moon, and the planets. The outermost sphere carried the so-called fixed stars, which maintain the same relative positions but appear to rotate across the sky.

This model with the Earth at the center was adopted by the Christian church. This view had the great attraction that it left plenty of room outside the sphere of the stars for Heaven and Hell, though how these were situated was never quite clear. The Earth-centered model of the universe remained in favor until the seventeenth century, when the observations of Galileo showed that it had to be replaced by the Copernican model, in which the Earth and the other planets orbit the Sun. Not only did this model get rid of the

spheres, but it also showed that the fixed stars must be at a very great distance because they did not show any apparent movement as the Earth went round the Sun (apart from that caused by the Earth's rotation about its axis).

With the Earth no longer at the center of the universe, it was fairly natural to postulate that the stars are other Suns like our own and that they are distributed roughly uniformly throughout an infinite universe. This uniform distribution, however, raised a problem: According to Newton's theory of gravity, published in 1687, each star would be attracted toward every other star in the universe. Why, then, did not the stars all fall together to a single point? Newton himself tried to argue that this would indeed happen for a bounded collection of stars; however, if the universe were infinite, the gravitational force on a star caused by the attraction of stars on one side of the star would be balanced by the force arising from stars on the other side. The net force on any star would therefore be zero and so the stars could remain motionless.

This argument is in fact an example of the fallacy one can fall into when one adds up an infinite number of quantities: By adding them up in different orders one can get different results. We now know that an infinite distribution of stars cannot remain motionless if they are all attracting each other; they will start to fall toward each other. The only way that one can have a static, infinite universe is if the force of gravity becomes repulsive at large distances. Even then the universe would be unstable, because if the stars got slightly nearer to each other, the attraction would win out over the repulsion and the stars would fall together. On the other hand, if they got slightly farther away from each other, the repulsion would win and they would move away from each other.

Despite these and other difficulties, nearly everyone in the eighteenth and nineteenth centuries believed that the universe was essentially unchanging in time. For such a universe the question of whether it had a beginning was metaphysical: One could equally well believe that it had existed forever or that it had been created in its present form a finite time ago. The belief in a static universe still persisted in 1915 when Einstein formulated his general theory of relativity, which modified Newton's theory of gravity to make it compatible with discoveries about the propagation of light. Einstein added a so-called cosmological constant, which produced a repulsive force between particles at a great distance. This repulsive force could balance the normal gravitational attraction and allow a static, uniform universe. This solution was unstable, but it had the interesting property that in it space was finite but unbounded, just as the surface of the Earth is finite in area but does not have any boundary or edge. Time in this solution, however, could be infinite.

Einstein's static model of the universe was one of the great missed opportunities of theoretical physics: If he had stuck to his original version of general relativity without the cosmological constant, he could have predicted that the universe ought to be either expanding or collapsing. As it happened, however, it was not realized that the universe was changing with time until astronomers like Slipher and Hubble began to observe the light from other galaxies.

Visible light is made up of waves—like radio waves but with a much shorter wavelength, or distance, between wave crests. If one passes the light through a prism, it is decomposed into its constituent wavelengths—colors like a rainbow. Slipher and Hubble found the same characteristic patterns of wavelengths or colors as for the light from stars in our own Galaxy, but the patterns were all shifted toward the red, or longer-wavelength, end of the rainbow or spectrum. The only reasonable explanation of this was that the galaxies are moving away from us. This effect, known as the Doppler shift, is used by the police to measure the speed of cars: As the cars speed toward the police, the distance between the wave crests is crowded up and the wavelength is reduced.

During the 1920s, Hubble observed the remarkable fact that the red shift is greater the further the galaxy is from our own. This meant that other galaxies are moving away from us at rates that are roughly proportional to their distance from us. The universe is not static, as had been previously thought, but is expanding. The rate of expansion is very low: It will take something like 20 billion years for the separation of the two galaxies to double, but this expansion completely changes the nature of the discussion about whether the universe has a beginning or an end. This is not just a metaphysical question

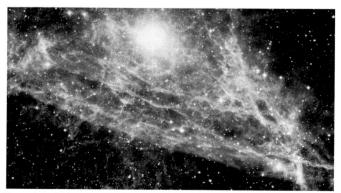

Part of the Vela supernova remnant.
(Anglo-Austrialian Telescope Board)

R I V U X G

as in the case of a static universe: As I shall describe, there may be a very real physical beginning or end of the universe.

The first model of an expanding universe that was consistent with Einstein's general theory of relativity and Hubble's observations of red shifts was proposed by the Russian physicist and mathematician Alexander Friedmann in 1922. However, it received very little attention until similar models were formulated by other people toward the end of the 1920s. In the Friedmann model and its later generalizations, the universe was assumed to be the same at every point in space and in every direction. This is obviously not a good approximation in our immediate neighborhood; there are local irregularities like the Earth and the Sun, and there are many more visible stars in the direction of the center of our Galaxy than in other directions. However, if we look at distant galaxies, we find that they are distributed roughly uniformly throughout the universe, the same in very direction. Thus, it does seem to be a good approximation on a large scale. Even better evidence comes from observations of the background of microwave radiation that was discovered in 1965 by two scientists at the Bell Telephone Laboratories. The universe is very transparent to radio waves with wavelengths of a few inches, so this radiation must have traveled to us from very great distances. Any large-scale irregularities in the universe would cause the radiation reaching us from different directions to have different intensities. Yet—to a very high degree of accuracy—the observed intensity is the same in every direction.

There are three kinds of generalized Friedmann models. In one of them, the galaxies are moving apart sufficiently slowly that the gravitational attraction between them will eventually stop their moving apart and start their approaching each other. The universe will expand to a maximum size and then recollapse. In the second model, the galaxies are moving apart so fast that gravity can never stop them and the universe expands forever. Finally, there is a third model in which the galaxies are moving apart at just the critical rate to avoid recollapse. In principle we could determine which model corresponds to our universe by comparing the present rate of expansion with the present average mass density. The mass of the matter in the universe that we can observe directly is not enough to stop the expansion. However, we have indirect evidence that there is more mass that we cannot see. Whether this "invisible" mass could be enough to stop the expansion eventually remains an open question.

In the Friedmann model that recollapses eventually, space is finite but unbounded, as in the Einstein static model. In the other two Friedmann models, which expand forever, space is infinite. Time, on the other hand, has a boundary or an edge. In all the models the expansion starts from a state of infinite density called the Big Bang singularity. In the model that recollapses there is another singularity, called the Big Crunch, at the end of the recollapse.

Singularities are places where the curvature of spacetime is infinite and the concepts of space and time cease to have any meaning. Scientific theories assume a background of space and time, so they all break down at a singularity. If there were events before the Big Bang, they would not enable one to predict the present state of the universe because predictability would break down at the Big Bang. Similarly, there is no way that one can determine what happened before the Big Bang from a knowledge of events after the Big Bang. This means that the existence or nonexistence of events before the Big Bang is purely metaphysical; they have no consequences for the present state of the universe. One might as well apply the principle of economy, known as Occam's razor, to cut such events out of the theory and say that time began at the Big Bang. Similarly, there is no way that we can predict or influence any events after the Big Crunch, so one might as well regard it as the end of time.

This beginning and possible end of time predicted by the Friedmann models are very different from earlier ideas. Prior to the Friedmann solutions, the beginning or end of time was something that had to be imposed from outside the universe; there was no necessity for a beginning or an end. In the Friedmann models, on the other hand, the beginning and end of time occur for dynamical reasons. One could still imagine the universe being created by an external agent in a state corresponding to some time after the Big Bang, but it would not have any meaning to say that the universe was created *before* the Big Bang. From the present rate of expansion of the universe, we can estimate that the Big Bang should have occurred between 10 and 20 billion years ago.

Many people disliked the idea that time has a beginning or an end because it smacks of divine intervention. There were, therefore, a number of attempts to avoid this conclusion. One of these was the steady-state model of the universe proposed in 1948 by Herman Bondi, Thomas Gold, and Fred Hoyle. In this model it was proposed that, as the galaxies move farther away from each other, new galaxies are formed in between out of matter that is being "constantly created." The universe would therefore look more or less the same at all times and the density would be roughly constant. This model had the great virtue that it made definite predictions that could be tested by observations. Unfortunately, observations of radio sources by Martin Ryle and his collaborators at Cambridge in the 1950s and early 1960s showed that the number of radio sources must have been greater in the past, contradicting the steady-state model. The final nail in the coffin of the steady-state theory was the discovery of microwave background radiation in 1965. There was no way this radiation could be accounted for in the model.

Another attempt to avoid a beginning of time was the suggestion that the singularity was simply a consequence of the high degree of symmetry of the Friedmann solutions: The relative motion of any two galaxies is restricted to the line joining them. It would therefore not be surprising that all galaxies collided with each other at some time. However, in the real universe, the galaxies also have some random velocities perpendicular to the line joining them. These "transverse velocities" could cause the galaxies to miss each other and allow the universe to pass from a contracting phase to an expanding one without the density ever becoming infinite. Indeed, in 1963 two Russian scientists claimed that this would happen in nearly every solution of the equations of general relativity. They based this claim on the fact that all the solutions with a singularity that they constructed had to satisfy some constraint or symmetry. They later realized, however, that there was a more general class of solutions with singularities that did not have to obey any constraint or symmetry.

This realization showed that singularities *could* occur in general solutions of general relativity, but it did not answer the question of whether they necessarily *would* occur. However, between 1965 and 1970, a number of theorems were proved that showed that any model of the universe must have a Big Bang singularity if the model obeys general relativity, satisfies a few other reasonable assumptions, and assumes as much matter as we observe in the universe. The same theorems predict that there will be a singularity that will be an end of time if the whole universe recollapses. Even if the universe is expanding too fast to collapse in its entirety, we nevertheless expect some localized regions, such as massive burnt-out stars, to collapse and form black holes. The theorems predict that the black holes will contain singularities that will be an end of time for anyone unfortunate or foolhardy enough to fall in.

Einstein's general theory of relativity is probably one of the two greatest intellectual achievements of the twentieth century. It is, however, incomplete because it is what is called a classical theory, that is, it does not incorporate the uncertainty principle of the other great discovery of this century, quantum mechanics. The uncertainty principle states that certain pairs of quantities, such as the position and velocity of a particle, cannot be predicted simultaneously with an arbitrarily high degree of accuracy. The more accurately one predicts the position of the particle, the less accurately one will be able to predict its velocity, and vice versa.

Quantum mechanics was developed in the early years of the twentieth century to describe the behavior of very small systems such as atoms or individual elementary particles. In particular, there was a problem with the structure of the atom, which was supposed to consist of a number of electrically charged particles called electrons orbiting around a central nucleus, as the planets orbit the Sun. The previous classical theory predicted that electrons would radiate light waves because of their motion. The waves would carry away energy and so would cause the electrons to spiral inward until they collided with the nucleus. However, such behavior is not allowed by quantum mechanics, because it would violate the uncertainty principle: If an electron were to sit on the nucleus, it would have both a definite position and a definite velocity. Instead, quantum mechanics predicts that the electron does not have a definite position; rather, the probability of finding it is spread out over some region around the nucleus, remaining finite even at the nucleus.

The prediction of classical theory of an infinite probability of finding the electron at the nucleus is rather similar to the prediction of classical general relativity that there should be a Big Bang singularity of infinite density. Thus, one might hope that if one were able to combine general relativity and quantum mechanics into a theory of quantum gravity, one would find that the singularities of gravitational collapse or expansion were smeared out as in the case of the collapse of an atom.

The first indication that singularities might be smeared out came with the discovery that black holes, formed by the collapse of localized regions such as stars, would not be completely black if one took into account the uncertainty principle of quantum mechanics. Instead, a black hole would radiate particles and radiation like a hot body. The radiation would carry away energy and so reduce the mass of the black

A spiral galaxy. (Anglo-Austrialian Telescope Board)   R I **V** U X G

hole. This in turn would increase the rate of emissions. Eventually, the black hole would disappear completely in a tremendous burst of emissions. All the matter that collapsed to form the black hole—and any astronaut who was unlucky enough to fall into the black hole—would disappear, at least from our region of the universe. However, the energy that corresponded to this mass (by Einstein's famous equation $E = mc^2$) would be emitted by the black hole in the form of radiation. Thus, the astronaut's mass-energy would be recycled to the universe. However, this would be rather a poor sort of immortality, because the astronaut's subjective concept of time would almost certainly come to an end, and the particles out of which he had been composed would not in general be the same as the particles that would be re-emitted by the black hole. Still, black hole evaporation did indicate that gravitational collapse might not lead to a complete end of time.

The real problem with spacetime having an edge or boundary at a singularity is that the laws of science do not determine the initial state of the universe at the singularity but only how it evolves thereafter. This problem would remain even if there were no singularity and time continued back indefinitely: The laws of science would not fix what the state of the universe was in the infinite past. In order to pick out one particular state for the universe from among all possible states that are allowed by the laws of science, one has to supplement the laws by boundary conditions that say what the state of the universe was at an initial singularity or in the infinite past. Many scientists are embarrassed at talking about the boundary conditions of the universe because they feel that it verges on metaphysics or religion. After all, they might say, the universe could have started off in a completely arbitrary state. That may be so, but in that case it could also have evolved in a completely arbitrary manner. Yet all the evidence that we have suggests it evolves in a well-determined way according to certain laws. It is therefore not unreasonable to suppose that there may also be simple laws that govern the boundary conditions and determine the state of the universe.

In the classical general theory of relativity, which does not incorporate the uncertainty principle, the initial state of the universe is a point of infinite density. It is very difficult to define where the boundary conditions of the universe should be at such a singularity. However, when quantum mechanics is taken into account, there is the possibility that the singularity may be smeared out and that space and time together may

form a closed four-dimensional surface without boundary or edge, like the surface of the Earth but with two extra dimensions. This would mean that the universe is completely self-contained and does not require boundary conditions. One would not have to specify the state in the infinite past, and there would not be any singularities at which the laws of physics break down. One could say that the boundary conditions of the universe are that it has no boundary.

It should be emphasized that this is simply a *proposal* for the boundary conditions of the universe. One cannot deduce them from some other principle: One can merely pick a reasonable set of boundary conditions, calculate what they predict for the present state of the universe, and see if they agree with observations. The calculations are very difficult and have been carried out so far only in simple models with a high degree of symmetry. However, the results are very encouraging. They predict that the universe must have started out in a fairly smooth and uniform state. It would have undergone a period of what is called exponential or "inflationary" expansion, during which its size would have increased by a very large factor but the density would have remained the same. The universe would then have become very hot and would have expanded to the state that we see today, cooling as it expanded. It would be uniform and the same in every direction on very large scales but would contain local irregularities that would develop into stars and galaxies.

What happened at the beginning of the expansion of the universe? Did spacetime have an edge at the Big Bang? The answer is that if the boundary conditions of the universe are that it has no boundary, time ceases to be well-defined in the very early universe just as the direction north ceases to be well defined at the north pole of the Earth. Asking what happened before the Big Bang is like asking for a point one mile north of the north pole. The quantity that we measure as time had a beginning, but that does not mean spacetime has an edge, just as the surface of the Earth does not have an edge at the north pole, or, at least, so I am told: I have not been there myself.

If spacetime is indeed finite but without boundary or edge, this would have important philosophical implications. It would mean that we could describe the universe by a mathematical model that was completely determined by the laws of science alone; they would not have to be supplemented by boundary conditions. We do not yet know the precise form of the laws: At the moment we have a number of partial laws governing the behavior of the universe under all but the most extreme conditions. However, it seems likely that these laws are all part of some unified theory that we have yet to discover. We are marking progress and there is a reasonable chance that we will discover it by the end of the century.

At first sight it might appear that such a theory would enable us to predict everything in the universe. However, our powers of prediction would be severely limited, first by the uncertainty principle that states that certain quantities cannot be exactly predicted but only their probability distribution, and second, and even more importantly, by the complexity of the equations, which makes them impossible to solve in any but very simple situations. Thus, we would still be a long way from omniscience.

Exploring the Early Universe

(Brookhaven National Laboratory)

1 n Chapter 28 we traced the history of our expanding universe back almost to its very birth in the Big Bang. Unfortunately, the early universe was so opaque that we cannot observe it directly. Can we nonetheless reconstruct those first 300,000 years? The answer is a tentative yes, because many aspects of today's universe are relics of the earliest events in the cosmos.

Immediately after the Big Bang, the universe was so dense and hot that particles were constantly colliding at nearly the speed of light. These violent collisions shattered particles down to the most fundamental building blocks of matter. Physicists today have learned to break particles apart in much the same way using enormous particle accelerators.

The accompanying image shows the result of a head-on collision between two gold nuclei, both of which were traveling at nearly the speed of light.

When the nuclei collided at the center of the image, much of the energy of their motion was transformed into a host of new particles, which made the tracks shown in the image. Experiments on the creation of matter from energy with particle accelerators may therefore hold important clues to the formation of our universe.

And so, cosmology, the study of the universe on the largest scale, is now intimately tied to *elementary particle physics*, the study of matter on the smallest scale. Advances in cosmology also benefit elementary particle physics. Not even the most powerful accelerators can match the energies that prevailed during the Big Bang. Because we cannot duplicate these energies ourselves, physicists cannot easily test many of their predictions. Instead, they look to the Big Bang as a laboratory in which we can explore the most deeply hidden secrets of nature.

As you read the sections of this chapter, look for the answers to the following questions.

29-1 Has the universe always expanded as it does today, or might it have suddenly "inflated"?

29-2 What is antimatter? How can it be created, and how is it destroyed?

29-3 Why is antimatter so rare today?

29-4 What materials in today's universe are remnants of nuclear reactions in the hot early universe?

29-5 How did the first galaxies form?

29-6 How were the fundamental forces of nature different during the first second after the Big Bang?

29-7 What exotic relics could have been left behind by a brief period of inflation?

29-8 Are scientists close to developing an all-encompassing "theory of everything"?

29-1 The newborn universe may have undergone a brief period of vigorous expansion

With the discovery of the cosmic microwave background, astronomers had direct evidence that the universe began with a hot Big Bang (Section 28-4). But the properties of the microwave background pose a substantial challenge to the standard theory of the expansion of the universe. The difficulty is that the microwave background is incredibly uniform, or *isotropic*, across the sky. If we subtract the effects of our own motion through the microwave background (Figure 28-8), we find that the temperature of the microwave background is the same in all parts of the sky to an accuracy of 1 part in 10,000.

To appreciate why this poses a problem, think about two opposite parts of the sky, labeled A and B in Figure 29-1. Why should the radiation that reaches us from A and B be so nearly the same? This radiation is left over from the primordial fireball, so the fireball at A must have had the same temperature as the fireball at B. But how? For two objects to be at the same temperature, they should have been in contact and able to exchange heat with each other. (A hot cup of coffee and a cold spoonful of cream reach the same temperature only after you pour the cream into the coffee.)

The problem is that the widely separated regions at A and B have absolutely no connection with each other. As Figure 29-1 shows, point A lies outside the cosmic particle horizon for point B, and point B lies outside the cosmic particle horizon for point A. Even traveling at the speed of light—the ultimate speed limit—no information of any kind has had time to travel between points A and B over the entire 14-billion-year history of the universe. How is it possible, then, that these unrelated parts of the universe have almost exactly the same temperature? This dilemma is called the **isotropy problem** or the **horizon problem**.

The density of our universe poses a second enigma called the *flatness problem*. To understand the origin of this problem, recall that the geometry of our universe depends on the density parameter Ω_0, which is the ratio of the combined average mass density in the universe (ρ_0) to the critical density (ρ_c) (see Section 28-6). Observations of temperature variations in the cosmic microwave background indicate that Ω_0 is very close to 1, which corresponds to a flat universe.

For the density parameter Ω_0 to be close to 1 today, it must have been *extremely* close to 1 during the Big Bang. In other words, the density of the early universe was almost

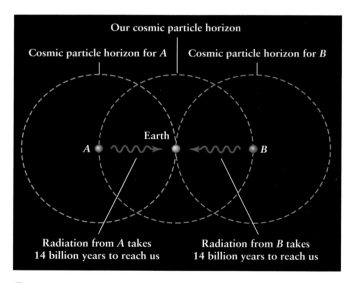

figure 29-1

The Isotropy Problem The temperature of radiation in the cosmic microwave background is almost exactly the same from all parts of the sky. This indicates that all parts of the cosmic fireball, from which this radiation comes, must have had almost exactly the same temperature. But regions A and B of the fireball, which now lie on our cosmic particle horizon and from which we receive radiation, are so far apart that they have never been in communication over the lifetime of the universe. The problem of explaining how such disconnected regions could have the same temperature is called the isotropy problem, or horizon problem.

exactly equal to the critical density. (The density was much higher than it is today, but the value of the critical density was also much higher. In a flat universe, the density of matter and the critical density decrease together as the universe expands, so that they remain equal at all times.)

The equations for an expanding universe show that any deviation from exact equality would have mushroomed within a fraction of a second. Had the average mass density been slightly less than ρ_c, the universe would have expanded so rapidly that matter could never have clumped together to form galaxies. Without galaxies, there would be no stars or planets, and humans would never have evolved. If, on the other hand, the density had been slightly greater than ρ_c, the universe would soon have become tightly packed with matter. Had this been the case, the entire cosmos would long ago have collapsed in a Big Crunch, and again humans would never have evolved.

In other words, immediately after the Big Bang the fate of the universe hung in the balance, like a pencil teetering on its point. The tiniest deviation from the precise equality $\rho_0 = \rho_c$ would have rapidly propelled the universe away from the special case of $\Omega_0 = 1$. Had there been such a deviation, the human race would not exist.

Happily, our universe is one in which galaxies, stars, planets, and humans do exist, the Big Crunch has not taken place, and Ω_0 is approximately 1. This tells us that the density of the universe immediately after the Big Bang must have been equal to the critical density to an incredibly high order of precision. In order for space to be as flat as it is today, the value of ρ_0 right after the Big Bang must have been equal to ρ_c to more than 50 decimal places!

What could have happened during the first few moments of the universe to ensure that $\rho_0 = \rho_c$ to such an astounding degree of accuracy? Because $\rho_0 = \rho_c$ means that space is flat, this enigma is called the **flatness problem**.

 In the early 1980s, a remarkable solution was proposed to both the isotropy problem and the flatness problem. Working independently, Alexei Starobinsky at the L. D. Landau Institute of Theoretical Physics in Moscow and Alan Guth at Stanford University suggested that the universe might have experienced a brief period of **inflation**, or extremely rapid expansion, shortly after the Planck time. (As we saw in Section 28-3, before the Planck time the known laws of physics do not properly describe the behavior of space, time, and matter.) During this **inflationary epoch**, which lasted for only about 10^{-24} s, the universe ballooned outward in all directions to become 10^{50} times its original size (Figure 29-2).

In the inflationary model, this violent expansion occurred because, for a brief interval, the cosmological constant (Λ) was huge. Recall that Einstein introduced the cosmological constant to hold up Newton's vision of a static universe (Section 28-1). Measurements of distant Type Ia supernovae suggest that the present-day value of Λ is slightly greater than zero (see Section 28-8). During the inflationary epoch, however, Λ was far larger—so large that it provided repulsive forces 10^{120} times larger than Einstein envisioned. This cosmic repulsion drove a violent expansion of space in all directions, literally producing the event we today call the Big Bang.

Inflation accounts for the isotropy of the microwave background. During the inflationary epoch, much of the material that was originally near our location was moved out to tremendous distances. Over the past 14 billion years, our cosmic particle horizon has expanded so that we can see radiation from these distant regions. Hence, when we examine microwaves from opposite parts of the sky, we are seeing radiation from parts of the universe that were originally in intimate contact with one another. This common origin is why all parts of the sky have almost exactly the same temperature.

 An inflationary epoch can also account for the flatness of the universe. To see why, think about a small portion of the Earth's surface, such as your backyard. For all practical purposes, it is impossible to detect the Earth's curvature over such a small area, and your backyard looks flat. Similarly, the observable universe is such a tiny fraction of the entire inflated universe that any overall curvature in it is

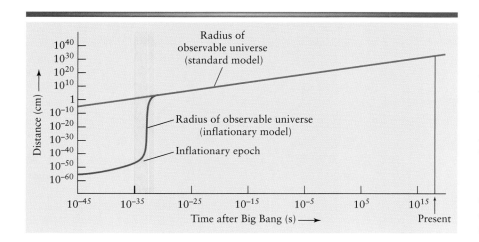

ƒigure 29-2

The Observable Universe With and Without Inflation According to the inflationary model (shown in red), the universe expanded by a factor of about 10^{50} shortly after the Big Bang. This growth in the size of the observable universe probably occurred during a very brief interval, as indicated by the vertical shaded area on the graph. For comparison, the projected size of the universe without inflation is shown in blue. (Adapted from A. Guth)

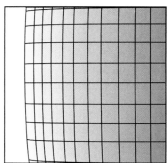

Figure 29-3

Inflation Solves the Flatness Problem This sequence of drawings shows how inflation can produce a locally flat geometry. In each successive frame, the sphere is inflated by a factor of 3 (and the number of grid lines on the sphere is increased by the same factor). Note how the curvature of the surface quickly becomes undetectable on the scale of the illustration. (Adapted from A. Guth and P. Steinhardt)

virtually undetectable (Figure 29-3). Like your backyard, the segment of space we can observe looks flat.

CAUTION! It is important to note that the concept of inflation does not violate Einstein's dictum that nothing can travel faster than the speed of light. Remember that the expansion of the universe is the expansion of space itself, not the motion of objects through space. During the inflationary epoch, the distances between particles increased enormously, but this was entirely the result of a sudden vigorous *expansion of space*. Particles did not move through space; space itself inflated.

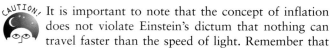

29-2 During inflation, all the mass and energy in the universe burst forth from the vacuum of space

Inflation was a brief but stupendous expansion of the universe immediately after the beginning of time. Physicists now realize that inflation helps explain where all the matter and radiation in the universe came from. To see how a violent expansion of space could create particles, we must first understand what quantum mechanics tells us about space.

WEB LINK 29-3 **Quantum mechanics** is the branch of physics that explains the behavior of nature on the atomic scale and smaller. For example, quantum mechanics tells us how to calculate the structure of atoms and the interactions between atomic nuclei. **Elementary particle physics** is the branch of quantum mechanics that deals with individual subatomic particles and their interactions.

The submicroscopic world of quantum mechanics is significantly different from the ordinary world around us. In the ordinary world we have no trouble knowing where things are.

You know where your house is; you know where your car is; you know where this book is. In the subatomic world of electrons and nuclei, however, you can no longer speak with this same confidence. A certain amount of fuzziness, or uncertainty, enters into the description of reality at the incredibly small dimensions of the quantum world.

To appreciate the reasons for this uncertainty, imagine trying to measure the position of a single electron. To find out where it is located, you must observe it. And to observe it, you must shine a light on it. However, the electron is so tiny and has such a small mass that the photons in your beam of light possess enough energy to give the electron a mighty kick. As soon as a photon strikes the electron, the electron recoils in some unpredictable direction. Consequently, no matter how carefully you try to measure the precise location of an electron, you necessarily introduce some uncertainty into the speed of that electron.

WEB LINK 29-4 These ideas are at the heart of the **Heisenberg uncertainty principle**, first formulated in 1927 by the German physicist Werner Heisenberg, one of the founders of quantum mechanics. This principle states that there is a reciprocal uncertainty between position and momentum (equal to the mass of a particle times its velocity). The more precisely you try to measure the position of a particle, the more unsure you become of how the particle is moving. Conversely, the more accurately you determine the momentum of a particle, the less sure you are of its location. These restrictions are not a result of errors in making measurements; they are fundamental limitations imposed by the nature of the universe.

There is an analogous uncertainty involving energy and time. You cannot know the energy of a system with infinite precision at every moment in time. Over short time intervals, there can be great uncertainty about the amounts of energy in the subatomic world. Specifically, let ΔE be the smallest possible uncertainty in energy measured over a short interval of time Δt. (Astronomers and physicists often use the capital Greek letter delta, Δ, as a prefix to denote a small quantity or a small change in a quantity.)

Heisenberg uncertainty principle for energy and time

$$\Delta E \times \Delta t = \frac{h}{2\pi}$$

ΔE = uncertainty in energy

Δt = time interval over which energy is measured

h = Planck's constant = 6.625×10^{-34} J s

This says that the shorter the time interval Δt, the greater the energy uncertainty ΔE must be in order to ensure that the product of ΔE and Δt is equal to $h/2\pi$.

We might look upon the Heisenberg uncertainty principle as merely an unfortunate limitation on our ability to know everything with infinite precision. But, in fact, this principle provides startling insights into the nature of the universe.

We have seen that one of the important conclusions of Einstein's special theory of relativity is the equivalence of mass and energy: $E = mc^2$ (see Section 18-1). There is nothing uncertain about the speed of light (c), which is an absolute constant. Therefore, any uncertainty in the energy of a physical system can be attributed to an uncertainty Δm in the mass. Thus,

$$\Delta E = \Delta m \times c^2$$

Combining this expression with the previous equation, we obtain

Heisenberg uncertainty principle for mass and time

$$\Delta m \times \Delta t = \frac{h}{2\pi c^2}$$

Δm = uncertainty in mass

Δt = time interval over which mass is measured

h = Planck's constant = 6.625×10^{-34} J s

c = speed of light = 3.00×10^8 m/s

This result is astonishing. It means that over a very brief interval Δt of time, we cannot be sure how much matter there is in a particular location, even in "empty space." During this brief moment, matter can spontaneously appear and then disappear. The greater the amount of matter (Δm) that appears spontaneously, the shorter the time interval (Δt) it can exist before disappearing into nothingness. This bizarre state of affairs is a natural consequence of quantum mechanics.

No particle can appear spontaneously by itself, however. For each particle created, so is a second, almost identical **antiparticle**. In other words, equal amounts of matter and **antimatter** come into existence and then disappear.

Despite its exotic name, there is actually nothing terribly mysterious about antimatter. A particle and an antiparticle are identical in almost every respect; their main distinction is that they carry opposite electric charges. For example, an ordinary electron (e^-) carries a negative charge; the corresponding antiparticle has the same mass but a positive charge, which is why it is called a **positron** (e^+). Because particles and antiparticles come and go in pairs, the total electric charge in the universe remains constant. Physicists often refer to this kind of balance between matter and antimatter as a **symmetry**.

A spontaneously created particle-antiparticle pair lasts for only an incredibly brief time. For example, consider an electron and a positron, each with a mass 9.11×10^{-31} kg. If we rewrite the Heisenberg uncertainty principle for mass and time to solve for Δt and then substitute the combined mass of $2 \times 9.11 \times 10^{-31}$ kg, we find that a spontaneously created electron-positron pair can last for a time

$$\Delta t = \frac{1}{\Delta m}\frac{h}{2\pi c^2} = \frac{1}{2 \times 9.11 \times 10^{-31}} \times \frac{6.625 \times 10^{-34}}{2\pi(3.00 \times 10^8)}$$

$$= 6.44 \times 10^{-22} \text{ s}$$

In other words, an electron and a positron can spontaneously appear and then disappear without violating any laws of physics—but they can remain in existence for no longer than 6.44×10^{-22} s. The more massive the particle, the shorter the time interval it can exist. For example, the proton has about 2000 times more mass than the electron. Pairs of protons and their antiparticles, called **antiprotons**, can appear and disappear spontaneously, but they exist for only 1/2000 as long as pairs of electrons and positrons do.

Spontaneous creation can and does happen absolutely anywhere and at any time, not just under the unusual conditions of the early universe. (It is happening right now in the space between this book and your eyes.) Quantum mechanics tells us that if a process is not strictly forbidden, then it must occur. Pairs of every conceivable particle and antiparticle are constantly being created and destroyed at every location across the universe. However, we have no way of observing these pairs directly without violating the uncertainty principle. For this reason, they are called **virtual pairs**. They do not "really" exist; they "virtually" exist.

Although virtual pairs of particles and antiparticles cannot be observed directly, their effects have nonetheless been detected. Imagine, for example, an electron in orbit about the nucleus of an atom, such as a hydrogen atom. Ideally, the electron should follow its orbit in a smooth, unhampered fashion. However, because of the constant brief appearance and disappearance of pairs of particles and antiparticles, miniscule electric fields exist for extremely short intervals of time. These tiny, fleeting electric fields cause the electron to jiggle slightly in its orbit. This jiggling produces slight changes in the energies of different electron orbits in the hydrogen atom, which manifest themselves as a miniscule shift in the wavelengths of the hydrogen spectral lines. (We discussed the connection between the energies of electron orbits in hydrogen and the hydrogen spectrum in Section 5-8.)

This shift was first detected in 1947 by the American physicists Willis Lamb and R. C. Retherford and today is

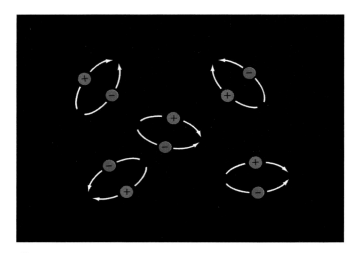

figure 29-4

Virtual Pairs Pairs of particles and antiparticles can appear and then disappear anywhere in space provided that each pair exists only for a very short time interval, as dictated by the uncertainty principle. In this sketch, electrons are shown in green and positrons are shown in red.

known as the **Lamb shift**. The Lamb shift provides powerful support for the idea that every point in space, all across the universe, is seething with virtual pairs of particles and anti-particles. In this sense, "empty space" is actually not empty at all. Figure 29-4 sketches the constant appearance and disappearance of virtual particles and antiparticles.

In some circumstances, virtual pairs can become *real* pairs of particles and antiparticles, a phenomenon called **pair production**. It has been known for years that highly energetic gamma rays (photons) can convert their energy into pairs of

particles and antiparticles. Quite simply, the gamma ray disappears upon colliding with a second photon, and a particle and an antiparticle appear in its place. These particles and antiparticles come from nature's ample supply of virtual pairs. The gamma rays provide a virtual pair with so much energy that the virtual particles can appear as real particles in the real world.

Pair production is routinely observed in high-energy nuclear experiments (see the image that opens this chapter). Indeed, it is one of the ways in which physicists manufacture exotic species of particles and antiparticles. The only requirement is that nature's balance sheet be satisfied. To create a particle and an antiparticle having a total mass M, the incoming gamma-ray photons must possess an amount of energy E that is greater than or equal to Mc^2. If the photons carry too little energy (less than Mc^2), pair production will not occur. Likewise, the more energetic the photons, the more massive the particles and antiparticles that can be manufactured. Figure 29-5a shows this process of pair production. The inverse process of **annihilation**, in which a particle and antiparticle collide with each other and are converted into high-energy gamma rays, is shown in Figure 29-5b.

Around 1980 physicists began applying these ideas to their thinking about the creation of the universe. During the inflationary epoch, space was expanding with explosive vigor. As we have seen, however, all space is seething with virtual pairs of particles and antiparticles. Normally, a particle and an antiparticle have no trouble getting back together in a time interval (Δt) short enough to be in compliance with the uncertainty principle. During inflation, however, the universe expanded so fast that particles were rapidly separated from their corresponding antiparticles. Deprived of the opportunity to recombine, these virtual particles had to become *real* particles in the real world. In this way, the universe was flooded with particles and antiparticles created by the violent expansion of space.

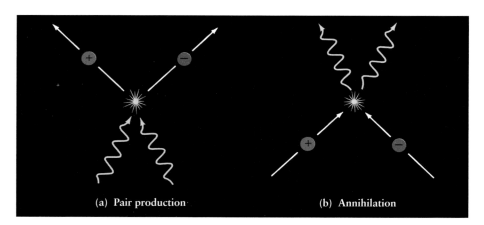

(a) Pair production   (b) Annihilation

figure 29-5

Pair Production and Annihilation (a) Pairs of virtual particles can be converted into real particles by high-energy gamma-ray photons. In this illustration, an electron (shown in green) and a positron (in red) are produced. This process can take place only if the combined energy of the two photons is no less than Mc^2, where M is the total mass of the electron and positron. (b) Conversely, a particle and an antiparticle can annihilate each other and be transformed into energy in the form of gamma rays.

29-3 As the primordial fireball cooled off, most of the matter and antimatter in the early universe annihilated each other

As soon as the flood of matter and antimatter appeared in the universe, collisions between particles and antiparticles began to produce numerous high-energy gamma rays. As these gamma rays collided, they promptly turned back into the particles and antiparticles from which they came. As a result, the rate of pair production soon equaled the rate of annihilation. For example, for every proton (p) and antiproton ($\overline{\text{p}}$) that annihilated each other to create gamma rays, two gamma rays collided elsewhere to produce a proton and an antiproton. In other words, reactions such as

$$\text{p} + \overline{\text{p}} \rightarrow \gamma + \gamma \text{ (proton-antiproton annihilation)}$$

and

$$\gamma + \gamma \rightarrow \text{p} + \overline{\text{p}} \text{ (proton-antiproton pair production)}$$

proceeded with equal vigor. In this state, called **thermal equilibrium**, precisely as many particles and antiparticles were being created as were being destroyed.

 Now let us think about what happened slightly later, soon after the Big Bang. As the universe continued to expand, all the gamma-ray photons became increasingly redshifted; as a result, the temperature of the radiation field fell. The temperature eventually became so low that the gamma rays no longer had enough energy to create particular kinds of particles and antiparticles. We say that the temperature dropped below the particles' **threshold temperature**. Collisions between particles and antiparticles continued to add photons to the cosmic-radiation background, but collisions between photons could no longer replenish the supply of particles and antiparticles.

When the universe was about 0.0001 second old (that is, at time $t = 10^{-4}$ s), the temperature of the radiation fell below 10^{13} K. At that moment, the universe became cooler than the threshold temperatures of both protons and neutrons. No new protons or neutrons appeared, but the annihilation of protons by antiprotons and of neutrons by antineutrons continued vigorously everywhere throughout space. This wholesale annihilation dramatically lowered the matter content (particles and antiparticles) of the universe, while simultaneously increasing the radiation (photon) content.

A little later, when the universe was about 1 second old, its temperature fell below 6×10^9 K, the threshold temperature for electrons and positrons. A similar annihilation of pairs of electrons and positrons further decreased the matter content of the universe while raising its radiation content. This radiation field, which fills all space, is the primordial fireball discussed in Section 28-5. This fireball, which dominated the universe for the next several hundred thousand years, therefore derived much of its energy from the annihilation of particles and antiparticles during the first second after the Big Bang.

Now we have a dilemma. If there had been perfect symmetry between particles and antiparticles, then for every proton there should have been an antiproton. For every electron, there should likewise have been a positron. Consequently, by the time the universe was 1 second old, every particle would have been annihilated by an antiparticle, leaving no matter at all in the universe.

Obviously, this did not happen. The planets, stars, and galaxies we see in the sky are made of matter, not antimatter. (If there were substantial amounts of antimatter in the universe, it would eventually collide with ordinary matter. We would then see copious amounts of gamma rays being emitted from the entire sky. While we do observe gamma-ray photons from various locations in the universe, they are neither numerous enough nor of the right energy to indicate the presence of much antimatter.) Thus, there must have been an excess of matter over antimatter immediately after the Big Bang. Physicists therefore say that somehow a **symmetry breaking** occurred during the earliest moments of the universe, meaning that the particles emerging from the Big Bang outnumbered the antiparticles.

We can estimate the extent of this symmetry breaking, because, as noted in Section 28-5, there are roughly a billion photons today in the microwave background for each proton and neutron in the universe. Thus, for every billion antiprotons, there must have been a billion plus one ordinary protons, leaving one surviving proton after annihilation. Similarly, for every billion positrons, there must have been a billion plus one ordinary electrons.

29-4 A background of neutrinos and most of the helium in the universe are relics of the primordial fireball

The early universe must have been populated with vast numbers of neutrinos (ν) and their antiparticles, the antineutrinos ($\overline{\nu}$). These particles have little or no mass, so their threshold temperature is quite low. These particles take part in the nuclear reactions that transform neutrons into protons and vice versa. For example, a neutron can decay into a proton by emitting an electron and an antineutrino:

$$\text{n} \rightarrow \text{p} + \text{e}^- + \overline{\nu}$$

This radioactive decay happens quickly (its half-life is about 10.5 minutes), which is why we do not find free neutrons floating around in the universe today. In the first 2 seconds after the Big Bang, however, neutrons were also created by collisions between protons and electrons:

$$\text{p} + \text{e}^- \rightarrow \text{n} + \nu$$

This reaction kept the number of neutrons approximately equal to the number of protons. This balance was maintained only as long as electrons were abundant, so that collisions between protons and electrons were frequent. But the number of electrons decreased precipitously as the temperature fell below 6×10^9 K and electrons and positrons annihilated each other (Section 29-3). By the time the universe was about 2 seconds old, no new neutrons were being formed, the natural tendency for neutrons to decay into protons took over, and the number of neutrons began to decline.

Before many of the neutrons could decay into protons, they began to combine with protons to form nuclei. Nuclei of helium, the first element more massive than hydrogen, consist of either two protons and two neutrons (^{4}He) or two protons and a single neutron (^{3}He). It is exceedingly improbable that two protons and one or two neutrons should all simultaneously collide with one another to form a helium nucleus. Instead, helium nuclei are built in a series of steps. The first step is to have a single proton and a single neutron combine to form deuterium (^{2}H), sometimes called "heavy hydrogen." A photon is emitted in this process, so we write this reaction as

$$p + n \rightarrow {}^2H + \gamma$$

Forming deuterium does not immediately lead to the formation of helium, however. The problem is that deuterium nuclei are easily destroyed, because a proton and a neutron do not stick together very well. Indeed, in the early universe, high-energy gamma rays easily broke deuterium nuclei back down into independent protons and neutrons. As a result, the synthesis of helium could not get beyond the first step. This block to the creation of helium is called the **deuterium bottleneck**.

When the universe was about 3 minutes old, the background radiation had cooled enough that its photons no longer had enough energy to break up the deuterium. By this time, most of the neutrons had decayed into protons, and protons outnumbered neutrons by about six to one. Because deuterium nuclei could now survive, the remaining neutrons combined with protons and rapidly produced helium. (Figure 18-2 depicts a similar sequence of reactions that produce helium from hydrogen in the core of the present-day Sun.)

The result was what we find in the universe today—about one helium atom for every ten hydrogen atoms. In addition to helium, nuclei of lithium (Li, which has three protons) and beryllium (Be, which has four) were also produced in small numbers. The process of building up nuclei such as deuterium and helium from protons and neutrons is called **nucleosynthesis** (Figure 29-6).

Because nuclei have positive electric charges, bringing them together to form more massive nuclei requires that they overcome their mutual electric repulsion. They are unable to do so if they are moving too slowly, which will be the case if the temperature is too low. As a result, by about 15 minutes after the Big Bang the universe was no longer hot enough for nucleosynthesis to take place. Only the four lightest elements (hydrogen, helium, lithium, and beryllium) were present in appreciable numbers. The heavier elements would be formed only much later, once stars had formed and nuclear reactions within those stars could manufacture carbon, nitrogen, oxygen, and all the other elements.

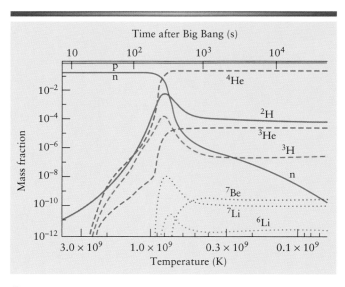

figure 29-6

Nucleosynthesis in the Early Universe This graph shows the production of nuclei between 10 seconds and 10 hours after the Big Bang. The vertical axis shows the fraction of the total mass that was in each type of particle or nucleus (p = proton, n = neutron, ^{2}H = deuterium, ^{3}He and ^{4}He = helium, ^{6}Li and ^{7}Li = lithium, ^{7}Be = beryllium). Very few nuclei were formed before the universe was 10 seconds old, thanks to the phenomenon of the deuterium bottleneck, which occurred at times earlier than those shown here. By about 10^3 seconds (roughly 15 minutes) after the Big Bang, the temperature had dropped below 4×10^8 K, no further nucleosynthesis was possible, and the relative amounts of different nuclei stabilized. The number of free neutrons declined rapidly as these particles decayed into protons, electrons, and antineutrinos. (Adapted from R. V. Wagoner).

While nuclei were being formed in the early universe, what happened to all those primordial neutrinos and antineutrinos that had interacted so vigorously with the protons and neutrons before the universe was 2 seconds old? And where are those neutrinos and antineutrinos today? As we saw in the discussion of the Sun (recall Section 18-4), neutrinos and antineutrinos are very difficult to detect because they interact only very weakly with matter. The Earth itself is virtually transparent to the neutrinos from the Sun.

By the time the universe was about 2 seconds old, matter was sufficiently spread out so that the universe had became transparent to neutrinos and antineutrinos. From that time on, neutrinos and antineutrinos could travel across the universe unimpeded. Consequently, a neutrino-antineutrino background should fill the universe much as the cosmic microwave background does. Indeed, these ancient neutrinos and antineutrinos may be about as populous today as the photons in the microwave background (of which there are 5.5×10^8 per cubic meter). The neutrino-antineutrino background should be slightly cooler than the photon background, which received extra energy from the electron-positron annihilations. Physicists estimate that the current temperature of the neutrino-antineutrino background is about 2 K, as opposed to 2.725 K for the microwave background.

29-5 Galaxies formed from density fluctuations in the early universe

The distribution of matter in the universe today is quite lumpy. Stars are grouped together in galaxies, galaxies into clusters, and clusters into superclusters that stretch across 50 Mpc (150 million ly) or more (see Section 26-6). Furthermore, galaxies seem to be concentrated along enormous sheets, which in turn surround voids measuring 30 to 120 Mpc (100 million to 400 million ly) across. These features, which characterize the large-scale structure of the universe, are seen in Figures 26-21 and 26-22.

Although there is a lot of lumpiness in the universe today, the early universe must have been exceedingly smooth. To see why, think back to the era of recombination that occurred approximately 300,000 years after the Big Bang (see Section 28-5). Before recombination, high-energy photons were constantly and vigorously colliding with charged particles throughout all space. After recombination, the universe became transparent, and these photons stopped interacting with the matter in the universe. Astronomers say that matter "decoupled" from radiation during the era of recombination. Because the cosmic microwave background is extremely isotropic, we can conclude that the matter with which these photons once collided so frequently must also have been spread smoothly across space.

The distribution of matter during the early universe must not have been *perfectly* uniform, however. If it had been, it would still have to be absolutely uniform today; there would now be only a few atoms per cubic meter of space, with no stars and no galaxies. Consequently, there must have been slight lumpiness, or **density fluctuations**, in the distribution of matter in the early universe. Through the action of gravity, these fluctuations eventually grew to become the galaxies and clusters of galaxies that we see today throughout the universe. As we saw in Section 28-5, the pattern of density fluctuations became imprinted on the cosmic background radiation during the era of recombination. Figure 29-7 shows a map of these fluctuations obtained from the MAXIMA microwave background experiment.

Our understanding of how gravity can amplify density fluctuations dates back to 1902, when the British physicist James Jeans solved a problem first proposed by Isaac Newton. Suppose that you have a gas with only very tiny fluctuations in density, as shown in Figure 29-8a. These regions of higher density will then gravitationally attract nearby material and thus gain mass. As this happens, however, the pressure of the gas inside these regions will also increase, which can make these regions expand and disperse. The question then becomes: Under what conditions does gravity overwhelm gas pressure so that a permanent object can form?

figure 29-7 R I V U X G

Variations in the Cosmic Microwave Background This high-resolution map of the cosmic background radiation shows details as small as 10 arcminutes across. Light-colored and dark-colored areas are respectively warmer and cooler than average. These data from the MAXIMA experiment, combined with other data from BOOMERANG (see Figure 28-14), help astrophysicists understand the origin of large-scale structure in the universe. (The MAXIMA Collaboration, Center for Particle Astrophysics, University of California, Berkeley)

Jeans proved that an object will grow from a density fluctuation provided that the fluctuation extends over a distance that exceeds the so-called **Jeans length** (L_J):

Jeans length for density fluctuations

$$L_J = \sqrt{\frac{\pi k T}{m G \rho_m}}$$

L_J = Jeans length

k = Boltzmann constant = 1.38×10^{-23} J/K

T = temperature of the gas (in kelvins)

m = mass of a single particle in the gas (in kilograms)

G = universal constant of gravitation
  = 6.67×10^{-11} N m² /kg²

ρ_m = average density of matter in the gas

Density fluctuations that extend across a distance larger than the Jeans length tend to grow, while fluctuations smaller than L_J tend to disappear (Figure 29-8b).

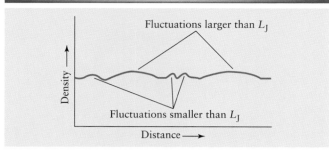

a   At an early time

b   At a later time

Figure 29-8

The Growth of Density Fluctuations **(a)** Small density fluctuations in the distribution of matter shortly after the era of recombination. **(b)** If the size of a fluctuation is greater than the Jeans length (L_J), it becomes gravitationally unstable and can grow in amplitude.

We can apply the Jeans formula to the conditions that prevailed during the era of recombination, when $T = 3000$ K and $\rho_m = 10^{-18}$ kg/m³. Taking m to be the mass of the hydrogen atom ($m = 1.67 \times 10^{-27}$ kg), we find that $L_J = 100$ light-years, the diameter of a typical globular cluster (Figure 29-9). Furthermore, the mass contained in a cube whose sides are 1 Jeans length in size (equal to the product of the density ρ_m and the volume of the cube $L_J{}^3$) is about 5×10^5 M$_\odot$, equal to the mass of a typical globular cluster.

As we saw in Section 21-3, globular clusters contain the most ancient stars we can find in the sky. For these reasons, Robert Dicke and P. J. E. Peebles at Princeton University proposed that globular clusters were among the first objects to form after matter decoupled from radiation. Objects the size of globular clusters may have merged to form still larger collections of matter. Over time, such mergers may have led to the population of galaxies we see today (see the discussion in Section 26-9, especially Figure 26-34).

If large-scale structure did indeed develop from density fluctuations, how did this development take place? One issue that complicates this problem is that 90% or more of the mass in the universe is in the form of dark matter, whose nature is not known (see Section 26-8). Researchers have hypothesized different types of dark matter in the hope of explaining the large-scale structure that we see. Neutrinos are an example of **hot dark matter**, so named because it consists of lightweight particles traveling at high speeds. **Cold dark matter**, on the other hand, consists of massive particles traveling at slow speeds. Examples include WIMPs, which we discussed in Section 25-4, as well as other, even more speculative exotic particles (discussed in Section 29-8).

Scientists use supercomputer simulations to see how different types of dark matter would influence the development of large-scale structure. Figure 29-10 shows the results of such a simulation in which the universe is dominated by *cold* dark matter. The simulation follows the motions of more than 2 million particles of cold dark matter in a box that expands as the universe expands. The box representing the present time (labeled $a = 0.60$, at the lower left of the figure) is about 50 Mpc on a side.

The simulation begins with an almost perfectly uniform distribution of particles, mimicking the tiny density fluctuations that must have been present just after inflation. A supercomputer then calculates how these particles move, based on Newton's laws in an expanding universe. At selected intervals, the program selects a narrow slice through the box and examines the distribution of particles within the slice. The slice through today's universe is about 6 Mpc thick.

Figure 29-10 shows nine such representative slices. Note how the tiny fluctuations barely visible in the first slice gradually become more pronounced as time passes. These features

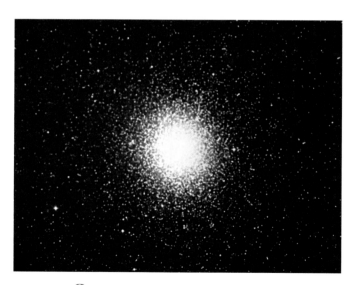

Figure 29-9   R I **V** U X G

A Globular Cluster A typical globular cluster contains 10^5 to 10^6 stars, each with an average mass of about 1 M$_\odot$, so the total mass of a typical cluster is 10^5–10^6 M$_\odot$. Cluster diameters range from about 6 to 120 pc (20 to 400 ly). Because these masses and diameters are comparable to the Jeans length (L_J) during the era of recombination, astronomers suspect that globular clusters were among the first objects to form in the universe. (NASA)

$a = 0.04$ $a = 0.06$ $a = 0.08$

$a = 0.10$ $a = 0.20$ $a = 0.40$

$a = 0.60$ $a = 0.80$ $a = 1.00$

figure 29-10

A Cold Dark Matter Simulation These nine views show slices through a large, box-shaped volume of space as the universe expands. The "expansion factor" a indicates the relative size of the box at the time when the slice was taken. For example, the actual dimensions of the box labeled $a = 0.10$ are one-tenth those of the box labeled $a = 1.00$. The first four views represent early epochs within the first few billion years after the Big Bang. The last five views are from much later times, with the $a = 0.60$ view possibly representing the present time. Galaxy formation probably occurs around the time of the $a = 0.20$ slice.

(Courtesy of J. M. Gelb and E. Bertschinger; Massachusetts Institute of Technology) merge to form larger structures as gravity continues to draw matter together, overcoming (at least partially) the expansion of the universe. You can see the development of both voids and filaments. Ordinary matter, which follows the course set by the cold dark matter, accumulates along density enhancements and forms galaxies and clusters of galaxies. The slice $a = 0.20$ seems to represent the era during which our Galaxy formed.

Similar simulations have been carried out using hot dark matter in the form of neutrinos. If a neutrino has no mass, it will always travel at the speed of light, just as a photon does. If, however, it has a little mass, it must slow down as the universe expands and cools. In this case, slow-moving neutrinos would accumulate over time within density fluctuations. The gravitational pull of these neutrinos on surrounding matter could eventually lead to the formation of clusters of galaxies.

A primary difference between simulations based on cold and hot dark matter is the way in which galaxies form. In calculations based on cold dark matter, galaxies form from the "bottom up." First, small clumps of matter appear, which then group together into galaxies; clusters and superclusters appear later. But in calculations based on hot dark matter, galaxies form from the "top down." Huge supercluster-sized sheets of matter form first and then fragment into galaxies. Recent observations of remote galaxies favor the "bottom up" scenario (see Figure 26-34), which in turn favors the model that dark matter is cold.

As is often the case in science, we need better observations to determine the true nature of dark matter. To this end, a new generation of spaceborne experiments will measure fluctuations in the cosmic background radiation with much finer detail than COBE, BOOMERANG, or MAXIMA. The essay by Professor John Ruhl that follows this chapter gives insight into what we may learn about the nature of dark matter and how large-scale structure developed in our universe.

29-6 In grand unified theories, all physical forces had the same strength immediately after the Big Bang

We have pushed our description of the early universe back to the first few minutes after the Big Bang, when neutrinos and helium nuclei formed. Remarkably, scientists can now describe events during the first second after the Big Bang, when the universe was a hot, dense sea of particles and antiparticles colliding with each other at high speeds. To discover what happened during that brief moment of time, we must understand how particles interact at very high energies.

Just *four* physical forces—gravity, electromagnetism, and the strong and weak forces—explain the interactions of everything in the universe. Of these forces, gravity is the most familiar. It is a long-range force that dominates the universe over astronomical distances. The electromagnetic force is also a long-range force, but it is intrinsically much stronger than the gravitational force. For example, the electromagnetic force between an electron and a proton is about 10^{39} times

stronger than the gravitational force between those two particles. That is why the electromagnetic force, not the gravitational force, is responsible for holding electrons in orbit about the nuclei in atoms.

We do not generally observe longer-range effects of the electromagnetic force, because there is usually a negative electric charge for every positive charge and a south magnetic pole for every north magnetic pole. Thus, over great volumes of space the effects of electromagnetism effectively cancel out. No similar canceling occurs with gravity because there is no equivalent "negative mass." That is why the force that holds the Moon in orbit about the Earth is gravitational, not electromagnetic.

Both the strong and the weak forces are said to be *short-range forces*. Their influence extends only over distances less than about 10^{-15} m, about the diameter of a proton. The **strong force** holds protons and neutrons together inside the nuclei of atoms. Without the strong force, nuclei would disintegrate because of the electromagnetic repulsion of the positively charged protons. In fact, the strong force overpowers the electromagnetic forces inside nuclei.

The **weak force** is at work in certain kinds of radioactive decay. An example is the transformation of a neutron (n) into a proton (p), in which an electron (e$^-$) and an antineutrino ($\bar{\nu}$) are released:

$$n \rightarrow p + e^- + \bar{\nu}$$

Numerous experiments in nuclear physics demonstrate that protons and neutrons are composed of more basic particles called **quarks**, the most common varieties being "up" (u) quarks and "down" (d) quarks. A proton is composed of two up quarks and one down quark, a neutron of two down quarks and one up quark.

In the 1970s the concept of quarks led to a breakthrough in our understanding of the strong and weak forces. The strong force holds quarks together, while the weak force is at work whenever a quark changes from one variety to another. For example, when a neutron decays into a proton, one of the neutron's down quarks changes into an up quark. Thus, the weak force is responsible for transformations such as

$$d \rightarrow u + e^- + \bar{\nu}$$

In the 1940s, physicists Richard P. Feynman and Julian S. Schwinger (working independently in the United States) and Sin-Itiro Tomonaga (in Japan) succeeded in developing a basic description of what we mean by force. Focusing their attention on the electromagnetic force, they tried to describe exactly what happens when two charged particles interact. According to their theory, now called **quantum electrodynamics**, charged particles interact by exchanging *virtual photons*. Virtual photons, like virtual particles, cannot be observed directly, because they exist for extremely short time intervals.

Quantum electrodynamics has proved the most successful theory in modern physics. It describes with remarkable accuracy many details of the electromagnetic interaction between

Force	Relative strength	Particles exchanged	Particles on which the force can act	Range	Example
Strong	1	gluons	quarks	10^{-15} m	holding protons, neutrons, and nuclei together
Electromagnetic	$\frac{1}{137}$	photons	charged particles	infinite	holding atoms together
Weak	10^{-4}	intermediate vector bosons	quarks, electrons, neutrinos	10^{-16} m	radioactive decay
Gravitational	6×10^{-39}	gravitons	everything	infinite	holding the solar system together

Table 29-1 — The Four Forces

charged particles. Inspired by these successes, physicists have tried to develop similar theories for the other three forces. In these theories, the weak force occurs when particles exchange **intermediate vector bosons**, the gravitational force occurs when particles exchange **gravitons**, and quarks stick together by exchanging **gluons**. Table 29-1 summarizes these features of the four physical forces.

In the 1970s, physicists made important progress in understanding the weak force. Steven Weinberg, Sheldon Glashow, and Abdus Salam proposed a theory with three types of intermediate vector bosons, which are exchanged in various manifestations of the weak force. These three particles were actually discovered in experiments in the 1980s, providing strong support for the theory.

A startling prediction of the Weinberg-Glashow-Salam theory is that the weak force and the electromagnetic force should be identical to each other for particles with energies greater than 100 GeV. (One GeV equals 10^9, or 1 billion, electron volts; see Section 5-5.) In other words, if particles are slammed together with a total energy greater than 100 GeV, then electromagnetic interactions become indistinguishable from weak interactions. We say that above 100 GeV the electromagnetic force and the weak force are "unified."

This unification occurs because the three types of intermediate vector bosons behave just like photons above 100 GeV. At such high energies, these three particles actually lose their mass, and the weak force becomes a long-range force with the same intrinsic strength as electromagnetism. Physicists describe this similarity by saying that "symmetry is restored" above 100 GeV.

In the world around us, however, the typical energies with which particles interact is very much lower, on the order of 1 eV or less. Below 100 GeV, intermediate vector bosons behave like massive particles, but photons are always massless. Because intermediate vector bosons and photons are not similar at low energies, we say that "symmetry is broken" below 100 GeV, which is why the electromagnetic and the weak forces behave so differently in the world around us.

In the 1970s, Sheldon Glashow, Howard Georgi, Jogesh Pati, and Abdus Salam proposed **grand unified theories** (or **GUTs**), which predict that the strong force becomes unified with the weak and electromagnetic forces at energies above 10^{14} GeV. In other words, if particles were to collide at energies greater than 10^{14} GeV, the strong, weak, and electromagnetic interactions would all be long-range forces and would be indistinguishable from each other.

Many physicists suspect that all four forces may be unified at energies greater than 10^{19} GeV (Figure 29-11). That is, if particles were to collide at these colossal energies, there would be no difference between the gravitational, electromagnetic,

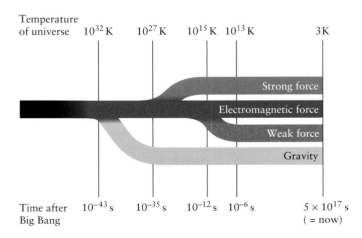

figure 29-11

Unification of the Four Forces The strength of the four physical forces depends on the energy of the particles that interact. As shown in this schematic diagram, the higher the energy, the more the forces resemble each other. Also included here are the temperature of the universe and the time after the Big Bang when the strengths of the forces are thought to have been equal.

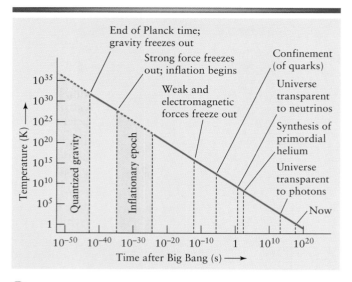

figure 29-12

The Early History of the Universe As the universe cooled, the four forces "froze out" of their unified state as a result of spontaneous symmetry breaking. Neutrons and protons froze out of the hot "quark soup" during a stage called confinement that occurred 10^{-6} second after the Big Bang. The inflationary epoch lasted from 10^{-35} to 10^{-24} s after the Big Bang. About 300,000 years after the Big Bang, the universe became transparent to light.

and nuclear forces. However, no one has yet succeeded in working out the details of such a **supergrand unified theory,** which is sometimes called a **theory of everything** (or **TOE**).

Physicists use particle accelerators to examine the unification of the weak and electromagnetic forces by slamming particles together at energies around 100 GeV. There is probably no hope, however, of ever constructing machines capable of making particles collide with energies approaching 10^{19} GeV. It may thus be impossible to test directly the grand and supergrand unified theories in a laboratory. However, the universe immediately after the Big Bang was so hot and its particles were moving so fast that they did indeed collide with energies on the order of 10^{19} GeV. The earliest moments of the universe have thus become a laboratory for testing some of the most elegant, sophisticated theories in physics.

Many of the ideas connecting elementary particle physics and cosmology are controversial and still quite speculative. Figure 29-12 summarizes our understanding so far. This figure shows how the average temperature of radiation has evolved with time, as well as other key events in the evolution of the universe.

Before the Planck time (from $t = 0$ s to $t = 10^{-43}$ s), particles collided with energies greater than 10^{19} GeV, and all four forces were unified. Because we do not yet have a TOE that properly describes the behavior of gravity, we remain ignorant of what was going on during the first 10^{-43} second of the universe's existence. We know, however, that by the end of the Planck time, the energy of particles in the universe had fallen

to 10^{19} GeV. At energies below this, gravity is thought not to be unified with the other three forces.

In the language of physics, at $t = 10^{-43}$ s there was a **spontaneous symmetry breaking** in which gravity was "frozen out" of the otherwise unified hot soup that filled all space. In such a "soup," energy (E) is related to temperature (T) by $E = kT$, where k is the Boltzmann constant (roughly 10^{-4} eV/K, or 10^{-13} GeV/K). Thus, the temperature of the universe was 10^{32} K when gravity emerged as a separate force.

At $t = 10^{-35}$ s, the energy of particles in the universe had fallen to 10^{14} GeV, equivalent to a temperature of 10^{27} K. At energies and temperatures below these, the strong force is no longer unified with the electromagnetic and weak forces. Thus, at $t = 10^{-35}$ s, there was a second spontaneous symmetry breaking, at which time the strong force "froze out." Calculations suggest that the inflationary epoch began at this point and lasted until about $t = 10^{-24}$ s. During this epoch the universe increased its size by a factor of roughly 10^{50}.

At $t = 10^{-12}$ s, the temperature of the universe had dropped to 10^{15} K, the energy of the particles had fallen to 100 GeV, and there was a final spontaneous symmetry breaking and "freeze-out" that separated the electromagnetic force from the weak force. From that moment on, all four forces have interacted with particles essentially as they do today.

The next significant event occurred at $t = 10^{-6}$ s, when the temperature was 10^{13} K and particles were colliding with energies of roughly 1 GeV. Prior to this moment, particles collided so violently that individual protons and neutrons could not exist, being constantly fragmented into quarks. After this time, appropriately called the period of **confinement,** quarks were finally able to stick together and became confined within individual protons and neutrons.

As we saw in Section 29-4, the universe became transparent to neutrinos at $t = 2$ s, and all the primordial helium was produced by $t = 3$ min. Finally, the universe became transparent to photons at $t = 3 \times 10^5$ yr.

29-7 Cosmic strings and other oddities may be relics of the early universe

We have seen how spontaneous symmetry breaking played an important role in the very early universe. This same phenomenon also explains how and why cosmic inflation may have taken place. The key idea is this: The vacuum of empty space is not truly empty, but in fact seethes with quantum-mechanical activity as virtual particles are created and destroyed.

The dynamic vacuum of space can have an excited state, just as an atom can be excited into a higher energy level. This excited or **false vacuum** would look identical to what we normally regard as empty space (a true vacuum) except that it would contain significantly more energy.

Alan Guth proposed that inflation occurred when the universe went from the high-energy, symmetric state of a false vacuum, in which the strong force was unified with the electromagnetic and weak forces, to the low-energy, symmetry-broken state of a true vacuum. The energy released during this

a   b   c

Figure 29-13

Symmetry Breaking and Cosmic Strings **(a)** The symmetry that existed in the early universe can be likened to orderly rows of pencils balanced on their points. **(b)** The state of broken symmetry that exists today is analogous to all the pencils having fallen over.

(c) A situation could arise, however, in which the falling pencils might topple toward each other so that one pencil is supported upright. Some physicists theorize that comparable structures, called cosmic strings, might be left over from the symmetry breaking that occurred during the earliest moments of the universe.

transition caused space to balloon outward in all directions. This is the source of the large value of Einstein's cosmological constant Λ during the inflationary epoch.

ANALOGY To visualize the transition from a symmetric false vacuum to an asymmetric true vacuum, imagine a pencil balanced on its point. For such a balanced pencil, no particular horizontal direction is singled out over any other. When the pencil falls over, however, the symmetry is broken, because the pencil then points toward some specific direction. That direction has become singled out over all others. Now imagine thousands of pencils precariously balanced on their points (Figure 29-13*a*), all of which soon fall over (Figure 29-13*b*). Symmetry has been broken everywhere, because every pencil points in some horizontal direction.

As soon as the universe cooled to 10^{27} K and the strong force froze out, the false vacuum that then filled space became unstable, like thousands of pencils balanced on their points. The rapid transition to a true vacuum is analogous to the pencils falling over. But note that something odd might occur. As thousands upon thousands of pencils knock each other over, it could happen that waves of falling pencils leave a single pencil upright, as shown in Figure 29-13*c*. For that pencil, symmetry would not have been broken. An analogous relic left over from the early universe, called a **cosmic string**, is a long, thin, massive wire of unbroken symmetry. Along a cosmic string, the strong, weak, and electromagnetic forces remain unified.

If cosmic strings exist, they would be incredibly massive (about 10^{12} tons per millimeter) but would emit little or no electromagnetic radiation. That makes them another candidate for cold dark matter. Ordinary matter would have collected around a cosmic string billions of years ago, forming a long chain of galaxies along the string's length.

Figure 29-14 shows some evidence that this may actually have taken place. This figure shows the locations of

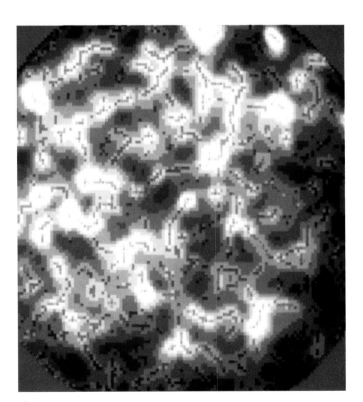

Figure 29-14

The Distribution of 400,000 Galaxies This map shows the locations of nearly half a million galaxies spanning 100° of the sky. The north pole of our Galaxy (in the constellation Coma Berenices) is at the center. Each small square covers 10 × 10 arcmin. White indicates a high density of galaxies, and darker reddish regions indicate few galaxies. Green squares emphasize filaments along which galaxies are concentrated, and the single red squares (near the centers of the white regions) indicate the points where galaxy counts reach a maximum. (Courtesy of E. L. Turner, J. E. Moody, and J. R. Gott)

400,000 galaxies across a region that covers approximately one-quarter of the sky. On this map, the sky is divided into tiny squares, each measuring about 10 arcminutes across; colors indicate the number of galaxies within each square. White indicates high densities of galaxies, whereas the darker colors denote low-density areas. The brown and black areas are voids that are nearly free of galaxies.

Note that many of the white regions in Figure 29-14 seem to be connected to one another by the green squares. These large-scale structures are typically 30 Mpc (100 million ly) long and contain up to 1 million galaxies with a combined mass of roughly 10^{16} $M_\odot$. Some physicists suspect that cosmic strings may be responsible for this distribution pattern.

29-8 Theories that attempt to unify the physical forces predict that the universe may have 11 dimensions

While we have a growing understanding of the early universe, there remains a veil obscuring the first 10^{-43} s after the Big Bang. These very *first* moments in the history of the universe, whose duration was the Planck time, determined everything that would come after. To understand this brief interval, we need to construct a quantum-mechanical theory that unifies gravity with the other physical forces of nature and that reconciles quantum mechanics with gravity. While this remains an unfinished task, remarkable progress has been made in recent years. One major breakthrough is that physicists have had to abandon the idea that there are only three dimensions of space.

In his special and general theories of relativity, Einstein combined time with the three dimensions of ordinary space, resulting in a four-dimensional combination called **spacetime**. In 1919, the Polish physicist Theodor Kaluza proposed the existence of a *fifth* dimension. Kaluza hoped to describe both gravity and electromagnetism in terms of the curvature of five-dimensional spacetime, just as Einstein had explained gravity by itself in terms of the curvature of four-dimensional spacetime (see Section 24-2).

A particle always follows the straightest possible path in the four space dimensions of Kaluza's theory. But in the three dimensions of ordinary space, the path appears curved. Hence, it appears to us that the particle has been deflected by gravitational and electromagnetic forces. Kaluza's hypothetical fifth dimension exists at every point in ordinary space but is curled up so tightly, like a very tiny loop, that it is not directly observable.

In 1926, the Swedish physicist Oskar Klein attempted to make Kaluza's five-dimensional theory compatible with quantum mechanics. While he was not successful, Klein discovered that particles of different masses could be identified with different vibrations of the tiny loop of Kaluza's fifth dimension. Today, any quantum-mechanical theory that uses more than four dimensions to provide a unified description of the forces of nature is called a **Kaluza-Klein theory**.

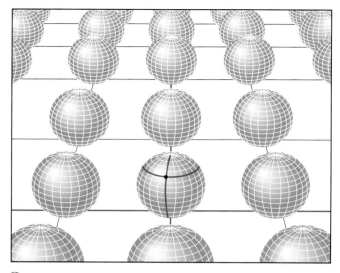

figure 29-15

Hidden Dimensions of Space Hidden dimensions of space might exist provided they are curled up so tightly that we cannot observe them. This drawing shows how an ordinary two-dimensional plane might contain two additional dimensions. At every point on the plane, there is a very tiny sphere so small that it cannot be seen. To pinpoint a particular location, you need to give not only a position on the plane but also a position on the sphere that is tangent to the plane at that point, as indicated by the red lines. (Adapted from D. Freedman and P. Van Nieuwenhuizen)

When Kaluza and Klein developed their theories, gravity and electromagnetism were the only known forces of nature. Today we know of four physical forces, which suggests that modern Kaluza-Klein theories should have even more than five dimensions. Edward Witten at Princeton University has argued that a geometric theory for describing all four forces would work best with 11 dimensions, ten of space and one of time. At every point in ordinary space and at every moment of time, the seven "extra" dimensions must be rolled up, like the loops of Kaluza and Klein, into compact structures far too tiny for us to detect (Figure 29-15).

While we do not yet have a definite 11-dimensional theory, physicists know something of how it would work. Particles would travel along the straightest possible paths in spacetime, but their paths in ordinary three-dimensional space would appear curved. From our perspective, we interpret these curved paths as the result of the four forces acting on the particles.

Theoretical physicists have shown that if there are indeed 11 dimensions, there must exist particles so massive that they have not yet been discovered. These speculative particles may be the dark matter that pervades the universe and holds together clusters of galaxies. Even more bizarre, the new theories no longer regard fundamental particles, such as electrons and quarks, as tiny points of mass. Instead, these particles may actually be multidimensional membranes, wrapped so tightly around the extra dimensions of space that they appear to us as points.

Andrew Strominger at Harvard University has shown that some of these membranes may fold on themselves in such a way that not even light can escape from them. In other words, these membranes may be black holes! This theory of membranes in 11 dimensions, or **M-theory**, may well explain the most exotic aspects of quantum mechanics, cosmology, and gravity.

At present, M-theory is only the speculative outline of a "theory of everything." We do not yet know how to describe a quark or an electron in terms of the higher dimensions of spacetime, nor can we yet explain the interactions among such particles. A full description of the first 10^{-43} s after the Big Bang is still far off. However, theoretical physicists around the world are devoting their efforts toward this goal.

We will first understand
How simple the universe is
When we realize
How strange it is.
—Anonymous

KEY WORDS

annihilation, p. 674

antimatter, p. 673

antiparticle, p. 673

antiproton, p. 673

cold dark matter, p. 678

confinement, p. 682

cosmic string, p. 683

density fluctuation, p. 677

deuterium bottleneck, p. 676

elementary particle physics, p. 672

false vacuum, p. 682

flatness problem, p. 671

gluon, p. 681

grand unified theory (GUT), p. 681

graviton, p. 681

Heisenberg uncertainty principle, p. 672

hot dark matter, p. 678

inflation, p. 671

inflationary epoch, p. 671

intermediate vector boson, p. 681

isotropy problem (horizon problem), p. 670

Jeans length, p. 677

Kaluza-Klein theory, p. 684

Lamb shift, p. 674

M-theory, p. 685

nucleosynthesis, p. 676

pair production, p. 674

positron, p. 673

quantum electrodynamics, p. 680

quantum mechanics, p. 672

quark, p. 680

spacetime, p. 684

spontaneous symmetry breaking, p. 682

strong force, p. 680

supergrand unified theory, p. 682

symmetry, p. 673

symmetry breaking, p. 675

theory of everything (TOE), p. 682

thermal equilibrium, p. 675

threshold temperature, p. 675

virtual pairs, p. 673

weak force, p. 680

KEY IDEAS

Cosmic Inflation: A brief period of rapid expansion, called inflation, is thought to have occurred immediately after the Big Bang. During a tiny fraction of a second, the universe expanded to a size many times larger than it would have reached through its normal expansion rate.

• Inflation may explain why the universe is nearly flat and the 2.725-K microwave background is almost perfectly isotropic.

Particles and Antiparticles: Heisenberg's uncertainty principle states that the amount of uncertainty in the mass of a subatomic particle increases as it is observed for shorter and shorter time periods.

• Because of the uncertainty principle, particle-antiparticle pairs can spontaneously form and disappear within a fraction of a second. These pairs, whose presence can be detected only indirectly, are called virtual pairs.

• A virtual pair can become a real particle-antiparticle pair when high-energy photons collide. In this process, called pair production, the photons disappear, and their energy is replaced by the mass of the particle-antiparticle pair. In the process of annihilation, a colliding particle-antiparticle pair disappear and high-energy photons appear.

The Primordial Fireball: Just after the inflationary epoch, the universe was filled with particles and antiparticles formed by pair production and with numerous high-energy photons formed by annihilation. A state of thermal equilibrium existed in this hot plasma.

• As the universe expanded, its temperature decreased. When the temperature fell below the threshold temperature required to produce each kind of particle, annihilation of that kind of particle began to dominate over production.

• Matter is much more prevalent than antimatter in the present-day universe. This is because particles and antiparticles were not created in exactly equal numbers just after the Planck time. This imbalance is an example of the process called symmetry breaking.

Nucleosynthesis: Helium could not have been produced until the cosmological redshift eliminated most of the high-energy photons. These photons created a deuterium bottleneck by breaking down deuterons before they could combine to form helium.

Density Fluctuations and the Origin of Galaxies: Galaxies are generally located on the surfaces of roughly spherical voids. Astronomers use supercomputers to simulate how this large-scale structure of the universe arose from primordial density fluctuations. Cold dark matter models give rough agreement with details of the large-scale structure.

The Four Forces and Their Unification: Four basic forces—gravity, electromagnetism, the strong force, and the weak force—explain all the interactions observed in the universe.

• Grand unified theories (GUTs) are attempts to explain three of the forces in terms of a single consistent set of physical laws. A supergrand unified theory would explain all four forces.

• GUTs suggest that all four physical forces were equivalent just after the Big Bang. However, because we have no satisfactory supergrand unified theory, we can as yet say nothing about the nature of the universe during this period before the Planck time ($t = 10^{-43}$ s after the Big Bang).

• At the Planck time, gravity froze out to become a distinctive force in a spontaneous symmetry breaking. During a second spontaneous symmetry breaking, the strong nuclear force became a distinct force. A final spontaneous symmetry breaking separated the electromagnetic force from the weak nuclear force; from that moment on, the universe behaved as it does today.

The Frontier of Knowledge: Many speculative aspects of the early universe remain the subject of active research.

• Cosmic strings—long, massive remnants of the spontaneous symmetry breaking that separated the strong nuclear force from the electromagnetic and weak forces—may yet be found to exist.

• The search for a theory that unifies gravity with the other physical forces suggests that the universe actually has 11 dimensions (ten of space and one of time), seven of which are folded on themselves so that we cannot see them.

REVIEW QUESTIONS

1. What aspects of the present-day universe can be explained by the idea of inflation?

2. In what ways is inflation similar to the present-day expansion of the universe? In what ways is it different?

3. The inflationary epoch lasted a mere 10^{-24} second. Why, then, is it worthy of so much attention by scientists?

4. What is the Heisenberg uncertainty principle? How does it lead to the idea that all space is filled with virtual particle-antiparticle pairs?

5. What is the difference between an electron and a positron?

6. Is it possible for a single hydrogen atom, with a positively charged proton and a negatively charged electron, to be created as a virtual pair? Why or why not?

7. Explain why antimatter was present in copious amounts in the early universe but is very rare today.

8. What is meant by the threshold temperature of a particle?

9. Explain the connection between the fact that humans exist and the phenomenon of symmetry breaking in the early universe.

10. Explain the connection between particles and antiparticles in the early universe and the cosmic microwave background that we observe today.

11. What is the deuterium bottleneck? Why was it important during the formation of nuclei in the early universe?

12. Why were only the four lightest chemical elements produced in the early universe?

13. Why is it reasonable to suppose that all space is filled with a neutrino background analogous to the cosmic microwave background?

14. Why do astronomers suspect that globular clusters were among the first objects to form in the history of the universe? Why not something larger and more massive?

15. What is the difference between hot and cold dark matter? How do astronomers decide which was more important in the formation of large-scale structures such as clusters of galaxies?

16. Describe an example of each of the four basic types of interactions in the physical universe. Do you think it possible that a fifth force might be discovered someday? Explain your answer.

17. Describe the observational evidence for (**a**) the Big Bang, (**b**) the inflationary epoch, (**c**) the era of recombination, and (**d**) the confinement of quarks.

18. Describe the large-scale structure of the universe as revealed by the distribution of clusters and superclusters of galaxies.

Problem-solving tips and tools

The mass of a proton is 1.67×10^{-27} kg, the mass of an electron is 9.11×10^{-31} kg, and the mass of the Sun is 1.99×10^{30} kg. As explained in Section 28-6, the critical density ρ_c is equal to $3H_0^2/8\pi G$; that is, ρ_c is proportional to the square of the Hubble constant H_0. If $H_0 = 70$ km/s/Mpc, the critical density is equal to about 9.2×10^{-27} kg/m^3. It is also useful to know that 1 m$^3 = 10^6$ cm^3, that 1 light-year $= 9.46 \times 10^{15}$ m, and that 1 GeV $= 10^3$ MeV $= 10^9$ eV.

19. An electron has a lifetime of 1.0×10^{-8} s in a given energy state before it makes a transition to a lower state. What is the uncertainty in the energy of the photon emitted in this process?

20. How long can a proton-antiproton pair exist without violating the principle of the conservation of mass?

 21. The mass of the intermediate vector boson W$^+$ (and of its antiparticle, the W$^-$) is 85.6 times the mass of the proton. The weak nuclear force involves the exchange of the W$^+$ and the W$^-$. **(a)** Find the rest energy of the W$^+$. Give your answer in GeV. **(b)** Find the threshold temperature for the W$^+$ and W$^-$. **(c)** From Figure 29-12, how long after the Big Bang did W$^+$ and W$^-$ particles begin to disappear from the universe? Explain.

22. Using the physical conditions present in the universe during the era of recombination ($T = 3000$ K and $\rho_m = 10^{-18}$ kg/m^3), show by calculation that the Jeans length for the universe at that time was about 100 ly and that the total mass contained in a sphere with this diameter was about 4×10^5 M$_\odot$.

23. Suppose that the average density of neutrinos throughout space is 100 neutrinos per cubic centimeter and that these neutrinos are responsible for giving the universe a density equal to the critical density. What range of neutrino masses corresponds to H_0 in the range from 65 to 85 km/s/Mpc? Give your answers in kilograms and as a fraction of the mass of the electron.

24. A typical dark nebula (see Figure 20-3) has a temperature of 30 K and a density of about 10^{-12} kg/m^3. **(a)** Calculate the Jeans length for such a dark nebula, assuming that the nebula is mostly composed of hydrogen.

Express your answer in meters and in light-years. **(b)** A typical dark nebula is several light-years across. Is it likely that density fluctuations within such nebulae will grow with time? **(c)** Explain how your answer to (b) relates to the idea that protostars form within dark nebulae (see Section 20-3).

DISCUSSION QUESTIONS

25. Some GUTs predict that the proton is unstable, although with a half-life far longer than the present age of the universe. What would it be like to live at a time when protons were decaying in large numbers?

26. If you hold an iron rod next to a strong magnet, the rod will become magnetized; one end will be a north pole and the other a south pole. But if you heat the iron rod to 1043 K (770°C = 1418°F) or higher, it will lose its magnetization. This demagnetization is said to be an example of *restoring* a spontaneously broken symmetry. Explain why.

 # WEB/CD-ROM QUESTIONS

27. Search the World Wide Web for information about the top quark. What kind of particle is it? How does it compare with the up and down quarks found in protons and neutrons? Why did physicists work so hard to try to find it?

28. Search the World Wide Web for information about primordial deuterium (that is, deuterium that was formed in the very early universe). Why are astronomers interested in knowing how abundant primordial deuterium is in the universe? What techniques do they use to detect it?

OBSERVING PROJECT

29. Use the *Starry Night* program to observe globular clusters. First turn off daylight (select **Daylight** in the **Sky** menu) and show the entire celestial sphere (select **Atlas** in the **Go** menu). Use the **Find...** command in the **Edit** menu to examine the following globular clusters. In each case, find the approximate angular diameter of the cluster. **(a)** M3; **(b)** M12; **(c)** M13. **(d)** Speculate on how these clusters would appear if you could see them at the same distance at the time of recombination.

JOHN RUHL

New Horizons in the Cosmic Microwave Background

Professor John Ruhl has been studying the cosmic microwave background for over a decade, using a variety of ground-based and balloon-borne instruments. His work has taken him to Antarctica six times, including many seasons at the "high and dry" South Pole, one of the world's best sites for millimeter-wave astronomy.

Professor Ruhl picked up the astrophysics bug as an undergraduate physics major at the University of Michigan, and did his graduate work in cosmology at Princeton University. After a postdoctoral fellowship at the University of Chicago, he moved to the University of California, Santa Barbara, in 1995. His research focuses on using the cosmic microwave background to learn about the early universe and to determine the global parameters of cosmology.

The BOOMERANG map of the cosmic microwave background radiation (CMB) in Figure 28-14 gives us a beautiful vision of the universe when it was only a few hundred thousand years old. This map was not—in any sense of the phrase—made overnight. We knew for a decade that measurements like these would be a wonderful tool for cosmology. Many independent groups toiled, and many years passed, before cutting-edge technology made these measurements possible.

The CMB contains an enormous amount of information that we have yet to mine. The next few years will be busy: Several instruments either already have data, or will have data soon, that will do a better job at measuring small angular scale features in the CMB. These features encode information about the matter in the universe—how much ordinary matter, and how much dark matter, is out there.

An important step forward is the Microwave Anisotropy Probe (MAP) satellite, launched by NASA in 2001. MAP will measure the whole sky from space during a two-year mission (as opposed to the few percent of the sky measured by BOOMERANG from a balloon). With this enormous increase in precision, we should be able to pin down parameters of the standard cosmological models such as Ω_0, Ω_Λ, and the Hubble constant to accuracies of a few percent. This is truly precision cosmology—which only a few years ago would have been an oxymoron!

While the temperature of the CMB (2.725 K) has been well measured, and its anisotropies are being attacked with vigor, a new frontier awaits. It turns out that we expect the CMB photons to be slightly polarized. The level of polarization is extremely small (a few percent of the temperature fluctuations) and very challenging to measure.

However, if we can detect and, eventually, precisely measure that polarization, it will be able to tell us a few very interesting things about the early universe. First, the prediction of polarization assumes that our current model of the very young universe is correct—specifically, that we understand the nature of the lumps that eventually collapsed into galaxies and clusters of galaxies, and that we understand the process of the decoupling of photons and matter that occurred when the universe was a few hundred thousand years old. We are always looking for ways to test this model, and the polarization provides a very strong one! Second, polarization information can be combined with the measured temperature anisotropies to improve our measurements of cosmological parameters. We are preparing now for another flight of BOOMERANG that will attempt to make better measurements of the smaller-scale information in the temperature anisotropies and detect the polarization of the CMB.

There is another, very long-shot—and long-term—goal of polarization measurements that, if successful, would lead to an incredible discovery. The simplest models of inflation predict that gravitational waves (see Section 24-2) were produced as inflation ends. These waves can impress their unique "fingerprints" in the pattern of CMB polarization. If inflation occurred at a sufficiently high energy scale (near the GUT scale of 10^{14} GeV), this gravity wave "fingerprint" may be strong enough for us to detect. But if inflation happened at a much lower energy scale, there would be too few gravitational waves for us to see the signature.

This measurement is very challenging, and not for the faint of heart. But if that signature can be detected, it would give us a fantastic probe of inflation!

The Search for Extraterrestrial Life

One of the most compelling questions in science is also one of the simplest: Are we alone? That is, is there life beyond the Earth? As yet, we have no definitive answer to this question. None of our spacecraft has found life elsewhere in the solar system, and radio telescopes have yet to detect signals of intelligent origin coming from space. Reports of aliens visiting our planet and abducting humans make compelling science fiction, but none of these reports has ever been verified.

Yet there are reasons to suspect that life might indeed exist beyond the Earth. One is that biologists find living organisms in some of the most "unearthly" environments on our planet. An example (shown here) is at the bottom of the Gulf of Mexico, where the crushing pressure and low temperature make methane—normally a gas—form into solid, yellowish mounds. Amazingly, these mounds teem with colonies of pink, eyeless, alien-looking worms the size of your thumb. If life can flourish here, might it not also flourish in the equally hostile conditions found on other worlds?

In this chapter we will look for places in our solar system where life may once have originated, and where it may exist today. We will see how scientists estimate the chances of finding life beyond our solar system, and how they search for signals from other intelligent species. And we will learn how a new generation of telescopes may make it possible to detect even single-celled organisms on worlds many light-years away.

(Dr. Charles Fisher, Eberly College of Science, Pennsylvania State University)   R I V U X G

As you read the sections of this chapter, look for the answers to the following questions.

30-1 What role could comets and meteorites have played in the origin of life on Earth?

30-2 Have spacecraft found any evidence for life elsewhere in our solar system?

30-3 Do meteorites from Mars give conclusive proof that life originated there?

30-4 How likely is it that other civilizations exist in our Galaxy?

30-5 How do astronomers search for evidence of civilizations on planets orbiting other stars?

30-6 Will it ever be possible to see Earthlike planets orbiting other stars?

30-1 The chemical building blocks of life are found throughout space

Suppose you were the first visitor to a new and alien planet. How would you recognize which of the strange objects around you were living, and which were inanimate? Questions such as these are central to **astrobiology** (also called **exobiology**), the study of life in the universe. Most astrobiologists suspect that if we find living organisms on other worlds, they will be "life as we know it"—that is, their biochemistry will be based on the unique properties of the carbon atom, as is the case for all terrestrial life.

Why carbon? The reason is that carbon has the most versatile chemistry of any element. Carbon atoms can form chemical bonds to create especially long and complex molecules. These carbon-based compounds, called **organic molecules,** include all the molecules of which living organisms are made.

Organic molecules can be linked together to form elaborate structures, such as chains, lattices, and fibers. Some of these structures are capable of complex, self-regulating chemical reactions. Furthermore, the primary constituents of organic molecules—carbon, hydrogen, nitrogen, oxygen, sulfur, and phosphorus—are among the most abundant elements in the universe. The versatility and abundance of carbon suggest that extraterrestrial life is also likely to be based on organic chemistry.

If life is based on organic molecules, then these molecules must initially be present on a planet in order for life to arise from nonliving matter. We now understand that many organic molecules originate from nonbiological processes in interstellar space. One such molecule is carbon monoxide (CO), which is made when a carbon atom and an oxygen atom collide and bond together. Carbon monoxide is found in abundance within giant interstellar clouds that lie along the spiral arms of our Milky Way Galaxy as well as other galaxies (see Figure 1-7). Carbon atoms have also combined with other elements to produce an impressive array of interstellar organic molecules, including ethyl alcohol (CH_3CH_2OH), formaldehyde (H_2CO), methyl cyanoacetylene (CH_3C_3N), and acetaldehyde (CH_3CHO). Radio astronomers have detected

these by looking for the telltale microwave emission lines of carbon-based chemicals in interstellar clouds.

The planets of our solar system coalesced from interstellar material (see Section 7-7), and some of the organic molecules in that material must have ended up on the planets' surfaces. Evidence for this comes from meteorites called **carbonaceous chondrites,** like the one in Figure 30-1. These are ancient meteorites that date from the formation of the solar system and that are often found to contain a variety of carbon-based molecules. The spectra of comets (see Section 7-5)—which are also among the oldest objects in the solar system—show that they, too, contain an assortment of organic compounds.

Comets and meteoroids were much more numerous in the early solar system than they are today, and they were correspondingly more likely to collide with a planet. These collisions would have seeded the planets with organic compounds from the very beginning of our solar system's history. Similar processes are thought to take place in other planetary systems,

Figure 30-1 R I **V** U X G

A Carbonaceous Chondrite Carbonaceous chondrites are primitive meteorites that date back to the very beginning of the solar system. Chemical analyses of newly fallen specimens disclose that they are rich in organic molecules, many of which are the chemical building blocks of life. This sample is a piece of the Allende meteorite, a large carbonaceous chondrite that fell in Mexico in 1969. (From the collection of Ronald A. Oriti)

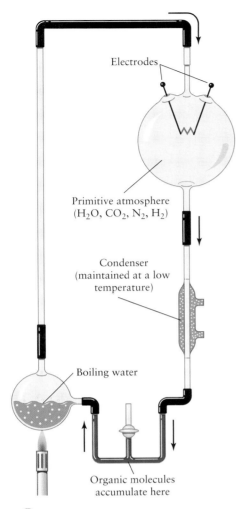

Electrodes

Primitive atmosphere
(H_2O, CO_2, N_2, H_2)

Condenser
(maintained at a low
temperature)

Boiling water

Organic molecules
accumulate here

ᶠigure 30-2

The Miller-Urey Experiment (Updated) Modern versions of this classic experiment prove that numerous organic compounds important to life can be synthesized from gases that were present in the Earth's primordial atmosphere. This experiment supports the hypothesis that life on the Earth arose as a result of ordinary chemical reactions.

which are thought to form in basically the same way as did our own (see Figure 7-13).

Comets and meteorites would not have been the only sources of organic material on the young planets of our solar system. In 1952, the American chemists Stanley Miller and Harold Urey demonstrated that under conditions that are thought to have prevailed on the primitive Earth, simple chemicals can combine to form the chemical building blocks of life. In a closed container, they prepared a sample of "atmosphere": a mixture of hydrogen (H_2), ammonia (NH_3), methane (CH_4), and water vapor (H_2O), the most common molecules in the solar system. Miller and Urey then exposed this mixture of gases to an elec-

tric arc (to simulate atmospheric lightning) for a week. At the end of this period, the inside of the container had become coated with a reddish-brown substance rich in amino acids and other compounds essential to life.

Since Miller and Urey's original experiment, most scientists have come to the conclusion that Earth's primordial atmosphere was composed of carbon dioxide (CO_2), nitrogen (N_2), and water vapor outgassed from volcanoes, along with some hydrogen. Modern versions of the Miller-Urey experiment (Figure 30-2) using these common gases have also succeeded in synthesizing a wide variety of organic compounds. The combination of comets and meteorites falling from space and chemical synthesis in the atmosphere could have made the chemical building blocks of life available in substantial quantities on the young Earth.

It is important to emphasize that scientists have *not* created life in a test tube. While organic molecules may have been available on the ancient Earth, biologists have yet to figure out how these molecules gathered themselves into cells and developed systems for self-replication. Nevertheless, because so many chemical components of life are so easily synthesized under conditions that simulate the primordial Earth, it seems reasonable to suppose that life could have originated as the result of chemical processes. Furthermore, because the molecules that combine to form these compounds are rather common, it seems equally reasonable that life could have originated in the same way on other planets.

Organic building blocks are commonplace throughout the universe, but this does not guarantee that life is equally commonplace. If a planet's environment is hostile, life may never get started or may quickly be extinguished. But we now have evidence that Jupiter-sized planets orbit other stars (see Section 7-9 and Geoff Marcy's essay "Alien Planets" at the end of Chapter 7) and that additional planetary systems are forming around young stars (see Section 7-7, especially Figure 7-13). It seems probable that there are Earthlike planets orbiting other stars, and that conditions on some of these worlds may be suitable for life as we know it.

30-2 Europa and Mars are promising places for life to have evolved

If life evolved on Earth from nonliving organic molecules, might the same process have taken place elsewhere in our solar system? Scientists have carefully scrutinized the planets and satellites in an attempt to answer this question, and most of the answers have been disappointing.

One major problem is that liquid water is essential for the survival of life as we know it. The water need not be pleasant by human standards—terrestrial organisms have been found in water that is boiling hot, fiercely acidic, or ice cold (see the image that opens this chapter)—but it must be liquid. In order for water on a planet's surface to remain liquid, the temperature cannot be too hot or too cold.

Furthermore, there must be a relatively thick atmosphere to provide enough pressure to keep liquid water from evaporating. Of all the worlds of the present-day solar system, only Earth has the right conditions for water to remain liquid on its surface. But there is now compelling evidence that Europa, one of the large satellites of Jupiter (see Table 7-2), has an ocean of water *beneath* its icy surface. As it orbits Jupiter, Europa is caught in a tug-of-war between Jupiter's gravitational influence and those of the other large satellites. This flexes the interior of Europa, and this flexing generates enough heat to keep subsurface water from freezing. Chunks of ice on the surface can float around on this underground ocean, rearranging themselves into a pattern that reveals the liquid water beneath.

No one knows whether life exists in Europa's ocean. But interest in this exotic little world is great, and scientists have proposed several missions to explore Europa in more detail.

The next best possibility for the existence of life is Mars. The present-day Martian atmosphere is so thin that water can exist only as ice or as a vapor. However, images made from Martian orbit show dried-up streambeds, flash flood channels, and sediment deposits. These features are evidence that the Martian atmosphere was once thicker and that water once coursed over the planet's surface. Could life have evolved on Mars during its "wet" period? If so, could life—even in the form of microorganisms—have survived as the Martian atmosphere thinned and the surface water either froze or evaporated?

In 1976, two spacecraft landed in different parts of Mars in search of answers to these questions. *Viking Lander 1* and *Viking Lander 2* each carried a scoop at the end of a mechanical arm to retrieve surface samples (Figure 30-3). These samples were deposited into a compact on-board biological laboratory that carried out three different tests for Martian microorganisms.

1. The *gas-exchange experiment* was designed to detect any processes that might be broadly considered as respiration. A surface sample was placed in a sealed container along with a controlled amount of gas and nutrients. The gases in the container were then monitored to see if their chemical composition changed.

2. The *labeled-release experiment* was designed to detect metabolic processes. A sample was moistened with nutrients containing radioactive carbon atoms. If any organisms in the sample consumed the nutrients, their waste products should include gases containing the telltale radioactive carbon.

3. The *pyrolytic-release experiment* was designed to detect photosynthesis, the biological process by which terrestrial plants use solar energy to help synthesize organic compounds from carbon dioxide. In the *Viking* experiments, a surface sample was placed in a container along with radioactive carbon dioxide and exposed to artificial sunlight. If plantlike photosynthesis occurred, microorganisms in the sample would take in some of the radioactive carbon from the gas.

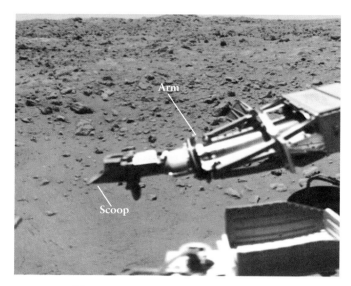

ƒigure 30-3   R I **V** U X G

Digging in the Martian Surface This view from the *Viking Lander 1* spacecraft shows the mechanical arm with its small scoop against the backdrop of the Martian terrain. The scoop was able to dig about 30 cm (12 in.) beneath the surface. (NASA)

The first data returned from these experiments caused great excitement, for in almost every case, rapid and extensive changes were detected inside the sealed containers. Further analysis of the data, however, led to the conclusion that these changes were due solely to nonbiological chemical processes. It appears that the Martian surface is rich in unstable chemicals that react with water to release oxygen gas. Because the present-day surface of Mars is bone-dry, these chemicals had nothing to react with until they were placed inside the moist interior of the *Viking Lander* laboratory.

At best, the results from the *Viking Lander* biological experiments were inconclusive. Perhaps life never existed on Mars at all. Or perhaps it did originate there, but failed to survive the thinning of the Martian atmosphere, the unstable chemistry of the planet's surface, and exposure to ultraviolet radiation from the Sun. (Unlike Earth, Mars has no ozone layer to block ultraviolet rays.) Another possibility is that Martian microorganisms have survived only in certain locations that the *Viking Lander*s did not sample, such as isolated spots on the surface or deep beneath the ground.

The exploration of Mars, including the search for microorganisms, is just beginning. (See the essay by Matthew Golombek that follows this chapter.) In 2003, a British spacecraft called *Beagle 2* will land on Mars and conduct an entirely different set of biological experiments (Figure 30-4). The spacecraft's name commemorates HMS *Beagle*, the surveying ship from which Charles Darwin made many of the observations that led to the theory of evolution.

Unlike the *Viking Lander*s, *Beagle 2* will test for the presence of either living or dead microorganisms. To do this, it will

Figure 30-4

The _Beagle 2_ Lander This lander (shown here in an artist's rendering) will be launched with the European Space Agency's _Mars Express_ spacecraft. On arrival at Mars, _Mars Express_ will go into orbit around the planet while _Beagle 2_ descends to the surface. A parachute and airbags will cushion the landing, after which the spacecraft will unfold to the configuration shown here. Each circular section is about the size of an extra-large pizza. (_Beagle 2_/European Space Agency)

bore into the interiors of rocks to gather pristine, undisturbed samples, then heat these samples in the presence of oxygen gas. All carbon compounds will decompose and form carbon dioxide (CO_2) when treated in this way, but biologically important molecules will signal their presence by decomposing at a lower temperature. As a further test for the chemicals of life, _Beagle 2_ will check to see how many of the CO_2 molecules contain the isotope ^{12}C (which appears preferentially in biological molecules) and how many contain ^{13}C (which does not). (See Box 5-5 for a description of isotopes.) If these experiments give a positive result, it will indicate that Martian rocks contain microorganisms that either survive to the present day or that died out at some point in the past.

Beagle 2 will also carry an experiment to search for traces of methane in the Martian atmosphere. Microorganisms on Earth can gain energy by converting carbon dioxide to methane, and presumably Martian microorganisms could do the same. Left to itself, methane rapidly decomposes in the Martian atmosphere. Hence, if _Beagle 2_ finds any methane, it will necessarily have been freshly formed—which would strongly suggest that life exists on Mars today.

In 1976, while the _Viking Landers_ were carrying out their biological experiments on the Martian surface, the companion _Viking Orbiter_ spacecraft photographed some surface features that could have been crafted by _intelligent_ life on Mars. The _Viking Orbiter 1_ image in Figure 30-5a shows what appears to be a humanlike face, perhaps the product of an advanced and artistic civilization. However, when the more advanced _Mars Global Surveyor_ spacecraft viewed the surface in 1998 using a superior camera (Figure 30-5b), it found no evidence for facial features.

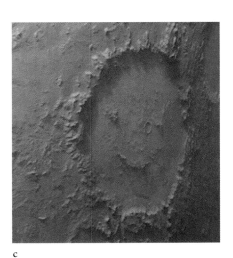

a   b   c

Figure 30-5   R I **V** U X G

A "Face" on Mars? **(a)** This 1976 image from _Viking Orbiter 1_ shows a Martian surface feature that resembles a human face. Some suggested that this feature might have been made by intelligent beings. **(b)** This 1998 _Mars Global Surveyor_ (MGS) image, made under different lighting conditions with a far superior camera, reveals the "face" to be just an eroded hill. **(c)** This MGS image shows features of natural origin within a 215-km (134-mi) wide crater. Can you see this "face"? (a: NSSDC/NASA and Dr. Michael H. Carr; b, c: Malin Space Science Systems/NASA)

Scientists are universally convinced that the "face" and other apparent patterns in the *Viking Orbiter* images were created by shadows on wind-blown hills. Microscopic life may once have existed on Mars, and may yet exist today, but there is no evidence that the red planet has ever been the home of intelligent beings.

30-3 Meteorites from Mars have been scrutinized for life-forms

While spacecraft can carry biological experiments to other worlds such as Mars, many astrobiologists look forward to the day when a spacecraft will return Martian samples to laboratories on Earth. (See the essay by Matthew Golombek that follows this chapter.) Until that day arrives, we have the next best thing: A dozen meteorites that appear to have formed on Mars have been found at a variety of locations on the Earth.

These meteorites are called **SNC meteorites** after the names given to the first three examples found (Shergotty, Nakhla, and Chassigny). What identifies SNC meteorites as having come from Mars is the chemical composition of trace amounts of gas trapped within them. This composition is very different from that of the Earth's atmosphere, but is a nearly perfect match to the composition of the Martian atmosphere found by the *Viking Landers*.

How could a rock have got from Mars to Earth? When a large piece of space debris collides with a planet's surface and forms an impact crater, most of the material thrown upward by the impact falls back onto the planet's surface. But some extraordinarily powerful impacts produce large craters—on Mars, roughly 100 km in diameter or larger. These tremendous impacts eject some rocks with such speed

that they escape the planet's gravitational attraction and fly off into space.

There are numerous large craters on Mars, so a good number of Martian rocks have probably been blasted into space over the planet's history. These ejected rocks then go into elliptical orbits around the Sun. A few such rocks will have orbits that put them on a collision course with the Earth, and these are the ones that scientists find as SNC meteorites. Using a technique called radioactive age-dating, scientists find that most SNC meteorites are between 200 million and 1.3 billion years old, much younger than the 4.6-billion-year age of the solar system. But one SNC meteorite, denoted by the serial number ALH 84001 and found in Antarctica in 1984, was discovered in 1993 to be 4.5 billion years old (Figure 30-6a). Thus, ALH 84001 is a truly ancient piece of Mars. Analysis of ALH 84001 suggests that it was fractured by an impact between 3.8 and 4.0 billion years ago, was ejected from Mars by another impact 16 million years ago, and landed in Antarctica a mere 13,000 years ago.

ALH 84001 is the only known specimen of a rock that was on Mars during the era when liquid water existed on the planet's surface. Scientists have therefore investigated its chemical composition carefully, in the hope that this rock may contain clues to the amount of water that once flowed on the Martian surface. One such clue is the presence of rounded grains of minerals called carbonates, which can form only in the presence of water.

In 1996, David McKay and Everett Gibson of the NASA Johnson Space Center, along with several collaborators, reported the results of a two-year study of the carbonate grains in ALH 84001. They made three remarkable findings. First, in and around the carbonate grains were large numbers of elongated, tubelike structures resembling fossilized microorganisms

a

R I **V** U X G   b

figure 30-6

A Meteorite from Mars (a) This 1.9-kg meteorite, known as ALH 84001, formed on Mars some 4.5 billion years ago. About 16 million years ago a massive impact blasted it into space, where it drifted in orbit around the Sun until landing in Antarctica 13,000 years ago. The small cube at lower right is 1 cm (0.4 in.) across. (b) This electron microscope image, magnified some 100,000 times, shows tubular structures about 100 nanometers (10^{-7} m) in length found within the Martian meteorite ALH 84001. One controversial interpretation is that these are the fossils of microorganisms that lived on Mars billions of years ago. (a: NASA Johnson Space Center; b: *Science*, NASA)

(Figure 30-6b). Second, the carbonate grains contain very pure crystals of iron sulfide and magnetite. These two compounds are rarely found together (especially in the presence of carbonates) but can be produced by certain types of bacteria. Indeed, about one-fourth of the magnetite crystals found in ALH 84001 are of a type that on Earth are formed only by bacteria. Third, the carbonates contain organic molecules—just the sort, in fact, that result from the decay of microorganisms.

McKay and Gibson concluded that the structures seen in Figure 30-6b are fossilized remains of microorganisms. If so, these organisms lived and died on Mars billions of years ago, during the era when liquid water was abundant.

Are McKay and Gibson's conclusions correct? Their claims of ancient life on Mars are extraordinary, and they require extraordinary proof. With only one rock like ALH 84001 known to science, however, such proof is hard to come by, and many scientists are skeptical. They argue that the structures found in ALH 84001 could have been formed in other ways that do not require the existence of Martian microorganisms. Studies by *Beagle 2* of rocks on Mars may help resolve the controversy.

For now, the existence of microscopic life on Mars in the distant past remains an open question. What is without question, however, is that ALH 84001 has fired the imagination of scientists and the public and generated new excitement about the search for extraterrestrial life.

30-4 The Drake equation helps scientists estimate how many civilizations may inhabit our Galaxy

We have seen that only a few locations in our solar system may have been suitable for the origin of life. But what about other planetary systems? The development of life on the Earth seems to suggest that extraterrestrial life, including intelligent species, might evolve on terrestrial planets around other stars, given sufficient time and hospitable conditions. How can we learn whether such worlds exist, given the tremendous distances that separate us from them? This is the great challenge facing the **search for extraterrestrial intelligence, or SETI.**

A tenet of modern folklore is the belief that alien civilizations do exist, and that their spacecraft have visited Earth. Indeed, surveys show that between one-third and one-half of all Americans believe in unidentified flying objects (UFOs). A somewhat smaller percentage believes that aliens have landed on Earth. But, in fact, there is *no* scientifically verifiable evidence of alien visitations. As an example, many UFO proponents believe that the U.S. government is hiding evidence of an alien spacecraft that crashed near Roswell, New Mexico, in 1947. However, the bits of "spacecraft wreckage" found near Roswell turn out to be nothing more than remnants of an unmanned research balloon. To find real evidence of the presence or absence of intelligent civilizations on worlds orbiting other stars, we must look elsewhere.

With our present technology, sending even a small unmanned spacecraft to another star requires a flight time of tens of thousands of years. Speculative design studies have been made for unmanned probes that could reach other stars within a century or less, but these are prohibitively expensive. Instead, many astronomers hope to learn about extraterrestrial civilizations by detecting radio transmissions from them. Radio waves are a logical choice for interstellar communication because they can travel immense distances without being significantly degraded by the interstellar medium, the thin gas and dust found between the stars (see Section 7-6).

Over the past several decades, astronomers have proposed various ways to search for alien radio transmissions, and several searches have been undertaken. In 1960, Frank Drake first used a radio telescope at the National Radio Astronomy Observatory in West Virginia to listen to two Sunlike stars, Tau Ceti and Epsilon Eridani, without success. More than 60 more extensive SETI searches have taken place since then, using radio telescopes around the world. Occasionally, a search has detected an unusual or powerful signal. But none has ever repeated, as a signal of intelligent origin might be expected to do. To date, we have no confirmed evidence of radio transmissions from another world.

Should we be discouraged by this failure to make contact? What are the chances that a radio astronomer might someday detect radio signals from an extraterrestrial civilization? The first person to tackle this issue was Frank Drake, who proposed that the number of technologically advanced civilizations in the Galaxy could be estimated by a simple equation. This is now called the **Drake equation:**

Drake equation

$$N = R_* \, f_p \, n_e \, f_l \, f_i \, f_c \, L$$

N = number of technologically advanced civilizations in the Galaxy whose messages we might be able to detect

R_* = the rate at which solar-type stars form in the Galaxy

f_p = the fraction of stars that have planets

n_e = the number of planets per solar system that are Earthlike (that is, suitable for life)

f_l = the fraction of those Earthlike planets on which life actually arises

f_i = the fraction of those life-forms that evolve into intelligent species

f_c = the fraction of those species that develop adequate technology and then choose to send messages out into space

L = the lifetime of that technologically advanced civilization

The Drake equation is enlightening because it expresses the number of extraterrestrial civilizations in a simple series of terms. We can estimate some of these terms from what we know about stars and stellar evolution. For example, the first two factors, R_* and f_p, can be determined by observation. In estimating R_*, we should probably exclude stars with masses greater than about 1.5 times that of the Sun. These more massive stars use up the hydrogen in their cores in 3 billion (3×10^9) years or less. On Earth, by contrast, human intelligence developed only within the last million years or so, some 4.6 billion years after the formation of the solar system. If that is typical of the time needed to evolve higher life-forms, then a star of 1.5 solar masses or more probably fades away or explodes into a supernova before creatures as intelligent as we can evolve on any of that star's planets.

Although stars less massive than the Sun have much longer lifetimes, they, too, seem unsuited for life because they are so dim. Only planets very near a low-mass star would be sufficiently warm for life as we know it, and a planet that close is subject to strong tidal forces from its star. We saw in Section 4-8 how the Earth's tidal forces keep the Moon locked in synchronous rotation, with one face continually facing the Earth. In the same way, a planet that orbits too close to its star would have one hemisphere that always faced the star, while the other hemisphere would be in perpetual, frigid darkness.

This leaves us with stars not too different from the Sun. (Like Goldilocks sampling the three bears' porridge, we must have a star that is not too hot and not too cold, but just right.) Based on statistical studies of star formation in the Milky Way, some astronomers estimate that roughly one of these Sunlike stars forms in the Galaxy each year, thus setting R_* at 1 per year.

As we saw in Sections 7-7 and 7-8, the planets in our solar system formed as a natural consequence of the birth of the Sun. We have also seen evidence suggesting that planetary formation may be commonplace around single stars (see Figure 7-13). Many astronomers suspect that most Sunlike stars probably have planets, and so they give f_p a value of 1.

Unfortunately, the rest of the terms in the Drake equation are very uncertain. Let's play with some hypothetical values. The chances that a planetary system has an Earthlike world suitable for life are not known. Were we to consider our own solar system as representative, we could put n_e at 1. Let's be more conservative, however, and suppose that one in ten solar-type stars is orbited by a habitable planet, making $n_e = 0.1$. From what we know about the evolution of life on the Earth, we might assume that, given appropriate conditions, the development of life is a certainty, which would make $f_l = 1$. This is an area of intense interest to astrobiologists.

For the sake of argument, we might also assume that evolution might naturally lead to the development of intelligence (a conjecture that is hotly debated) and also make $f_i = 1$. It's anyone's guess as to whether these intelligent extraterrestrial beings would attempt communication with other civilizations in the Galaxy, but were we to assume they would, f_c would be put at 1 also.

The last variable, L, involving the longevity of a civilization, is the most uncertain of all. Looking at our own

example, we see a planet whose atmosphere and oceans are increasingly polluted by creatures that possess nuclear weapons. If we are typical, perhaps L is as short as 100 years. Putting all these numbers together, we arrive at

$$N = \frac{1}{\text{year}} \times 1 \times 0.1 \times 1 \times 1 \times 1 \times 100 \text{ years} = 10$$

In other words, out of the hundreds of billions of stars in the Galaxy, we would estimate that there are only ten technologically advanced civilizations from which we might receive communications.

A wide range of values has been proposed for the terms in the Drake equation, and these various guesses produce vastly different estimates of N. Some scientists argue that there is exactly one advanced civilization in the Galaxy and that we are it. Others speculate that there may be hundreds or thousands of planets inhabited by intelligent creatures. If we wish to know whether our Galaxy is devoid of other intelligence, teeming with civilizations, or something in between, we must keep searching the skies.

30-5 Radio searches for alien civilizations are under way

Even if only a few alien civilizations are scattered across the Galaxy, we have the technology to detect radio transmissions from them. But if other civilizations are trying to communicate with us using radio waves, what frequency are they using? This is an important question, because if we fail to tune our radio telescopes to the right frequency, we might never know whether the aliens are out there.

A reasonable choice would be a frequency that is fairly free of interference from extraneous sources. SETI pioneer Bernard Oliver was the first to draw attention to a range of relatively noise-free frequencies in the neighborhood of the microwave emission lines of hydrogen (H) and hydroxide (OH) (Figure 30-7). This region of the microwave spectrum is called the **water hole**, because H and OH together make H_2O, or water.

In 1989, NASA began work on the High Resolution Microwave Survey (HRMS), an ambitious project to scan the entire sky at frequencies spanning the water hole from 10^3 to 10^4 MHz. HRMS would have observed more than 800 nearby solar-type stars over a narrower frequency range in the hope of detecting signals that were either pulsed (like Morse code) or continuous (like the carrier wave for a TV or radio broadcast). The sophisticated signal-processing technology of HRMS would have been able to sift through tens of millions of individual frequency channels simultaneously. It would even have been able to detect the minute Doppler shifts in a signal coming from an alien planet as that planet spun on its axis and moved around its star.

Sadly, just one year after HRMS began operation in 1992, the U.S. Congress imposed a mandate requiring that NASA no longer support HRMS or any other radio searches for extra-

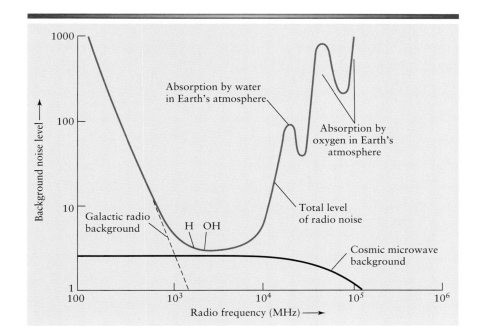

ƒigure 30-7

The Water Hole This graph shows the background noise level from the sky at various radio and microwave frequencies. The so-called water hole is a range of radio frequencies from about 1000 to 10,000 megahertz (Mhz) that happen to have relatively little cosmic noise. Some scientists suggest that this noise-free region would be well suited for interstellar communication. At lower frequencies there is substantial emission from interstellar gas (labeled "Galactic radio background"), and at higher frequencies radio waves tend to be absorbed by the Earth's atmosphere. Within the water hole itself, the principal source of noise is the afterglow of the Big Bang, called the cosmic background radiation. To put this graph in perspective, a frequency of 100 MHz corresponds to "100" on a FM radio, and 10^3 MHz is a frequency used for various types of radar. (Adapted from C. Sagan and F. Drake)

terrestrial intelligence. This decision, which was made on budgetary grounds, saved a few million dollars—an entirely negligible amount compared to the total NASA budget. Ironically, the senator who spearheaded this move was from the state of Nevada, where tax dollars have been spent to signpost a remote desert road as "The Extraterrestrial Highway."

 Even though NASA funding is no longer available, several teams of scientists remain actively involved in SETI programs. Funding for these projects has come from nongovernmental organizations such as the Planetary Society and from private individuals. Since 1995 the SETI Institute in California has been carrying out Project Phoenix, the direct successor to HRMS. When complete, this project will have surveyed a thousand Sunlike stars within 200 lightyears at millions of radio frequencies. At Harvard University, BETA (the *B*illion-channel *E*xtra*T*errestrial *A*ssay) is scanning the sky at even more individual frequencies within the water hole. Other multifrequency searches are being carried out under the auspices of the University of Western Sydney in Australia and the University of California.

 A major challenge facing SETI is the tremendous amount of computer time needed to analyze the mountains of data returned by radio searches. To this end, scientists at the University of California, Berkeley, have recruited more than 2 million personal computer users to participate in a project called SETI@home. Each user receives actual data from a detector called SERENDIP IV (*S*earch for *E*xtraterrestrial *R*adio *E*missions from *N*earby, *D*eveloped, *I*ntelligent *P*opulations) and a data analysis program that also acts as a screensaver. When the computer's screensaver is on, the program runs, the data are analyzed, and the results are reported via the Internet to the researchers

at Berkeley. The program then downloads new data to be analyzed. As of this writing, SETI@home users have provided as much computer time as a single computer working full-time for 400,000 years!

All current SETI projects make use of existing radio telescopes and must share telescope time with astronomical researchers. But by 2005, the SETI Institute plans to put into operation a radio telescope that will be dedicated solely to the search for intelligent signals. This telescope, called the Allen Telescope Array, will actually be hundreds of relatively small and inexpensive radio dishes working together. Perhaps this new array will be the first to detect a signal from a distant civilization.

30-6 Infrared telescopes in space will soon begin searching for Earthlike planets

Although no longer involved in SETI, NASA is planning a major effort to search for Earthlike planets suitable for the evolution of an advanced civilization. Such a search poses a major challenge. Planets the size of the Earth are too small to be detected by the indirect methods we described in Section 7-9. They are also too dim to be seen in visible light against the glare of their parent star.

Instead, an orbiting telescope called Terrestrial Planet Finder—currently targeted for a 2011 launch—will search for such planets by detecting their infrared radiation. The rationale is that stars like the Sun emit much less infrared radiation than visible light, while planets are relatively strong emitters

figure 30-8

The Spectrum of a Simulated Planet The image on the left is a simulation of what the Terrestrial Planet Finder infrared telescope might see when it is deployed around 2011. The white dot at the center is a nearby Sunlike star, and the smaller dots around it are planets orbiting the star. On the right is the simulated infrared spectrum of one of the planets, showing broad absorption lines of water vapor (H_2O), ozone (O_3), and carbon dioxide (CO_2). While all these molecules can be created by nonbiological processes, the presence of life will change the relative amounts of each molecule in the planet's atmosphere. Thus, the infrared spectrum of such planets will make it possible to identify worlds on which life may have evolved. (Jet Propulsion Laboratory)

of infrared. Hence, observing in the infrared makes it less difficult (although still technically challenging) to detect planets orbiting a star.

Over its planned six-year lifetime, Terrestrial Planet Finder will search for planets around the brightest 1000 stars within 15 parsecs (50 light-years) of the Sun. It will also analyze the infrared spectra of any planets that it finds, in the hope of seeing the characteristic absorption of atmospheric gases such as ozone, carbon dioxide, and water vapor (Figure 30-8). The relative amounts of these gases, as determined from a planet's spectrum, can reveal whether life is present on that planet.

WEB LINK 30.14 To achieve enough resolution to detect individual planets, Terrestrial Planet Finder will need to make use of interferometry. We discussed this technique for improving the resolution of telescopes in Section 6-6. By combining the light from four widely spaced 8-m dishes, Terrestrial Planet Finder will make the sharpest infrared images of any telescope in history. The European Space Agency is planning a similar planet-finding interferometry mission called Darwin.

A more speculative project using interferometry is Planet Imager. If funded, this will be an infrared telescope with sufficient resolution that some detail would be visible in the image of an extrasolar planet. One concept for such a mission would consist of five Terrestrial Planet Finder–type telescopes flying in a geometrical formation some 6000 km across (equal to the radius of the Earth). All five telescopes would collect light from the same extrasolar planet, then reflect it onto a single 8-m mirror. The combined light would go to detectors on board a sixth spacecraft (Figure 30-9).

Sometime in the next few decades, missions such as Terrestrial Planet Finder and Planet Imager may answer the question "Are there worlds like Earth orbiting other stars?" If the answer is yes, radio searches for intelligent signals will gain even more impetus.

The potential rewards from such searches are great. Detecting a message from an alien civilization could dramati-

cally change the course of our own civilization, through the sharing of scientific information with another species or an awakening of social or humanistic enlightenment. In only a few years our technology, industry, and social structure might advance the equivalent of centuries into the future. Such changes would touch every person on the Earth. Mindful of these profound implications, scientists push ahead with the search for extraterrestrial intelligence.

figure 30-9

Planet Imager This illustration shows the concept for Planet Imager, an infrared telescope array designed to create images of Earthlike planets orbiting other stars. Five Terrestrial Planet Finder–type telescopes—each of which has four 8-m mirrors—collect infrared light from a distant planet and reflect it onto a single 8-m mirror, which in turn reflects the combined light onto a spacecraft with infrared-sensitive detectors. To achieve the necessary resolution, the entire Planet Imager array will need to be about 6000 km across, very much larger than shown here. The technology to build Planet Imager does not yet exist. (Jet Propulsion Laboratory) astrobiology, p. 690

carbonaceous chondrite, p. 690

Drake equation, p. 695

exobiology, p. 690

organic molecules, p. 690

search for extraterrestrial intelligence
 (SETI), p. 695

SNC meteorite, p. 694

water hole, p. 696

KEY IDEAS

Organic Molecules in the Universe: All life on Earth, and presumably on other worlds, depends on organic (carbon-based) molecules. These molecules occur naturally throughout interstellar space.

• The organic molecules needed for life to originate were probably brought to the young Earth by comets or meteorites. Another likely source for organic molecules is chemical reactions in the Earth's primitive atmosphere. Similar processes may occur on other worlds.

Life in the Solar System: Besides Earth, only two worlds in our solar system—the planet Mars and Jupiter's satellite Europa—may have had the right conditions for the origin of life.

• Mars once had liquid water on its surface, though it has none today. Life may have originated on Mars during the liquid water era.

• The *Viking Lander* spacecraft searched for microorganisms on the Martian surface, but found no conclusive sign of their presence. The *Beagle 2* mission to Mars will carry out a different set of biological experiments on samples taken from the interiors of rocks.

• An ancient Martian rock that came to Earth as a meteorite shows circumstantial evidence that microorganisms once existed on Mars. Additional rock samples are needed to provide corroboration.

• Europa appears to have extensive liquid water beneath its icy surface. Future missions may search for the presence of life there.

Radio Searches for Extraterrestrial Intelligence: Astronomers have carried out a number of searches for radio signals from other stars. No signs of intelligent life have yet been detected, but searches are continuing and using increasingly sophisticated techniques.

• The Drake equation is a tool for estimating the number of intelligent, communicative civilizations in our Galaxy.

Telescope Searches for Earthlike Planets: A new generation of orbiting infrared telescopes may be able to detect terrestrial planets around nearby stars. If such planets are found, their spectra may reveal the presence or absence of life.

REVIEW QUESTIONS

1. What is meant by "life as we know it"? Why do astrobiologists suspect that extraterrestrial life is likely to be of this form?

2. How have astronomers discovered organic molecules in interstellar space? Does this discovery mean that life of some sort exists in the space between the stars?

3. Mercury, Venus, and the Moon are all considered unlikely places to find life. Suggest why this should be.

4. Summarize the differences in philosophy between the biological experiments on board the *Viking Lander*s and those that will be carried on *Beagle 2*.

5. Suppose someone brought you a rock that he claimed was a Martian meteorite. What scientific tests would you recommend be done to test this claim?

6. Why are most searches for extraterrestrial intelligence made using radio telescopes? Why are most of these carried out at frequencies between 10^3 MHz and 10^4 MHz?

7. Explain why Terrestrial Planet Finder and Planet Imager, both of which are planned to be infrared telescopes, need to be placed in space.

ADVANCED QUESTIONS

Problem-solving tips and tools

The small-angle formula, discussed in Box 1-1, will be useful. Section 5-2 gives the relationship between wavelength and frequency, while Section 5-9 and Box 5-6 discuss the Doppler effect. Section 6-3 gives the relationship between the angular resolution of a telescope, the telescope diameter, and the wavelength used. You will find useful data about the planets in Appendix 1.

8. In 1802, when it seemed likely to many scholars that there was life on Mars, the German mathematician Karl Friedrich Gauss proposed that we signal the Martian

inhabitants by drawing huge geometric patterns in the snows of Siberia. His plan was never carried out. (a) Suppose patterns had been drawn that were 1000 km across. What minimum diameter would the objective of a Martian telescope need to have to be able to resolve these patterns? Assume that the observations are made at a wavelength of 550 nm, and assume that Earth and Mars are at their minimum separation. (b) Ideally, the patterns used would be ones that could not be mistaken for natural formations. They should also indicate that they were created by an advanced civilization. What sort of patterns would you have chosen?

9. Assume that all the terms in the Drake equation have the values given in the text, except for N and L. (a) If there are 1000 civilizations in the Galaxy today, what must be the average lifetime of a technological civilization? (b) What if there are a million such civilizations?

10. (a) Of the visually brightest stars in the sky listed in Appendix 5, which might be candidates for having Earthlike planets on which intelligent civilizations have evolved? Explain your selection criteria. (b) Repeat part (a) for the nearest stars, listed in Appendix 4.

11. It has been suggested that extraterrestrial civilizations would choose to communicate at a wavelength of 21 cm. Hydrogen atoms in interstellar space naturally emit at this wavelength, so astronomers studying the distribution of hydrogen around the Galaxy would already have their radio telescopes tuned to receive extraterrestrial signals. (a) Calculate the frequency of this radiation in megahertz. Is this inside or outside the water hole? (b) Discuss the merits of this suggestion.

12. Imagine that a civilization in another planetary system is sending a radio signal toward Earth. As our planet moves in its orbit around the Sun, the wavelength of the signal we receive will change due to the Doppler effect. This gives SETI scientists a way to distinguish stray signals of terrestrial origin (which will not show this kind of wavelength change) from interstellar signals. (a) Use the data in Appendix 1 to calculate the speed of the Earth in its orbit. For simplicity, assume the orbit is circular. (b) If the alien civilization is transmitting at a frequency of 3000 MHz, what wavelength (in meters) would we receive if the Earth were moving neither toward nor away from their planet? (c) The maximum Doppler shift occurs if the Earth's orbital motion takes it directly toward or directly away from the alien planet. How large is that maximum wavelength shift? Express your answer both in meters and as a percentage of the unshifted wavelength you found in (b). (d) Discuss why it is important that SETI radio receivers be able to measure frequency and wavelength to very high precision.

13. Suppose that Planet Imager has an effective diameter of 6000 km and uses infrared radiation with a wavelength of 10 μm. If it is used to observe an Earthlike planet orbiting the star Epsilon Eridani, 3.22 parsecs (10.5 light-years) from Earth, what is the size of the smallest detail that Planet Imager will be able to resolve on the face of that planet? Give your answer in kilometers.

DISCUSSION QUESTIONS

14. Suppose someone told you that the *Viking Landers* failed to detect life on Mars simply because the tests were designed to detect terrestrial life-forms, not Martian life-forms. How would you respond?

15. Science-fiction television shows and movies often depict aliens as looking very much like humans. Discuss the likelihood that intelligent creatures from another world would have (a) a biochemistry similar to our own, (b) two legs and two arms, and (c) about the same dimensions as a human.

16. The late, great science-fiction editor John W. Campbell exhorted his authors to write stories about organisms that think as well as humans, but not *like* humans. Discuss the possibility than an intelligent being from another world might be so alien in its thought processes that we could not communicate with it.

17. If a planet always kept the same face toward its star, just as the Moon always keeps the same face toward Earth, most of the planet's surface would be uninhabitable. Discuss why.

18. How do you think our society would respond to the discovery of intelligent messages coming from a civilization on a planet orbiting another star? Explain your reasoning.

19. What do you think will set the limit on the lifetime of our technological civilization? Explain your reasoning.

20. The first of all Earth spacecraft to venture into interstellar space were *Pioneer 10* and *Pioneer 11*, which were launched in 1972 and 1973, respectively. Their missions took them past Jupiter and Saturn and eventually beyond the solar system. Both spacecraft carry a metal plaque with artwork (reproduced here) that shows where the spacecraft is from and what sort of creatures designed it. If an alien civilization were someday to find one of these spacecraft, which of the features on the plaque do you think would be easily understandable to them? Explain.

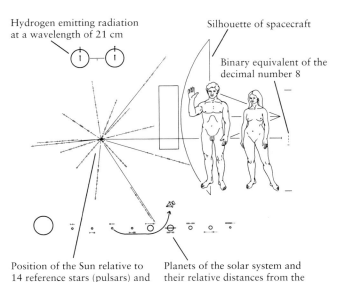

Hydrogen emitting radiation at a wavelength of 21 cm

Silhouette of spacecraft

Binary equivalent of the decimal number 8

Position of the Sun relative to 14 reference stars (pulsars) and the center of the Galaxy

Planets of the solar system and their relative distances from the Sun (shown in binary code)

21. Any living creatures in the subsurface ocean of Europa would have to survive without sunlight. Instead, they might obtain energy from Europa's inner heat. Search the World Wide Web for information about "black smokers," which are associated with high-temperature vents at the bottom of Earth's oceans. What kind of life is found around black smokers? How do these life-forms differ from the more familiar organisms found in the upper levels of the ocean?

22. Like other popular media, the World Wide Web is full of claims of the existence of "extraterrestrial intelligence"—namely, UFO sightings and alien abductions. **(a)** Choose a web site of this kind and analyze its content using the idea of *Occam's razor*, the principle that if there is more than one viable explanation for a phenomenon, one should choose the simplest explanation that fits all the observed facts. **(b)** Read what a skeptical web site has to say about UFO sightings. A good example is the web site of the Committee for the Scientific Investigation of Claims of the Paranormal, or CSICOP. After considering what you have read on both sides of the UFO debate, discuss your opinions about whether aliens really have landed on Earth.

23. The Drake Equation. Access the Active Integrated Media Module "The Drake Equation" in Chapter 30 of the *Universe* web site or CD-ROM. **(a)** For each of the terms in the Drake equation, choose a value that seem reasonable to you. How did you choose these values? Using the module, what do you find for the number of civilizations in our Galaxy? From your calculation, are civilizations common or uncommon in our Galaxy? **(b)** Using the module, choose a set of values that give $N = 10^6$ (a million civilizations). What values did you use? Which of these seem reasonable to you, and why?

OBSERVING PROJECT

24. Use the *Starry Night* program to view the Earth as it might be seen by a visiting spacecraft. First select **Viewing Location…** in the **Go** menu and set the viewing location to your city or town. (You use the list of cities provided, or you can use the mouse to click on your approximate position on the world map. Then click the **Set Location** button.) In the Control Panel at the top of the window, set the local time to 12:00:00 P.M. (noon). To see the Earth from space, use the elevation buttons (the ones that look like a rocket landing or taking off) in the Control Panel to raise yourself off the surface until you can see the entire Earth. (You may want to adjust your field of view so that you are looking directly at the Earth.) **(a)** Describe any features you see that suggest life could exist on Earth. Explain your reasoning. **(b)** Using the controls at the right-hand end of the Control Panel, zoom in to show more detail around your city or town. The amount of detail is comparable to the view from a spacecraft a few million kilometers away. Can you see any evidence that life does exist on Earth? **(c)** From a distance of a few million kilometers, are there any measurements that a spacecraft could carry out to prove that life exists on Earth? Explain your reasoning.

MATTHEW GOLOMBEK

Exploring Mars

Dr. Matt Golombek, chief scientist of the *Mars Pathfinder* mission, is a research geologist at the Jet Propulsion Laboratory, California Institute of Technology, NASA's lead center for planetary exploration. He received his undergraduate degree from Rutgers University and his master's degree and Ph.D. from the University of Massachusetts, Amherst. Dr. Golombek was a postdoctoral fellow at the Lunar and Planetary Institute, Houston, before joining JPL.

Dr. Golombek has conducted research in the structural geology of the Earth, planets, and satellites. He is presently NASA's Mars Exploration Program Landing Site Scientist,

leading the effort to select landing sites for Mars missions.

Dr. Golombek cowrote the book *Mars: Uncovering the Secrets of the Red Planet* for the National Geographic Society and has served as editor and associate editor of professional journals. He has received numerous awards, including the NASA Exceptional Scientific Achievement Medal. In his honor, asteroid 6456 was named Golombek.

Mars is the most Earthlike planet in our solar system. It is the first planet humans will visit and the only planet with abundant water that can support life—including people, perhaps—in the future. The geologic record on Mars suggests a warmer early climate in which liquid water (a requirement for life) may have been stable. Further, meteorites from Mars may contain evidence of primitive life. Mars is also a unique terrestrial planet with evidence of major climatic change and a geologic record of rocks on the surface that spans the entire history of the solar system. An exploration program to the red planet will allow not only the investigation of fundamental geologic, climatologic, and exobiologic topics but can also address one of the most compelling scientific questions: "Are we alone in the universe?"

The underlying questions that drive the exploration of Mars are: Did life ever develop on Mars? What is the current climate and how has it changed with time? How have the surface and interior (geology) of the planet evolved through time and what resources are available for future use? Water plays a central role in all these questions.

Robotic spacecraft are among the most complex devices built by humans, and launching and operating them in space are somewhat risky and difficult endeavors. Because of the orbital motions of Mars and Earth, spacecraft can be launched to Mars along a favorable trajectory only once every 26 months. The spacecraft must withstand a nine-month interplanetary journey, observe the planet with carefully designed instruments, and send data back to Earth. These spacecraft typically take three to ten years to design and build, require highly coordinated work by hundreds of scientists and engineers, cost hundreds of millions of dollars to complete, and have a few percent chance of exploding during launch on even the most reliable launch vehicles available.

Future missions are designed to provide an understanding of Mars as a whole. Missions such as NASA's *Mars Surveyor*

orbiter being readied for launch in 2001, Europe's *Mars Express* mission to be launched in 2003, Japan's *Nozomi* orbiter departing for Mars in 2003, and other NASA missions under study will obtain imaging data to investigate the geological processes and history of the surface, spectral imaging data directed at understanding the chemistry and mineralogy of surface materials, sounder data to probe the subsurface distribution of water, atmospheric data to study the present state and circulation of the atmosphere, and other data directed at understanding the interaction of the solar wind and other particles with the uppermost atmosphere of Mars.

Remotely sensed data from orbit will always carry an inherent ambiguity or uncertainty, so information collected from a probe on the Martian surface is required to provide "ground truth." NASA's two *Mars Exploration Rovers* being readied for launch in 2003, Britain's *Beagle 2* lander to be launched on *Mars Express* in 2003, and France's four *Netlanders* scheduled for launch in 2005 will provide such "ground truth" and make simultaneous measurements of global atmospheric and interior processes.

An extremely high-priority mission being studied is a "sample-return" mission to bring rock, soil, and atmosphere samples of Mars to the Earth. Much more sophisticated tests can be done with samples in Earth laboratories than by the limited instruments that can be miniaturized and placed on landed spacecraft. Analyses of these samples will allow extensive testing of our knowledge about the evolution of Mars. These studies will also provide information for the design of eventual human exploration missions.

Was Mars once warm and wet like the Earth? Did life begin on Mars? If it did, what happened to it? Could it have come to the Earth in chunks of rocks ejected from Mars? Could we all be Martians? Is life common or rare in the universe? These are the questions we hope to address by exploring our fascinating neighbor.

Appendices

Appendix 1 | The Planets: Orbital Data

Planet	Semimajor axis (AU)	Semimajor axis (10^6 km)	Sidereal period (years)	Sidereal period (days)	Synodic period (days)	Average orbital speed (km/s)	Orbital eccentricity	Inclination of orbit to ecliptic(°)
Mercury	0.3871	57.91	0.2408	87.969	115.88	47.9	0.206	7.00
Venus	0.7233	108.2	0.6152	224.70	583.92	35.0	0.0068	3.39
Earth	1.0000	149.60	1.0000	365.256	—	29.79	0.017	0.00
Mars	1.5236	227.93	1.8808	686.98	779.94	24.1	0.093	1.85
Jupiter	5.2026	778.30	11.856		398.9	13.1	0.048	1.30
Saturn	9.5719	1431.9	29.369		378.1	9.64	0.053	2.48
Uranus	19.194	2877.4	84.099		369.7	6.83	0.043	0.77
Neptune	30.066	4497.8	164.86		367.5	5.5	0.010	1.77
Pluto	39.537	5914.7	248.60		366.7	4.7	0.250	17.12

Appendix 2 | The Planets: Physical Data

Planet	Equatorial diameter (km)	Equatorial diameter (Earth = 1)	Mass (kg)	Mass (Earth = 1)	Average density (kg/m³)	Rotation period (days)*	Inclination of equator to orbit (°)	Surface gravity (Earth = 1)	Albedo	Escape speed (km/s)
Mercury	4880	0.383	3.302×10^{23}	0.0553	5430	58.646	0.5	0.38	0.12	4.3
Venus	12,104	0.949	4.868×10^{24}	0.8150	5243	243.01^R	177.4	0.91	0.59	10.4
Earth	12,756	1.000	5.974×10^{24}	1.000	5515	1.000	23.45	1.000	0.39	11.2
Mars	6794	0.533	6.418×10^{23}	0.1074	3934	1.026	25.19	0.38	0.15	5.0
Jupiter	142,984	11.209	1.899×10^{27}	317.8	1326	0.414	3.12	2.36	0.44	60.2
Saturn	120,536	9.449	5.685×10^{26}	95.16	687	0.444	26.73	1.1	0.47	35.5
Uranus	51,118	4.007	8.682×10^{25}	14.53	1318	0.718^R	97.86	0.92	0.56	21.3
Neptune	49,528	3.883	1.024×10^{26}	17.15	1638	0.671	29.56	1.1	0.51	23.5
Pluto	2300	0.180	1.31×10^{22}	0.002	2000	6.387^R	122.52	0.07	0.5	1.2

* For Jupiter, Saturn, Uranus, and Neptune, the internal rotation period is given. A superscript R means that the rotation is retrograde (opposite the planet's orbital motion).

Appendix 3 | Satellites of the Planets

Planet	Satellite	Discoverer(s)	Average distance from center of planet (km)	Orbital (sidereal) period (days)*	Orbital eccentricity	Diameter of satellite (km)	Mass (kg)
EARTH	Moon	—	384,400	27.322	0.0549	3476	7.349×10^{22}
MARS	Phobos	Hall (1877)	9378	0.319	0.01	$28 \times 23 \times 20$	1.1×10^{16}
	Deimos	Hall (1877)	23,460	1.263	0.00	$16 \times 12 \times 10$	1.8×10^{15}
JUPITER**	Metis	Synott (1979)	127,960	0.2948	0.00	40	1×10^{17}
	Adrastea	Jewitt et al. (1979)	128,980	0.2983	0 (?)	$24 \times 20 \times 16$	1.9×10^{16}
	Amalthea	Barnard (1892)	181,300	0.4981	0.00	$270 \times 200 \times 155$	7.2×10^{18}
	Thebe	Synott (1979)	221,900	0.6745	0.01	100	8×10^{17}
	Io	Galileo (1610)	421,600	1.769	0.00	3642	8.932×10^{22}
	Europa	Galileo (1610)	670,900	3.551	0.01	3120	4.791×10^{22}
	Ganymede	Galileo (1610)	1,070,000	7.155	0.00	5268	1.482×10^{23}
	Callisto	Galileo (1610)	1,883,000	16.689	0.01	4800	1.077×10^{23}
	Leda	Kowal (1974)	11,094,000	238.72	0.15	16	6×10^{15}
	Himalia	Perrine (1904)	11,480,000	250.57	0.16	180	1×10^{19}
	Lysithea	Nicholson (1938)	11,720,000	259.22	0.11	40	8×10^{16}
	Elara	Perrine (1905)	11,737,000	259.65	0.21	80	8×10^{17}
	Ananke	Nicholson (1951)	21,200,000	631[R]	0.17	30	4×10^{16}
	Carme	Nicholson (1938)	22,600,000	692[R]	0.21	44	1×10^{17}
	Pasiphae	Melotte (1908)	23,500,000	735[R]	0.38	70	2×10^{17}
	Sinope	Nicholson (1914)	23,700,000	758[R]	0.28	40	8×10^{16}
	S/1999 J1	Spacewatch (2000)	24,200,000	768[R]	0.12	10	(?)
SATURN***	Pan	Showalter (1990)	133,570	0.573	0.00	20	(?)
	Atlas	Terrile (1980)	137,640	0.602	0 (?)	$20 \times 20 \times 20$	(?)
	Prometheus	Collins et al. (1980)	139,350	0.613	0.00	$140 \times 100 \times 80$	3×10^{17}
	Pandora	Collins et al. (1980)	141,700	0.629	0.00	$110 \times 90 \times 70$	2×10^{17}
	Epimetheus	Walker (1966)	151,422	0.694	0.01	$140 \times 120 \times 100$	6×10^{17}
	Janus	Dolfuss (1966)	151,472	0.695	0.01	$220 \times 200 \times 160$	2×10^{18}
	Mimas	Herschel (1789)	185,520	0.942	0.02	392	3.8×10^{19}
	Enceladus	Herschel (1789)	238,020	1.370	0.00	500	7.3×10^{19}
	Tethys	Cassini (1684)	294,660	1.888	0.00	1060	6.2×10^{20}
	Calypso	Smith et al. (1980)	294,660	1.888	0 (?)	$34 \times 28 \times 26$	(?)
	Telesto	Smith et al. (1980)	294,660	1.888	0 (?)	$24 \times 22 \times 22$	(?)
	Dione	Cassini (1684)	377,400	2.737	0.00	1120	1.1×10^{21}
	Helene	Laques et al. (1980)	377,400	2.737	0.01	$36 \times 32 \times 30$	(?)
	Rhea	Cassini (1672)	527,040	4.518	0.00	1530	2.3×10^{21}
	Titan	Huygens (1655)	1,221,850	15.945	0.03	5150	1.4×10^{23}
	Hyperion	Bond (1848)	1,481,000	21.277	0.10	$410 \times 260 \times 220$	2×10^{19}
	Iapetus	Cassini (1671)	3,561,300	79.331	0.03	1460	1.6×10^{21}
	Phoebe	Pickering (1898)	12,952,000	550.48[R]	0.16	220	4×10^{18}

Planet	Satellite	Discoverer(s)	Average distance from center of planet (km)	Orbital (sidereal) period (days)*	Orbital eccentricity	Diameter of satellite (km)	Mass (kg)
URANUS	Cordelia	*Voyager 2* (1986)	49,752	0.335	0 (?)	26	(?)
	Ophelia	*Voyager 2* (1986)	53,763	0.376	0 (?)	32	(?)
	Bianca	*Voyager 2* (1986)	59,166	0.435	0 (?)	44	(?)
	Cressida	*Voyager 2* (1986)	61,767	0.464	0 (?)	66	(?)
	Desdemona	*Voyager 2* (1986)	62,658	0.474	0 (?)	58	(?)
	Juliet	*Voyager 2* (1986)	64,358	0.493	0 (?)	84	(?)
	Portia	*Voyager 2* (1986)	66,097	0.513	0 (?)	110	(?)
	Rosalind	*Voyager 2* (1986)	69,927	0.558	0 (?)	58	(?)
	1986 U10	Karkoschka (1999)	75,000	0.62	0 (?)	40	(?)
	Belinda	*Voyager 2* (1986)	75,260	0.624	0 (?)	67	(?)
	Puck	*Voyager 2* (1986)	86,010	0.762	0 (?)	154	(?)
	Miranda	Kuiper (1948)	129,780	1.413	0.00	473	6.6×10^{19}
	Ariel	Lassell (1851)	191,240	2.520	0.00	1160	1.4×10^{21}
	Umbriel	Lassell (1851)	265,790	4.144	0.00	1170	1.2×10^{21}
	Titania	Herschel (1787)	435,840	8.706	0.00	1580	3.5×10^{21}
	Oberon	Herschel (1787)	582,600	13.463	0.00	1520	3.0×10^{21}
	Caliban	Gladman (1997)	7,168,900	579^R	0 (?)	60	(?)
	Stephano	Gladman (1999)	7,942,000	676^R	0 (?)	30	(?)
	Sycorax	Nicholson (1997)	12,213,000	1289^R	0 (?)	120	(?)
	Prospero	Holman (1999)	16,114,000	1953^R	0 (?)	50	(?)
	Setebos	Kavelaars (1999)	18,205,000	2345^R	0 (?)	40	(?)
NEPTUNE	Naiad	*Voyager 2* (1989)	48,227	0.294	0 (?)	58	(?)
	Thalassa	*Voyager 2* (1989)	50,075	0.312	0 (?)	80	(?)
	Despina	*Voyager 2* (1989)	52,526	0.335	0 (?)	150	(?)
	Galatea	*Voyager 2* (1989)	61,953	0.429	0 (?)	160	(?)
	Larissa	*Voyager 2* (1989)	73,548	0.555	0 (?)	190	(?)
	Proteus	*Voyager 2* (1989)	117,647	1.122	0 (?)	420	(?)
	Triton	Lassell (1846)	354,760	5.877^R	0.00	2706	2.15×10^{22}
	Nereid	Kuiper (1949)	5,513,400	360.14	0.75	340	(?)
PLUTO	Charon	Christy (1978)	19,640	6.387	0.00	1190	1.9×10^{21}

* *A superscript R means that the satellite orbits in a retrograde direction (opposite to the planet's rotation).*

** *Eleven additional small satellites of Jupiter (S/2000 J1 through S/2000 J11)were discovered in 2000. Their average distances from Jupiter range from 7 to 24 million km, and their sizes range from 3 to 8 km. Nine are in retrograde orbits. The discoveries of these satellites had not been confirmed as of this writing (early 2001), so they do not appear in the above table.*

*** *In 2000, ten additional small satellites of Saturn (S/2000 S1 through S/2000 S10) were discovered beyond the orbit of Phoebe. While the existence of these satellites has been confirmed, their orbits are still tentative as of this writing (early 2001) and so do not appear in this table.*

Appendix 4 | The Nearest Stars

Name*	Parallax (arcsec)	Distance (light-years)	Spectral type	Radial velocity** (km/s)	Proper motion (arcsec/year)	Apparent visual magnitude	Absolute visual magnitude	Luminosity (Sun = 1)
Sun			G2 V			−26.7	+4.83	1.00
Proxima Centauri	0.772	4.22	M5.5 V	−22	3.853	+11.01	+15.45	8.2×10^{-4}
Alpha Centauri A	0.742	4.40	G2 V	−25	3.710	−0.01	+4.34	1.77
Alpha Centauri B	0.742	4.40	K0 V	−21	3.724	+1.35	+5.70	0.55
Barnard's Star	0.549	5.94	M4 V	−111	10.358	+9.54	+13.24	3.6×10^{-3}
Wolf 359	0.418	7.80	M6 V	+13	4.689	+13.45	+16.6	3.5×10^{-4}
Lalande 21185	0.392	8.32	M2 V	−84	4.802	+7.49	+10.46	0.023
L 726-8 A	0.381	8.56	M5.5 V	+29	3.366	+12.41	+15.3	9.4×10^{-4}
L 726-8 B	0.381	8.56	M6 V	+32	3.366	+13.2	+16.1	5.6×10^{-4}
Sirius A	0.379	8.61	A1 V	−9	1.339	−1.44	+1.45	26.1
Sirius B	0.379	8.61	white dwarf	−9	1.339	+8.44	+11.3	2.4×10^{-3}
Ross 154	0.336	9.71	M3.5 V	−12	0.666	+10.37	+13.00	4.1×10^{-3}
Ross 248	0.316	10.32	M5.5 V	−78	1.626	+12.29	+14.8	1.5×10^{-3}
Epsilon Eridani	0.311	10.49	K2 V	+17	0.977	+3.72	+6.18	0.40
Lacaille 9352	0.304	10.73	M1.5 V	+10	6.896	+7.35	+9.76	0.051
Ross 128	0.300	10.87	M4 V	−31	1.361	+11.12	+13.50	2.9×10^{-3}
L 789-6	0.294	11.09	M5 V	−60	3.259	+12.33	+14.7	1.3×10^{-3}
61 Cygni A	0.287	11.36	K5 V	−65	5.281	+5.20	+7.49	0.16
61 Cygni B	0.285	11.44	K7 V	−64	5.172	+6.05	+8.33	0.095
Procyon A	0.286	11.40	F5 IV–V	−4	1.259	+0.40	+2.68	7.73
Procyon B	0.286	11.40	white dwarf	−4	1.259	+10.7	+13.0	5.5×10^{-4}
BD +59° 1915 A	0.281	11.61	M3 V	−1	2.238	+8.94	+11.18	0.020
BD +59° 1915 B	0.281	11.61	M3.5 V	+1	2.313	+9.70	+11.97	0.010
Groombridge 34 A	0.280	11.65	M1.5 V	+12	2.918	+8.09	+10.33	0.030
Groombridge 34 B	0.280	11.65	M3.5 V	+11	2.918	+11.07	+13.3	3.1×10^{-3}
Epsilon Indi	0.276	11.82	K5 V	−40	4.704	+4.69	+6.89	0.27
GJ 1111	0.276	11.82	M6.5 V	−5	1.288	+14.79	+17.0	2.7×10^{-4}
Tau Ceti	0.274	11.90	G8 V	−17	1.922	+3.49	+5.68	0.62
GJ 1061	0.270	12.08	M5.5 V	−20	0.836	+13.03	+15.2	1.0×10^{-3}
L 725-32	0.269	12.12	M4.5 V	+28	1.372	+12.10	+14.25	1.7×10^{-3}
BD +05° 1668	0.263	12.40	M3.5 V	+18	3.738	+9.84	+11.94	0.011
Kapteyn's Star	0.255	12.79	M0 V	+246	8.670	+8.86	+10.89	0.013
Lacaille 8760	0.253	12.89	M0 V	+28	3.455	+6.69	+8.71	0.094
Krüger 60 A	0.250	13.05	M3 V	−33	0.990	+9.85	+11.9	0.010
Krüger 60 B	0.250	13.05	M4 V	−32	0.990	+11.3	+13.3	3.4×10^{-3}

This table, compiled from the Hipparcos General Catalogue *and from data reported by the Research Consortium on Nearby Stars, lists all known stars within 4.00 parsecs (13.05 light-years).*

* *Stars that are components of binary systems are labeled A and B.*

** *A positive radial velocity means the star is receding; a negative radial velocity means the star is approaching.*

Name	Designation	Distance (light-years)	Spectral type	Radial velocity (km/s)*	Proper motion (arcsec/year)	Apparent visual magnitude	Apparent visual brightness (Sirius = 1)**	Absolute visual magnitude	Luminosity (Sun = 1)
Sirius A	α CMa A	8.61	A1 V	−9	1.339	−1.44	1.000	+1.45	26.1
Canopus	α Car	313	F0 I	+21	0.031	−0.62	0.470	−5.53	8.2×10^{-4}
Arcturus	α Boo	36.7	K2 III	−5	2.279	−0.05	0.278	−0.31	190
Alpha Centauri A	α Cen A	4.4	G2 V	−25	3.71	−0.01	0.268	+4.34	1.77
Vega	α Lyr	25.3	A0 V	−14	0.035	+0.03	0.258	+0.58	61.9
Capella	α Aur	42.2	G8 III	+30	0.434	+0.08	0.247	−0.48	180
Rigel	α Ori A	773	B8 Ia	+21	0.002	+0.18	0.225	−6.69	7.0×10^{5}
Procyon	α CMi A	11.4	F5 IV–V	−4	1.259	+0.4	0.184	+2.68	7.73
Achernar	α Eri	144	B3 IV	+19	0.097	+0.45	0.175	−2.77	5250
Betelgeuse	α Ori	427	M2 Iab	+21	0.029	+0.45	0.175	−5.14	4.1×10^{4}
Hadar	β Cen	525	B1 II	−12	0.042	+0.61	0.151	−5.42	8.6×10^{4}
Altair	α Aql	16.8	A7 IV–V	−26	0.661	+0.76	0.132	+2.2	11.8
Aldebaran	α Tau A	65.1	K5 III	+54	0.199	+0.87	0.119	−0.63	370
Spica	α Vir	262	B1 V	+1	0.053	+0.98	0.108	−3.55	2.5×10^{4}
Antares	α Sco A	604	M1 Ib	−3	0.025	+1.06	0.100	−5.28	3.7×10^{4}
Pollux	β Gem	33.7	K0 III	+3	0.627	+1.16	0.091	+1.09	46.6
Fomalhaut	α PsA	25.1	A3 V	+7	0.368	+1.17	0.090	+1.74	18.9
Deneb	α Cyg	3230	A2 Ia	−5	0.002	+1.25	0.084	−8.73	3.2×10^{5}
Mimosa	β Cru	353	B0.5 III	+20	0.05	+1.25	0.084	−3.92	3.4×10^{4}
Regulus	α Leo A	77.5	B7 V	+4	0.249	+1.36	0.076	−0.52	331

Data in this table were compiled from the Hipparcos General Catalogue.

* *A positive radial velocity means the star is receding; a negative radial velocity means the star is approaching.*

** *This is the ratio of the star's apparent brightness to that of Sirius, the brightest star in the night sky.*

Note: Acrux, or α Cru (the brightest star in Crux, the Southern Cross), appears to the naked eye as a star of apparent magnitude +0.87, the same as Aldebaran, but it does not appear in this table because Acrux is actually a binary star system. The blue-white component stars of this binary system have apparent magnitudes of +1.4 and +1.9, and thus they are dimmer than any of the stars listed here.

Appendix 6 | Some Important Astronomical Quantities

Astronomical unit	$1\ \text{AU} = 1.4960 \times 10^{11}\ \text{m}$ $= 1.4960 \times 10^{8}\ \text{km}$
Light-year	$1\ \text{ly} = 9.4605 \times 10^{15}\ \text{m}$ $= 63{,}240\ \text{AU}$
Parsec	$1\ \text{pc} = 3.2616\ \text{ly}$ $= 3.0857 \times 10^{16}\ \text{m}$ $= 206{,}265\ \text{AU}$
Year	$1\ \text{y} = 365.2564\ \text{days}$ $= 3.156 \times 10^{7}\ \text{s}$
Solar mass	$1\ M_\odot = 1.989 \times 10^{30}\ \text{kg}$
Solar radius	$1\ R_\odot = 6.9599 \times 10^{8}\ \text{m}$
Solar luminosity	$1\ L_\odot = 3.90 \times 10^{26}\ \text{W}$

Appendix 8 | Some Useful Mathematics

Area of a rectangle of sides a and b	$A = ab$
Volume of a rectangular solid of sides a, b, and c	$V = abc$
Hypotenuse of a right triangle whose other sides are a and b	$c = \sqrt{a^2 + b^2}$
Circumference of a circle of radius r	$C = 2\pi r$
Area of a circle of radius r	$A = \pi r^2$
Surface area of a sphere of radius r	$A = 4\pi r^2$
Volume of a sphere of radius r	$V = \tfrac{4}{3}\pi r^3$
Value of π	$\pi = 3.1415926536$

Appendix 7 | Some Important Physical Constants

Speed of light	$c = 2.9979 \times 10^{8}\ \text{m/s}$
Gravitational constant	$G = 6.6726 \times 10^{-11}\ \text{N m}^2/\text{kg}^2$
Planck's constant	$h = 6.6261 \times 10^{-34}\ \text{J s}$ $= 4.1357 \times 10^{-15}\ \text{eV s}$
Boltzmann constant	$k = 1.3807 \times 10^{-23}\ \text{J/K}$ $= 8.6174 \times 10^{-5}\ \text{eV/K}$
Stefan-Boltzmann constant	$\sigma = 5.6705 \times 10^{-8}\ \text{W m}^{-2}\ \text{K}^{-4}$
Mass of electron	$m_\text{e} = 9.1094 \times 10^{-31}\ \text{kg}$
Mass of proton	$m_\text{p} = 1.6726 \times 10^{-27}\ \text{kg}$
Mass of neutron	$m_\text{n} = 1.6749 \times 10^{-27}\ \text{kg}$
Mass of hydrogen atom	$m_\text{H} = 1.6735 \times 10^{-27}\ \text{kg}$
Rydberg constant	$R = 1.0968 \times 10^{7}\ \text{m}^{-1}$
Electron volt	$1\ \text{eV} = 1.6022 \times 10^{-19}\ \text{J}$

Glossary

You can find more information about each term in the indicated chapter or chapters.
See the Key Words section in each chapter for the specific page on which the meaning of a given term is discussed.

A ring   One of three prominent rings encircling Saturn. (Chapter 15)

absolute magnitude   The apparent magnitude that a star would have if it were at a distance of 10 parsecs. (Chapter 19)

absolute zero   A temperature of –273°C (or 0 K), at which all molecular motion stops; the lowest possible temperature. (Chapter 5)

absorption line spectrum   Dark lines superimposed on a continuous spectrum. (Chapter 5)

acceleration   The rate at which an object's velocity changes due to a change in speed, a change in direction, or both. (Chapter 4)

accretion   The gradual accumulation of matter in one location, typically due to the action of gravity. (Chapter 7, Chapter 20)

accretion disk   A disk of gas orbiting a star or black hole. (Chapter 27)

active galactic nucleus (AGN)   The center of an active galaxy. (Chapter 27)

active galaxy   A galaxy that is emitting exceptionally large amounts of energy; a Seyfert galaxy, radio galaxy, or quasar. (Chapter 27)

active optics   A technique for improving a telescopic image by altering the telescope's optics to compensate for variations in air temperature or flexing of the telescope mount. (Chapter 6)

adaptive optics   A technique for improving a telescopic image by altering the telescope's optics in a way that compensates for distortion caused by the Earth's atmosphere. (Chapter 6)

aerosol   Tiny droplets of liquid dispersed in a gas. (Chapter 15)

AGB star   See *asymptotic giant branch star.*

AGN   *See active galactic nucleus.*

albedo   The fraction of sunlight that a planet, asteroid, or satellite reflects. (Chapter 8)

alpha particle   The nucleus of a helium atom, consisting of two protons and two neutrons. (Chapter 21)

amino acids   The chemical building blocks of proteins. (Chapter 17)

andesite   Low-density, silicon-rich volcanic rocks found in the young northern lowlands of Mars. (Chapter 12)

angle   The opening between two lines that meet at a point. (Chapter 1)

angular diameter   The angle subtended by the diameter of an object. (Chapter 1)

angular distance   The angle between two points in the sky. (Chapter 1)

angular measure   The size of an angle, usually expressed in degrees, arcminutes, and arcseconds. (Chapter 1)

angular momentum   See *conservation of angular momentum.*

angular resolution   The angular size of the smallest feature that can be distinguished with a telescope. (Chapter 6)

angular size   See *angular diameter.*

anisotropic   Possessing unequal properties in different directions; not isotropic. (Chapter 26)

annihilation   The process by which the masses of a particle and antiparticle are converted into energy. (Chapter 29)

annular eclipse   An eclipse of the Sun in which the Moon is too distant to cover the Sun completely, so that a ring of sunlight is seen around the Moon at mid-eclipse. (Chapter 3)

anorthosite   Rock commonly found in ancient, cratered highlands on the Moon. (Chapter 9)

Antarctic Circle   A circle of latitude 23½° north of the Earth's South Pole. (Chapter 2)

antimatter   Matter consisting of antiparticles such as antiprotons, antineutrons, and positrons. (Chapter 29)

antiparticle   A particle with the same mass as an ordinary particle but with other properties, such as electric charge, reversed. (Chapter 29)

antiproton   A particle with the same mass as a proton but with a negative electric charge. (Chapter 29)

aphelion   The point in its orbit where a planet is farthest from the Sun. (Chapter 4)

apogee   The point in its orbit where a satellite or the Moon is farthest from the Earth. (Chapter 3)

apparent brightness   The flux of a star's light arriving at the Earth. (Chapter 19)

apparent magnitude   A measure of the brightness of light from a star or other object as measured from Earth. (Chapter 19)

apparent solar day   The interval between two successive transits of the Sun's center across the local meridian. (Chapter 2)

apparent solar time Time reckoned by the position of the Sun in the sky. (Chapter 2)

arcminute One-sixtieth (1/60) of a degree, designated by the symbol ′. (Chapter 1)

arcsecond One-sixtieth (1/60) of an arcminute or 1/3600 of a degree, designated by the symbol ″. (Chapter 1)

Arctic Circle A circle of latitude 23½° south of the Earth's North Pole. (Chapter 2)

asteroid One of tens of thousands of small, rocky, planetlike objects in orbit about the Sun. Also called "minor planet." (Chapter 7, Chapter 17)

asteroid belt A region between the orbits of Mars and Jupiter that encompasses the orbits of many asteroids. (Chapter 7, Chapter 17)

asthenosphere A warm, plastic layer of the mantle beneath the lithosphere of the Earth. (Chapter 8)

astrobiology The study of life in the universe. (Chapter 30)

astrometric method A technique for detecting extrasolar planets by looking for stars that "wobble" periodically. (Chapter 7)

astronomical unit (AU) The semimajor axis of the Earth's orbit; the average distance between the Earth and the Sun. (Chapter 1)

asymptotic giant branch A region on the Hertzsprung-Russell diagram occupied by stars in their second and final red-giant phase. (Chapter 22)

asymptotic giant branch star (AGB star) A red-giant star that has ascended the giant branch for the second and final time. (Chapter 22)

atmosphere (atm) A unit of atmospheric pressure. (Chapter 8)

atmospheric pressure The force per unit area exerted by a planet's atmosphere. (Chapter 8)

atom The smallest particle of an element that has the properties characterizing that element. (Chapter 5)

atomic number The number of protons in the nucleus of an atom of a particular element. (Chapter 5, Chapter 7)

AU See *astronomical unit.*

aurora (plural **aurorae**) Light radiated by atoms and ions in the Earth's upper atmosphere, mostly in the polar regions. (Chapter 8)

aurora australis Aurorae seen from southern latitudes; the southern lights. (Chapter 8)

aurora borealis Aurorae seen from northern latitudes; the northern lights. (Chapter 8)

autumnal equinox The intersection of the ecliptic and the celestial equator where the Sun crosses the equator from north to south. Also used to refer to the date on which the Sun passes through this intersection. (Chapter 2)

average density The mass of an object divided by its volume. (Chapter 7)

average density of matter In cosmology, the average density of all the material substance in the universe. (Chapter 28)

B ring One of three prominent rings encircling Saturn. (Chapter 15)

Balmer line An emission or absorption line in the spectrum of hydrogen caused by an electron transition between the second and higher energy levels. (Chapter 5)

Balmer series The entire pattern of Balmer lines. (Chapter 5)

Barnard object One of a class of dark nebulae discovered by E. E. Barnard. (Chapter 20)

barred spiral galaxy A spiral galaxy in which the spiral arms begin from the ends of a "bar" running through the nucleus rather than from the nucleus itself. (Chapter 26)

baseline In interferometry, the distance between two telescopes whose signals are combined to give a higher-resolution image. (Chapter 6)

belt A dark band in Jupiter's atmosphere. (Chapter 13)

Big Bang An explosion of all space that took place roughly 15 billion years ago and from which the universe emerged. (Chapter 1, Chapter 28)

Big Crunch The fate of the universe if it is bounded and ultimately collapses in upon itself. (Chapter 28)

binary star (also **binary star system** or **binary**) Two stars orbiting about each other. (Chapter 19)

biosphere The layer of soil, water, and air surrounding the Earth in which living organisms thrive. (Chapter 8)

bipolar outflow Oppositely directed jets of gas expelled from a young star. (Chapter 20)

BL Lac object See *blazar.*

black hole An object whose gravity is so strong that the escape speed exceeds the speed of light. (Chapter 1, Chapter 24)

black hole evaporation The process by which black holes emit particles. (Chapter 24)

blackbody A hypothetical perfect radiator that absorbs and re-emits all radiation falling upon it. (Chapter 5)

blackbody curve The intensity of radiation emitted by a blackbody plotted as a function of wavelength or frequency. (Chapter 5)

blackbody radiation The radiation emitted by a perfect blackbody. (Chapter 5)

blazar A type of active galaxy whose nucleus has a featureless spectrum without spectral lines. (Chapter 27)

blueshift A decrease in the wavelength of photons emitted by an approaching source of light. (Chapter 5)

Bohr orbits In the model of the atom described by Niels Bohr, the only orbits in which electrons are allowed to move about the nucleus. (Chapter 5)

Bok globule A small, roundish, dark nebula. (Chapter 20)

bounded universe A universe whose expansion slows until it eventually stops, after which the universe contracts. (Chapter 28)

bright terrain (Ganymede) Young, reflective, relatively crater-free terrain on the surface of Ganymede. (Chapter 14)

brightness See *apparent brightness*.

brown dwarf A starlike object that is not massive enough to sustain hydrogen burning in its core. (Chapter 7, Chapter 19)

brown ovals Elongated, brownish features usually seen in Jupiter's northern hemisphere. (Chapter 13)

C ring One of three prominent rings encircling Saturn. (Chapter 15)

capture theory The hypothesis that the Moon was gravitationally captured by the Earth. (Chapter 9)

carbon burning The thermonuclear fusion of carbon to produce heavier nuclei. (Chapter 22)

carbon star A peculiar red-giant star whose spectrum shows strong absorption lines of carbon and carbon compounds. (Chapter 22)

carbonaceous chondrite A type of meteorite that has a high abundance of carbon and volatile compounds. (Chapter 17, Chapter 30)

Cassegrain focus An optical arrangement in a reflecting telescope in which light rays are reflected by a secondary mirror to a focus behind the primary mirror. (Chapter 6)

Cassini division An apparent gap between Saturn's A and B rings. (Chapter 15)

CCD See *charge-coupled device*.

celestial equator A great circle on the celestial sphere 90° from the celestial poles. (Chapter 2)

celestial poles The points about which the celestial sphere appears to rotate. (Chapter 2) See also *north celestial pole; south celestial pole*.

celestial sphere An imaginary sphere of very large radius centered on an observer; the apparent sphere of the sky. (Chapter 2)

center of mass The point between a star and a planet, or between two stars, around which both objects orbit. (Chapter 7, Chapter 19)

central bulge A spherical distribution of stars around the nucleus of a spiral galaxy. (Chapter 25)

Cepheid variable A type of yellow, supergiant, pulsating star. (Chapter 21)

Cerenkov radiation The radiation emitted by a particle traveling through a substance at a speed greater than the speed of light in that substance. (Chapter 22)

Chandrasekhar limit The maximum mass of a white dwarf. (Chapter 22)

charge-coupled device (CCD) A type of solid-state device designed to detect photons. (Chapter 6)

chemical composition A description of which chemical substances make up a given object. (Chapter 7)

chemical differentiation The process by which the heavier elements in a planet sink toward its center while lighter elements rise toward its surface. (Chapter 7)

chemical element See *element*.

chondrule A glassy, roughly spherical blob found within meteorites. (Chapter 7)

chromatic aberration An optical defect whereby different colors of light passing through a lens are focused at different locations. (Chapter 6)

chromosphere A layer in the atmosphere of the Sun between the photosphere and the corona. (Chapter 18)

circumpolar A term describing a star that neither rises nor sets but appears to rotate around one of the celestial poles. (Chapter 2)

circumstellar accretion disk An accretion disk that surrounds a protostar. (Chapter 20)

close binary A binary star system in which the stars are separated by a distance roughly comparable to their diameters. (Chapter 21)

closed universe A universe with positive curvature, so that its geometry is analogous to that of the surface of a sphere. (Chapter 28)

cluster of galaxies A collection of galaxies containing a few to several thousand member galaxies. (Chapter 26)

cluster of stars A group of stars that formed together and that have remained together due to their mutual gravitational attraction. (Chapter 20)

CNO cycle A series of nuclear reactions in which carbon is used as a catalyst to transform hydrogen into helium. (Chapter 18)

cocoon nebula The nebulosity surrounding a protostar. (Chapter 20)

co-creation theory The hypothesis that the Earth and the Moon formed at the same time from the same material. (Chapter 9)

cold dark matter Slowly moving, weakly interacting particles presumed to constitute the bulk of matter in the universe. (Chapter 29)

collisional ejection theory The hypothesis that the Moon formed from material ejected from the Earth by the impact of a large asteroid. (Chapter 9)

color-magnitude diagram A plot of the apparent magnitudes of stars (that is, apparent brightnesses) of stars in a cluster against their color indices. (Chapter 21)

color ratio The ratio of the apparent brightness of a star measured in one spectral region to its brightness measured in a different region. (Chapter 19)

coma (of a comet) The diffuse gaseous component of the head of a comet. (Chapter 17)

coma (optical) The distortion of off-axis images formed by a parabolic mirror. (Chapter 6)

combined average mass density The average density in the observable universe of all forms of matter and energy, measured in units of mass per volume. (Chapter 28)

comet A small body of ice and dust in orbit about the Sun. While passing near the Sun, a comet's vaporized ices give rise to a coma and tail. (Chapter 7, Chapter 17)

condensation temperature The temperature at which a particular substance in a low-pressure gas condenses into a solid. (Chapter 7)

conduction The transfer of heat by directly passing energy from atom to atom. (Chapter 18)

confinement The permanent bonding of quarks to form particles like protons and neutrons. (Chapter 29)

conic section The curve of intersection between a circular cone and a plane; this curve can be a circle, ellipse, parabola, or hyperbola. (Chapter 4)

conjunction The geometric arrangement of a planet in the same part of the sky as the Sun, so that the planet is at an elongation of 0°. (Chapter 4)

conservation of angular momentum A law of physics stating that in an isolated system, the total amount of angular momentum—a measure of the amount of rotation—remains constant. (Chapter 7)

constellation A configuration of stars in the same region of the sky. (Chapter 2)

contact binary A binary star system in which both members fill their Roche lobes. (Chapter 21)

continuous spectrum A spectrum of light over a range of wavelengths without any spectral lines. (Chapter 5)

convection The transfer of energy by moving currents of fluid or gas containing that energy. (Chapter 8, Chapter 18)

convection cell A circulating loop of gas or liquid that transports heat from a warm region to a cool region. (Chapter 8)

convection current The pattern of motion in a gas or liquid in which convection is taking place. (Chapter 8)

convective zone The region in a star where convection is the dominant means of energy transport. (Chapter 18)

core (of the Earth) The iron-rich inner region of the Earth's interior. (Chapter 8)

core helium burning The thermonuclear fusion of helium at the center of a star. (Chapter 21, Chapter 22)

core hydrogen burning The thermonuclear fusion of hydrogen at the center of a star. (Chapter 21)

corona (of the Sun) The Sun's outer atmosphere, which has a high temperature and a low density. (Chapter 18)

coronal hole A region in the Sun's corona that is deficient in hot gases. (Chapter 18)

coronal mass ejection An event in which billions of tons of gas from the Sun's corona is suddenly blasted into space at high speed. (Chapter 8, Chapter 18)

cosmic background radiation See *cosmic microwave background.*

cosmic censorship, law of See *law of cosmic censorship.*

cosmic microwave background An isotropic radiation field with a blackbody temperature of about 2.7 K that permeates the entire universe. (Chapter 28)

cosmic particle horizon An imaginary sphere, centered on the Earth, whose radius equals the distance light has traveled since the Big Bang. (Chapter 28)

cosmic singularity The Big Bang. (Chapter 28)

cosmic string A hypothetical long, thin, wirelike concentration of matter left over from the Big Bang. (Chapter 29)

cosmological constant (Λ) In the equations of general relativity, a quantity signifying a pressure throughout all space that helps the universe expand; a type of dark energy. (Chapter 28)

cosmological principle The assumption that the universe is homogeneous and isotropic on the largest scale. (Chapter 28)

cosmological redshift A redshift that is caused by the expansion of the universe. (Chapter 28)

cosmology The study of the structure and evolution of the universe. (Chapter 28)

coudé focus An optical arrangement with a reflecting telescope. A series of mirrors is used to direct light to a remote focus away from the moving parts of the telescope. (Chapter 6)

crater A circular depression on a planet or satellite caused by the impact of a meteoroid. (Chapter 9)

critical density The combined average mass density throughout the universe at which space would be flat. (Chapter 28)

crust (of a planet) The surface layer of a terrestrial planet. (Chapter 8)

crustal dichotomy (Mars) The contrast between the young northern lowlands and older southern highlands on Mars. (Chapter 12)

crystal A material in which atoms are arranged in orderly rows. (Chapter 8)

current sheet A broad, flat region in Jupiter's magnetosphere that contains an abundance of charged particles. (Chapter 13)

D ring One of several faint rings encircling Saturn. (Chapter 15)

dark energy A form of energy that appears to pervade the universe but has no discernible gravitational effect. (Chapter 28)

dark energy density parameter (Ω_Λ) The ratio of the average mass density of dark energy in the universe to the critical density. (Chapter 28)

dark-energy-dominated universe A universe in which the mass density of dark energy exceeds both the average density of matter and the mass density of radiation. (Chapter 28)

dark matter Nonluminous matter that appears to be quite abundant in galaxies and throughout the universe. (Chapter 25)

dark-matter problem The enigma that most of the matter in the universe seems not to emit radiation of any kind and is detectable by its gravity alone. (Chapter 26)

dark nebula A cloud of interstellar gas and dust that obscures the light of more distant stars. (Chapter 20)

dark terrain (Ganymede) Older, heavily cratered, dark-colored terrain on the surface of Ganymede. (Chapter 14)

decametric radiation Radiation from Jupiter whose wavelength is about 10 meters. (Chapter 13)

deceleration parameter (q_0) A number whose value indicates the rate at which the expansion of the universe is slowing down (if $q_0 > 0$) or speeding up (if $q_0 < 0$). (Chapter 28)

decimetric radiation Radiation from Jupiter whose wavelength is about a tenth of a meter. (Chapter 13)

declination Angular distance of a celestial object north or south of the celestial equator. (Chapter 2)

deferent A stationary circle in the Ptolemaic system along which another circle (an epicycle) moves, carrying a planet, the Sun, or the Moon. (Chapter 4)

degeneracy The phenomenon, due to quantum mechanical effects, whereby the pressure exerted by a gas does not depend on its temperature. (Chapter 21)

degenerate electron (neutron) pressure The pressure exerted by degenerate electrons (neutrons). (Chapter 21, Chapter 23)

degree A basic unit of angular measure, designated by the symbol °. (Chapter 1)

degree Celsius A basic unit of temperature, designated by the symbol °C and used on a scale where water freezes at 0° and boils at 100°. (Chapter 5)

degree Fahrenheit A basic unit of temperature, designated by the symbol °F and used on a scale where water freezes at 32° and boils at 212°. (Chapter 5)

density See *average density*.

density fluctuation A variation of density from one place to another. (Chapter 29)

density parameter (Ω_0) The ratio of the average density of matter and energy in the universe to the critical density. (Chapter 28)

density wave In a spiral galaxy, a localized region in which matter piles up as it orbits the center of the galaxy; this is a proposed explanation of spiral structure. (Chapter 25)

detached binary A binary star system in which neither star fills its Roche lobe. (Chapter 21)

deuterium bottleneck The situation during the first 3 minutes after the Big Bang when deuterium (an isotope of hydrogen whose nucleus contains one proton and one neutron) inhibited the formation of heavier elements. (Chapter 29)

differential rotation The rotation of a nonrigid object in which parts adjacent to each other at a given time do not always stay close together. (Chapter 13, Chapter 18)

differentiated asteroid An asteroid in which chemical differentiation has taken place, so that denser material is toward the asteroid's center. (Chapter 17)

differentiation See *chemical differentiation*.

diffraction The spreading out of light passing through an aperture or opening in an opaque object. (Chapter 6)

diffraction grating An optical device, consisting of thousands of closely spaced lines etched in glass or metal, that disperses light into a spectrum. (Chapter 6)

direct motion The apparent eastward movement of a planet seen against the background stars. (Chapter 4)

disk (of a galaxy) The disk-shaped distribution of Population I stars that dominates the appearance of a spiral galaxy. (Chapter 25)

distance ladder The sequence of techniques used to determine the distances to very remote galaxies. (Chapter 26)

distance modulus The difference between the apparent and absolute magnitudes of an object. (Chapter 19)

diurnal motion Any apparent motion in the sky that repeats on a daily basis, such as the rising and setting of stars. (Chapter 2)

Doppler effect The apparent change in wavelength of radiation due to relative motion between the source and the observer along the line of sight. (Chapter 5)

double radio source An extragalactic radio source characterized by two large regions of radio emission, typically located on either side of an active galaxy. (Chapter 27)

double star A pair of stars located at nearly the same position in the night sky. Some, but not all, double stars are binary stars. (Chapter 19)

Drake equation An equation used to estimate the number of intelligent civilizations in the Galaxy with which we might communicate. (Chapter 30)

dredge-up The transporting of matter from deep within a star to its surface by convection. (Chapter 22)

dust devil Whirlwind found in dry or desert areas on both Earth and Mars. (Chapter 12)

dust grain A microscopic bit of matter found in interplanetary or interstellar space. (Chapter 20)

dust tail The tail of a comet that is composed primarily of dust grains. (Chapter 17)

dwarf elliptical galaxy A low-mass elliptical galaxy that contains only a few million stars. (Chapter 26)

E ring A very broad, faint ring encircling Saturn. (Chapter 15)

earthquake A sudden vibratory motion of the Earth's surface. (Chapter 8)

eccentricity A number between 0 and 1 that describes the shape of an ellipse. (Chapter 4)

eclipse The cutting off of part of or all the light from one celestial object by another. (Chapter 3)

eclipse path The track of the tip of the Moon's shadow along the Earth's surface during a total or annular solar eclipse. (Chapter 3)

eclipse year The interval between successive passages of the Sun through the same node of the Moon's orbit. (Chapter 3)

eclipsing binary A binary star system in which, as seen from Earth, the stars periodically pass in front of each other. (Chapter 19)

ecliptic The apparent annual path of the Sun on the celestial sphere. (Chapter 2)

Eddington limit A constraint on the rate at which a black hole can accrete matter. (Chapter 27)

electromagnetic radiation Radiation consisting of oscillating electric and magnetic fields. Examples include gamma rays, X rays, visible light, ultraviolet and infrared radiation, radio waves, and microwaves. (Chapter 5)

electromagnetic spectrum The entire array of electromagnetic radiation. (Chapter 5)

electromagnetism Electric and magnetic phenomena, including electromagnetic radiation. (Chapter 5)

electron A subatomic particle with a negative charge and a small mass, usually found in orbits about the nuclei of atoms. (Chapter 5)

electron volt (eV) The energy acquired by an electron accelerated through an electric potential of one volt. (Chapter 5)

element A chemical that cannot be broken down into more basic chemicals. (Chapter 5)

elementary particle physics The branch of quantum mechanics that deals with individual subatomic particles and their interactions. (Chapter 29)

ellipse A conic section obtained by cutting completely through a circular cone with a plane. (Chapter 4)

elliptical galaxy A galaxy with an elliptical shape and no conspicuous interstellar material. (Chapter 26)

elongation The angular distance between a planet and the Sun as viewed from Earth. (Chapter 4)

emission line spectrum A spectrum that contains bright emission lines. (Chapter 5)

emission nebula A glowing gaseous nebula whose spectrum has bright emission lines. (Chapter 20)

Encke gap A narrow gap in Saturn's A ring. (Chapter 15)

energy flux The rate of energy flow, usually measured in joules per square meter per second. (Chapter 5)

energy level In an atom, a particular amount of energy possessed by an atom above the atom's least energetic state. (Chapter 5)

energy-level diagram A diagram showing the arrangement of an atom's energy levels. (Chapter 5)

epicenter The location on the Earth's surface directly over the focus of an earthquake. (Chapter 8)

epicycle A moving circle in the Ptolemaic system about which a planet revolves. (Chapter 4)

epoch The date used to define the coordinate system for objects on the sky. (Chapter 2)

equal areas, law of See *Kepler's second law.*

equilibrium resurfacing hypothesis (Venus) A proposed explanation for the young age of Venus's surface. It hypothesizes that volcanic eruptions on Venus are continually covering up craters at about the same rate that craters are formed by impact. (Chapter 11)

equinox One of the intersections of the ecliptic and the celestial equator. Also used to refer to the date on which the Sun passes through such an intersection. (Chapter 2) See also *autumnal equinox; vernal equinox.*

equivalence principle In the general theory of relativity, the principle that in a small volume of space, the downward pull of gravity can be accurately and completely duplicated by an upward acceleration of the observer. (Chapter 24)

era of recombination The moment, approximately 300,000 years after the Big Bang, when the universe had cooled sufficiently to permit the formation of hydrogen atoms. (Chapter 28)

ergoregion The region of space immediately outside the event horizon of a rotating black hole where it is impossible to remain at rest. (Chapter 24)

escape speed The speed needed by an object (such as a spaceship) to leave a second object (such as a planet or star) permanently and to escape into interplanetary space. (Chapter 7)

eV See *electron volt*.

event horizon The location around a black hole where the escape speed equals the speed of light; the surface of a black hole. (Chapter 24)

evolutionary track The path on an H-R diagram followed by a star as it evolves. (Chapter 20)

excited state A state of an atom, ion, or molecule with a higher energy than the ground state. (Chapter 5)

exobiology The biology of life on other planets. (Chapter 30)

exponent A number placed above and after another number to denote the power to which the latter is to be raised, as n in 10^n. (Chapter 1)

extinction (interstellar) See *interstellar extinction*.

extrasolar planet A planet orbiting a star other the Sun. (Chapter 7)

eyepiece lens A magnifying lens used to view the image produced at the focus of a telescope. (Chapter 6)

F ring A thin, faint ring encircling Saturn just beyond the A ring. (Chapter 15)

false vacuum Empty space with an abnormally high energy density that possibly filled the universe immediately after the Big Bang. (Chapter 29)

far-infrared The part of the infrared spectrum most different in wavelength from visible light. (Chapter 25)

far side (of the Moon) The side of the Moon that faces perpetually away from the Earth. (Chapter 9)

favorable opposition An opposition of Mars that affords good Earth-based views of the planet. (Chapter 12)

filament A portion of the Sun's chromosphere that arches to high altitudes. (Chapter 18)

first quarter moon The phase of the Moon that occurs when the Moon is 90° east of the Sun. (Chapter 3)

fission theory The hypothesis that the Moon was pulled out of a rapidly rotating proto-Earth. (Chapter 9)

flare See *solar flare*.

flat space Space that is not curved; space with zero curvature. (Chapter 28)

flatness problem The dilemma posed by the fact that the combined average mass density of the universe is very nearly equal to the critical density. (Chapter 29)

flocculent spiral galaxy A spiral galaxy with fuzzy, poorly defined spiral arms. (Chapter 25)

fluorescence A process in which high-energy ultraviolet photons are absorbed and the absorbed energy is radiated as lower-energy photons of visible light. (Chapter 20)

focal length The distance from a lens or mirror to the point where converging light rays meet. (Chapter 6)

focal plane The plane in which a lens or mirror forms an image of a distant object. (Chapter 6)

focal point The point at which a lens or mirror forms an image of a distant point of light. (Chapter 6)

focus (of an ellipse) (plural **foci**) One of two points inside an ellipse such that the combined distance from the two foci to any point on the ellipse is a constant. (Chapter 4)

focus (of a lens or mirror) The point to which light rays converge after passing through a lens or being reflected from a mirror. (Chapter 6)

force A push or pull that acts on an object. (Chapter 4)

frequency The number of crests or troughs of a wave that cross a given point per unit time. Also, the number of vibrations per unit time. (Chapter 5)

full moon A phase of the Moon during which its full daylight hemisphere can be seen from Earth. (Chapter 3)

fundamental plane A relationship among the size of an elliptical galaxy, the average motions of its stars, and how the galaxy's brightness appears distributed over its surface. (Chapter 26)

fusion crust The coating on a stony meteorite caused by the heating of the meteorite as it descended through the Earth's atmosphere. (Chapter 17)

G ring A thin, faint ring encircling Saturn. (Chapter 15)

galactic cannibalism A collision between two galaxies of unequal mass and size in which the smaller galaxy seems to be absorbed by the larger galaxy. (Chapter 26)

galactic cluster See *open cluster*.

galactic nucleus The center of a galaxy. (Chapter 25)

galaxy A large assemblage of stars, nebulae, and interstellar gas and dust. (Chapter 1, Chapter 25)

Galilean satellites The four large moons of Jupiter. (Chapter 14)

gamma-ray bursters Objects found in all parts of the sky that emit intense bursts of high-energy radiation. (Chapter 27)

gamma rays The most energetic form of electromagnetic radiation. (Chapter 5)

general theory of relativity A description of gravity formulated by Albert Einstein. It explains that gravity affects the geometry of space and the flow of time. (Chapter 24)

geocentric model An Earth-centered theory of the universe. (Chapter 4)

giant A star whose diameter is typically 10 to 100 times that of the Sun and whose luminosity is roughly that of 100 Suns. (Chapter 19)

giant elliptical galaxy A large, massive, elliptical galaxy containing many billions of stars. (Chapter 26)

giant molecular cloud A large cloud of interstellar gas and dust in which temperatures are low enough and densities high enough for atoms to form into molecules. (Chapter 20)

glitch A sudden speedup in the period of a pulsar. (Chapter 23)

global catastrophe hypothesis (Venus) A proposed explanation for the young age of Venus's surface. It hypothesizes that the entire planet was resurfaced over a short period with fresh lava, covering up any older craters. (Chapter 11)

global warming The upward trend of the Earth's average temperature caused by increased amounts of greenhouse gases in the atmosphere. (Chapter 8)

globular cluster A large spherical cluster of stars, typically found in the outlying regions of a galaxy. (Chapter 21, Chapter 25)

gluon A particle that is exchanged between quarks. (Chapter 29)

grand-design spiral galaxy A galaxy with well-defined spiral arms. (Chapter 25)

grand unified theory (GUT) A theory that describes and explains the four physical forces. (Chapter 29)

granulation The rice grain–like structure found in the solar photosphere. (Chapter 18)

granule A convective cell in the solar photosphere. (Chapter 18)

grating See *diffraction grating.*

gravitational force See *gravity.*

gravitational lens A massive object that deflects light rays from a remote source, forming an image much as an ordinary lens does. (Chapter 26)

gravitational radiation (gravitational waves) Oscillations of space produced by changes in the distribution of matter. (Chapter 24)

gravitational redshift The increase in the wavelength of a photon as it climbs upward in a gravitational field. (Chapter 24)

graviton The particle that is responsible for the gravitational force. (Chapter 29)

gravity The force with which all matter attracts all other matter. (Chapter 4)

Great Attractor A huge concentration of matter toward which many galaxies, including the Milky Way, appear to be moving. (Chapter 28)

Great Dark Spot A prominent high-pressure system that once existed in Neptune's southern hemisphere. (Chapter 16)

Great Red Spot A prominent high-pressure system in Jupiter's southern hemisphere. (Chapter 13)

greatest eastern elongation The configuration of an inferior planet at its greatest angular distance east of the Sun. (Chapter 4, Chapter 10)

greatest western elongation The configuration of an inferior planet at its greatest angular distance west of the Sun. (Chapter 4, Chapter 10)

greenhouse effect The trapping of infrared radiation near a planet's surface by the planet's atmosphere. (Chapter 8)

greenhouse gas A substance whose presence in a planet's atmosphere enhances the greenhouse effect. (Chapter 8)

ground state The state of an atom, ion, or molecule with the least possible energy. (Chapter 5)

group (of galaxies) A poor cluster of galaxies. (Chapter 26)

GUT See *grand unified theory.*

H I Neutral, unionized hydrogen. (Chapter 25)

H II region A region of ionized hydrogen in interstellar space. (Chapter 20)

halo (of a galaxy) A spherical distribution of globular clusters and Population II stars that surround a spiral galaxy. (Chapter 25)

head-tail source A radio source consisting of a bright "head" and a long, dim "tail." (Chapter 27)

Heisenberg uncertainty principle A principle of quantum mechanics that places limits on the precision of simultaneous measurements. (Chapter 24, Chapter 29)

heliocentric model A Sun-centered theory of the universe. (Chapter 4)

helioseismology The study of the vibrations of the Sun as a whole. (Chapter 18)

helium burning The thermonuclear fusion of helium to form carbon and oxygen. (Chapter 21)

helium flash The nearly explosive beginning of helium burning in the dense core of a red-giant star. (Chapter 21)

helium shell flash A brief thermal runaway that occurs in the helium-burning shell of a red supergiant. (Chapter 22)

Herbig-Haro object A small, luminous nebula associated with the end point of a jet emanating from a young star. (Chapter 20)

Hertzsprung-Russell (H-R) diagram A plot of the luminosity (or absolute magnitude) of stars against their surface temperature (or spectral type). (Chapter 19)

high-velocity star A star traveling at an exceptionally high speed relative to the Sun. (Chapter 25)

highlands (on Mars) See *southern highlands*.

highlands (on the Moon) See *lunar highlands*.

Hirayama family A group of asteroids with nearly identical orbits about the Sun. (Chapter 17)

homogeneous Having the same property in one region as in every other region. (Chapter 28)

horizon problem See *isotropy problem*.

horizontal branch A group of stars on the color-magnitude diagram of a typical globular cluster that lie near the main sequence and have roughly constant luminosity. (Chapter 22)

horizontal-branch star A low-mass, post–helium-flash star on the horizontal branch. (Chapter 21)

hot dark matter Dark matter consisting of particles moving at high speeds. (Chapter 29)

hot spot (on Jupiter) An unusually warm and cloud-free part of Jupiter's atmosphere. (Chapter 13)

hot-spot volcanism Volcanic activity that occurs over a hot region buried deep within a planet. (Chapter 11)

H-R diagram See *Hertzsprung-Russell diagram*.

Hubble classification A method of classifying galaxies as spirals, barred spirals, ellipticals, or irregulars according to their appearance. (Chapter 26)

Hubble constant (H_0) In the Hubble law, the constant of proportionality between the recessional velocities of remote galaxies and their distances. (Chapter 26)

Hubble flow The recessional motions of remote galaxies caused by the expansion of the universe. (Chapter 26)

Hubble law The empirical relationship stating that the redshifts of remote galaxies are directly proportional to their distances from Earth. (Chapter 26)

hydrocarbon Any one of a variety of chemical compounds composed of hydrogen and carbon. (Chapter 15)

hydrogen burning The thermonuclear conversion of hydrogen into helium. (Chapter 18)

hydrogen envelope A huge, tenuous sphere of gas surrounding the head of a comet. (Chapter 17)

hydrostatic equilibrium A balance between the weight of a layer in a star and the pressure that supports it. (Chapter 18)

hyperbola A conic section formed by cutting a circular cone with a plane at an angle steeper than the side of the cone. (Chapter 4)

hyperbolic space Space with negative curvature. (Chapter 28)

hypothesis An idea or collection of ideas that seems to explain a specified phenomenon; a conjecture. (Chapter 1)

ice rafts (Europa) Segments of Europa's icy crust that have been moved by tectonic disturbances. (Chapter 14)

ices Solid materials with low condensation temperatures, including ices of water, methane, and ammonia. (Chapter 7)

ideal gas A gas in which the pressure is directly proportional to both the density and the temperature of the gas; an idealization of a real gas. (Chapter 21)

igneous rock A rock that formed from the solidification of molten lava or magma. (Chapter 8)

imaging The process of recording the image made by a telescope of a distant object. (Chapter 6)

impact breccia A type of rock formed from other rocks that were broken apart, mixed, and fused together by a series of meteoritic impacts. (Chapter 9)

impact crater A crater formed by the impact of a meteoroid. (Chapter 9)

inertia, law of See *Newton's first law of motion*.

inferior conjunction The configuration when an inferior planet is between the Sun and Earth. (Chapter 4)

inferior planet A planet that is closer to the Sun than the Earth is. (Chapter 4)

inflation A sudden expansion of space. (Chapter 29)

inflationary epoch A brief period shortly after the Big Bang during which the scale of the universe increased very rapidly. (Chapter 29)

infrared radiation Electromagnetic radiation of wavelength longer than visible light but shorter than radio waves. (Chapter 5)

inner core (of the Earth) The solid innermost portion of the Earth's iron-rich core. (Chapter 8)

inner Lagrangian point The point between the two stars constituting a binary system where their Roche lobes touch; the point across which mass transfer can occur. (Chapter 21)

instability strip A region of the H-R diagram occupied by pulsating stars. (Chapter 21)

interferometry A technique of combining the observations of two or more telescopes to produce images better than one telescope alone could make. (Chapter 6)

intermediate-period comet A comet with an orbital period between 20 and 200 years. (Chapter 17)

intermediate vector boson The particle that is responsible for the weak force. (Chapter 29)

internal rotation period The period with which the core of a Jovian planet rotates. (Chapter 13)

interstellar extinction The dimming of starlight as it passes through the interstellar medium. (Chapter 20, Chapter 25)

interstellar medium Gas and dust in interstellar space. (Chapter 7, Chapter 20)

interstellar reddening The reddening of starlight passing through the interstellar medium as a result of blue light being scattered more than red. (Chapter 20)

intracluster gas Gas found between the galaxies that make up a cluster of galaxies. (Chapter 26)

inverse-square law The statement that the apparent brightness of a light source varies inversely with the square of the distance from the source. (Chapter 19)

Io torus A doughnut-shaped ring of gas circling Jupiter at the distance of Io's orbit. (Chapter 14)

ion tail The relatively straight tail of a comet produced by the solar wind acting on ions. (Chapter 17)

ionization The process by which a neutral atom becomes an electrically charged ion through the loss or gain of electrons. (Chapter 5)

iron meteorite A meteorite composed primarily of iron. (Chapter 17)

irregular cluster (of galaxies) A sprawling collection of galaxies whose overall distribution in space does not exhibit any noticeable spherical symmetry. (Chapter 17)

irregular galaxy An asymmetrical galaxy having neither spiral arms nor an elliptical shape. (Chapter 26)

isotope Any of several forms for the same chemical element whose nuclei all have the same number of protons but different numbers of neutrons. (Chapter 5)

isotropic Having the same property in all directions. (Chapter 26, Chapter 28)

isotropy problem The dilemma posed by the fact that the cosmic microwave background is isotropic; also called the horizon problem. (Chapter 29)

Jeans length The smallest scale over which a density fluctuation in a medium will contract to form a gravitationally bound object. (Chapter 29)

jet An extended line of fast-moving gas ejected from the vicinity of a star or a black hole. (Chapter 7)

joule (J) A unit of energy. (Chapter 5)

Jovian planet Any of the four largest planets: Jupiter, Saturn, Uranus, or Neptune. (Chapter 7)

Jupiter-family comet A comet with an orbital period of less than 20 years. (Chapter 17)

Kaluza-Klein theory A theory that attempts to describe the four physical forces in terms of the geometry of spacetime with more than four dimensions. (Chapter 29)

kelvin (K) A unit of temperature on the Kelvin temperature scale, equivalent to a degree Celsius. (Chapter 5)

Kelvin-Helmholtz contraction The contraction of a gaseous body, such as a star or nebula, during which gravitational energy is transformed into thermal energy. (Chapter 7)

Kepler's first law The statement that each planet moves around the Sun in an elliptical orbit with the Sun at one focus of the ellipse. (Chapter 4)

Kepler's second law The statement that a planet sweeps out equal areas in equal times as it orbits the Sun; also called the law of equal areas. (Chapter 4)

Kepler's third law A relationship between the period of an orbiting object and the semimajor axis of its elliptical orbit. (Chapter 4)

kiloparsec (kpc) One thousand parsecs; about 3260 light-years. (Chapter 1)

kinetic energy The energy possessed by an object because of its motion. (Chapter 7)

Kirchhoff's laws Three statements about circumstances that produce absorption lines, emission lines, and continuous spectra. (Chapter 5)

Kirkwood gaps Gaps in the spacing of asteroid orbits, discovered by Daniel Kirkwood. (Chapter 17)

Kuiper belt A region that extends from around the orbit of Pluto to about 500 AU from the Sun, where many icy objects orbit the Sun. (Chapter 17)

Lamb shift A tiny shift in the spectral lines of hydrogen caused by the presence of virtual particles. (Chapter 29)

last quarter moon The phase of the Moon that occurs when the Moon is 90° west of the Sun. (Chapter 3)

lava Molten rock flowing on the surface of a planet. (Chapter 8)

law of cosmic censorship The hypothesis that all singularities must be surrounded by an event horizon. (Chapter 24)

law of equal areas See *Kepler's second law.*

law of inertia See *Newton's first law of motion.*

law of universal gravitation A formula deduced by Isaac Newton that expresses the strength of the force of gravity that two masses exert on each other. (Chapter 4)

laws of physics A set of physical principles with which we can understand natural phenomena and the nature of the universe. (Chapter 1)

length contraction In the special theory of relativity, the shrinking of an object's length along its direction of motion. (Chapter 24)

lenticular galaxy A galaxy with a central bulge and a disk but no spiral structure; an S0 galaxy. (Chapter 26)

libration An apparent rocking of the Moon whereby an Earth-based observer can, over time, see slightly more than one-half the Moon's surface. (Chapter 9)

light curve A graph that displays how the brightness of a star or other astronomical object varies over time. (Chapter 19)

light-gathering power A measure of the amount of radiation brought to a focus by a telescope. (Chapter 6)

light pollution Light from cities and towns that degrades telescope images. (Chapter 6)

light scattering The process by which light bounces off particles in its path. (Chapter 5, Chapter 15)

light-year (ly) The distance light travels in a vacuum in one year. (Chapter 1)

limb darkening The phenomenon whereby the Sun looks darker near its apparent edge, or limb, than near the center of its disk. (Chapter 18)

line of nodes The line where the plane of the Earth's orbit intersects the plane of the Moon's orbit. (Chapter 3)

liquid metallic hydrogen Hydrogen compressed to such a density that it behaves like a liquid metal. (Chapter 13)

lithosphere The solid, upper layer of the Earth; essentially the Earth's crust. (Chapter 8)

Local Bubble A large cavity in the interstellar medium within which the Sun and nearby stars are located. (Chapter 25)

Local Group The cluster of galaxies of which our Galaxy is a member. (Chapter 26)

local meridian See *meridian*.

long-period comet A comet that takes hundreds of thousands of years or more to complete one orbit of the Sun. (Chapter 17)

long-period variable A variable star with a period longer than about 100 days. (Chapter 21)

lookback time How far into the past we are looking when we see a particular object. (Chapter 28)

Lorentz transformations Equations that relate the measurements of different observers who are moving relative to each other at high speeds. (Chapter 24)

lower meridian The half of the meridian that lies below the horizon. (Chapter 2)

lowlands (on Mars) See *northern lowlands*.

luminosity The rate at which electromagnetic radiation is emitted from a star or other object. (Chapter 5, Chapter 18, Chapter 19)

luminosity class A classification of a star of a given spectral type according to its luminosity. (Chapter 19)

luminosity function The numbers of stars of differing brightness per cubic parsec. (Chapter 19)

lunar eclipse An eclipse of the Moon by the Earth; a passage of the Moon through the Earth's shadow. (Chapter 3)

lunar highlands Ancient, high-elevation, heavily cratered terrain on the Moon. (Chapter 9)

lunar month See *synodic month*.

lunar phase The appearance of the illuminated area of the Moon as seen from Earth. (Chapter 3)

Lyman series A series of spectral lines of hydrogen produced by electron transitions to and from the lowest energy state of the hydrogen atom. (Chapter 5)

M-theory A theory that attempts to describe fundamental particles as 11-dimensional membranes. (Chapter 29)

MACHO See *massive compact halo object*.

magma Molten rock beneath a planet's surface. (Chapter 8)

magnetic axis A line connecting the north and south magnetic poles of a planet or star possessing a magnetic field. (Chapter 16)

magnetic-dynamo model A theory that explains the solar cycle as a result of the Sun's differential rotation acting on the Sun's magnetic field. (Chapter 18)

magnetic resonance imaging (MRI) A technique for viewing the interior of the human body that makes use of how protons respond to magnetic fields. (Chapter 25)

magnetogram An artificial picture of the Sun that shows regions of different magnetic polarity. (Chapter 18)

magnetopause That region of a planet's magnetosphere where the magnetic field counterbalances the pressure from the solar wind. (Chapter 8)

magnetosphere The region around a planet occupied by its magnetic field. (Chapter 8)

magnification The factor by which the apparent angular size of an object is increased when viewed through a telescope. (Chapter 6)

magnifying power See *magnification*.

magnitude scale A system for denoting the brightnesses of astronomical objects. (Chapter 19)

main sequence A grouping of stars on the Hertzsprung-Russell diagram extending diagonally across the graph from hot, luminous stars to cool, dim stars. (Chapter 19)

main-sequence lifetime The total time that a star spends burning hydrogen in its core, and hence the total time that it will spend as a main-sequence star. (Chapter 21)

main-sequence star A star whose luminosity and surface temperature place it on the main sequence on an H-R diagram; a star that derives its energy from core hydrogen burning. (Chapter 19)

major axis (of an ellipse) The longest diameter of an ellipse. (Chapter 4)

mantle (of a planet) That portion of a terrestrial planet located between its crust and core. (Chapter 8)

mare (plural **maria**) Latin for "sea"; a large, relatively crater-free plain on the Moon. (Chapter 9)

mare basalt A type of lunar rock commonly found in the mare basins. (Chapter 9)

marginally bounded universe A universe in which the rate of expansion continually decreases but never stops, so that it just barely manages to continue expanding forever. (Chapter 28)

maser An interstellar cloud in which water molecules emit intensely at microwave wavelengths. (Chapter 26)

mass A measure of the total amount of material in an object. (Chapter 4)

mass density of radiation The energy possessed by a radiation field per unit volume divided by the square of the speed of light. (Chapter 28)

mass loss A process by which a star gently loses matter. (Chapter 21)

mass-luminosity relation A relationship between the masses and luminosities of main-sequence stars. (Chapter 19)

mass-radius relation A relationship between the masses and radii of white dwarf stars. (Chapter 22)

mass transfer The flow of gases from one star in a binary system to the other. (Chapter 21)

massive compact halo object (MACHO) A dim star or low-mass black hole that may constitute part of the unseen dark matter. (Chapter 25)

matter density parameter (Ω_m) The ratio of the average density of matter in the universe to the critical density. (Chapter 28)

matter-dominated universe A universe in which the average density of matter exceeds both the mass density of radiation and the mass density of dark energy. (Chapter 28)

mean solar day The interval between successive meridian passages of the mean Sun; the average length of a solar day. (Chapter 2)

mean Sun A fictitious object that moves eastward at a constant speed along the celestial equator, completing one circuit of the sky with respect to the vernal equinox in one tropical year. (Chapter 2)

medium (plural **media**) A material through which light travels. (Chapter 6)

medium, interstellar See *interstellar medium*.

megaparsec (Mpc) One million parsecs. (Chapter 1)

melting point The temperature at which a substance changes from solid to liquid. (Chapter 8)

meridian (or local meridian) The great circle on the celestial sphere that passes through an observer's zenith and the north and south celestial poles. (Chapter 2)

meridian transit The crossing of the meridian by any astronomical object. (Chapter 2)

mesosphere A layer in a planet's atmosphere above the stratosphere. (Chapter 8)

metal In reference to stars and galaxies, any elements other than hydrogen and helium; in reference to planets, a material such as iron, silver, and aluminum that is a good conductor of electricity and of heat. (Chapter 19)

metal-poor star A star that, compared to the Sun, is deficient in elements heavier than helium; a Population II star. (Chapter 21)

metal-rich star A star whose abundance of heavy elements is roughly comparable to that of the Sun; a Population I star. (Chapter 21)

metamorphic rock A rock whose properties and appearance have been transformed by the action of pressure and heat beneath the Earth's surface. (Chapter 8)

meteor The luminous phenomenon seen when a meteoroid enters the Earth's atmosphere; a "shooting star." (Chapter 17)

meteor shower Many meteors that seem to radiate from a common point in the sky. (Chapter 17)

meteorite A fragment of a meteoroid that has survived passage through the Earth's atmosphere. (Chapter 7, Chapter 17)

meteoritic swarm A collection of meteoroids moving together along an orbit about the Sun. (Chapter 17)

meteoroid A small rock in interplanetary space. (Chapter 17)

microlensing A phenomenon in which a compact object such as a MACHO acts as a gravitational lens, focusing the light from a distant star. (Chapter 25)

microwaves Short-wavelength radio waves. (Chapter 5)

Milky Way Galaxy Our Galaxy; the band of faint stars seen from the Earth in the plane of our Galaxy's disk. (Chapter 25)

millisecond pulsar A pulsar with a period of roughly 1 to 10 milliseconds. (Chapter 23)

mineral A naturally occurring solid composed of a single element or chemical combination of elements, often in the form of crystals. (Chapter 8)

minor planet See *asteroid*.

minute of arc See *arcminute*.

model A hypothesis that has withstood experimental or observational tests; the results of a theoretical calculation that gives the values of temperature, pressure, density, and so forth throughout the interior of an object such as a planet or star. (Chapter 1)

molecule A combination of two or more atoms. (Chapter 7)

moonquake   Sudden, vibratory motion of the Moon's surface. (Chapter 9)

MRI   See *magnetic resonance imaging.*

nanometer (nm)   One billionth of a meter: 1 nm = 10^{-9} meter = 10^{-6} millimeter = 10^{-3} μm. (Chapter 5)

neap tide   An ocean tide that occurs when the Moon is near first-quarter or third-quarter phase. (Chapter 4)

near-Earth object (NEO)   An asteroid whose orbit lies wholly or partly within the orbit of Mars. (Chapter 17)

near-infrared   The part of the infrared spectrum closest in wavelength to visible light. (Chapter 25)

nebula   A cloud of interstellar gas and dust. (Chapter 1)

nebulosity   See *nebula.*

negative curvature   The curvature of a surface or space in which parallel lines diverge and the sum of the angles of a triangle is less than 180°. (Chapter 28)

negative hydrogen ion   A hydrogen atom that has acquired a second electron. (Chapter 18)

NEO   See *near-Earth object.*

neon burning   The thermonuclear fusion of neon to produce heavier nuclei. (Chapter 22)

neutrino   A subatomic particle with no electric charge and little or no mass, yet one that is important in many nuclear reactions. (Chapter 18)

neutron   A subatomic particle with no electric charge and with a mass nearly equal to that of the proton. (Chapter 5)

neutron capture   The buildup of neutrons inside nuclei. (Chapter 22)

neutron star   A very compact, dense star composed almost entirely of neutrons. (Chapter 23)

new moon   The phase of the Moon when the dark hemisphere of the Moon faces the Earth. (Chapter 3)

Newtonian mechanics   The branch of physics based on Newton's laws of motion. (Chapter 1, Chapter 4)

Newtonian reflector   A reflecting telescope that uses a small mirror to deflect the image to one side of the telescope tube. (Chapter 6)

Newton's first law of motion   The statement that a body remains at rest, or moves in a straight line at a constant speed, unless acted upon by a net outside force; the law of inertia. (Chapter 4)

Newton's form of Kepler's third law   A relationship between the period of two objects orbiting each other, the semimajor axis of their orbit, and the masses of the objects. (Chapter 4)

Newton's second law of motion   A relationship between the acceleration of an object, the object's mass, and the net outside force acting on the mass. (Chapter 4)

Newton's third law of motion   The statement that whenever one body exerts a force on a second body, the second body exerts an equal and opposite force on the first body. (Chapter 4)

no-hair theorem   A statement of the simplicity of black holes. (Chapter 24)

noble gas   An element whose atoms do not combine into molecules. (Chapter 13)

node   See *line of nodes.*

nonthermal radiation   Radiation other than that emitted by a heated body. (Chapter 13, Chapter 27)

north celestial pole   The point directly above the Earth's North Pole where the Earth's axis of rotation, if extended, would intersect the celestial sphere. (Chapter 2)

northern lights   See *aurora borealis.*

northern lowlands (on Mars)   Relatively young and crater-free terrain in the Martian northern hemisphere. (Chapter 12)

nova (plural novae)   A star that experiences a sudden outburst of radiant energy, temporarily increasing its luminosity roughly a thousandfold. (Chapter 23)

nuclear density   The density of matter in an atomic nucleus; about 4×10^{17} kg/m^3. (Chapter 22)

nucleosynthesis   The process of building up nuclei such as deuterium and helium from protons and neutrons. (Chapter 29)

nucleus (of an atom)   The massive part of an atom, composed of protons and neutrons, about which electrons revolve. (Chapter 5)

nucleus (of a comet)   A collection of ices and dust that constitutes the solid part of a comet. (Chapter 17)

nucleus (of a galaxy)   See *galactic nucleus.*

OB association   A grouping of hot, young, massive stars, predominantly of spectral types O and B. (Chapter 20)

OBAFGKM   The temperature sequence (from hot to cold) of spectral types. (Chapter 19)

objective lens   The principal lens of a refracting telescope. (Chapter 6)

objective mirror   The principal mirror of a reflecting telescope. (Chapter 6)

oblate   Flattened at the poles. (Chapter 13)

oblateness   A measure of how much a flattened sphere (or spheroid) differs from a perfect sphere. (Chapter 13)

observable universe   That portion of the universe inside the cosmic particle horizon. (Chapter 28)

Occam's razor   The notion that a straightforward explanation of a phenomenon is more likely to be correct than a convoluted one. (Chapter 4)

occultation The eclipsing of an astronomical object by the Moon or a planet. (Chapter 14, Chapter 16)

oceanic rift A crack in the ocean floor that exudes lava. (Chapter 8)

Olbers's paradox The dilemma associated with the fact that the night sky is dark. (Chapter 28)

1-to-1 spin-orbit coupling See *synchronous rotation*.

Oort cloud A presumed accumulation of comets and cometary material surrounding the Sun at distances of roughly 50,000 to 100,000 AU. (Chapter 17)

open cluster A loose association of young stars in the disk of our Galaxy; a galactic cluster. (Chapter 20)

open universe A universe with negative curvature, so that its geometry is analogous to that of a saddle-shaped hyperbolic surface. (Chapter 28)

opposition The configuration of a planet when it is at an elongation of 180° and thus appears opposite the Sun in the sky. (Chapter 4)

optical double star Two stars that lie along nearly the same line of sight but are actually at very different distances from us. (Chapter 19)

optical telescope A telescope designed to detect visible light. (Chapter 6)

optical window The range of visible wavelengths to which the Earth's atmosphere is transparent. (Chapter 6)

organic molecules Molecules containing carbon, some of which are the molecules of which living organisms are made. (Chapter 30)

outer core (of the Earth) The outer, molten portion of the Earth's iron-rich core. (Chapter 8)

outgassing The release of gases into a planet's atmosphere by volcanic activity. (Chapter 8)

overcontact binary A close binary system in which the two stars share a common atmosphere. (Chapter 21)

oxygen burning The thermonuclear fusion of oxygen to produce heavier nuclei. (Chapter 22)

ozone A type of oxygen whose molecules contain three oxygen atoms. (Chapter 8)

ozone hole A region of the Earth's atmosphere over Antarctica where the concentration of ozone is abnormally low. (Chapter 8)

ozone layer A layer in the Earth's upper atmosphere where the concentration of ozone is high enough to prevent much ultraviolet light from reaching the surface. (Chapter 8)

P wave One of three kinds of seismic waves produced by an earthquake; a primary wave. (Chapter 8)

pair production The creation of a particle and its antiparticle from energy. (Chapter 23, Chapter 29)

parabola A conic section formed by cutting a circular cone at an angle parallel to the side of the cone. (Chapter 4)

parallax The apparent displacement of an object due to the motion of the observer. (Chapter 4, Chapter 19)

parsec (pc) A unit of distance; 3.26 light-years. (Chapter 1, Chapter 19)

partial lunar eclipse A lunar eclipse in which the Moon does not appear completely covered. (Chapter 3)

partial solar eclipse A solar eclipse in which the Sun does not appear completely covered. (Chapter 3)

Paschen series A series of spectral lines of hydrogen produced by electron transitions between the third and higher energy levels. (Chapter 5)

Pauli exclusion principle A principle of quantum mechanics stating that no two electrons can have the same position and momentum. (Chapter 21)

penumbra (of a shadow) (plural **penumbrae**) The portion of a shadow in which only part of the light source is covered by an opaque body. (Chapter 3)

penumbral eclipse A lunar eclipse in which the Moon passes only through the Earth's penumbra. (Chapter 3)

perigee The point in its orbit where a satellite or the Moon is nearest the Earth. (Chapter 3)

perihelion The point in its orbit where a planet or comet is nearest the Sun. (Chapter 4)

period (of a planet) The interval of time between successive geometric arrangements of a planet and an astronomical object, such as the Sun. (Chapter 4)

period-luminosity relation A relationship between the period and average density of a pulsating star. (Chapter 21)

periodic table A listing of the chemical elements according to their properties, invented by Dmitri Mendeleev. (Chapter 5)

permafrost Frozen soil that lies just underneath the surface. (Chapter 12)

photodisintegration The breakup of nuclei by high-energy gamma rays. (Chapter 22)

photoelectric effect The phenomenon whereby certain metals emit electrons when exposed to short-wavelength light. (Chapter 5)

photometry The measurement of light intensities. (Chapter 6, Chapter 19)

photon A discrete unit of electromagnetic energy. (Chapter 5)

photosphere The region in the solar atmosphere from which most of the visible light escapes into space. (Chapter 18)

photosynthesis A biochemical process in which solar energy is converted into chemical energy, carbon dioxide and water are absorbed, and oxygen is released. (Chapter 8)

physics, laws of See *laws of physics.*

pixel A picture element. (Chapter 6)

plage A bright region in the solar atmosphere as observed in the monochromatic light of a spectral line. (Chapter 18)

Planck time A fundamental interval of time defined by basic physical constants. (Chapter 28)

Planck's law The relationship between the energy of a photon and its wavelength or frequency; $E = hc/\lambda = h\nu$. (Chapter 5)

plane of the ecliptic The plane in which the Earth orbits the Sun. (Chapter 3)

planetary nebula A luminous shell of gas ejected from an old, low-mass star. (Chapter 22)

planetesimal A small body of primordial dust and ice from which the planets formed. (Chapter 7)

plasma A hot ionized gas. (Chapter 13, Chapter 18, Chapter 28)

plastic The attribute of being nearly solid yet able to flow. (Chapter 8)

plate A large section of the Earth's lithosphere that moves as a single unit. (Chapter 8)

plate tectonics The motions of large segments (plates) of the Earth's surface over the underlying mantle. (Chapter 8)

polarized radiation Electromagnetic waves whose electric fields are aligned in a particular direction. (Chapter 27)

polymer A long molecule consisting of many smaller molecules joined together. (Chapter 15)

poor cluster (of galaxies) A cluster of galaxies with very few members; a group of galaxies. (Chapter 26)

Population I star A star whose spectrum exhibits spectral lines of many elements heavier than helium; a metal-rich star. (Chapter 21)

Population II star A star whose spectrum exhibits comparatively few spectral lines of elements heavier than helium; a metal-poor star. (Chapter 21)

positional astronomy The study of the apparent positions of the planets and stars and how those positions change. (Chapter 2)

positive curvature The curvature of a surface or space in which parallel lines converge and the sum of the angles of a triangle is greater than 180°. (Chapter 28)

positron An electron with a positive rather than negative electric charge; the antiparticle of the electron. (Chapter 18, Chapter 29)

power of ten The exponent n in 10^n. (Chapter 1)

powers-of-ten notation A shorthand method of writing numbers, involving 10 followed by an exponent. (Chapter 1)

precession (of the Earth) A slow, conical motion of the Earth's axis of rotation caused by the gravitational pull of the Moon and Sun on the Earth's equatorial bulge. (Chapter 2)

precession of the equinoxes The slow westward motion of the equinoxes along the ecliptic due to precession of the Earth. (Chapter 2)

primary mirror See *objective mirror.*

prime focus The point in a telescope where the objective focuses light. (Chapter 6)

primitive asteroid See *undifferentiated asteroid.*

primordial black hole A type of black hole that may have formed in the Big Bang. (Chapter 24)

primordial fireball The extremely hot gas that filled the universe immediately following the Big Bang. (Chapter 28)

principle of equivalence See *equivalence principle.*

progenitor star A star that later explodes into a supernova. (Chapter 22)

prograde orbit An orbit of a satellite around a planet that is in the same direction as the rotation of the planet. (Chapter 14)

prograde rotation A situation in which an object (such as a planet) rotates in the same direction that it orbits around another object (such as the Sun). (Chapter 11)

prominence Flamelike protrusions seen near the limb of the Sun and extending into the solar corona. (Chapter 18)

proper distance See *proper length.*

proper length A length measured by a ruler at rest with respect to an observer. (Chapter 24)

proper motion The angular rate of change in the location of a star on the celestial sphere, usually expressed in arcseconds per year. (Chapter 19)

proper time A time interval measured with a clock at rest with respect to an observer. (Chapter 24)

proplyd See *protoplanetary disk.*

proton A heavy, positively charged subatomic particle that is one of two principal constituents of atomic nuclei. (Chapter 5)

proton-proton chain A sequence of thermonuclear reactions by which hydrogen nuclei are built up into helium nuclei. (Chapter 18)

protoplanet A Moon-sized object formed by the coalescence of planetesimals. (Chapter 7)

protoplanetary disk (proplyd) A disk of material encircling a protostar or a newborn star. (Chapter 7, Chapter 20)

protostar A star in its earliest stages of formation. (Chapter 20)

protosun The part of the solar nebula that eventually developed into the Sun. (Chapter 7)

Ptolemaic system The definitive version of the geocentric cosmology of ancient Greece. (Chapter 4)

pulsar A pulsating radio source thought to be associated with a rapidly rotating neutron star. (Chapter 1, Chapter 23)

pulsating variable star A star that pulsates in size and luminosity. (Chapter 21)

pulsating X-ray source An object that emits pulses of X rays at regular intervals of a few seconds. (Chapter 23)

quantum electrodynamics A theory that describes details of how charged particles interact by exchanging photons. (Chapter 29)

quantum mechanics The branch of physics dealing with the structure and behavior of atoms and their constituents as well as their interaction with light. (Chapter 5, Chapter 29)

quark One of several particles thought to be the internal constituents of certain heavy subatomic particles such as protons and neutrons. (Chapter 29)

quasar A starlike object with a very large redshift. (Chapter 1, Chapter 27)

radial velocity That portion of an object's velocity parallel to the line of sight. (Chapter 5, Chapter 19)

radial velocity curve A plot showing the variation of radial velocity with time for a binary star or variable star. (Chapter 19)

radial velocity method A technique used to detect extrasolar planets. (Chapter 7)

radiant (of a meteor shower) The point in the sky from which meteors of a particular shower seem to originate. (Chapter 17)

radiation darkening The darkening of methane ice by electron impacts. (Chapter 16)

radiation-dominated universe A universe in which the mass density of radiation exceeds both the average density of matter and the mass density of dark energy. (Chapter 28)

radiation pressure Pressure exerted on an object by radiation falling on the object. (Chapter 17)

radiative diffusion The random migration of photons from a star's center toward its surface. (Chapter 18)

radiative zone A region within a star where radiative diffusion is the dominant mode of energy transport. (Chapter 18)

radio galaxy A galaxy that emits an unusually large amount of radio waves. (Chapter 27)

radio lobe A region near an active galaxy from which significant radio radiation emanates. (Chapter 27)

radio telescope A telescope designed to detect radio waves. (Chapter 6)

radio waves The longest-wavelength electromagnetic radiation. (Chapter 5)

radio window The range of radio wavelengths to which the Earth's atmosphere is transparent. (Chapter 6)

recombination The process in which an electron combines with a positively charged ion. (Chapter 20)

red giant A large, cool star of high luminosity. (Chapter 19, Chapter 21)

red-giant branch The region of an H-R diagram occupied by stars in their first red-giant phase. (Chapter 22)

reddening (interstellar) See *interstellar reddening.*

redshift The shifting to longer wavelengths of the light from remote galaxies and quasars; the Doppler shift of light from a receding source. (Chapter 5, Chapter 26)

reflecting telescope A telescope in which the principal optical component is a concave mirror. (Chapter 6)

reflection The return of light rays by a surface. (Chapter 6)

reflection nebula A comparatively dense cloud of dust in interstellar space that is illuminated by a star. (Chapter 20)

reflector A reflecting telescope. (Chapter 6)

refracting telescope A telescope in which the principal optical component is a lens. (Chapter 6)

refraction The bending of light rays when they pass from one transparent medium to another. (Chapter 6)

refractor A refracting telescope. (Chapter 6)

refractory element An element with high melting and boiling points. (Chapter 9)

regolith The layer of rock fragments covering the surface of the Moon. (Chapter 9)

regular cluster (of galaxies) A spherical cluster of galaxies. (Chapter 26)

relativistic cosmology A cosmology based on the general theory of relativity. (Chapter 28)

residual polar cap An ice-covered polar region on Mars that does not completely evaporate during the Martian summer. (Chapter 12)

respiration A biological process that produces energy by consuming oxygen and releasing carbon dioxide. (Chapter 8)

retrograde motion The apparent westward motion of a planet with respect to background stars. (Chapter 4)

retrograde orbit An orbit of a satellite around a planet that is in the direction opposite to which the planet rotates. (Chapter 14)

retrograde rotation A situation in which an object (such as a planet) rotates in the direction opposite to which it orbits around another object (such as the Sun). (Chapter 11)

rich cluster (of galaxies) A cluster of galaxies containing many members. (Chapter 26)

rift valley A feature created when a planet's crust breaks apart along a line. (Chapter 12)

right ascension A coordinate for measuring the east-west positions of objects on the celestial sphere. (Chapter 2)

ring particles Small particles that constitute a planetary ring. (Chapter 15)

ringlet One of many narrow bands of particles of which Saturn's ring system is composed. (Chapter 15)

Roche limit The smallest distance from a planet or other object at which a second object can be held together by purely gravitational forces. (Chapter 15)

Roche lobe A teardrop-shaped volume surrounding a star in a binary inside which gases are gravitationally bound to that star. (Chapter 21)

rock A mineral or combination of minerals. (Chapter 8)

rotation curve A plot of the orbital speeds of stars and nebulae outward from the center of a galaxy. (Chapter 25)

RR Lyrae variable A type of pulsating star with a period of less than one day. (Chapter 21)

runaway greenhouse effect A greenhouse effect in which the temperature continues to increase. (Chapter 11)

runaway icehouse effect A situation in which a decrease in atmospheric temperature causes a further decrease in temperature. (Chapter 12)

S wave One of three kinds of seismic waves produced by an earthquake; a secondary wave. (Chapter 8)

Sagittarius A A grouping of several powerful radio sources near the center of our Galaxy. (Chapter 25)

saros A particular cycle of similar eclipses that recur about every 18 years. (Chapter 3)

scarp A line of cliffs formed by the faulting or fracturing of a planet's surface. (Chapter 10)

scattering of light See *light scattering*.

Schwarzschild radius (R_{Sch}) The distance from the singularity to the event horizon in a nonrotating black hole. (Chapter 24)

scientific method The basic procedure used by scientists to investigate phenomena. (Chapter 1)

seafloor spreading The separation of plates under the ocean due to lava emerging in an oceanic rift. (Chapter 8)

search for extraterrestrial intelligence (SETI) The scientific search for evidence of intelligent life on other planets. (Chapter 30)

second of arc See *arcsecond*.

sedimentary rock A rock that is formed from material deposited on land by rain or winds, or on the ocean floor. (Chapter 8)

seeing disk The angular diameter of a star's image. (Chapter 6)

seismic wave A vibration traveling through a terrestrial planet, usually associated with earthquake-like phenomena. (Chapter 8)

seismograph A device used to record and measure seismic waves, such as those produced by earthquakes. (Chapter 8)

self-propagating star formation The process by which the formation of stars in one location in a galaxy stimulates the formation of stars in a neighboring location. (Chapter 25)

semidetached binary A binary star system in which one star fills its Roche lobe. (Chapter 21)

semimajor axis One-half of the major axis of an ellipse. (Chapter 4)

SETI See *search for extraterrestrial intelligence*.

Seyfert galaxy A spiral galaxy with a bright nucleus whose spectrum exhibits emission lines. (Chapter 27)

shell helium burning The thermonuclear fusion of helium in a shell surrounding a star's core. (Chapter 22)

shell hydrogen burning The thermonuclear fusion of hydrogen in a shell surrounding a star's core. (Chapter 21)

shepherd satellite A satellite whose gravity restricts the motions of particles in a planetary ring, preventing them from dispersing. (Chapter 15)

shield volcano A volcano with long, gently sloping sides. (Chapter 11)

shock wave An abrupt, localized region of compressed gas caused by an object traveling through the gas at a speed greater than the speed of sound. (Chapter 8)

SI units The International System of Units (*Système international des unités*), based on the meter (m), the second (s), and the kilogram (kg). (Chapter 1)

sidereal clock A clock that measures sidereal time. (Chapter 2)

sidereal day The interval between successive meridian passages of the vernal equinox. (Chapter 2)

sidereal month The period of the Moon's revolution about the Earth with respect to the stars. (Chapter 3)

sidereal period The orbital period of one object about another as measured with respect to the stars. (Chapter 4)

sidereal time Time reckoned by the location of the vernal equinox. (Chapter 2)

sidereal year The orbital period of the Earth about the Sun with respect to the stars. (Chapter 2)

silicon burning The thermonuclear fusion of silicon to produce heavier elements, especially iron. (Chapter 22)

singularity A place of infinite spacetime curvature; the center of a black hole. (Chapter 24)

small-angle formula A relationship between the angular and linear sizes of a distant object. (Chapter 1)

SNC meteorite A meteorite that came to Earth from Mars. (Chapter 30)

solar constant The average amount of energy received from the Sun per square meter per second, measured just above the Earth's atmosphere. (Chapter 5)

solar corona Hot, faintly glowing gases seen around the Sun during a total solar eclipse; the uppermost regions of the solar atmosphere. (Chapter 3)

solar cycle See *22-year solar cycle.*

solar eclipse An eclipse of the Sun by the Moon; a passage of the Earth through the Moon's shadow. (Chapter 3)

solar flare A sudden, temporary outburst of light from an extended region of the solar surface. (Chapter 18)

solar nebula The cloud of gas and dust from which the Sun and solar system formed. (Chapter 7)

solar neutrino A neutrino emitted from the core of the Sun. (Chapter 18)

solar neutrino problem The discrepancy between the predicted and observed numbers of solar neutrinos. (Chapter 18)

solar system The Sun, planets and their satellites, asteroids, comets, and related objects that orbit the Sun. (Chapter 1)

solar transit The passage of an object in front of the Sun. (Chapter 10)

solar wind An outward flow of particles (mostly electrons and protons) from the Sun. (Chapter 7, Chapter 18)

south celestial pole The point directly above the Earth's South Pole where the Earth's axis of rotation, if extended, would intersect the celestial sphere. (Chapter 2)

southern highlands (on Mars) Older, cratered terrain in the Martian southern hemisphere. (Chapter 12)

southern lights See *aurora australis.*

space velocity How fast and in what direction a star moves through space. (Chapter 19)

spacetime A four-dimensional combination of time and the three dimensions of space. (Chapter 24, Chapter 29)

special theory of relativity A description of mechanics and electromagnetic theory formulated by Albert Einstein, which explains that measurements of distance, time, and mass are affected by the observer's motion. (Chapter 24)

spectral analysis The identification of chemical substances from the patterns of lines in their spectra. (Chapter 5)

spectral class A classification of stars according to the appearance of their spectra. (Chapter 19)

spectral line In a spectrum, an absorption or emission feature that is at a particular wavelength. (Chapter 5)

spectral type A subdivision of a spectral class. (Chapter 19)

spectrograph An instrument for photographing a spectrum. (Chapter 6)

spectroscopic binary A binary star system whose binary nature is deduced from the periodic Doppler shifting of lines in its spectrum. (Chapter 19)

spectroscopic parallax The distance to a star derived by comparing its apparent brightness to a luminosity inferred from the star's spectrum. (Chapter 19)

spectroscopy The study of spectra and spectral lines. (Chapter 5, Chapter 6, Chapter 7)

spectrum (plural spectra) The result of dispersing a beam of electromagnetic radiation so that components with different wavelengths are separated in space. (Chapter 5)

spectrum binary A binary star whose binary nature is deduced from the presence of two sets of incongruous spectral lines. (Chapter 19)

speed Distance traveled divided by the time elapsed to cover that distance. (Chapter 4)

spherical aberration The distortion of an image formed by a telescope due to differing focal lengths of the optical system. (Chapter 6)

spherical space Space with positive curvature. (Chapter 28)

spicule A narrow jet of rising gas in the solar chromosphere. (Chapter 18)

spin The intrinsic angular momentum possessed by certain particles. (Chapter 25)

spin-flip transition A transition in the ground state of the hydrogen atom, which occurs when the orientation of the electron's spin changes. (Chapter 25)

spin-orbit coupling See *1-to-1 spin-orbit coupling; 3-to-2 spin-orbit coupling.*

spiral arms Lanes of interstellar gas, dust, and young stars that wind outward in a plane from the central regions of a galaxy. (Chapter 25)

spiral galaxy A flattened, rotating galaxy with pinwheel-like spiral arms winding outward from the galaxy's nucleus. (Chapter 26)

spontaneous symmetry breaking A symmetry breaking that occurs suddenly. (Chapter 29)

spring tide An ocean tide that occurs at new moon and full moon phases. (Chapter 4)

stable Lagrange points Locations along Jupiter's orbit where the combined gravitational effects of the Sun and Jupiter cause asteroids to collect. (Chapter 17)

standard candle An astronomical object of known intrinsic brightness that can be used to determine the distances to other galaxies. (Chapter 26)

starburst galaxy A galaxy that is experiencing an exceptionally high rate of star formation. (Chapter 26)

star cluster See *cluster of stars.*

stationary absorption line An absorption line in the spectrum of a binary star that does not show the same Doppler shift as other lines, indicating that it originates in the interstellar medium. (Chapter 20)

Stefan-Boltzmann law A relationship between the temperature of a blackbody and the rate at which it radiates energy. (Chapter 5)

stellar association A loose grouping of young stars. (Chapter 20)

stellar evolution The changes in size, luminosity, temperature, and so forth that occur as a star ages. (Chapter 20)

stellar parallax The apparent displacement of a star due to the Earth's motion around the Sun. (Chapter 19)

stony iron meteorite A meteorite composed of both stone and iron. (Chapter 17)

stony meteorite A meteorite composed of stone. (Chapter 17)

stratosphere A layer in the atmosphere of a planet directly above the troposphere. (Chapter 8)

strong force The force that binds protons and neutrons together in nuclei. (Chapter 29)

subduction zone A location where colliding tectonic plates cause the Earth's crust to be pulled down into the mantle. (Chapter 8)

subtend To extend over an angle. (Chapter 1)

summer solstice The point on the ecliptic where the Sun is farthest north of the celestial equator. Also used to refer to the date on which the Sun passes through this point. (Chapter 2)

sunspot A temporary cool region in the solar photosphere. (Chapter 18)

sunspot cycle The semiregular 11-year period with which the number of sunspots fluctuates. (Chapter 18)

sunspot maximum/minimum That time during the sunspot cycle when the number of sunspots is highest/lowest. (Chapter 18)

supercluster A collection of clusters of galaxies. (Chapter 26)

superconductivity The phenomenon whereby a flowing electric current does not experience any electrical resistance. (Chapter 23)

superfluidity The phenomenon whereby a fluid flows without experiencing any viscosity. (Chapter 23)

supergiant A very large, extremely luminous star of luminosity class I. (Chapter 19, Chapter 22)

supergrand unified theory A complete description of all forces and particles, as well as the structure of space and time; a "theory of everything" (TOE). (Chapter 29)

supergranule A large convective feature in the solar atmosphere, usually outlined by spicules. (Chapter 18)

superior conjunction The configuration of a planet being behind the Sun as viewed from the Earth. (Chapter 4)

superior planet A planet that is more distant from the Sun than the Earth is. (Chapter 4)

superluminal motion Motion that appears to involve speeds greater than the speed of light. (Chapter 27)

supermassive black hole A black hole with a mass of a million or more Suns. (Chapter 24, Chapter 27)

supernova (plural **supernovae**) A stellar outburst during which a star suddenly increases its brightness roughly a millionfold. (Chapter 1, Chapter 17, Chapter 22)

supernova remnant The gases ejected by a supernova. (Chapter 20, Chapter 22)

supersonic Faster than the speed of sound. (Chapter 20)

surface wave A type of seismic wave that travels only over the Earth's surface. (Chapter 8)

symmetry Any situation in which certain properties of particles or forces are equivalent. (Chapter 29)

symmetry breaking A process by which certain symmetries in the mathematics of particle physics are altered to produce new particles and forces. (Chapter 29)

synchrotron radiation A type of nonthermal radiation emitted by charged particles moving through a magnetic field. (Chapter 13, Chapter 23)

synchronous rotation The rotation of a body with a period equal to its orbital period; also called 1-to-1 spin-orbit coupling. (Chapter 3, Chapter 9)

synodic month The period of revolution of the Moon with respect to the Sun; the length of one cycle of lunar phases. Also called the lunar month. (Chapter 3)

synodic period The interval between successive occurrences of the same configuration of a planet. (Chapter 4)

T Tauri stars Young variable stars associated with interstellar matter that show erratic changes in luminosity. (Chapter 20)

T Tauri wind A flow of particles away from a T Tauri star. (Chapter 7)

tail (of a comet) Gas and dust particles from a comet's nucleus that have been swept away from the comet's head by the radiation pressure of sunlight and the solar wind. (Chapter 17)

tangential velocity That portion of an object's velocity perpendicular to the line of sight. (Chapter 19)

temperature See *degree Celsius; degree Fahrenheit; kelvin.*

terminator The line dividing day and night on the surface of the Moon or a planet; the line of sunset or sunrise. (Chapter 9)

terrae Cratered lunar highlands. (Chapter 9)

terrestrial planet Mercury, Venus, Earth, and Mars; the classification sometimes also includes the Galilean satellites and Pluto. (Chapter 7)

theory A hypothesis that has withstood experimental or observational tests. (Chapter 1)

theory of everything See *supergrand unified theory.*

thermal equilibrium A balance between the input and outflow of heat in a system. (Chapter 18, Chapter 29)

thermal pulses Brief bursts in energy output from the helium-burning shell of an aging low-mass star. (Chapter 22)

thermal radiation The radiation naturally emitted by any object that is not at absolute zero. Blackbody radiation is an idealized case of thermal radiation. (Chapter 13, Chapter 27)

thermonuclear fusion The combining of nuclei under conditions of high temperature in a process that releases substantial energy. (Chapter 18)

thermosphere A region in the Earth's atmosphere between the mesosphere and the exosphere. (Chapter 8)

3-to-2 spin-orbit coupling The rotation of Mercury, which makes three complete rotations on its axis for every two complete orbits around the Sun. (Chapter 10)

threshold temperature The temperature above which photons spontaneously produce particles and antiparticles of a particular type. (Chapter 29)

tidal force A gravitational force whose strength and/or direction varies over a body and thus tends to deform the body. (Chapter 4, Chapter 15)

tidal heating The heating of the interior of a satellite by continually varying tidal stresses. (Chapter 14)

time dilation The slowing of time due to relativistic motion. (Chapter 24)

time zone A region on the Earth where, by agreement, all clocks have the same time. (Chapter 2)

TOE See *supergrand unified theory.*

total lunar eclipse A lunar eclipse during which the Moon is completely immersed in the Earth's umbra. (Chapter 3)

total solar eclipse A solar eclipse during which the Sun is completely hidden by the Moon. (Chapter 3)

totality (lunar eclipse) The period during a total lunar eclipse when the Moon is entirely within the Earth's umbra. (Chapter 3)

totality (solar eclipse) The period during a total solar eclipse when the disk of the Sun is completely hidden. (Chapter 3)

transit See *meridian transit; solar transit.*

triple alpha process A sequence of two thermonuclear reactions in which three helium nuclei combine to form one carbon nucleus. (Chapter 21)

Trojan asteroid One of several asteroids that share Jupiter's orbit about the Sun. (Chapter 17)

Tropic of Cancer A circle of latitude 23½° north of the Earth's equator. (Chapter 2)

Tropic of Capricorn A circle of latitude 23½° south of the Earth's equator. (Chapter 2)

tropical year The period of revolution of the Earth about the Sun with respect to the vernal equinox. (Chapter 2)

troposphere The lowest level in the Earth's atmosphere. (Chapter 8)

Tully-Fisher relation A correlation between the width of the 21-cm line of a spiral galaxy and the total luminosity of that galaxy. (Chapter 26)

tuning fork diagram A diagram that summarizes Edwin Hubble's classification scheme for spiral, barred spiral, and elliptical galaxies. (Chapter 26)

turnoff point The point on an H-R diagram where the stars in a cluster are leaving the main sequence. (Chapter 21)

21-cm radio emission Radio radiation emitted by neutral hydrogen atoms in interstellar space. (Chapter 25)

22-year solar cycle The semiregular 22-year interval between successive appearances of sunspots at the same latitude and with the same magnetic polarity. (Chapter 18)

Type I Cepheid A metal-rich Cepheid variable. (Chapter 21)

Type Ia supernova A supernova whose spectrum lacks hydrogen lines but has a strong absorption line of ionized silicon. (Chapter 22)

Type Ib supernova A supernova whose spectrum lacks hydrogen and silicon lines but has a strong helium absorption line. (Chapter 22)

Type Ic supernova A supernova whose spectrum is almost devoid of emission or absorption lines. (Chapter 22)

Type II Cepheid A metal-poor Cepheid variable. (Chapter 21)

Type II supernova A supernova with hydrogen emission lines in its spectrum, caused by the explosion of a massive star. (Chapter 22)

UBV photometry A system for determining the surface temperature of a star by measuring the star's brightness in the ultraviolet (U), blue (B), and visible (V) spectral regions. (Chapter 19)

ultramafic lava A type of lava enriched in magnesium and iron. These give the lava a higher melting temperature. (Chapter 14)

ultraviolet radiation Electromagnetic radiation of wavelengths shorter than those of visible light but longer than those of X rays. (Chapter 5)

umbra (of a shadow) (plural **umbrae**) The central, completely dark portion of a shadow. (Chapter 3)

unbounded universe A universe that continues to expand forever. (Chapter 28)

undifferentiated asteroid An asteroid within which chemical differentiation did not occur. (Chapter 17)

universal constant of gravitation (**G**) The constant of proportionality in Newton's law of gravitation. (Chapter 4)

upper meridian The half of the meridian that lies above the horizon. (Chapter 2)

Van Allen belts Two doughnut-shaped regions around the Earth where many charged particles (protons and electrons) are trapped by the Earth's magnetic field. (Chapter 8)

velocity The speed and direction of an object's motion. (Chapter 4)

vernal equinox The point on the ecliptic where the Sun crosses the celestial equator from south to north. Also used to refer to the date on which the Sun passes through this intersection. (Chapter 2)

very-long-baseline interferometry (**VLBI**) A method of connecting widely separated radio telescopes to make very high-resolution observations. (Chapter 6)

virtual pair A particle and antiparticle that exist for such a brief interval that they cannot be observed. (Chapter 24, Chapter 29)

visible light Photons detectable by the human eye. (Chapter 5)

visual binary A binary star in which the two components can be resolved through a telescope. (Chapter 19)

VLBI See *very-long-baseline interferometry.*

void A large volume of space, typically 100 to 400 million light-years in diameter, that contains very few galaxies. (Chapter 26)

volatile element An element with low melting and boiling points. (Chapter 9)

waning crescent moon The phase of the Moon that occurs between third quarter and new moon. (Chapter 3)

waning gibbous moon The phase of the Moon that occurs between full moon and third quarter. (Chapter 3)

water hole A region in the microwave spectrum suitable for interstellar radio communication. (Chapter 30)

watt A unit of power, equal to one joule of energy per second. (Chapter 5)

wavelength The distance between two successive wave crests. (Chapter 5)

wavelength of maximum emission The wavelength at which a heated object emits the greatest intensity of radiation. (Chapter 5)

waxing crescent moon The phase of the Moon that occurs between new moon and first quarter. (Chapter 3)

waxing gibbous moon The phase of the Moon that occurs between first quarter and full moon. (Chapter 3)

weak force The short-range force that is responsible for transforming certain particles into other particles, such as the decay of a neutron into a proton. (Chapter 29)

weakly interacting massive particle (**WIMP**) A hypothetical massive particle that may constitute part of the unseen dark matter. (Chapter 25)

weight The force with which gravity acts on a body. (Chapter 4)

white dwarf A low-mass star that has exhausted all its thermonuclear fuel and contracted to a size roughly equal to the size of the Earth. (Chapter 19, Chapter 22)

white hole A black hole from which matter and radiation emerge. (Chapter 28)

white ovals Round, whitish feature usually seen in Jupiter's southern hemisphere. (Chapter 13)

Widmanstätten patterns Crystalline structure seen in certain types of meteorites. (Chapter 17)

Wien's law A relationship between the temperature of a blackbody and the wavelength at which it emits the greatest intensity of radiation. (Chapter 5)

WIMP See *weakly interacting massive particle.*

winding dilemma The problem of why the spiral arms of a galaxy like the Milky Way do not disappear. (Chapter 25)

winter solstice The point on the ecliptic where the Sun reaches its greatest distance south of the celestial equator. Also used to refer to the date on which the Sun passes through this point. (Chapter 2)

X-ray burster A nonperiodic X-ray source that emits powerful bursts of X rays. (Chapter 1, Chapter 23)

X rays Electromagnetic radiation whose wavelength is between that of ultraviolet light and gamma rays. (Chapter 5)

ZAMS See *zero-age main sequence.*

Zeeman effect A splitting or broadening of spectral lines due to a magnetic field. (Chapter 18)

zenith The point on the celestial sphere directly overhead an observer. (Chapter 2)

zero-age main sequence The main sequence of young stars that have just begun to burn hydrogen at their cores. (Chapter 21)

zero-age main-sequence star A newly formed star that has just arrived on the main sequence. (Chapter 21)

zero curvature The curvature of a surface or space in which parallel lines remain parallel and the sum of the angles of a triangle is exactly 180°. (Chapter 28)

zodiac A band of 12 constellations around the sky centered on the ecliptic. (Chapter 2)

zonal winds The pattern of alternating eastward and westward winds found in the atmospheres of Jupiter and Saturn. (Chapter 13)

zone A light-colored band in Jupiter's atmosphere. (Chapter 13)

Answers to Selected Questions

Answers to Selected Questions

Quiz: How to Get the Most from UNIVERSE
1. Paragraphs labeled by the "caution" icon **2.** Paragraphs labeled by the "analogy" icon **3.** These indicate references to sections, figures, or tables in previous chapters **4.** "Tools of the Astronomer's Trade," "The Heavens on the Earth" **5.** On the *Universe* web site; in addition, all but the "Web Links" can be found on the CD-ROM that accompanies this book **6.** A planetarium program; on the CD-ROM that accompanies this book; it comes with the book at no additional cost **7.** It shows whether the image was made with **R**adio waves, **I**nfrared radiation, **V**isible light, **U**ltraviolet light, **X** rays, or **G**amma rays **8.** On the *Universe* web site or on the CD-ROM that accompanies this book **9.** (**a**) page A-6 (**b**) page A-1 (**c**) page A-3 (**d**) page A-4 **10.** (**a**) page 403 (**b**) page 123 (**c**) page 84 (**d**) page 75 **11.** 656.9 nm **12.** At the back of the book, following the Index

 Only mathematical answers are given in the following, not answers that require interpretation or discussion. Your instructor will expect you to show the steps required to calculate each mathematical answer.

Part I
Chapter 1
19. About 3×10^{36} times larger **20.** 8.94×10^{56} hydrogen atoms **21.** 8.7×10^3 km **22.** 2.9×10^7 Suns **23.** (**a**) 1.58×10^{-5} ly (**b**) 4.84×10^{-6} pc **24.** 4.99×10^2 s **25.** 4.3×10^9 km **26.** (**a**) 9.95×10^{13} km (**b**) 10.5 years **27.** 4.7×10^{17} s **29.** (**a**) 1.1 m (**b**) 69 m (**c**) 4100 m **30.** 3.4×10^3 km **31.** 3.7 km **32.** 6.9 m **33.** 0.292 arcmin

Chapter 2
23. Around 11:28 P.M. **28.** (**b**) About 10 hours **32.** 26½° **36.** October 25, 1917 **39.** 50° **41.** (**a**) 6:00 P.M. (**b**) September 21 **42.** 3:50 A.M. local time

Chapter 3
23. (**a**) 0.91 hour (**b**) 11° **24.** 49 arcsec **35.** (**a**) July 2, 2019; southern Pacific and South America (**b**) July 24, 2055

Chapter 4
15. Semimajor axis = 0.25 AU, period = 0.125 year **16.** Average = 25 AU, farthest = 50 AU **18.** 12 N, 6 m/s² **21.** 16 times stronger **22.** Earth exerts 1.98×10^{20} N on the Moon **29.** 87.97 days **33.** (**a**) 4 AU (**b**) 8 years **34.** (**a**) 9.0 AU (**b**) 0.5 AU **35.** (**a**) 0.86 AU (**b**) 146 days **36.** ⅓ as much as on Earth **37.** Forces are approximately the same, Earth's acceleration is 100 times greater **38.** 67 newtons, 0.14 **39.** (**a**) 24 hours (**b**) 42,300 km **40.** 0.5 year **41.** 119 minutes **43.** (**a**) 5.00 years (**b**) 2.92 AU **44.** (**a**) 3.43×10^{-5} N (**b**) 3.21×10^{-5} N (**c**) 2.2×10^{-6} N

Chapter 5
1. 1.28 s **5.** 2.83×10^{14} Hz **6.** 0.336 m **11.** 753 nm **12.** 24,000 K **20.** 9980°F **21.** 9400 nm **22.** About 10 μm **23.** 5.1 times as much **24.** (**a**) 4890 nm = 4.89 μm (**b**) 540 times as much **25.** (**a**) 5.74×10^8 W/m² (**b**) 10,000 K **26.** 2.9 nm **28.** 2.43×10^{-3} nm **29.** (**a**) 1005 nm **30.** 372.2 nm **33.** Possible transitions: energy = 1 eV, λ = 1240 nm; energy = 2 eV, λ = 620 nm; energy = 3 eV, λ = 414 nm **34.** Coming toward us at 13.0 km/s **35.** 656.9 nm **36.** 8,600 km/s

Chapter 6
21. ¹/₂₅ = 0.04 **22.** Subaru has 12 times the light-gathering power **23.** (**a**) 222× (**b**) 100× (**c**) 36×; angular resolution = 0.75 arcsec **25.** 300 km (Hubble Space Telescope, Jupiter's moons); 110 km (human eye, our Moon) **26.** (**a**) 34 ly (**b**) 37 km **28.** (**a**) 0.18 m (**b**) 5.9×10^{-4} arcsec **29.** (**a**) 5.39×10^{-4} m (**c**) 0.56 m

Part II
Chapter 7
19. Mass = 6.4×10^{23} kg, average density = 3900 kg/m³ **22.** (**a**) 3.3×10^{21} J (**b**) Equivalent to 40 million Hiroshima-type weapons **23.** (**a**) 1.2 m/s **24.** 12 km/s **25.** (**a**) 617 km/s **26.** (**a**) 2.02 km/s (**b**) 19.4 km/s **27.** (**a**) About 50 times Earth's actual mass **29.** (**a**) 1.76 years **31.** (**a**) About 180 AU **32.** 2.9×10^7 AU = 140 parsecs = 460 light-years **33.** (**a**) About 600 AU (**b**) About 5×10^{40} cubic meters (**c**) About 10^{55} atoms (**d**) About 3×10^{14} atoms per cubic meter **35.** 2.03×10^{30} kg = 1.02 times the mass of the Sun **36.** (**a**) 12 m/s (**b**) 1.3×10^{-3} arcsec (**c**) 9.0×10^{-5} arcsec

Chapter 8
20. (**a**) 6.8×10^{16} W (**b**) 1.07×10^{17} W (**c**) 209 W (**d**) 246 K = −27°C **21.** 4 km **22.** Core: 17%; mantle: 82%; crust: 1% **23.** 0.020 (2% of the total mass) **24.** About 15,000 kg/m³ **25.** 2.2×10^8 years ago (based on the present-day separation of 6600 km) **29.** (**a**) 2.225 eV **33.** Pressure is 0.001 of sea-level value at 55 km altitude

Chapter 9
20. 130 N on the Moon, 780 N on Earth **21.** (**a**) 5.78×10^{-5} newton (**b**) 4.15×10^{-5} newton (**c**) 1.39 **26.** In 100 years, probability is 8.3×10^{-8} (one chance in 12 million); in 10^6 years, probability is 8.3×10^{-4} (one chance in 1200) **27.** 2.51 s **28.** 5.5×10^5 km **30.** (**b**) 10^6 times greater

Chapter 10
18. 730 km **21.** For $T = 430°C$, $\lambda_{max} = 4.1$ μm **23.** (**a**) 3.03 m/s (**b**) 1.26 nm **24.** (**a**) About 30 **25.** 300 N on Mercury, 130 N on the Moon, 780 N on Earth

Chapter 11
24. 0.23 nm   25. For T = 460°C, λ_{max} = 4.0 μm   32. 920 m

Chapter 12
25. Earth completes 2.14 orbits, Mars completes 1.14 orbits
27. (a) 0.16 AU   28. (a) 270 km (b) 27 km   29. 2.0 cm
32. (a) 3770 km (b) 370 km   46. (a) Radius = 20,400 km, altitude = 17,000 km

Chapter 13
24. 12.7 km/s   26. 8.5×10^{53} hydrogen atoms, 7.1×10^{52} helium atoms   27. Roughly 600 km/h   28. 127 K   29. (a) About 160 K = –113°C = –171°F (b) About 6 atm (c) About 50 m
30. 59.5 km/s, or 5.3 times the escape speed from the Earth
31. 8300 newtons   32. 71,000 kg/m^3

Chapter 14
28. 760 km   29. 0.046 arcsec   32. 2.8×10^{11} (280 billion) years
33. About 8.1 minutes   34. (b) 159,000 km

Chapter 15
21. (a) 14.4 hours for inner edge of A ring, 7.9 hours for inner edge of B ring   25. Escape speed = 2.6 km/s; mass of molecule = 2.0×10^{-26} kg, corresponding to a molecular weight of 12
28. 0.09 arcsec   29. (a) 65.7 hours (b) 27.6 arcsec

Chapter 16
23. Sun-Uranus force = 1.39×10^{21} N, Neptune-Uranus force = 2.24×10^{17} N; Neptune reduces the sunward gravitational pull on Uranus by 1.61×10^{-4}, or 0.0161%   24. (a) 2900 kg/m^3
28. (a) 1.13×10^{-3} as bright as on Earth (b) 4.10×10^{-4} as bright as on Earth (c) 2.77 times brighter   29. About 7200 km
31. 0.95 arcsec

Chapter 17
25. 2.82 AU   26. 30 km   28. (a) 3.6×10^{12} kg (b) 0.83 m/s
30. About 4%   31. (a) Period = 350 years, lifetime = 3.5×10^4 years (b) Period = 1.1×10^4 years, lifetime = 1.1×10^6 years (c) Period = 3.5×10^5 years, lifetime = 3.5×10^7 years (d) Period = 1.1×10^7 years, lifetime = 1.1×10^9 years
34. (a) 10^{15} kg; (b) 10^{-16} kg/m^3

Part III
Chapter 18
24. (a) 1.8×10^{-9} J (b) 9.0×10^{16} J (c) 5.4×10^{41} J
25. (a) 4.6×10^{-36} s (b) 2.3×10^{-10} s (c) 1.4×10^{15} s = 4.4×10^7 years   26. 0.048 (4.8%) of the Sun's mass will be converted from hydrogen to helium; chemical composition of Sun (by mass) will be 69% hydrogen, 30% helium   27. (a) 8.89×10^{28} kg of hydrogen consumed, mass lost = 6.23×10^{26} kg
28. (a) 1.64×10^{-13} J (b) 2.43×10^{-3} nm   29. 1.4×10^{13} kilograms per second   33. 98,600 kg/m^3 34. 1.9×10^{-7} nm
35. (a) 1700 nm   37. For the photosphere, 500 nm; for the chromosphere, 58 nm; for the corona, 1.9 nm   39. For the umbra, 670 nm; for the penumbra, 580 nm   40. (a) (Flux from patch of penumbra)/(flux from patch of photosphere) = 0.55 (b) (Flux from patch of penumbra)/(flux from patch of umbra) = 1.8

Chapter 19
24. Average Sun-Pluto distance = 1.92×10^{-4} pc; Proxima Centauri is 6780 times farther away   25. (a) 9.7 pc
(b) 0.10 arcsec   26. 3.92 pc   27. (a) 161 km/s (b) 294 km/s
28. Distance = 105 pc; tangential velocity would have to be about 4340 km/s   29. 110 km/s   30. (a) +59.4 km/s (c) 486.23 nm
32. The Sun is 1.11×10^{-3} as bright on Neptune as on Earth
33. Star A is 16 times more luminous than star B   34. Star D appears 9 times brighter than star C   35. The farther star is 144 times more luminous than the closer star   36. 37.0 AU
37. 6.8 $L_\odot$ 38. 0.38 pc = 7.9×10^4 AU   39. (a) +13.8 (b) (Luminosity of HIP 72509)/(luminosity of Sun) = 2.4×10^{-4}
40. +17   41. 6300 pc   42. (a) Brightest: M = –2.37; dimmest: M = +4.32 (b) M = +0.79   45. (b) $m_B - m_V$ = –0.23 (Bellatrix), +0.68 (Sun), +1.86 (Betelgeuse)   46. 99 $R_\odot$ 47. Radius increases by a factor of 2, luminosity increases by a factor of 64
50. T = 10,000 K, R = 3.8 $R_\odot$ 51. L = 35 $L_\odot$, so distance = 14 pc
52. (b) 0.26 $R_\odot$ = 1.8×10^5 km   53. (a) 5.0 pc (b) 22.5 AU
(c) 1.5 $M_\odot$ 54. (a) 40 $M_\odot$ (b) M_1 = 32 $M_\odot$, M_2 = 8 $M_\odot$
56. 2500 pc, 125 times greater volume

Chapter 20
25. 0.34%   26. 3.4×10^4 atoms per cm^3 30. 1100 $R_\odot$ = 7.4×10^8 km = 4.9 AU   31. (a) 250 AU = 3.7×10^{10} km
(b) 4000 years (c) 1000 AU; 24 years   32. 1.3×10^{31} m^3
33. (a) 2.9° (b) 1300 km/s (c) 180 years

Chapter 21
26. 657,000 km   27. (a) 618 km/s (b) 61.8 km/s   28. (a) 12.0 km/s
(b) 9.3 km/s   29. 1.9×10^{21} kg, or 12% of the original mass of hydrogen   30. 4.1×10^7 years (b) 3.2×10^{11} years   32. About 1900 K (1600°C, or 2900°F)   33. 3.4×10^7 years   36. About 650 pc   37. About 370 pc   39. (a) 1.65×10^6 km

Chapter 22
22. 0.11 $R_\odot$ 24. 7000 years ago   25. (a) 97 nm
26. (a) 1.8×10^9 kg/m^3 (b) 6400 km/s   28. (a) 1.7×10^{30} kg = 0.84 $M_\odot$ (b) 1.1×10^{12} newtons (c) 1.5×10^8 m/s, or 0.5 of the speed of light   29. (a) 6.4×10^5 m/s   30. 3.24×10^9 km = 21.7 AU   31. (Maximum luminosity of SN 1993J)/(maximum luminosity of SN 1987A) = 4.5   33. (a) 1.3×10^{-7} $b_\odot$ (b) It would be 130 times brighter than Venus   34. 1.3×10^8 pc = 130 Mpc
35. 1100 km/s, or 3.8×10^{-3} of the speed of light

Chapter 23
21. About 5400 B.C.   23. It will take about 3100 years, or until about A.D. 5100 (note that "expansion rate" is the rate at which the surface of the nebula moves)   24. (a) Distance = 2200 pc = 7200 ly; estimate that explosion occurred around 6100 B.C.
25. (b) Maximum correction = 10^{-4} of the pulsar period
26. (a) Density of matter in a neutron = 4×10^{17} kg/m^3
29. (a) 0.097 nm (b) Luminosity of neutron star = 5.8×10^{31} W = 1.5×10^5 $L_\odot$ 30. (b) 14 km

Chapter 24
19. 0.98 of the speed of light   20. 0.99995 of the speed of light
21. (a) 25 years (b) 20 light-years as measured by an Earth

observer, 12 light-years as measured by the astronaut
23. $2.8\ M_\odot$ **24.** 5.7×10^8 years **25.** (a) 2.01×10^6 km
26. 0.20 year = 70 days **27.** 2.8 m **28.** (a) R_{Sch} = 8.86 mm,
density = $2.05 \times 10^{30}\ kg/m^3$ (b) R_{Sch} = 2.95 km, density =
$1.85 \times 10^{19}\ kg/m^3$ (c) R_{Sch} = 8.85×10^9 km = 59.1 AU,
density = $2.06\ kg/m^3$ **29.** $1.9 \times 10^{17}\ kg/m^3$ **30.** 2.7×10^{38} kg =
$1.4 \times 10^8\ M_\odot$

Part IV

Chapter 25
22. (a) 1.2×10^{12} cubic parsecs (b) 1.1×10^8 cubic parsecs
(c) Probability = 9.6×10^{-5}; we can expect to see a supernova
within 300 pc once every 350,000 years **24.** 9.5×10^{-25} J; it takes
3.2×10^5 such photons to equal the energy of one H_α photon
25. (b) 5700 pc **26.** 21 times **28.** (a) 3.1×10^8 years
(b) $7.4 \times 10^{11}\ M_\odot$ **29.** A 10% error in radius results in a
10% error in mass; a 10% error in velocity results in a 20%
error in mass **31.** $2.7 \times 10^{11}\ M_\odot$ **34.** (a) 8.9×10^6 km =
0.059 AU (b) 7.4×10^{-6} arcsec **35.** (a) 2300 AU for S0-1,
950 AU for S0-2 (b) 0.29 arcsec for S0-1, 0.12 arcsec for S0-2
36. $2.5 \times 10^6\ M_\odot$

Chapter 26
21. (a) 6.9 Mpc (b) 8.3 Mpc (c) 1.4 Mpc = 1400 kpc **22.** 9.5 Mpc
23. (a) 1.1×10^{10} km = 70 AU (b) 7.0 Mpc **24.** (a) 154 Mpc =
5.03×10^8 light-years **25.** 54 km/s/Mpc **26.** (a) 1.2×10^{70} atoms
(b) 3.6×10^{-6} atoms per cm^3 **27.** $1.2 \times 10^{12}\ M_\odot$ **28.** Period =
4.4×10^8 years; mass = $3.6 \times 10^{11}\ M_\odot$

Chapter 27
22. 13.1 **23.** 2.87×10^5 km/s = 0.958 of the speed of light
24. 9.7×10^4 km/s = 3.5×10^8 km/h = 0.32 of the speed
of light **25.** 2.93×10^5 km/s = 0.976 of the speed of light
28. (a) 1.80×10^5 km/s = 0.600 of the speed of light (b) 134 hours
(c) 970 AU **29.** (a) 30 years (b) 2006 (c) 5/3 of the speed of light
30. (a) Maximum luminosity = $9 \times 10^{13}\ L_\odot$ **31.** $1.5 \times 10^8\ M_\odot$
32. 2.9×10^9 km = 20 AU

Chapter 28
25. 771 nm **26.** (a) 20 billion years (b) 13 billion years
(c) 9.7 billion years **27.** 1.6×10^8 km/s/Mpc **28.** 65 times denser
29. 23.6 K **30.** (a) $4.6 \times 10^{-23}\ kg/m^3$ **31.** (a) $9.5 \times 10^{-18}\ kg/m^3$
(b) $4.8 \times 10^{-4}\ kg/m^3$ (c) $1.3 \times 10^{-7}\ kg/m^3$ **32.** (a) $4.7 \times$
$10^{-27}\ kg/m^3$ (b) $1.9 \times 10^{-26}\ kg/m^3$ **33.** (a) 79 Mpc
(b) $5 \times 10^{-25}\ kg/m^3$ (c) T = 17.3 K, ρ_{rad} = $7.5 \times 10^{-28}\ kg/m^3$
34. Ω_Λ = $1/3$ **35.** (a) −0.11 (b) 0.16 **36.** $1.8 \times 10^{-52}\ m^{-2}$

Chapter 29
19. 1.1×10^{-26} J **20.** 3.5×10^{-25} s **21.** (a) 80.5 GeV
(b) 9.34×10^{14} K **23.** For H_0 = 65 km/s/Mpc, mass =
7.9×10^{-35} kg = 8.7×10^{-5} of the mass of the electron; for
H_0 = 85 km/s/Mpc, mass = 1.4×10^{-34} kg = 1.5×10^{-4} of the
mass of the electron **24.** (a) 1.1×10^{14} m = 0.011 light-year

Chapter 30
8. (a) 3.7 cm **9.** (a) 10,000 years (b) 10 million years
11. (a) 1430 MHz **12.** (a) 29.8 km/s (b) 0.10 m (c) 9.9×10^{-9} m,
0.0000099% of unshifted wavelength

Index

Page numbers in *italic* indicate figures.

E ring, 328, *329*, 336
E0 to E7 galaxies, 593
Eagle (lunar module), 211
Eagle Nebula, 466
Earth, 151, 179–201
  albedo of, 180, 181
  asthenosphere of, 253
  atmosphere of, 140, 180, 182,
    194–197, 199, 250, 271–272, 691
  average density of, 153, 154, 181, 183
  axis of rotation of, 29–33
  carbon dioxide on, 250, 271–272
  circumference of, 55–58
  climate of and solar activity, 411
  core of, 183–186, 234, 235
  crust of, 186–189
  curvature of, *10*
  data table, 181
  day length, 218
  diameter of, *10*, 32, 57–58, 152, 153,
    181
  equatorial plane of, 33
  escape speed of, 160, 181
  formation of, 183
  geologic processes on, 253, *254*
  and global warming, 200
  and gravitational force of Moon, 85,
    86
  and gravitational force of the Sun, 81
  heliocentric model of universe, 66
  human population of, 198, 199
  inclination of, 181
  interior structure of, 183–186
  life on, 390
  and line of nodes, 49
  location of in Galaxy, 131, 565–566
  and lunar eclipses, 49–*51*
  magnetic field of, 191–193, 350, *351*,
    235
  magnetosphere of, 193, 297
  mass of, 153, 181
  and meteoroids, 371
  and night sky, 23–24
  oblateness of, 295
  orbit of, 24, 29, 34, 151, 153, 181,
    198, *262*
  ozone layer of, 200–201
  plant life on, *198*
  and precession, 32–33
  radiation emitted from, 180–181
  relative size of, 57–58
  revolution of about Sun, 29
  rotation of, 23–24, 26, 32, 181
  rotation axis of, 198
  seasons on, 29–32
  semimajor axis of, 76
  shadow of, 47, 49–52
  sidereal period of, 68, 76
  surface of, 180, 182
  surface temperature of, 180–182, 198
  and tidal effects of Sun and Moon,
    85–86
  ultramafic lavas on, 309
  umbra of, 50–52
  volcanoes on, 250
  water on, 180, *181*

  weather patterns on, 290
Earth-Jupiter distance, 285
Earth-Mars distance, 261, 263
Earth-Moon distance, 44, 58, 206, 207,
  214–216, 218
Earth-Saturn distance, 325
Earth-Sun distance, *4, 9, 10,* 12, 29, 58,
  68, 153, 181, 391
earthquakes, 184, 185, 213–214
eccentricity (of an ellipse), 75
Eckart, Andreas, 581
eclipse path, 53–55
eclipse series, 56
eclipse year, 56
eclipses, 49–50
  Galilean satellites, 304–305
  predicting, 50, 55, 56
  see also lunar eclipses; solar eclipses;
    total lunar eclipse; total solar
    eclipse
eclipsing binary star systems, 445–447,
  494, 526, 534
ecliptic, 30–34, 44, 66
Eddington, Sir Arthur, 392, 492, 629
Eddington limit, 629–630
Egyptian pyramids, 21
Egyptians, ancient, 18
Einstein Observatory, 143, *624, 626*
Einstein, Albert, 84, 104, 113, 392, 394,
  531, 542, 545–546, 557, 641–642,
  644, 656, 666, 667, 671
  see also relativity, general theory of;
    relativity, special theory of
Einstein-Rosen bridge, 557
Eistla Regio (Venus), *254*
El Niño phenomenon, 198
Elara, 317
electric field, 95–97
electricity, 95
  see also electromagnetism
electromagnetic radiation, 96–101, 244,
  635
electromagnetic spectrum, 97
electromagnetism, 2, 95–96, 680–682,
  684
  Maxwell's theory of, 109–110
electron-positron pairs, 675
electron transitions, 114–115
electron-type neutrinos, 418
electron volt (eV), 105
electrons, 104, 110–111, 393, 568–569
  and Bohr orbits, 112–114
  and spin, 570
  in early universe, 675–676
  see also positrons
elementary particle physics, 669, 672
elements, *see* chemical elements; heavy
    elements; radioactive elements;
    refractory elements; volatile elements
eleven-dimensional theory, 684, 685
Elliot, James L., 351
ellipse, 83, *84*
elliptical (E) galaxies, 591, 593–595,
  597–599, 603, 610, 615
  see also blazars; giant elliptical
    galaxies; radio galaxies

Elmegreen, Debra and Bruce, 579
elongation (of planets), 67
Eltanin (γ Draconis), 498
Elysium Mons, 267
emission line spectrum, 107, 113–114
  of chromosphere, 402–403
  of corona, 404
  of iron vapor, 109
  of NGC 2363 nebula, 109
  of planetary nebula, 504–505
emission lines
  from hydrogen, 621–622
  of quasars, 618, 619
  of protostars, 463
  on spectrogram, *136*
emission nebula, 455, *456*, 472
empty space, 673, 674, 682
Enceladus, 328, 335, 336
Encke, Johann Franz, 327
Encke gap, 327, 330
energy flux (*F*), 102
energy-level diagram, 114, 115
energy levels (in atom), 114–115
energy transport, 394–396, 503
Enif (ε Pegasi), 498
epicenter (of earthquake), 184
epicycle, 65–66
Epimetheus, *329*, 335
epoch, 33–34
Epsilon Eridani, 425, 695
Epsilon Indi, 429
equatorial bulge, 32–33
equilibrium resurfacing hypothesis, 255
equinox, 30, 31
equinoxes, precession of the, 33
equivalence principle, 546
era of recombination, 651–652, 677, 678
Eratosthenes, 55–57, 58
ergoregion, 555
Eridanus (constellation), 426
Eros (asteroid), *158*, 369
error bars, *599*
escape speed (of a planet), 160
Esposito, Larry W., 247
Eta Aquarids, 383
ethane, 334
Eurasia, 187
Eurasian plate, 188, *189*
Europa, 155, *157*, 302–305, 307,
  311–313, 317, 692
  data table, 304
European Southern Observatory, 134
European Space Agency, 140, 143, 172,
  236, 278, 337, *376*, 380, 405, 422,
  693, 698
evening star
  Mercury as, 226
  Venus as, 241
event horizon, 553–556
evolutionary tracks, 492–493
  of high-mass stars, 485, 508
  of post–main-sequence stars, 501–502
  of protostars, 460–461
  of white dwarfs, 507
Ewen, Harold, 569
excited states, 114

solar system, 3
origin and evolution of, 4, 163–169
see also Milky Way Galaxy
solar transits, 228
of Mercury, 228, *229*, 243
of Venus, 243
solar wind, 169, 192–193, 377, 405
on Jupiter, 297
and Mercury, 235–236
solids, 165
atoms in, 115
Solomon, Philip, 469
sound waves, and Doppler effect, 116
South America, 186–188
South American plate, 189
south celestial pole, 27, 28, 33
South Pole–Aitken Basin, *210*, 212, 219
Southern Cross (Crux), 7, 20, 23
southern highlands (Mars), 265, 266, *268*, *269*
southern lights (aurora australis), 193, *194*
Southern Wall (of galaxies), 604
Soviet Union, and space missions, 211–212, 245
Space Infrared Telescope Facility (SIRTF), 141
space, and relativity, 542, 666
space velocity, 423
spacecraft
orbits of, 82
see also individual entries, e.g., Pioneer spacecraft, *Voyager* spacecraft
spacetime, 542, 546, *547*, 666, 668, 684
and black holes, 548, *549*, 554
five-dimensional, 684
special theory of relativity, *see* relativity, special theory of
spectra, Kirchhoff's laws for, 106–107
spectral analysis, 106
spectral classes, 432–434, 435
spectral lines, 105–106, 109
Lyman series, 114
splitting of, 408–409
wavelengths of, and relative motion 115–117
spectral types, 432, 439, 440
and Hertzsprung-Russell diagram, 437
spectrogram, *136*
spectrograph, 136–137
spectroscopic binaries, 444–445
spectroscopic parallax, 425, 440, 441, 595
spectroscopy, 109, 136–136, 155–156, 444
see also stellar spectroscopy
spectrum, spectra, 94–95
see also absorption line spectrum; continuous spectrum; emission line spectrum
spectrum binary, 444
speed, 78, 542
measurement of, 13

see also light, speed of
spherical aberration, 130, *131*
Hubble Space Telescope, 142, *143*
Spica (α Virginis), 25, 435
spicules, 403, 411, 412
spin, 568–569
spin-flip transition, 569, 570
spin-orbit coupling, 228–231
spiral arms (of galaxies), 6, 568, 571, 572, 576–579, 592
and star formation, 469
density-wave model of, 578–579
formation of, 606–607
spiral density waves, 579
spiral galaxies, *459*, *572*, 579, 591–592, 594–595, 603, 609–610, 615, *668*
and tidal forces, 86–87
see also Seyfert galaxies
spiral nebulae, 588–589
spokes (in Saturn's rings), 339
spontaneous symmetry breaking, 682–683
spring tides, 85
spring (season), 29–31
stable Lagrange points, 370
stades, 56
standard candles, 596
star catalogs, 34
star charts, 34
star clusters, 465–467, 485–489, *607*
H-R diagram, *467*
see also globular clusters
star death, 5, 161, *500*, *504*, 506–507, 510–511, 537
and black holes, 548, 554
supernovae, 510–511
star formation, 4–5, 162, 453–472, 610
in elliptical galaxies, 593
in spiral arms, 571, 578
in spiral galaxies, 592
self-propagating, 579
star trails, *27*
starburst galaxies, 605
Stardust spacecraft, 378
starlight, 92
Starobinsky, Alexei, 671
stars, 419–447
absorption lines of, 434
apparent motion of, 422
atmospheres of, 447
and Balmer absorption lines, 111
blackbody spectrum of, 435
chemical composition of, 434
circumpolar, 27
colors of, 430–431
determining surface temperature of, 102
ejected matter from, 161–162
estimating distances to, 440
evolution of, 6
heavy elements in, 161–162, 163
on Hertzsprung-Russell diagram, 437–439
high-mass, 508–511
high-velocity, 567

in galactic center, *582*
light from, 44
low-mass, 501–502, 506–507
luminosity of, 420, 435–437
metal poor, Population II, 489–490
metal rich, Population I, 489–490
naming of, 23
navigation by, 20
progenitor, 512–513
pulsating, 490–493
radii of, 435, 436, 447
shining of, 4
sizes and types of, 434–436
surface temperatures of, 430–431, 433–437
wobble of, 170, 172
see also binary stars; double stars; giants; globular clusters; horizontal-branch stars; main-sequence stars; Population I stars; Population II stars; red giants; supergiant stars; T Tauri stars; variable stars; white dwarfs; *and entries under* stellar
stationary absorption lines, 456
steady-state model of universe, 667
Stefan, Josef, 102
Stefan-Boltzmann law, 102–104, 406, 435, 436
stellar association, 467, 471
stellar corpses, 524, 526, 536
stellar distances, 420–422
and luminosity and brightness, 424–426
stellar evolution, 454, 478–495, 500–519
and young star clusters, 465–467
stellar-mass black holes, 552, 556
stellar masses, 440–443, *444*
and star formation, 463–465
stellar motions, 423
stellar nurseries, *5*, 460, 461, 471, 472
stellar parallax, 73, 420–422, 436, 437, 440
stellar spectra, 432–434, 436, 437
and star types, 439–440
stellar spectroscopy, 432
stellar winds, 470, 471, 503
Stephano, 354
Stern, Alan, 358
Stjerneborg observatory, Denmark, 73
Stonehenge, 21
stones, *see* stony meteorites
stony asteroid, 382, *383*
stony iron meteorites, 373, 374
stony meteorites, 373
storms
on Neptune, 347–348
on Saturn, 330
on Uranus, 346
stratosphere, *196*, 197
Strom, Robert G., 255
Strominger, Andrew, 685
strong nuclear force, 111, 537, 680–682
Subaru telescope, 131, 134, 359

Northern Hemisphere Star Charts

The following set of star charts, one for each month of the year, is from *Griffith Observer* magazine. It is useful in the northern hemisphere only. For a set of star charts suitable for use in the southern hemisphere, see the *Universe* CD-ROM or web site.

To use these charts, first select the chart that best corresponds to the date and time of your observations. Hold the chart vertically and turn it so that the direction you are facing shows at the bottom.

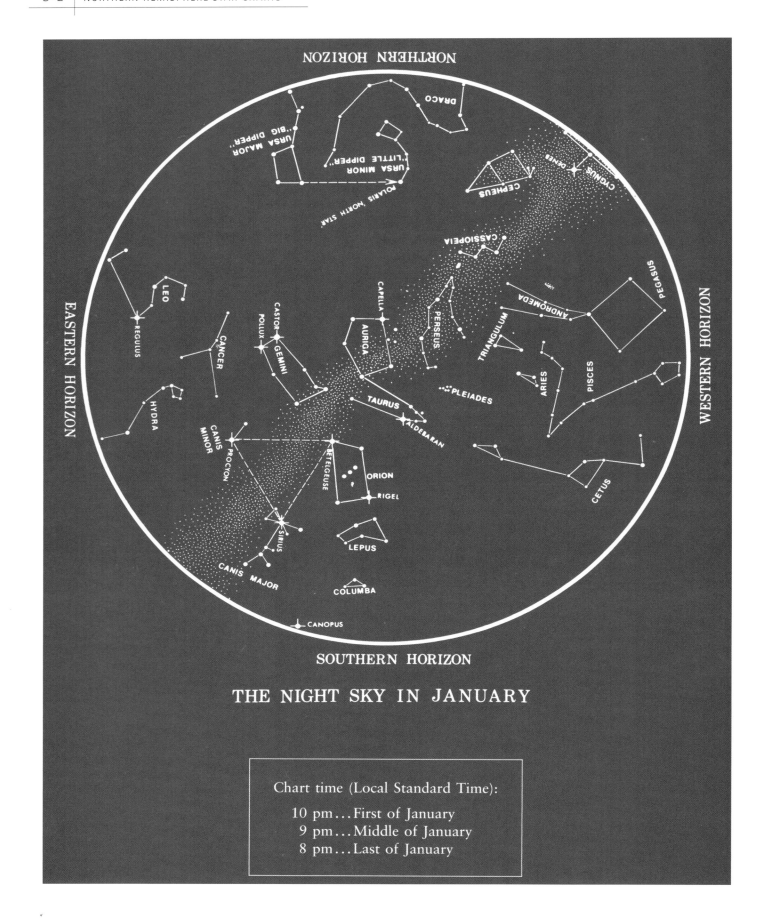

THE NIGHT SKY IN JANUARY

Chart time (Local Standard Time):

10 pm...First of January
9 pm...Middle of January
8 pm...Last of January

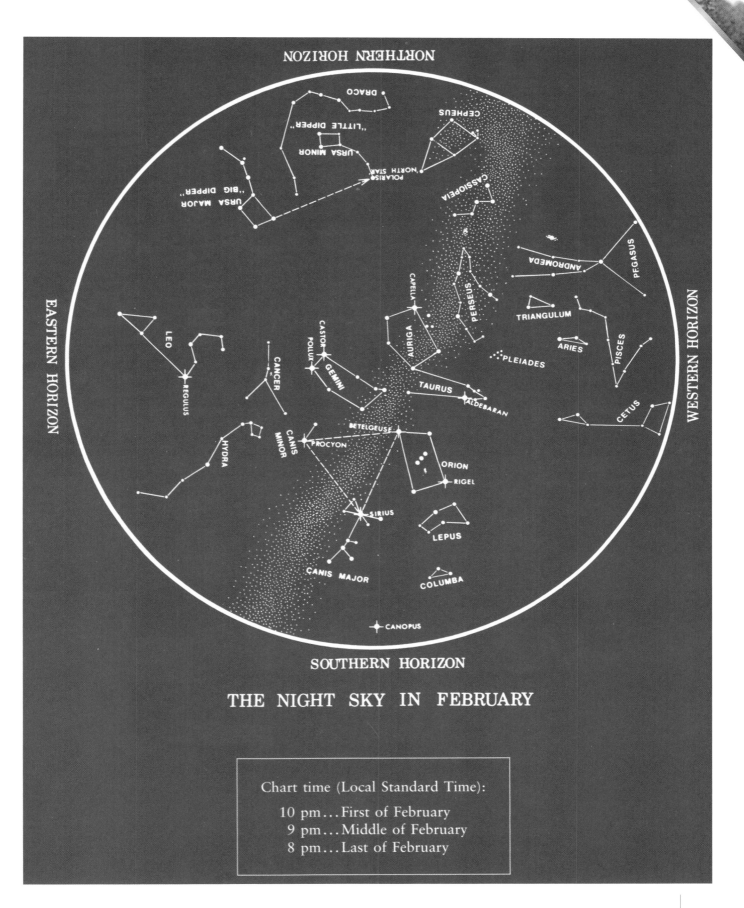

NORTHERN HORIZON

DRACO

"LITTLE DIPPER"

CEPHEUS

URSA MINOR

POLARIS "NORTH STAR"

CASSIOPEIA

"BIG DIPPER"

URSA MAJOR

ANDROMEDA

PEGASUS

CAPELLA

PERSEUS

TRIANGULUM

LEO

AURIGA

POLLUX

CASTOR

CANCER

GEMINI

ARIES

PISCES

PLEIADES

REGULUS

TAURUS

ALDEBARAN

CETUS

HYDRA

CANIS MINOR

PROCYON

BETELGEUSE

ORION

RIGEL

SIRIUS

LEPUS

CANIS MAJOR

COLUMBA

CANOPUS

EASTERN HORIZON

WESTERN HORIZON

SOUTHERN HORIZON

THE NIGHT SKY IN FEBRUARY

Chart time (Local Standard Time):

10 pm...First of February
9 pm...Middle of February
8 pm...Last of February

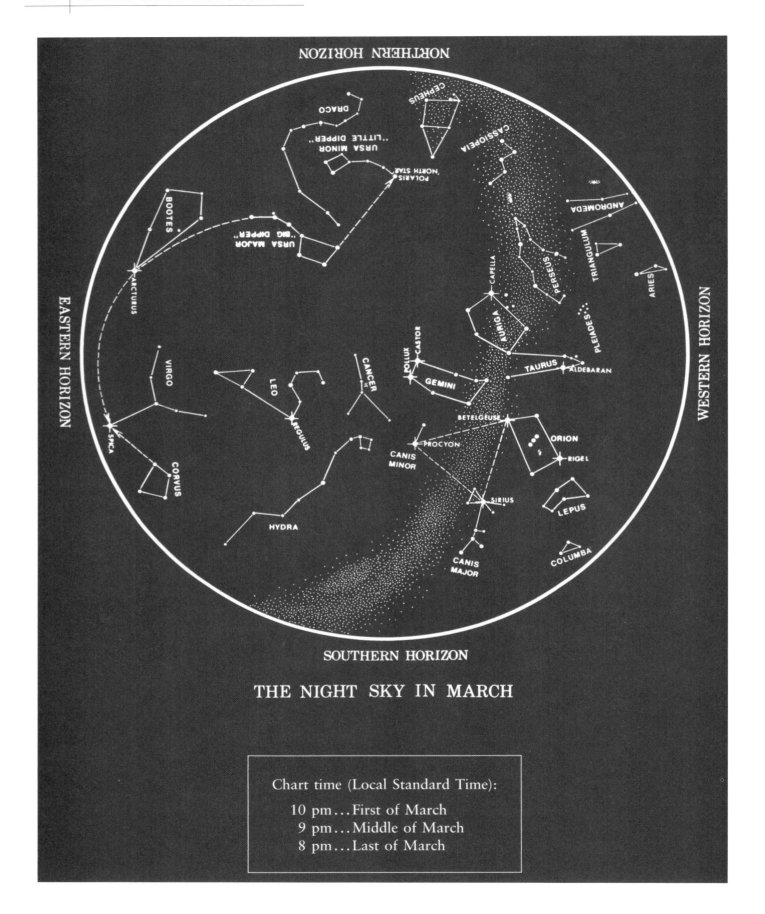

THE NIGHT SKY IN MARCH

Chart time (Local Standard Time):

10 pm...First of March
9 pm...Middle of March
8 pm...Last of March

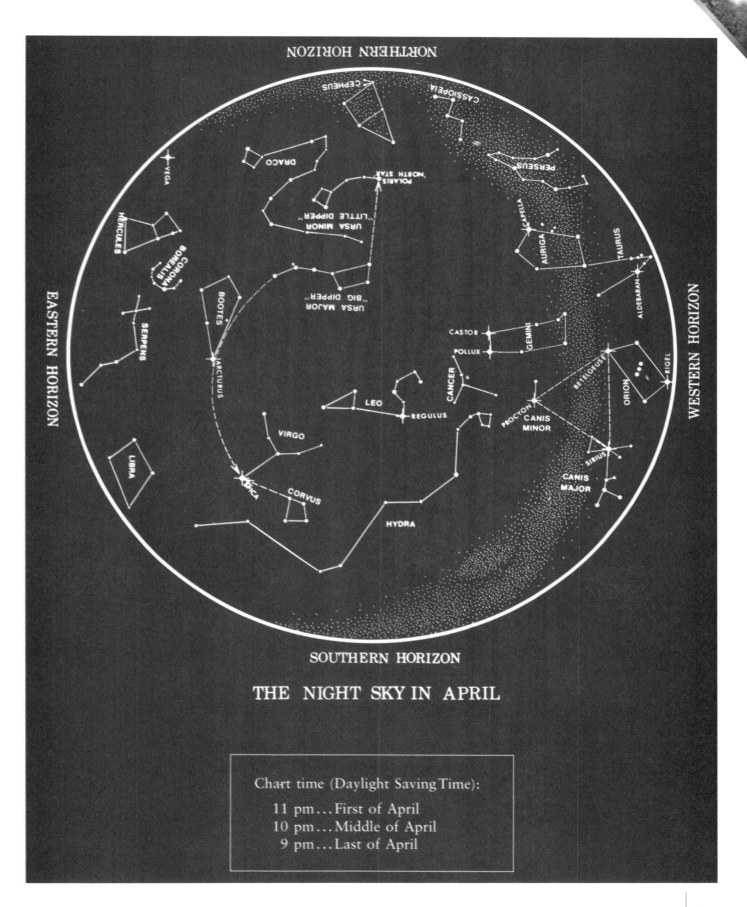

THE NIGHT SKY IN APRIL

Chart time (Daylight Saving Time):

11 pm...First of April
10 pm...Middle of April
9 pm...Last of April

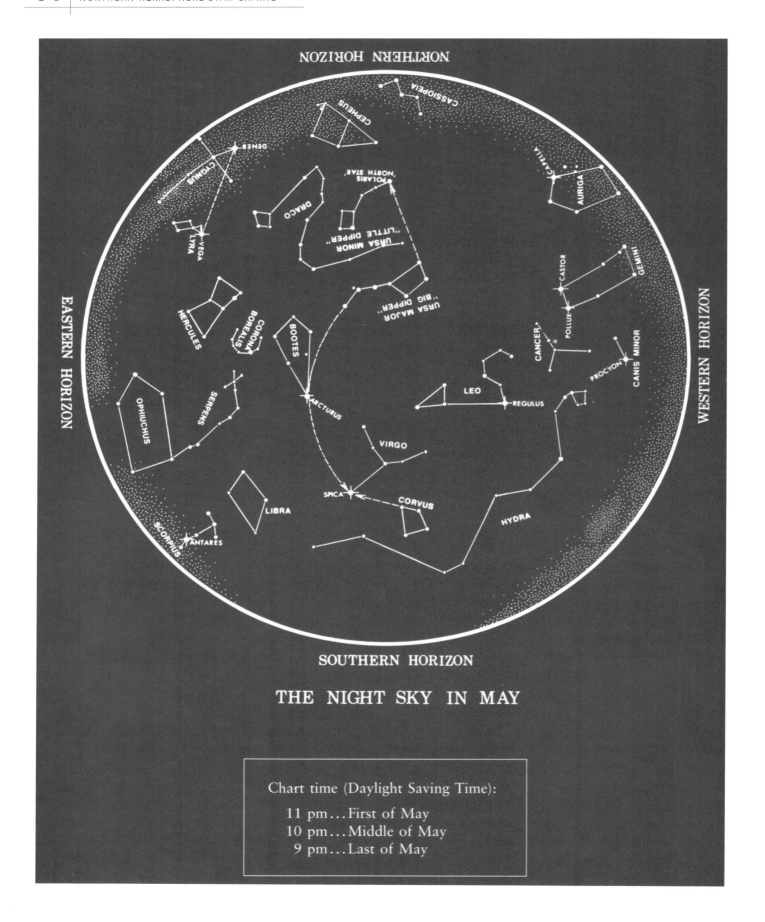

THE NIGHT SKY IN MAY

Chart time (Daylight Saving Time):

11 pm...First of May
10 pm...Middle of May
9 pm...Last of May

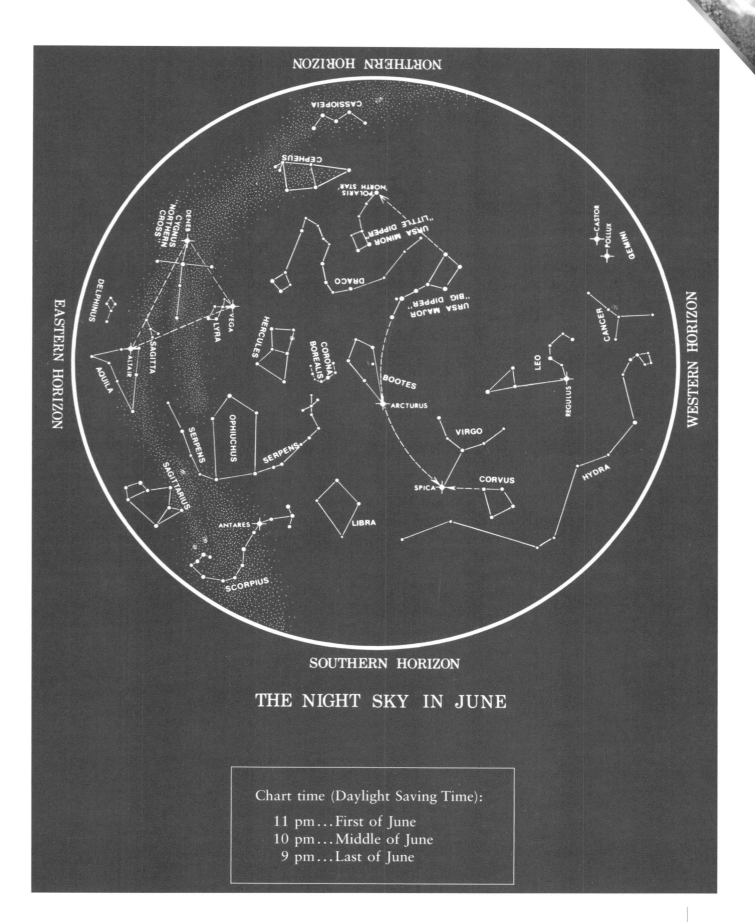

THE NIGHT SKY IN JUNE

Chart time (Daylight Saving Time):

11 pm... First of June
10 pm... Middle of June
9 pm... Last of June

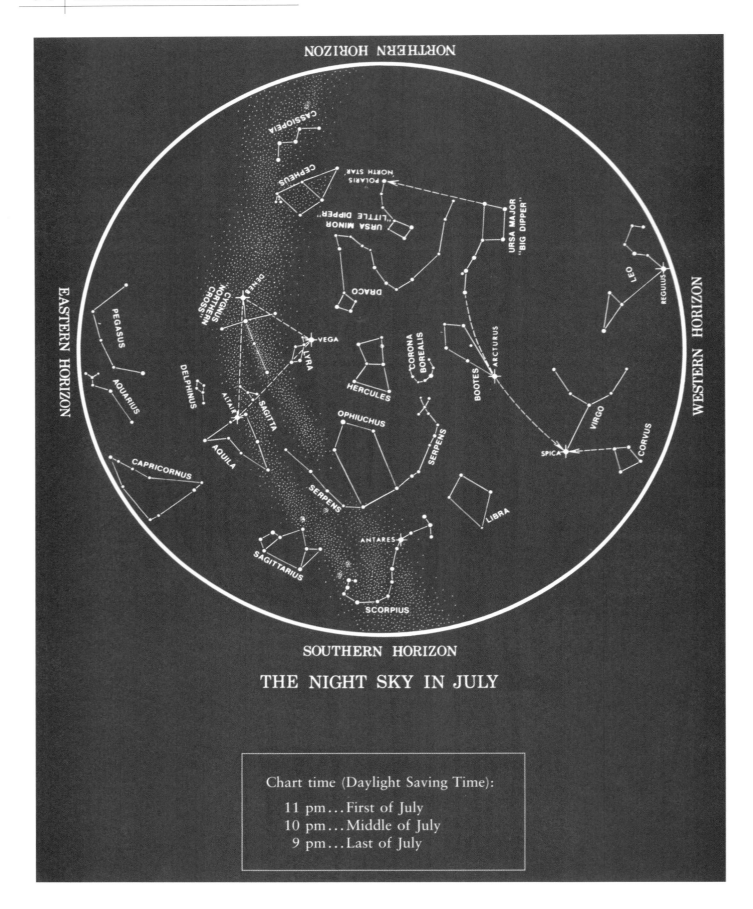

SOUTHERN HORIZON

THE NIGHT SKY IN JULY

Chart time (Daylight Saving Time):

11 pm...First of July
10 pm...Middle of July
9 pm...Last of July

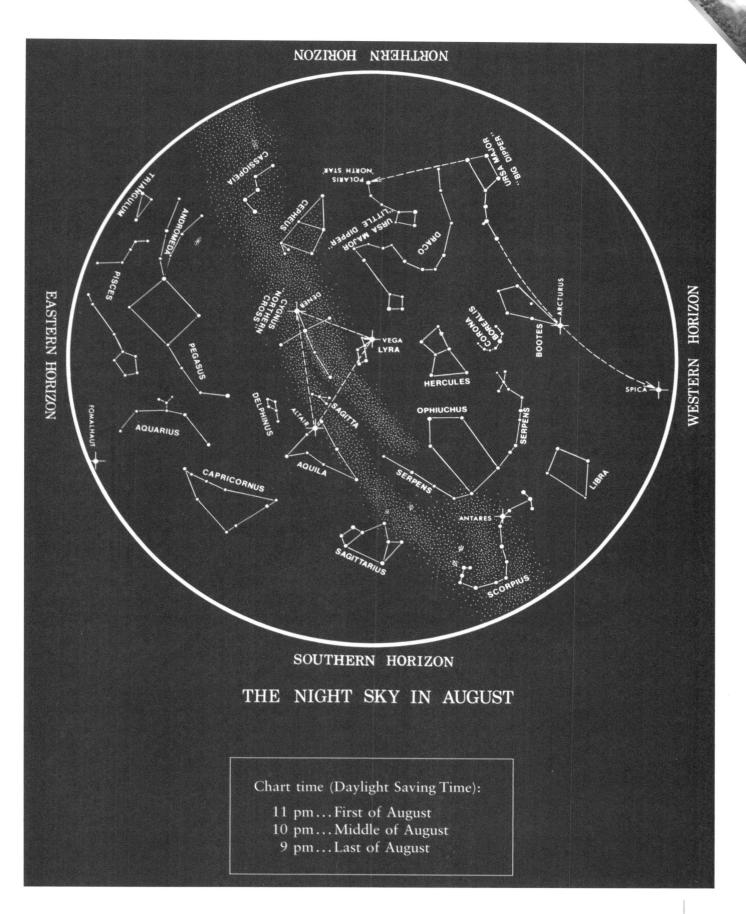

THE NIGHT SKY IN AUGUST

Chart time (Daylight Saving Time):

11 pm…First of August

10 pm…Middle of August

9 pm…Last of August

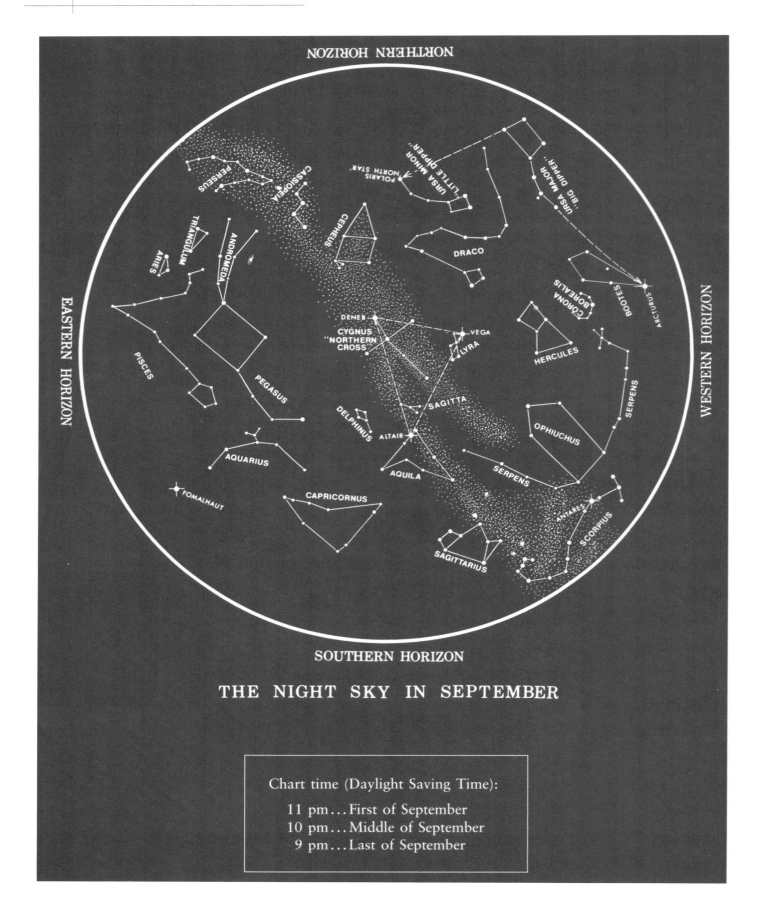

THE NIGHT SKY IN SEPTEMBER

Chart time (Daylight Saving Time):

11 pm...First of September
10 pm...Middle of September
9 pm...Last of September

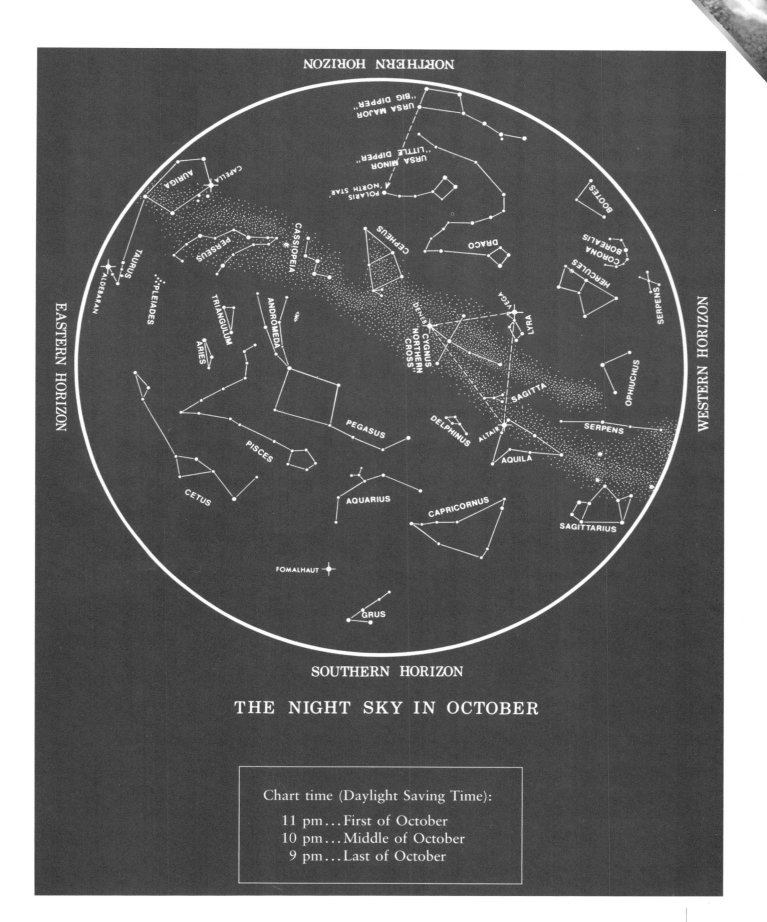

THE NIGHT SKY IN OCTOBER

Chart time (Daylight Saving Time):

11 pm...First of October
10 pm...Middle of October
9 pm...Last of October

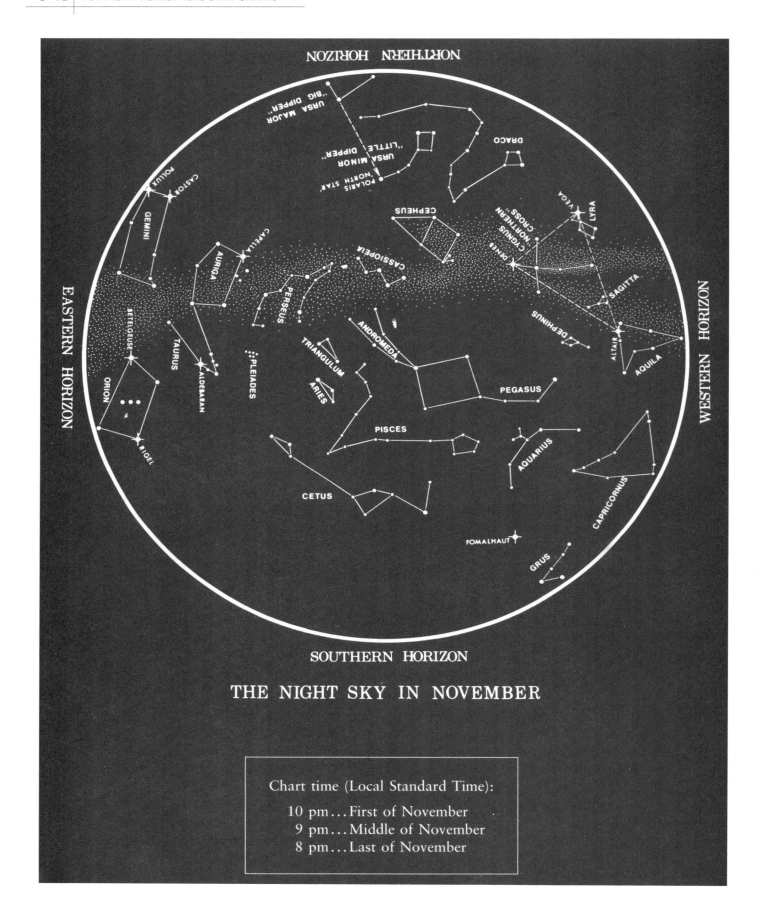

THE NIGHT SKY IN NOVEMBER

Chart time (Local Standard Time):

10 pm...First of November
9 pm...Middle of November
8 pm...Last of November